Handbook of Pest Management in Organic Farming

Handbook of Pest Management in Organic Farming

Edited by

Vincenzo Vacante

Mediterranean University, Reggio Calabria, Italy

and

Serge Kreiter

SupAgro, Montpellier, France

CABI

CABI is a trading name of CAB International

CABI
Nosworthy Way
Wallingford
Oxfordshire OX10 8DE
UK

Tel: +44 (0)1491 832111
Fax: +44 (0)1491 833508
E-mail: info@cabi.org
Website: www.cabi.org

CABI
745 Atlantic Avenue
8th Floor
Boston, MA 02111
USA

Tel: +1 (617)682-9015
E-mail: cabi-nao@cabi.org

A catalogue record for this book is available from the British Library, London, UK.

Library of Congress Cataloging-in-Publication Data

Names: Vacante, Vincenzo, editor. | Kreiter, Serge, editor.
Title: Handbook of pest management in organic farming / edited by Vincenzo Vacante and Serge Kreiter.
Description: Boston, MA : CABI, [2017] | Includes bibliographical references and index.
Identifiers: LCCN 2017023564 (print) | LCCN 2017024620 (ebook) | ISBN 9781780645001 (ePDF) | ISBN 9781786393807 (ePub) | ISBN 9781780644998 (hbk : alk. paper)
Subjects: LCSH: Agricultural pests--Control. | Organic farming.
Classification: LCC SB950 (ebook) | LCC SB950 .H343 2017 (print) | DDC 628.9/6--dc23
LC record available at https://lccn.loc.gov/2017023564

ISBN-13: 978 1 78064 499 8

Commissioning editor: Rachael Russell
Editorial assistant: Emma McCann
Production editor: Shankari Wilford

Typeset by SPi, Pondicherry, India
Printed and bound in the UK by CPI Group (UK) Ltd, Croydon, CR0 4YY, UK

Contents

Contributors

Babin, Régis, International Centre of Insect Physiology and Ecology (icipe), PO Box 30772-00100, Nairobi, Kenya. E-mail: regis.babin@cirad.fr

Bàrberi, Paolo, Institute of Life Sciences, Scuola Superiore Sant'Anna, Piazza Martiri della Libertà 33, 56127 Pisa PI, Italy. E-mail: paolo.barberi@santannapisa.it

Barloggio, Guendalina, Research Institute of Organic Agriculture, Forschungsinstitut für biologischen Landbau (FiBL), Ackerstrasse 113, CH-5070 Frick, Switzerland. E-mail: guendalina.barloggio@fibl.org

Barrera, Juan F., Departamento de Agricultura, Sociedad y Ambiente, El Colegio de la Frontera Sur (ECOSUR), Carretera Antiguo Aeropuerto km 2.5, Tapachula, Chiapas, CP 30700 Mexico. E-mail: jbarrera@ecosur.mx

Baspinar, Hüseyin, Department of Plant Protection, Adnan Menderes University, Agricultural Faculty, 09100 Aydın, Turkey. E-mail: hbaspinar@adu.edu.tr

Benuzzi, Massimo, CBC (Europe) S.r.l. – BIOGARD Division, Area Tecnica, Via Calcinaro 2085/7, 47521 Cesena (FC), Italy. E-mail: mbenuzzi@cbceurope.it

Bonsignore Carmelo Peter, Laboratorio di Entomologia ed Ecologia Applicata, Dipartimento PAU, Mediterranean University, 89100 Salita Melissari, sn, Reggio Calabria, Italy. E-mail: cbonsignore@unirc.it

Bordat, Dominique, [Ex CIRAD Vegetables Expert in Entomology] 222 Chemin de la qualité, 34980 Montferrier, France. E-mail: dominique.bordat@cirad.fr

Buamas, Chamaiporn, Plant Protection Research and Development Office, Department of Agriculture, Bangkok, Thailand. E-mail: c.buamas@gmail.com

Carrillo, Daniel, University of Florida, Tropical Research and Education Center, Homestead, FL 33031, USA. E-mail: dancar@ufl.edu

Collier, Rosemary, Warwick Crop Centre, School of Life Sciences, The University of Warwick, Wellesbourne, Warwick, CV35 9EF, UK. E-mail: rosemary.collier@warwick.ac.uk

Costanzo, Ambrogio, Institute of Life Sciences, Scuola Superiore Sant'Anna, Piazza Martiri della Libertà 33, 56127 Pisa PI, Italy. Present address: Organic Research Centre, Elm Farm, Hamstead Marshall, Newbury RG20 0HR, UK. E-mail: ambrogio.c@organicresearchcentre.com

Cruz, Maribel, Fondo Latinoamericano para Arroz de Riego (FLAR), Cali, Colombia. E-mail: maribel.cruz@cgiar.org

Daniel, Claudia, Research Institute of Organic Agriculture, Forschungsinstitut für biologischen Landbau (FiBL), Ackerstrasse 113, CH-5070 Frick, Switzerland. E-mail: claudia.daniel@fibl.org

Delrio, Gavino, Agriculture Department, University of Sassari, via Enrico de Nicola, 1, 07100 Sassari, Italy. E-mail: gdelrio@uniss.it

Doll, David, University of California Cooperative Extension, Merced County, 2145 Wardrobe Avenue, Merced, CA 95341-6445 (209) 385-7403, USA. E-mail: dadoll@ucanr.edu

Faber, Ben, UC Cooperative Extension 669 County Square Dr., Suite 100 Ventura CA 93003, USA. E-mail: bafaber@ucanr.edu

Garcia, Anamaria, Food and Agriculture Organization of the United Nations (FAO), Regional Office for Latin America and the Caribbean, Santiago, Chile. E-mail: anmagadi@gmail.com

Gaterol, Eduardo J., Fondo Latinoamericano para Arroz de Riego (FLAR), Cali, Colombia. E-mail: e.j.gaterol@cgiar.org

Graziosi, Ignazio, International Center for Tropical Agriculture (CIAT) Asia regional office, Hanoi, Vietnam. Present address: Department of Entomology, University of Kentucky, S-225 Agricultural Science Center North, Lexington, KY 40546-0091, USA. E-mail: i.graziosi@uky.edu

Grissa-Lebdi, Kaouthar, Plant Protection Department, National Agronomic Institute of Tunisia (INAT), 43 Avenue Charles Nicolle, 1082 Tunis, Tunisia. E-mail: grissak@yahoo.fr

Karagöz, Mehmet, Department of Plant Protection, Adnan Menderes University, Agricultural Faculty, 09100 Aydın, Turkey. E-mail: Mkaragoz@52@hotmail.com

Koller, Martin, Research Institute of Organic Agriculture, Forschungsinstitut für biologischen Landbau (FiBL), Ackerstrasse 113, CH-5070 Frick, Switzerland. E-mail: martin.koller@fibl.org

Kreiter, Serge, Montpellier SupAgro Unité Mixte de Recherche (UMR) Centre de Biologie pour la Gestion des Populations (CBGP) (SupAgro/CIRAD/INRA/IRD), 755 avenue du Campus Agropolis (Baillarguet) CS 30016, 34988 Montferrier-sur-Lez cedex, France. E-mail: serge.kreiter@supagro.fr

Ladurner, Edith, CBC (Europe) S.r.l. – BIOGARD Division, Area Tecnica, Via Calcinaro 2085/7, 47521 Cesena (FC), Italy. E-mail: eladurner@cbceurope.it

Lentini, Andrea, Agriculture Department, University of Sassari, via Enrico de Nicola, 1, 07100 Sassari, Italy. E-mail: lentini@uniss.it

Luka, Henryk, Research Institute of Organic Agriculture, Forschungsinstitut für biologischen Landbau (FiBL), Ackerstrasse 113, CH-5070 Frick, Switzerland. E-mail: henryk.luka@fibl.org

Matray, Silvia, Research Institute of Organic Agriculture, Forschungsinstitut für biologischen Landbau (FiBL), Ackerstrasse 113, CH-5070 Frick, Switzerland. E-mail: silviamatray@gmx.de

Niggli, Urs, Research Institute of Organic Agriculture, Forschungsinstitut für biologischen Landbau (FiBL), Ackerstrasse 113, CH-5070 Frick, Switzerland. E-mail: urs.niggli@fibl.org

Ozman-Sullivan, Sebahat K., Ondokuz Mayis University, Faculty of Agriculture, Department of Plant Protection, 55139 Samsun, Turkey. E-mail: sozman@omu.edu.tr; ozman.sullivan@gmail.com

Pantoja, Alberto, Food and Agriculture Organization of the United Nations (FAO), Regional Office for Latin America and the Caribbean, Santiago, Chile. Present address: United States Department of Agriculture (USDA), Agricultural Research Service, 1815 North University Street, Peoria, IL 61604, USA. E-mail: Alberto.pantoja@ars.usda.gov *and* eldelavegabaja@gmail.com

Peña, Jorge E., University of Florida, Tropical Research and Education Center, Homestead, FL 33031, USA. E-mail: jepena@ufl.edu

Prado, Gustavo A., Centro Internacional de Agricultura Tropical (CIAT), Cali, Colombia. E-mail: gaprado66@yahoo.com

Rijal, Jhalendra, University of California Cooperative Extension, Stanislaus County, 3800 Cornucopia Way, Suite A, Modesto, CA 95358 (209)525-6800, USA. E-mail: jrijal@ucanr.edu

Sahraoui, Hajer, Citrus Technical Center, Plant Protection Department, 8099 Beni-Khalled (Zaouiet Jedidi), BP 318, Tunisia. E-mail: hajersahraoui@yahoo.fr

Stoeckli, Sibylle, Research Institute of Organic Agriculture, Forschungsinstitut für biologischen Landbau (FiBL), Ackerstrasse 113, CH-5070 Frick, Switzerland. E-mail: sibylle.stoeckli@fibl.org

Tien Dung, Ngo, Plant Protection Department (PPD), Ministry of Agriculture, Hanoi, Vietnam. E-mail: ipmppd@fpt.vn

Tixier, Marie-Stéphane, Montpellier SupAgro Unité Mixte de Recherche (UMR) Centre de Biologie pour la Gestion des Populations (CBGP) (SupAgro/CIRAD/INRA/IRD), 755 avenue du Campus Agropolis (Baillarguet) CS 30016, 34988 Montferrier-sur-Lez cedex, France. E-mail: marie-stephane.tixier@supagro.fr

Torres, Edgar A., Centro Internacional de Agricultura Tropical (CIAT), Cali, Colombia. E-mail: mejoarroz@gmail.com

Vacante, Vincenzo, Laboratorio di Entomologia ed Ecologia Applicata, Dipartimento PAU, Mediterranean University, 89100 Salita Melissari, sn, Reggio Calabria, Italy. E-mail: vvacante@unirc.it

Wongtiem, Prapit, Department of Agriculture, Rayong Field Crops Research Center, Rayong, Thailand. E-mail: wongtiem_prapit@yahoo.com

Wyckhuys, Kris A.G., International Center for Tropical Agriculture (CIAT) Asia regional office, Hanoi, Vietnam. E-mail: k.wyckhuys@cgiar.org

Xiao, Qiang, Tea Research Institute, Chinese Academy of Agricultural Sciences, Hangzhou, 310008, China. E-mail: xqtea@vip.163.com

Preface

Conventional agriculture provides the use of different technical means, including fertilizers and pesticides. Artificial fertilizers were discovered during the 18th century, first as superphosphates and then as ammonia-based fertilizers mass produced according to the 'Haber-Bosch process' developed during World War I. Likewise, important pesticides were discovered as molecules at the end of the 1800s and similar innovations occurred in the 1940s, to the point that the decade is reported as the 'pesticide era' (Horne and Page, 2008), and within a century the consumption of these substances has become highly significant. In this regard, the Food and Agriculture Organization of the United Nations (FAO) Fertilizer Archives (2013) reported for 2002 a consumption of fertilizers in Eastern European countries of about 140,000,000 t, followed by other European countries (about 130,000,000 t), Oceania (120,000,000 t), Asia (110,000,000 t) and the Americas (30,000,000 t). Inter alia, the world consumption of nitrogen, phosphate and potash in 2009 was 69.3 kg/ha, 25.8 kg/ha and 14.8 kg/ha, respectively (FAOSTAT, 2015). As regards pesticides, Tilman *et al.* (2002) reported for 2000 a global production (herbicides, insecticides, fungicides and others) of 3×10^6 t, and global imports of about US$12 billion for 1996. The estimated worldwide annual sales of pesticides for 1999 has been about US$35 billion for herbicides, over US$15 billion for insecticides, over US$5 billion for fungicides, and less than US$5 billion for other products (Agrios, 2005). Approximately 3 billion t of pesticides are applied each year in the world (Pimentel, 2009a), and annually US$30 billion is spent on chemical pest control (CropLife International, 2009). In addition, the cost of new agrochemical product discovery, including development and registration, associated with the environmental, social and economic costs highlight the emblematic status of pesticides. Currently > 3.5 million active ingredients are being tested, with a success ratio of 1:140,000, and developmental costs of US$256 million over a developmental time of 10 years. The benefit:cost ratio is 2:1, with large risks of pest resistance, limited specificity and large numbers of harmful side effects (Pimentel *et al.*, 1980; Pimentel, 2009b; McDougall, 2010).

Despite these efforts heavy agricultural reliance on synthetic chemical fertilizers and pesticides produce serious impacts on public health and the environment (Pimentel *et al.*, 2005a). These new technical possibilities offer advantages in the short term, but in the longer term cause serious side effects such as soil compaction, soil erosion and overall decline in soil fertility, along with health problems following the use of toxic chemicals entering the food supply (Stinner, 2007).

The solution to this problem is based on the implementation of integrated pest management (IPM). However, despite this strategy representing an ideal solution, its proposal is in some way derived from the failure (or limits) of biological control and it tends to mediate between biological control on the one hand and chemical control (as a treatment in the first instance) on the other. Numerous reports confirm its wide application in different crops, but we cannot deny at the same time the failures and the high negative ecotoxicological impact of some active ingredients considered to be fundamental to particular IPM strategies (e.g. the use of cyhexatin on strawberry, azynphos-methyl on orchards and methyl parathion on vineyards, now all banned from such crops). Hokkanen (2015) recently analysed the factors that affect the development and implementation of the method, highlighting the role of scientific knowledge, politics and business.

'Organic agriculture' is a solution to the problems referred to above, reducing reliance on agrochemical inputs as well as making agriculture environmentally and economically healthy. Historically, this technique relates to studies on soil biology carried out in the late 1800s and early 1900s, especially to the search for new biological methods to resolve the impact of side effects of agricultural practices, while at the same time maintaining higher levels of production. The concept of organic agriculture, initially called 'biodynamic agriculture', was formulated for the first time in 1924 by the Austrian philosopher and esotericist Rudolf Steiner, founder of the so-called 'anthroposophy'. In a series of eight lectures in Koberwitz he delivered his agriculture course about the health of the earth and the optimization of its fertility in order to improve the quality of food (Paull, 2011a). However, Steiner's approach was mainly spiritual and unscientific, and only over time has its proposal been better enunciated by various authors. In fact, organic agriculture was developed in the late 1930s and early 1940s by the botanists Albert Howard and his wife Gabrielle Howard, whose vision was influenced by their scientific approach and experience with traditional biodynamic farming methods carried out in India (Paull, 2006). So, Sir A. Howard is widely acknowledged as the 'father of organic farming', having been the first to apply scientific knowledge and principles to traditional and more natural methods (Stinner, 2007). At the same time in the 1940s in the USA, J.I. Rodale founded 'The Rodale Institute', a working organic farm for the study of organic agriculture, and the 'Rodale Press' to spread information about organic farming to the wider public. Other contributors include Lady Eve Balfour working in the UK, and others authors across the world.

From an etymological point of view we do not know why the term 'organic' was applied to agriculture. The term 'organic agriculture' could be due to Lord Northbourne, who in 1940 adhered to the philosophy of Steiner. In this sense, the term organic identifies the farm as a living organism and derives from Steiner's unscientific anthroposophy (Paull, 2011b). Conversely, the term 'organic' could have been derived from the work of early soil biologists, the so-called 'humus farming'. The etymology of the term could then originate from the organic matter used to improve the humus content of soils. This concept was spread by Howard and Rodale, and since the early 1940s both terms have merged (Diver, 2014; Nayler, 2014).

According to the above recollections, the aim of organic agriculture is to increase ecological processes that facilitate plant nutrition and conserve soil and water resources. Pimentel *et al.* (2005b) reported that various organic technologies have been employed for about 6000 years to make agriculture sustainable while conserving soil, water, energy and biological resources. Advantages of organic technologies include: (i) higher soil organic matter and nitrogen; (ii) lower fossil energy inputs; (iii) yields similar to those of conventional systems; and (iv) conservation of soil moisture and water resources (especially advantageous under drought conditions). These aspects would seem to highlight the need for conventional agriculture to become more sustainable and ecologically healthy by adopting some traditional organic farming technologies.

From a practical point of view, organic systems do not allow the use of agrochemicals and external inputs to improve the environment and farm economics, and rely on techniques such as crop rotation, green manure, compost and biological pest control. Natural fertilizers and pesticides (herbicides, insecticides and fungicides) such as bonemeal from animals or pyrethrins from chrysanthemum flowers may be applied, but the use of other means, including synthetic petrochemical fertilizers and pesticides, plant growth regulators such as hormones, antibiotic application in livestock, genetically modified organisms, human sewage sludge and nanomaterials is excluded or strongly limited (Paull, 2011c; Directorate General for Agriculture and Rural Development of the European Commission, 2015).

In recent decades increasing environmental awareness has modified the supply-driven organic proposal to a demand-driven one. Premium prices and government subsidies have involved growers. Many growers have adopted traditional methods, comparable to organic farming, but are not certified, or have converted to modern organic agriculture for economic reasons (Paull, 2007). In less than a century the land that is organically farmed in the world has been growing and the latest data published in 2015 by FiBL (Research Institute of Organic Agriculture) and IFOAM (International Federation of Organic Agriculture Movements) are very encouraging. This indicates a total of 43 million ha in the world is managed under organic farming, including 17.3 million ha in Oceania, 11.5 million ha in Europe, 6.6 million ha in Latin America, 3.4 million ha in Asia, 3.0 million ha in North America and 1.2 million ha in Africa.

In the last 70 years the principles and guidelines of organic agriculture in different crops have been discussed by eminent specialists. None the less, there is no single updated contribution on pest control of the main crops in the world. This shortage has suggested a need for this book, which is divided into 21 chapters, each dedicated to a specific crop, with exception of the cultivation of field vegetables, which is treated in two chapters according to the wide distribution of these crops and the strong peculiarities occurring in different areas of the world. Each chapter was entrusted to eminent specialists in the pest control of the particular crop, and includes a brief presentation of the crop and its pests, followed by a discussion on the bio-ecology and current methods of control that are known to be used according to the organic farming technique.

The aims of the book, the complexity of issues and the limited available space have meant that only the main aspects of the topic are covered. Although the project consists of a rational approach managed by qualified specialists of pest control, we cannot completely rule out some risk of approximations, and we invite interested specialists to contact us to indicate correction of any unintended errors. With the enormity of the subject which is constantly evolving the work cannot be considered to be concluded.

Vincenzo Vacante
Mediterranean University, Reggio Calabria, Italy
and
Serge Kreiter
SupAgro, Montpellier, France

References

Agrios, G.N. (2005) *Plant Pathology*, 5th edn. Academic Press, New York, 952 pp.

CropLife International (2009) *Facts and Figures – the Status Of Global Agriculture (2008–2009)*. CropLife International, Brussels, p. 12.

Directorate General for Agriculture and Rural Development of the European Commission (2015) What is Organic Farming? Available at: http://ec.europa.eu/agriculture/organic/index_en.htm (accessed 11 July 2015).

Diver, S. (2014) Controlled Microbial Composting and Humus Management: Luebke Compost. Available at: http://www.ibiblio.org/steved/Luebke/Luebke-compost2.html (accessed 19 July 2015).

FAOSTAT (2015) Food and Agriculture Organization of the United Nations Statistic Division. Available at: http://faostat3.fao.org/home/E (accessed 28 June 2015).

Food and Agriculture Organization of the United Nations (FAO) (2013) *World Food and Agriculture. FAO Statistical Yearbook*. Food and Agriculture Organization of the United Nations, Rome, p. 289.

Forschungsinstitut für biologischen Landbau (FiBL) and International Federation of Organic Agriculture Movements (IFOAM) (2015) The World of Organic Agriculture. Statistics & Emerging Trends 2015. Available at: https://www.fibl.org/fileadmin/documents/shop/1663-organic-world-2015.pdf (accessed 19 July 2015).

Hokkanen, H.M.T. (2015) Integrated pest management at the crossroads: science, politics, or business (as usual)? *Arthropod-Plant Interactions* 9, 543–545.

Horne, P.A. and Page, J. (2008) *Integrated Pest Management for Crops and Pastures*. Commonwealth Scientific and Industrial Research Organisation (CSIRO), Landlinks Press, Melbourne, Australia, p. 119.

McDougall, P. (2010) *The Cost of New Agrochemical, Product Discovery, Development & Registration in 1995, 2000 and 2005–8. R&D Expenditure in 2007 and Expectations for 2012*. A Consultancy Study for Crop Life America and the European Crop Protection Association, Saughland, Pathhead, Midlothian EH7 5XP, UK, 22 pp.

Nayler, J. (2014) Second Thoughts About Organic Agriculture. Soil and Health Library. Available at: http://www.soilandhealth.org/01aglibrary/Second.Thoughts.pdf (accessed 19 July 2015).

Paull, J. (2006) The Farm as Organism: The Foundational Idea of Organic Agriculture Elementals. *Journal of Bio-Dynamics Tasmania* 83, 14–18.

Paull, J. (2007) China's organic revolution. *Journal of Organic Systems* 2(1), 1–11.

Paull, J. (2011a) Attending the first organic agriculture course: Rudolf Steiner's agriculture course at Koberwitz, 1924. *European Journal of Social Sciences* 21(1), 64–70.

Paull, J. (2011b) The Betteshanger summer school: missing link between biodynamic agriculture and organic farming. *Journal of Organic Systems* 6(2), 13–26.

Paull, J. (2011c) Nanomaterials in food and agriculture: the big issue of small matter for organic food and farming. In: *Proceedings of the Third Scientific Conference of ISOFAR* (International Society of Organic Agriculture Research), 28 September–1 October, Namyangju, Korea, 2, pp. 96–99.

Pimentel, D. (2009a) Pesticides and pest control. In: Rajinder, P. and Dhawan, A.K. (eds) *Integrated Pest Management: Innovation–Development Process*, Vol. 1. Springer, Dordrecht, The Netherlands, pp. 83–87.

Pimentel, D. (2009b) Environmental and economic costs of the application of pesticides primarily in the United States. In: Peshin, R. and Dhawan, A.K. (eds) *Integrated Pest Management: Innovation–Development Process*, Vol. 1. Springer, Dordrecht, The Netherlands, pp. 89–111.

Pimentel, D., Andow, D., Dyson-Hudson, R., Gallahan, D., Jacobson, S., Irish, M., Kroop, S., Moss, A., Schreiner, I., Shephard, M., Thompson, T. and Vinzant, B. (1980) Environmental and social costs of pesticides: a preliminary assessment. *Oikos* 34, 126–140.

Pimentel, D., Hepperly, P., Hanson, J., Siedel, R. and Douds, D. (2005a) Organic and conventional farming systems: environmental and economic issues. *Environmental Biology* p. 52. Available at: http://dspace.library.cornell.edu/bitstream/1813/2101/1/pimentel_report_05-1.pdf (accessed 17 July 2015).

Pimentel, D., Hepperly, P., Hanson, J. Douds, D. and Seidel, R. (2005b) Environmental, energetic, and economic comparisons of organic and conventional farming systems. *BioScience* 55(7), 573–582.

Stinner, D.H. (2007) The science of organic farming. In: Lockeretz, W. (ed.) *Organic Farming: An International History*. CAB International, Wallingford, UK, pp. 40–72.

Tilman, D., Cassman, K.G., Matson, P.A., Naylor, R. and Polasky, S. (2002) Agricultural sustainability and intensive production practices. *Nature* 418, 671–677.

Acknowledgements

The editors extend special thanks to Rachael Cutts, Product Manager, e-books at CABI, who with her suggestions has allowed the realization of the book project. Thanks also to Rachael Russell, Commissioning Editor at CABI, for valuable assistance in the development of the project. Moreover, they sincerely thank all colleagues who have agreed to participate in the initiative.

1 Management of Crops to Prevent Pest Outbreaks

Claudia Daniel,* Guendalina Barloggio, Sibylle Stoeckli, Henryk Luka and Urs Niggli
Research Institute of Organic Agriculture, Forschungsinstitut für biologischen Landbau (FiBL), Frick, Switzerland

Introduction

Organic farmers face the same potentially severe pest problems as their colleagues in integrated pest management (IPM) and conventional farming systems. However, approaches to manage the pest insects are different because the aim of organic farming is a holistic system perspective rather than simple reductionist control approaches. Organic cropping systems are designed to prevent damaging levels of pests, thus minimizing the need for direct and curative pest control (Peacock and Norton, 1990). Within this chapter, we will briefly explain the standards for organic farming, which also set the framework for pest control. We present a conceptual model for pest control in organic farming and describe the influence of functional agrobiodiversity and conservation biological control on pest management. We focus on the use of preventive strategies and cultural control methods. The system approach is illustrated with examples in organic *Brassica* vegetable and oilseed rape production, because these economically important crops (Ahuja *et al.*, 2010) are attacked by a broad range of different pest insects (Smukler *et al.*, 2008; Ahuja *et al.*, 2010) and show different levels of tolerance. Economic thresholds for pests on oilseed rape are usually higher than on vegetables. Therefore, less control is used in oilseed rape which might lead to the build-up of large pest populations, threatening nearby vegetable fields. With the increasing area of oilseed rape production, pest problems in these crops are likely to increase.

Standards for Organic and IPM Production: Similarities and Differences

Organic farming

Organic farming is regulated by international and national organic production standards, such as the IFOAM (International Federation of Organic Agriculture Movements) Norms (IFOAM, 2012), *Codex Alimentarius* (FAO and WHO, 2007), or European Union (EU) regulation (EC, 2007). Organic standards all have the same principal norms for plant production as described in the *Codex Alimentarius*:

> Organic agriculture is a holistic production management system which promotes and enhances agroecosystem health, including biodiversity, biological cycles, and soil biological activity. It emphasizes the use of

* Corresponding author: claudia.daniel@fibl.org

> management practices in preference to the use of off-farm inputs, taking into account that regional conditions require locally adapted systems. This is accomplished by using, where possible, cultural, biological, biotechnical, physical and mechanical methods, as opposed to using synthetic materials, to fulfil any specific function within the system.
>
> (FAO and WHO, 2007)

Thus, the maintenance of plant health primarily relies on preventative measures, such as: (i) the choice of appropriate species and varieties resistant to pests and diseases; (ii) appropriate crop rotations, cultivation techniques, mechanical and physical methods; and (iii) the protection of natural enemies of pests. In the case of an established threat to a crop, plant protection products may only be used if they have been authorized for use in organic production. Within the EU, products authorized for organic farming are listed in Annex II of the implementation rule 889/2008 (EC, 2008). Substances used for plant protection should be of plant, animal, microbial or mineral origin. Genetically modified organisms (GMOs) and products produced from or by GMOs, as well as mineral nitrogen fertilizers are not allowed. Chemically synthesized products are only allowed if they are not available in sufficient quantities in their natural form (e.g. pheromones) and if conditions for their use do not result in contact of the product with the edible parts of the crop (e.g. application in dispensers).

IPM

IPM standards were developed and defined by the International Organisation for Biological and Integrated Control (IOBC) (Boller *et al.*, 2004). With the Sustainable Use Directive (EC, 2009), IPM has become the main part of the European crop protection policy. Central goals of IPM are the prevention and suppression of harmful organisms, as well as the preference of non-chemical methods with few side effects on non-targets (Kogan, 1998). In addition, monitoring of pest insects, economic action thresholds and anti-resistance strategies are centrepieces of IPM strategies. Nevertheless, pest management in IPM is still dominated by the use of synthetic pesticides. In particular the strong focus on economic thresholds leads to a reductionist view of the systems (El-Wakeil, 2010). Environmental considerations and the presence or absence of beneficial insects are mostly not included in the economic thresholds (El-Wakeil, 2010). According to Ehler (2006), this perpetuates a 'quick-fix mentality', where symptoms are treated instead of causes. IPM principles are only reluctantly implemented by the farmers due to higher costs, and higher risk of failures of non-chemical control methods, as well as lack of experience with these methods (Gruys, 1982). Incentives for farmers to use alternative methods are missing, because the advantage of using sustainable and preventive measures 'is at the social and environmental level and on the long-term, rather than at the private economic level and on the short-term' (Gruys, 1982). In addition, the low price for synthetic pesticides does not reflect the true ecological costs. Thus, for the individual farmer it is often more economical to use a curative pesticide instead of preventive measures. The use of pesticides is more regulated in organic farming systems: only naturally derived substances are allowed. As availability and efficacy of these substances is limited and most of them are considerably more expensive than synthetic pesticides, organic farmers have a stronger incentive to consequently apply preventive measures.

Conceptual Model for Pest Management in Organic Farming

A conceptual model for pest management in organic farming (Fig. 1.1) was proposed by Wyss *et al.* (2005), refined by Zehnder *et al.* (2007) and complemented by Luka (2012, cited in Forster *et al.*, 2013). The fundamental first step of this holistic approach is the benefits of nature conservation measures: ecosystem diversity is increased through habitat management, extensification of land uses, establishment of non-crop habitats and biotope

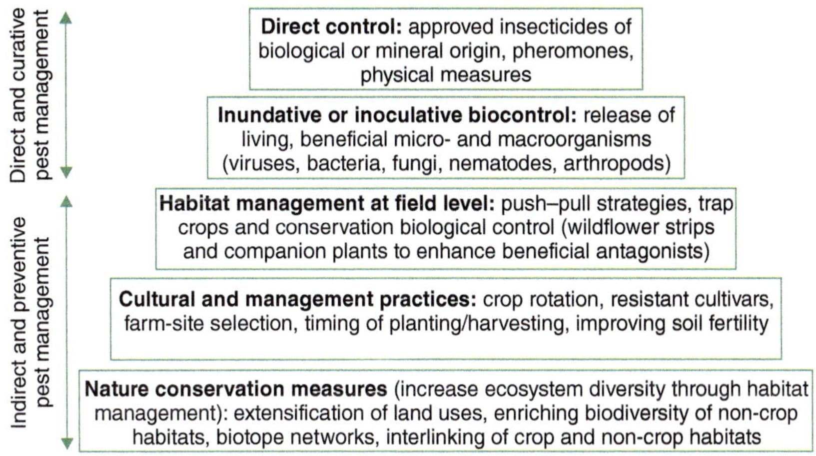

Fig. 1.1. Conceptual model for pest management in organic farming. (Adapted and supplemented based on Luka, 2012, cited in: Wyss *et al.*, 2005; Zehnder *et al.*, 2007; Forster *et al.*, 2013.)

networks. The second step of the pyramidal model are cultural practices applied by the farmers in order to avoid pest damage (Peacock and Norton, 1990). These practices include crop rotation, increasing crop diversity, timely planting and harvesting, transplanting, weed management, choice of resistant varieties and avoiding areas with high pest presence on the farm level. These practices go hand in hand with the third step which is habitat management at the field level (i.e. companion plants, tailored wildflower strips, push–pull strategies) which aims at interlinking crop and non-crop habitats. These first three steps create a broad and solid basis for healthy plant development. Direct control methods based on biocontrol organisms or bioinsecticides are the fourth and fifth steps of the model. However, these methods can have side effects on beneficial arthropods and thus adversely affect ecosystem services needed for pest prevention. Thus, direct control measures should only be applied in case of threatening pest outbreaks and selective methods should be preferred. The use of non-selective biopesticides should be limited to a minimum. Within this chapter we will focus on the use of preventive strategies (the first three steps in the multi-level model). The last two steps (biocontrol and organically approved insecticides) are only briefly mentioned here and discussed in detail in Chapters 2 and 3 of this volume, respectively.

Nature Conservation Measures: the Basis for Biodiversity and Ecosystem Services

According to the *Convention on Biological Diversity* of Rio de Janeiro in 1992, biodiversity encompasses the variety of life on earth ranging from genes, through species, to entire ecosystems (United Nations, 1992). Ecosystem diversity covers the diversity of habitats or patches within a landscape and includes the diversity of farming systems, ratio of arable land to other land uses as well as interactions between agricultural land and nearby natural biotopes. Ecosystem diversity and diversified cropping systems have a range of benefits, both short term (e.g. by increase in crop yield and quality due to improved pest control) and long term (e.g. by re-establishing agroecosystem sustainability), on the agronomic level (e.g. biotic and abiotic stress resistance, production of cultivated ecosystems), as well as on the societal and ecological level (e.g. by landscape aesthetics, water and soil quality and flora and fauna conservation, including endangered species, existence of typical habitats with particular species) (Clergue *et al.*, 2009; Malézieux *et al.*, 2009).

Integrating biodiversity conservation into production systems

Agricultural ecosystems comprise productive areas (managed fields), as well as semi-natural and natural habitats (Moonen and Bàrberi, 2008). The productive areas can have a negative impact on biodiversity: monocultures treated with broad-spectrum pesticides to prevent pest outbreaks (Landis *et al.*, 2000) decrease the natural enemies' diversity, reduce species richness, abundance and effectiveness (Naranjo and Ellsworth, 2009; Winqvist *et al.*, 2012). This can start a negative loop where the decrease in the natural enemy populations is followed by an increase in pest populations which necessitate an increase in pesticide applications, which once again negatively impact natural enemy populations (Sandhu *et al.*, 2008; Geiger *et al.*, 2010; Krauss *et al.*, 2011). This negative loop, where practical protection of the rapeseed yield also ensures the highest possible pest population of *Meligethes aeneus* (Fabricius) for the next year, has been described by Hokkanen (2000). Contrary to productive areas, semi-natural and natural habitats are expected to have a positive impact on biodiversity which also benefit the productive areas, for example through biological control or pollination (Sandhu *et al.*, 2008). The *Millennium Ecosystem Assessment* (World Resources Institute, 2005) distinguishes the following ecosystem functions: (i) supporting services; (ii) provisioning services (e.g. food, pollination); (iii) regulating services (e.g. pest and disease control); and (vi) cultural services. The value of ecosystem services to agriculture is enormous and often underappreciated (Tscharntke *et al.*, 2012; Power, 2014). The consequent use of functional agrobiodiversity might not only break the negative loop but even induce a positive loop (Krauss *et al.*, 2011) where reduction of pesticides leads to an increase in antagonists which in turn leads to further reductions of pesticides.

However, there is still a debate how to integrate biodiversity conservation into production systems and how to best achieve the multiple objectives in agriculture. Balmford *et al.* (2012) describe the two main approaches, land sharing and land sparing. Ecosystem schemes in most European countries (EC, 2005) aim at conserving and promoting general biodiversity in order to mitigate the adverse impact of intense farming on nature (Aviron *et al.*, 2009; Birrer *et al.*, 2014). This 'land-sparing concept' implies that biodiversity is functionally negligible for production systems (Tscharntke *et al.*, 2012). In addition, there is limited interaction between conservation practitioners and agronomists which leads to a large gap in translation of ecosystem services into economical yield increase (Letourneau and Bothwell, 2008; Shanker *et al.*, 2012). Other ecosystem schemes directly aim at shaping and influencing biodiversity within the productive area ('land-sharing concept') with the purpose of providing ecological functions which positively influence agricultural production (functional agrobiodiversity) (Ratnadass *et al.*, 2012; Balmer *et al.*, 2013, 2014). Herzog and Schüepp (2013) underlined the value of nature reserves for the protection of highly sensitive species (land sparing), but also pointed out the relevance of semi-natural habits within production fields (land sharing). With the promotion of outcome-oriented agri-environmental schemes being common in agricultural policy, the implementation of nature conservation measures may be supported (Birrer *et al.*, 2014). Outcome-oriented agri-environmental schemes are directly bound to the outcome of a desired ecosystem service in contrast to action-oriented agri-environmental schemes which prescribe a defined set of management actions (Derissen and Quaas, 2013). Such schemes provide more flexibility for land management decisions and underline the importance of the integration of farmers in nature conservation measures (Birrer *et al.*, 2014).

Influence of landscape complexity and farming system

Structurally complex landscapes with a high plant diversity maintain and preserve high levels of arthropod species and provide a spillover of these species towards crop fields (Molina *et al.*, 2014). Many studies show that herbivore density and crop damage decreases with increasing proportions of non-crop habitats in the landscape (Wezel *et al.*, 2014). Heterogeneous landscapes can sustain higher parasitoid densities than homogeneous production areas (Landis *et al.*, 2000; Tscharntke *et al.*, 2007; Chaplin-Kramer *et al.*, 2011; Rusch *et al.*, 2012). In complex landscapes, higher parasitism rates of pollen beetle *M. aeneus* and lower crop damage were observed than in simple landscapes (Thies and Tscharntke, 1999). Predation and parasitism of *Mamestra brassicae* (Linnaeus) were also found to be related to landscape variables (Bianchi *et al.*, 2005). Interlinking biotope networks with crop and non-crop habitats has a positive effect on abundance and diversity of epigeic predators, such as carabid beetles or spiders, or birds (Pfiffner and Luka, 2000; Östman *et al.*, 2001; Weibull *et al.*, 2003). Approaches to manage non-production areas to create a more biodiverse set of habitats and greater landscape heterogeneity and finally to increase ecosystem services are used by farmers in the USA ('farmscaping'): habitat enhancement through farmscaping increased both biodiversity (particularly plants) and multiple ecosystem functions of agricultural interest (Smukler *et al.*, 2010).

The positive effect of a complex landscape is reinforced by organic farming practices (Östman *et al.*, 2001; Pfiffner and Luka, 2003; Winqvist *et al.*, 2011, 2012): differences in farm structure, pesticide and fertilization regimes, rotations, historical removal of particular landscape elements and differing management strategies (MacFadyen *et al.*, 2009; Puech *et al.*, 2014) result in an increase in conservation biological control on organic farms and a subsequently reduced pest incidence (Östman *et al.*, 2001; Birkhofer *et al.*, 2008; Crowder *et al.*, 2010; Meyling *et al.*, 2013). In particular, the ban of herbicides on organic farms leads to a higher weed biodiversity compared with conventional farms, which also alters species richness and food-web structure (Pfiffner and Luka, 2003; MacFadyen *et al.*, 2009). Organic farming fosters biodiversity of birds, mammals, invertebrates, arable flora (Hole *et al.*, 2005), microbial and faunal decomposers (Birkhofer *et al.*, 2008), and especially beneficial arthropods (MacFadyen *et al.*, 2009; Gomiero *et al.*, 2011; Krauss *et al.*, 2011; Puech *et al.*, 2014) such as spiders and carabid beetles (Pfiffner and Luka, 2003) or parasitoid wasps (MacFadyen *et al.*, 2009). According to Nentwig (2003), a combination of organic farming and semi-natural habitats is important for the conservation and enhancement of species-rich assemblages in an agricultural landscape. Thus, organic farms are harbouring a treasure of high biodiversity and should take special care to conserve it. The preservation and transmission of traditional ecological knowledge is therefore of key importance (Berkes *et al.*, 2000).

Influence of Organic Farming Practices on Abundance of Pest Insects and their Antagonists

Farming practices and cropping systems have their roots in traditional farming and are among the oldest techniques for pest, disease, weed and soil fertility management. They need to be adapted to crops, local climate and soil conditions. Cropping systems – in conventional as well as in organic farming – range from large-scale commercial production in monocultures to highly diversified intercropping systems of subsistence farming (Bajawa and Kogan, 2004). Applied cultural practices therefore vary among different cropping systems. Cultural control practices aim at prevention, avoidance or suppression of pests by creating conditions that are detrimental to the pest or favourable to natural enemies (Hill, 2014). Optimal and expedient implementation of cultural practices requires in-depth knowledge of pest biology and careful long-term planning. Bajawa and Kogan (2004) give a very comprehensive overview on cultural practices for pest control which include: (i) crop rotation; (ii) sanitation; (iii) the use of healthy seed and planting material; (iv) the choice of adapted/resistant/tolerant cultivars; (v) agronomic measures aimed at soil quality and functioning (minimum tillage, animal and green manure, compost); (vi) agronomic measures favouring healthy plant development (irrigation, optimal nutrition, weed management, row spacing); and (vii) adapted timing for planting or harvest in order to disrupt the crop–pest phenological synchrony. Farming practices, such as crop rotation, soil cultivation and fertilization, also have an effect on below-ground functional biodiversity: Differences in rhizobia strains associated with soybean plants were shown to influence honeydew composition of aphids feeding on those plants (Whitaker *et al.*, 2014). This altered honeydew composition can in turn influence the whole above-ground food web of aphid antagonists and mutualists.

Crop rotation

The yield-stabilizing effect of crop rotation has been known for thousands of years: it was practised during the Han dynasty of China, as well as by the Romans and Greeks (Karlen *et al.*, 1994). Many factors, processes and mechanisms contribute to the yield-stabilizing effect of crop rotations: influence of crop rotation on biotic and abiotic soil properties seem most important, but effects on weed control, soil-borne diseases or decreased insect pressure are also contributing factors (Karlen *et al.*, 1994). Crop rotation drastically changes the above and below ground environment and thus increases temporal diversity in an agricultural landscape which again promotes biodiversity. In organic farming, a diverse crop rotation is still a standard cultural practice and an essential part of organic philosophy. Crop rotation for pest control is useful against pests which have a narrow host range and a limited dispersal ability (Karlen *et al.*, 1994). For instance, maize rootworm (*Diabrotica* spp.) is efficiently controlled by a 3-year rotation (Francis and Porter, 2011). Crop rotation and isolation is also an important control method for the cabbage pest *Contarinia nasturtii* (Kieffer), which overwinters in the soil of the previous crop and migrates less than 100 m. In addition, there are indirect effects of crop rotation on pest incidence: legumes in a crop rotation are an important source of nitrogen and nitrogen availability influences susceptibility of plants to pest damage.

Fertilization

Level and source of nitrogen fertilization also have an effect on pest abundance and can promote crop-plant resistance to insect pests (Culliney and Pimentel, 1986) as well as tri-trophic interactions (Banfield-Zanin *et al.*, 2012). In cabbage production, lower densities of flea beetles, aphids and caterpillars were observed on organically manured plants compared with chemically fertilized and unfertilized plants (Culliney and Pimentel, 1986; Arancon *et al.*, 2005). Data indicate that leaf nitrogen, water content, glucosinolate content and plant size may have influenced insect populations (Eigenbrode and Pimentel, 1988; Staley *et al.*, 2009). Glucosinolates content in plants can also affect higher trophic levels, due to reduced host quality and because specialist herbivores may use glucosinolates for their own defence (Hopkins *et al.*, 2009). This might also be one explanation for the observations of Stafford *et al.* (2012) who found that specialist cabbage aphid, *Brevicoryne brassicae* (Linnaeus) performed better on organically fertilized cabbage plants, whereas the generalist green peach aphid, *Myzus persicae* (Sulzer) had a lower performance on organically fertilized plants. Positive effects of organic fertilization were also observed in other crops: in potato production, Colorado potato beetle densities were lower in organically manured fields due to altered mineral content of potato leaves (Alyokhin *et al.*, 2005). Synthetic fertilizers were found to increase sap-feeding insects (aphids, mites, whiteflies) due to increased availability of nitrogen (Garratt *et al.*, 2011). Tri-trophic interactions are also influenced by source of nitrogen: in a meta-analysis, Garratt *et al.* (2011) showed a significant positive effect of organic fertilizers on natural enemy responses. Similar results were obtained by Banfield-Zanin *et al.* (2012), who observed that mortality of ladybird beetle larvae was 10% higher if they fed on aphids on conventionally fertilized compared with aphids on organically fertilized *Brassica* plants. Thus, in organic farming systems, natural enemies may have a higher efficacy than in conventional farming systems. The positive impact of organic fertilizers on natural enemies might be one factor explaining the higher number of natural enemies observed in organic systems (Garratt *et al.*, 2011). Even though higher yields might be possible using mineral fertilizers, this comes at the cost of higher insect levels which necessitate insecticide applications. In view of a system approach, the ban of mineral fertilizers in organic farming seems appropriate. However, it is sometimes challenging for organic farmers to synchronize soil nutrient supply and release in the rhizosphere with the crop nutrient demand (Dorais, 2007), especially if cultivars bred under and adapted for conventional conditions are used.

Tillage and soil cultivation

Minimum tillage is an agronomic measure aimed at soil quality and functioning (Gadermaier *et al.*, 2012). Reduced tillage and organic farming practices have synergistic positive effects on soil biota (Kuntz *et al.*, 2013). Minimum tillage and no tillage (direct seeding) help to reduce erosion, subsoil compaction, nitrate leaching to groundwater, and energy consumption, while increasing soil biota activity, soil organic matter, and thus carbon sequestration (Dorais, 2007; Palm *et al.*, 2014; Wezel *et al.*, 2014). High organic matter and an active soil biology are essential for good soil fertility. Crops growing in these conditions generally show lower abundance of several insect herbivores (Altieri and Nicholls, 2003). Phelan *et al.* (1995) showed that females of European corn borer, *Ostrinia nubilalis* (Hübner) preferred plants in conventional soil for oviposition. Thus, soil-management practices can significantly affect the susceptibility of crops to pests (Lenardis *et al.*, 2014). While minimum tillage seems preferable based on soil quality and pest susceptibility, tillage is often necessary for weed control as well as to accelerate decomposition of crop residues (Dorais, 2007). The destruction of cabbage roots and harvest residues immediately after harvest is a key method to prevent pupation of cabbage root fly larvae, *Delia radicum* (Linnaeus) or lepidopteran pests (*M. brassicae*, *Plutella xylostella* (Linnaeus), *Pieris* sp.). Soil cultivation after harvest and removal of volunteer oilseed rape plants is important to reduce population levels of swede midge (*C. nasturtii*), especially in regions were oilseed rape is grown in close vicinity to production of *Brassica* vegetables. In addition, soil cultivation reduces the risk of diseases that survive on infected debris such as phoma stem canker and light leaf spot. However, there is clearly a conflict of strategies: no tillage is recommended to avoid the spread of clubroot, another major oilseed rape disease, as well as to protect parasitoids of pollen beetles which overwinter as pupae in the soil of previous oilseed rape fields and which are destroyed by ploughing (Nilsson, 2010). All soil cultivation measures (ploughing, non-inversion tillage, superficial soil loosening, mechanical weed control and grass cutting) potentially disturb epigeic predators and lead to an increased mortality and emigration of these insects. Spiders were found to be most vulnerable, but carabid and staphylinid beetles were also reduced (Kromp, 1999; Thorbek and Bilde, 2004; Legrand *et al.*, 2011). However, mechanical weed control in organic farming doesn't perturb the flora like herbicide-using farming systems, which generally leads to a higher weed density and diversity on organic farms. The increased weed density was shown to have a positive effect on carabid beetles in organic wheat fields (Diehl *et al.*, 2012). In addition, a higher weed density interferes with host plant location of specialized pest insects such as *D. radicum*: plants in bare soil are more heavily attacked than plants growing in diverse backgrounds (Finch and Collier, 2000). Thus, possible positive and negative effects of tillage and soil cultivation require a balanced decision based on the observed situation and pest pressure in the field. Potentially negative effects of soil cultivation can be mitigated by refuge areas adjacent to fields and by maintaining crop and landscape diversity.

Host plant resistance/cultivar choice

Cultivar choice has a huge impact on the outbreak of insect pests. However, first criteria for cultivar choice are often market demands and product quality (appearance, taste, nutritional value and health compounds, shelf life and shipping tolerance) (Dorais, 2007). The second criterion is often the resistance to plant pathogens. Pest insect resistance or tolerance usually only play a subordinate role for cultivar choice and is rarely addressed in breeding programmes. This can be partly explained by the fact that pest attacks often occur infrequently and artificial infestation (as often applied in disease screening) is often too time-consuming. Moreover, the plant reaction to pest attack is influenced by very complex interactions and often inherited in a quantitative manner. In general, partial resistance or tolerance might be more effective in the long term than complete resistance, because tolerant cultivars pose a lower selection pressure on pests for forming adapted biotypes (van Emden, 1991) and they can support a certain level of pests and thus maintain antagonist activity. In *Brassica* vegetables and oilseed rape, glucosinolates are produced in the leaves and play a major role in insect–plant interactions (Giamoustaris and Mithen, 1995; Hopkins *et al.*, 2009). Huge differences in glucosinolate content are observed between different cultivars. While high levels of glucosinolates serve as a feeding deterrent for generalist herbivores, they often act as a feeding attractant for specialized herbivores and their predators or parasitoids (Hopkins *et al.*, 2009). Other traits like leaf colour, thickness of wax layer or wax composition can influence susceptibility to pest attack and be used for indirect selection. Voorrips *et al.* (2008) could show that tolerance of cabbage to thrips was related to earliness, Brix and leaf surface wax. In contrast, higher levels of damage caused by lepidopteran pests were reported for *Brassica* genotypes with light green leaves and a reduced wax layer (Eigenbrode and Espelie, 1995). A strong selection for pest tolerance or resistance could result in unintended changes in flavour and taste. In many cases, even more complex defence mechanisms and chemical cues are mediating insect–plant interactions (Bottrell *et al.*, 1998): semiochemicals emitted by plants after damage by herbivores can directly affect the herbivores due to toxic or repellent properties as well as indirectly by attracting natural enemies (Simpson *et al.*, 2013). In addition they can also act as plant-to-plant signals, warning their neighbouring plants. In a comprehensive review, Cortesero *et al.* (2000) summarizes how 'plant attributes influence natural enemy efficiency by providing shelter, mediating host/prey accessibility, providing host/prey finding cues, influencing host/prey suitability, mediating host/prey availability, and providing supplemental food sources for natural enemies'. However, the active contribution of plants for the efficacy of natural enemies has rarely been addressed in breeding programmes. Breeding for conventional farming focuses on increasing the yield under optimized conditions with large external inputs of fertilizers and pesticides which can result in loss of ability to attract natural enemies. This was shown by Degenhardt *et al.* (2009) for maize varieties: modern North American varieties have lost the ability to emit (*E*)-β-caryophyllene which attracts entomopathogenic nematodes that infect and kill the western corn rootworm. Thus, these varieties receive little protection from the nematodes. Currently, organic farming still largely depends on varieties bred by conventional breeders (Lammerts van Bueren *et al.*, 2002). Varieties that fit in the system perspective of organic farming are still lacking. This is a very vulnerable point of the whole system approach. Plant traits especially important for organic farming systems include: (i) adaptation to organic fertilization and crop protection; (ii) a better root system; (iii) ability to interact with beneficial soil microorganisms; (iv) the ability to suppress weeds; and (v) the ability to tolerate pests and diseases (Lammerts van Bueren *et al.*, 2002). Unravelling the underlying genetic and physiological mechanisms for pest tolerance is just at the beginning. Broekgaarden *et al.* (2010) tested two cabbage cultivars for their herbivore community composition throughout the season and found significant differences in resistance level which could be attributed to a high level of RNA expression of potential defence genes. Jyoti *et al.* (2001) tested wild-crop relatives in order to identify genetic resources with improved tolerance against cabbage maggot. Breeding programmes focusing on these traits are urgently needed to fill this gap in the whole system approach. In addition, different cultivar types might be considered for efficient pest control: Instead of cultivating homogeneous F_1 hybrids, open-pollinated populations or cultivar mixtures with different traits and tolerance levels against various pests might be a promising strategy.

Other agronomic measures

Other different agronomic measures are used in order to reduce or avoid pest damage. Certified seed and planting material are a prerequisite for healthy plant development. Adapted timing for planting or harvest can disrupt the crop–pest phenological synchrony: in areas with high pressure of swede midge (*C. nasturtii*), broccoli is produced mainly in spring and autumn instead of summer. During summer, cauliflower, which is less susceptible to swede midge, is produced as a substitute. Damage by autumn oilseed rape pests, such as flea beetles, *Psylloides chrysocephala* (Linnaeus) or *Athalia rosae* (Linnaeus) is diminished by early sowing and by creating conditions favourable for rapid plant development. Measures to create favourable growing conditions and healthy plant development include adjusted irrigation, drainage, optimal nutrition, weed management, or adapted row spacing. Overhead irrigation during evening hours instead of drip irrigation was shown to reduce infestation with *P. xylostella* by more than 85% (McHugh and Foster, 1995), but this strategy is only possible in areas with low pressure of fungal diseases. Increased irrigation – overhead or drip irrigation – can also mitigate damage caused by flea beetles whereas a reduction in irrigation can reduce damage of cabbage fly *D. radicum* because its eggs are highly sensitive to drought. Thus, an overall pest and disease risk assessment is necessary to select suitable agronomic measures for pest prevention. As cultural practices can have opposing effects on different pests and diseases, they need to be adapted according to local pest and disease pressure. This requires a lot of attention and knowledge of the farmers. Adapted cultural practices can also stimulate compensatory plant growth after pest infestation: in cabbage production, seedlings are planted deeper and are earthed up after transplanting in order to stimulate secondary root growth to compensate for damage caused by *D. radicum*. In oilseed rape, favourable growing conditions can stimulate compensatory growth of side shoots after bud damage by pollen beetle (*M. aeneus*) on the main shoot. This can even result in an overcompensation leading to higher yields in fields with moderate pollen beetle incidence compared with fields with low or no pollen beetle incidence (Wahmhoff, 2000). Mechanical weed control can also reduce pest incidence: in oilseed rape hoeing in autumn reduces not only the weeds but also removes the oldest oilseed rape leaves with the highest infestation of flea beetle larvae from the plants (Wahmhoff, 2000). Hoeing in spring has a positive effect on soil temperature and thus on nitrogen mineralization which creates favourable conditions for compensatory growth after pollen beetle infestations (Wahmhoff, 2000). In cabbage production, machines for mechanical weed control can reduce damage by *D. radicum* to a certain extent. Since cultural control practices do not result in eradication of pest insects, they allow the conservation of natural enemies.

Habitat Management at Field Level

The cultivation of crop plants necessarily leads to a simplification of nature's biodiversity and creates artificial ecosystems which need constant human interventions (Altieri, 2007). Habitat management at field level restores a certain level of biodiversity within crop fields and aims at creating conditions favourable to natural enemies (conservation biological control) or detrimental to the pest (push–pull strategy). Apart from positive effects on pest control, habitat management at field level can also improve other ecosystem functions, such as weed control, mitigation of soil erosion, and nutrient cycling (e.g. by fixing atmospheric nitrogen in legume plants) (Simpson *et al.*, 2013).

A vast variety of measures and strategies are used for habitat management at field level (Malézieux *et al.*, 2009; Parolin *et al.*, 2012), for example:

- Intercropping and mixed cropping stands for the simultaneous growing of different harvested crop species in one field.
- Under-sowing crops, often clover, are sown with or after the main crop and are not harvested; their most intensive growth occurs before covering by the main crop or after harvest of the main crop.
- Companion plants are non-crop plants grown within the fields for different purposes: (i) attraction and maintenance of natural enemies by providing pollen and nectar (insectary plants); (ii) repellence and/or interception pest insects (repellent plants); and (iii) influence on nutrition and/or chemical defence of the crop plants (Parolin *et al.*, 2012).
- Banker plants, mainly used in greenhouse production, are a mini-rearing system for natural enemies (Huang *et al.*, 2011). The banker plants supply a non-pest prey (e.g. aphids which infest the banker plant but not the crop plant) and thus sustain the natural enemies within the greenhouse.
- Beetle banks – grass-covered earth banks in the middle of the field – are shelter habitats which provide suitable overwintering sites for predatory carabid and staphylinid beetles or spiders (Jonsson *et al.*, 2008).
- Cover crops are sown after harvest of the main crop before sowing of the new crop mainly to prevent nitrogen leaching and soil erosion.
- Flowering strips usually consist of insectary plants sown at field margins and are aimed to attract natural enemies by providing food and shelter.
- Barrier plants are also sown at field margins and are aimed at intercepting immigrating pest insects (Parolin *et al.*, 2012).
- Trap crops or trap plants are of a preferred growth stage, cultivar or species and thus attract, divert, intercept and/or retain targeted insects because they are more attractive than the main crop (Parolin *et al.*, 2012). Trap crops serve as a sink for insects, preventing the movement of insects to the main crop (Shelton and Nault, 2004). Dead-end trap crops are plants highly attractive to insects, but unsuitable for their reproduction (Shelton and Badenes-Perez, 2006).

All these approaches are applied in different combinations to address different pest problems. A maximum of spatial diversity is created in permaculture or agroforestry with the idea to confront pests with a diverse array of non-host vegetation and thus prevent build-up of pest populations (Francis and Porter, 2011).

Intercropping and cover cropping

In cabbage production, intercropping and cover cropping is implemented as an efficient strategy for *D. radicum* prevention: oviposition of *D. radicum* is significantly reduced in cabbage fields intercropped with clover, because non-host plants interfere with host-plant location of this specialist cabbage pest (Finch and Collier, 2000; Meyling *et al.*, 2013). Reduced pest attacks were reported for cabbage intercropped with onion or tomato (Asare-Bediako *et al.*, 2010). Disruption of host location resulted from the green leaves of the non-host plants, and not from their odours and/or tastes (Finch *et al.*, 2003). The higher weed density observed in organic farming can have a similar effect: plants in bare soil are more heavily attacked by specialist insect pests than plants growing in diverse backgrounds (Finch and Collier,

2000). Similar observations were made by Andow *et al.* (1986) for the specialist cabbage pests *Phyllotreta cruciferae* (Goeze) and *B. brassicae*: cabbage growing in living mulches resulted in lower pest populations than cabbage growing in bare soil. Cover cropping also provides habitat refuges for predators between seasons until the time of cabbage establishment. In addition cover crops prevent soil erosion and help to control weed problems.

Push–pull strategy

This strategy is based on the behavioural manipulation of pest insects: repellent or deterrent companion or intercrop plants within the field 'push' the pest insect from the crop and attractive trap crop plants around the field 'pull' them from the crop (Cook *et al.*, 2007; Khan *et al.*, 2010; Ratnadass *et al.*, 2012). The most important example of a successful application of the push–pull strategy is the stem borer management developed by the International Centre of Insect Physiology and Ecology for African subsistence maize and sorghum production (Khan *et al.*, 2010): Napier grass, (*Pennisetum purpureum* Schumach.) and Sudan grass (*Sorghum vulgare* Pers. var. *sudanense* Hitchc.) are highly attractive for egg laying and pull adult pest insects away from the main crop. At the same time, larval development of the stem borer is very poor in Napier grass, resulting in low survival rates. Legumes of the genus *Desmodium* and molasses grass (*Melinis minutiflora* P. Beauv.) are used as the push component within the fields. Apart from repelling the stem borer, *Desmodium* also suppresses the main weed *Striga hermonthica* (Delile) Benth. by an allelopathic mechanism and has a positive effect on soil quality, whereas molasses grass increased parasitism of stem borer larvae by *Cotesia sesamiae* (Cameron). The economic benefit from the application of this strategy results from an increase in yield by at least 2 t/ha/year but is also due to the fact that the push and pull plants can be used for animal fodder (Cook *et al.*, 2007; Khan *et al.*, 2010).

Trap crops are an important part of push–pull strategies, but can also be used as a single measure (Hokkanen, 1991). In particular, highly mobile insects which cause damage immediately upon immigration in the field (e.g. flea beetles *Phyllotreta* sp., pollen beetles *M. aeneus* in cabbage and oilseed rape production) are good targets for trap-crop strategies (Shelton and Badenes-Perez, 2006). Turnip rape (*Brassica rapa* Linnaeus) sown as a perimeter trap crop around oilseed rape fields is used as a trap crop for pollen beetles. Due to its advanced growth stage, the olfactory and visual cues of turnip rape plants are more attractive to pollen beetles than oilseed rape plants (Cook *et al.*, 2006). Perimeter turnip rape trap crops can significantly reduce pollen beetle populations in the centre of the oilseed rape fields (Büchi, 1989; Büchs and Katzur, 2004; Cook *et al.*, 2004). Currently, a push–pull strategy for pollen beetle is being developed based on this trap crop in combination with within-field application of repellents: different essential oils, such as lavender oil (Mauchline *et al.*, 2013), lemongrass oil or cornmint oil (Daniel, 2014) were shown to have a repellent effect on pollen beetles, but there are still several open questions concerning formulation and application of these oils. Silicate rock dusts also showed a significant repellent effect on pollen beetles (Daniel *et al.*, 2013), but further research is needed to bring a push–pull strategy for pollen beetle control into practice. Turnip rape trap crops are also used around cauliflower and broccoli fields in order to prevent immigration of pollen beetles from neighbouring oilseed rape fields to cauliflower fields. Because immigration occurs shortly before harvest, the use of insecticides is not possible and trap crops are the only option for control (Hokkanen, 1991).

Conservation biological control

Conservation biological control is another major focus of habitat management at field level. Eilenberg *et al.* (2001) defined conservation biological control as 'Modification of the environment or existing practices to protect and enhance specific natural enemies or other organisms to reduce the effect of pests'. Many natural enemies depend on non-host food during parts of their life cycle, for example parasitoids of Lepidoptera need nectar as food during the adult life stage. Flowering strips at field margins, within-field companion plants, intercropping or cover crops provide plant-based food sources (nectar, pollen), alternative hosts and/or honeydew to the natural enemies (Jonsson *et al.*, 2008). In addition, these structures also provide favourable microclimatic conditions, shelter, habitats for hibernation or aestivation and refuge from disturbance caused by agricultural practices (Jonsson *et al.*, 2008). Many authors have shown the benefits of planting flowers near crop production sites for increasing parasitoid densities (Jervis *et al.*, 1996; Heimpel *et al.*, 2004; Lavandero *et al.*, 2005; Winkler *et al.*, 2006; Bianchi and Wäckers, 2008; Pfiffner *et al.*, 2009). The concept of conservation biological control has been readily accepted and implemented by many organic farmers, because it absolutely coincides with organic farming principles, such as biodiversity and biological cycles, and because it is supported by lower fertilization levels, reduced insecticide applications and a higher tolerance to pest infestations (Simpson *et al.*, 2013). Organic farming practices and conservation biological control thus support each other: for instance Ponti *et al.* (2007) observed that both intercropping with buckwheat and mustard and the use of compost instead of mineral fertilizer decreased abundance of the cabbage aphid *B. brassicae* in broccoli. However, vegetation diversification does not necessarily reduce pest insect incidence, because polyphagous pests are able to use a wide range of host plants (Ratnadass *et al.*, 2012). In addition, intraguild predators might also be enhanced by habitat management and can sometimes disrupt biological control (Straub *et al.*, 2008). However, in the majority of cases, conservation of natural enemy biodiversity and biological control are compatible or even complementary goals (Straub *et al.*, 2008), but in-depth knowledge of the biology and requirements of the pest as well as of the antagonists and hyperparasitoids is necessary to develop tailored measures of in-field habitat management. Resources that selectively benefit key natural enemies are needed. Within the next section, we present a case study on the development of conservation biological control in Swiss cabbage production.

Case Study: Development of Conservation Biological Control for Swiss Cabbage Production

Different Lepidoptera larvae (*M. brassicae*, *Pieris brassicae* (Linnaeus), *Pieris rapae* (Linnaeus) and *P. xylostella*) are among the key pests of *Brassica* vegetables (Peacock and Norton, 1990; Cartea *et al.*, 2009; Ahuja *et al.*, 2010). As part of the concept on ecological compensation areas, Swiss farmers established 3500 ha of flower strips (Aviron *et al.*, 2009) using seed mixtures of about 25 species (Pfiffner and Wyss, 2004). This approach is very broad, benefitting biodiversity in general, but it is not focused on the species of agronomic interest (Ratnadass *et al.*, 2012). The effects of these species-rich wildflower strips growing next to or in close vicinity to cabbage fields on the parasitation rate of Lepidoptera larvae was monitored in 2001 and 2002 (Pfiffner *et al.*, 2003). The most abundant parasitoid species were *Microplitis mediator* (Haliday) in *M. brassicae*, *Cotesia rubecula* (Marshall) in *P. rapae* and *Diadegma semiclausum* (Helen) in *P. xylostella* (Pfiffner *et al.*, 2003; Lauro *et al.*, 2005). However, the wildflower strips did not consistently improve the control of *P. rapae* and *M. brassicae* (Pfiffner *et al.*, 2009) because only a few of the 24 plant species in the mixture (e.g. *Centaurea cyanus* Linnaeus, *Fagopyrum esculentum* Moench and *Daucus carota* Linnaeus) might have benefitted the target parasitoids. Other authors have shown that flower strips can increase the reproductive lifespan of *Diadegma* sp. (Winkler *et al.*, 2006; Lee and Heimpel, 2008) as well as parasitation rates in neighbouring cabbage fields (Lee and Heimpel, 2005; Lavandero *et al.*, 2006; Winkler *et al.*, 2006, 2009). Thus, an improvement of conservation biological control might be achieved by targeted selection of flowering species. The most important features of flower species are the attractiveness to parasitoids, nectar accessibility and food quality (Wyss and Pfiffner, 2008).

In order to select plants most suitable for *M. mediator*, several laboratory experiments were conducted: Olfactory attractiveness of five different flowers (bishop's weed *Ammi majus* Linnaeus, cornflower *C. cyanus*, buckwheat *F. esculentum*, candytuft *Iberis amara* Linnaeus, and oregano *Origanum vulgare* Linnaeus) was tested in laboratory Y-tube olfactometer experiments (Belz *et al.*, 2013). *C. cyanus*, *F. esculentum* and *I. amara* were found to be particularly attractive and might therefore be able to recruit *M. mediator*. In addition to a high attractiveness, a suitable flower must also provide accessible nectar in a utilizable quality for the parasitoid. Effects of different nectar sources on fecundity and longevity of *M. mediator* were tested in laboratory experiments (Géneau *et al.*, 2012): nectar from *F. esculentum*, *C. cyanus* and *Vicia sativa* Linnaeus significantly increased fecundity (parasitation rate) and longevity of *M. mediator*, whereas *A. majus* increased only longevity but not fecundity. In addition, *F. esculentum*, *C. cyanus* and *V. sativa* also had a positive effect on longevity of *Diadegma fenestrale* (Holmgren), a generalist parasitoid of lepidopteran pests (Géneau *et al.*, 2012). *F. esculentum* is also known to support *D. semiclausum* (Lavandero *et al.*, 2006). The experiments also showed that *M. mediator* can use the extra-floral nectar of *C. cyanus* and *V. sativa* as a food source (Géneau *et al.*, 2012). The fact that extra-floral nectaries usually produce nectar for a much longer period than floral nectaries makes this two flower species especially interesting for the enhancement of *M. mediator* (Géneau *et al.*, 2012, 2013).

In order to avoid enhancement of the pest insect, the effect of nectar sources on fecundity and longevity of *M. brassicae* was tested in another laboratory experiment: none of the flowers positively influenced fecundity and longevity of *M. brassicae* (Géneau *et al.*, 2012). *P. rapae* is also unlikely to benefit from flowers, because it can only access nectar from very few plant species: during its flight period under sunny and dry conditions, nectar concentration and viscosity is too high for exploitation by Lepidoptera (Winkler *et al.*, 2009). The contrary is true for *P. xylostella* which is mainly active at dusk when relative humidity is high: all plant species suitable for the parasitoid *D. semiclausum* also benefitted *P. xylostella* (Winkler *et al.*, 2009). Different observations were made by Lavandero *et al.* (2006), who observed *F. esculentum* to selectively benefit *D. semiclausum* but not its host *P. xylostella*.

Based on these experiences, *F. esculentum* and *C. cyanus* have been selected for the composition of a tailored wildflower strip. *V. sativa* and *A. majus* were added to the seed mixture in order to have floral and extra-floral nectar available from the end of May until the end of September. Although *A. majus* does not benefit the parasitoids, it is a valuable plant in the mixture because it ensures a soil covering and weed suppression during the summer, as well as a nectar supply for a broad spectrum of beneficials (e.g. hover flies) (Balmer *et al.*, 2013, 2014). In addition to the tailored wildflower strips, cornflowers (*C. cyanus*) were established as companion plants within the cabbage fields in order to provide nectar in closest vicinity to the hosts. The parasitation of *M. brassicae* larvae by *M. mediator* was significantly higher in the presence of within-field companion plants, whereas the distance to the tailored wildflower strip did not affect parasitation of Lepidoptera larvae (Balmer *et al.*, 2013, 2014). Parasitation of *M. brassicae* eggs was significantly increased in the vicinity of the wildflower strip, whereas within-field companion plants had little influence. Larval parasitoids responded more readily to the provision of nectar resources, because their larger body size enables a target-oriented flight, whereas egg parasitoids cover only short distances by active dispersal (Pfiffner *et al.*, 2009). For the reduction of crop damage, however, egg parasitation and egg predation seems more important than larval parasitation, because parasitized larvae still cause a certain amount of damage. The main egg parasitoid observed in the experiments was *Telenomus* sp. (Pfiffner *et al.*, 2009; Balmer *et al.*, 2013, 2014). A laboratory mass rearing and subsequent release of this insect (augmentative biological control) seems therefore an interesting approach to further improve the system. The cabbage yield in the field experiments was positively, but not significantly, influenced by the within-field companion plants: the weight per cabbage head was increased by 18% in the presence of cornflowers.

Egg predation was significantly increased by within-field companion plants but remained unaffected by the distance to the tailored wild flower strip (Balmer *et al.*, 2013, 2014). This is contrary to the observations of Pfiffner *et al.* (2009), who observed a higher egg predation in the vicinity of multi-species flower strips. Wildflower strips are known to harbour many epigeic polyphagous arthropods, such as carabid beetles and spiders (Pfiffner and Luka, 2000; Ditner *et al.*, 2013). Within-field companion plants were also able to shift diversity and species composition of epigeic predators resulting in differences compared with cabbage monocultures (Ditner *et al.*, 2013). Less-specific interventions, like a certain level of weeds, is already beneficial for the predator community (Balmer *et al.*, 2013). The stomach contents of captured predators were analysed by molecular gut analyses (Traugott *et al.*, 2006) in order to identify the main prey species: pest Lepidoptera belonged to the prey spectrum, whereas parasitoid DNA was rarely detected in predator guts (Balmer *et al.*, 2013). This indicates that carabids, staphylinids and spiders do not substantially interfere with parasitoid biocontrol. Hyperparasitoids might be another factor disrupting efficient biological control: Lee and Heimpel (2005) observed hyperparasitation of *Diadegma insulare* (Cresson) by *Conura side* (Walker) in field experiments but the presence of buckwheat did not increase hyperparasitism rates.

Thus, every intervention needs to be tailored and adapted to local situations in order to selectively promote the desired antagonist. If crops are attacked by a complex of insect pests, the implementation of trap cropping and tailored conservation biological control becomes more difficult (Shelton and Badenes-Perez, 2006). In these situations, the use of insecticides which control more than one pest and which is less knowledge intensive and less complex based on agronomical manipulations becomes more attractive. As habitat management strategies do not result in a marketable product, such as an insecticide, research funding is often limited (Shelton and Badenes-Perez, 2006). There are still huge gaps in knowledge, because complex interactions between species, environments and management practices are difficult to research in traditional factorial experimental approaches. A system approach in research and an understanding of the dynamic interactions is needed.

Direct Control Measures

The last two steps in the pyramidal model of organic pest control are inundative biological control (mass release of antagonists or application of biocontrol products) and the use of physical control measures, pheromones or approved insecticides (Fig. 1.1). Contrary to cultural practices and implementation of within-field habitat management, the application of biocontrol agents or bioinsecticides provides the farmers with methods for rapid reactions.

Biological control

Biological control agents are described in detail in Chapter 3 of this volume. They are often used in *Brassica* vegetables: the use of *Bacillus thuringiensis* Berliner var. *kurstaki* and *aizawai* is a very effective direct method against lepidopteran larvae without causing side effects on natural enemies. In the cabbage example above, the rearing and mass release of the egg parasitoid *Telenomus* sp. also belongs to the third step of the pyramidal model.

Physical pest control

Physical methods of pest control include nets, fences, particle films or inert dusts

(Vincent *et al.*, 2003). Crop netting is used in cabbage production against *C. nasturtii*, *D. radicum*, Lepidoptera or flea beetles *Phyllotreta* sp. Although this method is highly efficient, it has the disadvantage of excluding natural enemies from the crop. In particular, problems with cabbage whitefly, *Aleurodes proletella* (Linnaeus) can increase under net covering due to missing antagonists. Crop netting also affects disease outbreaks by increasing humidity. In order to avoid this problem, exclusion fences were developed for *C. nasturtii* and *D. radicum*, two very low-flying insects (Vernon and Mackenzie, 1998). Fences of 1.4 m height with an overhang could reduce damage caused by *C. nasturtii* in broccoli and kohlrabi by 78% (Wyss and Daniel, 2004). The use of inert dusts is also considered to be a physical control method. There are many different kinds of inert dusts: lime, common salt, sand, kaolin, paddy husk ash, wood ash, clays, and diatomaceous earths (Vincent *et al.*, 2003). Silicon compounds are used to strengthen plants and to constitute a barrier against insect feeding (Simpson *et al.*, 2013). In addition, silicon compounds can boost plant volatile production after herbivore infestation which attracts natural enemies (Simpson *et al.*, 2013). In oilseed rape production, the good efficacy of inert dusts (i.e. clinoptilolithe) against pollen beetles was shown to increase yield by 23% (Daniel *et al.*, 2013). Kaolin particle film technology has been developed for fruit production (Daniel *et al.*, 2005) but was recently registered for pollen beetle control in Switzerland (Dorn *et al.*, 2014). The use of sounds and vibrations is another physical pest control method, but examples for efficient applications are still rare. Sound traps as a part of an attract-and-kill strategy are used for mole crickets (Parkman and Frank, 1993). Field efficacy of disruptive vibrational signals for mating disruption was demonstrated for the leafhopper, *Scaphoideus titanus* Ball on grapevine plants (Eriksson *et al.*, 2012).

Natural insecticides

Insecticides for organic farming must meet the standards for organic farming and are therefore of natural origin. They are described in detail in Chapter 2 of this volume. In *Brassica* vegetable production, potassium soap, horticultural oils and pyrethrum are used against aphids. Pyrethrum is also applied against flea beetles. Neem can be used against *A. proletella*, but the efficacy is only sufficient if drop-leg technology for under-leaf application is used. Spinosad is used against different Lepidoptera larvae, thrips, *C. nasturtii* and *D. radicum*. Most natural products (pyrethrum, neem oil, rotenone, nicotine) have a long history of use as insecticides (Isman, 2006; Rosell *et al.*, 2008; Gerwick and Sparks, 2014). Their main advantage lies in their lack of persistence and bioaccumulation in the environment, because they generally degrade faster in sunlight, air and moisture than synthetic products (Grdiša and Gršić, 2013). Compared with conventional pesticides, they are usually more selective to non-target insects (Grdiša and Gršić, 2013). However, some insecticides used in organic farming (such as spinosad, pyrethrum and rotenone) can have detrimental side effects on non-target organisms (Jansen *et al.*, 2010). After application of spinosad against *C. nasturtii* or Lepidoptera, side effects on aphid parasitoids often lead to an increase in aphid infestation (Hommes and Herbst, 2014). Parasitoids of Lepidoptera are also negatively influenced. Thus, all efforts to establish conservation biological control can be annihilated. Crop netting against pests can also have this destabilizing effect by excluding antagonists, but side effects of crop netting are more restricted in time and space than the side effects of spinosad. In order to avoid the negative impact of direct control measures on ecosystem functionality, selective methods for pest control should be preferred and the necessity of applications should be carefully assessed. To date, the limiting factors for organic production of *Brassica* vegetables are *C. nasturtii* and *D. radicum* because no efficient preventive control strategy is available which often makes the application of spinosad necessary.

Economic thresholds to determine the necessity of direct interventions in IPM are often blindly accepted for organic farming systems. However, these thresholds do not reflect the system approach. The presence of antagonists, different fertilization levels, influence of cultivar, and interaction between different pests are not taken into account. In particular, potential negative effects on beneficial insects are missing (i.e. if the treatment of a primary pest leads to extinction of antagonists and thus to the necessity of treatments against secondary pests, a higher economic threshold for the primary pest seems appropriate – from the economic as well as from the ecological point of view). Most economic thresholds are not based on sound experimental data, but are mainly estimations based on expert opinion (El-Wakeil, 2010). In oilseed rape production, the economic threshold for pollen beetle control in the UK is tenfold higher than in Germany or Switzerland (Wahmhoff, 2000). Because of the uncertainties concerning insect population growth rates, most thresholds are rather conservative in order to prevent damage. The challenge of the forthcoming years will be to develop economic thresholds that consider multiple factors but that are still simple to use (El-Wakeil, 2010).

Outlook and Conclusions

According to Francis and Porter (2011), 'Sustainability means preserving economic productivity while taking seriously the ecological foundation and social implications and impacts of farming. It includes designing systems that are resilient and can endure for the indefinite future.' In order to achieve sustainability, the positive impacts of functional agrobiodiversity and of conservation biological control need to be fully exploited. Therefore, agricultural practice needs to be adapted at crucial points: most important is the reduction of non-selective insecticides in order to avoid side effects on beneficial arthropods. This insecticide reduction can only be achieved if robust and adapted cultivars are planted. However, cultivars that fit in the system perspective of organic farming are still lacking which is a very vulnerable point of the whole system approach.

The preventive approach of crop protection requires system-based research, which integrates biological, chemical, physical, ecological, economic and social sciences in a comprehensive way (Lichtfouse *et al.*, 2009) in order to describe relationships and interactions between soil, microorganisms, plants and insects, as well as the influence of agronomic measures on these relationships. Currently research is exploring many interactions, such as plant–microbes–insect, and knowledge is rapidly growing. The challenge in the coming years will be to translate this knowledge into efficient, environmentally safe and economic ways for pest control.

The redesign and development of sustainable cropping systems is very knowledge intensive and requires highly educated and attentive farmers. So far, many farmers are reluctant to change their usual practices, mainly because short-term benefits seem too low and are not easily measurable. As pest problems do not end at farm gates, a closer collaboration between neighbouring farmers could tackle pest problems at a region-wide scale and might increase the impact of conservation biological control and cultural measures. Region-wide control approaches, especially for highly mobile pests, will play a bigger role in future pest control.

Acknowledgements

Many thanks to Martin Koller, Maike Krauss, Monika Messmer and Carolin Möller for help and comments on the manuscript.

References

Ahuja, I., Rohloff, J. and Bones, A.M. (2010) Defence mechanisms of *Brassicaceae*: implications for plant–insect interactions and potential for integrated pest management. A review. *Agronomy for Sustainable Development* 30, 311–348.

Altieri, M.A. (2007) Fatal harvest: old and new dimensions of the ecological tragedy of modern agriculture. In: Nemetz, P.N. (ed.) *Sustainable Resource Management: Reality or Illusion?* Edward Elgar Publishing, Cheltenham, UK, pp. 189–213.

Altieri, M.A. and Nicholls, C.I. (2003) Soil fertility management and insect pests: harmonizing soil and plant health in agroecosystems. *Soil and Tillage Research* 72, 203–211.

Alyokhin, A., Porter, G., Groden, E. and Drummond, F. (2005) Colorado potato beetle response to soil amendments: a case in support of the mineral balance hypothesis? *Agriculture, Ecosystems & Environment* 109, 234–244.

Andow, D.A., Nicholson, A.G., Wien, H.C. and Willson, H.R. (1986) Insect populations on cabbage grown with living mulches. *Environmental Entomology* 15, 293–299.

Arancon, N.Q., Galvis, P.A. and Edwards, C.A. (2005) Suppression of insect pest populations and damage to plants by vermicomposts. *Bioresource Technology* 96, 1137–1142.

Asare-Bediako, E., Addo-Quaye, A.A. and Mohammed, A. (2010) Control of diamondback moth (*Plutella xylostella*) on cabbage (*Brassica oleracea* var. *capitata*) using intercropping with non-host crops. *American Journal of Food Technology* 5, 269–274.

Aviron, S., Nitsch, H., Jeanneret, P., Buholzer, S., Luka, H., Pfiffner, L., Pozzi, S., Schüpbach, B., Walter, T. and Herzog, F. (2009) Ecological cross compliance promotes farmland biodiversity in Switzerland. *Frontiers in Ecology and the Environment* 7, 247–252.

Bajawa, I. and Kogan, M. (2004) Cultural practices: springboard to IPM. In: Koul, O., Dhaliwal, G.S. and Cuperus, G.W. (eds) *Integrated Pest Management: Potential, Constraints and Challenges*. CABI International, Wallingford, UK, pp. 21–38.

Balmer, O., Pfiffner, L., Schied, J., Willareth, M., Leimgruber, A., Luka, H. and Traugott, M. (2013) Noncrop flowering plants restore top-down herbivore control in agricultural fields. *Ecology and Evolution* 3, 2634–2646.

Balmer, O., Géneau, C.E., Belz, E., Weishaupt, B., Förderer, G., Moos, S., Ditner, N., Juric, I. and Luka, H. (2014) Wildflower companion plants increase pest parasitation and yield in cabbage fields: experimental demonstration and call for caution. *Biological Control* 76, 19–27.

Balmford, A., Green, R. and Phalan, B. (2012) What conservationists need to know about farming. *Proceedings of the Royal Society B* 279, 2714–2724.

Banfield-Zanin, J.A., Rossiter, J.T., Wright, D.J., Leather, S.R. and Staley, J.T. (2012) Predator mortality depends on whether its prey feeds on organic or conventionally fertilised plants. *Biological Control* 63, 56–61.

Belz, E., Kölliker, M. and Balmer, O. (2013) Olfactory attractiveness of flowering plants to the parasitoid *Microplitis mediator*: potential implications for biological control. *BioControl* 58, 163–173.

Berkes, F., Colding, J. and Folke, C. (2000) Rediscovery of traditional ecological knowledge as adaptive management. *Ecological Applications* 10, 1251–1262.

Bianchi, F.J.J.A. and Wäckers, F.L. (2008) Effects of flower attractiveness and nectar availability in field margins on biological control by parasitoids. *Biological Control* 46, 400–408.

Bianchi, F.J.J.A., van Wingerden, W.K.R.E., Griffioen, A.J., van der Veen, M., van der Straten, M.J.J., Wegman, R.M.A. and Meeuwsen, H.A.M. (2005) Landscape factors affecting the control of *Mamestra brassicae* by natural enemies in Brussels sprout. *Agriculture, Ecosystems & Environment* 107, 145–150.

Birkhofer, K., Bezemer, T.M., Bloem, J., Bonkowski, M., Christensen, S., Dubois, D., Ekelund, F., Fließbach, A., Gunst, L., Hedlund, K., Mäder, P., Mikola, J., Robin, C., Setälä, H., Tatin-Froux, F., van der Putten, W.H. and Scheu, S. (2008) Long-term organic farming fosters below and aboveground biota: implications for soil quality, biological control and productivity. *Soil Biology & Biochemistry* 40, 2297–2308.

Birrer, S., Zellweger-Fischer, J., Stoeckli, S., Korner-Nievergelt, F., Balmer, O., Jenny, M. and Pfiffner, L. (2014) Biodiversity at the farm scale: a novel credit point system. *Agriculture Ecosystems & Environment* 197, 195–203.

Boller, E.F., Avilla, J., Joerg, E., Malavolta, C., Wijnands, F.G. and Esbjerg, P. (2004) Integrated production: principles and technical guidelines. *IOBC (International Organisation for Biological and Integrated Control) WPRS Bulletin* 27, 1–54.

Bottrell, D.G., Barbosa, P. and Gould, F. (1998) Manipulating natural enemies by plant variety selection and modification: a realistic strategy? *Annual Review of Entomology* 43, 347–367.

Broekgaarden, C., Poelman, E.H., Voorrips, R.E., Dicke, M. and Vosman, B. (2010) Intraspecific variation in herbivore community composition and transcriptional profiles in field-grown *Brassica oleracea* cultivars. *Journal of Experimental Botany* 61, 807–819.

Büchi, R. (1989) Modelle für die Verteilung des Rapsglanzkäfers, *Meligethes aeneus* F., auf Raps und Rübsenpflanzen. *Zeitschrift für angewandte Entomologie* 108, 363–371.

Büchs, W. and Katzur, K. (2004) Cultivation techniques as means to control pests in organic oilseed rape production. *IOBC (International Organisation for Biological and Integrated Control) WPRS Bulletin* 27, 225–236.

Cartea, M.E., Padilla, G., Vilar, M. and Velasco, P. (2009) Incidence of the major Brassica pests in northwestern Spain. *Journal of Economic Entomology* 102, 767–773.

Chaplin-Kramer, R., Rourke, M.E., Blitzer, E.J. and Kremen, C. (2011) A meta-analysis of crop pest and natural enemy response to landscape complexity. *Ecology Letters* 14, 922–932.

Clergue, B., Amiaud, B., Pervanchon, F., Lasserre-Joulin, F. and Plantureux, S. (2009) Biodiversity: function and assessment in agricultural areas: a review. In: Lichtfouse, E., Navarrete, M., Debaeke, P., Souchère, V. and Alberola, C. (eds) *Sustainable Agriculture*. Springer, Dordrecht, The Netherlands, pp. 309–327.

Cook, S.M., Watts, N.P., Hunter, F., Smart, L.E. and Williams, I.H. (2004) Effects of a turnip rape trap crop on the spatial distribution of *Meligethes aeneus* and *Ceutorhynchus assimilis* in oilseed rape. *IOBC (International Organisation for Biological and Integrated Control) WPRS Bulletin* 27, 199–206.

Cook, S.M., Smart, L.E., Martin, J.L., Murray, D.A., Watts, N.P. and Williams, I.H. (2006) Exploitation of host plant preferences in pest management strategies for oilseed rape (*Brassica napus*). *Entomologia Experimentalis et Applicata* 119, 221–229.

Cook, S.M., Khan, Z.R. and Pickett, J.A. (2007) The use of push–pull strategies in integrated pest management. *Annual Review of Entomology* 52, 375–400.

Cortesero, A.M., Stapel, J.O. and Lewis, W.J. (2000) Understanding and manipulating plant attributes to enhance biological control. *Biological Control* 17, 35–49.

Crowder, D.W., Northfield, T.D., Strand, M.R. and Snyder, W.E. (2010) Organic agriculture promotes evenness and natural pest control. *Nature* 466, 109–112.

Culliney, T.W. and Pimentel, D. (1986) Ecological effects of organic agricultural practices on insect populations. *Agriculture, Ecosystems & Environment* 15, 253–266.

Daniel, C. (2014) Olfactometer screening of repellent essential oils against the pollen beetle (*Meligethes* spp.). *IOBC (International Organisation for Biological and Integrated Control) WPRS Bulletin* 104, 79–83.

Daniel, C., Pfamatter, W., Kehrli, P. and Wyss, E. (2005) Processed kaolin as an alternative insecticide against the European pear sucker, *Cacopsylla pyri* (L.). *Journal of Applied Entomology* 129, 363–367.

Daniel, C., Dierauer, H. and Clerc, M. (2013) The potential of silicate rock dust to control pollen beetles (*Meligethes* spp.). *IOBC (International Organisation for Biological and Integrated Control) WPRS Bulletin* 96, 47–55.

Degenhardt, J., Hiltpold, I., Köllner, T.G., Frey, M., Gierl, A., Gershenzon, J., Hibbard, B.E., Ellersieck, M.R. and Turlings, T.C.J. (2009) Restoring a maize root signal that attracts insect-killing nematodes to control a major pest. *Proceedings of the National Academy of Sciences USA* 106, 13213–13218.

Derissen, S. and Quaas, M.F. (2013) Combining performance-based and action-based payments to provide environmental goods under uncertainty. *Ecological Economics* 85, 77–85.

Diehl, E., Wolters, V. and Birkhofer, K. (2012) Arable weeds in organically managed wheat fields foster carabid beetles by resource- and structure-mediated effects. *Arthropod–Plant Interactions* 6, 75–82.

Ditner, N., Balmer, O., Beck, J., Blick, T., Nagel, P. and Luka, H. (2013) Effects of experimentally planting non-crop flowers into cabbage fields on the abundance and diversity of predators. *Biodiversity and Conservation* 22, 1049–1061.

Dorais, M. (2007) Organic production of vegetables: state of the art and challenges. *Canadian Journal of Plant Science* 87, 1055–1066.

Dorn, B., Jossi, W., Humphrys, C. and Hiltbrunner, J. (2014) Screening of natural products in the laboratory and the field for control of pollen beetles. *Journal of Applied Entomology* 138, 109–119.

Ehler, L.E. (2006) Integrated pest management (IPM): definition, historical development and implementation, and the other IPM. *Pest Management Science* 62, 787–789.

Eigenbrode, S.D. and Espelie, K.E. (1995) Effects of plant epicuticular lipids on insect herbivores. *Annual Review of Entomology* 40, 171–194.

Eigenbrode, S.D. and Pimentel, D. (1988) Effects of manure and chemical fertilizers on insect pest populations on collards. *Agriculture, Ecosystems & Environment* 20, 109–125.

Eilenberg, J., Hajek, A. and Lomer, C. (2001) Suggestions for unifying the terminology in biological control. *BioControl* 46, 387–400.

El-Wakeil, N.E. (2010) Insect economic levels in relation to crop production. *Archives of Phytopathology and Plant Protection* 43, 1711–1745.

Eriksson, A., Anfora, G., Lucchi, A., Lanzo, F., Virant-Doberlet, M. and Mazzoni, V. (2012) Exploitation of insect vibrational signals reveals a new method of pest management. *PLoS One* 7, e32954.

European Commission (EC) (2005) *Agri-environment Measures – Overview on General Priciples, Types of Measures and Application*. European Comission, Directorate General for Agriculture and Rural Development, 24 pp.

European Commission (EC) (2007) European communities council regulation (EC) No. 834/2007 of 28 June 2007 on organic production and labelling of organic products and repealing Regulation (EEC) No. 2092/91. *Official Journal of the European Union* L 189, 1–23.

European Commission (EC) (2008) European communities commission regulation (EC) No. 889/2008 of 05 September 2008 laying down detailed rules for the implementation of Council Regulation (EC) No. 834/2007 on organic production and labelling of organic products with regard to organic production, labelling and control. *Official Journal of the European Union* L 250, 1–84.

European Commission (EC) (2009) Directive 2009/128/EC of the European Parliament and of the Council of 21 October 2009 establishing a framework for Community action to achieve the sustainable use of pesticides. *Official Journal of the European Union* L 309, 71–86.

Finch, S. and Collier, R.H. (2000) Host-plant selection by insects – a theory based on 'appropriate/inappropriate landings' by pest insects of cruciferous plants. *Entomologia Experimentalis et Applicata* 96, 91–102.

Finch, S., Billiald, H. and Collier, R.H. (2003) Companion planting – do aromatic plants disrupt host-plant finding by the cabbage root fly and the onion fly more effectively than non-aromatic plants? *Entomologia Experimentalis et Applicata* 109, 183–195.

Food and Agriculture Organization of the United Nations (FAO) and World Health Organization (WHO) (2007) *Codex Alimentarius: Organically Produced Foods*, 3rd edn. FAO, Rome, 63 pp.

Forster, D., Adamtey, N., Messmer, M.M., Pfiffner, L., Baker, B., Huber, B. and Niggli, U. (2013) Organic agriculture-driving innovations in crop research. In: Bhullar, G.S. and Bhullar, N.K. (eds) *Agricultural Sustainability – Progress and Prospects in Crop Research*. Elsevier Academic Press, London, pp. 21–45.

Francis, C.A. and Porter, P. (2011) Ecology in sustainable agriculture practices and systems. *Critical Reviews in Plant Sciences* 30, 64–73.

Gadermaier, F., Berner, A., Fließbach, A., Friedel, J.K. and Mäder, P. (2012) Impact of reduced tillage on soil organic carbon and nutrient budgets under organic farming. *Renewable Agriculture and Food Systems* 27(1), 68–80.

Garratt, M.P.D., Wright, D.J. and Leather, S.R. (2011) The effects of farming system and fertilisers on pests and natural enemies: a synthesis of current research. *Agriculture, Ecosystems & Environment* 141, 261–270.

Geiger, F., Bengtsson, J., Berendse, F., Weisser, W., Emmerson, M., Morales, M., Ceryngier, P., Liira, J., Tscharntke, T. and Winqvist, C. (2010) Persistent negative effects of pesticides on biodiversity and biological control potential on European farmland. *Basic and Applied Ecology* 11, 97–105.

Géneau, C.E., Wäckers, F.L., Luka, H., Daniel, C. and Balmer, O. (2012) Selective flowers to enhance biological control of cabbage pests by parasitoids. *Basic and Applied Ecology* 13, 85–93.

Géneau, C.E., Wäckers, F.L., Luka, H. and Balmer, O. (2013) Effects of extrafloral and floral nectar of *Centaurea cyanus* on the parasitoid wasp *Microplitis mediator*: olfactory attractiveness and parasitization rates. *Biological Control* 66, 16–20.

Gerwick, B.C. and Sparks, T.C. (2014) Natural products for pest control: an analysis of their role, value and future. *Pest Management Science* 70, 1169–1185.

Giamoustaris, A. and Mithen, R. (1995) The effect of modifying the glucosinolate content of leaves of oilseed rape (*Brassica napus* ssp. *oleifera*) on its interaction with specialist and generalist pests. *Annals of Applied Biology* 126, 347–363.

Gomiero, T., Pimentel, D. and Paoletti, M.G. (2011) Environmental impact of different agricultural management practices: conventional vs. organic agriculture. *Critical Reviews in Plant Sciences* 30, 95–124.

Grdiša, M. and Gršić, K. (2013) Botanical insecticides in plant protection. *Agriculturae Conspectus Scientificus* 78, 85–93.

Gruys, P. (1982) Hits and misses. The ecological approach to pest control in orchards. *Entomologia Experimentalis et Applicata* 31, 70–87.

Heimpel, G.E., Lee, J.C., Wu, Z., Weiser, L., Wäckers, F. and Jervis, M.A. (2004) Gut sugar analysis in field-caught parasitoids: adapting methods originally developed for biting flies. *International Journal of Pest Management* 50, 193–198.

Herzog, F. and Schüepp, C. (2013) Are land sparing and land sharing real alternatives for European agricultural landscapes? *Aspects of Applied Biology* 121, 109–116.

Hill, S.B. (2014) Considerations for enabeling the ecological redesign of organic and conventional agriculture: a social ecology and psychological perspective. In: Bellon, S. and Penvern, S. (eds) *Organic Farming, Prototype for Sustainable Agricultures*. Springer, Dordrecht, The Netherlands, pp. 401–422.

Hokkanen, H.M. (1991) Trap cropping in pest management. *Annual Review of Entomology* 36, 119–138.
Hokkanen, H.M.T. (2000) The making of a pest: recruitment of *Meligethes aeneus* onto oilseed *Brassicas*. *Entomologia Experimentalis et Applicata* 95, 141–149.
Hole, D.G., Perkins, A.J., Wilson, J.D., Alexander, I.H., Grice, P.V. and Evans, A.D. (2005) Does organic farming benefit biodiversity? *Biological Conservation* 122, 113–130.
Hommes, M. and Herbst, M. (2014) Supervised control of aphids and caterpillars on white cabbage and impact of flower strips. *IOBC (International Organisation for Biological and Integrated Control) WPRS Bulletin* 107, 43–49.
Hopkins, R.J., van Dam, N.M. and van Loon, J.J.A. (2009) Role of glucosinolates in insect–plant relationships and multitrophic interactions. *Annual Review of Entomology* 54, 57–83.
Huang, N., Enkegaard, A., Osborne, L.S., Ramakers, P.M., Messelink, G.J., Pijnakker, J. and Murphy, G. (2011) The banker plant method in biological control. *Critical Reviews in Plant Sciences* 30, 259–278.
International Federation of Organic Agriculture Movements (IFOAM) (2012) *The IFOAM Norms for Organic Production and Processing*. IFOAM, Bonn, Germany, 134 pp.
Isman, M.B. (2006) Botanical insecticides, deterrents and repellents in modern agriculture and an increasingly regulated world. *Annual Review of Entomology* 51, 45–66.
Jansen, J.P., Defrance, T. and Warnier, A.M. (2010) Effects of organic-farming-compatible insecticides on four aphid natural enemy species. *Pest Management Science* 66, 650–656.
Jervis, M.A., Kidd, N.A.C. and Heimpel, G.E. (1996) Parasitoid adult feeding behaviour and biocontrol – a review. *Biocontrol News and Information* 17, 11–26.
Jonsson, M., Wratten, S.D., Landus, D.A. and Gurr, G.M. (2008) Recent advances in conservation biological control of arthropods by arthropods. *Biological Control* 45, 172–175.
Jyoti, J.L., Shelton, A.M. and Earle, E.D. (2001) Identifying sources and mechanims of resistance in crucifers for control of cabbage maggot (Diptera: Anthomyiidae). *Journal of Economic Entomology* 94, 942–949.
Karlen, D.L., Varvel, G.E., Bullock, D.G. and Cruse, R.M. (1994) Crop rotations for the 21st century. In: Donald, L.S. (ed.) *Advances in Agronomy* 53, Academic Press, London, pp. 1–45.
Khan, Z.R., Midega, C.A.O., Bruce, T.J.A., Hooper, A.M. and Pickett, J.A. (2010) Exploiting phytochemicals for developing a 'push–pull' crop protection strategy for cereal farmers in Africa. *Journal of Experimental Botany* 61, 4185–4196.
Kogan, M. (1998) Integrated pest management: historical perspectives and contemporary developments. *Annual Review of Entomology* 43, 243–270.
Krauss, J., Gallenberger, I. and Steffan-Dewenter, I. (2011) Decreased functional diversity and biological pest control in conventional compared to organic crop fields. *PLoS One* 6, e19502.
Kromp, B. (1999) Carabid beetles in sustainable agriculture: a review on pest control efficacy, cultivation impacts and enhancement. *Agriculture Ecosystems & Environment* 74, 187–228.
Kuntz, M., Berner, F., Gattinger, A., Scholberg, J.M., Mäder, P. and Pfiffner, L. (2013) Influence of reduced tillage on earthworm and microbial communities under organic arable farming. *Pedobiologia* 56, 251–260.
Lammerts van Bueren, E.T., Struik, P.C. and Jacobsen, E. (2002) Ecological concepts in organic farming and their consequences for an organic crop ideotype. *NJAS – Wageningen Journal of Life Sciences* 50, 1–26.
Landis, D.A., Wratten, S.D. and Gurr, G.M. (2000) Habitat management to conserve natural enemies of arthropod pests in agriculture. *Annual Review of Entomology* 45, 175–201.
Lauro, N., Kuhlmann, U., Mason, P.G. and Holliday, N.J. (2005) Interaction of a solitary larval endoparasitoid, *Microplitis mediator*, with its host, *Mamestra brassicae*: host acceptance and host suitability. *Agriculture and Agri-Food Canada* 129, 567–573.
Lavandero, B., Wratten, S.D., Shishehbor, P. and Worner, S. (2005) Enhancing the effectiveness of the parasitoid *Diadegma semiclausum* (Helen): movement after use of nectar in the field. *Biological Control* 34, 152–158.
Lavandero, B., Wratten, S.D., Didham, R.K. and Gurr, G. (2006) Increasing floral diversity for selective enhancement of biological control agents: a double-edged sward? *Basic and Applied Ecology* 7, 236–243.
Lee, J.C. and Heimpel, G.E. (2005) Impact of flowering buckwheat on Lepidopteran cabbage pests and their parasitoids at two spatial scales. *Biological Control* 34, 290–301.
Lee, J.C. and Heimpel, G.E. (2008) Floral resources impact longevity and oviposition rate of a parasitoid in the field. *Journal of Animal Ecology* 77, 565–572.
Legrand, A., Gaucherel, C., Baudry, J. and Meynard, J.M. (2011) Long-term effects of organic, conventional, and integrated crop systems on Carabids. *Agronomy for Sustainable Development* 31, 515–524.
Lenardis, A.E., Szpeiner, A. and Ghersa, C.M. (2014) Arthropod assemblage related to volatile cues in flowering wheat: interaction between aphid herbivory and soil conditions as induction factors. *Environmental Entomology* 43, 448–457.

Letourneau, D.K. and Bothwell, S.G. (2008) Comparison of organic and conventional farms: challenging ecologists to make biodiversity functional. *Frontiers in Ecology and the Environment* 6, 430–438.

Lichtfouse, E., Navarrete, M., Debaeke, P., Souchère, V., Alberola, C. and Ménassieu, J. (2009) Agronomy for sustainable agriculture. A review. *Agronomy for Sustainable Development* 29, 1–6.

MacFadyen, S., Gibson, R., Polaszek, A., Morris, R.J., Craze, P.G., Planqué, R., Symondson, W.O.C. and Memmott, J. (2009) Do differences in food web structure between organic and conventional farms affect the ecosystem service of pest control? *Ecology Letters* 12, 229–238.

Malézieux, E., Crozat, Y., Dupraz, C., Laurans, M., Makowski, D., Ozier-Lafontaine, H., Rapidel, B., de Tourdonnet, S. and Valantin-Morison, M. (2009) Mixing plant species in cropping systems: concepts, tools and models: a review. In: Lichtfouse, E., Navarrete, M., Debaeke, P., Souchère, V. and Alberola, C. (eds) *Sustainable Agriculture*. Springer, Dordrecht, The Netherlands, pp. 329–353.

Mauchline, A.L., Cook, S.M., Powell, W. and Osborne, J.L. (2013) Effects of non-host plant odour on *Meligethes aeneus* during immigration to oilseed rape. *Entomologia Experimenatlis et Applicata* 146, 313–320.

McHugh, J.J. and Foster, R.E. (1995) Reduction of diamondback moth (Lepidoptera: Plutellidae) infestation in head cabbage by overhead irrigation. *Journal of Economic Entomology* 88, 162–168.

Meyling, N.V., Navntoft, S., Philipsen, H., Thorup-Kristensen, K. and Eilenberg, J. (2013) Natural regulation of *Delia radicum* in organic cabbage production. *Agriculture Ecosystems & Environment* 164, 183–189.

Molina, G.A., Poggio, S.L. and Ghersa, C.M. (2014) Epigeal arthropod communities in intensively farmed landscapes: effects of land use mosaics, neighbourhood heterogeneity, and field position. *Agriculture Ecosystems & Environment* 192, 135–143.

Moonen, A.-C. and Bàrberi, P. (2008) Functional biodiversity: an agroecosystem approach. *Agriculture, Ecosystems & Environment* 127, 7–21.

Naranjo, S.E. and Ellsworth, P.C. (2009) The contribution of conservation biological control to integrated control of *Bemisia tabaci* in cotton. *Biological Control* 51, 458–470.

Nentwig, W. (2003) Management of biodiversity in agroecosystems. *Basic and Applied Ecology* 4, 105–106.

Nilsson, C. (2010) Impact of soil tillage on parasitoids of oilseed rape pests. In: Williams, I.H. (ed.) *Biocontrol-based Integrated Management of Oilseed Rape Pests*. Springer, Dordrecht, The Netherlands, pp. 305–312.

Östman, Ö., Ekbom, B., Bengtsson, J. and Weibull, A.C. (2001) Landscape complexity and farming practice influence the condition of polyphagous carabid beetles. *Ecological Applications* 11, 480–488.

Palm, C., Blanco-Canqui, H., DeClerck, F., Gatere, L. and Grace, P. (2014) Conservation agriculture and ecosystem services: an overview. *Agriculture, Ecosystems & Environment* 187, 87–105.

Parkman, J.P. and Frank, J.H. (1993) Use of a sound trap to inoculate *Steinernema scapterisci* (Rhabditida, Steinernematidae) into pest mole cricket populations (Orthoptera, Gryllotalpidae). *Florida Entomologist* 76, 75–82.

Parolin, P., Bresch, C., Desneux, N., Brun, R., Bout, A., Boll, R. and Poncet, C. (2012) Secondary plants used in biological control: a review. *International Journal of Pest Management* 58, 91–100.

Peacock, L. and Norton, G.A. (1990) A critical analysis of organic vegetable crop protection in the UK. *Agriculture, Ecosystems & Environment* 31, 187–197.

Pfiffner, L. and Luka, H. (2000) Overwintering of arthropods in soils of arable fields and adjacent semi-natural habitats. *Agriculture, Ecosystems & Environment* 78, 215–222.

Pfiffner, L. and Luka, H. (2003) Effects of low-input farming systems on carabids and epigeal spiders – a paired farm approach. *Basic and Applied Ecology* 4, 117–127.

Pfiffner, L. and Wyss, E. (2004) Use of sown wildflower strips to enhance natural enemies of agricultural pests. In: Gurr, G.M., Wratten, S.D. and Altieri, M.A. (eds) *Ecological Engineering for Pest Management – Advances in Habitat Manipulation for Arthopods*. CSIRO Publishing, Collingwood, Australia, pp. 165–186.

Pfiffner, L., Merkelbach, L. and Luka, H. (2003) Do sown wildflower strips enhance the parasitism of lepidopteran pests in cabbage crops? *IOBC (International Organisation for Biological and Integrated Control) WPRS Bulletin* 26, 111–116.

Pfiffner, L., Luka, H., Schlatter, C., Juen, A. and Traugott, M. (2009) Impact of wildflower strips on biological control of cabbage lepidopterans. *Agriculture, Ecosystems & Environment* 129, 310–314.

Phelan, P.L., Mason, J.F. and Stinner, B.R. (1995) Soil-fertility management and host preference by European corn borer, *Ostrinia nubilalis* (Hübner), on *Zea mays* L.: a comparison of organic and conventional chemical farming. *Agriculture, Ecosystems & Environment* 56, 1–8.

Ponti, L., Altieri, M.A. and Gutierrez, A.P. (2007) Effects of crop diversification levels and fertilization regimes on abundance of *Brevicoryne brassicae* (L.) and its parasitization by *Diaeretiella rapae* (M'Intosh) in broccoli. *Agricultural and Forest Entomology* 9, 209–214.

Power, A.G. (2014) Ecosystem services and agriculture: tradeoffs and synergies. *Philosophical Transactions of the Royal Society B* 365, 2959–2971.

Puech, C., Baudry, J., Joannon, A., Poggi, S. and Aviron, S. (2014) Organic vs. conventional farming dichotomy: does it make sense for natural enemies? *Agriculture, Ecosystems & Environment* 194, 48–57.

Ratnadass, A., Fernandes, P., Avelino, J. and Habib, R. (2012) Plant species diversity for sustainable management of crop pests and diseases in agroecosystems: a review. *Agronomy for Sustainable Development* 32, 273–303.

Rosell, G., Quero, C., Coll, J. and Guerrero, A. (2008) Biorational insecticides in pest management. *Journal of Pesticide Science* 33, 103–121.

Rusch, A., Valantin-Morison, M., Roger-Estrade, J. and Sarthou, J.P. (2012) Using landscape indicators to predict high pest infestations and successful natural pest control at the regional scale. *Landscape and Urban Planning* 105, 62–73.

Sandhu, H.S., Wratten, S.D., Cullen, R. and Case, B. (2008) The future of farming: the value of ecosystem services in conventional and organic arable land. An experimental approach. *Ecological Economics* 64, 835–848.

Shanker, C., Katti, G., Padmakumari, A.P., Padmavathi, C. and Sampathkumar, M. (2012) Biological control, functional biodiversity and ecosystem services in insect pest management. In: Venkateswarlu, B., Shanker, A.K., Shanker, C. and Maheswari, M. (eds) *Crop Stress and Its Management: Perspectives and Strategies*. Springer, Dordrecht, The Netherlands, pp. 471–495.

Shelton, A.M. and Badenes-Perez, F.R. (2006) Concepts and applications of trap cropping in pest management. *Annual Review of Entomology* 51, 285–308.

Shelton, A.M. and Nault, B.A. (2004) Dead-end trap cropping: a technique to improve management of the diamondback moth, *Plutella xylostella* (Lepidoptera: Plutellidae). *Crop Protection* 23, 497–503.

Simpson, M., Read, D.M.Y. and Gurr, G.M. (2013) Application of chemical cues in arthropod pest management for organic crops. In: Wajnberg, E. and Colazza, S. (eds) *Chemical Ecology of Insect Parasitoids*. Wiley-Blackwell, Oxford, pp. 266–281.

Smukler, S.M., Jackson, L.E., Murphree, L., Yokota, R., Koike, S.T. and Smith, R.F. (2008) Transition to large-scale organic vegetable production in the Salinas Valley, California. *Agriculture Ecosystems & Environment* 126, 168–188.

Smukler, S.M., Sanchez-Moreno, S., Fonte, S.J., Ferris, H., Klonsky, K., O'Geen, A.T., Scow, K.M., Steenwerth, K.L. and Jackson, L.E. (2010) Biodiversity and multiple ecosystem functions in an organic farmscape. *Agriculture Ecosystems & Environment* 139, 80–97.

Stafford, D.B., Tariq, M., Wright, D.J., Rossiter, J.T., Kazana, E., Leather, S.R., Ali, M. and Staley, J.T. (2012) Opposing effects of organic and conventional fertilizers on the performance of a generalist and a specialist aphid species. *Agricultural and Forest Entomology* 14, 270–275.

Staley, J.T., Stewart-Jones, A., Poppy, G.M., Leather, S.R. and Wright, D.J. (2009) Fertilizer affects the behaviour and performance of *Plutella xylostella* on brassicas. *Agricultural and Forest Entomology* 11, 275–282.

Straub, C.S., Finke, D.L. and Snyder, W.E. (2008) Are the conservation of natural enemy biodiversity and biological control compatible goals? *Biological Control* 45, 225–237.

Thies, C. and Tscharntke, T. (1999) Landscape structure and biological control in agroecosystems. *Science* 285, 893–895.

Thorbek, P. and Bilde, T. (2004) Reduced numbers of generalist arthropod predators after crop management. *Journal of Applied Ecology* 41, 526–538.

Traugott, M., Zangerl, P., Juen, A., Schallhart, N. and Pfiffner, L. (2006) Detecting key parasitoids of lepidopteran pests by multiplex PCR. *Biological Control* 39, 39–46.

Tscharntke, T., Bommarco, R., Clough, Y., Crist, T.O., Kleijn, D., Rand, T.A., Tylianakis, J.M., van Nouhuys, S. and Vidal, S. (2007) Conservation biological control and enemy diversity on a landscape scale. *Biological Control* 43, 294–309.

Tscharntke, T., Clough, Y., Wanger, T.C., Jackson, L., Motzke, I., Perfecto, I., Vandermeer, J. and Whitbread, A. (2012) Global food security, biodiversity conservation and the future of agricultural intensification. *Biological Conservation* 151, 53–59.

United Nations (UN) (1992) *Convention on Biological Diversity*. UN, Rio de Janeiro.

van Emden, H.F. (1991) The role of host plant resistance in insect pest mismanagement. *Bulletin of Entomological Research* 81, 123–126.

Vernon, S. and Mackenzie, J.R. (1998) The effect of exclusion fences on the colonization of rutabagas by cabbage flies (Diptera: Anthomyidae). *The Canadian Entomologist* 130, 153–162.

Vincent, C., Hallman, G., Panneton, B. and Fleurat-Lessard, F. (2003) Management of agricultural insects with physical control methods. *Annual Review of Entomology* 48, 261–281.

Voorrips, R.E., Steenhuis-Broers, G., Tiemens-Hulscher, M. and Lammerts van Bueren, E.T. (2008) Plant traits associated with resistance to *Thrips tabaci* in cabbage (*Brassica oleracea* var. *capitata*). *Euphytica* 163, 409–415.

Wahmhoff, W. (2000) *Integrierter Rapsanbau: Untersuchungen zur Entwicklung integrierter Produktionsverfahren am Beispiel des Winterrapses (*Brassica napus *L.)*. Erich Schmidt Verlag, Berlin, 284 pp.

Weibull, A.C., Östman, Ö. and Granqvist, Å. (2003) Species richness in agroecosystems: the effect of landscape, habitat and farm management. *Biodiversity and Conservation* 12, 1335–1355.

Wezel, A., Casagrande, M., Celette, F., Vian, J.F., Ferrer, A. and Peigné, J. (2014) Agroecological practices for sustainable agriculture. A review. *Agronomy for Sustainable Development* 34, 1–20.

Whitaker, M.R.L., Katayama, N. and Ohgushi, T. (2014) Plant–rhizobia interactions alter aphid honeydew composition. *Arthropod–Plant Interactions* 8, 213–220.

Winkler, K., Wäckers, F., Bukovinszkine-Kiss, G. and van Lenteren, J. (2006) Sugar resources are vital for *Diadegma semiclausum* fecundity under field conditions. *Basic and Applied Ecology* 7, 133–140.

Winkler, K., Wäckers, F.L., Kaufman, L.V., Larraz, V. and van Lenteren, J.C. (2009) Nectar exploitation by herbivores and their parasitoids is a function of flower species and relative humidity. *Biological Control* 50, 299–306.

Winqvist, C., Bengtsson, J., Aavik, T., Berendse, F., Clement, L.W., Eggers, S., Fischer, C., Flohre, A., Geiger, F., Liira, J., Pärt, T.T., Thies, C., Tscharntke, T., Weisser, W.W. and Bommarco, R. (2011) Mixed effects of organic farming and landscape complexity on farmland biodiversity and biological control potential across Europe. *Journal of Applied Ecology* 48, 570–579.

Winqvist, C., Ahnstrom, J. and Bengtsson, J. (2012) Effects of organic farming on biodiversity and ecosystem services: taking landscape complexity into account. *Annals of the New York Academy of Sciences* 1249, 191–203.

World Resources Institute (2005) *Millennium Ecosystem Assessment 2005, Ecosystems and Human Well-Being*. Island Press, Washington, DC.

Wyss, E. and Daniel, C. (2004) The effect of exclusion fences on the colonization of broccoli and kohlrabi by the swede midge, *Contarinia nasturtii* (Diptera: Cecidomyiidae). *Mitteilungen der Deutschen Gesellschaft für Allgemeine und Angewandte Entomologie* 14, 387–390.

Wyss, E. and Pfiffner, L. (2008) Biodiversity in organic horticulture – an indicator for sustainability and a tool for pest management. *Acta Horticulturae* 767, 75–80.

Wyss, E., Luka, H., Pfiffner, L., Schlatter, C., Uehlinger, G. and Daniel, C. (2005) Approaches to pest management in organic agriculture: a case study in European apple orchards. Presented at Symposium 'IPM in Organic Systems', XXII International Congress of Entomology, Brisbane, Australia, 16 August 2004, 33N–36N. CAB International, Organic-Research.com May 2005. Available at: https://www.researchgate.net/publication/262002057 (accessed 29 May 2017).

Zehnder, G., Gurr, G.M., Kühne, S., Wade, M.R., Wratten, S.D. and Wyss, E. (2007) Arthropod pest management in organic crops. *Annual Review of Entomology* 52, 57–80.

2 Plant Protection Tools in Organic Farming

Massimo Benuzzi* and Edith Ladurner
CBC (Europe) S.r.l. – BIOGARD Division, Cesena, Italy

Introduction

Organic plant protection tools are all those biological control tools and products of natural origin which control pests and diseases in agriculture, and are not synthetized via chemical processes. We will see later on that this definition is not truly correct, because it is not always easy to exactly define the actual difference between a product of 'natural origin' and a product obtained by chemical synthesis. In addition, frequently certifying bodies and even organic growers themselves do not take into account the importance of a formulated product allowed in organic farming being registered and authorized for a specified use within the country, in the same way as a 'normal' conventional plant protection product. Usually this is obviously linked with operational limits and may not be accepted by everybody, especially those who produce organically more for the philosophy behind it than for the aim of combining ecology and economy. Nevertheless, if organic farming should be considered the 'most controlled and secure' farming method, how can any competitive advantage be claimed if the production tools themselves are not equally controlled and certified? For this reason, in this chapter, for straightforwardness, we will stick to the more generic definition of 'organic plant protection tool' given by the International Federation of Organic Agriculture Movements (IFOAM), which we consider to be correct, and that is: *Organic production and processing systems are based on the use of natural, biological, renewable, and regenerative resources.*

A clear legislative definition of methods and tools allowed in organic production systems has not yet been identified, and differences among countries exist. In fact, something certified as organic in the USA may not be certified as such according to the regulations of the European Union (EU) or other countries, for example Switzerland, and may not be in compliance with the *IFOAM Norms for Organic Production and Processing* (IFOAM, 2014). In short, no written law, which applies everywhere, does exist. No 'worldwide' organic production exists! For this reason, we will consider 'organic' as what is organic according to EU regulations. Where deemed necessary, we will indicate the differences among countries by referring to the specific national regulations. In theory, according to EU regulations, identifying the plant protection products allowed for use in

* Corresponding author: mbenuzzi@cbceurope.it

organic farming systems is extremely simple. In fact, all the active substances which are approved for plant protection in organic farming within the EU are listed in Annex II of Reg. (EC) No. 889/2008, laying down detailed rules for the implementation of Reg. (EU) No. 834/2007, and subsequent amendments (last revision: Commission Implementing Regulation (EU) 2016/673 of 29 April 2016).

Each active substance must obviously comply also with the national laws of each Member State, and therefore each plant protection product must be authorized for the intended use also on a national level, except where specific indications exist that this requirement does not apply. For example, in some Member States products based on active substances, such as beeswax and quartz sand, are not considered plant protection products. In some other Member States, such as Spain, Germany and Italy, specific laws exist, which allow for the use of so-called 'plant strengtheners'. Finally, we would like to draw attention to an incongruity at the basis of the above-mentioned EU regulation. This regulation contains a sentence, which cannot be considered applicable by any plant protection expert, particularly if specialized in organic farming. Article 12 of Reg. (EC) No. 834/2007 states: 'the prevention of damage caused by pests, diseases and weeds shall rely primarily on the protection by natural enemies, the choice of species and varieties, crop rotation, cultivation techniques and thermal processes'. However, it is definitely extremely difficult to respect the subsequent recommendation of using organic plant protection products only 'in the case of an established threat to a crop'. Biocontrol experts know very well that most of the active substances listed in Annex II of Reg. (EC) No. 889/2008 (and also those listed by IFOAM) cannot be applied only once the pest or disease has already reached high levels of infestation/infection, that is once it has become 'an established threat to the crop'. Almost none of the active substances allowed in organic farming have a so-called 'curative' action. Most of these substances, instead, have a 'preventative' action, or should be applied 'at the very first appearance of the pest or disease on the crop', and definitely not only in case of 'an established threat to the crop'. It appears as if organic farming systems, confounding all ecological principles, can be considered a sort of 'desert island' where no human interference is deemed necessary, because nature is always able to 'restore an equilibrium'. In other words, it seems as if it would be sufficient that organic growers wait for nature taking its course, because sooner or later the organically cultivated field will restore its 'natural equilibrium'. However, this conflicts with some of the basic principles of ecology: any cultivated field (irrespective of whether it is being cultivated under organic or conventional production systems) as such constitutes a fragile ecosystem, subjected to the eco-resistance of the initially present systems, usually represented by outbreaks of pests, diseases and weeds. It is easily forgotten that without human interference a rich biocenosis, consisting of numerous plant and animal species actively taking part in complex food chains, and guaranteeing the maintenance of long-lasting and silent natural balances of the involved species at low demographic levels, would be present in that field. In short, human choices not easily combine with those of nature: the former necessarily tend to simplify processes, while the latter found survival on the complexity of systems. Agricultural production systems therefore require 'external inputs', capable of restoring an equilibrium which has been destabilized by human actions. These inputs primarily consist of products for plant nutrition and plant protection.

In conventional agriculture, chemically synthetized substances were apparently able to solve those crop protection problems that growers face on a day-by-day basis, but these substances may have undesired side effects. In organic production systems a similar need exists: several problems must still be solved and not always valuable solutions are available. For this reason, the actions that are taken differ in their basic principles, but are nevertheless in a certain way similar to those taken in conventional agriculture: after having taken a series of agronomic measures (e.g. selection of the site, choice of species and varieties, timing of cultivation, adequate plant nutrition), if deemed necessary, appropriate plant protection products are applied. Plant protection in organic farming systems must not be considered a mere substitution process of a 'chemical product' with a 'biological product', but a novel way of conceiving the process of agricultural production based on a series of unique and brave decisions, capable of combining adequate agronomic measures, cultivar selection, plant nutrition, and – above all – on a change of the grower's mindset. Unfortunately, for reasons ranging from philosophical to economic, many growers have engaged in organic agriculture without the adequate technical know-how, even though very often the technical know-how of organic growers must be superior to that of conventional growers. Very often organic growers do not know the novel organic control measures which are now available, or do not take them into consideration, because they are used to part of their production being systematically damaged by pests and/or diseases.

This chapter aims at filling this gap by listing the number and type of currently available organic plant protection tools, and by providing information on how they should be used.

The Different Categories of Plant Protection Products Allowed in Organic Farming

There are some general principles which define the plant protection products allowed in organic farming. The active substances allowed for use in organic farming must not be produced from or by genetically modified organisms (GMOs), and thus '*Naturally occurring plants, animals, fungi, bacteria and other organisms are generally allowed*' (IFOAM, 2014). For 'non-renewable' resources deriving from mines (e.g. some active substances of inorganic origin) an evaluation is required and frequently compositional requirements and conditions for use exist. Furthermore, the environmental impact of the technologies used during the production process may be evaluated (IFOAM, 2014). In general, substances produced from non-renewable resources of synthetic origin are not allowed in organic farming systems, with the exception of substances that are not available in nature in sufficient quantities. In this case, these synthetic substances must be chemically well-defined substances, analogous to their natural form, for example insect pheromones. However, pheromones can be used only in traps and dispensers, and can never get in direct contact with the crop. For this reason, liquid sprayable formulations of pheromones (which are applied to the crop) are not allowed in organic farming neither in the USA nor in Europe. Furthermore, it may be useful to point out that in the USA, the Organic Materials Review Institute (OMRI) allows for the use of organic (i.e. based on substances of natural origin or microbials) herbicides in organic farming, while within the EU only mechanical (mowing and cutting, tillage, mulching) and physical (flame and thermal weeding) practices are allowed for weed control.

In order to facilitate the description of the biocontrol tools, we have divided them into six categories based on the active substance that they contain:

1. products based on microorganisms (or substances produced by microorganisms);
2. products based on plant extracts;
3. products based on pheromones;
4. beneficial organisms;
5. products based on substances from traditional use in organic farming; and
6. other substances.

Products Based on Microorganisms

Plant protection products based on microorganisms are formulations which contain a microorganism (fungi, bacteria, viruses, etc.) as the active substance, provided that the microorganism is not a GMO. Viruses have been included in this category for several years, even though they cannot be considered microorganisms *sensu stricto*. Microbial-based plant protection products are active against a broad range of targets in agriculture: insecticides, nematicides, fungicides, bactericides and even herbicides based on microorganisms are available on the market (as already mentioned above, with the latter not being allowed in organic farming within the EU).

Regulation (EC) No. 889/2008 allows for the use of all microorganisms provided that they are not produced from or by GMOs, and the same requirement is made by IFOAM and OMRI in the USA. OMRI and FiBL (Forschungsinstitut für biologischen Landbau, Research Institute of Organic Agriculture) in Switzerland were the first to include a substance produced by microorganisms (i.e. spinosad) in the list of substances allowed in organic farming. In Europe, the use of spinosad in organic farming has been allowed since 2008. Also IFOAM lists spinosad among the allowed substances, but erroneously catalogues it as a fungus. Furthermore, IFOAM does not refer to substances produced by microorganisms. According to Reg. (EC) No. 889/2008, no conditions for use exist for microorganisms provided that they are non-GMO, while for substances produced by microorganisms measures must be taken to minimize the risk to key parasitoids and to minimize the risk of development of resistance. In short, an additional evaluation is required for the latter in order to be listed as allowed in organic farming. For this reason, dead microorganisms, which show activity because of metabolites contained in the technical material or formulation or because of the dead microorganisms themselves acting as elicitors, are not automatically included among the substances allowed in organic farming.

Within the broad 'biological arsenal' that nature has selected for, over the course of evolution, there are numerous microorganisms that may be successfully used for plant protection, while others are still in line for a possible application in the future. Unfortunately, when moving from theory to practice, a strong selection also occurs among the possible candidates. Frequently a microorganism showing high efficacy in the laboratory is then not used in practice not 'only' because of excessive production costs, but also for more technical reasons, such as difficulties in obtaining a stable formulation with a shelf life that can be considered acceptable for common commercial distribution channels. There is also the 'hurdle' of registration. Plant protection products containing a microorganism and not a chemical substance as the active ingredient must be authorized for the intended use. Although the costs for registration of a microorganism are much lower than those for a conventional plant protection product, registration costs in compliance with the new EU Regulation amount to several millions of euros, and very often the time necessary to achieve approval and related costs are prohibitive for small companies operating in this sector. Finally, 'biopesticides' may also end up being placed on the market as fertilizers, with the complacency of some certifying bodies and without the guarantees provided by the registration process.

Plant protection with microbial control agents has many advantages. For example, microbial control agents can easily be applied with conventional spray equipment. Thus, the growers do not need to change their habits, and application remains cost-effective. Furthermore, microbial control agents usually have a very short pre-harvest interval (and sometimes, no restrictions at all exist) and a favorable toxicological and ecotoxicological profile. They can be successfully inserted into integrated plant protection strategies and they can help to reduce the risk of the development of resistance to chemical insecticides and fungicides in pest populations and of undesired residues in the final production. The latter has actually been the driving force for their development. This may sound contradictory, but biological plant protection products can help to increase the life of conventional pesticides.

However, there are also numerous problems that must be faced in order to apply a microbial control agent successfully. They must be stored properly under controlled temperature conditions (especially those based on fungi), as should be reported on the label. They often have a limited shelf life, which must be respected in order to achieve acceptable efficacy. Unfortunately, distribution channels and retailers of plant protection products frequently are not equipped for the handling of pesticides requiring cold storage. Therefore, in addition to the lack of technical knowledge on the use of such innovative products, the end-user very often has no assurance that the product he/she acquires has been stored properly prior to application.

Numerous records on insect 'diseases' can be found in the literature, but insect pathology became a real science only in 1835 due to the findings of Agostino Bassi (Porter, 1973). This scientist demonstrated that a microorganism (a fungus, afterwards named *Beauveria bassiana* (Bals.-Criv.) Vuill. in his honour) was the causal agent of the white muscardine disease of the silkworm, *Bombix mori* (Linnaeus), and that this disease can be transmitted from one insect to the other. In addition, Bassi was also the first to assume that these microorganisms may be used for pest control on cultivated crops. From these early days on, a lot of progress has been made from both an applied and a scientific point of view. Microbial control consists of causing disease outbreaks (epizootics in the case of animals) in pest or pathogen populations in order to control these populations. For a long time, the only formulated products available on the market containing a microorganism as the active substance were those based on the bacterium *Bacillus thuringiensis* (Berliner), but now several products based on microbial control agents are commercially available.

A brief description of the microorganisms currently most commonly applied in agriculture is provided below.

Viruses

Viruses are ultramicroscopic, metabolically inert infectious agents that cannot grow or reproduce outside a living cell. All organisms can be infected by viruses. Currently approximately 1100 viruses are known to cause disease in insects (Eberle *et al.*, 2012). The viruses used in crop protection are insecticides, but recently also so-called attenuated virus isolates, which are able to reduce the virulence of virus pathogens, have started to appear on the market. The insect pathogenic viruses belong to different virus families, but only those belonging to the family *Baculoviridae* have entered the market up to now. They have been isolated exclusively from invertebrates, and are therefore considered safe to mammals and other more complex life forms. Numerous records on virus-based microbial pest control can be found in the literature. Viruses are applied for the control of many lepidopteran pests, especially noctuid moths (*Mamestra brassicae* L., *Spodoptera exigua* (Hubner), *Helicoverpa armigera* Hubner, *Agrotis ipsilon* (Hufnagel), etc.), but there are also reports on the successful use of viruses against forest pests, such as *Neodiprion sertifer* Geoffroy and *Orgyia pseudotsugata* (McDonnough). The latter have almost exclusively been developed in North America within governmental projects. Baculoviruses are double-stranded DNA viruses, and are divided into four genera: (i) *Alphabaculovirus* (Lepidopteran-specific nucleopolyhedrovirus (NPV)); (ii) *Bctabaculovirus* (Lepidopteran-specific granulovirus (GV)); (iii) *Gammabaculovirus* (Hymenopteran-specific NPV); and (iv) *Deltabaculovirus* (Dipteran-specific NPV) (Jehle *et al.*, 2006). The formulated products currently on the market therefore all contain viruses belonging to the genera *Alphabaculovirus* and *Betabaculovirus* as the active substance, and are, respectively, nucleopolyhedroviruses (NPVs) and granuloviruses (GVs) (Table 2.1).

NPVs are irregularly spheroidal in shape. They sometimes have a tetrahedral structure but more frequently a polyhedral structure, with their diameter varying from 0.5 μm to 15 μm. Each NPV can contain many virions, with a single nucleocapsid (SNPV) or multiple nucleocapsids per virion (MNPV). Viral replication occurs in the cell nucleus. After nucleocapsids are replicated in the nucleus of the midgut epithelial cells, they need to exit the cell to spread the NPV infection. GVs, instead, are spheroidal, and their diameter ranges from 0.2 μm to 0.5 μm. Each GV contains only one single virion per occlusion body. As in NPVs, also in GVs viral replication starts in the cell

Table 2.1. Virus-based active substances commercially available for the control of arthropod pests in Europe. (From BCPC, 2014.)

Active substance[a]	Virus	Target pest
Adoxophyes orana GV (AoGV)	Granulovirus	*Adoxophyes orana*
Autographa californica NPV (AcNPV)	Nucleopolyhedrovirus	*Trichoplusia ni*
Cydia pomonella GV (CpGV)	Granulovirus	*Cydia pomonella*
Helicoverpa armigera NPV (HearNPV)	Nucleopolyhedrovirus	*Helicoverpa armigera*
Helicoverpa zea NPV (HezeNPV)	Nucleopolyhedrovirus	*Helicoverpa zea* and *Heliothis virescens*
Lymantria dispar NPV (LdNPV)	Nucleopolyhedrovirus	*Lymantria dispar*
Neodiprion abietis NPV (NaNPV)	Nucleopolyhedrovirus	*Neodiprion abietis*
Spodoptera exigua NPV (SpexNPV)	Nucleopolyhedrovirus	*Spodoptera exigua*
Spodoptera littoralis NPV (SpliNPV)	Nucleopolyhedrovirus	*Spodoptera littoralis*

[a]GV, Granulovirus; NPV, nucleopolyhedrovirus.

nucleus, but later in the infection the nuclear membrane appears to disintegrate and the nucleoplasm and cytoplasm merge, while in the case of NPV infection the nuclear membrane remains intact (Rohrmann, 2013). Some GV infections are limited to the midgut, whereas others cause systemic infections and can replicate in a wide variety of tissues similar to NPVs. Others still, appear to spread to and are limited to replicating only in fat body tissues.

The major advantages of virus applications are: (i) their high species specificity; (ii) their ease of use (they can be applied with conventional spray equipment); and (iii) their safety for non-target organisms. However, their high specificity is also one of their major disadvantages that has hampered their commercial development, except for *Cydia pomonella* granulovirus (CpGV) and more recently *Helicoverpa armigera* nucleopolyhedrovirus (HearNPV). Furthermore, many viruses have a slow action (the insect dies several days after infection), which is not what growers want. They are susceptible to ultraviolet (UV) degradation, and production is laborious because viruses produce new copies of themselves only in live host cells. Therefore, the host must also be mass-produced (frequently on an artificial diet), which results in an increase in production costs.

CpGV was originally isolated in Mexico and described by Tanada (1964). This CpGV isolate, replicated directly on insect host larvae, has been used in experimental applications for several years (Falcon *et al.*, 1968). As for other entomopathogenic viruses, infection of codling moth larvae occurs mainly through ingestion of food sources contaminated with viral granules. The granule thus constitutes a sort of natural micro-encapsulation. After ingestion, the viral occlusion body protein dissolves in the highly alkaline pH of the midgut (> 9), and occlusion-derived virions (ODVs) are released. These subsequently negotiate through the peritrophic membrane lining the midgut, and start the infection in the midgut columnar cells. Nucleocapsids (NCs) are released into the cytoplasm of the midgut cells, and enter the nucleus for replication. In the first stages of viral infection, NCs are transported to the cytoplasm, and emerge as budded viruses (BVs). The infection is spread by these non-occluded virions to other host tissues, causing the death of the larva within a few days. In the very late stages of infection, NCs are again occluded in the polyhedral-shaped protein matrix. New occlusion bodies, which serve to protect virions from the external environment and spread the infection to other hosts, are produced. The dead host larva turns flaccid and whitish in colour. The larva therefore does not die due to the action of toxins (as in the case of *B. thuringiensis*), but due to the disruption of the activity of vital organs. Considering that CpGV acts by ingestion and that it is highly virulent (one single virus particle may be sufficient to kill a newly hatched larva), but rapidly degraded by UV radiation, the importance of correct timing of CpGV applications becomes evident. The monitoring of the target population is of great importance, because CpGV must be applied close to *C. pomonella* egg hatching in order to increase the likelihood of the newly hatched larva ingesting the virus.

It is well known that especially eggs of the first *C. pomonella* generation are primarily laid on leaves (up to 90%), and therefore there is increased likelihood that the larvae hatching from these eggs ingest one or more virus particles while 'wandering' from the leaves to the fruit seeking a penetration site. Eggs of subsequent generations are mainly laid on fruits (only 15–20% on leaves), and therefore the likelihood of these small larvae ingesting the virus before penetrating into the fruit is lower, but still possible. CpGV therefore works best against first generation larvae.

CpGV infection does not result in an immediate interruption of the feeding activity of the larva (as in the case of *B. thuringiensis*). Despite the relatively rapid speed of kill, exposed larvae live long enough to damage fruit. Because larvae must ingest the virus, contact with the fruit and shallow entry points (stings) are inevitable. Most infected larvae die just below the surface of the fruit (Falcon *et al.*, 1968). The stings, which may be observed, correspond to the attempts of already-virus-infected larvae to enter the fruit. These stings, however, do not have any effect on fruit conservation and quality.

Another factor considered to be a major drawback by growers, is the persistence of CpGV in the field. Exposure to solar radiation (UVB, 280–320 nm) is the most critical factor limiting the persistence of entomopathogenic viruses in the field. When applied at the recommended field rates, most CpGV products currently on the market have a half-life of eight sunny days, and thus re-application after eight sunny days is recommended. Under conditions of overcast sky, the half-life in days of CpGV increases, and the time interval between applications may be increased.

CpGV can be used as stand-alone product for codling moth control, and in this case at least three to four applications are necessary for an efficient control of the first codling moth generation. As a long-term codling moth population management tool, CpGV is extensively deployed in conventional orchards in conjunction with insecticides and pheromone-mediated mating disruption, and with mating disruption only in organic orchards (Lacey *et al.*, 2008).

Other viruses that have already entered the crop protection market are *Adoxophyes orana* granulovirus (AoGV), almost exclusively used in organic orchards. Its mode of action can be considered identical to that of CpGV, but the application strategies must be adapted to the life cycle of the target pest *Adoxophyes orana* Fischer von Rösslerstamm.

On vegetables, especially in protected crops, three NPVs are increasingly successfully used for the control of the noxious Noctuid moths *H. armigera*, *S. exigua* and *Spodoptera littoralis* (Boisduval). Also in this case, repeated applications (two to three at weekly intervals) are deemed necessary for efficient pest control, and the first application should be conducted preferably at egg hatching or when the very first symptoms of damage appear on the crop. Since all virus-based products act only when ingested, thorough crop coverage is of great importance.

Fungi

Fungi can be used for both pest and disease control. Up to recent times, fungi were primarily studied for their activity against insect pests, and more than 800 species of entomopathogenic fungi have been described worldwide.

The species of fungi that have been commercially developed for pest and disease control are listed in Table 2.2. The list only includes fungi that are available for placing on the market in Europe, because the registration process is more complex in Europe than in non-European countries. Furthermore, in Europe data on field efficacy must be provided.

Antagonistic fungi are relatively easy to produce (at least in comparison to viruses) because they can be obtained via solid state

Table 2.2. Fungus-based active substances commercially available in Europe. (From BCPC, 2014.)

Fungus species	Activity	Major target(s)
Ampelomyces quisqualis	Fungicide	Powdery mildews
Aureobasidium pullulans	Fungicide/bactericide	Grey mould and fire blight
Beauveria bassiana[a]	Insecticide/acaricide	Rhynchota, Thysanoptera, Coleoptera, Diptera and tetranychid mites
Coniothyrium minitans	Fungicide	*Sclerotinia* spp.
Gliocladium catenulatum	Fungicide	Grey mould and several soil-borne diseases
Isaria fumosorosea	Insecticide/acaricide	Rhynchota, Coleoptera and tetranychid mites
Lecanicillium muscarium	Insecticide/acaricide	Whiteflies, thrips and tetranychid mites
Metarhizium anisopliae[a]	Insecticide	Coleoptera, Lepidoptera, scales and mealy bugs, termites and thrips
Paecilomyces fumosoroseus	Insecticide	Whiteflies and aphids
Paecilomyces lilacinus (= *Purpureocillium lilacinus*)	Nematicide	Several plant parasitic nematodes, especially root-knot nematodes
Phlebiopsis gigantea	Fungicide	Diseases of forest trees
Pythium oligandrum	Fungicide	Grey mould, *Sclerotinia* spp., *Alternaria* spp.
Trichoderma asperellum[a]	Fungicide	Soil-borne disease
Trichoderma atroviridae[a]	Fungicide	*Eutypa lata* and soil-borne diseases
Trichoderma gamsii[a]	Fungicide	Soil-borne diseases
Trichoderma hamatum[a]	Fungicide	Soil-borne diseases
Trichoderma harzianum[a]	Fungicide	Soil-borne diseases
Trichoderma polysporum[a]	Fungicide	Soil-borne and stem diseases
Trichoderma viridae[a]	Fungicide	Soil-borne diseases

[a]Various strains.

fermentation by seeding the fungus on flat-beds of a solid culture substrate (usually grain seeds), or via liquid state fermentation in tanks on an adequate substrate. Entomopathogenic fungi primarily act by contact (penetration into the insect through its cuticle). They can control insects that are usually difficult to control with conventional means, such as Rhynchota with sucking mouthparts (whiteflies, aphids and scale insects; in general they do not move a lot and have sedentary stages) or soil pests. These pests may easily develop populations resistant to synthetic plant protection products, and therefore additional alternative control tools, such as microbial control agents, are necessary. However, the activity of fungi may be impaired by biotic factors (e.g. the fungal strain, its physiology, the defence mechanisms of the host), abiotic factors (temperature and humidity) and by compatibility of the fungal strain with conventional fungicides.

Among the fungi used for disease control, many formulated products based on strains of *Trichoderma* spp. are commercially available. Also this fungus, as mentioned above for entomopathogenic fungi, can be applied against targets which are usually difficult to control with conventional means, that is soil-borne diseases.

Below a brief description of the fungal control agents, which can most commonly be found on the market, is provided.

The ascomycete fungus *B. bassiana* belonging to the family Clavicipitaceae has been studied and applied successfully for many years. The fungus grows naturally in soils throughout the world, and has been isolated from more than 700 arthropod species. It thus parasitizes a wide range of arthropod hosts and numerous records of susceptible insect orders, such as Coleoptera, Lepidoptera, Orthoptera, Rhynchota, Homoptera, Hemiptera and Thysanoptera, can be found in the literature. Currently

B. bassiana is the most studied entomopathogenic fungus, and it is also widely used for crop protection.

When the spores of *B. bassiana* come into contact with the cuticle of an insect host under conditions of adequate humidity and temperature, they germinate and form an appressorium that is used to remain attached to and infect the host. A penetration hypha emerges from the appressorium and is driven through the host's cuticle. If the insect moults during this early stage the infection cycle is interrupted, otherwise the fungus will continue to grow and proliferate inside the host (BCPC, 2014). Having penetrated the cuticle, the fungus alters its growth morphology and produces blastospores or hyphal bodies, which circulate in the haemolymph and proliferate by budding. Infected host insects usually die within 3–5 days. While proliferating in the host, several *B. bassiana* strains also produce toxic secondary metabolites, such as beauvericin and bassianolide. The *B. bassiana* strains available on the market as plant protection products do not produce any toxic secondary metabolite. Their insecticidal activity is mainly due to mechanical and enzymatic degradation of the cuticle by penetration hyphae producing different chitinolytic enzymes, and due to physical rupture of internal organs by vegetative growth of the fungus (Ortiz-Urquiza *et al.*, 2010). Under conditions of very high relative humidity (> 90%), the fungus may also reproduce asexually on the host by formation of conidia on aerial conidiophores. The dead insect may thus serve as a source of spores for secondary spread of the fungus.

Recent studies evidenced also other, additional modes of action of *B. bassiana* strains, such as the oviposition deterrent activity of *B. bassiana* strain ATCC 74040 against fruit flies (Ortu *et al.*, 2009; Ruiu *et al.*, 2013) and the ability of many entomopathogenic *B. bassiana* strains to endophytically colonize plants (Quesada-Moraga *et al.*, 2013). These mechanisms, if adequately exploited, could help to keep pest populations below the action threshold.

Insects vary in susceptibility to different *B. bassiana* strains. The *B. bassiana* strains currently on the market are primarily used for the control of whiteflies, thrips, aphids and tetranychid mites (Ladurner *et al.*, 2013). They can be applied with conventional spray equipment, and generally do not have any pre-harvest interval restriction.

Recently in Europe several novel strains of entomopathogenic fungi, such as *Isaria fumosorosea* (Wize) (formerly *Paecilomyces fumosoroseus*), *Metarhizium anisopliae* (Metchnikoff)(= *M. brunneum*), and *Lecanicillium muscarium* (Petch) Zare et Gams, have been granted approval because they are compliant with the requirements of Reg. (EC) No. 1107/2009. This regulation states that:

> substances should only be included in plant protection products where it has been demonstrated that they present a clear benefit for plant production and they are not expected to have any harmful effect on human or animal health or any unacceptable effects on the environment.

I. fumosorosea is an ascomycete fungus with nearly worldwide distribution belonging to the family Trichocomaceae. It controls whiteflies, especially in protected crops, and also shows some activity against aphids and thrips. This active substance was the first biological insecticide that was approved by the new European approval process. As for most entomopathogenic fungi, *I. fumosorosea* infects its host by breaching the cuticle and dispersing through the haemocoel. As the fungus continues to grow inside the insect, the insect will die. Under optimum conditions white fungal growth may appear outside the dead insect, and more spores are released to infect other insects. The fungus is most effective under conditions of high relative humidity (> 80%), and therefore its use (by label information) is limited to protected crops.

L. muscarium is the approved name of an entomopathogenic fungus species that was previously widely known as *Verticillium lecanii*. It belongs to the same family as *B. bassiana* and has a worldwide distribution. It is used for the control of whitefly larvae, with a significant side effect on thrips larvae and spider mites. After germination, the spores produce hyphae that penetrate the body cavity where they proliferate, destroying the tissues. The strain Ve 6 is currently marketed in many European countries.

M. anisopliae (= *M. brunneum*) is another ascomycete fungus belonging to the family Clavicipitaceae. This fungus has a wide host range. It has been observed to infect over 200 insect pest species (Cloyd, 1999) belonging to different orders (e.g. Lepidoptera, Coleoptera, Rhynchota). While in *B. bassiana* and *I. fumosorosea* the mould growing on infected cadavers remains white, in *M. anisopliae* it soon turns green as spores are produced. A large number of *M. anisopliae* strains that are adapted to certain groups of insects exist. Therefore, since most strains show activity against a limited number of insect hosts, various products containing different strains as active substance are available on the market, especially in South America. Throughout the world several strains of *M. anisopliae* are commercialized as plant protection products.

Fungi can also be successfully used for the control of fungal plant diseases. As an example of successful transfer from the laboratory to the field, from science to practice, it is definitely worth mentioning the many fungal species belonging to the genus *Trichoderma*. The most common species are listed in Table 2.2. However, since it is not the aim of this chapter to provide a detailed description of each single species and/or formulated product, they are all referred to as *Trichoderma* spp. hereafter.

Trichoderma spp. are ascomycete fungi (family Hypocreaceae), which have been developed as biocontrol agents of fungal diseases of plants in several countries (Lorito *et al.*, 2010). The genus has worldwide distribution, and is present in all soils, where *Trichoderma* spp. are the most prevalent culturable fungi. Cultures are typically fast growing on agar, and with conidia production they soon turn yellowish-green or dark green in colour. The optimum temperature for growth is usually around 25°C, but *Trichoderma* spp. also grow at temperatures ranging from 10°C to above 30°C. Some *Trichoderma* spp. strains can even grow at temperatures close to 0°C and above 35°C.

The mode of action of *Trichoderma* spp. is complex and varies depending on the strain and on the application site (soil or phylloplane). In the soil, the antagonist colonizes the rhizosphere, and acts as a 'barrier' against the attack from soil pathogens. Some species and strains are rhizosphere competent (i.e. able to grow on roots as they develop). They may also produce chitinolytic enzymes, which interact with and may cause irreversible damage to fungal cell walls of soil-borne pathogens. Another mechanism of antagonism is the competition for space and nutrients. If conidia of *Trichoderma* spp. germinate on plant surfaces before the fungal plant pathogen has become established, the antagonistic fungus can occupy space, use nutrients and create conditions which are adverse to the development of fungal pathogens such as *Botrytis cinerea* Pers., frequently present as a saprophyte on the vegetation during the early stages of infection.

Other mechanisms of antagonism consist of protease production and resistance induction. For example, it has been shown that *Trichoderma* spp. strains produce proteolytic enzymes, which inhibit or degrade pectinases and other enzymes that are essential for plant-pathogenic fungi, such as *B. cinerea*, to penetrate leaf surfaces (Harman *et al.*, 2004). De Meyer *et al.* (1998) showed the ability of soil-applied *Trichoderma harzianum* strain T 39 to induce a plant-mediated effect against *B. cinerea* on pepper and lettuce. Also when applied as foliar spray, *T. harzianum* appears to be capable of inducing the production of proteins and hormone-like metabolites in plants and thus induce preventative plant-mediated effects against plant pathogens (Palmieri *et al.*, 2012).

In conclusion, *Trichoderma* spp. have the potential to control of a wide range of plant pathogens. Most commercially developed *Trichoderma* strains are used for the control of soil-borne pathogens (e.g. *Pythium* spp., *Phytophthora* spp., *Fusarium* spp. and others), but some have also been developed for the control of pathogens affecting aerial plant parts, such as esca (a grapevine trunk disease), *Eutypa lata* (Pers.: Fr.) Tul. (dead arm of grapevine) and *B. cinerea* (grey mould).

Ampelomyces quisqualis Cesati is an ascomycete fungus belonging to the family Phaeosphaeriaceae. Pycnidia, sexual fruiting bodies of *A. quisqualis*, are commonly found in the cells of the hyphae, conidiophores and immature ascomata of powdery mildew fungi worldwide (Falk *et al.*, 1995a, b). The microorganism overwinters as pycnidia, but the relative importance of pycnidia within different host structures (i.e. hyphae versus conidiophores versus chasmothecia) as survival structures is unknown. Conidia of *A. quisqualis* are probably rain dispersed from pycnidia in spring. They germinate on the young leaves and, if powdery mildew colonies are present, they penetrate the hyphae and invade the mycelia of powdery mildews. Free water is required for infection, and may occur in less than 24 h under favourable conditions. Optimal temperatures for growth of *A. quisqualis* range from 20°C to 30°C. The mycoparasite can invade hyphae, mycelia, conidiophores and immature chasmothecia of powdery mildews directly through the cell walls. As *Ampelomyces* acts against powdery mildews through mycoparasitism without producing antifungal compounds, it destroys the invaded powdery mildew colonies only slowly, in 5–10 days, depending on the ambient temperature, relative humidity and other abiotic factors (Kiss, 2008). Infected cells generally die soon after pycnidial formation begins. The life cycle of the antagonist starts again when pycnidia are mature, and the released conidia can contribute to the spread of the infection. *A. quisqualis* can also spread over long distances as hyphal fragments in parasitized and detached powdery mildew conidia (Kiss *et al.*, 2004).

A. quisqualis kills the parasitized powdery mildew cells by causing a rapid degeneration of the cytoplasm. Powdery mildew colonies parasitized by *A. quisqualis* are off-white to grey in colour, and powdery mildew spore production is reduced or absent in the parasitized areas. Uninfected hyphae and conidiophores of powdery mildews are transparent, but turn translucent-white soon after infection. Once the mycoparasite has begun to produce pycnidia, the hyphae and conidiophores swell to several times their normal diameter, and the amber colour of the pycnidial wall of the mycoparasite may be noticed through the cell walls of the host.

Since the microorganism commonly occurs in all ecosystems, applications of *A. quisqualis* can contribute to enhance the activity of naturally occurring populations. The recent studies conducted by Caffi *et al.* (2012) confirmed what had already been assumed based on previous studies (Haas *et al.*, 2005; Zanzotto *et al.*, 2005): in grapevine, pre-harvest and/or postharvest applications of *A. quisqualis* that reduce powdery mildew chasmothecia formation, delay disease onset and epidemic development the following spring, and also result in reduced disease severity the following spring. The antagonist is primarily applied to control powdery mildew of grapevine, cucurbits, solanaceous crops and strawberry.

Coniothyrium minitans Campb. is an ascomycete fungus belonging to the family Leptosphaeriaeceae acting in nature almost exclusively against sclerotia (a certain type of resting stage) of species of the genus *Sclerotinia*, such as *Sclerotinia sclerotiorum* (Lib.) de Bary and *Sclerotinia trifoliorum* Erikss. The mycoparasite can commonly be found in soils worldwide. In humid soil, the spores of the antagonist germinate at temperatures ranging from 5°C to 25°C. If hyphae or pycnidiospores of *C. minitans* get in contact with sclerotia of certain *Sclerotinia* species, the latter are invaded. Following penetration of the hyphal walls of the sclerotium by hyphae of *C. minitans*, the cytoplasm of the host disintegrates, the walls collapse and the host is killed, or its ability to germinate is drastically reduced. Sclerotial exudates and exudates of *Sclerotinia*-infected plants seem to stimulate spore germination and directional growth of the mycelium of the mycoparasite. Sporulation of *C. minitans* occurs on the surface of and inside the sclerotium. Under ideal conditions the fungus forms pycnidiospores within 14 days. These spores, if spread (e.g. by water or soil mesofauna) to other sclerotia can cause new infections. The biocontrol agent is thus used primarily against sclerotia. It is applied by spraying the soil surface or crop residues at the end of the crop cycle with conventional spray equipment, and then incorporating the product into the upper soil layer. The antagonist should be applied at least 3 months prior to the onset of the disease to allow for the active ingredient to reduce inoculum loads of sclerotia in the soil and control disease outbreaks.

Pythium oligandrum Dreschler is a fungus that occurs worldwide, living in the soil where it colonizes the rhizosphere and parasitizes several fungi that cause plant diseases. *P. oligandrum* has four modes of action: (i) mycoparasitism; (ii) enhancement of plant resistance; (iii) competition for space; and (iv) competition for nutrients (BCPC, 2014). The fungicide can be used either against or for the control of several diseases (Verticillium wilt, grey mould, leaf spot and *Sclerotinia* spp.) on field crops such as oilseed rape and sunflower, but also on protected vegetable crops.

Recently a new product enlarging the product portfolio of organic growers has been placed on the market. It contains the yeast-like fungus *Aureobasidium pullulans* (De Bary) as the active ingredient, and is used for the control of grey mould and fire blight.

Bacteria

Bacteria can also be used for disease and pest control. In recent years, several new active substances based on different strains of *Bacillus subtilis* (Erhenberg), *Bacillus amyloliquefaciens* Fukumoto and *Bacillus pumilus* Meyer and Gottheil have been placed on the market, and are increasingly applied in practice. Records of at least 100 different species of bacteria being pathogenic to arthropods can be found in the literature (Thacker, 2002), with *B. thuringiensis* (Berliner) being most commonly used as a commercial biological pesticide (Table 2.3). Its efficacy against larvae of Lepidoptera, Diptera and Coleoptera (e.g. the Colorado potato beetle, *Leptinotarsa decemlineata* Say) is well known.

B. thuringiensis was first discovered by Ishiwata Shigetane in Japan in 1901 as the cause of a disease in the silkworm. The biologist called the bacterium '*Sotto Bacillus*', because 'sotto' in Japanese means 'collapse'. *B. thuringiensis* was then rediscovered in Thuringia (Germany) in 1911 by Ernst Berliner, who isolated it as the cause of a disease called *Schlaffsucht* in flour moth, *Ephestia kuehniella* Zeller. For many years, the two bacteria have been considered two distinct species, and only in the 1950s they were identified as the same species.

B. thuringiensis is a Gram-positive, soil-dwelling bacterium, occurring naturally worldwide in soils and insect-rich environments, such as flour mills and grain-storage

Table 2.3. Bacterium-based active substances commercially available in Europe. (From BCPC, 2014.)

Species	Activity	Target
Bacillus amyloliquefaciens[a]	Fungicide/ bactericide	*Botrytis*, powdery mildews, *Sclerotinia* spp. and bacterial diseases
Bacillus firmus	Nematicide	Several plant parasitic nematodes, especially root-knot nematodes
Bacillus pumilus	Fungicide	Powdery mildews
Bacillus subtilis[a]	Fungicide/ bactericide	*Botrytis*, powdery mildews, *Sclerotinia* spp. and bacterial diseases
Bacillus thuringiensis subsp. *kurstaki*	Insecticide	Lepidopteran larvae
B. thuringiensis subsp. *aizawai*	Insecticide	Lepidopteran larvae, especially noctuid larvae
B. thuringiensis subsp. *israeliensis*	Insecticide	Larvae of Diptera, especially Culicidae and Simulidae
B. thuringiensis subsp. *tenebrionis*	Insecticide	Larvae of Coleoptera, especially Chrysomelidae
Pseudomonas chlororaphis	Fungicide for seed treatment	Wheat and barley foliar diseases
Streptomyces griseoviridus	Fungicide	Soil-borne diseases
Streptomyces lydicus	Fungicide	Soil-borne diseases

[a]Various strains.

facilities, but also on leaf surfaces, aquatic environments and animal faeces. Its life cycle consists of two phases, a phase of vegetative cell division and a phase of spore development (Ibrahim *et al.*, 2010). In the presence of abundant nutrients, the bacterium multiplies rapidly via vegetative cell division. In the case of a lack of nutrients or adverse environmental conditions, the bacterium, instead, starts to sporulate. The spores are metabolically inactive and are able to survive in the environment for a long time until conditions become again favourable to vegetative growth. Upon sporulation, *B. thuringiensis* forms crystals of proteinaceous δ-endotoxins (Cry toxins) with insecticidal activity. These crystal proteins actually are pro-toxins, because they are activated only once ingested by the insect at high pH levels, which can be found in the gut of many lepidopteran larvae (Caroli *et al.*, 2000). When insects ingest toxin crystals, their alkaline digestive tracts denature the insoluble crystals, making them soluble and thus amenable to being cut with proteases found in the insect gut, which liberate the toxin from the crystal. The toxin binds to receptors in the insect gut cell membrane, and is then inserted into the membrane, paralysing the digestive tract and forming a pore. The insect stops feeding and therefore does not cause damage to plants any more almost immediately, and starves to death within a few days. Live bacteria may also colonize the insect which can contribute to death. *B. thuringiensis* acts by ingestion, and only insects with chewing mouthparts may ingest the crystals and/or spores applied on to the vegetation. The activity may vary depending on gut pH of the target insect, the presence of digestive enzymes and/or specific δ-endotoxin receptors (which vary among insects), and the developmental stage of the target pest (early-instar larvae are more susceptible than late-instar larvae).

Each *B. thuringiensis* strain produces δ-endotoxins, with their activity being often limited primarily to one insect species or a well-defined group of species. Different parasporal crystals are made either of single or multiple Cry proteins, and usually a single *B. thuringiensis* strain produces from two to five different Cry toxins. Cry toxins are encoded by *cry* genes found mainly on large plasmids. However, the genes may be integrated into the chromosome. In the classification proposed by Crickmore *et al.* (1998) the *cry* genes are divided into 51 groups and subgroups and the Cry toxins are separated into six major classes according to their insect host specificities and include:

- group 1-lepidopteran (Cry1, Cry9 and Cry15);
- group 2-lepidopteran and dipteran (Cry2);
- group 3-coleopteran (Cry3, Cry7 and Cry8);
- group 4-dipteran (Cry4, Cry10, Cry11, Cry16, Cry17, Cry19 and Cry20);
- group 5-lepidopteran and coleopteran (Cry1I); and
- group 6-nematodes (Cry6).

The Cry1I, Cry2, Cry3, Cry10 and Cry11 toxins (73–82 kDa) are unique because they appear to be natural truncations of the larger Cry1 and Cry4 proteins (130–140 kDa).

Due to its proven effectiveness, its short pre-harvest interval and re-entry time, and no MRL (maximum residue level) restrictions, *B. thuringiensis* can be used in any crop protection strategy. *B. thuringiensis*-based products have been and are being successfully inserted in pest control strategies, for example for the control of tortricid moths on grapevine, and on pome and stone fruits, especially close to harvest to avoid the risk of undesired residues in fruit. On vegetables, *B. thuringiensis* has been increasingly applied only recently, mainly due to the development of innovative integrated crop protection strategies and due to its selectivity to beneficial insects and mites.

Currently in Europe many formulated products based on *B. thuringiensis* (*B. thuringiensis* subsp. *israelensis* excluded) are commercially available. Most of these products contain the well-known *B. thuringiensis* subsp. *kurstaki* (henceforth *Btk*) strain HD1 as the active ingredient, which produces five different Cry toxins. Additional new formulations based on other *Btk* strains, such as SA11, SA12, EG 2348, etc., have been subsequently developed for their – sometimes improved – efficacy against certain target pests and longer persistence in the field.

More recent is the placement on the market of products containing *B. thuringiensis* subsp. *aizawai* (referred to as *Bta*) strains (e.g. ABTS 1857and GC 91).

B. thuringiensis strains are highly effective insecticides. In fact, multinational corporations producing genetically engineered seed have tried to exploit the characteristics of *B. thuringiensis* (*Bt*) toxins. They have developed crops that are genetically modified to produce toxins from *Bt*, in order to prolong the insecticidal effect of the toxins throughout the entire crop cycle. However, the target pest may rapidly develop resistance to the *Bt* toxin produced by *Bt* crops due to strong selective pressure, because genetically modified (GM) *Bt*-containing plants produce the Cry toxins continuously. There are a lot of controversies around GMOs on several levels, especially whether making them is ethical and not against nature, and debates on the effects of GM crops on the ecosystem are still ongoing.

B. subtilis and *B. amyloliquefaciens* (family *Bacillaceae*) are Gram-positive bacteria with worldwide distribution. They can be found in soil, but also in the gastrointestinal tract of ruminants and humans. *B. subtilis* was first described in 1835 and originally named *Vibrio subtilis*, and renamed *B. subtilis* in 1872. *B. amyloliquefaciens* was discovered in soil in Japan in 1943 by a Japanese scientist, who gave the bacterium its name because it produced (*faciens*) a liquifying (*lique*) amylase (*amylo*).

The cells of both bacteria are rod shaped. Both species can form endospores, which are able to survive extreme environmental conditions. Several scientists labelled *B. amyloliquefaciens* as a strain or variety of *B. subtilis* up to 1967, when, based on their studies, Welker and Campbell (1967) concluded that *B. amyloliquefaciens* should be considered a valid species.

In nature these microrganisms compete for space and nutrients with other microrganisms. In addition, several strains of these species produce cyclic lipopeptides with antifungal and antibacterial activity (Ongena and Jacques, 2007; Cawoy *et al.*, 2015) in order to protect their ecological niche, to inhibit the growth of and eventually outcompete potential microbial competitors, fungal and bacterial plant pathogens included. The strains of *B. subtilis/amyloliqeufaciens* currently authorized as plant protection products all produce lipopeptides, and, as many other microbial agents, they show a preventative action against plant pathogens. They can inhibit plant pathogen spore germination, disrupt germ-tube growth, and interfere with the attachment of the pathogen to the plant. They are also reported to be capable of inducing resistance in plants against bacterial pathogens (Ongena and Jacques, 2007). The *B. subtilis/amyloliquefaciens* strains currently available on the market have been registered for the control of both air- and soil-borne fungal diseases, and of several bacterial diseases, such as fire blight (*Erwinia amylovora* (Burril)), *Xanthomonas* spp. and *Pseudomonas* spp.

Due to their mode of action, these microorganisms should be applied preventatively. They can be applied by themselves as well as in alternation and/or in combination with conventional fungicides and bactericides in integrated crop protection strategies. The likelihood exists, that within a few years they may become the microbial active substances most commonly used as plant protection products. It should though always be kept in mind that *B. subtilis/amyloliquefaciens* strains differ in the type and amount of lipopeptides they produce, in their response to climatic conditions (e.g. growth rate under different temperature conditions), and thus in their activity against plant diseases. Furthermore, improved formulation technology may result in increased field performance of a certain strain. The formulated products of *B. subtilis/amyloliquefaciens* strains contain spores, which allow the bacterium to remain dormant under hostile conditions, and therefore the formulations usually have a long shelf life at room temperature (up to 2–3 years).

Bacillus firmus Bredemann and Werner (family *Bacillaceae*) is a naturally occurring soil bacterium with nematicidal activity. It is effective against several plant parasitic nematodes, especially root-knot nematodes, *Meloidogyne* spp. Together with the fungus *Paecilomyces lilacinus*, it is one of the few microorganisms that has been commercially developed for nematode control. *B. firmus* inhibits hatching of nematode eggs, reduces nematode motility and gall formation on roots, and stimulates plant root development. The formulated product must be applied before sowing/transplanting by incorporating the product into the soil.

Pseudomonas chlororaphis (Guignard and Sauvageau) (family *Pseudomonadaceae*) is another bacterium species used for crop protection, and it is the only bacterium currently authorized in Europe for seed treatment of cereals. The authorized strain is effective against several types of seed-borne diseases in cereals, such as Helminthosporium diseases (*Drechslera graminea* (Rabenh. ex Schltdl.), *Drechslera teres* (Sacc.), *Drechslera avenae* (Eidam) Scharif), stinking smut (*Tilletia caries* (DC.) Tul. and C. Tul), loose smut of oats and covered smut of barley (*Ustilago avenae* (Pers.) Rostr., *Ustilago hordeii* (Pers.) Lagerh.), *Septoria nodorum* (Berk.) blotch and *Bipolaris sorokiniana* (Sacc.).

P. chlororaphis competes with the plant pathogens for space and nutrients on the seeds and induces a resistance reaction in the plant (Tombolini *et al.*, 1999; European Commission, 2004). Once in contact with the roots, the bacteria also produce small amounts of antifungal substances. The bacterial cells enter a resting stage on the seeds after treatment, are activated when seed germinates, and remain active up to the stage when three to four leaves have unfolded (Hökeberg, 2006; Strasser *et al.*, 2007).

The authorized strain of *P. chlororaphis* belongs to a group of commonly occurring soil bacteria, is a good spermosphere, but a poor rhizosphere and phyllosphere colonizer. Applied to seed it does not colonize the shoot and roots. In the spermosphere, the strain is out-competed to non-detectable levels in a couple of weeks. The effect is very restricted both in time and in space, and the effect on the environment is thus regarded acceptable.

Protozoa

Protozoa are a taxonomically diverse group of insect pathogens, but only one species, *Nosema locustae* Canning, has been commercially developed for pest control, in particular for the control of grasshoppers and Mormon crickets in the USA (EPA, 2000). For use in pesticide products, the spores are mixed with bait, which is then applied to soil as a solid or liquid. When a target insect ingests the bait, the spores become active, the microbe grows and replicates in the insect's digestive system and the insect soon dies. The bait is most effective if used when the insects are still in their immature nymph stage, before they become adults. However, field performance is not always optimal, because infections may be chronic instead of acute (Lange and Cigliano, 2005).

Products Based on Plant Extracts

Products based on plant extracts are products that contain one or more plant extracts of natural origin as the active substance. Among these, those based on pyrethrins and azadirachtin (both approved in Europe) are the most well-known products (Table 2.4).

Quassia, even though listed in Annex II of Reg. (EC) No. 889/2008, has not been approved yet at EU level as an active substance for plant protection. However, it may be listed among the approved basic substances in the near future.

Rotenone, listed in Annex II of Reg. (EC) No. 889/2008, and nicotine, listed in previous EU regulations concerning organic production, have not been included in Annex I of Directive 91/414/EEC following the re-registration process due to ecotoxicological issues for rotenone and toxicological issues for nicotine. The authorizations for plant protection products containing these active substances have thus been withdrawn. Nicotine has been withdrawn from the market also in the USA.

Plant oils are listed among the active substances authorized for plant protection within the EU as well as by IFOAM and OMRI. All refer to a generic category named plant or horticultural oils (IFOAM, 2014). However, within the EU each single plant oil (e.g. citronella oil, spearmint oil, rapeseed oil) must be approved in order to be listed as active substance authorized for plant protection,

Table 2.4. Primary plant extracts available on the market worldwide. (From BCPC, 2014.)

Active substance[a]	Activity	Status
Azadirachtin	Insecticide	Approved in Europe, listed in Reg. EC 889/2008, and allowed in organic farming worldwide
Garlic extract	Nematicide/insecticide	Approved in Europe, not yet listed in Reg. EC 889/2008
Ryania extract	Insecticide	Not approved in Europe
Sabadilla seed extract	Insecticide	Not approved in Europe
Extract from tea tree	Fungicide	Approved in Europe
Extract from *Chenopodium ambrosioides*	Insecticide	Not approved in Europe
Reynoutria sachalinensis extract	Fungicide/resistence inducer	Approval pending in Europe
Quillaja saponaria extract	Nematicide	Not approved in Europe
Laminarin	Resistance inducer	Approved in Europe
Nicotine	Insecticide	Withdrawn in European Union and the USA, may potentially still be used in other countries (according to IFOAM)[b]
Fennel oil	Fungicide	Not approved in Europe (available in Switzerland)
Spearmint oil	Suppresses potato sprouting	Allowed in organic farming worldwide
Clove oil	Postharvest fungicide	Allowed in organic farming worldwide
Capsaicin/hot pepper oil	Insecticide/repellent	Not allowed in Europe
Rapeseed oil	Insecticide/acaricide	Not registered in Europe as an insecticide
Sesame oil	Nematicide/insecticide	Not approved in Europe
Thyme oil	Insecticide/fungicide	Not approved in Europe
Orange oil	Insecticide/fungicide	Approved in Europe
Pyrethrins	Insecticide	Allowed in organic farming worldwide
Quassia	Insecticide	Not approved in Europe but listed in Reg. EC 889/2008, allowed in organic farming worldwide
Rotenone	Insecticide	Not approved and withdrawn in Europe, may potentially be used in other countries (according to IFOAM)[b]

[a]All products based on plant oils are allowed in organic farming in Europe (after approval).
[b]IFOAM, International Federation of Organic Agriculture Movements.

while IFOAM and OMRI are much less restrictive, and any plant oil may be used for plant protection provided that it is of non-synthetic origin (and not nicotine-based!). In addition to the category 'plant oils', the IFOAM list of crop protectants also contains an extremely generic category named 'plant preparations' in addition to different plant extracts listed one by one, such as pyrethrins, neem (*Azadirachta indica* A.Juss.), quassia, rotenone, ryania and sabadilla.

The registration and commercial development of a product based on a plant extract is much more complicated, time-consuming and expensive than that of a product based on a microorganism, because, at least at an EU level, data requirements are very similar to those for synthetic active substances, residue studies included. Unfortunately, this also favours the illegal commercialization of plant extracts as fertilizers and/or plant growth stimulants, and listing the most commonly used plant extracts is difficult. We will therefore provide a detailed description only of those plant extracts that have actually been approved as active substances at an EU level.

Pyrethrins

Pyrethrins are a class of insecticidal organic compounds naturally occurring in the seeds of the flowers of plants belonging to the

Compositae familiy, normally *Chrysanthemum cinerariaefolium* (Vis.). Pyrethrins are probably the oldest biopesticide. It is the only plant extract that has 'survived' World War II, it has been used effectively as a botanical insecticide around the world for centuries, and it is one of the most commonly used allowed non-synthetic insecticides in certified organic agriculture (Crosby, 1995).

Pyrethrins have been identified in antiquity in China. They spread west to Persia probably via the Silk Roads during the Middle Ages. Dried powdered flower heads were known as 'Persian powder'. In the early 19th century Persian powder was introduced to Dalmatia, France, the USA and Japan. Dalmatia then became the main production area up to World War I, and dried powdered flower heads were also called 'Dalmatian pellitory'. Current production is mainly from Kenya, Tanzania, Rwanda, Australia, New Guinea and Ecuador.

Pyrethrins are extracted as an oil or dry powder shortly after the flower blooms. The flower contains about 1–2% pyrethrins relative to its dry weight, but approximately 94% of the total yield is concentrated in the seeds of the flower (Crosby, 1995). The standardized pyrethrin solution is composed of six esters which are insecticidal: pyrethrin I and II, cinerin I and II, and jasmolin I and II. Pyrethrins are thus a mixture of esters of two acids, chrysanthemic acid and pyrethric acid, with three alcohols, pyrethrolone, cinerolone and jasmolone (Gunasekara, 2004).

Pyrethrins affect the nervous system of insects, causing paralysis and a 'knockdown' effect. They bind to sodium channels of nerve cells, prolonging their opening, and thereby causing possible death (Tomlin, 2000). However, exposure to pyrethrins does not always cause the insect's death. A 'knockdown dose' does not mean a killing dose, because natural pyrethrins are swiftly detoxified by enzymes in the insect. Thus, some pests will recover. To delay detoxification and enhance the potency of pyrethrins, synergists such as piperonyl butoxide (not available any more on the market in Europe) are added to the formulation.

Pyrethrins act by contact and do not have any systemic action. Therefore, thorough and full coverage of the vegetation is of great importance to obtain optimal efficacy. In agriculture, the major targets are aphids, whiteflies, leafhoppers, thrips, psyllids and lace bugs. Pyrethrins can also be used to control insect pests in stored grain and food. On the vegetation, pyrethrins are rapidly degraded by sunlight, and thus persist only for a short period of time. Under dark conditions, there is little degradation of pyrethrins over time, however, in light there is rapid degradation from 100% to less than 1% within 5 h (Crosby, 1995). Therefore beneficial insects which may have been affected by the biopesticide are able to re-colonize treated crops, and beneficial insects may be released just 24–36 h after application. In organic farming systems, pyrethrins can be applied in a tank mixture with mineral oil or potassium soap to increase field performance.

Azadirachtin

Azadirachtin is a compound found in the neem tree (*A. indica*) which belongs to the mahogany family *Meliaceae* and is native to India and the Indian subcontinent. Products made from neem trees have been used in India for over two millennia for their medicinal properties (Schmutterer, 1995). In 1959 Heinrich Schmutterer, a German Scientist, noticed during a locust attack in Sudan that neem trees were the only plants that remained green and healthy while all other vegetation was completely destroyed by the locust plague. Swarms of locusts also settled on the neem trees but left without feeding. At that moment Schmutterer may have grasped what incredible possibilities this tree could offer in the area of plant protection. He decided to study this unusual phenomenon in depth. During the following 35 years he, his students, associates and scores of scientists throughout the world have studied the properties of the compounds of neem and their mode of action. However, a real breakthrough was only possible after knowledge on insect behaviour and physiology had improved greatly during the last decades and after sophisticated instruments became available, which could identify very small amounts of compounds and elucidate their structural formulae. Among these compounds, azadirachtin, a very complex tetranortriterpenoid obtained from the seed kernels *of A. indica* and in low concentrations from tissue culture, was shown to be one of the most promising plant ingredients for insect control.

Azadirachtin is very often erroneously identified as neem oil. However, neem oil is a plant oil pressed from the fruits and seeds of the neem tree, while azadirachtin is a mixture of isomeric compounds contained in neem oil, with its content varying from 300 ppm to over 2500 ppm depending on the extraction technology and quality of the neem seeds crushed. Neem oil contains an array of compounds with insecticidal, acaricidal and fungicidal activity, chemically classed as triterpenoids or limonoids (the term 'limonoid' is derived from limonin, which was first identified as the bitter constituent of *Citrus* seeds in 1841), but only the 13 isomers of azadirachtin are able to exert a real insecticidal effect. Azadirachtin A is usually the major constituent among these isomers, and commercial formulations of azadirachtin usually contain a stated amount of azadirachtin A. Neem oil contains other substances (more than 60) belonging to the same chemical class with different – and sometimes not clear – activity against insects, such as salannin, salannol, nimbin and meliantrol.

Neem oil extraction is done by mechanical pressing of seeds followed by filtering and steam pressure or solvent extraction to obtain neem oil extract, and therefore the content of azadirachtin and other limonoids in the extract varies depending on the quality of the raw material (seeds must be harvested at the right moment in order to have high azadirachtin content) and the extraction process used. Different methods of extraction produce different compositions of oil. Whether the kernels are pressed or extracted using solvents, it is questionable whether the extraction processes will ever produce the exact same mix of components twice. The technical-grade material, instead, used to manufacture commercial azadirachtin-based insecticides, is produced using a standardized extraction and concentration process, and manufacturers are required to comply with a minimum purity of azadirachtin A (FAO, 2006; European Commission, 2011). Manufacturers of commercial products use this technical-grade material to produce the insecticide formulations containing 1 5% of azadirachtin A. Different formulated products are available on the market.

Azadirachtin has a unique and multiple mode of action (National Research Council, 1992; Gilbert and Gill, 2010). It can be considered a botanical insect growth regulator. Because of its structural resemblance to the natural insect moulting hormone ecdysone, azadirachtin interrupts moulting, metamorphosis and development of the female reproductive system. Immature insects exposed to azadirachtin may moult prematurely or die before they can complete a properly timed moult. Those insects that survive a treatment are likely to develop into a deformed adult incapable of feeding, dispersing or reproducing.

Azadiractin acts primarily by ingestion, and has a systemic action, which was first discovered by Gill and Lewis (1971). It has no knockdown effect, and is considered relatively safe to beneficial insects. In the main, this is because neem products must be ingested to be effective (National Research Council, 1992). Thus, insects that feed on plant tissues succumb, while those that feed on nectar or other insects rarely contact significant concentrations of neem products. In some species, such as the Colorado potato beetle *L. decemlineata*, azadirachtin has been shown to be able to reduce fecundity in females (Schmutterer, 1995).

Azadirachtin is a broad-spectrum insecticide. The major targets are aphids, such as the rosy apple aphid (*Dysaphis plantaginea* Pass.), several lepidopteran species (e.g. *Phyllocnistis citrella* Stainton and *Tuta absoluta* (Meirick)), thrips, whiteflies, leafhoppers, some dipteran leafminers, and the Colorado potato beetle (National Research Council, 1992). Caroppo *et al.* (2002) observed that azadirachtin could be effectively used also for the control of the root-knot nematode *Meloidogyne incognita* (Kofoid and White).

The application method depends on the type of pest which must be controlled and on its biology. For example, the rosy apple aphid overwinters in the egg stage. The overwintering eggs in spring give rise to only female aphids known as stem mothers, which give birth to living nymphs. Therefore, before flowering, the majority of the aphid population is present in the nymph stage, and one single application of azadirachtin may be sufficient to control the target insect. In other pest populations (e.g. thrips and whiteflies), instead, several overlapping generations and different developmental stages may be present contemporaneously. Since the active substance acts primarily against larval stages, repeated applications (two to three at 7–14-day intervals) may be necessary to control the pest.

Other plant extracts

The plant extracts that have been placed on the market recently are laminarin, spearmint oil and clove oil. The uses supported by available data of the latter two substances provided as postharvest applications are sprouting control on stored potato for spearmint oil, and *Gloeosporium* spp. and *Penicillium* spp. control on stored apples, pears and peaches for clove oil.

Laminarin is a polysaccharide extracted from brown seaweed (*Laminaria digitata* (Huds) Lamouroux). It is not a fungicide, but stimulates the plants natural defence mechanisms against plant pathogens by inducing SAR (systemic acquired resistance) (Klarzynski *et al.*, 2000). The substance must therefore be applied preventatively prior to the appearance of disease symptoms on the crop. The product can be used for powdery mildew control on strawberry, fire blight control on pome fruit, and scab control on apple. In the EU, laminarin is not listed as a plant extract, but among 'other substances' (see Reg. (EC) No. 354/2014; European Commission, 2014).

In addition to ryania and sabadilla seed extract, already mentioned above, a countless number of plant extracts and plant oils have been studied for the control of both pests and plant pathogens, and several have shown promising results. Extracts of quassia (*Quassia amara* L.) wood or bark, with quassin being the main active ingredient, have been known to act as a natural insecticide for a long time. However, as for all other active substances and formulations already available on the market, a thorough evaluation of the quality of the raw material, the extraction and production process as well as of the formulated product is essential.

Products Based on Pheromones

It could be argued that insects live in an environment where odours lead the way: in fact, communication via chemical compounds is extremely widespread among insects, and, due to evolution, so is the spectrum of odour compounds produced and the range of insect organs capable of perceiving them. At present, these chemical messages are called 'semiochemicals', and they are subdivided into allelochemicals and pheromones depending on whether the interactions are inter- or intraspecific. Allelochemicals are used to exchange messages with species different from the source species (both plant and animal species), while pheromones are released by one member of a species to cause a specific interaction with another member of the same species.

The most well-known compounds among these chemical messages are sex pheromones. They are usually emitted by females, and are used to communicate between sexes and to favour encounters between individuals of the opposite sex. Among all these chemical messages, the sex pheromones of insects are those of particular interest to agriculture.

Up to some decades ago, entomologists were surprised to see how males of extremely rare lepidopteran species were unerringly able to find their females even under unfavourable conditions and over vast areas. The studies conducted already in the 19th century by Fabre on large Saturnidae moths, had shown that the males did not locate the females simply by chance. Now, after new methods of analysis and increasingly sophisticated equipment have become available, it is well known that usually the male is guided towards the female by the chemical compound (i.e. the sex pheromone) emitted by the female. The chemical structure of the first sex pheromone (the sex pheromone of the silkworm moth, *B. mori*) was identified in 1959 by the German scientist Butenandt.

Since this time many different sex pheromones have been identified and synthesized and they are now commonly used in pest management programmes for easy and cost-effective pest detection and monitoring. In addition, sex pheromones can also be used directly as plant protection tools, in particular for:

- mating disruption;
- mass trapping; and
- attract and kill.

Mating disruption

Mating disruption may be accomplished in two principle ways: (i) male confusion; or (ii) false trail following (Welter *et al.*, 2005; Flint and Doane, 2013).

Male confusion is thought to be the result of ambient pheromone concentrations sufficient to hide the trails of calling females. The high ambient concentration of the species-specific pheromone (several milligrams per day, more than a hundred times higher than that of a calling female) is emitted by different types of point-source dispensers or microcapsules. The number of dispensers that must be deployed per unit area depends on the insect species that must be controlled and the type of dispenser used. The effect is the adaptation of antennal receptor sites and/or habituation of the male's central nervous system. The receptor site becomes unresponsive and the male is not able to locate the calling female any more, which finally results in reduced mating and thus in a reduction of the population density of the following generation. Recently, a new male-confusion technology consisting of aerosol formulations of pheromones has been developed and placed on the market. The 'dispenser' in this case consists of an aerosol can containing the pheromone inserted into a programmable electromechanical device, which allows the pheromone emission frequency to be programmed over a selected time period during the day (usually evening/night hours when moths actually fly and mate). With this technology, the required number of emission points (can + device) per hectare is very low (usually between two and four/ha), thus allowing for a considerable reduction in labour costs/time for dispenser deployment. However, this technology should preferably be used only over large areas, even larger than those recommended as the minimum size areas for point-source dispensers.

False trail following results from placing many more point sources of pheromone per unit area than the anticipated numbers of females in the crop (usually 2000–3000 dispensers/ha). The odds of males finding females at the end of the pheromone trail must be greatly reduced. Emission of pheromone from each dispenser in this case is low compared with that from male-confusion dispensers. A downwind trail of pheromone is created, which is not lost in a background of released pheromone. Males following these trails are thought to spend their mating energies in pursuit of artificial pheromone sources instead of calling females. However, usually two to three applications per season are necessary to cover the entire flight period of the insect pests.

Currently mating disruption is commonly used in many countries all over the world not only in organic farming, but also in integrated crop production systems in order to reduce the pest populations. Resistance to synthetic insecticides has probably created the necessary economic impetus for the inclusion of mating disruption in integrated pest management (IPM) strategies. The overall area covered by mating disruption is estimated to exceed 750,000 ha (Miller and Gut, 2015), but it may be that these figures are underestimated. In Europe, mating disruption was introduced in the 1990s, and it is mainly used for the control of carpophagous pests in fruit crops. The most important insect pests, for which one or more mating-disruption products are commercially available, are listed in Table 2.5.

As for other plant protection tools, mating disruption must be applied by following certain rules, and it must be kept in mind that its efficacy may vary depending on several factors, such as initial population density, shape and slope of orchard, and exposure to winds. The most important recommendations are:

- In order to work, mating-disruption products should be applied in homogeneous and large fields of at least 1–2 ha in size for isolated fields, and possibly 5 ha for non-isolated fields. In non-isolated fields, in particular, additional dispensers should be applied along the borders as reinforcement against pheromone depletion due to wind, especially along the upwind border. Aerosol mating-disruption products should be applied over areas of at least 10–15 ha in size.
- Mating-disruption products may be used by themselves for pest control only in fields with low-to-medium population density; in case of high population density, the likelihood of males locating females, even just by chance, is high, and mating disruption should be used in combination with insecticide sprays. For example on pome fruit, especially in organic orchards, codling moth can be controlled by using mating disruption in combination with applications of *C. pomonella* granulovirus-based products.
- Hand-applied dispensers must be deployed uniformly throughout the field (for border reinforcement see above). Also the placement of aerosol units should be as uniform as possible (by taking into account the typical wind direction), and accurately planned, for example by using a GPS (global positioning system) device.
- Dispensers (aerosol units included) must be deployed in the field before the beginning of the first flight of the target pest.
- Other factors that may impair the efficacy of mating-disruption products are high temperatures and extremely windy conditions, which result in a higher and faster release of pheromone from hand-applied dispensers.

The instructions provided by the manufacturer should always be read carefully and followed, because different dispenser types vary in their properties.

Table 2.5. Most important insect pests for which one or more mating disruption products are commercially available.

Insect	Common name	Crop
Cydia pomonella	Codling moth	Pome fruit and walnut
Grapholita molesta[a]	Oriental fruit moth	Stone and pome fruit
Adoxophyes orana, Archips spp., *Pandemis* spp.	Tortricid moths	Pome fruit
Anarsia lineatella	Peach twig borer	Stone fruit
Lobesia botrana	European grapevine moth	Grapevine
Eupoecilia ambiguella	European grape berry moth	Grapevine
Chilo suppressalis	Asiatic rice borer	Rice
Ostrinia nubilati	European corn borer	Pepper
Lymantria dispar	Gypsy moth	Forests
Synanthedon spp.	Clearwing	Currant
Tuta absoluta	Tomato leafminer	Solanaceous crops (under permanent protection)
Zeuzera pyrina	Leopard moth	Fruit crops, olive
Keiferia lycopersicella	Tomato pinworm	Tomato

[a]The dispensers active against *G. molesta* may be effectively used also against *Cydia funebrana* on plum.

Mass trapping and attract and kill

Mass trapping (place enough traps, catch enough males, and leave the females of the species without mates, according to Flint and Doane, 2013) may successfully be used to control large lepidopterans, such as the cossid moths *Zeuzera pyrina* Linnaeus and *Cossus cossus* Linnaeus, by deploying eight to ten large cone traps baited with pheromone lures/ha. These moths usually are present at low population levels. Furthermore, the females that do not mate within 24–48 hours after emergence are not able to lay fertile eggs, and therefore the removal of even just part of the male population over several years may result in a considerable population reduction. Mass trapping by deploying water traps baited with pheromone lures is also used to mass trap males of *S. littoralis* and *T. absoluta*. However, no scientific evidence exists that this actually results in effective population reduction.

Several attract-and-kill devices (combination of an attractant with a contact insecticide; Welter *et al.*, 2005) have been commercially developed for the control of the Mediterranean fruit fly, *Ceratitis capitata* Wiedmann and the olive fruit fly, *Bactrocera oleae* Gmelin. The outer surface of the attract-and-kill device (usually an envelope-like trap) is coated with an insecticide. A specific attractant is used to attract the flies to the insecticide-treated surface. According to EU regulations, in organic production systems only two pyrethroids, deltamethrin and lambda-cyhalothrin, are allowed to be used in traps. These insecticides, even though synthetic, are tolerated because they do not get in direct contact with the vegetation, and will most likely be substituted with biological insecticides, as soon as the latter become available. Direct contact between the attract-and-kill devices and the vegetation must be avoided. Depending on the product, crop and pest level, the devices should be deployed at a rate of a few dozen to a few hundred per hectare. The attractants may be: (i) synthetic food-grade attractants, such as ammonium acetate; (ii) parapheromones (antagonistic compounds, agonists, pheromone mimics and synergists, broadly grouped under the term parapheromones, according to Renou and Guerrero, 2000) used to attract and kill food-seeking females before their eggs mature; and/or (iii) pheromones. As for mating disruption products, attract-and-kill strategies only work if they are applied over large areas (at least 2–5 ha). Also available on the market are attract-and-kill systems that consist of traps that only contain food attractants such as hydrolysed proteins. These systems are used to attract and kill food-seeking females before their eggs mature, and they, too, must be applied over large areas in order to work. Table 2.6 lists mass trapping and attract-and-kill systems that are available on the market.

Beneficial Organisms

The release of beneficial insects and mites as well as the application of entomopathogenic nematodes for crop protection is not expressly mentioned in Reg. EC 889/2008,

Table 2.6. Mass trapping and attract-and-kill systems available on the market.

Insect	System/crop	Comments
Anthonomus grandis	Mass trapping/cotton	Both males and females attracted by pheromone
Bactrocera oleae	Attract and kill/olive	Males attracted by pheromone, females attracted by food attractant
Ceratitis capitata	Attract and kill/citrus and other fruit	Males attracted by pheromone, females attracted by food attractant
Cossus cossus	Mass trapping	Males attracted by pheromone
Spodoptera littoralis	Mass trapping	Males attracted by pheromone
Zeuzera pyrina	Mass trapping	Males attracted by pheromone

but it is specified in the *IFOAM Norms for Organic Production and Processing* (IFOAM, 2014). The release of beneficials is perfectly in line with the principles of organic production listed in Reg. (EC) No. 834/2007 and by IFOAM (IFOAM, 2014): organic production shall preserve natural resources (thus natural populations of beneficials are included) and shall include biological, cultural and mechanical mechanisms to manage pests, weeds and diseases, including the release of natural enemies, such as predators and parasites. Techniques and methods for mass production of beneficial insects, mites and nematodes have been developed by different companies since the 1960s (if not earlier), and currently several dozen species are commercially available (Tables 2.7 and 2.8). They can be applied in both vegetable and fruit crops. Interestingly, the release of beneficial insects and mites is becoming increasingly important in integrated rather than in organic protected crops. This might seem strange, but in integrated farming systems the steadily increasing spread of the appropriate technical knowledge, of great importance for this 'sophisticated' technology, is apparently beginning to bear fruit. Very often, instead, communication or transfer of information among players (from companies to retailers, and finally to growers) is missing, and the effectiveness of proven biocontrol techniques, such as the use of the predator *Phytoseiulus persimilis* Dosse against the two-spotted spider mite *Tetranychus urticae* Koch or the use of the parasitoid *Diglyphus isaea* (Walker) against leafminers, is still being questioned.

Companies producing beneficials put a lot of effort and research not only in the improvement of their mass-production techniques, but also in the development of adequate packaging. The biology and behaviour of each single species is investigated thoroughly in order to identify the most appropriate storage and transport conditions, prerequisites for good performance in the field.

The use of natural enemies for biological control started in the 1960s. Crop protection in European greenhouses became strongly chemically oriented shortly after World War II in the 1950s. However, this excellent climate for fast reproduction of pests and diseases demanded high spray frequencies and, thus, resulted in quick development of resistance against pesticides. This initiated a search for alternatives to chemical pesticides. One of the first efficient candidates identified was the predatory mite *P. persimilis* for the control of *T. urticae* (Scopes, 1985). The first natural enemies for control of pests in European greenhouse vegetables became available in the 1960s. From then on, mass-production and application techniques were developed for numerous beneficial insects and mites, and biocontrol with natural enemies spread from Northern Europe to the Mediterranean zone, first to Sicily, Italy, in the 1970s and then also to other countries. A change from chemical control to very advanced IPM took place over a time span of about 20 years, especially from 2007/2008 onwards in Spain. The change to IPM was not based on idealism about a cleaner or healthier environment, but at least in part was due to an analysis made and published by Greenpeace Germany's consumer organization in 2005 on fruit and vegetable samples coming from Spain. This found that the legal residue limits for pesticides had been exceeded in one out of every five samples of fruit and vegetables, and in 27 instances illegal pesticide residues were discovered, and was mainly due to the fact that chemical control had become increasingly complicated because of resistance issues, especially on bell pepper. Growers in Spain were 'forced' to change their crop protection strategies, and the use of natural enemies provided clear advantages. It is estimated that currently biocontrol with natural enemies is used on at least 10,000 ha of greenhouse crops, with most of them being integrated and not organic production systems.

Products Based on Substances from Traditional Use in Organic Farming

Copper

Currently copper-based fungicides are among the most widely used products for fungal and bacterial disease control not only in organic,

Table 2.7. Main beneficial insect and mite species commercially available in Europe. (From: BCPC, 2014.)

Beneficial organism	Target	Parasitoid (PS)/ predator (PR)	Mainly used on
Adalia bipunctata	Aphids	PR	Vegetables and ornamentals
Amblydromalus limonicus	Thrips and whiteflies	PR	Vegetables and ornamentals
Amblyseius andersoni	Tetranychid, eriophyid and tarsonemid mites	PR	Grapevine, small fruits, vegetables and ornamentals
Amblyseius barkeri	Tarsonemid mites	PR	Strawberry
Amblyseius cucumeris	Thrips and mites	PR	Vegetables and ornamentals
Amblyseius degenerans	Thrips	PR	Vegetables and ornamentals
Amblyseius fallacis	Tetranychid and eriophyid mites	PR	Pome fruit, vegetables and ornamentals
Amblyseius montdorensis	Whiteflies, thrips and tarsonemid mites	PR	Vegetables and ornamentals
Amblyseius swiskii	Whiteflies, thrips and mites	PR	Vegetables and ornamentals
Anagrus atomus	Leafhoppers	PS	Vegetables and ornamentals
Anagyrus pseudococci	*Planococcus* spp.	PS	Citrus, grapevine
Anisopteromalus calandrae	Pests of stored grain	PS	Stored grain
Anthocoris nemoralis	*Cacopsilla pyri*	PR	Pear
Aphelinus abdominalis	*Macrosiphum euphorbiae, Aulacorthum solani*	PS	Vegetables and ornamentals
Aphidius colemani	Aphids	PS	Vegetables and ornamentals
Aphidius ervi	Aphids	PS	Vegetables and ornamentals
Aphidius matricariae	Aphids	PS	Vegetables and ornamentals
Aphidoletes aphidimyza	Aphids	PR	Vegetables and ornamentals
Aphytis melinus	California red scale; oleander scale	PS	Citrus
Atheta coriata	Fungus gnats and pupae of soil insects	PR	Vegetables and ornamentals
Chrysoperla carnea	Aphids	PR	Vegetables and ornamentals
Cryptolaemus montrouzieri	Mealybugs	PR	Citrus and ornamentals
Cydnodromus califirnicus	Thrips and mites	PR	Vegetables and ornamentals
Dacnusa sibirica	Agromizid leafminers	PS	Vegetables and ornamentals
Delphastus catalinae	Whiteflies	PR	Vegetables and ornamentals
Delphastus pusillus	Whiteflies	PR	Vegetables and ornamentals
Diglyphus isaea	Agromizid leafminers	PS	Vegetables and ornamentals
Encarsia formosa	Whiteflies	PS	Vegetables and ornamentals
Ephedrus cerasicola	Aphids	PS	Vegetables and ornamentals
Epysirphus balteatus	Aphids	PR	Vegetables and ornamentals
Eretmocerus eremicus	Whiteflies	PS	Vegetables and ornamentals
Eretmocerus mundus	Whiteflies	PS	Vegetables and ornamentals
Euseius gallicus	Spider mites, thrips and whiteflies	PR	Roses and ornamentals
Feltiella acarisuga	Tetranychid mites	PR	Vegetables and ornamentals
Gaeolaelaps aculeifer	Fungus gnats and pupae of soil insects	PR	Vegetables and ornamentals
Habrobracon hebetor	*Ephestia* spp., *Plodia* spp., *Sitotroga cerealella*	PS	Storehouses and flour mills
Leptomastix algirica	Mealybugs	PS	Citrus, vegetables and ornamentals
Leptomastix dactylopii	*Planococcus citri*	PS	Citrus and ornamentals
Macrocheles robustulus	Fungus gnats and other soil pests	PR	Ornamentals
Macrolophus pygmeus	Whiteflies	PR	Vegetables and ornamentals

Continued

Table 2.7. Continued.

Beneficial organism	Target	Parasitoid (PS)/ predator (PR)	Mainly used on
Nesidiocoris tenuis	Whiteflies	PR	Vegetables and ornamentals
Orius insidiosus	Thrips	PR	Vegetables and ornamentals
Orius laevigatus	Thrips	PR	Vegetables and ornamentals
Orius majusculus	Thrips	PR	Vegetables and ornamentals
Phytoseiulus persimilis	Spidermites	PR	Vegetables and ornamentals
Podisus maculiventris	Colorado potato beetle, noctuid moths	PR	Potato, vegetables and ornamentals
Praon volucre	Aphids	PS	Strawberry
Stratiolaelaps scimitus	Fungus gnats and other soil pests	PR	Ornamentals
Trichogramma brassicae	European corn borer and other Lepidoptera	PS	Maize and vegetables
Trichogramma evanescens	Stored grain Lepidoptera	PS	Stored grain

Table 2.8. Main entomopathogenic nematode species commercially available in Europe. (From BCPC, 2014.)

Nematode	Target	Mainly used on
Heterorhabditis bacteriophora	*Otiorhynchus* spp. (black wine weevil, etc.)	Vegetables, berries and ornamentals
Heterorhabditis megidis	*Otiorhynchus* spp. (black wine weevil, etc.)	Vegetables, berries and ornamentals
Phasmarhabditis hermaphrodita	Slugs	Vegetables and ornamentals
Steinernema carpocapsae	Lepidoptera, European mole cricket, flat-headed root borer	Stone and pome fruit, ornamentals, vegetables
Steinernema feltiae	Fungus gnats and Lepidoptera	Stone and pome fruit, vegetables and ornamentals
Steinernema kraussei	*Otiorhynchus* spp. (black wine weevil, etc.)	Vegetables, berries and ornamentals

but also in conventional farming (BCPC, 2006). Copper can be used against a wide range of fungal and bacterial plant pathogens, powdery mildews excluded. In fact, for over a century (its disease control properties were discovered in 1882) copper has been representing a milestone on which farmers can rely on for crop protection. Copper is an inorganic compound, thus it does not break down like organic compounds and consequently copper can accumulate in soil. Therefore, in recent years, concerns about its environmental fate have been raised. This stimulated research for improvements in formulation and application technology of copper-based fungicides in order to reduce the amount of metallic copper applied per hectare per year, and for alternatives to copper, particularly in organic farming. Possible alternatives under investigation are microorganisms (fungi and bacteria), natural compounds (plant extracts), and minerals (acidified clays).

Copper-based fungicides can be described as insoluble compounds, yet their action as fungicides and bactericides is due to the release of small amounts of copper ions (Cu^{++}), when in contact with water. Following absorption into the fungus or bacterium, the copper ions will link to various chemical groups present in many proteins, and disrupt the function of these proteins. Proteins are denatured, and enzymes that are critical for cell functioning are destroyed. Thus, the mode of action of copper fungicides is the non-specific denaturation (disruption) of cellular proteins, resulting in alteration of cell membrane permeability, inhibition of respiration and interruption of spore germination.

Different formulations have different physical-chemical properties, and the release of copper ions varies among formulations. Furthermore, copper can also cause phytotoxic effects. Copper can inhibit photosynthesis, promote iron chlorosis, and an excess of copper in plants may result in stunted growth. It can also cause fruit russeting and fruit necrosis, for example on apple and pear.

EU regulations clearly state which copper compounds are allowed in organic farming, namely those in the form of copper hydroxide, copper oxychloride, copper oxide, Bordeaux mixture and tribasic copper sulfate. These compounds are in compliance with Reg. (EC) No. 354/2014 of 8 April 2014, amending and correcting Reg. (EC) No. 889/2008, which lays down detailed rules for the implementation of Reg. (EC) No. 834/2007, the regulation on organic production and labelling of organic products, repealing Reg. (EC) No. 2092/91. Reg. (EC) 354/2014 furthermore poses the following conditions for use: only uses of copper as bactericide and fungicide up to 6 kg copper/ha/year are allowed. For perennial crops, Member States may, by derogation, allow the 6 kg copper limit to be exceeded in a given year provided that the average quantity actually used over a 5-year period consisting of that year and of the preceding 4 years does not exceed 6 kg. Risk mitigation measures shall be taken to protect water and non-target organisms such as by using buffer zones.

In the (near) future it will be necessary to find alternatives to the use of copper to continue to grow crops organically because copper compounds have been included in the EU list of pesticides for which national authorities need to find substitute plant protection products, including non-chemical methods. The aim is to encourage more sustainable crop protection.

Sulfur

Sulfur, together with copper, is the most well-known and widely used fungicide, and its use as a fungicide may be traced back many centuries (Large, 1958). Early in agricultural history, the Greeks recognized its efficacy against rust diseases on wheat. Sulfur is primarily used for the control of powdery mildews, scab and rust diseases, but it also controls mites (particularly eryophid mites) on a range of crops (BCPC, 2014). Currently sulfur is still one of the most important fungicides, especially in organic farming, where only a limited number of disease control tools are available at the moment.

Up to several decades ago, sulfur was almost exclusively used in the form of a more-or-less coarse powder, suitable to be applied in a dry state (dusting sulfur). Subsequently, formulations consisting of very fine particles (e.g. wettable sulfur, micronized sulfur and colloidal sulfur) and finally water dispersible granule (WDG) formulations, which can be suspended in water and applied as an aqueous suspension, were placed on the market. Unlike copper, at the moment there are no restrictions or limitations in the EU concerning the use of sulfur in organic farming. However, issues were raised whether elemental sulfur is elemental sulfur regardless of where it comes from. Currently, the major part (if not all) of the elemental sulfur contained in sulfur-based fungicides comes from oil refining, and not from mines. Discussions are ongoing whether mined sulfur should be considered natural and oil-processed sulfur not.

Despite sulfur having being used as a fungicide for a very long time its mode of action remains unclear. It is considered a non-systemic protectant fungicide with direct contact and indirect vapour action of its breakdown product hydrogen sulfide (Thind, 2012; BCPC, 2014). The mode of interaction with plant tissues is primarily binding of sulfur to cuticular waxes, with higher doses resulting in sulfur sitting on sulfur. There is also a low level of sulfur penetrating into plant tissues (Emmett *et al.*, 2003). Sulfur, as well as copper, has multisite activity (FRAC group M), that is it interferes with several of the fungus vital life functions. Sulfur is reported to inhibit respiration, disrupt protein function and inhibit biochemical pathways by chelation of metals important for normal enzyme function (Hewitt, 1998). Sulfur is considered a pre-infection fungicide. Some authors have stated that sulfur has no curative (post-infection) activity. Actual contact of the sulfur particle with the fungus seems to be necessary before fungicidal action can occur, and there is still some debate as to whether vapour inhibits spore germination. However, vapour activity is reported to be the primary source of activity against *Erysiphe necator* (Schwein.) Burrill when the temperature is above 15°C, while below 15°C vapour activity is negligible (Emmett *et al.*, 2003).

What is known is that particle size, temperature and relative humidity influence fungicidal activity of sulfur. The optimal temperature range for hydrogen sulfide production (and thus vapour activity) is 25–30°C (Thind, 2012). At temperatures greater than 30°C, sulfur may pose a risk of phytotoxicity to the crop, and the extent of the phytotoxic symptoms may vary depending on particle size, applied rate and type of formulation. Manufacturers of sulfur-based fungicides usually recommend to reduce rates or avoid applications during periods of high temperatures. Phytotoxic effects are most likely to occur on some apple (e.g. 'Golden Delicious', 'Imperatore', 'Commercio', 'Jonathan', 'Rome Beauty', 'Stayman') and pear (e.g. 'William', 'Decana del Comizio', 'Kaiser') cultivars, on peach, on almost all apricot cultivars, and on cucurbits. Furthermore, caution is recommended when spray programmes require applications of both mineral oils and sulfur. Together they are phytotoxic to plants, and thus should not be applied in a tank mixture. A spray interval of at least 10–14 days is recommended. As far as the selectivity of sulfur to beneficials such as parasitoids and phytoseiid mites is concerned, different considerations can be found in the literature (Auger *et al.*, 2003; Thomson, 2012). In our opinion, it is also important to take into account the application rate and the type of sulfur formulation used for disease control: the likelihood exists that applications of 20–30 kg/ha of elemental sulfur dusted on plants have a greater impact on beneficials than the more recent WDG and liquid formulations.

Lime sulfur (calcium polysulfide)

Lime sulfur is believed to be the earliest synthetic chemical used as a pesticide for both pest and disease control. In the 1840s it was first used in France to control grapevine powdery mildew (*E. necator*), and in 1886 it was first used in California to control San Jose scale. Lime sulfur (calcium polysulfide) used to be listed in Reg. (EC) No. 889/2008 as a fungicide, insecticide and bactericide, but it is now listed only as a fungicide in Reg. (EC) No. 354/2014, which amends and corrects Reg. (EC) No. 889/2008. Lime sulfur is formed by reacting calcium hydroxide with sulfur, which results in the formation of several compounds (mono-, bi-, tri-, tetra-, pentasulfides, thiosulfates, sulfates and sulfites). This cocktail of sulfur compounds has both an insecticidal (especially against scale insects) and fungicidal action. It is a deep-orange liquid, which in the environment degrades to calcium hydroxide and sulfur (Paranjape *et al.*, 2014). It acts directly by contact against insects by forming hydrogen sulfide (toxic to insects) and causing a caustic action on the integument of insects, occlusion of respiratory spiracles and inhibition of respiration (BCPC, 2006). It softens the wax of scale insects and prevents the establishment of young nymphs on the vegetation (BCPC, 2006). Under favourable environmental conditions, its insecticidal effect may persist for a long time, up to 1 month. High concentrations of lime sulfur can be phytotoxic. Therefore, up to a few years ago, lime-sulfur-based products were primarily applied as dormant season products. By reducing application rates, these products are now also used during periods of vegetative growth, especially in organic farming. Lime sulfur is used for the control of several diseases, such as scab, peach leaf curl, powdery mildews, *Coryneum* blight, brown rot (*Monilinia* spp.), peacock leaf spot and sooty moulds.

Paraffin oils (dormant oils and narrow-range summer oils)

Petroleum-derived mineral oils have been used for pest control for more than 100 years. They used to be applied primarily for cool-season control of pests that overwinter on trees and shrubs. The first applications date back to the 1880s and were based on the use of a kerosene–soap emulsion, and later of distillates in the range between kerosene and lighter lubricating oils (Agnello, 2002). These oils were effective against insect pests, but were too phytotoxic. In 1915, E. de Ong showed that tree injury was caused by substances (aromatics and other unsaturated components) in oils that were removable with sulfuric acid. Further improvements of the formulation were then made by developing aqueous emulsions, which enabled increased crop coverage. After World War II, the development of chemical pesticides resulted in a decrease in the use of the so-called 'alternative means', which included paraffin oils, and synthetic pesticides quickly became dominant in insect and mite control. It was not until the discovery of insect and mite resistance in the 1970s and the increasing concerns of environmental pollution that the use of oils really began to develop (Agnello, 2002). In the meantime, considerable improvements in oil refinement technology and the availability of highly refined, lighter weight, lower impurity oils with sunscreens to reduce phytotoxicity extended the use of paraffin oil products to control pests on crops in full foliage (Agnello, 2002). Paraffin oils are derived from crude petroleum, which is separated into fractions by heat. The lighter parts such as gasoline, kerosene or diesel evaporate at lower temperatures, while paraffin waxes, bitumen and lubricating fat are collected at higher temperatures.

Although crude petroleum oils are complex mixtures of a large number of compounds, they consist primarily of hydrocarbons. The main components can be categorized as paraffin chains (straight or branched), naphthene rings, aromatic rings, unsaturated hydrocarbons or asphaltic material, and the proportion and type of compounds generally varies among crude oils (Agnello, 2002). The unsaturated hydrocarbons contain one or more double bonds between carbon atoms. These compounds are chemically highly reactive compounds, which are easily oxidized and can form phytotoxic substances. Most of the aromatic compounds are saturated during the refining process, while resins (asphaltic substances containing oxygen or sometimes sulfur) and compounds containing nitrogen (e.g. quinolone) are removed with sulfuric acid (Davidson *et al.*, 1991; Agnello, 2002).

The so-called 'summer' oils or narrow-range oils are highly refined petroleum oils with a mid-boiling (distillation) point from 211°C to 224°C under controlled vacuum conditions. They exhibit both pesticidal efficacy and safety to plants, because phytotoxic components are 'excluded' due to the narrow range of distillation. Traditional spray oils (dormant oils), instead, were taken between a median boiling point from 190°C to 274°C under a vacuum of 10 mm Hg (range: more than 80°C). Narrow range oils contain a high proportion of paraffins (more than 60% of the total hydrocarbons), while the content in (potentially phytotoxic) aromatic structures and (not insecticidal) naphthenic compounds is considerably reduced. They may therefore also be called paraffinic oils or superior oils, a term originated by P.J. Chapman in 1967 to categorize summer-use oils that met certain specifications, including a high proportion of paraffinic hydrocarbons, and purification that allowed year-round use without phytotoxicity.

Both dormant and summer oils act primarily by contact, causing pests to die from asphyxiation within a few hours by blocking the spiracles and respiratory system through which insects breathe. In some cases, oils may also act as poisons, interacting with the fatty acids of the insect and interfering with normal metabolism (Taverner, 2002). Oils may also act as repellents by disrupting oviposition and how an insect feeds (Stansly *et al.*, 2002), a feature that is particularly important in the transmission of some plant viruses by aphids. The application recommendations for summer oils differ from those of dormant oils. The former can be applied at high rates (from 0.6–0.8 l/100 l up to 1.5–2 l/100 l) directly on to the vegetation. Paraffin oils can be used against a wide range of pests, for example tetranychid mites, scale insects and mealybugs, aphids, whiteflies, psyllids, and to a reduced extent also against thrips, lepidopteran and coleopteran eggs, and leafminers. Once primarily used for the control of San Jose scale on citrus, paraffin oils, or better narrow-range oils, are now commonly used also on pome and stone fruit, and especially on vegetables and ornamentals.

Mineral oils are often added to other pesticide products to improve efficacy. In this sense, they are considered spray adjuvants, even though they may have pesticidal activity on their own. In organic crop protection systems they are often added as adjuvants to pyrethrins- and azadirachtin-based products, and to products containing microorganisms as the active ingredient. In addition to high efficacy, narrow-range oils have several other benefits in comparison with conventional mineral oils, such as reduced side effects on beneficials (beneficials can re-colonize the crop as soon as the spray deposit has dried on the vegetation) and negligible risk of development of resistance in target populations due to their mode of action.

Fatty acid potassium soap (soft soap)

Soaps are water-soluble salts of fatty acids. They are made by saponification from fats and oils by treating them chemically with a strong alkali, such as sodium hydroxide and potassium hydroxide. Only fatty acid potassium soaps (soft soaps) are allowed in organic farming for insect control, while sodium soaps (hard soaps) are not (see Reg. (EC) No. 354, 2014; European Commission, 2014). Chemically soft soaps can be defined as potassium salts of long-chain carboxylic acids. These salts have the ability to solubilize non-polar organic substances, such as fats. They can therefore be used as both detergents and pest control tools. The insecticidal activity of soft soaps depends on the structure of the lipid component, usually oleic or linoleic acid obtained from plant seeds, for example soy, sunflower and rape seeds. Other plant oils, such as palm oil, instead, have no insecticidal action. Soft soaps are most effective against soft-bodied insects (aphids, psyllids, whiteflies, lace bugs, etc.) and tetranychid mites (Osborne and Petitt, 1985). Furthermore, insecticidal soaps are also an effective leaf wash to remove honeydew from leaves.

Insecticidal soaps work only on direct contact. The fatty acids of the soap disrupt the structure and permeability of the insect's cell membrane, the cell contents leak from the damaged cells, and they remove protective waxes that cover the insect, resulting in dehydration. Under direct sunlight the insect quickly dies. The same effect is obtained by using insecticidal soaps as a leaf wash to remove honeydew. Honeydew is produced by juvenile stages of pests, such as pear psylla, also to protect them from prolonged exposure to direct sunlight. Removing the honeydew results in increased mortality of early-instar larvae. However, insecticidal soaps have no residual activity once the spray application has dried.

Currently, formulated products containing fatty acid potassium soap as the active ingredient are registered in Europe for plant protection.

Ethylene

Chemically ethylene is an unsaturated hydrocarbon, a gas, which is also an important plant hormone. It is used in agriculture to force the ripening of fruits. Ethylene promotes fruit ripening by triggering several changes that lead to the conversion of starch into free sugar, by inducing the synthesis of pectolytic enzymes that degrade pectic substances in cell walls, which result in pulp and peel softening. In addition, ethylene promotes the accumulation of coloured pigments and the degradation of chlorophyll (Nath *et al.*, 2006). According to Reg. (EC) No. 354/2014, the use of ethylene in organic farming systems is only allowed for: (i) degreening of bananas, kiwis and kakis; (ii) degreening of citrus fruit only as part of a strategy for the prevention of fruit fly damage in citrus; (iii) flower induction of pineapple; and (iv) sprouting inhibition in potatoes and onions. As a plant growth regulator only indoor uses may be authorized, and authorizations shall be limited to professional users.

Quartz sand

Quartz is the most important sand-forming mineral, and quartz sand consists mainly of silica glass (silicon dioxide). It has been notified as a repellent for use on deciduous and coniferous trees in forestry (EFSA, 2011).

Repellents by smell of animal or plant origin/sheep fat

These repellents are applied on to the plants and work by their repellant odour in order to protect the plants from being eaten by herbivorous animals, such as red deer, roe deer and chamois. Their use is allowed only on non-edible parts of the crop and where crop material is not ingested by sheep or goats.

Other Substances

Ferric phosphate

Ferric phosphate is listed as a molluscicide in Reg. (EC) No. 889/2008. Iron phosphate minerals, such as beraunite, dufrenite, phosphosiderite and strengite, occur in nature in solid form. Iron phosphate is a stomach poison in slugs and snails. It damages their digestive tissue. When the ferric phosphate pellet is ingested by the slug or snail feeding ceases immediately. The iron phosphate then interferes with calcium metabolism within the slug and eventually causes the slug's death (Horgan, 2006). In Europe, granular bait formulation containing ferric phosphate has been placed on the market some time ago, and is still commercially available. In organic farming, ferric phosphate can be used if surface-spread between cultivated plants; it must not get into direct contact with the plants. With exception of the entomopathogenic nematode species *Phasmarhabditis hermaphrodita*, ferric phosphate is currently the only molluscicide allowed in organic farming in Europe.

Beeswax

In organic farming beeswax, produced by honeybees, is used to seal and help healing of pruning cuts and grafts, as it promotes callusing. In Europe beeswax it is not considered a plant protection product, but a pruning agent. As for many other natural products, beeswax presents a variable composition. It consists mainly of esters of fatty acids and long-chain alcohols, such as myricyl palmitate (33%), myricyl cerotate (12%), myricyl palmitoleate (12%), myricyl hydroxypalmitate (6%) (Pickthall, 1955) and others. Furthermore, it contains free acids, hydrocarbons and other substances.

Lecithin

Lecithin is a generic term that describes the group of yellow-brownish fatty substances occurring in animal and plant tissues composed of phosphoric acid, choline, fatty acids, glycerol, glycolipids, triglycerides and phospholipids. It is used by food manufacturers as an emulsifier and is a normal component of cell membranes in plant tissues. In agriculture it is mainly used as adjuvant, for its anti-drift properties. It is listed in Reg. (EC) 354/2014 as a fungicide. It acts by direct contact, and its activity seems to be related to the disturbance of hyphal penetration into the host plant (Misato *et al.*, 1983).

Potassium hydrogen carbonate (aka potassium bicarbonate)

Potassium hydrogen carbonate is a fungicide of traditional use in organic farming. The substance is listed as a fungicide and insecticide in Reg. (EC) 354/2014, and as a fungicide by IFOAM and OMRI. The same applies to sodium hydrogen carbonate (sodium bicarbonate; $NaHCO_3$), which shows the same mode of action as potassium hydrogen carbonate. Recently sodium bicarbonate has been approved at EU level as a basic substance to be used for plant protection, and therefore it could be used in organic farming.

The mode of action of bicarbonate salts is linked to the perturbation of pH, osmotic pressure and the bicarbonate/carbonate ion balance of sensitive fungi. Potassium bicarbonate acts by contact with fungi in aqueous solution and inhibits the development of fungal mycelium and spores. It is reported to damage the cell wall membrane and to disrupt cell physiology. It has a multi-site mode of action, and the risk of resistance development can be considered low (Jamar *et al.*, 2007; Milling *et al.*, 2012; Marku *et al.*, 2014). It is mainly applied against powdery mildews on grapevine, fruit and vegetable crops, and it also has successfully been tested against *B. cinerea* on grapevine, *Venturia inaequalis* (Cooke) Wint. on apple, and *Leveillula taurica* (Lev.) G. Arnaud on tomatoes and peppers (Milling *et al.*, 2012). Potassium bicarbonate may be phytotoxic, especially on certain varieties and at high application rates. The product, especially when used by itself against destructive diseases such as apple scab, may not provide optimal disease control. In these cases, the substance can be used either in a tank mixture or in combination with other control tools (e.g. sulfur) in order to obtain better field performance.

Diatomaceous earth

Diatomaceous earth (henceforth DE), also known as diatomite or Kieselgur, is a naturally occurring, soft, siliceous sedimentary rock that is easily crumbled into a fine powder. It consists of millions of fossilized remains of diatoms, a type of hard-shelled algae. Diatoms are a major group of algae, and are among the most common types of phytoplankton. A unique feature of diatom cells is that they are enclosed within a cell wall or skeleton made of silica (hydrated silicon dioxide) called a frustule. These algae grow in both lacustrine and marine environments, where they absorb the silicon dioxide dissolved in the water to form their frustules. DE forms by the accumulation of the amorphous silica remains of dead diatoms in lacustrine or marine sediments, in which the frustules have been naturally fossilized over millions of years. DE deposits accumulate very slowly over time and can be extracted in several areas around the world. The commercial deposits of DE are restricted to Tertiary or Quaternary periods (Cummins, 1960).

The mining and processing of DE is delicate and complicated. DE is usually mined in open-pit, surface mines (although some operations do use underground extraction methods). The stockpiled material is then hauled to the processing plant for crushing, drying, milling, etc. Drying and thermal processing is conducted to remove moisture, and to adapt particle size to operational requirements. The final product is a soft white talc-like powder. DE consisting of amorphous silica containing less than 0.1% crystalline silica (in the form of quartz or cristobalite) can be considered toxicologically safe to mammals (EFSA, 2012). Inhaled crystalline silica is a human carcinogen and unless it can be demonstrated that the levels of crystalline silica are below 0.1%, the substance would require the classification 'May cause cancer by inhalation' according to Reg. (EC) No. 1272/2008.

DE is used in many sectors, for example as a filter medium for swimming pools, in beer and wine production, in cosmetics (e.g. as very mild abrasive in toothpaste and facial scrubs), and even as a food additive. For centuries stored grain has been protected from insect attack by adding some form of powder or dust to it. China was the first country in which powders were used for this purpose (approximately 4000 years ago). The natives in India and South Africa as well as American Indians used ashes to protect stored grain against insect pests. Although patents for DE formulations were issued in the USA already in the late 1800s, it was not until the 1950s that the first commercial formulations of it became available. From then on a series of studies were conducted, which resulted in an increased use of DE in agriculture. Also in the case of stored grain protection, the reawakened interest in alternative control tools was primarily due to the development of resistance by insects, contamination of foodstuffs with residues, and exposure of users to toxic chemicals, rather than due to a request for solutions from organic farming.

DE formulations used for stored grain protection consist mainly of silicon dioxide. It acts through absorption of the lipid layer of arthropod's chitin and abrasion which leads to desiccation of the pest (Subramanyam and Roesli, 2000). The target insects usually die within a few days, but DE can also show a long-lasting residual insecticidal effect provided that humidity and temperature conditions are adequate (effectiveness decreases as humidity increases, and increases as temperatures increase) (Arthur, 2000; Ciesla and Guéry, 2014).

DE-based products are mainly applied as insecticides and/or acaricides to stored grain, such as wheat, barley, maize, rice, sunflower, soybean, beans, chickpeas, lentils, peas and other grains. They can be applied in all places where dry grain is stored or transported, such as silos, containers, railcars, ships, trucks, boxes, cans and bags. Before applying the treatment, the dry grain should be checked whether it is already infested by insects. The product should preferably be mixed with the grain prior to transfer into the storage room/container. It is important to provide for an equal and uniform distribution of DE within the grain, especially when DE is applied during filling of containers or silos. DE can successfully be used to control numerous storage pests, such as the rice weevil, grain weevil and maize weevil (*Sitophilus oryzae* (Linnaeus), *Sitophilus granarius* (Linnaeus) and *Sitophilus zeamais* (Motschulsky), respectively), the mealworm beetle (*Tenebrio molitor* Linnaeus), the red flour beetle and confused flour beetle (*Tribolium castaneum* Herbst. and *Tribolium confusum* Du Val, respectively), the pea weevil (*Bruchus pisorum* Linnaeus), the bean weevil (*Acanthoscelides obtectus* Say), the sawtoothed grain beetle (*Oryzaephilus surinamensis* Linnaeus), the lesser grain borer (*Rhyzopertha dominica* (Fabricius)), the rusty grain beetle (*Cryptolestes ferrugineus* Stephens), the Indian meal moth (*Plodia interpunctella* (Hubner)), the Mediterranean flour moth (*E. kuehniella*) and other Lepidopteran pests (Quarles and Winn, 1996). The use of DE can result in a reduction of the specific weight of the treated grain, but it does not affect its properties. DE is listed among the active substances allowed in organic farming according to IFOAM and OMRI, but it was only recently listed in the EU.

References

Agnello, A. (2002) Petroleum-derived spray oils: chemistry, history, refining and formulation. In: Beattie, G.A.C., Watson, D.M., Stevens, M.L., Ray, D.J. and Spooner-Hart, R.N. (eds) *Spray Oils Beyond 2000*. University of Western Sydney Press, Sydney, Australia, pp. 2–18.

Arthur, F.H. (2000) Toxicity of diatomaceous earth to red flour beetles and confused flour beetles (Coleoptera: Tenebrionidae): effects of temperature and relative humidity. *Journal of Economic Entomology* 93(2), 526–532.

Auger, P., Guichou, S. and Kreiter, S. (2003) Variations in acaracidal effect of wettable sulfur on *Tetranychus urticae* (Acari: Tetranychidae): effect of temperature, humidity and life stage. *Pest Management Science* 59, 559–565.

British Crop Production Council (BCPC) (2006) *The Pesticide Manual*, 14th edn. Edited by C.D.S. Tomlin. BCPC, Farnham, UK.

British Crop Production Council (BCPC) (2014) *Manual of Biocontrol Agents*, 5th edn. Edited by Leonard G. Copping. BCPC, Farnham, UK.

Caffi, T., Legler, S.E., Bugiani, R. and Rossi, V. (2012) Combining sanitation and disease modelling for control of grapevine powdery mildew. *European Journal of Plant Pathology* 135(4), 817–829.

Caroli, L., Valentini, F. and Pasqualini, E. (2000) *Bacillus thuringiensis* ssp. *kurstaki* e *aizawai*. In: Nicoli, G. and Radeghieri, P. (coordinators) *Gli Ausiliari nell'agricoltura Sostenibile*. CALDERINI Edagricole, Bologna, Italy, pp. 277–289.

Caroppo, S., Ambrogioni, L. and Capella, A. (2002) Valutazione dell'efficacia dell'azadiractina A (32 g/l)nel controllo di *Meloidogyne incognita* (Kofoid *et* White) Chitwood su pomodoro in ambiente protetto. *Redia* LXXXV, 2002, 121–130.

Cawoy, H., Debois, D., Franzil, L., De Pauw, E., Thonart, P. and Ongena, M. (2015) Lipopeptides as main ingredients for inhibition of fungal phytopathogens by *Bacillus subtilis/amyloliquefaciens*. *Microbial Biotechnology* 8(2), 281–295.

Chapman, P.J. (1967) Petroleum oils for the control of orchard pests. *New York State Agricultural Experiment Station Bulletin* 814, 1–24.

Ciesla, Y. and Guéry, B. (2014) Efficacy of diatomaceous earth against the rice weevil *Sitophilus oryzae* (L.) after a preventive treatment of wheat. *IOBC (International Organisation for Biological and Integrated Control) WPRS Bulletin* 98, 183–191.

Cloyd, R.A. (1999) The entomopathogenic fungus *Metarhizium anisopliae*. *Midwest Biological Control News* VI(7).

Crickmore, N., Zeigler, D.R., Feitelson, J., Schnepf, E., van Rie, J. and Lereclus, D. (1998) Revision of the nomenclature for the *Bacillus thuringiensis* pesticidal crystal proteins. *Microbiology and Molecular Biology Reviews* 62, 807–813.

Crosby, D.G. (1995) Environmental fate of pyrethrins. In: Casida, J.E. and Quistad, G.B. (eds) *Pyrethrum Flowers: Production, Chemistry, Toxicology, and Uses*. Oxford University Press, New York, pp. 194–213.

Cummins, A.B. (1960) Diatomite. In: *Industrial Minerals and Rocks*, 3rd edn. American Institute of Mining, Metallurgical, and Petroleum Engineers, New York, pp. 303–319.

Davidson, N.A., Dibble, J.E., Flint, M.L., Marer, P.J. and Guye, A. (1991) *Managing Insects and Mites with Spray Oils*. University of California (Publication 3347), Berkeley, California.

De Meyer, G., Bigirama, J., Elad, Y. and Hofte, M. (1998) Induced resistance in *Trichoderma harzianum* T39 biocontrol of *Botrytis cinerea*. *European Journal of Plant Pathology* 104, 279–286.

Eberle, K.E., Jehle, J.A. and Huber, J. (2012) Microbial control of crop pests using insect virus. In: Abral, D.P. and Shankor, V. (eds) *Integrated Pest Management: Principles and Practice*. CAB International, Wallingford, UK, pp. 281–297.

Emmett, B., Wicks, T. and Magarey, P. (2003) Sulphur formulations, particle size and activity – a review. In: *Strategic use of Sulphur in Integrated Pest and Disease Management (IPM) Programs for Grapevines*. Department of Primary Industries, Mildura, Victoria, Australia, pp. 52–57.

Environmental Protection Agency (EPA) (2000) *Nosema locustae* (117001) Fact Sheet. p. 2. Available at: http://www.epa.gov/pesticides/chem_search/reg_actions/registration/fs_PC-117001_01-Oct-00.pdf (accessed 22 June 2017).

European Commission (2004) Review Report for the Active Substance *Pseudomonas chlororaphis*. *Pseudomonas chlororaphis* 4204/VI/98-Final, 14 pp. Available at: http://ec.europa.eu/food/plant/pesticides/eu-pesticides-database/public/?event=activesubstance.detail&language=EN&selectedID=1786 (accessed 22 June 2017).

European Commission (2011) Review Report for the Active Substance Azadirachtin SANCO/10311/2011 Final, 8 pp. Available at: http://ec.europa.eu/food/plant/pesticides/eu-pesticides-database/public/?event=activesubstance.detail&language=EN&selectedID=976 (accessed 22 June 2017).

European Commission (2014) Commission Implementing Regulation (EU) No. 354/2014 of 8 April 2014 amending and correcting Regulation (EC) No. 889/2008 laying down detailed rules for the implementation of Council Regulation (EC) No. 834/2007 on organic production and labelling of organic products with regard to organic production, labelling and control. *Official Journal of the European Union* L 106, 7–14.

European Commission (2016) Commission Implementing Regulation (EU) No. 2016/673 of 29 April 2016.

European Food Safety Authority (EFSA) (2011) Conclusion on the peer review of the pesticide risk assessment of the active substance quartz sand. *EFSA Journal* 9(7), 2300 (37 pp.).

European Food Safety Authority (EFSA) (2012) Conclusion on the peer review of the pesticide risk assessment of the active substance kieselgur (diatomaceous earth). *EFSA Journal* 10(7), 2797 (35 pp.).

Falcon, L.A., Kane, W.R. and Bethell, R.S. (1968) Preliminary evaluation of a granulosis virus for control of the codling moth. *Journal of Economic Entomology* 61, 1208–1213.

Falk, S.P., Gadoury, D.M., Cortesi, P., Pearson, R.C. and Seem, R.C. (1995a) Parasitism of *Uncinula necator* ascomata by the mycoparasite *Ampelomyces quisqualis*. *Phytopathology* 85, 794–800.

Falk, S.P., Gadoury, D.M., Pearson, R.C. and Seem, R.C. (1995b) Partial control of grape powdery mildew by the mycoparasite *Ampelomyces quisqualis*. *Plant Disease* 79, 483–490.

Flint, H.M. and Doane, C.C. (2013) Understanding semiochemicals with emphasis on insect sex pheromones in integrated pest management programs. In: Radcliffe, E.B., Hutchison, W.D. and Cancelado, R.E. (eds) *Radcliffe's IPM World Textbook*. University of Minnesota, St Paul, Minnesota. Available at: http://ipmworld.umn.edu (accessed 22 June 2017).

Food and Agriculture Organization of the United Nations (FAO) (2006) FAO Specifications and Evaluations for Azadirachtin. 23 pp. Available at: http://www.fao.org/fileadmin/templates/agphome/documents/Pests_Pesticides/Specs/azadirachtin2006.pdf (accessed 22 June 2017).

Gilbert, S.I. and Gill, S.S. (eds) (2010) *Insect Control: Biological and Synthetic Agents*. Academic Press, San Diego, California, 470 pp.

Gill, G.S. and Lewis, C.T. (1971) Systemic action of an insect feeding deterrent. *Nature* 232(5310), 402–403.

Greenpeace (2005) *Essen ohne Pestizide: Einkaufsratgeber für Obst und Gemüse*. Greenpeace, Hamburg, Germany, 25 pp.

Gunasekara, A.S. (2004) Environmental fate of pyrethrins. California Department of Pesticide Regulation, p. 22. Available at: http://www.cdpr.ca.gov/docs/emon/pubs/fatememo/pyrethrin_efate2.pdf (accessed 22 June 2017).

Haas, E., Roschatt, C. and Schweigkofler, W. (2005) L'oidio in Alto Adige nel 2004: attacco e difesa. *Frutta e Vite* 3, 77–81.

Harman, G.E., Howell, C.R., Viterbo, A., Chet, I. and Lorito, M. (2004) *Trichoderma* species – opportunistic, avirulent plant symbionts. *Nature Reviews Microbiology* 2, 43–56.

Hewitt, H.G. (1998) *Fungicides in Crop Protection*. CAB International, Wallingford, UK.

Hökeberg, M. (2006) *Pseudomonas chlororaphis* MA 342, active organism in the Lantmännen BioAgri products Cedomon® and Cerall®. Presented at Rebeca Workshop, 18–22 September 2006, Salzau, Germany, p. 2.

Horgan, A.R. (2006) The potential for slug control with ferric phosphate. *Aspects of Applied Biology* 79, 225–226.

Ibrahim, M.A., Griko, N., Junker, M. and Bulla, L.A. (2010) *Bacillus thuringiensis*. A genomics and proteomics perspective. *Bioengineered Bugs* 1(1), 31–50.

International Federation of Organic Agriculture Movements (IFOAM) (2014) *The IFOAM Norms for Organic Production and Processing*. Die Deutsche Bibliothek, 134 pp. Available at: http://infohub.ifoam.bio/sites/default/files/ifoam_norms_version_july_2014.pdf (accessed 22 June 2017).

Jamar, L., Lefrancq, B. and Lateur, M. (2007) Control of apple scab (*Venturia inaequalis*) with bicarbonate salts under controlled environment. *Journal of Plant Diseases and Protection* 114(5), 221–227.

Jehle, A., Blissard, G.W., Bonning, B.C., Cory, J.S., Hernlou, E.A., Rohrmann, G.F., Theilmann, D.A., Thiem, S. M. and Vlak, J.M. (2006) On the classification and nomenclature of baculoviruses: a proposal for revision. *Archives of Virology* 151, 1257–1266.

Kiss, L. (2008) Intracellular mycoparasites in action: interactions between powdery mildew fungi and *Ampelomyces*. In: *British Mycological Society Symposia Series*. The British Mycological Society, Elsevier Ltd, Dordrecht, The Netherlands, pp. 37–52.

Kiss, L., Russell, J.C., Szentiványi, O., Xu, X. and Jeffries, P. (2004) Biology and biocontrol potential of *Ampelomyces* mycoparasites, natural antagonists of powdery mildew fungi. *Biocontrol Science and Technology* 14(7), 635–651.

Klarzynski, O., Plesse, B., Joubert, J. M., Yvin, J.-C., Kopp, M., Kloareg, B. and Fritig, B. (2000) Linear β-1,3 glucans are elicitors of defense responses in tobacco. *Plant Physiology* 124(3), 1027–1038.

Lacey, L.A., Thomson, D., Vincent, C. and Arthurs, S.P. (2008) Codling moth granulovirus: a comprehensive review. *Biocontrol Science and Technology* 18, 639–663.

Ladurner, E., Benuzzi, M., Braggio, A. and Franceschini, S. (2013) Microbial control of European red spider mite (*Panonychus ulmi*) with *Beauveria bassiana* strain ATCC 74040. *IOBC (International Organisation for Biological and Integrated Control) WPRS Bulletin* 90, 51–55.

Lange, C.E. and Cigliano, M.M. (2005) Overview and perspectives on the introduction and establishment of the grasshopper (Orthoptera: Acridoidea) biocontrol agent *Paranosema locustae* (Canning) (microsporidia) in the Western pampas of Argentina. *Vedalia* 12(1), 61–84.

Large, E.C. (1958) *The Advance of the Fungi*. Jonathan Cape, London.

Lorito, M., Woo, S.L., Harman, G.E. and Monte, E. (2010) Translation research on *Trichoderma*: from 'omics to the fields. *Annual Review of Phytopathology* 48, 1–23.

Marku, L., Vrapi, H. and Hasani, M. (2014) Effect of potassium bicarbonate (Armicarb) on the control of apple scab (*Venturia inequalis*) in the region of Puka in Albania. *International Refereed Journal of Engineering and Science* 3(6), 25–30.

Miller, J.R. and Gut, L.R. (2015) Mating disruption for the 21st century: matching technology with mechanism. *Environmental Entomology* 44(3), 427–453.

Milling, R., Laffranque, J.-P. and Orpella, M. (2012) Armicarb® – a new bio-fungicide for use in organic and conventional fruit growing in Europe. In: *Proceedings of the 14th International Conference on Organic Fruit-Growing Eco-Fruit*. Fördergemeinschaft Ökologischer Obstbau e.V., Weinsberg. Germany, pp. 264–267.

Misato, T., Homma, Y. and Ko, K. (1983) The development of a natural fungicide: soybean lecithine. *Netherlands Journal of Plant Pathology* 83(suppl.1), 395–402.

Nath, P., Trivedi, P.K., Sane, V.A. and Sane, A.P. (2006) Role of ethylene in fruit ripening. In: Khan N.A. (ed.) *Ethylene Action in Plants*. Springer, Berlin, pp. 151–184.

National Research Council (1992) *Neem: A Tree For Solving Global Problems*. National Academy Press, Washington, DC.

Ongena, M. and Jacques, P. (2007) *Bacillus* lipopeptides: versatile weapons for plant disease control. *Trends in Microbiology* 16(3), 115–125.

Ortiz-Urquiza, A., Riveiro-Miranda, L., Santiago-Álvarez, C. and Quesada-Moraga, E. (2010) Insect-toxic secreted proteins and virulence of the entomopathogenic fungus *Beauveria bassiana*. *Journal of Invertebrate Pathology* 105, 270–278.

Ortu, S., Cocco, A. and Dau, R. (2009) Evaluation of the entomopathogenic fungus *Beauveria bassiana* strain ATCC 74040 for the management of *Ceratitis capitata*. *Bulletin of Insectology* 62, 245–252.

Osborne, L.S. and Petitt, F.L. (1985) Insecticidal soap and the predatory mite, *Phytoseiulus persimilis* (Acari: Phytoseiidae), used in management of the two-spotted spider mite (Acari: Tetranychidae) on greenhouse grown foliage plants. *Journal of Economic Entomology* 78, 687–691.

Palmieri, M.C., Perazzolli, M., Matafora, V., Moretto, M., Bachi, A. and Pertot, I. (2012) Proteomic analysis of grapevine resistance induced by *Trichoderma harzianum* T39 reveals specific defence pathways activated against downy mildew. *Journal of Experimental Botany* 63(17), 6237–6251.

Paranjape, K., Gowariker, V., Krishnamurthy, V.N. and Gowariker, S. (2014) *The Pesticide Encyclopedia*. CAB International, Wallingford, UK.

Pickthall, J. (1955) The effects upon emulsions of the hydroxy compounds in beeswax. *Journal of the Society of Cosmetic Chemists* 6(4), 263–275.

Porter, J.R. (1973) Agostino bassi bicentennial. *Bacteriological Reviews* 37(3), 284–288.

Quarles, W. and Winn, P.S. (1996) Diatomaceous earth and stored product pests. *The IPM Practitioner* XVIII (5/6), 1–9.

Quesada-Moraga, E., Landa del Castillo, B.B. and López-Díaz, C. (2013) Vertical transmission of an endophytic strain of *Beauveria bassiana* (Ascomycota; Hypocreales) colonizing opium poppy *Papaver somniferum*. *IOBC (International Organisation for Biological and Integrated Control) WPRS Bulletin* 90, 68.

Renou, M. and Guerrero, A. (2000) Insect parapheromones in olfaction research and semiochemical-based pest control strategies. *Annual Review of Entomology* 45, 605–630.

Rohrmann, G.F. (2013) *Baculovirus Molecular Biology*, 3rd edn. National Center for Biotechnology Information, Bethesda, Maryland.

Ruiu, L., Falchi, G. and Ladurner, E. (2013) *Beauveria bassiana* strain ATCC 74040 interferes with oviposition behavior of Mediterranean fruit fly. *IOBC (International Organisation for Biological and Integrated Control) WPRS Bulletin* 90, 43–46.

Schmutterer, H. (1995) *The Neem Tree*. VCH Publishers, New York.

Scopes, L.E.A. (1985) Red spider mites and the predator *Phytoseiulus persimilis*. In: Hussey, N.W. and Scopes, L.E.A. (eds) *Biological Pest Control. The Glasshouse Experience*. *Bulletin OILB-SROP* [*Cette Organisation Internationale de Lutte Biologique (OILB)-Section régionale ouest-paléarctique (SROP)*] 1985, pp. 43–52.

Stansly, P.A., Liu, T.X. and Schuster, D.J. (2002) Effects of horticultural mineral oils on a polyphagous whitefly, its plant hosts and its natural enemies. In: Beattie, G.A.C., Watson, D.M., Stevens, M.L., Ray, D.J. and Spooner-Hart, R.N. (eds) *Spray Oils Beyond 2000*. University of Western Sydney Press, Sydney, Australia, pp. 120–133.

Strasser, H., Strauch, O., Ehlers, R.-U. and Hauschield, R. (2007) Positive list of 'low risk' candidates. Deliverable n. 12, REBECA [Registration of Biological Control Agents], Project no. SSPE-CT-2005-022709, 37 pp. Available at: http://www.rebeca-net.de/downloads/report/deliverable%2012.pdf (accessed 22 June 2017).

Subramanyam, B. and Roesli, R. (2000) Inert dusts. In: Subramanyam, B. and Hagstrum, D.W. (eds) *Alternatives to Pesticides in Stored-Product IPM*. Kluwer, Boston, Massachusetts, pp. 321–380.

Tanada, J. (1964) A granulosis-virus of the codling moth *Carpocapsa pomonella* L. (Olethreitidae, Lepidoptera). *Journal of Insect Pathology* 6, 378–380.

Taverner, P. (2002) Drowning or just waving? A perspective on the ways petroleum-based oils kill arthropod pests of plants. In: Beattie, G.A.C., Watson, D.M., Stevens, M.L., Ray, D.J. and Spooner-Hart, R.N. (eds) *Spray Oils Beyond 2000*. University of Western Sydney Press, Sydney, Australia, pp. 78–87.

Thacker, J.R.M. (2002) *An Introduction to Arthropods Pest Control*. Cambridge University Press, Cambridge.

Thind, T.S. (2012) *Fungicide Resistance in Crop Protection: Risk and Management*. CAB International, Wallingford, UK, 285 pp.

Thomson, L. (2012) Pesticide impacts on beneficial species. Factsheet May 2012. Australian Government, Grape and Wine Research and Development Corporation, Adeleide, Australia, 7 pp.

Tombolini, R., van der Gaag, D.J., Gerhardson, B. and Jansson, J.K. (1999) Colonization pattern of the biocontrol strain *Pseudomonas chlororaphis* MA 342 on barley seeds visualized by using green fluorescent protein. *Applied and Environmental Microbiology* 65(8), 3674–3680.

Tomlin, C.D.S. (2000) *The Pesticide Manual*, 12th edn. British Crop Production Council, Farnham, UK, pp. 178–179.

Welker, N.E. and Campbell, L.L. (1967) Unrelatedness of *Bacillus amyloliquefaciens* and *Bacillus subtilis*. *Journal of Bacteriology* 94(4), 1124–1130.

Welter, S.C., Pickel, C., Millar, J., Cave, F., van Steenwyk, R.A. and Dunley, J. (2005) Pheromone mating disruption offers selective management options for key pests. *California Agriculture* 59(1), 16–22.

Zanzotto, A., Bellotto, D. and Borgo, M. (2005) Efficacia antioidica di AQ10 ed effetti sulla formazione di cleistoteci in vigneto. *Informatore Fitopatologico* 7–8, 68–76.

3 Natural Enemies and Pest Control

Vincenzo Vacante* and Carmelo Peter Bonsignore
Mediterranean University, Salita Melissari, Italy

Introduction

A number of invertebrates, mainly nematodes, mites and insects, directly or indirectly affect the human economy. According to ecological conditions and their evolutionary adaptations, some of these may damage human health and that of domestic and wild animals while others offend our sense of aesthetics or cause severe economic damage to cultivated plants. There is a wide spectrum of different types of damage caused and these have been variously studied. Phytophagous species living on cultivated plants and/or their products are commonly referred to as pests; the damage they cause induces a deleterious effect on the physiology of the host plant and/or on the quality and quantity of their products, and the resulting loss can be defined as the amount of negative economic impact measured in terms of utility or production of the plant species attacked. The relationship between pests and cultivated plants can be approached from various viewpoints, for example taxonomic, morphological, bio-ecological and economic.

Usually, outbreaks of pest populations are associated with changes in the ecological stability of ecosystems, due to the action of ecological perturbations that are able to influence different factors, among which are the populations of natural enemies of pests. To all this is added the introduction of pests from other areas of origin, that are able to spread rapidly with serious consequences for cultivated plants and for biodiversity. In some cases the pest is considered to be at low-pest density but is still of concern as the species may be a vector of a plant, animal or human disease. Knowledge of the ecological factors that govern ecosystems helps to rationalize the pest control, highlighting the most suitable means and strategies, including the priority action of natural enemies (that may be indigenous and/or introduced).

The use of these natural resources enables the biological control of pests, a strategy that has been known historically for centuries in different countries (e.g. China and ancient Egypt), but has technically established itself during the last century when the scientific assessment of its economic and ecotoxicological benefits began. The relationship between pests and their natural enemies (and the possible application of this in the field) involves various ecological factors. Of these, the flows of matter and energy within ecosystems, their stability and the perturbations

* Corresponding author: vvacante@unirc.it

that may occur within them play a key role. Understanding these factors make it possible to examine the role of natural enemies in ecosystems and their possible use in the field. Unfortunately, there is not space here in the current book to remind the reader of these concepts and it is suggested that the interested reader consults specific texts that cover the applied ecology of production systems. Here we simply treat the natural enemies of pests and their implications, including their contribution to organic farming.

Natural Enemies and Biological Control

In each ecosystem (wild and/or artificial) the containment of pest populations is strongly related to the maintainance of its stability and integrity. Predation, parasitism and parasitoidism of pests by natural enemies play a key role and have occurred since the evolution of the first terrestrial ecosystems some 500 million years ago. It acts in all of the world's ecosystems without any human intervention, and from an economic point of view biological control has made a considerable contribution to agriculture (Waage and Greathead, 1988). Control of pests by natural enemies living in nature or through their inoculation (classical biological control), conservation and/or augmentation in the agroecosystems defines 'biological control'. A more 'artificial' but less restrictive concept of biological control includes among the biological means also some 'non-living means', consisting of extracts of various natural substrates (animal, vegetable, etc.). In the 1990s the European Union (EU) regulated biological control with specific legislation (Annex II B, EEC Regulation 2092/91, repealed in June 2007 by Reg. No. 834/2007), which also included various biological substances of plant origin (e.g. azadirachtin, natural pyrethrins) commonly known as biopesticides, as well as certain chemicals (petroleum oils, narrow range oils) which are distillates from petroleum. For a detailed discussion on biopesticides see Grant *et al.* (2010) and Benuzzi and Ladurner (Chapter 2, this volume). Despite this last technical possibility, according to the aspects mentioned above biological control *sensu stricto* exploits the action of different organisms belonging to various taxa (viruses, bacteria, fungi, nematodes, spiders, mites, molluscs, insects, reptiles, birds and mammals). Among these, the arthropods play the most significant role and their 'manipulation' represents a strong scientific challenge involving plant protection, ecological and socio-economic implications. The topic has been extensively investigated in the last century and is supported by a rich literature that is difficult to summarize. Here we outline the main aspects, and invite the reader who is interested in further insights to consult specialist contributions.

Common pathogens of pests include viruses, bacteria, fungi and microspora. The frequent association of arthropods with microorganisms is well documented (van der Geest, 1985, 2010; Tanada and Kaya, 1993; McCoy, 1996; Boucias and Pendland, 1998; Poinar and Poinar, 1998; van der Geest *et al.*, 2000; Bruin and van der Geest, 2009); some pathogens contamine the pests externally, and exhibit a variable typology with the local environment (Lacey and Kaya, 2007). Viruses have been investigated as potential control agents of insect pests, especially the Lepidoptera (Hunter-Fujita *et al.*, 1998; Miller and Ball, 1998) but less is known about viruses that infect mites and ticks (Hoy, 2011). Bacteria and fungi are common pathogens of arthropods and most species belong to the division Eumycotina. Among the family *Bactericeae*, *Bacillus thuringiensis* Berliner (and other species) is a well-known sporigenous species that is nowadays produced by industry and its different strains are sold for the control of various pest insects, such as moths (Lepidoptera), beetles (Coleoptera) and flies (Diptera). The most important pathogenic fungi are ascribed to the subdivisions Deuteromycotina, Mastigomycotina and Ascomycotina (Samson *et al.*, 1988). The Deuteromycotina include pathogens of pest mites, such as eriophyiids and spider mites and insect pests (Coccoidea, Aleyrodidae, Aphididae). Common species of this taxon are *Beauveria bassiana* Vuillemin, *Metharhizium anisopliae* (Metschnikoff) Sorokin, *Paecilomyces* sp., *Verticilium lecanii* (Zimmerman) Viegas, *Aschersonia aleirodidis*

Webber (Samson and McCoy, 1983) and *Hirsutella thompsonii* Fisher (Deuteromycota or mitosporic fungi) (Bell, 1974). About 150 species in the genus *Entomophthora* (Mastigomycotina) are pathogens of mites and insects. Different species of the Exobasidiomycetidae (*Meira argovae* Boekhout *et al.*, *Meira geulakonigii* Boekhout *et al.*, *Acaromyces ingoldii* Boekhout *et al.*) develop on citrus rust mite, *Phyllocoptruta oleivora* (Ashmead), and some spider mites (Paz *et al.*, 2007). About 58 fungal pathogens have been found infecting at least 73 species of mites in the Acaridida, Oribatida, Thrombidiforma, Gamasida or Ixodida in either the field or the laboratory (Chandler *et al.*, 2000). Most fungal pathogens are in the Entomophthorales (Entomophthoromycotina, formerly Zygomycota), Ascomycota and the imperfect (or Deuteromycota) fungi, which are all transmitted horizontally. A number of species, such as *H. thompsonii* and *Neozygites floridana* (Weiser *et* Muma) (Entomophthorales), are specific to mites, while other fungal species may kill both mites and insects (Gerson *et al.*, 1979; Chandler *et al.*, 2000). The topic has been extensively covered by Evans and Prior (1990).

Most of the important taxa of arthropods involved in biological control include the Araneae, Acari and insects. The Araneae are generalist predators affecting the stability of pest populations and are also commonly used as biological indicators of the integrity of the ecosystems; they are called spiders, and include about 35,000 species in 80 families. Araneae live worldwide in different ecosystems, for example in the citrus of the Mediterranean region where dozens of families and species are known (Benfatto *et al.*, 1992; Ribeiro *et al.*, 2006; Monzó *et al.*, 2008). The Acari are commonly referred to as mites and include approximately 55,000 described species (the group will grow to number over 1,000,000 species) (Walter and Proctor, 1999). They number about 5500 genera and 550 families (Krantz, 2009), and almost three dozen of these families are involved in the biological control of pests. Of these, some families (Phytoseiidae, Cheyletidae, Stigmaeidae, Cunaxidae and Hemisarcoptidae) are very important in the control of mite pest populations (Gerson *et al.*, 2003; Gerson, 2014; Vacante, 2015). Other species such as centipedes (Chilopoda), scorpions and false scorpions (Pseudoscorpions) are important natural enemies. Despite the great importance in the biological control of the groups referred to above, the most important natural enemies belong to the insects and the orders Hemiptera (Anthocoridae, Miridae), Neuroptera (Chrysopidae, Conioterygidae), Diptera (Cecidomyiidae, Muscidae, Syrphidae), Coleoptera (Alleculidae, Anthribidae, Cantharidae, Coccinellidae, Cybocephalidae, Endomychidae, Nitidulidae, Staphylinidae and Tenebrionidae) and Hymenoptera (Braconidae, Platygastridae, Pteromalidae, Encyrtidae, Eulophidae, Aphelinidae). The Coleoptera include about 300,000 species and 150 families (Balduf, 1935; Drea, 1990), and the Hymenoptera more than 200,000 species (Gauld and Bolton, 1988), most of which are important in pest control. Some species exhibit an ambiguous feeding habit, for example the mirid bug *Nesidiocoris tenuis* (Reuter), which on tomato plants both preys on the whitefly populations, *Trialeurodes vaporariorum* (Westwood), and feeds on various organs of the plants (Vacante and Tropea Garzia, 1994). Other insects adopt different nutritional strategies with reference to the different life stages of development. This happens in staphylinid beetles that are parasitoids when juvenile and predatory as adults.

When crop producers are able to have the support of natural enemies as predators or parasitoids for insect pest management, it is clearly important not to destroy these natural enemies using insecticides to contain the parasite. The effects of pesticides on the natural enemies of pests are well documented. So today it is increasingly desirable to intervene with insecticidal compounds that can act in harmony with predators and parasitoids, or at least reduce the effects against natural enemies.

Implementation of Biological Control

As discussed above, the implementation of biological control exploits the action of natural enemies living in the field using one or more of the three above-mentioned techniques, that is inoculation (or classical biological control), conservation or augmentative biological control. These may be integrated with various technical options (agronomic, etc.). Some natural enemies may be absent in crops grown in new areas or may be unable to survive in the new habitat. This lack of natural enemies calls for research into the natural enemies of pests in the areas where the host plants are native or to areas where the natural enemies of the pests are reported. For example, specific programmes of introduction and acclimatization of predatory mites are being conducted worldwide in different crops. The topic is articulated in the phases of introducing, culturing and establishing of biological agents, and has been extensively discussed by Gerson *et al.* (2003). A review of experiences carried out on citrus in the Mediterranean region was produced by Vacante (2012). Many other cases of different natural enemies have been discussed by different specialists.

The above options are similar to those provided by integrated pest management (IPM), with exception of the very restricted use or absence of the threshold concept, resulting in the prevalent preventative measures of choice. Implementation of biological control needs the availability of a complete control programme for each crop. Moreover, it is necessary that the total costs of this programme are similar to the costs of chemical control, and the extension service is reliable, independent and not compromised by the pesticide industry (van Lenteren, 2012). In addition, the implementation of a biological control programme provides the realization of four steps: (i) prevention of pest introduction; (ii) identification of pests and natural enemies; (iii) monitoring of pests and natural enemies; and (iv) use of biological means for pest control.

Prevention of pest introduction is basically to prevent the economic damage associated with the development of exotic pests in new environments. Specific means (regulatory, technical, etc.) and facilities enable management of such an introduction and its prevention. The problem is current and involves both pathogical and entomological aspects, for example the introduction of the red palm weevil (Vacante, 2013) on palms, or the accidental introduction of the *Citrus* tristeza virus (CTV) in Mediterranean citriculture (Bar-Joseph and Catara, 2012).

Identification of pests and natural enemies is essential for the implementation of a biological control programme (Bonsignore and Vacante, 2012a). Various examples testify the risk associated with misidentification. A recent interesting case concerns the identification of *Cales noacki* Howard (Aphelinidae), parasitoid of the woolly whitefly, *Aleurothrixus floccosus* (Maskell) (Aleyrodidae). Ongoing studies seem to indicate that *C. noacki* exists as a 'cryptic species complex', that is a group of individual organisms which are virtually morphologically uniform in appearance but each is in fact a distinct biological species (Mottern and Heraty, 2014). This event, apart from the practical point of view, opens very questionable scenarios, and points out the expensive and almost useless release programme of the parasitoid in Italian citriculture, *C. noacki* being already largely present in the field and able to develop spontaneously. Other significant examples of misidentification of natural enemies (and the negative impact on biological control) are discussed from time to time in the literature (DeBach, 1974; Rosen and DeBach, 1979; Rosen, 1986).

Monitoring is a fundamental step to assess the dynamics of pests and natural enemy populations in the field, in order to estimate their presence, possible damage caused by pests, and population development of natural enemies. Each pest–natural enemies complex needs the implementation of specific protocols, according to the crop(s) and environmental conditions.

Implementation of natural, inoculative and conservative biological control depends largely on public funding; conversely, augmentative biological control, with the exception of some countries, is largely related to economic activities and needs the mass production and release of natural enemies. Moreover, biological control includes other 'non-living means'. In some cases the use of some botanical pesticides and nematodes against an insect pest could be uneconomical and possibly disadvantageous, in particular where there is an established predator–prey population (Bonsignore and Vacante, 2012b).

Natural biological control varies according to the ecosystem, and in addition to the action of indigenous fauna may exploit the actions of a number of exotic natural enemies variously introduced over time with classical biological programmes, especially in perennial crops. Different examples confirm the high value of natural biological control in extremely specialized agroecosystems, for example protected crops of the Mediterranean and European temperate regions. Some examples help to better explain this topic. In these 'artificial' environments natural populations of *Diglyphus isaea* (Walker) (Eulophidae) produce a serious control of different leafminers (Fig. 3.1), and this represents one of the most successful cases of natural biological control known to date. Similarly, the natural development of populations of the minute pirate bug, *Orius laevigatus* (Fieber) (Anthocoridae) may produce a strong control of western flower thrips, *Frankliniella occidentalis* (Pergande) (Fig. 3.2), and in particular conditions even control by the phytoseiid mite, *Phytoseiulus persimilis* Athias-Henriot, may suppress the development of populations of the two-spotted spider mite, *Tetranychus urticae* Koch (Tetranychidae) (Fig. 3.3). Moreover, populations of *T. urticae* are often suppressed by the natural development of the cecidomyid *Feltiella acarisuga* (Vallot) (= *Therodiplosis persicae* (Kieffer)) (Fig. 3.4). Among the cases documented in the literature, that of the plant bug *N. tenuis* (= *Cyrtopeltis tenuis* Reuter) (Miridae) is probably the most emblematic. The role of this mirid in the control of greenhouse whitefly, *T. vaporariorum* was reported by Arzone *et al.* (1990) and successively evaluated on tomato in the Mediterranean cold greenhouses of Sicily (Italy) by Vacante and Tropea Grazia (1994)

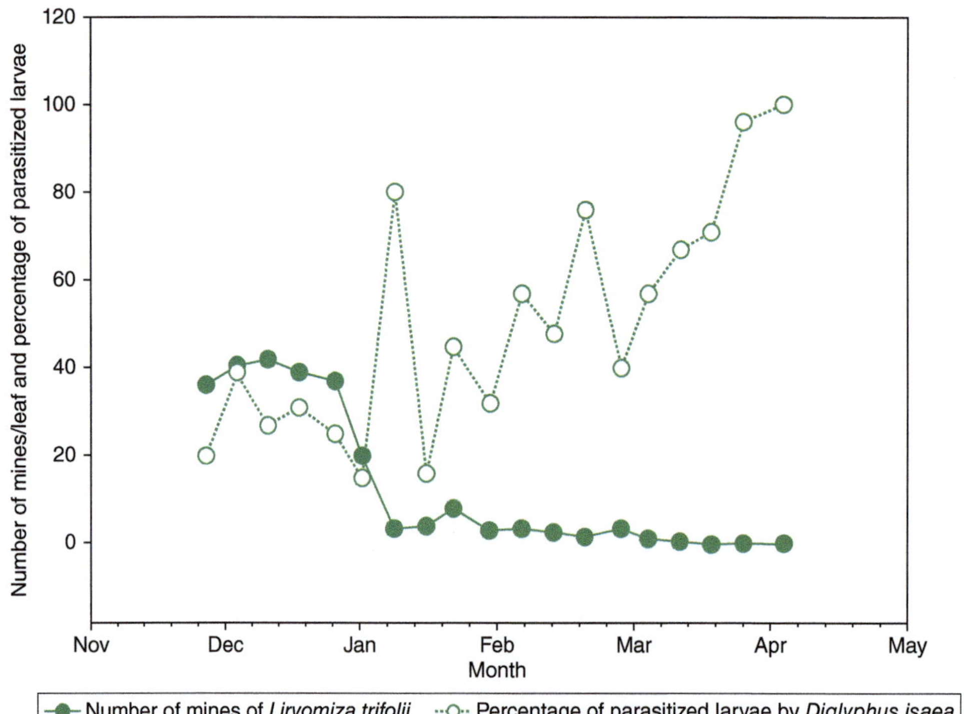

Fig. 3.1. Development of a natural population of the eulophid *Diglyphus isaea* (Walker) on aubergine infested by the leafminer (*Liryomiza trifolii* Burgess) in a cold greenhouse in Sicily (Italy). (Modified from Vacante, 2000.)

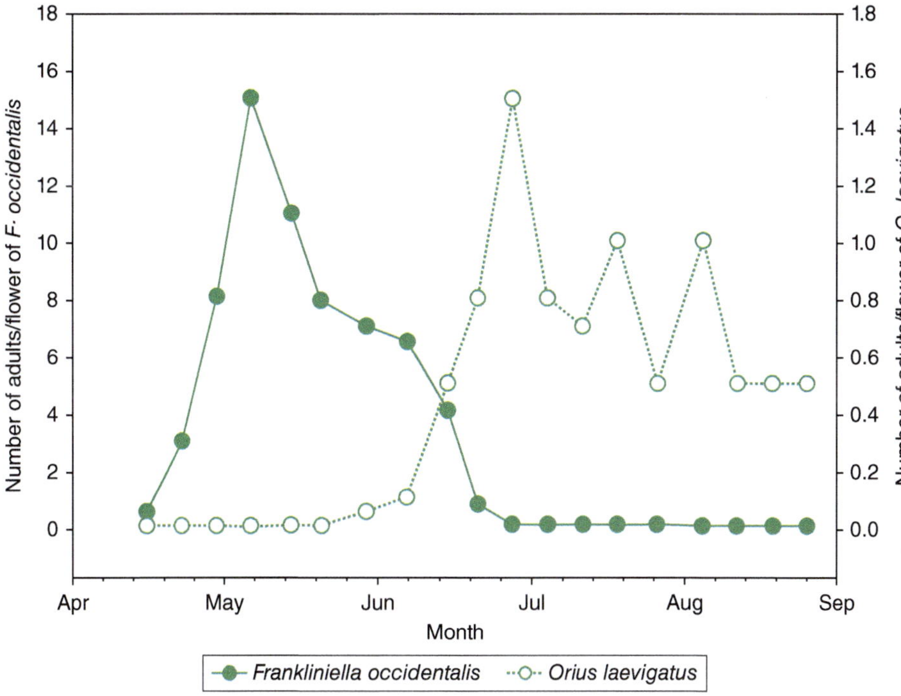

Fig. 3.2. Development of a natural population of the minute pirate bug, *Orius laevigatus* (Fieber), on a pepper infested by western flower thrips, *Frankliniella occidentalis* (Pergande) in a cold greenhouse in Sicily (Italy). (Modified from Vacante, 2000.)

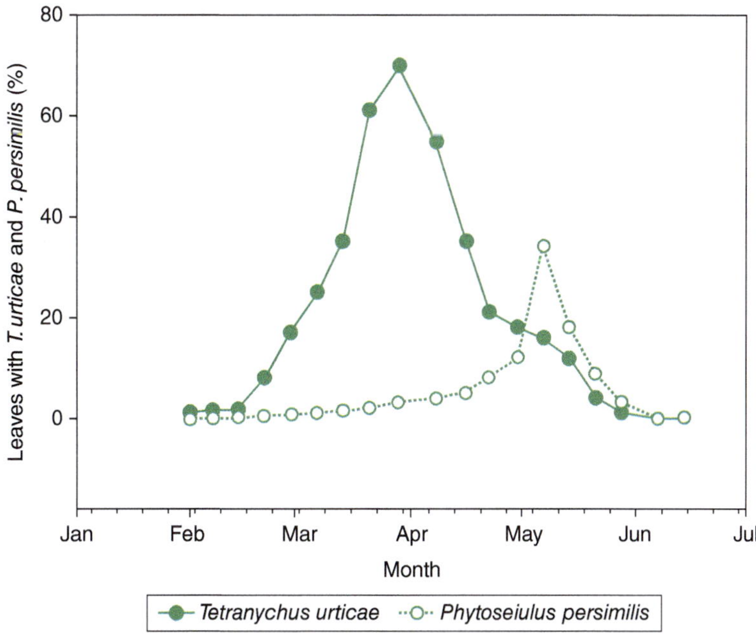

Fig. 3.3. Development of a natural population of the phytoseiid mite, *Phytoseiulus persimilis* Athias Henriot, on pepper infested by the two-spotted spider mite, *Tetranychus urticae* Koch in a cold greenhouse in Sicily (Italy). (Modified from Vacante, 2000.)

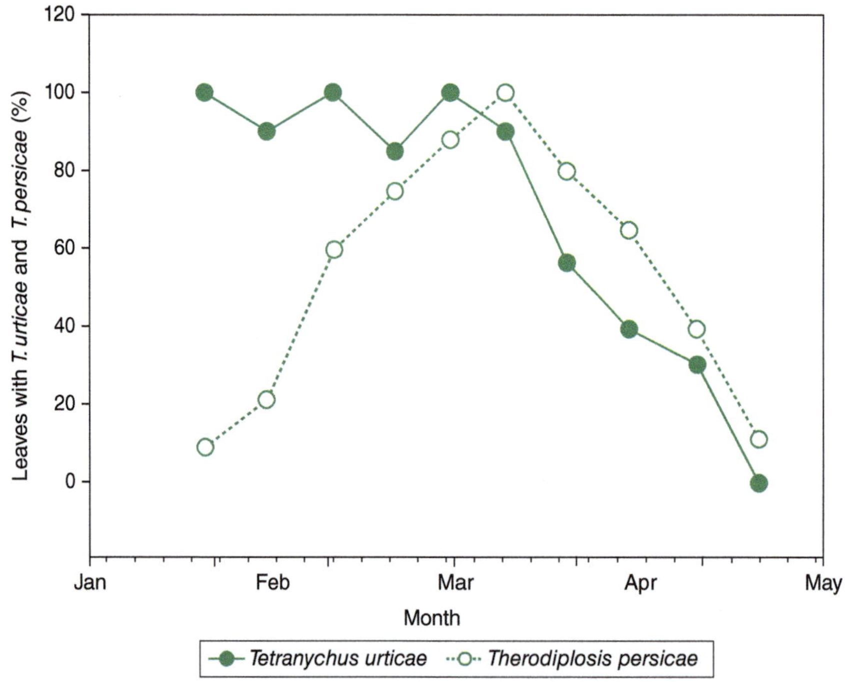

Fig. 3.4. Development of a natural population of the cecidomyid *Therodiplosis persicae* (Kieffer) on rose infested by *Tetranychus urticae* Koch in a cold greenhouse in Sicily (Italy). (Modified from Vacante, 1981.)

(Fig. 3.5). However, the singular adaptation of the mirid to phytophagy and the risk of its damage to host plants (Malausa and Henao, 1988; Malausa, 1989) suggested for some decades the release of other natural enemies, for example *Macrolophus caliginosus* (Wagner) (Miridae) or *Encarsia formosa* Gahan (Aphelinidae), reporting a general failure in the control of the greenhouse whitefly in the Western Mediterranean. With exception of the writers, very few other specialists have taken into account that the populations of the mirid constantly develop when the use of insecticides is suppressed in conjunction with the biological control programme with other natural enemies (or in the absence of biological control). Recently, an intensive research study has enabled better understanding of the bio-ecology of the mirid and the impact of its population with greenhouse crops. So, Sánchez and Lacasa (2008), to estimate the density thresholds, predicted the variation in truss weight as a function of fruit weight and aborted fruit. The two latter variables were expressed as a function of the cumulative number of *N. tenuis* (CNN). A maximum of 15% truss weight overcompensation was predicted at 15.8% of fruit abortion. Yield reduction was predicted at fruit abortion rates > or = 27.7%, which corresponded to 566 CNN per plant or 32.11 CNN per leaf. *N. tenuis* may be considered a useful predator of small pests in tomato crops if kept under these thresholds. Mathematical models predict a yield increase and fruit upgrade that overcompensates for the reduction in the number of fruit below the density threshold. Calvo *et al.* (2009) investigated the behaviour of *N. tenuis* in the control of the sweet potato whitefly, *Bemisia tabaci* (Gennadius) (Aleyrodidae) and concluded that mirid predators respond differently to different plant types as well. For instance, another mirid, *Dicyphus tamaninii* Wagner (Miridae) damages tomato but not cucumber (Gabarra *et al.*, 1995; Castañé *et al.*, 1996). Thus, the balance between plant injury and biological control is determined by both the quantity and the quality of plant and prey types (Eubanks and Denno, 1997;

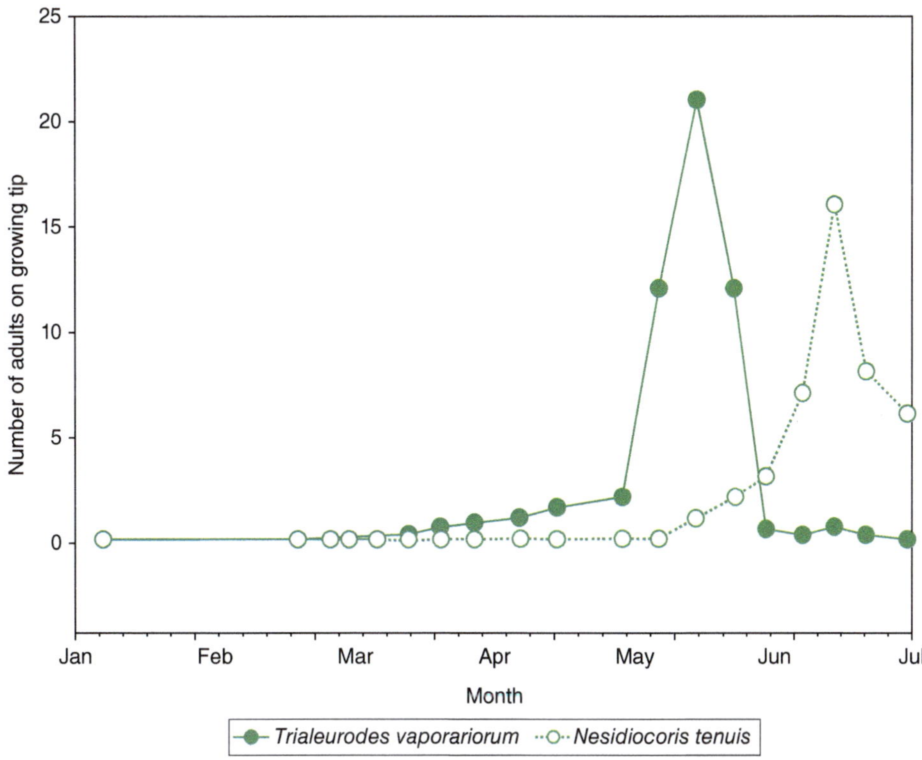

Fig. 3.5. Development of a natural population of the plant bug *Nesidiocoris tenuis* (Reuter) on tomato infested by *Trialeurodes vaporariorum* (Westwood) in a cold greenhouse in Sicliy (Italy). (Modified from Vacante and Tropea Grazia, 1994.)

Agrawal *et al.*, 1999). As a consequence, the status of a mirid species such as *N. tenuis* as pest or biological control agent will depend on crop, pest complex, and possibly other circumstances. Today the mass rearing of *N. tenuis* occurs in Spain and Morocco. Another 'ambiguous' case concerns the killer fly or hunter fly, *Coenosia attenuata* Stein (Muscidae), effective predator of little arthropods including whiteflies that are distributed worldwide (Vacante, 2000; Gerling *et al.*, 2001; Couri and Salas, 2010; Bonsignore, 2016), and largely underestimated until now. Many other natural enemies are known worldwide in various crops.

Classical biological control, so called because it was the first type of biological control applied widely, is based on the introduction of one or more natural enemies from the areas of the pest's origin, or from other regions with similar climates (DeBach, 1964, 1974; DeBach and Rosen, 1991). A typical example from the 19th century concerns the vedalia beetle, *Rodolia cardinalis* (Mulsant) (Coccinellidae), a natural enemy of cottony cushion scale, *Icerya purchasi* (Maskell) (Margarodidae), and native of Australia. Many successful examples are reported in the literature, for example 26 species of entomophagous insects have been successfully introduced into Mediterranean citriculture over the last century, of which at least half are now acclimatized and distributed in the region. Most of the citrus pests in this region are introduced species, which for various reasons are more amenable to biological control by the importation of their natural enemies (DeBach *et al.*, 1971). Citrus scale insect although having serious impact on their host species, doesn't reproduce quickly and different species tend to disperse from tree to tree slowly so that the natural enemies can have more opportunities to act against them. Even for the California red scale, *Aonidiella aurantii* Maskell (Diaspididae) the biological control has had some success in different areas of the world by releases of its parasitoid *Aphytis melinus* DeBach (Aphelinidae). Among the three biological control strategies, classical biological control is the most advantageous, as it can guarantee, when bolstered by a concrete success, sustainable pest control at low economic costs. Classical biological control may be based on the introduction of the largest possible number of natural antagonists or on that of the best natural antagonist (Huffaker *et al.*, 1976; Zeddies *et al.*, 2001). The advantages of one or more of the other strategies are not always clear. The best solution may occur through the integration of cultural practices including intercropping and the use of varieties resistant to pests, all integrated with classical biological control. Application of classical biological control is usually entrusted to universities, research centres and/or technical extension service structures (Vacante and Bonsignore, 2012).

Conservation biological control uses the implementation of different techniques to protect and stimulate the performance of naturally occurring natural enemies (Gurr and Wratten, 2000), as it includes: (i) the periodical introduction of natural enemies (the inoculative method); (ii) undertaking measures to preserve their existence (protective method); and (iii) the containment of levels of pest attacks. Horticultural practices and rational use of selective pesticides are of considerable practical importance. For example, the negative impact of unselective or partially selective chemicals can be mitigated on citrus (and on other crops) by not treating some plants per hectare in order to preserve populations in these plants of important natural enemies (*C. noacki*, *R. cardinalis* and others) (Vacante, 2009; Vacante and Bonsignore, 2012).

Augmentative biological control uses the mass distribution of natural enemies, and is a common technique of control, aimed at having a rapid, direct impact on the pest populations. The majority of natural enemies used in the implementation of this technique belong to the Arthropoda (219 out of 230 species), followed by one mollusc and ten nematodes. Only four taxonomic groups of arthropods play a significant role: (i) the Hymenoptera with 120 species; (ii) the Acari with 30 species; (iii) the Coleoptera with 28 species; and (iv) the Heteroptera with 19 species (van Lenteren, 2012). About 3500 natural enemies have been tested, with a success ratio of 1:10, and developmental costs of US$2 million over a development time of 10 years. The benefit:cost ratio is 2.5–20:1, with a small or no risk of resistance, large specificity and few or no side effects (Pimentel *et al.*, 1980; Bale *et al.*, 2008; Cock *et al.*, 2009, 2010; Pimentel, 2009).

According to crops and environment conditions, it may also provide seasonal inoculative interventions, meant to integrate with the activities of any species present or with those introduced earlier. Augmentative biological control aims to prevent, or to inhibit, the development of pest populations throughout the growing season; it does not, however, preclude any positive effects in subsequent years. The technique is more expensive than classical biological control and requires more than the mass production of natural enemies (e.g. storage, transport, technical assistance). The commercial mass production and sale of natural enemies has occurred for approximately the last 120 years, and in different crops the implementation of this method may represent an alternative to chemical pest control (van Lenteren and Bueno, 2003). According to Cock *et al.* (2010) over 170 species of natural enemies (fundamentally invertebrates) were used in augmentative biological control programmes in Europe, and in 2010 about 230 species of natural enemies included ten taxonomic groups that were applied in pest management worldwide (van Lenteren, 2012).

Regardless of the aspects discussed above, in general the impact of natural enemies on the population of insects is not easy to identify. The different approaches (observations, experiments and models) (Hunter, 2001) used to study the efficacy do not always allow the identification of real effectiveness of such biological control.

The observation of the abundance of the insects and their potential natural enemies is essential to find out what natural enemies are important in reducing the number of a particular species of insects. In various ecosystems different species of polyphagous predators are often more abundant than monospecific predators (Pons *et al.*, 2005). Since the first surveys on the effectiveness of a natural enemy it has been found that this may vary according to traits expressed by the host plant on which the herbivore is feeding (Price *et al.*, 1980). To locate their insect prey on a plant, the natural enemies can use the plant's chemical cues, so the efficacy on predation or parasitism could depend on plant traits (Brown *et al.*, 1995). The physical structure of the cuticle of plants can influence both the location as well as the capture of prey, as in the parasitoid *E. formosa* which has greater difficulty in finding *T. vaporariorum* in cucumber varieties with hairy leaves (van Lenteren *et al.*, 1995). The interaction among natural enemies and intraguild predation often affect in an unpredictable way the effects of natural enemies on the population of the insect herbivore (Rosenheim *et al.*, 1993).

Factors Affecting the Use of Biological Control

Various factors affect the implementation of biological control. The most important of these factors are discussed here, taken from different works and in particular from the comprehensive reviews of van Lenteren (2012) and Zalucki *et al.* (2015).

According to van Lenteren (2012) the factors that positively affect the implementation of biological control include the reduction of pesticide application, with a consequent increase in food security and quality, and the improvement of human health; moreover, biological control helps to control invasive exotic pests, protects biodiversity and maintains ecosystem services (Cock *et al.*, 2010). Implementation of biological control is directly related to: (i) the resistance of arthropods to various insecticides; (ii) the residue demands by food retailers and supermarket chains; (iii) the attitude of consumers; and (iv) the change in attitude of governmental institutions. However, some of these factors appear to be not well documented, for example the tendencies of consumers to favour biological control (McNeil *et al.*, 2010). In addition to the factors above, governmental institutions can stimulate the use of biological control, and the case of the European Commission (EC) is eloquent (EU Directive 2009/128/EC; EC, 2010). Interventions to reduce the number of active ingredients (750 of 1000 active ingredients are being phased out in the coming years), the development of a national action programme for IPM by 2014, and the application of IPM including biological control have provided an important incentive for biological control (van Lenteren, 2012).

Factors negatively affecting the implementation of biological control include the attitude of the pesticide industry, farmers and governmental institutions (van Lenteren, 2012). Moreover, guidelines and regulations, and the attitude of the biological control community play an important role. The specificity of many natural enemies, the impossibility to patent them or to store them for any length of time has dampened the interest of the pesticide industry in biological control; often, the use of different natural enemies cannot be integrated with chemical control and needs special training of sales personnel and farmers. Nevertheless, taking into account that chemical control alone does not control all pests, biological control became necessary and some chemical companies started to produce and/or even commercialize natural enemies (Merino-Pachero, 2007). Many farmers believe that pest control cannot renounce the use of pesticides. National or international policies to encourage the use of sustainable solutions for pest control are often unclear. Governments can effect change by enforcing the use of non-chemical pest control, but often there is a lack of funding, long-term support for research and implementation of biological control. The increasing number of guidelines and regulations is a factor limiting the application of biological control. Some of these regulations, for example the *Guidelines for the Export, Shipment, Import and Release of Biological Control Agents and Other Beneficial Organisms* (IPPC, 2005), the guidelines for environmental risk assessment and/or the national regulations for import and release of biological control agents, may delay the implementation of biological control (Bolckmans, 1999).

Conversely, Zalucki *et al.* (2015) reviewed the factors hampering the implementation of the strategy according to results of studies carried out by various authors. The most important of these factors are: (i) insufficient knowledge of natural enemies (i.e. the relationship between natural enemies and pest mortality, examination of trophic interactions between predators and their prey, etc.); (ii) the deficient application of alternative methods; (iii) the uncertainty of growers on the application of insecticides and on the impact of natural enemies; (iv) better understanding of the ecological implications of arthropod predators feeding on various types of prey; and (v) the assessment of natural enemy impact in relation to the needs of farmers and to specific pest problems.

Commonly, the natural enemies are rarely investigated, and their role in agroecosystems is largely ignored from a practical point of view. Despite the considerable amount of scientific reports on natural enemies, especially on their role in the control of pests in the field (Wiedenmann and Smith, 1997), most studies are based on confined laboratory conditions, small field plots, or small-scale interventions (Naranjo, 2001; Macfadyen *et al.*, 2014). Actually, few investigations address the effectiveness and impact of natural enemies on the pest density at a scale significant to farmers (MacFadyen *et al.*, 2015). Usually, farmers underestimate the natural enemies in the field, and the approach with 'selective' insecticides continues until they fail due to resistance (Zhao *et al.*, 2002) or the occurrence of large climate-driven population outbreaks (Maelzer *et al.*, 1996; Maelzer and Zalucki, 1999, 2000; Rochester *et al.*, 2002; Zalucki and Furlong, 2005).

The tactical approach to alternative methods of control as the 'ecosystem services' potential of natural enemies is not completely exploited to reduce the risk of pest population outbreaks. The design and build of agricultural landscapes represent a challenge, in order to decrease both the mean pest survival and its variance within the fields and across the agricultural landscape (Gurr *et al.*, 2004; Schellhorn *et al.*, 2008, 2014, 2015).

The uncertainty of growers, either real or perceived, associated with the impact of natural enemies on pest populations leads to the use of pesticides in order to relate the natural enemies to pest mortality. Furlong and Zalucki (2010) report that there have been relatively few investigations into the contribution of natural enemies (especially predators) to pest mortality and a restricted number of studies examined this impact under different pest management regimes. Moreover, MacFadyen *et al.* (2015) note that predators consume their prey and usually leave no direct evidence of their activity, other than the suppressive effects on populations; the phenomenon is more difficult to study than the effects of parasitoids, which, by comparison, can be sampled from host populations in a relatively easy way.

To improve the application of natural enemies, and get over the simple increases in 'biodiversity' it is necessary to relate natural enemies to pest mortality. The methodology necessary to achieve this aim has been discussed in detail in various different works. Most of their contents are not new, but have simply not been applied (Zalucki *et al.*, 2015). Furlong (2015) reported that ecological experiments associated with advances in molecular techniques to identify predator diets, and the diffusion of organic agriculture, provide both the mechanisms and a platform upon which many predator–prey interactions in pest management can be examined at a significant scale. This opportunity may only produce benefits if current approaches to research are modified and relevant ecological data are collected at appropriate scales. DNA-based and antibody-based techniques may be applied in studies on arthropod predation. These methodologies have been developed subsequent to the precipitin test first applied to arthropod predation 60 years ago (Dempster, 1960), and allow the investigation of the trophic interactions between predators and their prey. Unfortunately, only in a restricted number of research studies have these techniques been used in experiments to evaluate the impact of the pest predators in field populations, although Dempster (1960, 1967) and Ashby (1974) have pioneered research into predator gut contents analysis to understand the ecological implications of arthropod predators feeding on different types of prey.

MacFadyen *et al.* (2015) reported that many studies on the presence and/or abundance of pests and natural enemies deduce the impact qualitatively, providing findings to address ecological questions and a critical evaluation of impact to lead to pest-management decision making. Sometimes, very simple techniques can be applied to estimate the natural enemy impact, for example: (i) cages or barriers to restrict natural enemy or prey movement; (ii) direct observation of natural enemy attack; and (iii) direct observation of prey to evaluate mortality. Assessment of a natural enemy's impact must be related to the needs of farmers and to the specific pest problems, for example the magnitude of mortality attributed to natural enemies may be less important than the timing and consistency of mortality occurring between the seasons. The best impact assessments produce general information that can be used to aid decision making regarding how to conserve natural enemies and how these natural enemies may be best used to control pest populations.

Finally, pest control thresholds can be established according to the different crop development stages to measure the impact of natural enemies on pests. Incorporation of knowledge on model pest-and-natural-enemies systems and the integration of area-wide landscape management with in-field pest management enable pest control to be achieved.

World Spread of Biological Control

It has been estimated that natural biological control extends over 89.5 billion ha of the world's ecosystems (land with vegetation), and 44.4 billion ha of this surface concerns some form of agricultural activity, including forestry and grassland. Natural and inoculative biological control contribute to control of the development of indigenous and exotic pest populations in natural and agro-ecosystems, and both possibilities define the 'ecosystem service', whose annual value was estimated to be about US$400 billion (Costanza *et al.*, 1997).

Inoculative or classical biological control is applied on 10% of land under cultivation (Bale *et al.*, 2008) and, over the last 120 years, 165 pests have been brought under long-term control (Cock *et al.*, 2010). The most widely used natural enemies in inoculative insect control (e.g. *Aphelinus mali* (Haldeman), *Aphytis lingnanensis* Compère, *R. cardinalis* and other species) have been introduced in more than 20 countries/regions worldwide, where they assure a permanent control of different pests (Cock *et al.*, 2009, 2010). Obviously, these data are only of general value (e.g. the authors of this chapter believe that the data on the surface area treated by biological control is not credible because taking, for example, the relevant surfaces of citrus and other crops treated in Italy with biological control probably depends on the phase of the crop for economic reasons and there may be false findings). However, the values do provide an indication of the widespread use of biological control methods.

As regards augmentative biological control, Cock *et al.* (2010) estimated that worldwide 170 species of natural enemies (fundamentally invertebrate) are reared and sold globally for the control of over 100 pests on about 0.4% of land under cultivation. Augmentative biological control is operated by state-funded or commercial insectaries (van Lenteren and Bueno, 2003). Conversely to natural and inoculative biological control, augmentative biological control extends on a reduced surface, estimated to be about 16 million ha and corresponding to 0.4% of cultivated land with crops. Worldwide, 30 important commercial producers are known (Bolckmans, 2008), 20 of which are active in Europe, and employ more than ten workers. In addition to these important companies, it is estimated that about 500 small commercial companies are active. Fewer than five companies employ more than 50 workers. The most important company employed about 600 workers during 2011. Different associations organize the producers, for example in Europe it is the International Biocontrol Manufacturers Association (IBMA), in North America the Association of Natural Biocontrol Producers (ANBP), in Australia the Australasian Biological Control (ABC) and in Brazil the Brazilian Association of Biological Control (ABCbio) (van Lenteren, 2012).

The 25 species of natural enemies most often released make up more than 90% of the approximately €300 million of the total world market at end-user level (Bolckmans, 2008; Cock *et al.*, 2010). Greenhouse crops in The Netherlands, the UK, France and Spain, followed by the USA are the most important commercial markets (expressed in sales volume) for natural enemies, and together they constitute about two-thirds of the total market (Bolckmans, 1999). Nevertheless, Africa, Asia and Latin America represent significant and growing markets. The commercial market for field crops is smaller than the greenhouse market (van Lenteren, 2012).

The choice of natural enemies may be affected by various factors, including socio-economic interests. So, more than 90% of the natural enemies commercially suggested for use in Africa derive from material collected in and – initially mass reared on – other continents, as well as in Canada, Japan, Mexico and South Korea. In Australia, New Zealand and the USA almost equal numbers of indigenous and exotic natural enemies are used. In South and Central America (Argentina, Brazil and Cuba), most of the natural enemies released in augmentative biological control are indigenous. In Europe nine exotic species have been substituted by indigenous species, for example the replacement of *Eretmocerus eremicus* Howard by *Eretmocerus mundus* (Mercet) and of *Orius insidiosus* (Say) by *O. laevigatus* (van Lenteren, 2012).

Despite the above findings, there are conflicting opinions on the validity of augmentative biological control. So, while van Lenteren (2012) praises the effectiveness of the method, Collier and van Steenwyk (2004) published a critical evaluation of augmentative biological control, mainly referring to studies carried out in the USA. Obviously, other contributions are of the opposite opinion. In this context, it is likely that the discussion is influenced by economic interests related on the one hand to the chemical industry, interested to diminish the importance of biological control, and on the other hand to the mass production of natural enemies, animated by opposing interests.

Biological Control and Organic Farming

Aspects discussed before may induce the conclusion that biological control of pests identifies itself with organic farming alone. Although the action of natural enemies of pests is a crucial step of organic farming alone, it does not always solve the problem of pest outbreaks and usually it must be integrated with other important technical tools, such as crop management (see Daniel *et al.*, Chapter 1, this volume). The topic has been well investigated by different authors, for example by Altieri and Nicholls (2003), who studied soil fertility management and insect pests, in order to harmonize soil and plant health in agroecosystems. From a practical point of view, Wyss (2011) of the Research Institute of Organic Agriculture, FiBL (Switzerland) reported that organic pest, disease and weed management can be seen as a three-step approach with multiple tools. The first step consists of providing good growing conditions for plants to enhance their resilience and resistance. The second step consists of encouraging natural control mechanisms through promotion of natural enemies, and the third step involves application of direct control measures to kill the pests, diseases or weeds in a way that has minimum residual effect on the ecosystem. The tools to manage pests and diseases include cultural practices, habitat management to control pests and diseases, and direct control measures in pest management.

The last includes the biological control of pests by natural enemies and the use of insecticides of vegetable or mineral origin, pheromones, repellents, traps for mass trapping of pests, and physical barriers against pests. The protocol suggested by Caldwell *et al.* (2013) of Cornell University and other American institutions basically includes the use of cultural controls and various materials approved for organic production. Many other references confirm this.

Conclusions

The lack of comprehensive bio-ecological knowledge on natural enemies suggests that such studies should be intensified. In particular, it is necessary to know the structure of agroecosystems, especially the ecological interrelationships between the different trophic levels. Regardless of this aspect, one may wonder what role biological control plays within an organic farming strategy, and if so what biological control technique is best from a practical point of view. As regards the first point, there is considerable scientific evidence that confirms that biological control represents a key aspect of organic farming (and of other agricultural practices) – although its performance needs to be improved. As regards the second point, between the different options natural biological control is superior with respect to other technical possibilities; often its action is decisive and usually it cannot be underestimated. Usually, there is an integration between natural and classical biological control, and they are less expensive and more long lasting than other control strategies. When and where possible, both strategies must be encouraged. Augmentative biological control is applied and integrated with natural biological control, and although it may be effective at controlling pest populations, its application often exhibits insurmountable limits related to the cost and quality of the biological material used. The application of this strategy must be concretely examined under economic and ecological profiles. In augmentative biological control the technical choices are often imposed by economic designs that are strangers to the real needs of the specific environments. The case of *N. tenuis* irrefutably confirms this aspect. From a practical point of view usually biological control of pests in organic farming must be integrated with other technical means which do not produce a negative impact on ecosystems. From this point of view we can conclude that populations of natural enemies can exercise strong practical effects on populations of pests infesting crops, and their action is the cornerstone of biological control.

References

Agrawal, A.A., Kobayashi, C. and Thaler, J.S. (1999) Influence of prey availability and induced host-plant resistance on omnivory by western flower thrips. *Ecology* 80, 1713–1723.

Altieri, M.A. and Nicholls, C.I. (2003) Soil fertility management and insect pests: harmonizing soil and plant health in agroecosystems. *Soil and Tillage Research* 72, 203–211.

Arzone, A., Alma, A. and Tavella, L. (1990) Role of Mirids (Rhynchota Heteroptera) in the control of *Trialeurodes vaporariorum* Westw. (Rhynchota Aleyrodidae). *Bollettino di Zoologia Agraria e di Bachicoltura* 22, 43–51.

Ashby, J.W. (1974) A study of arthropod predation of *Pieris rapae* L. using serological and exclusion techniques. *Journal of Applied Ecology* 11, 419–425.

Balduf, W.V. (1935) *The Bionomics of Entomophagous Coleoptera*. John S. Swift Co., Inc., St Louis, Missouri, p. 220.

Bale, J., van Lenteren, J.C. and Bigler, F. (2008) Biological control and sustainable food production. *Philosophical Transactions of the Royal Society of London B* 363(1492), 761–776.

Bar-Joseph, M. and Catara, A. (2012) Endemic and emerging vector-borne Mediterranean citrus diseases and their epidemiological consequences. In: Vacante, V. and Gerson, U. (eds) *Integrated Control of Citrus Pests in the Mediterranean Region*. Bentham Science Publishers, London, pp. 137–155.

Bell, J.V. (1974) Mycoses. In: Cantwell, G.E. (ed.) *Insect Diseases*. Marcel Dekker, New York, pp. 185–236.

Benfatto, D., Di Franco, F. and Vacante, V. (1992) Spiders of Italian citrus groves. In: Tribulato, E., Gentile, A. and Reforgiato, G. (eds) *Proceedings of the International Society of Citriculture*, Vol. 3. Acireale, Scicily, Italy, 8–13 March, pp. 938–942.

Bolckmans, K.J.F. (1999) Commercial aspects of biological pest control in greenhouses. In: Albajes, R., Gullino, M.L., van Lenteren, J.C. and Elad, Y. (eds) *Integrated Pest and Disease Management in Greenhouse Crops*. Kluwer Publishers, Dordrecht, The Netherlands, pp. 310–318.

Bolckmans, K.J.F. (2008) De insectenfabriek. In: Osse, J., Schoonhoven, L., Dicke, M. and Buiter, R. (eds) *Natuur als bondgenoot: biologische bestrijding van ziekten en plagen*. Bio-Wetenschappen en Maatschappij, Den Haag, The Netherlands, pp. 51–52.

Bonsignore, C.P. (2016) Environmental factors affecting the behavior of *Coenosia attenuata* Stein, a predator of *Trialeurodes vaporariorum* (Westwood) in tomato greenhouses. *Entomologia Experimentalis et Applicata* 158, 87–96.

Bonsignore, C.P. and Vacante, V. (2012a) Natural enemies. In: Vacante, V. and Gerson, U. (eds) *Integrated Control of Citrus Pests in the Mediterranean Region*. Bentham Science Publishers, London, pp. 66–87. eISBN: 978-1-60805-294-3.

Bonsignore, C.P. and Vacante, V. (2012b) Influences of botanical pesticides and biological agents on *Orius laevigatus – Frankliniella occidentalis* dynamics under greenhouse conditions. *Journal of Plant Protection Research* 52, 15–23.

Boucias, D.G. and Pendland, J.C. (1998) *Principles of Insect Pathology*. Kluwer Academic Publishers, Boston, Massachusetts, p. 548.

Brown, J.M., Abrahamson, W.G., Packer, R.A. and Way, P.A. (1995) The role of natural-enemy escape in a gallmaker host-plant shift. *Oecologia* 104, 52–60.

Bruin, J. and van der Geest, L. (eds) (2009) *Diseases of Mites and Ticks*. Reprinted from *Experimental and Applied Acarology* 46(1–4), VI, p. 350.

Caldwell, B., Sideman, E., Seaman, A., Shelton, A. and Smart, C. (2013) *New York State Agricultural Experiment Station (NYSAES)*. Cornell UniversityArnold Printing Corp, Ithaca, New York, p. 210.

Calvo, J., Bolckmans, K., Stansly, P.A. and Urbaneja, A. (2009) Predation by *Nesidiocoris tenuis* on *Bemisia tabaci* and injury to tomato. *BioControl* 54, 237–246.

Castañé, C., Alomar, O. and Riudavets, J. (1996) Management of western flower thrips on cucumber with *Dicyphus tamaninii* (Heteroptera: Miridae). *Biological Control* 7, 114–120.

Chandler, D., Davidson, G., Pell, J.L., Ball, B.V., Shaw, K. and Sunderland, K.D. (2000) Fungal biocontrol of Acari. *Biocontrol Science and Technology* 10, 357–384.

Cock, M.J.W., van Lenteren, J.C., Brodeur, J., Barratt, B.I.P., Bigler, F., Bolckmans, K., Cônsoli, F.L., Haas, F., Mason, P.G. and Parra, J.R.P. (2009) *The Use and Exchange of Biological Control Agents for Food and Agriculture*. Food and Agriculture Organization of the United Nations (FAO) Background Study No. 47. FAO, Rome, 88 pp.

Cock, M.J.W., van Lenteren, J.C., Brodeur, J., Barratt, B.I.P., Bigler, F., Bolckmans, K., Consoli, F.L., Haas, F., Mason, P.G. and Parra, J.R.P. (2010) Do new access and benefit sharing procedures under the convention on biological diversity threaten the future of biological control? *BioControl* 55, 199–218.

Collier, T. and van Steenwyk, R. (2004) A critical evaluation of augmentative biological control. *Biological Control* 31, 245–256.

Costanza, R., d'Arge, R., de Groot, R., Farber, S., Grasso, M., Hannon, B., Limburg, K., Naeem, S., O'Neill, R.V., Paruelo, J., Raskin, R.G., Sutton, P. and van den Belt, M. (1997) The value of the world's ecosystem services and natural capital. *Nature* 387, 253–260.

Couri, M.S. and Salas, C. (2010) First record of *Coenosia attenuate* Stein (Diptera, Muscidae) from Chile, with biological notes. *Revista Brasileira de Entomologia* 54(1). Available at: http://dx.doi.org/10.1590/S0085-56262010000100020 (accessed 9 June 2016).

DeBach, P. (1964) The scope of biological control. In: DeBach, P. (ed.) *Biological Control of Insect Pests and Weeds*. Chapman and Hall, London, pp. 3–20.

DeBach, P. (1974) *Biological Control by Natural Enemies*. Cambridge University Press, Cambridge, p. 323.

DeBach, P. and Rosen, D. (1991) *Biological Control by Natural Enemies*, 2nd edn. Cambridge University Press, New York, p. 440.

DeBach, P., Rosen, D. and Kennett, C.E. (1971) Biological control of coccids by introduced natural enemies. In: Huffaker, C.B. (ed.) *Biological Control*. Plenum Press, New York, pp. 165–194.

Dempster, J.P. (1960) A quantitative study of the predators on the eggs and larvae of the broom beetle, *Phytodecta olivacea* Forster, using the precipitin test. *Journal of Animal Ecology* 29, 149–167.

Dempster, J.P. (1967) The control of *Pieris rapae* with DDT. I. The natural mortality of the young stages of *Pieris*. *Journal of Applied Ecology* 4, 485–500.
Drea, J.J. (1990) Other Coleoptera. In: Rosen, D. (ed.) *Armored Scale Insects: Their Biology, Natural Enemies and Control*, Vol. B. Elsevier, Amsterdam, pp. 41–49.
Eubanks, M.D. and Denno, R.F. (1997) The effects of plant quality and alternative prey on the suppression of aphid population by an omnivore. *Bulletin of the Ecological Society of America* 90, 155–161.
European Commission (EC) (2010) Directive 2009/128/EC of the European Parliament and of the Council of 21 October 2009 establishing a framework for Community action to achieve the sustainable use of pesticides. Available at: http://eur-lex.europa.eu/eli/dir/2009/128/2009-11-25 (accessed 16 June 2017).
Evans, H.C. and Prior, C. (1990) Entomopathogenic fungi. In: Rosen, D. (ed.) *Armored Scale Insects: Their Biology, Natural Enemies and Control*, Vol. B. Elsevier, Amsterdam, pp. 3–17.
Furlong, M.J. (2015) Knowing your enemies: integrating molecular and ecological methods to assess the impact of arthropod predators on crop pests. *Insect Science* 22, 6–19.
Furlong, M.J. and Zalucki, M.P. (2010) Exploiting predators for pest management: the need for sound ecological assessment. *Entomologia Experimentalis et Applicata* 135, 225–236.
Gabarra, R., Castañé, C. and Albajes, R. (1995) The mirid bug *Dicyphus tamaninii* as a greenhouse whitefly and western flower thrips predator on cucumber. *Biocontrol Science and Technology* 5, 475–488.
Gauld, I. and Bolton, B. (1988) *The Hymenoptera*. Oxford University Press, Oxford and the Natural History Museum, London, p. 332.
Gerling, D., Alomar, O. and Arnó, J. (2001) Biological control of *Bemisia tabaci* using predators and parasitoids. *Protection* 20, 779–799.
Gerson, U. (2014) Pest control by mites (Acari): present and future. *Acarologia* 54(4), 371–394.
Gerson, U., Kenneth, R. and Muttath, T.I. (1979) *Hirsutella thompsonii*, a fungal pathogen of mites. II. Host–pathogen interactions. *Annals of Applied Biology* 91, 29–40.
Gerson, U., Smiley, R.L. and Ochoa, R. (2003) *Mites (Acari) for Pest Control*. Blackwell Publishing, Oxford, p. 539.
Grant, W.P., Greaves, J., Tatchell, M., Chandler, D., Prince, G. and Bailey, A. (2010) *Biopesticides. Pest Management and Regulation*. CAB International, Wallingford, UK, p. 232.
Gurr, G.M. and Wratten, S.D. (eds) (2000) *Biological Control: Measures of Success*. Kluwer Academic Publishers, Dordrecht, The Netherlands, p. 429.
Gurr, G.M., Wratten, S.D. and Altieri, M.A. (eds) (2004) *Ecological Engineering: Advances in Habitat Manipulation for Arthropods*. CSIRO Publishing, Melbourne, Australia, p. 232.
Hoy, M.A. (2011) *Agricultural Acarology: Introduction to Integrated Mite Management*. CRC Press, Boca Raton, Florida, p. 430.
Huffaker, C.B., Simmonds, F.J. and Laing, J.E. (1976) The theoretical and empirical basis of biological control. In: Huffaker, C.B. and Messenger, P.S. (eds) *Theory and Practice of Biological Control*. Academic Press, New York, pp. 41–78.
Hunter, M.D. (2001) Multiple approaches to estimating the relative importance of top-down and bottom-up forces on insect populations: experiments, life tables, and time-series analysis. *Basic and Applied Ecology* 2, 295–309.
Hunter-Fujita, F.R., Entwistle, P.F., Evans, H.F. and Cook, N.E. (1998) *Insect Viruses and Pest Management*. Wiley, Chichester, UK, p. 235.
International Plant Protection Convention (IPPC) (2005) *International Standards for Phytosanitary Measures (ISPM) 3. Guidelines for the Export, Shipment, Import and Release of Biological Control Agents and Other Beneficial Organisms*. IPPC, Food and Agriculture Organization of the United Nations (FAO), Rome.
Krantz, G.W. (2009) Introduction. In: Krantz, G.W. and Walter, D.E. (eds) *Acarology*. Texas Tech University Press, Lubbock, Texas, pp. 1–2.
Lacey, L.A. and Kaya, H.K. (eds) (2007) *Field Manual of Techniques in Invertebrate Pathology*, 2nd edn. Springer, Dordrecht, The Netherlands, p. 870.
Macfadyen, S., Hardie, D.C., Fagan, L., Stefanova, K., Perry, K.D., DeGraaf, H.E., Holloway, J., Spafford, H. and Umina, P.A. (2014) Reducing insecticide use in broad-acre grains production: an Australian study. *PLoS One* 9, e89119.
Macfadyen, S., Davies, A.P. and Zalucki, M.P. (2015) Assessing the impact of arthropod natural enemies on crop pests at the field scale. *Insect Science* 22, 20–34.
Maelzer, D.A. and Zalucki, M.P. (1999) Analysis and interpretation of long term light trap data for *Helicoverpa armigera* (Lepidoptera: Noctuidae) in Australia: the effect of climate and crop host plants. *Bulletin of Entomological Research* 89, 455–464.

Maelzer, D.A. and Zalucki, M.P. (2000) Long range forecasts of the *Helicoverpa punctigera* and *H. armigera* (Noctuidae: Lepidoptera) in Australia using the SOI and the SST. *Bulletin of Entomological Research* 90, 133–146.

Maelzer, D.A., Zalucki, M.P. and Laughlin, R. (1996) An analysis of historic light trap data for *Helicoverpa punctigera*: forecasting the size of pest population. *Bulletin of Entomological Research* 86, 547–557.

Malausa, J.C. (1989) Lutte intégrée sous serre: les punaises prédatrices Mirides dans les cultures de Solanacées du sud-est de la France. *PHM Revue Horticole* 298, 39–43.

Malausa, J.C. and Henao, B. (1988) First observations in France of *Cyrtopeltis* (*Nesidiocoris*) *tenuis* Reuter, 1895 (Het. Miridae). *Nouvelle Revue d'Entomologie* 5, 180.

McCoy, C.W. (1996) Pathogens of eriophyoid mites. In: Lindquist, E.E., Sabelis, M.W. and Bruin, J. (eds) *Eriophyoid Mites. Their Biology, Natural Enemies and Control.* World Crop Pests, Vol. 6. Elsevier Science B.V., Amsterdam, pp. 481–490.

McNeil, J.N., Cotnoir, P.A., Leroux, T., Laprade, R. and Schwartz, J.L. (2010) A Canadian national survey on the public perception of biological control. *BioControl* 55, 445–454.

Merino-Pachero, M. (2007) Almeria finally forced to turn green. *Fruit and Vegetable Technology* 7, 23–25.

Miller, L.K.J. and Ball, L.A. (eds) (1998) *The Insect Viruses*. Plenum Press, New York, p. 413.

Monzó, H.C., Mollà, O., Montón, H., Melic, A., Castañera, P. and Urbaneja, A. (2008) Ground-dwelling spiders (Araneae) in citrus orchards in Spain. *IOBC (International Organisation for Biological and Integrated Control) WPRS Bulletin* 38, 267.

Mottern, J.L. and Heraty, J.M. (2014) Revision of the *Cales noacki* species complex (Hymenoptera, Chalcidoidea, Aphelinidae). *Systematic Entomology* 39, 354–379. DOI: 10.1111/syen.12060.

Naranjo, S.E. (2001) Conservation and evaluation of natural enemies in IPM systems for *Bemisia tabaci*. *Crop Protection* 20, 835–852.

Paz, Z., Gerson, U. and Sztejnberg, A. (2007) Assaying three new fungi against citrus mites in the laboratory, and field trial. *BioControl* 52, 855–862.

Pimentel, D. (2009) Environmental and economic costs of the application of pesticides primarily in the United States. In: Peshin, R. and Dhawan, A.K. (eds) *Integrated Pest Management: Innovation-Development Process*. Springer, Dordrecht, The Netherlands, pp. 89–111.

Pimentel, D., Andow, D., Dyson-Hudson, R., Gallahan, D., Jacobson, S., Irish, M., Kroop, S., Moss, A., Schreiner, I., Shephard, M., Thompson, T. and Vinzant, B. (1980) Environmental and social costs of pesticides: a preliminary assessment. *Oikos* 34,126–140.

Poinar, G.O. Jr and Poinar, R. (1998) Parasites and pathogens of mites. *Annual Review of Entomology* 43, 449–469.

Pons, X., Núñez, E., Lumbierres, B. and Albajes, R. (2005) Epigeal aphidophagous predators and the role of alfalfa as a reservoir of aphid predators for arable crops. *European Journal of Entomology* 102, 519–525.

Price, P.W., Bouton, C.E., Gross, P., McPheron, B.A., Thompson, J.N. and Weiss, A.E. (1980) Interaction among three trophic levels: influence of plants on interactions between insect herbivores and natural enemies. *Annual Review of Ecology and Systematics* 11, 41–65.

Ribeiro, C., Cardoso, P. and Franco, J.C. (2006) Abundance and diversity of spiders in lemon orchards with different weed management systems. *IOBC (International Organisation for Biological and Integrated Control) WPRS Bulletin* 29(3), 167–178.

Rochester, W.A., Zalucki, M.P., Ward, A., Miles, M. and Murray, D.A.H. (2002) Testing insect movement theory: empirical analysis of pest data routinely collected from agricultural crops. *Computers and Electronics in Agriculture* 35, 139–149.

Rosen, D. (1986) The role of taxonomy in effective biological control programs. *Agriculture, Ecosystems and Environment* 15, 121–129.

Rosen, D. and DeBach, P. (1979) *Species of* Aphytis *of the World (Hymenoptera: Aphelinidae)*. Israel University Press, Jerusalem and Dr. W. Junk, The Hague, The Netherlands, p. 801.

Rosenheim, J.A., Wilhoit, L.R. and Armer, C.A. (1993) Influence of intraguild predation among generalist insect predators on the suppression of an herbivore population. *Oecologia* 96, 439–449.

Samson, R.A. and McCoy, C.W. (1983) *Aschersonia aleyrodis*, a fungal pathogen of whitefly I. Scanning electron microscopy of the development on the citrus whitefly. *Zeitschrift Fur Angewandte Entomologie* [*Journal of Applied Entomology*] 96(1–5), 380–386.

Samson, R.A., Evans, H.C. and Latge, J.P. (1988) *Atlas of Entomopathogenic Fungi*. Springer, Heidelberg, Germany, p. 187.

Sánchez, J.A. and Lacasa, A. (2008) Impact of the zoophytophagous plant bug *Nesidiocoris tenuis* (Heteroptera: Miridae) on tomato yield. *Journal of Economic Entomology* 101(6), 1864–1870.

Schellhorn, N.A., Pierce, S., Bianchi, F.J.J.A., Williams, D. and Zalucki, M.P. (2008) Designing landscapes for multiple outcomes in broad-acre environments. *Australian Journal of Experimental Agriculture* 48, 1549–1559.

Schellhorn, N.A., Bianchi, F.J.J.A. and Hsu, C.L. (2014) Movement of entomophagous arthropods in agricultural landscapes: links to pest suppression. *Annual Review of Entomology* 59, 559–581.

Schellhorn, N.A., Parry, H.R., Macfadyen, S., Wang, Y.M. and Zalucki, M.P. (2015) Connecting scales: achieving in-field pest control from areawide and landscape ecology studies. *Insect Science* 22, 35–51.

Tanada, Y. and Kaya, H.K. (1993) *Insect Pathology*. Academic Press, San Diego, California, p. 666.

Vacante, V. (1981) Notizie sulla presenza di *Therodiplosis persicae* Kieffer (Diptera, Cecidomyiidae) in serra su piante orticole e floreali, attaccate da *Tetranychus urticae* Koch (Acarina, Tetranychidae). *Tecnica Agricola, Catania* XXXIII(5), 303–312. (in Italian)

Vacante, V. (2000) Animali dannosi alle ortive da serra. In: Baccetti, B., Barbagallo, S., Süss, L. and Tremblay, E. (eds) *Manuale di Zoologia Agraria*. Delfino, A. Editore, Rome, pp. 429–448.

Vacante, V. (2009) Mezzi e strategie di lotta contro i fitofagi degli agrumi. In: Vacante, V. and Calabrese, F. (eds) *Citrus, Trattato di Agrumicoltura*. Edagricole, Bologna, Italy, pp. 291–314.

Vacante, V. (2012) *The History of IPM in the Mediterranean Citriculture*. In: *Integrated Control of Citrus Pests in the Mediterranean Region*, Vacante, V. and Gerson, V. (eds). Bentham Science Publishers, London, pp. 19–27.

Vacante, V. (2013) Punteruolo rosso delle palme: aggiornamenti sulle tecniche di lotta. *Protezione Delle Colture* 5, 38–51.

Vacante, V. (2015) *Handbook of Mites of Economic Plants*. CAB International, Wallingford, UK, p. 872.

Vacante, V. and Bonsignore, C. (2012) Implementation of IPM in citriculture. In: Vacante, V. and Gerson, U. (eds) *Integrated Control of Citrus Pests in the Mediterranean Region*. Bentham Science Publishers, London, pp. 28–55.

Vacante, V. and Tropea Garzia, G. (1994) Investigations on the role of *Nesidiocoris tenuis* (Reuter) (Hemiptera: Miridae) on tomato in unheated greenhouses in the Ragusa area. *Informatore Fitopatologico* 44, 45–48.

van der Geest, L.P.S. (1985) Pathogens of spider mites. In: Helle, W. and Sabelis, M.W. (eds) *Spider Mites. Their Biology, Natural Enemies and Control*. World Crop Pests, Vol. 1B. Elsevier Science B.V., Amsterdam, pp. 247–258.

van der Geest, L.P.S. (2010) IPM potentials of microbial pathogens and diseases of mites. In: Ciancio, A. and Mukerji, K.G. (eds) *Integrated Management of Arthropod Pests and Insects Borne Diseases*, Vol. 5. Springer Science+Business Media B.V., Dordrecht, The Netherlands, pp. 249–308.

van der Geest, L.P.S., Elliot, S.L., Breeuwer, J.A.J. and Beerling, E.A.M. (2000) Diseases of mites. *Experimental and Applied Acarology* 24, 497–560.

van Lenteren, J.C. (2012) The state of commercial augmentative biological control: plenty of natural enemies, but a frustrating lack of uptake. *BioControl* 57, 1–20.

van Lenteren, J.C. and Bueno, V.H.P. (2003) Augmentative biological control of arthropods in Latin America. *BioControl* 48, 123–139.

van Lenteren, J.C., Hua, L.Z., Kamerman, J.W. and Rumei, X (1995) The parasite–host relationship between *Encarsia formosa* (Hym., Aphelinidae) and *Trialeurodes vaporariorum* (Hom., Aleyrodidae) XXVI. Leaf hairs reduce the capacity of *Encarsia* to control greenhouse whitefly on cucumber. *Journal of Applied Entomology* 119, 553–559.

Waage, J.K. and Greathead, D.J. (1988) Biological control: challenges and opportunities. *Philosophical Transactions of the Royal Society of London, Series B* 318, 111–128.

Walter, D.E. and Proctor, H.C. (1999) *Mites – Ecology, Evolution and Behaviour*. University of New South Wales Press, Sydney, Australia and CAB International, Wallingford, UK, 322 pp.

Wiedenmann, R.N. and Smith, J.W.J. (1997) Attributes of natural enemies in ephemeral crop habitats. *Biological Control* 10, 16–22.

Wyss, E. (FiBL) (2011) *African Organic Agriculture Training Manual Module 04 Pest, Desease and Weed Management*. FiBL, Research Institute of Organic Agriculture, Switzerland, p. 33.

Zalucki, M.P. and Furlong, M.J. (2005) Forecasting *Helicoverpa* populations in Australia: a comparison of regression based models and a bioclimatic based modelling approach. *Insect Science* 12, 45–56.

Zalucki, M.P., Furlong, M.J., Schellhorn, N.A., Macfadyen, S. and Davies, A.P. (2015) Assessing the impact of natural enemies in agroecosystems: toward 'real' IPM or in quest of the Holy Grail? *Insect Science* 22, 1–5.

Zeddies, J., Schaab, R.P., Neuenschwander, P. and Herren, H.R. (2001), Economics of biological control of cassava mealybug in Africa. *Agricultural Economics* 24, 209–219. DOI: 10.1111/j.1574-0862.2001.tb00024.x.

Zhao, J.Z., Li, Y.X., Collins, H.L., Gusukuma-Minuto, L., Mau, R.F.L., Thompson, G.D. and Shelto, A.M. (2002) Monitoring and characterization of diamondback moth (Lepidoptera: Plutellidae) resistance to spinosad. *Journal of Economic Entomology* 95, 430–436.

4 Pest Control in Organic Citrus Groves

Kaouthar Grissa-Lebdi[1]* and Hajer Sahraoui[2]
[1]*National Agronomic Institute of Tunisia (INAT), Tunisia;* [2]*Citrus Technical Center, Beni-Khalled, Tunisia*

Introduction

Citrus production takes place throughout the tropical and subtropical countries of the world (Spreen, 2012). About 140 countries grow citrus in the world and it is estimated that the total harvested surface is 9.6 million ha (FAO, 2015), with a total worldwide production of fresh citrus fruit amounting to 121,273.2 thousand t (FAO, 2015). The leading citrus-fruit-producing countries are China (29,567 thousand t), Brazil (18,966 thousand t), the USA (9394 thousand t), Mexico (7503 thousand t), India (7400 thousand t) and Spain (6512.6 thousand t), representing close to two-thirds of global production (FAO, 2015). Citrus production can be divided among four primary groups: (i) sweet oranges; (ii) mandarins (also known as tangerines); (iii) grapefruit; and (iv) lemons and limes (Spreen, 2012).

Citrus fruits grown organically cover almost 75,000 ha, which constituted 0.8% of the world's total citrus area in 2014 (Lernoud and Willer, 2016). The largest producer is Italy with nearly 230,000 ha, followed by Mexico (about 12,000 ha) and China (nearly 8000 ha). Nevertheless, no crop details are available for the organic area of some of the world's leading citrus producers, including Brazil (0.92 million ha), Nigeria (0.73 million ha) and India (0.69 million ha) (Lernoud and Willer, 2016). Crop details that are available show that oranges were grown in 44% of the organic citrus fruit area, pomelos and grapefruit followed with 5% of the area, lemons and limes with 4%, tangerine with 1% and others with 46% of the citrus area (Lernoud and Willer, 2016). The most significant exporters of fresh citrus are Italy, Spain, Argentina, the USA and Greece; while the main exporters of organic citrus juices are Brazil, Israel, Costa Rica, the USA, Italy, Mexico and Cuba (FAO, 2015).

Organic citrus farming is facing several difficulties, and among them, pest attack is one of the most important problems, which hinder the diversification of citriculture in the world. The components of a pest management programme in organic farming system outline basic arthropod monitoring procedures, and specific practices used in sustainable production to prevent the growth of pest populations beyond economic thresholds (Spreen, 2012). In pest management strategies for organic crops, priority is given to preventative strategies, followed by more direct measures: (i) conservation biological

* Corresponding author: grissak@yahoo.fr

control; (ii) inundation and inoculation biocontrol; (iii) cultural practices; (iv) host plant resistance; (v) approved insecticides of biological and mineral origin; (vi) mating disruption; (vii) pheromones; (viii) attract-and-kill techniques; and (ix) repellent agents as physical barriers.

There are a large number of pests which attack citrus and their relative importance varies among countries and changes throughout the years. This chapter deals with key and minor pests (insects and mites) that affect citrus production and they are identified as key pests if they were considered to be in that category by many authors, and if they have a worldwide distribution. Minor pests are those that are sporadic or important only locally.

I Diptera

I-1 Tephritidae

This is a large family of fruit flies, which includes more than 4500 described species (White and Elson-Harris, 1992). These pests are tropical and subtropical species with a broad range of host plants (Liquido *et al.*, 1991). Some species from this family attack citrus fruit in the most important growing areas in the world, among these only a few species are considered important, belonging mostly to the genera *Ceratitis*, *Anastrepha* and *Bactrocera* (Cavalloro and Di Martino, 1986).

Species and their distribution

The Mediterranean fruit fly, ***Ceratitis capitata*** (Wiedemann), is the most important species because of its worldwide distribution and highly polyphagous behaviour (Liquido *et al.*, 1991; De Meyer *et al.*, 2002; Duyck and Quilici, 2002). *C. capitata* was eradicated from North America (Florida, California and Texas) and New Zealand (Klassen and Curtis, 2005). Other citrus fruit fly species are important but still almost confined to their areas of origin (Delrio, 1986). The American fruit fly, ***Anastrepha fraterculus*** (Wiedemann), occurs in subtropical areas of Central America, the Caribbean and South America (Silva *et al.*, 2006). The Mexican fruit fly, ***Anastrepha ludens*** (Loew), is located in Mexico and Central America. The Caribbean fruit fly, ***Anastrepha suspensa*** (Loew), is distributed in Central America, the Caribbean and Florida. The oriental fruit fly, ***Bactrocera dorsalis*** (Hendel), occurs in South-east Asia, Hawaii, the Mariana Islands and Tahiti. The Queensland fruit fly, ***Bactrocera tryoni*** (Froggatt), is locted in Australia. The Japanese orange fruit fly, ***Bactrocera tsuneonis*** (Miyake), occurs in Japan, China, Taiwan and Vietnam. The peach fruit fly, ***Bactrocera zonata*** (Saunders), occurs from Iran to South-east Asia, through its native areas (India and Pakistan) and the south-east Mediterranean region. The Natal fruit fly, ***Ceratitis rosa*** Karsch, is distributed in southern and eastern Africa and the islands of Mauritius and Reunion (EPPO, 2015).

Bio-ecology

The basic life cycles of tropical fruit flies are roughly similar: the female fruit fly uses its telescopic ovipositor to insert the eggs beneath the skin of citrus fruits, the emerging larvae or 'maggots' will feed on the fruit pulp and grow. At the completion of the third instar, the larvae will leave the fruit and pupate in the soil. At the end of the process of metamorphosis, adults exit the puparium. Tephritid adults need to consume protein in order to reach sexual maturity. The duration of the different developmental stages of fruit flies is dependent on the fruit fly species and environmental conditions (Christenson and Foote, 1960; Delrio, 1986).

Damage

Injury to fruits occurs through oviposition punctures and larval development. Fungal and bacterial infections take place through puncture scars and as a result of which fruits become rotten and fall (Delrio, 1986). In addition to heavy losses in fruit production, fruit flies are phytosanitary barriers to the export of fresh fruits due to severe quarantine regulations designed by importing countries. Major control and eradication programmes have been developed in various parts of the world to combat these devastating species (Ayers, 1957; De Woskin, 1981).

Organic controls

Control programmes should be directed at preventing fruit flies from laying eggs in the fruit. As with the control of other pest species, a single control method by itself is often not sufficient to effectively control the fruit flies and best results are gained from a combination of methods found in this section.

MONITORING. When applied against small fruit fly populations, control and eradication programmes are most effective (Papadopoulos *et al.*, 2001). Traps and attractants are used to detect fruit fly species and to determine their flight periods and their abundance. In organic citrus orchards two types of attractant can be used for monitoring. Jackson traps baited with para-pheromone capture the males (Niccoli *et al.*, 1991). The para-pheromone trimedlure (TML) captures *C. capitata* and *C. rosa*; the para-pheromones methyl eugenol (ME) and cuelure (CUE) capture *Bactrocera* species (Beroza *et al.*, 1961; Cunningham, 1989; FAO/IAEA, 2003; Tan *et al.*, 2014). McPhail trap types (McPhail, 1939) baited with liquid protein capture both females and males, with a higher percentage of females being captured. These traps are used for monitoring of several tephritid species (Epsky *et al.*, 1999) and are useful especially for the detection of *Anastrepha* species which do not respond to any known sex attractant (Robacker and Moreno 1988; Felix *et al.*, 2009; Epsky *et al.*, 2011). Fruit sampling can also be used for population monitoring (Papadopoulos, 1999).

PHYSICAL AND CULTURAL CONTROLS. *Sanitation* is important for the control of citrus fruit flies (IPPC, 2005). It prevents fruit fly larvae from developing or young emerging adult flies from returning to the crop to reproduce (Klungness *et al.*, 2005). All dropped and prematurely ripe fruits and fruits suspected of being infested should be rapidly collected and destroyed to break the cycle of infection. The collected fruits may be boiled in hot water or collected in a sealed, black plastic bag placed in the sun, so the larvae will be destroyed by the heat. Infected fruit may also be buried 0.5–1 m (2–4 feet) below the soil surface so that adult flies will not be able to emerge (Panhwar, 2005). In Pakistan, it is advisable to plough the field two to three times to kill larvae and pupae by exposure to sun and air (Panhwar, 2005).

A structure called an 'augmentarium' can be used. It is a tent-like structure developed by researchers in Hawaii (Klungness *et al.*, 2005) with the purpose of enclosing fruits and/or vegetables infested with fruit fly larvae. By having a fine mesh, this structure sequesters adult flies emerging from infested fruit while allowing the parasitoids to escape. More recent research with augmentoria has been conducted in Reunion Island (Deguine *et al.*, 2011). Augmentoria have proved effective at reducing fruit fly populations when compared with trapping. Sanitation using augmentoria may become a major component of agroecological protection against fruit flies and it could be considered as a potential and significant tool of biological control by releasing parasitoids, lowering fruit fly populations and producing compost.

Physical exclusion is where fruits are enclosed in an appropriately sized bag tied around the fruit stem (Leblanc and Mararuai, 1999). This method of control is more adapted to small plantings. In some situations, whole-orchard netting may be justified where protection is required against a number of other important pests.

Use of a *particle film barrier* is where processed kaolin clay is used to cover the fruit. This has the potential for reducing damage caused by fruit flies in organic citrus orchards and is commercially available. There is some evidence that it can be an effective deterrent to oviposition of some species of fruit flies (*C. capitata*, *C. rosa*, etc.) (Caleca *et al.*, 2010; LoVerde *et al.*, 2011). The mode of action is hypothesized to be non-recognition of the host because of changes in the colour and texture of the fruit that result from the kaolin clay application (Glenn, 2012).

MASS TRAPPING. High densities of protein-baited traps can capture enough fruit flies (especially females) and thus significantly reduce the fruit damage. Mass trapping of fruit flies is being implemented successfully in different countries, especially in the Mediterranean (Lez *et al.*, 2008; Martínez-Ferrer *et al.*, 2011). Liquid traps (food traps baited with hydrolysed proteins and diammonium phosphate) are suitable for use in organic orchards and are available commercially (FAO/IAEA, 2003). To be effective, the traps should be placed before the beginning of fruit ripening. The attractiveness of food lures extends over just a few metres, so these traps should ideally be no more than 5 m or 6 m apart and are best placed from 1.5 m to 2 m above the ground in shady areas (FAO/IAEA, 2003).

ATTRACT AND KILL: PROTEIN BAITING/MALE ANNIHILATION. Organically acceptable fruit fly baits containing a food attractant and the insecticide spinosad (natural substance made by a soil bacterial species *Saccharopolyspora spinosa* Mertz & Yao) are commercially available. The bait is applied as a coarse spot spray on to tree foliage throughout the orchard. After feeding on the bait, the flies are killed by the insecticide spinosad. Baiting should begin when fruit flies are detected in traps and the fruit is susceptible to egg laying (FAO/IAEA, 2003).

The male annihilation technique is achieved through use of a high density of bait stations of male lures mixed with spinosad and replaced every 8 weeks, to reduce the male population of fruit flies. Ready-to-use formulations are commercially available. Commercial products are used for the control of *Bactrocera* species, whereas TML captures *C. capitata* (Tan *et al.*, 2014; Vargas *et al.*, 2014).

BIOLOGICAL CONTROL. Parasitoids, *Diachasmimorpha longicaudata* (Ashmead) and *Diachasmimorpha tryoni* (Cameron) are already proven and available as augmentative releases against *Ceratitis* and *Anastrepha* species and have been used with some success in Latin America and the southern USA (Wong *et al.*, 1984; Wong and Ramadan 1987; Sivinski *et al.*, 2000; Ovruski *et al.*, 2003). Currently, *D. longicaudata* is considered as one of the most important biological control agents against *Anastrepha* species and *C. capitata* in tropical and subtropical America. *Biosteres longicaudatus* (Ashmead) and *Aceratoneuromyia indica* (Silvestri) were introduced in Mexico against *A. ludens* and they became established. It was claimed that *A. indica* accounted for up to 80% parasitism (Clausen, 1978; Aluja *et al.*, 1990). *Opinus longicaudatus* (Ashmead), *B. longicaudatus* and *Fopius arisanus* (Sonan) were found parasitizing *B. zonata* in Pakistan (Syed *et al.*, 1970; Agarwal and Kapoor, 1986; Rousse *et al.*, 2007). Pupal predation by ants is stated by several authors (Bateman, 1968; Newell and Haramoto, 1968; Thomas, 1993, 1995). In Hawaii, *Pheidole megacephala* (Fabricius) attacks *B. dorsalis* larvae in fallen fruit, causing 36% mortality (Newell and Haramoto, 1968). Similar results showed that ants are responsible for 10% of the mortality of the soil-inhabiting stages of *B. tryoni* in Australia (Bateman, 1968).

Entomopathogenic nematodes are of potential interest for control of fruit flies focusing on the soil stages. Laboratory experiments showed good efficacy of four nematode species, *Steinernema feltiae*, *Steinernema carpocapsae*, *Heterorhabditis bacteriophora* and *Heterorhabditis baujardi*, to control tephritids (*C. capitata*, *B. zonata*, *B. tyroni*) (Lindegren *et al.*, 1990; Langford *et al.*, 2014; Nouh and Hussein, 2014; Minas *et al.*, 2016).

STERILE INSECT TECHNIQUE (SIT). This control method involves the mass rearing, sterilization and then release of large numbers of male flies over an infested area of wild flies. Gradually, the wild flies can only find sterile flies with which to mate, consequently no offspring are produced and the wild population is eradicated (Knipling, 1955; Dyck *et al.*, 2005). The SIT is used currently in Mexico, Texas and California to maintain a fly-free zone of *C. capitata*. Attempts to control *C. capitata* with SIT have been applied in some countries arround the world (Argentina, Australia, Brazil, Costa Rica, Egypt, Italy, Israel, Jordan, Morocco, Nicaragua, Palestine, Peru, Portugal, Spain, South Africa and Tunisia) (Hendrichs, 2000).

MICROBIOLOGICAL CONTROL. *Bacillus thuringiensis* Berliner used to control *A. ludens* and *C. capitata* showed adult mortality (Toledo *et al.*, 2006). The fungal pathogen *Metarhizium anisopliae* (Metchnikoff) Sorokin is another biopesticide with activity against larvae and pupae of *Anastrepha* and *Ceratitis* species in soil (Toledo *et al.*, 2006).

II Lepidoptera

Lepidopteran adults feed on nectar, while larvae can consume various plant parts, mainly leaves but also stems, flowers, fruits, roots or seeds. Many species are pests of economic interest (Vincent and Carde, 2009).

II-1 Gracillariidae

This family has about 1800 described species but in citrus crops it is represented by one major citrus pest: ***Phyllocnistis citrella*** Stainton (citrus leafminer, CLM).

Distribution

This species has a worldwide distribution (Waterhouse and Sands, 2001; Karamaouna *et al.*, 2008; Muniappan *et al.*, 2012).

Bio-ecology

Females lay eggs individually on the underside of young, expanding leaf flushes (Rogers *et al.*, 2016). Upon emerging, larvae immediately enter the leaf and begin feeding on epidermal cells inside the young leaf; larvae are rarely found on fruits (Kerns *et al.*, 2001a). Pupation is within the mine at the leaf margin, under a slight curl of the leaf. Depending on the environmental conditions, the total generation time ranges from 13 to 52 days (French, 2002). In Japan, the CLM is reported to pass the winter in the adult stage (Hoy and Nguyen, 1997). In India, the CLM overwinters as larvae and pupae (Hoy and Nguyen, 1997).

Damage

The CLM is most easily detected by its silver meandering larval mine, usually on the ventral side of the leaf (French, 2002). Leafminers may attack the young shoots and cause them to wilt. Severe infestations result in twisted and damaged leaves that dry out and may defoliate (Lo Genco *et al.*, 2008). Larval damage increases the susceptibility of plant leaves and trees to pathogens, including citrus canker bacterium *Xanthomonas axopodis* pv. *citri* Hasee (Rogers *et al.*, 2016). CLM is particularly harmful to nursery plantings and in young groves, which can be severely stunted with delayed maturity of 1 to 2 years (Waterhouse and Sands, 2001; Prakash, 2012). Mature trees can tolerate leaf damage without appreciable harm to the tree or its fruit (Elekcioglu and Uygun, 2013).

Organic controls

MONITORING. Pheromone traps can be used to monitor adult male CLMs, and determine their abundance (Kerns *et al.*, 2001a). However, control decisions are based on sampling for active larvae. The number of larvae required to cause a level of damage was estimated to be 0.74 larvae per tender leaf in southern China or when 30% of the leaves have active mines with live larvae (Kerns *et al.*, 2001a). In Thailand, treatments for the CLM are recommended when more than 50% of the flushes in pomelo orchards are infested (Hoy and Nguyen, 1997).

CULTURAL CONTROL. Orange trees have three flushes: (i) in spring (February–April); (ii) in summer (June–July); and (iii) in autumn (mid-August–October). The shoots of spring are the most significant (60% of the foliage) as they are primarily responsible for supporting fruit development. Also they are only attacked to a small extent by the leaf miner compared with the other two flushes, which can be 100% infested. CLM populations could be partially suppressed if tree flushing patterns could be modified by: (i) fertilizing and pruning in late winter to promote flush growth in spring when the pest is either absent or relatively scarce (Hoy and Nguyen, 1997); and (ii) limiting flush growth in late summer and autumn by not overfertilizing or overirrigating (but avoiding drought) during summer and autumn to keep trees in a healthy state, thus improving their resistance to attack by *P. citrella*. This restricted growth is used worldwide – in the Mediterranean area (Spain, Italy, Morocco and Tunisia), the USA, Australia and Asia (India and China) (Prakash, 2012). Suckers, the vigorous shoots that grow from the trunk below the graft union, should always be removed. Zhou *et al.* (1994) found that planting *Ageratum conyzoides* L. as ground cover, in hilly citrus orchards in China, increased relative humidity in summer and temperature in winter and, also, decreased summer temperatures in the canopy and soil, and thus improved conditions for the survival of *P. citrella*'s natural enemies.

HOST PLANT RESISTANCE. Batra *et al.* (1992) tested 134 citrus species/cultivars for resistance. Nineteen species/cultivars were fairly resistant, 27 slightly susceptible, 53 moderately susceptible and 25 highly susceptible. 'Cleopatra', a promising rootstock for sweet orange, was slightly susceptible whereas the commercial rootstock 'Jatti Khatti' (*Citrus jambhiri*) was highly susceptible. Padmanaban (1994) evaluated 31 species or varieties; lowest infestation levels were observed in the sweet orange variety 'Ruby Malta' (0.33%) and in the hybrid variety 'Kinnow Mandarin' (8.34%). The variation in susceptibility could be due to the availability of tender flushes and seasonal fluctuation in the leafminer population.

PETROLEUM OIL SPRAYS. Oil sprays do not kill CLM, but instead affect the behaviour of the female moth, which tends to lay fewer eggs on the oil-sprayed leaves and also the movement of moths between and within trees is affected (Beattie and Hardy, 2005). Multiple low concentration (0.25–0.5%) oil sprays applied in early summer (every 5–14 days) to the young shoots prevent the moth establishing large populations; the effectiveness of mineral oils for the control of *P. citrella* has been demonstrated in Australia, India and China (Beattie and Hardy, 2005). Spraying mature trees should only be considered if trees are moderately to severely infested and when they may be an important source of infestation on adjacent immature trees.

BIOLOGICAL CONTROL. In Iran, mortality of *P. citrella* because of natural enemies can be as high as 89%, it is about 40–70% in China, up to 80% in India (Prakash, 2012), and 25–92% in organic pomelo orchards in Thailand (Nanta *et al.*, 1996).

The most important and abundant parasitoid that has significant effect for natural control of CLM is *Citrostichus phyllocnistoides* (Narayanan), signalled in China, Japan (Hoy and Nguyen, 1997), India and Turkey (Prakash, 2012; Elekçioğlu and Uygun, 2013). In Spain, *C. phyllocnistoides* become the most abundant parasitoid in all the orchards (99.4%) by displacing native and other introduced parasitoids (Karamaouna *et al.*, 2008). The establishment, dispersal and high rates of parasitism (>80%) of *C. phyllocnistoides* were also observed in Israel (Argov and Rossler, 1996), Italy (Siscaro *et al.*, 1999), Morocco, Cyprus (Orphanides *et al.*, 1996), Turkey (Elekçioğlu and Uygun, 2006) and Greece (Kalaitzaki *et al.*, 2011). *Semilacer petiolatus* Girault was also established in the Mediterranean region as in Israel (Argov and Rossler, 1996), Italy (Lo Genco *et al.* 2008), Morocco (Abbassi *et al.*, 1996), Syria, Spain, Cyprus (Orphanides *et al.*, 1996), Tunisia, Turkey (Elekçioğlu and Uygun, 2006) and Greece (Kalaitzaki *et al.*, 2011); but with lower numbers, less frequency and a lower rate of parasitism compared with *C. phyllocnistoides*. This fact leads to the conclusion that *C. phyllocnistoides* and *S. petiolatus* are effective and well adapted in the Mediterranean climate with a great tolerance to low relative humidities. *Ageniaspis citricola* Logvinovskaya was the most important parasitoid for natural control of CLM in Taiwan, Thailand and Australia (Waterhouse and Sands, 2001; Muniappan *et al.*, 2012). Introduced to Florida, *A. citricola* established and became the dominant parasitoid causing up to 90% mortality of larvae and pupae in action with other natural enemies (Hoy and Nguyen, 1997); it is also present in the Mediterranean Basin (Jacas *et al.*, 2010).

Predators, also, appear to be important biological control agents of CLMs. Lacewings are generally associated with heavy infestations; in southern China, *Ancylopteryx octopunctata* and *Chrysopa boninensis* controlled the population of CLM with predacious mites, bedbugs, ants and spiders (Chen *et al.*, 1989). In Alabama, predation by spiders (*Hibana* sp.) was the single most important mortality element (50–70% of mortality), predation by ants (*Solenopis invicta* Buren) was second (10–19% of mortality) and predation by predatory insect larvae (*Chrysoperla* sp. and *Harmonia axyridis* Pallas) accounted for 3–27% of all mortalities (Xiao and Fadamiro, 2010).

MASS TRAPPING. The pheromone lure (sex pheromone) is active for a period of 8–12 weeks, at a rate of four traps/ha (Lapointe *et al.*, 2006). Trap placement in the middle and upper parts of the citrus plant gives better results for *P. citrella* capture. The models and colours of the traps do not interfere with *P. citrella* capture efficiency (Parra-Pedrazzoli *et al.*, 2009). However, during high infestations traps alone are not recommended without additional measures (Parra-Pedrazzoli *et al.*, 2009).

MATING DISRUPTION. The CLM sex pheromone can also be used in the field without a trap, as a form of mating disruption to control and manage the pest. The technique is based on the use of SPLAT™ (ISCA Technologies) which can provide successful disruption of *P. citrella* in commercial citrus plantations over several weeks (Lapointe *et al.*, 2006; Stelinski *et al.*, 2008, 2010; Lapointe *et al.*, 2015). A single application may be capable of effective mating disruption for an entire growing season and should lead to economic and effective control of the leafminer (Lapointe *et al.*, 2006, 2015).

MICROBIOLOGICAL CONTROL. *B. thuringiensis* is active against the leafminer and because CLM is protected inside the mine, it is suggested that mineral oils are used to act as surfactant by reducing the surface tension. This increases the penetration of *B. thuringiensis* suspension through the epidermis of the citrus leaf and kills the larvae (Amiri-Besheli, 2007). *B. thuringiensis* has been found to cause 80–97% mortality in *P. citrella* (Niu *et al.*, 2014). Spinosad shows some efficacy against larvae; however, it might need to be reapplied every 7–14 days (Kerns *et al.*, 2001a).

BOTANICAL CONTROL. The neem formulations can be used as follow-up sprays under heavy infestation and as prophylactic sprays during periods of new flush emergence. Azadiractin was potentially useful for controlling CLM; however, it might need to be reapplied every 7–14 days (Perovic and Hrnčić, 2008).

II-2 Yponomeutidae

In citrus crops, the Yponomeutidae family is represented by one major species; ***Prays citri*** Milliere, communly known as citrus flower moth (CFM).

Distribution

The CFM is a citrus pest in the Mediterranean region, in many parts of Asia (India, Indonesia, Malaysia, Pakistan, Philippines, Japan), in some African countries (Reunion, South Africa, Zimbabwe) and in Oceania (Australia and some Pacific islands) (EFSA, 2008).

Bio-ecology

It is reported specially on *Citrus limon* (lemon) and *Citrus aurantiifolia* (lime); (Ibrahim and Shahateh, 1984; El-Metwally *et al.*, 2010). All stages of the insect may be found throughout the year. The first attacks occur in spring and they are significant when the trees are in bloom. Generally, the eggs are laid on the flowers, the buds and sometimes on the young fruits (Ibrahim and Shahateh, 1984). The emerging larvae bore into these organs, wherein they feed and produce abundant webbing that ties the flowers together. Cocoons may be found on fruits, flowers and leaves. The number of generations varies from three to 16, depending on climatic conditions: three in Greece, 11 in Italy, between eight and ten in Israel and between nine and 11 in Tunisia and Egypt (El-Metwally *et al.*, 2010; Karamaouna *et al.*, 2010).

Damage

The damage is more severe in lemons and citrons. The moths feed both internally and externally on flowers, young buds, young shoots and small fruits; more rarely they can cause damage to mature fruits (El-Metwally *et al.*, 2010). The damaged flowers and buds dry out and drop; the small fruits fall down. Damage of economic importance causes up to 90% loss in flower production in Spain and Tunisia, and 15–70% flower reduction in Portugal (Garrido *et al.*, 1984; EFSA, 2008).

Organic controls

MONITORING. Monitoring the infestation is possible by sampling and controlling flowers and newly formed fruits. Monitoring the male flights is also possible by pheromone traps fitted with synthetic sex pheromone. Pheromone traps placed on the trees during spring may serve as an early warning system. The beginning of trapping should precede the mass blossoming of citrus (Karamaouna *et al.*, 2010). Biological control is most effective when timed according to captures and performed when the majority of the young larvae hatch from the eggs.

CULTURAL CONTROL. Control of the pest depends largely on cultural practices and methods directed against the adult moth, since the larvae mining within the flower tissues are less vulnerable to biopesticides and insect pathogens. Water stress is considered to be a key factor to manage CFM in *C. limon* (lemon) groves. In Sicily, orchards submitted to water stress, resulted in total infestation in buds, flowers and set fruits below the economic threshold (Mineo, 1993; Ben Salah *et al.*, 1996). However, in orchards with adequate water, the total registered infestation surpassed the economic threshold in the same period. In Greece and in twice-flowering lemons, good control *of P. citri* can be obtained by forcing early flowering in spring and summer, before the peaks of adult pest flight in each season are observed (Karamaouna *et al.*, 2010).

BIOLOGICAL CONTROL. The most important parasitoid of *P. citri* is *Ageniaspis fuscicollis* (Dalman) subsp. *praysincola* Silvestri (Hymenoptera: Encyrtidae), signalled in the Mediterranean region (Greece, Turkey, Italy, Spain, Tunisia) (Karamaouna *et al.*, 2010) and used as biological control in citrus in India and Reunion. Some other parasitoids can be useful, for example *Elasmus flabellatus* Boyer de Fonscolombe (Eulophidae) in Greece and *Bracon laetus* (Braconidae) in Spain (Jacas *et al.*, 2010). The egg parasitoids, *Trichogramma evanescens* (Westwood) and *Trichogramma cacoeciae* Marchal, are suitable candidates for the control of *P. citri* in lime orchards in Egypt and lemon orchards in Tunisia, respectively (Abo-Sheaesha and Agamy, 2004). Some predators feed on the eggs, such as *Aeolothrips tenuicornis* Bagnall and *Galendromus occidentalis*; *Chrysoperla carnea* feeds on the larvae.

MASS TRAPPING AND MATING DISRUPTION. Mass trapping by placing 120 traps/ha in a citrus orchard can be efficient and reduce the flower infestation by the larvae (Sternlicht *et al.*, 1990). The synthetic sex pheromone showed a high degree of selectivity against *P. citri*. Studies have shown good results with pheromone mating disruption in lemon in Portugal with over 95% reductions in male captures (Silva *et al.*, 2006).

MICROBIOLOGICAL CONTROL. *B. thuringiensis* and *Bacillus sphaericus* are two safe biological control agents (EFSA, 2008). In Egypt, an application of *B. thuringiensis* reduced larval infestations by about 60–75%. Two sprays of Bactospeine are adequate to effectively reduce the population of *P. citri* in lime orchards during the main flowering period of trees. Also *Beauveria bassiana* ((Bals.-Criv.) Vuill.) can be effective to reduce the population of *P. citri* (El-Metwally *et al.*, 2010). In Turkey, two or three applications of *B. thuringiensis* at 10-day intervals are usually sufficient.

II-3 Pyralidae

Most of these small moths are inconspicuous and many are economically important pests. The carob moth, ***Ectomyelois ceratoniae*** (Zeller), is considered a minor pest of citrus crops.

Distribution

It is found in the Mediterranean Basin, Iran, South Africa, Australia and the Americas.

Bio-ecology

During summer, the pest oviposits on citrus fruit, preferring grapefruits and navel orange, especially when infested with mealybugs, or on fruits that touch each other (Serghiou, 1983). The hatched larvae enter into any available openings or cracks in the fruit, wherein they feed without harming the seeds. They often remain there even after harvest, thus invading storage facilities. The carob moth remains active in nature for about 6–7 months; which makes it accomplish four to five annual generations from May to November in Tunisia and Turkey (Öztürk *et al.*, 2011). Fallen fruits and leaves serve as overwintering sites, within which the larvae hibernate.

Damage

Damage is due to larval burrowing around the calyx and the moths tunnel into grapefruits. The attack causes artificial ripening of the fruit that turn prematurely yellow, exude gum from the entry wounds and drop prematurely. Losses may reach 10–20% (Argov and Gerson, 2012).

Organic controls

MONITORING. Sex pheromone traps are placed in citrus orchards to follow the pest's populations.

CULTURAL CONTROL. In Turkey and Tunisia, mechanical control can decrease the infestation by 60–80%, by collecting fruits dropped every 4 days and burying them in the ground (Uygun and Satar, 2008). As the pest develops on *Acacia farnesiana* pods all year round, the planting of these trees as fences around citrus orchards should be discouraged.

MICROBIOLOGICAL CONTROL. In Turkey, *B. thuriengiensis* is 95–99% effective; treatments should be started in early summer and repeated at 20-day intervals (Uygun and Satar, 2008). *B. subtilis* SPB1 could also be effective against carob moth (Mnif *et al.*, 2013).

BIOLOGICAL CONTROL. In Tunisia, the release of *T. cacoeciae* in the field (300 eggs/tree) (25,000 parasitoids/ha) can succeed if the time of releases coincide with the pest oviposition periods. It lowered the infestation level to ≈ 8% (Dhouibi *et al.*, 2016). In Turkey, the parasitoid, *Phanerotoma flavitestacae* Fisher, had controlled the carob moth after release (Uygun and Satar, 2008).

MASS TRAPPING. The density of ten traps/ha seems quite effective to control *E. ceratoniae.* These traps capture the maximum number of adult males consequently reducing the population level of the pest. Treatment with mass trapping lowered the infestation level to 8% (Dhouibi *et al.*, 2016).

MATING DISRUPTION. In Tunisia, SPLAT™ (EC) was applied as discrete points at the dose of 350g/ha, distributed on the branches of the trees. Results seem promising by reducing the rate of infestation by *E. ceratoniae* in citrus orchards. In fact, after 13 days of the treatment, the reduction of the infestation rate reached 100% (Dhouibi *et al.*, 2016).

ATTRACT AND KILL. This control method is based on a combination between a carob moth sex pheromone (0.08%), an insecticide (Permethrin: 6%) and inert ingredients (93.93%). In the organic field, we put two drops of the product on a branch of every selected tree with a gun (400 drops/ha). This product attracts the males *of E. ceratoniae,* and when they contact the product, the insecticide kills them (Dhouibi *et al.*, 2016).

BOTANICAL CONTROL. Azaderachtin (compound of neem used at the dose 0.25 l/ha) is an approved insecticide in organic agriculture and periodically utilized to reduce damaging pest populations. Azaderachtin works as an insect growth regulator and prevents immature stages from moulting, reducing insect infestations.

SIT (STERILE INSECT TECHNIQUE). This approach was conducted in Tunisia and Iran in order to control the moth attacking pomegranates and citrus. The results of fecundity, fertility, egg hatch and adult emergence of the date moth suggest that the most effective radiation dose to sterilize females and partially sterilize males is 250 Gy in Tunisia. This dose allowed no emergence of adults for the F_1 (Chakroun *et al.*, 2015). In Iran, the result shows that the best controlling doses of young and old pupa are 120 Gy and 160 Gy, respectively (Zolfagharieh *et al.*, 2009).

II-4 Tortricidae

Many of the tortrix moths or leafroller moths are economically important pests. The citrus leafroller, ***Homona coffearia*** Nietner, is considered as a minor pest to citrus.

Distribution

This species is widespread in Asia, from India to Japan.

Bio-ecology

Adult moths are active during the night and shelter in the tree during the day. A single female can deposit approximately 50 eggs during her lifetime, usually on the upper surface of the leaf near the midrib. The newly hatched larvae spin a thin silk web on the leaf surface, and stay beneath it to feed. As the larvae grow, they begin to fold the leaf longitudinally and feed within the roll. When the larvae are mature, they pupate inside a cocoon (Muniappan *et al.*, 2012).

Damage

Damage is mainly from the loss of young leaves, which reduces photosynthesis and depresses the yield.

Biological control

In India, massive release of *Trichogramma dendrolimi* Matsumurain on a large scale, significantly caused the reduction of *H. coffearia* (Prakash, 2012).

II-5 Papilionidae

The family includes the largest butterflies in the world; though the majority are tropical, members of the family inhabit every continent. The citrus butterfly or citrus dog, ***Papilio demoleus*** Linnaeus, is considered as a minor pest to citrus.

Distribution

This species is distributed in Asia (India, Pakistan, Bangladesh, Sri Lanka, Iran, China, Taiwan, Indonesia and the Philippines), Australia, Europe (only Portugal), Central America and the Caribbean (Cuba, Jamaica and the Dominican Republic) (Sarada *et al.*, 2014).

Bio-ecology

Adults are active fliers. A female lays about 180 eggs during 2–6 days. The larva feeds on leaves and when disturbed it contracts and emits a repulsive smell. The last instar is a voracious eater, consuming several leaves each day. After that the larva wanders and selects a suitable space for pupation. In different seasons, the life cycle varies from 20 to 100 days. There may be five to six overlapping generations each year and it hibernates at the chrysalis stage (Mineo, 1986).

Damage

The pest attacks mandarin and acid lime plantations almost throughout the year but attacks are particularly serious during the summer. Caterpillars cause severe defoliation as they are voracious feeders. It is a highly destructive pest of citrus plants in India (Prakash, 2012).

Organic controls

CULTURAL CONTROL. Low-level infestations can be controlled by hand-picking and destroying larvae (Prakash, 2012).

BOTANICAL CONTROL. The pest can be controlled by foliar sprays at an early larval stage (Prakash, 2012). Some plant extracts such as pipal (*Ficus religiosa*), carrot grass (*Parthenium hysterophorus*), neem (*Azadirachta indica*) and datura (*Datura stramoneum*), can significantly reduce the *P. demoleus* larval population (by 52.8%). The leaf extracts of *Eucalyptus globulus* and *Ageratum conyzoides* are effective against butterfly larvae. An aqueous extract of neem seeds, used twice at 8 days interval, effectively checks the pest population as it has strong antifeedant and repellent activities (Pandey *et al.*, 2011).

BIOLOGICAL CONTROL. *Apanteles flavipes* Cameron (braconid) and *Melalophacharops* sp. (ichneumonid) are the predominant larval parasitoids and *Pteromalus puparium* Linnaeus (chalcid) and *Holcojoppa coelopyga* Morley (ichneumonid) are the major pupal parasitoids which could be utilized for effective biological control of the pest (Prakash, 2012). In India, parasitoids known to parasitize larvae of citrus butterfly are *Apanteles papilionis* Viereck and *Bracon hebetor* Say. The release of *Trichogramma chilonis* Ishii (500 adults per tree) can be effective with the egg parasitism rate as high as 75.9%. In Thailand, the larval parasite, *Erycia nymphalidophaga* Baranov (tachinid), and the egg parasite, *Ooencyrtus malayensis* Ferriere (encyrtid), were found active against *P. demoleus* (Sarada *et al.*, 2014).

MICROBIOLOGICAL CONTROL. The entomopathogens, *B. thuringiensis* and *Beauveria bassiana* give good control of the pest in Nepal and India (Prakash, 2012; Sarada *et al.*, 2014).

III Hemiptera

III-1 Pentatomidae-Tessaratomidae

Most of the bug species occurring in citrus production areas are generalist feeders (e.g. the southern green stink bug ***Nezara viridula*** (Linnaeus)) (Drake, 1920); other species are locally serious pests to citrus, such as the spined citrus bug, ***Biprorulus bibax*** (Breddin), and the bronze orange bug, ***Musgraveia sulciventris***. These species are confined to Australia and would pose a serious threat in other citrus-producing areas of the world if accidently introduced (Panizzi *et al.*, 2000).

Species and their distribution

The generalist *N. viridula* is a highly polyphagous feeder. Its distribution includes the tropical and subtropical regions of Europe, Asia, Africa and the Americas (Panizzi, 2008). In autumn and early winter, this species may become a serious pest in citrus groves, feeding especially upon the fruit, young seedlings and young shoots of older trees (Drake, 1920). The spined citrus bug, *B. bibax*, has emerged as a major citrus pest in Australia (Panizzi *et al.*, 2000). *B. bibax* feeds on the fruits of lemons, mandarins and oranges, causing drying and brown staining of the fruit segments, gumming on the skin and premature fruit drop (James, 1989; Panizzi *et al.*, 2000).

Bio-ecology

The female oviposits on leaves, fruit or twigs. Adults and nymphs of this species attack fruits, young leaves and shoots by sucking the sap (Drake, 1920; Panizzi *et al.*, 2000). These bugs are common in natural habitats surrounding orchards and on weeds and cover crops within orchards (Mundinger and Chapman, 1932; Beers *et al.*, 1993); migrating adults invade orchards and feed on the fruit.

Damage

Bugs feed by inserting their stylets into the plant tissue and sucking up nutrients (Drake, 1920). During the feeding process, they also may transmit plant pathogens (bacterial and fungal transmissions), which increase their damage potential (Mitchell, 2004). They cause injury to plant tissues, resulting in plant wilt and, in many cases, abortion of fruits and seeds. The damage on fruit from the punctures is hard brownish or black spots. These punctures affect the fruit's edible qualities and decidedly lower its market value. Young fruit growth is retarded and the fruit often withers and drops from the plant (Mundinger and Chapman, 1932; McPherson, 1982).

Organic controls

CULTURAL CONTROL. Weed management is a basic agronomic practice used to suppress these pests. The potential of trap cropping to control bugs was demonstrated by Rea *et al.* (2002), where the density of the southern green stink bug, *N. viridula*, was reduced when black mustard, *Brassica nigra*, was grown around the perimeter of fields. The trap crop should be ploughed under before the developing bugs become adults to prevent them from migrating to the main crop. Also it has been demonstrated that hand-collection with large nets can be done successfully and profitably. Using large nets, the work should be done in the early morning or on cool days (Drake, 1920).

BIOLOGICAL CONTROL. Parasitoids, usually wasps and flies, provide biological control of *N. viridula*. In Florida, *Trichopoda pennipes* Fabricius parasitizes adults and nymphs, and *Trissolcus basalis* Wollaston parasitizes eggs (Buschman, 1980; Jones, 1988; Corrêa-Ferreira and Moscardi, 1995). These two parasites have also been introduced as biological control agents in other areas, such as Australia and Hawaii. *T. basalis* has been used in California in an effort to control its southern green stink bug population (Davis, 1967; Jones, 1988).

B. bibax eggs are parasitized by various wasp species. The main parasitic wasps are: *Trissolcus oenone* Dodd, *Trissolcus ogyges* Dodd, *Anastatus biproruli* Girault, *Acroclisoides tectacorisi* Girault and *Centrodora darwini* Girault. Parasitism can be as high as 60–100%, with most parasitism occurring during the spring to early summer. Predators of *B. bibax* nymphs and adults include: spiders, predatory bugs, praying mantis and the assassin bug (*Pristhesancus plagipenis*). Ants and lacewing larvae consume significant numbers of *B. bibax* eggs (Panizzi *et al.*, 2000).

BOTANICAL CONTROL. The use of a natural insecticide, such as the extract of neem seeds, decreased the scars caused by the feeding of *N. viridula* (Seymour *et al.*, 1995; Riba *et al.*, 2003).

MICROBIOLOGICAL CONTROL. The entomopathogenic fungi, *Beauveria bassiana*, *Metarhizium anisopliae* and *Paecilomyces lilacinus* (Thom) Samson have been evaluated for control of the black bug of rice, *Scotinophara coarctata*, with some success (Rombach *et al.*, 1986).

III-2 Cicadellidae

Leafhoppers belonging to the family Cicadellidae are plant feeders that suck sap from grasses, shrubs or trees. Some species have a cosmopolitan distribution, or occur throughout the temperate and tropical regions. Some are pests or vectors of plant viruses and phytoplasmas (Stiller, 2009).

Circulifer tenellus (Baker), the beet leafhopper is considered as a key pest to citrus; while ***Circulifer haematoceps*** (Mulsant & Rey) is a minor pest.

Distribution

C. tenellus has a wide geographical distribution, including North America (the USA), the Caribbean (Jamaica), Europe (France, Greece, Italy and Spain), Africa (Angola, Egypt, Morocco, South Africa, Sudan and Tunisia) and Asia (India, Iran, Israel, Saudi Arabia and Turkey) (EFSA, 2015).

C. haematoceps has a restricted distribution including Europe (Cyprus, France, Greece, Italy, Portugal and Spain), North Africa (Egypt, Morocco and Tunisia) and the Middle East (Iraq, Israel, Syria, Turkey and Iran) (EFSA, 2015).

Bio-ecology

Available data suggest that the two species, *C. tenellus* and *C. haematoceps*, have a similar biology. Development time, from egg to adult, is strongly dependent on temperature and varies from 19 to 119 days for *C. tenellus*. They have a variable number of generations per year depending on the geographical area (from one to six) and overwintering is sustained by mated females, largely in uncultivated areas (EFSA,

2015). Each female can lay from one to 200 eggs in the leaf veins and petioles of the host plants (Meyerdirk and Moratorio, 1987). Both *C. tenellus* and *C. haematoceps* are highly polyphagous, feeding on a variety of herbaceous plants (weeds and cultivated) and shrubs. Neither species is harmful by itself, but both are vectors of *Spiroplasma citri*, the causal agent of the citrus stubborn disease (Rana *et al.*, 1975). After acquiring *S. citri* by feeding on infected plants, the midgut, haemocoel and salivary glands of leafhoppers are colonized and they transmit *S. citri* in a persistent way through infected saliva (Liu *et al.*, 1983).

Damage

There are no reports of direct damage due to the feeding activity of *C. haematoceps* and *C. tenellus* on cultivated host plants (EFSA, 2015). The economic importance is always associated with their role in spreading plant-pathogenic organisms such as citrus stubborn disease of citrus. The disease affects both the quality and the yield of fruits. Data from Cyprus indicate, on two cultivars of navel oranges, yield reductions from 19% to 34%, with impact on reduction in fruit size, weight and quality (Mello *et al.*, 2010). Visual surveys indicated that the proportion of trees infected with stubborn disease in affected orchards in California and Morocco ranged from less than 1% to over 50% (EFSA, 2015).

Organic controls

CULTURAL CONTROL. Management procedures such as proper weed control or the plantation of weeds preferred by these Cicadellidae species in rows between the citrus trees to attract the pest individuals are encouraged.

BIOLOGICAL CONTROL. *Anagrus epos* Girault and *Anagrus nigriventris* Girault, parasitoids of beet leafhopper eggs, which successfully control their host, are proposed as biological control agents. The parasitism rate is about 71.4% (Niu *et al.*, 2014).

III-3 Psylloidea

Jumping plant lice or psyllids are small plant-feeding insects that tend to be very host specific (monophagous) or they feed on a few closely related plants (oligophagous) (Burckhardt and Ouvrard, 2012). Two psyllid species are considered in citrus: the Asian citrus psyllid, ***Diaphorina citri*** Kuwayama, is considered as a major pest, while the African citrus psyllid, ***Trioza erytreae*** (Del Guercio) is considered as a minor pest.

III-3-1 Liviidae

Diaphorina citri Kuwayama, Asian citrus psyllid (ACP)

Distribution

D. citri is widely distributed, in Asia (India, Pakistan, China, the Philippines, Taiwan, Indonesia and Malaysia) (Muniappan, 2012), North America (the USA), South America (Brazil and Uruguay), Central America (Mexico and the Caribbean) and Europe (Spain). The ACP is the natural vector of the greening disease or Huanglongbing (HLB) responsible for the citrus decline (Manjunath *et al.*, 2008), which is already present in a number of countries in Asia (India, Burma, Malaysia, Indonisia, China and the Philippines), Africa (Tanzania and Ethiopia), North America (the USA), Central America (Jamaica) and South America (in Brazil but absent in Uruguay). HLB has not been reported in Australia or in the Mediterranean Basin.

Bio-ecology

New flush is required for psyllid females to lay eggs as well as for subsequent development of the psyllid nymphs (French, 2002). When a suitable flush is not available for egg laying, psyllids may remain on a tree feeding on mature leaves until a new flush is available for reproduction (Rogers *et al.*, 2016). The ACP is the vector of the greening disease (HLB) which is caused by the phloem bacteria *Candidatus Liberibacter asiaticus* (Hoddle and Pandey, 2014). This Asian form is heat tolerant and symptoms of the disease can develop at temperatures of up to 35°C. Acquisition of the pathogen is usually through psyllid nymphal feeding while adults are responsible for transmission from tree to tree. Transmission of the pathogen is thought to occur through salivary secretions, requiring 1–7 h of feeding for successful transmission (Polek *et al.*, 2007).

Damage

Psyllid feeding damage is limited to new growth resulting in curling and distortion of the young leaves (Rogers *et al.*, 2016). The psyllids also produce wax and honeydew, which allows the growth of sooty mould. Trees affected by greening disease bear small, asymmetrical fruit which are partially green and which are unsaleable because of their poor size and quality (Muniappan *et al.*, 2012). Wherever the disease has appeared, citrus production has been compromised with the loss of millions of trees. In India, the incidence of greening disease was more on sweet orange and mandarin varieties compared with other cultivars such as acid lime and lemon (Prakash, 2012).

Organic controls

Reducing psyllid populations not only slows the rate of HLB spread but also reduces the severity of the disease once established. However, eliminating HLB from an area is difficult to achieve with vector control alone.

CULTURAL CONTROL. Management practices used within a grove can affect psyllid populations, especially those that promote new flushes such as hedging, topping and fertilization (Rogers *et al.*, 2016). Alternate host plants such as orange jasmine (*Murraya paniculata*), curry leaf (*Murraya koengi*) and box orange (*Severinia buxifolia*) in the vicinity of the grove can serve as sources of psyllids for infestation and they should be removed from areas surrounding citrus groves. However, the possibility of using these plants as a 'trap crop' may be used (Prakash, 2012). Also, non-commercial plants neighbouring managed areas, maintained without control sprays against psyllid, are probably the main source of inoculum for HLB spread to managed citrus areas. Only clean and healthy plants should be transported from nursery stock. In areas with low incidence of greening disease, any infected trees should be removed to prevent them from being reservoirs of the pathogen.

BIOLOGICAL CONTROL. Two host-specific parasitoids have shown efficacy in Asia (Taiwan, China, India and the Philippines) for controlling the ACP: *Tamarixia radiata* (Waterston) and *Diaphorencyrtus aligarhensis* (Shafee, Alam, & Agarwal) (Aubert, 1987; Hall, 2008; Muniappan *et al.*, 2012). Classical biological control projects have been conducted to establish these two parasitoids in a number of countries invaded by *D. citri*, but only *T. radiata* was successfully established and it is generally considered as the better of these two parasitoids against *D. citri* (Hall, 2008; Hoddle and Pandey, 2014). *T. radiata* is now widely distributed (the USA, Venezuela, Uruguay, Brazil, the Caribbean, Guadeloupe, India, Taiwan, Mauritius and Reunion Islands) and it is an efficient agent for the inundative type of biological control with an ideal parasitoid:host ratio of 1:10 to obtain an optimal offspring size (Hall, 2008). In India, the nymphal parasitoid *T. radiata* has been quite effective causing as high as 30–40% and even up to 90% parasitism at certain locations (Prakash, 2012). *Tetrastichus radiatus* Waterston is reported to control *D. citri* in Pakistan and India (Baloch, 1996). No egg parasitoids of *D. citri* have been recorded and adults seem to be fairly free from natural enemies (Hall, 2008). *T. radiata* in India and other areas in Asia are attacked by a complex of hyperparasitoids.

The syrphid fly, *Allograpta oblique* (Say), has been found attacking *D. citri* in Reunion, Nepal (Aubert, 1987) and Florida (Michaud, 2002). Chrysopids reported to attack *D. citri* in Florida, India, China, Reunion and Nepal are *Chrysopa boninensis* Okamoto, *Chrysoperla rufilabris* Burmeister and *Apertochrys acrassinervis*; with possible release of *Mallada boninensis* Okamoto (Michaud, 2004; Prakash, 2012). At least 12 species of coccinellids (*Brumus suturalis* Fabricius, *Coccinella rependox* Thunberg, *Coccinella septempunctata* L., *Cheilomenessex maculate* Fabricius, *Harmonia axyridis* Pallas, *Chilocorus nigrita* (Fab.), *Hyppodamia convergens* Guérin-Méneville, *Adelia bipunctata* L., etc.) that prey upon nymphal psylla and can be used for biological control are reported in India, China, Reunion and Nepal (Aubert, 1987; Hall, 2008). Certain spider species (*Hibana velox* Becker) may be important predators of *D. citri* (Hall, 2008; Michaud, 2002, 2004). The phytoseiid mite, *Amblyseius swirskii* Athias-Henriot, can be used as a biological control agent.

MICROBIOLOGICAL CONTROL. In Brazil, extract of *Burkholderia rinojensis* strain A396 (a bacterium isolated from a Japanese soil sample) is effective as a microbiological insecticide in reducing populations of *D. citri* and had no phytotoxic effect on the plants. Spinosad effectively controls ACP and is recommended for use in organic production (Florida). Entomopathogenic fungi are important biological control agents of ACP and often cause epizootics that reduce host populations dramatically. A number of species of entomopathogenic fungi have been reported to infect *D. citri* worldwide including: *Isaria fumosorosea* Wize (syn. *Paecilomyces fumosoroseus*), *Hirsutella citriformis* Speare, *Beauveria bassiana* and *Lecanicillium lecanii* (Zimm.) Zare & W. Gams (Hall, 2008; Guizar-Guzman and Sanchez-Pena, 2013). Spinosad provided 87% mortality accumulated over 72 h (Qureshi *et al.*, 2013).

HOST PLANT RESISTANCE. *D. citri* shows an oviposition preference for Duncan grapefruit, rough lemon and sweet orange, while Rangpur lime, 'Eureka' lemon and citron were the least preferred genotypes. On another side, sweet orange and mandarin orange are highly susceptible to the HLB disease; while sour orange, grapefruit and lemons are moderately susceptible (Baloch, 1996). In the USA, some hybrids of *Poncirus trifoliata*, when used as rootstocks for commercial citrus cultivars, produce trees with reduced HLB symptoms and higher fruit production. The choice of rootstocks to be used in commercial plantings expected to be affected by HLB can help reducing the infestation (Polek *et al.*, 2007).

III-3-2 Triozidae

The African citrus psyllid, ***Trioza erytreae*** (Del Guercio) (AfCP)

Distribution

This species is distributed in sub-Saharan Africa, Asia (India, the Philippines, China, Taiwan, Indonesia), South America (Brazil) and Europe (Portugal and Spain) (CABI/EPPO, 2006). AfCP is a vector of the African citrus greening disease or HLB caused by '*Candidatus*' *Liberibacter africanus*. However, the distribution of *T. erytreae* is wider than that of citrus greening bacterium, which has not been recorded in Spain and Portugal.

Bio-ecology

T. erytreae is confined to the family *Rutaceae*, especially limes (*C. aurantiifolia*) and lemons (*C. limon*). The AfCP as well as the citrus greening agent (the African form, *L. africanus*) are sensitive to hot dry conditions and will not develop at temperatures exceeding 25°C. By contrast the ACP and the citrus greening agent (the Asian form, *L. asiaticus*) are heat tolerant and symptoms of the disease can develop at temperatures of up to 35°C (Polek *et al.*, 2007). The AfCP adult is able to acquire HLB and transmit the pathogen within 24 h of initial feeding; in contrast, the ACP does not transmit the disease until approximately 24 days after initial feeding. *T. erytreae* is, also, able experimentally to transmit the Asian form and in Mauritius and Reunion, where both forms occur, *T. erytreae* probably transmits both. Nymphs, as well as adults are capable of transmitting the greening agent to citrus. The population fluctuation of *T. erytreae* is correlated with the flushing rhythm of the citrus host. A female psyllid can lay up to 2000 eggs during 4–7 weeks on the new-growth leaves and the nymphal development takes 17–43 days. Both periods are inversely related to mean temperature and directly related to the nutritional value of the leaves (Van den Berg *et al.*, 1991). Nine to ten generations may occur per year.

Damage

The psyllid itself severely distorts leaves, which become stunted and galled. Honeydew excreted by psyllids covers the outside of fruits and leaves and promotes the growth of sooty mould fungus that inhibits photosynthesis, weakens the plant and makes fruit unattractive (Polek *et al.*, 2007). Nevertheless, the main economic importance of AfCP is as vector of the citrus greening which causes: (i) mottling and yellowing (chlorosis) of the leaves; and (ii) misshapen and bitter taste of fruits. Trees are frequently stunted and have partial defoliation. Twig dieback, leaf and fruit drop, and off-season blooming are other common symptoms (Aubert, 1987).

Organic controls

CULTURAL CONTROL. Importation of citrus from countries where the citrus greening bacterium or either of its vectors occur should be prohibited. *T. erytreae* enters orchards from indigenous hosts in surrounding vegetation, so it is recommended to remove them. Citrus material (budwoods, grafted trees, rootstocks) from infected areas can carry eggs and/or nymphs over longer distances. It is possible to fumigate citrus budwood material against *T. erytreae* (Van den Berg *et al.*, 1991).

BIOLOGICAL CONTROL. In Reunion Island, *T. erytreae* has been successfully controlled and eliminated by the introduction of *Tamarixia dryi* (Waterston) from South Africa (Aubert, 1987). Two primary parasitoids are associated with *T. erytreae* in Zimbabwe: *Terrastichus radiatus* is an external parasitoid whereas *Psyllaephagus pulvinatus* (Waterston) is an internal one.

III-4 Superfamily: Coccoidea

Coccoidea (scales and mealybugs) contains nearly 8000 species of plant-feeding hemipterans comprising 32 known families (Gullan and Cook, 2007). Species attacking citrus trees belong essentially to four families: (i) Diaspididae Targioni-Tozzetti (armoured scales); (ii) Coccidae Fallen (soft scales); (iii) Pseudococcidae Heymons (mealybugs); and (iv) Monophlebidae Signoret (giant scales). These species are mostly sap-sucking small insects feeding on plant tissues and living under waxy covers.

III-4-1 Diaspididae

The Diaspididae (armoured scales) is a large family of nearly 2500 described species, and among these, some species are considered as major pests of citrus crops (Miller and Davidson, 1990).

Species and their distribution

California red scale, ***Aonidiella aurantii*** (Maskell), is a major pest of citrus with a worldwide distribution (EPPO, 2015). Florida red scale (or circular scale), ***Chrysomphalus aonidum*** (Linnaeus), is a polyphagous scale widespread in tropical and subtropical areas of North, Central and South America, the Mediterranean, southern Africa, Australia, Asia and the Pacific Islands (Kennett *et al.*, 1999). Black scale, ***Parlatoria ziziphi*** Lucas, is widely disseminated throughout the world, especially in the tropics but also in temperate zones (EPPO, 2015). This species has been reported as the most important pest of citrus in Egypt (Coll and Abd-Rabou, 1998). Foldi (2001) listed this species as an economically important pest in France. Purple scale, ***Lepidosaphes beckii*** (Newman), is a major pest of citrus in southern California, Florida, Iran, South Africa, the Mediterranean area and most of the citrus-growing countries of South America (EPPO, 2015). Citrus snow scale, ***Unaspis citri*** (Comstock), is one of the principal pests of citrus in many citrus-growing regions of the world, especially in the tropics (Miller and Davidson, 1990). Chaff scale, ***Parlatoria pergandii*** (Comstock), is a cosmopolitan and polyphagous pest. Miller and Davidson (1990) listed this insect as a serious and widespread pest. It was reported to be a very important pest in Spain, Japan, Italy, Turkey, Lebanon, Israel, South-east Asia, Central America, Mexico, Florida and Texas (Talhouk, 1975). Dictyospermum scale, ***Chrysomphalus dictyospermi*** Morgan, has a worldwide distribution and is known mainly as a serious pest of citrus (Miller and Davidson, 1990; Danzig and Pellizzari, 1998; Foldi, 2001). Other species have been reported on citrus but are considered as sporadic or locally important, such as the Japanese citrus scale, ***Unaspis yanonensis*** (Kuwana), which is established in China (Ebeling, 1959), Korea, Japan, Italy and southern France (Commeau and Sola, 1964) where it is now a major pest of citrus (Panis, 1982). Greedy scale, ***Hemiberlesia rapax*** (Comstock), was originally described from host plants in California and Florida, USA, and has since been found in countries in Africa, Central and South America, Europe and southern Asia (EPPO, 2015). It is an occasional pest of citrus, considered of little economic importance (Jeppson, 1989). Glover's scale, ***Lepidosaphes gloverii*** (Packard), is widely distributed throughout the tropical and subtropical regions of the world (Nakahara, 1982). Due to the natural enemies and especially parasitoids, it is now of less importance. However, it is still occasionally a serious pest that requires control (Maddison, 1976; Chua and Wood, 1990; Danzig and Pellizzari, 1998).

Bio-ecology

When the crawlers (first instar) settle and start feeding, they begin to secrete the armour. The second and third nymphal stages develop beneath this armour (Miller and Davidson, 1990). The number of generations per year depends on the temperature and humidity. Usually, armoured scales have several overlapping generations a year: (i) California red scale raise three to seven annual generations; (ii) *U. citri* produces up to nine in Texas; (iii) Florida red scale produce four to six generations per year; and (iv) purple scales may produce three or more generations each year (Miller and Davidson, 1990).

Damage

At feeding, these pests inject toxins into the leaves, which causes them to yellow and drop (Beardsley and Gonzalez, 1975). On fruits, the type of damage depends on the scale species. Species such as California red scale create a yellow halo around the scale (Miller and Davidson, 1990); other species, like *L. gloverii* can cause a delay in the development of colour in maturing fruit because the area around the scale insect remains green (Bruwer, 1998). Large infestations may drastically affect plant vigour, defoliate the tree, causes dieback of twigs, lower fruit production and occasionally cause death of the tree (Jeppson, 1989). The citrus snow scale feeds primarily on the trunk and tree limbs of young trees. With heavy infestations, the normal growth of the bark is prevented and it may become split. This split may allow access for other insect pathogens to invade the damaged tree (Russo and Longo, 2004).

Organic controls

MONITORING. Scale infestation can be visually monitored to detect the appearance of the scale armour on stems, leaves and fruits of hosts. High populations are easy to detect. The crawlers can be detected by the use of sticky traps (Miller and Davidson, 1990).

CULTURAL CONTROL. Plants with proper cultural care (pruning, good fertilization and irrigation) are more resistant to scale damage. Pruning to open up canopies can reduce populations of scales by increasing scale mortality from exposure to heat (Ortu *et al.*, 2004). If they are limited to a few parts of small plants, heavily infested twigs and branches can be eliminated. High-pressure water sprays (of 10–20 l/tree at 300 psi) may reduce scale populations on the trunk and main branches by over 60% within 3 months, and by about 90% after 1 year (Cesnik and Medina, 1995).

BIOLOGICAL CONTROL. The most important natural enemies of scales are parasitic wasps, including species of *Aphytis*, *Coccophagus*, *Encarsia* and *Metaphycus* (Flint and Dreistadt, 1998).

Classical biological control against California red scale was initiated in California with *Aphytis lingnanesis* (Compere) and *Aphytis melinus* (DeBach), which were established in the area (DeBach and White, 1960; Luck *et al.*, 1997). These two species were also imported and successfully established in many other citrus-growing countries (Mellado, 2011). In Japan, two parasitoids (*Aphytis yanonensis* Rosen and *Coccobius fulvus* Compere & Annecke) were introduced from China in the 1980s to control arrowhead scale (Furuhashi and Nishino, 1983). A parasitic wasp, *Aphytis holoxanthus* (DeBach), was introduced from Israel in 1959 to control Florida red scale. The wasp became established and is now the major biological control agent for Florida red scale on Texas citrus (Maltby *et al.*, 1968; Selhime *et al.*, 1969; Dean, 1982). The ladybird beetle, *Chilocorus circumdatus* Gyllenhal is effective to control citrus snow scale (Rogers, 2012). Two wasp parasites are effective natural enemies for chaff scale: (i) *Aphytis hispanicus* (Mertect); and (ii) *Prospaltella fisciata* (Malenotti) (French, 2002). The most effective purple scale parasite is *Aphytis lepidosaphes* (Compere) (Dreistadt, 2012). Biological control agents of *P. ziziphi* include fungi of the genera *Aschersonia*, parasites of the genera *Aspidiotiphagus* and *Aphytis*, and the predators *Chilocorus nigrita*, *Lindorus lophanthae* and *Orcus chalybeus* (Dekle, 1976). Browning (1994) discusses effective biological control of Glover's scale in Florida, USA, and provides a list of natural enemies found there.

ORGANIC INSECTICIDES. Mineral oils and soap insecticide are used to control armoured scales (Miller and Davidson, 1990).

III-4-2 Coccidae

This is a family of scale insects that are commonly known as soft scales, wax scales or tortoise scales. The females are flat with elongated oval bodies and a smooth integument which may be covered with wax (Allen, 1995). Among this family, ***Saissetia oleae*** (Olivier) is considered as a major pest, while ***Coccus hesperidum*** Linnaeus and ***Ceroplastes floridensis*** Comstock are treated as minor pests to citrus.

Saissetia oleae (Olivier), olive scale or black scale

Distribution

This species has a worldwide distribution.

Bio-ecology

The female lays 1000–4000 eggs in a cavity under her body, where they are protected for the 16–40 days they take to hatch (Gill, 1988). The first instar (crawler) migrates to leaves and petioles to locate a feeding site. As scales mature, they gradually move back to twigs and interior branches. Overwintering takes place at the intermediate immature stage. In culture, a complete generation takes 70–90 days. High temperatures and low humidity affect the immature stages adversely and influence their distribution on the plant by colonizing the shaded and more humid parts (Gill, 1988).

Damage

The young black scale excretes sticky and shiny honeydew on leaves of infested trees. At first, affected trees and leaves glisten and then become sooty and black in appearance as sooty mould fungus grows on the honeydew (Waterhouse and Sands, 2001). Infestations reduce vigour and productivity of the tree. At higher infestation levels, females infest the outer canopy where the honeydew they secrete may cover the fruits and these become covered in sooty mould. Continued feeding causes defoliation which reduces the bloom in the following year. High infestations are most common in young groves and may cause twig dieback.

Organic controls

CULTURAL CONTROL. Pruning to provide open airy trees discourages black scale infestation.

BIOLOGICAL CONTROL. The primary biological control agent in Florida against *S. oleae* is the egg parasitoid *Scutellista caerulea* (Fonscolombe), which lays its egg under the female scale from which a grub hatches to feed on the scale eggs (Browning 1994). *Metaphycus helvolus* (Compere) was an effective introduced natural enemy of the scale in coastal areas of California. From California, *M. helvolus* has been introduced successfully to Greece, Crete, Chile and Iran to control *S. oleae*. However, since the introduction of *Metaphycus bartletti* (syn. *Metaphycus lounsburyi*) Annecke and Mynhardt (Hymenoptera: Encyrtidae), the latter species has now become the most important control agent in California, some European countries, Israel and Egypt (Argov and Rossler, 1993; Abd-Rabou, 2011). The most abundant and widely distributed parasitoids of black scale in citrus crops in Spain are *S. caerulea* and *Metaphycus flavus* (parasitoid of young instars). *M. helvolus* and *M. lounsburyi*, considered the main parasitoids in other citrus areas of the world, had a limited incidence in Spain (Tena *et al.*, 2008). In Cyprus, the most abundant natural enemies were *S. caerulea*, *M. lounsburyi* and *M. flavus* (Davarci, 1996). In Australia, control of *S. oleae* has been attributed mainly to the parasitoids *M. lounsburyi*, *M. helvolus*, *Metaphycus anneckei*, *S. caerulea* and *Moranila californica* (Howard) (Waterhouse and Sands, 2001). Their effects are complemented by several species of coccinellids, important predators of immature stages, including *Cryptolaemus montrouzieri* Mulsant, *Chilocorus bipustulatus* Linnaeus, *Rhyzobius ventralis* Erichson *and Rhyzobius forestieri* Mulsant (Waterhouse and Sands, 2001; Jacas *et al.*, 2010). However, ants attracted to the scales by honeydew may deter natural enemies from attacking them (Gill, 1988). Several species of ants that feed on the honeydew produced by *S. oleae* stop predators from attacking the scales (Waterhouse and Sands, 2001).

MICROBIOLOGICAL CONTROL. A fungal pathogen, *Verticillum lecanii*, destroys high densities of *S. oleae* in late summer and autumn when conditions are humid (Waterhouse and Sands, 2001).

ORGANIC INSECTICIDES. Only early instars are susceptible to treatment by organic pyrethrum sprays or authorized insecticides. To achieve the best control, the oil spray should be timed to coincide with the greatest number of young stages. Spot spraying heavily infested trees rather than whole blocks is also an option for these scales.

Coccus hesperidum Linnaeus, brown soft scale

Distribution

The brown soft scale has a worldwide distribution (Waterhouse and Sands, 2001).

Bio-ecology

Brown soft scale eggs are retained in the female's body until hatching and each female lays about 75 eggs (French, 2002). Females give birth to live crawlers that remain under the protective cover of the female's body for several days. Once leaving their mother, the tiny crawlers find a feeding site where they remain until maturity. The threshold of development is at 13°C and 515 degree days are necessary for the completion of a generation. They are most noticeable on plant stems and branches, but also will settle on leaves (French, 2002). In warmer environments, crawlers are present throughout the year and a generation may be completed in 40–60 days (Waterhouse and Sands, 2001).

Damage

Brown soft scale is a piercing-sucking insect that sucks plant sap at all stages of its development, causing leaves to yellow and die.

However, the main damage is indirect, through the production of honeydew that is colonized by sooty mould fungi, which can be more damaging than the scale itself (Waterhouse and Sands, 2001). Sooty mould covers leaves, flowers and fruits with a thick black mass, which decreases photosynthesis activity and vigour of trees. When the sooty mould occurs on fruits, it can reduce their size and cause them to be unmarketable or of a lower grade, as the fungus is difficult to wash off (French, 2002). This pest was considered a major pest of citrus in several countries before being completely controlled by its natural enemies.

Organic controls

MONITORING. Early signs of infestation can easily be overlooked until blackening of leaves by sooty mould becomes noticeable. Check the level of parasitism by looking for parasite exit holes and for developing parasites within the scale body (Stansly and Rogers, 2016).

CULTURAL CONTROL. Prune and destroy infected branches and plant parts.

BOTANICAL CONTROL. Control should be achieved with insecticidal soap, horticultural oil, mineral white oils or neem oil and this should be thoroughly applied to all infested plant surfaces (Beattie and Hardy, 2005; Stansly and Rogers, 2016). At least three treatments are needed to control an infestation repeated every 6–7 days until scales have been eradicated. The timing of the oil spray is critical, and needs to coincide with the presence of the young, recently hatched stages (crawlers) which generally occur on the upper leaf surfaces. Soft brown scales are more difficult to control because they have overlapping generations (French, 2002).

MICROBIOLOGICAL CONTROL. Entomopathogenic fungi, especially *Verticillium lecanii* (Zimmerman) Viegas can offer good control against brown soft scale in the field, under suitable conditions of relatively high humidity.

BIOLOGICAL CONTROL. In Texas citrus, parasitoids such as *Coccophagus lycimnia* (Walker) and *Microterys flavus* (Howard), generally hold the brown soft scale at below damaging levels (French, 2002). A complex of *Metaphycus* spp. parasites attack brown soft scale; the most common of these is *Metaphycus angustifrons* Compere in southern California. In Iran, *C. lycimnia* is effective on *C. hesperidum* (Davarci, 1996). The brown soft scale is controlled by *M. flavus* (Howard) and *Coccophagus scutellaris* Dalman in Cyprus (Davarci, 1996). *M. flavus* (Encyrtidae) is sensitive to ant presence; whereas *Coccophagus* spp. (Aphelinidae) are dominant in ant-attended scale colonies. The rarity of *C. hesperidum* in Peruvian citrus orchards was due to the parasitoids *M. luteolus* Timberlake and *Coccophagus quaestor* Girault. In South Africa, *C. hesperidum* is attacked by more than 25 species of chalcidoids (Prinsloo, 1984). In Australia, the most abundant parasitoids to control *C. hesperidum* are *Coccophagus ceroplastae* Howard, *C. lycimnia*, *Encyrtus infelix* Embleton, *Diversinervis elegans* Silvestri, *Microterys nietneri* Motschulsky and *Metaphycus anneckei* Guerrieri and Noyes (Waterhouse and Sands, 2001; French, 2002). In California, the ladybird beetles *Rhyzobius* (*Lindorus*) *lophanthae* Blaisdell, *Chilocorus orbus* Casey and *Chilocorus cacti* Linneatus prey on brown soft scales. In Australia, predators include the coccinellids *Cryptolaemus montrouzieri*, *Diomus notescens*, *Harmonia conformis* Boisduval, *Parapriasus australasiae* Boisduval, *Rhyzobius lophanthae* Blaisdell and *Rhyzobius ventralis* Erichson, and the chrysopid *Micromus tasmaniae* Walker (Waterhouse and Sands, 2001).

Ceroplastes floridensis Comstock, Florida wax scale (FWS)

Distribution

This species has a worldwide distribution (Davarci, 1996; Hodges *et al.*, 2001).

Bio-ecology

Two to three generations per year are common throughout its global range. Each generation lasts about 3–4 months. Each female can lay 80 eggs at maturity. Upon hatching, first instars (crawlers) emerge from underneath the female, disperse and settle on other leaves, stems and twigs to begin feeding and secreting wax around their bodies. Older nymphs can move around within the same plant to search for new flushes of growth on which to feed. FWSs can also overwinter as newly mature females (Drees *et al.*, 2005).

Damage

The damage caused by the FWS includes both direct and indirect damage. Direct damage (for the purpose of feeding) can discolour leaves, cause premature leaf drop and branch dieback. Indirect damage is due to honeydew, which favours sooty mould growth (Waterhouse and Sands, 2001). Sooty mould can cause a significant reduction in photosynthesis and aesthetic value (Hodges *et al.*, 2001).

Organic controls

CULTURAL CONTROL. When purchasing any plant material for installation, it is necessary to be certain that each plant is pest free. Otherwise, you have to prune off and destroy any infested plant parts. It is important to plant them in a location that is suitable in terms of light and soil-type needs for the duration of the plant's growth.

BIOLOGICAL CONTROL. Three parasitoids are known to attack FWSs in the USA. They are *C. lycimnia* (Aphelinidae), *Metaphycus eruptor* Howard (Encyrtidae) and *Scutellista cynea* Motschulsky (Pteromalidae) (Drees *et al.*, 2005). In Australia, several parasitoids and predators are associated with *C. floridensis*: the egg predators *Scutellista caerulea* and *Moranila californica*, and the parasitoids *Coccophagus ceroplastae*, *Diversinervis elegans* and *Microterys neitneri*, are the most important (Waterhouse and Sands, 2001).

MICROBIOLOGICAL CONTROL. The fungus *Verticillum lecanii* sometimes infects immature scales during humid conditions (Waterhouse and Sands, 2001).

III-4-3 Pseudococcidae

Mealybugs are slow-moving insects with flat and oval bodies covered with a thin layer of white wax. One mealybug species is economically important to citrus: ***Planococcus citri*** (Risso) (Panis, 1977; Williams, 1985). Other species occur in citrus orchards but are considered as minor.

Distribution

P. citri has a worldwide distribution (CABI/EPPO, 1999).

Bio-ecology

Eggs are deposited as white, cottony masses (ovisacs) giving the appearance of cotton spread on the plant. Depending on the season, egg hatch may occur after 6–10 days or several weeks (Kerns *et al.*, 2001b). Nymphs emerge from the ovisacs and settle along midribs and veins on the underside of leaves, young twigs and fruit buttons. They can also be found where two fruits are touching each other. The nymphs take 6–10 weeks to reach maturity (Griffiths and Thompson, 1957). Several overlapping generations of mealybug occur in a year and are most common during the spring and early summer (Gill *et al.*, 2013).

Damage

The citrus mealybug extracts plant sap and reduces tree vigour. Its feeding results in wilted, distorted and yellowed chlorotic leaves, premature leaf drop, stunted growth and occasional death of infested plants or plant parts (Gill *et al.*, 2013). The mealybug secretes honeydew, on which sooty mould develops and this reduces the photosynthetic capacity of leaves and degrades the fruit quality leading to commercially unacceptable appearance of fruits (Gausman and Hart, 1974).

Organic controls

MONITORING. Sex pheromone traps were found to be effective for surveying and testing the population density of citrus mealybugs (Kerns *et al.*, 2001b). Wax and honeydew secreted by crawlers are visible indicators of infestations (Griffiths and Thompson, 1957). The presence of ants is also an indicator of the presence of mealybugs or other sap-sucking insects.

CULTURAL CONTROL. Hedging citrus groves to reduce contact between trees and thorough cleaning of equipment and harvest materials are useful for reducing the spread of this insect. Some ornamental hosts near citrus groves should be avoided or monitored and treated for infestations to prevent mealybug spread (Kerns *et al.*, 2001b).

BIOLOGICAL CONTROL. Several natural enemies have been identified that are effective at controlling citrus mealybug: *Leptomastidea abnormis* (Girault), *Leptomastix dactylopii* Howard, *Chrysoplatycerus splendens* Howard and *Anagyrus pseudococci* (Girault) are common wasps parasitic on second and third instar nymphs (Griffiths and Thompson, 1957; Davarci, 1996). Lacewing *Sympherobius barberi* (Banks) and *Chrysopa lateralis* Guérin are common predators of citrus mealybug. Classical biolocal control using *Cryptolaemus montrouzieri* is used successfully in augmentation programmes against the mealybugs infesting citrus around the world (Kairo *et al.*, 2013).

MINERAL OILS. Mealybug populations may be reduced by spraying applications of petroleum oils (Jeppson, 1989). Timing of the oil spray is critical and needs to coincide with the presence of the crawlers. After the spring flush, sprays should be applied immediately after most eggs have hatched (Griffiths and Thompson, 1957).

III-4-4 Margarodidae Monophlebidae

Icerya purchasi (Maskell), the cottony cushion scale is a major species in citrus.

Distribution

I. purchasi has a worldwide distribution (French, 2002).

Bio-ecology

The cottony cushion scale gets its name from the female's white fluted egg sac, which contains from 600 to 800 eggs, and is usually found on twigs (French, 2002). The heavily laden egg sac may become two or three times longer than the body of the female. Mature females are often found on younger plant parts. The eggs hatch in 2 days during warmer months but will incubate as long as 2 months in winter. After leaving the egg sac, the crawlers settle along the midribs and veins of the leaves. The next two instars migrate to the larger twigs and branches. The cottony-cusion scale retains its legs and remains mobile throught its life (French, 2002).

Damage

Most damage is caused by the feeding of the immature stages on the foliage. Older nymphs migrate to the larger twigs and the adults to the larger branches and trunk (French, 2002). Decreased vitality, fruit drop and defoliation result from scale feeding. Heavy infestations can result in defoliation and dieback of twigs. Economic damage is also caused by sooty mould, which disfigures the fruit and reduces photosynthesis when it occurs on leaves.

Organic controls

CULTURAL CONTROL. Pruning and hedging trees prevent touching between trees and helps prevent within-grove spread of infestations. Additionally, pruning will aid in opening up the canopy to maximize light penetration. Providing plants with good growing conditions and proper cultural care, especially appropriate irrigation, is important so they are more resistant to scale damage.

BIOLOGICAL CONTROL. The cottony cushion scale was controlled by the vedalia beetle, *Rodolia cardinalis* (Mulsant) which was the first successful use of classical biological control, in which a beneficial organism was introduced and complete control was achieved in the USA (French, 2002). *R. cardinalis* has a high reproduction rate and sufficient ability of adaptation. It has successfully settled and spread in citrus-growing areas of China, Cyprus, Egypt, Iran, Japan and Turkey where it has inhibited the frequent massive occurrence of *I. purchasi* (Davarci, 1996; Takagi, 2003; Niu *et al.*, 2014). The parasitoid *Cryptochaetum iceryae* has also proved to be effective in regulating *I. purchasi* populations. Adult *C. iceryae* are sensitive to heat and aridity and are most effective in regulating cottony cushion scale populations in cooler coastal areas; in more arid areas, vedalia beetles are more effective. Ants attending *I. purchasi* infestations to collect honeydew may defend the scales from attack by their natural enemies; it may be useful to control the ants to help the natural enemies bring the scale population under control.

III-5 Aphididae

The Aphididae is a large cosmopolitan family of nearly 4000 species. Worldwide, about 25 aphid species have been recorded in citrus crops throughout the world (Barbagallo and Patti, 1986; Blackman and Eastop, 1994; Stoetzel, 1994). Some of these species are considered of economic importance, either by direct or indirect damage, and others are of occasional importance in some citrus areas (Barbagallo and Patti, 1986).

Species and their distribution

The most widespread and harmful species are the green citrus aphid ***Aphis spiraecola*** (syn. *citricida*) Patch, the cotton aphid ***Aphis gossypii*** Glover, the black citrus aphid ***Toxoptera aurantii*** (Boyer de Fonscolombe) and the most efficient vector of citrus tristeza virus (CTV), the oriental citrus aphid ***Toxoptera citricida*** (Kirkaldy). *A. spiraecola*, the green peach aphid ***Myzus persicae*** (Sulzer) and *A. gossypii* are widely distributed on citrus in temperate and tropical regions. *T. citricida* has been known to be widely distributed on citrus in Asia, India, New Zealand, Australia, the Pacific Islands, Africa, Madagascar, the Indian Ocean Islands and South America. The Mediterranean region except north-west Spain and northern Portugal remains free of this species (EPPO, 2015). Other aphid species can also be important, including *M. persicae*, the cowpea aphid ***Aphis craccivora*** Koch, the black bean aphid ***Aphis fabae*** Scopoli and the potato aphid ***Macrosiphum euphorbiae*** (Thomas).

Bio-ecology

Aphids pierce through the host-plant tissue, inject saliva that dissolves the sap and then suck the sap. This feeding system enables aphids to transmit plant viruses and other pathogens from diseased to healthy plants (Jeppson, 1989). Aphids have complicated life cycles, often alternating between amphigenic and parthenogenetic generations, with different apterous and winged adult forms. Nymphs appear in the spring and become adult apterous females that reproduce parthenogenetically, giving birth to living nymphs. Winged females appear after several generations in order to migrate to other plants of the same species, called the primary host (monoecious cycle), or to other host plants, called the secondary host (heteroecious cycle). By autumn female sexuparae appear and produce winged females and males that return to the primary host, where they copulate and the females lay the over wintering eggs (Komazaki, 1987; Blackman and Eastop, 1994; Uygun *et al.*, 2012). The life cycle of some aphid species, like the brown citrus aphid, is much less complex than that of most aphids. These species are permanently anholocyclic: all individuals throughout the year are viviparous parthenogenetic females (Komazaki, 1987; Halbert and Brown, 1998). In the Mediterranean region, the aphid populations show two main peaks synchronized with emergence of tender shoots, during spring and early autumn, and sometimes with another minor population increase during summer. The aphids complete one or two generations before the flush hardens off and then alate (winged) aphids are produced (Uygun *et al.*, 2012).

Damage

These insects affect citrus trees directly by sucking the sap. As a result, the foliage may become chlorotic and die prematurely. Their feeding also causes distortion and leaf curling, hindering photosynthetic capacity of the plant (Blackman and Eastop, 1994). In addition, they excrete honeydew, which provides a substrate for black sooty mould fungi, like *Aspergillus niger* Van Tieghem, which cover the leaf surfaces reducing respiration and photosynthesis. The major damage caused by aphids is through transmission of plant viruses (Jeppson, 1989). The brown citrus aphid, *T. citricida*, is one of the world's most serious pests of citrus because of its efficient transmission of CTV (Yokomi, 1995). One of the most devastating citrus crop losses ever reported followed the introduction of brown citrus aphid into Brazil and Argentina: 16 million citrus trees on sour orange rootstock were killed by CTV (Carver, 1978).

Organic controls

Generally, aphids are not a problem on citrus except on young trees or continually flushing varieties. Management of virus inoculum is the most important control strategy because spread of severe strains of CTV is the major problem associated with aphids and especially *T. citricida* (Garnsey *et al.*, 1998).

MONITORING. High numbers are found on the leaf surfaces during the period of flushing. Yellow traps are commonly used for population monitoring.

CULTURAL CONTROL. Aphids can be managed by encouraging natural diversity around and within the orchard using plants that encourage the 'no-citrus aphids' (Rodrigues *et al.*, 2006).

BIOLOGICAL CONTROL. Predators, parasites and fungal diseases attack aphids and occur naturally in the orchards (Grafton-Cardwell *et al.*, 2015). Van Emden *et al.* (1969) and Volkl *et al.* (2007) provide a good list of beneficial organisms, most are general predators. The honeydew produced by the aphids provides a good food source for many natural enemies. The wasp parasite, *Aphidius testaceipes* (Cresson), is commonly found to be beneficial, and sometimes causes up to 99% parasitism. *Lipolexis oregmae* (Gahan) attacks *T. aurantii*, *A. spiraecola*, *A. gossypii* and *A. craccivora* on citrus and other crops in Florida (Hoy and Nguyen, 2000).

MICROBIOLOGICAL CONTROL. Entomopathogenic fungi attack Aphididae species, but a critical requirement for efficacy of such fungi is high humidity. This was confirmed when using *Beauveria bassiana*, *Paecilomyces fumosoroseus* and *Metarhizium anisopliae* against *T. citricida* (Poprawski *et al.*, 1999).

BOTANICAL CONTROL. Oil seems to be most effective when the amount of pest is at a low level. In heavily infested plants, neem sprays can be applied around the aphid populations, especially around new shoots and under the leaves (Lowery *et al.*, 1993; Ulrichs *et al.*, 2001; Tang *et al.*, 2002).

III-6 Aleyrodidae

The family Aleyrodidae (whiteflies) is a taxon of nearly 161 genera placed in about 1556 species (Martin and Mound, 2007). Several species of whiteflies are well documented and considered as major pests of citrus whereas others occasionally feed on citrus.

Major species and their distribution

Citrus whitefly, ***Dialeurodes citri*** (Ashmead), has a cosmopolitan distribution in Southeast Asia, the Middle East, the Mediterranean region, the USA, and Central and South America (EPPO, 2015). Orange spiny whitefly, ***Aleurocanthus spiniferus*** (Quaintance) is distributed in: Asia (China, India, Pakistan, the Philippines, Thailand, Vietnam), Africa (Kenya, Tanzania, South Africa), Australia, Europe (Italy), the Caribbean and the Pacific

Islands (EPPO, 2015). Citrus blackfly, ***Aleurocanthus woglumi*** Ashby occurs in Asia (China, India, Pakistan, the Philippines), Africa (Kenya, Tanzania), the USA (Florida), and Central and South America (Nguyen *et al.*, 1998; EPPO, 2015). Cloudy winged whitefly ***Singhiella*** (***Dialeurodes***) ***citrifolii*** (Morgan) is widely distributed in Central America and the Caribbean, the USA and South America, Asia, Europe (France, Greece, Italy, Spain, Turkey), Commonwealth of Independent States (Azerbaijan, Russia, Tajikistan, Uzbekistan) and Algeria (Fasulo and Weems, 1999).

These pests are now generally found under effective biological control and pest populations rarely require treatment (Nguyen and Sailer, 1979; Nguyen *et al.*, 1983; Davarci, 1996; Stansly and Rogers, 2016).

Bio-ecology

The duration of the life cycle and the number of generations per year vary among species and are greatly influenced by the prevailing climate. A mild temperature with high relative humidity provides ideal conditions for growth and development (Byrne and Bellows, 1991).

Damage

Whiteflies typically feed on the undersides of plant leaves and affect citrus trees by sucking the sap, causing a general weakening of the infested trees (Argov *et al.*, 2012). They also cause indirect damage by the transmission of viral diseases (Jones, 2003) and the production of honeydew leading to black fungus accumulation. This black fungus may cover the leaves and fruit completely and negatively affect photosynthesis. The fruit often must be washed before it is put on the market (Grafton-Cardwell *et al.*, 2015).

Organic controls

MONITORING. Yellow sticky traps are useful for monitoring and detecting whiteflies and are commercially available (Grafton-Cardwell *et al.*, 2015). In cases of strong infestation, control measures against these pests can be required.

BIOLOGICAL CONTROL. Classical biological control against whiteflies was initiated in many countries around the world and often maintains these pests at acceptable levels. It started against *A. floccosus* by introducing the specific parasitoid, *Cales noacki* Howard, in the Mediterranean region (Morocco, Tunisia), and it was a success while the pest was under economic levels in these countries (Davarci, 1996). An introduced wasp parasite (*Encarsia lahorensis* Howard) is established in Florida's citrus regions and is reducing *D. citri* populations. Current established biological control options that are effective against citrus blackfly in Florida include *Encarsi perplexa* (Huang & Polaszek) and *Amitus hesperidium* Silvestri. Possibly these existing biological controls would also control orange spiny whitefly (Nguyen and Sailer, 1979; Nguyen *et al.*, 1983). In Turkey, the release of *E. lahorensis* and *Serangium parcesetosum* Sicard (predator coccinellid) has been undertaken (Davarci, 1996). In Japan, the parasitoids *Prospaltella smithi* Silvestri and *Cryptognatha* sp. introduced from China achieved satisfactory control by killing more than 74% of the *A. spiniferus* population (Kuwana and Ishii 1927). In Guam, the use of *P. smithi* and *Amitus hesperidium* Silvestri achieved 80–95% parasitism (Peterson, 1955). Classical biological control of *A. spiniferus* brought significant reduction within 8 months of release of *Encarsia* cf. *smithi* (Hymenoptera: Aphelinidae) in the commercial orchards of southern Africa with a mean parasitism rate of about 72.9% parasitism (Van den Berg and Greenland, 1997). Biological control against citrus blackfly in Florida is effective using *Encarsi perplexa* Huang & Polaszek and *Amitus hesperidum* (Silvestri) (Nguyen *et al.*, 1983).

CULTURAL CONTROL. A regularly maintained programme of hedging and topping can help avoid whitefly problems. Excessive use of nitrogen fertilizer will increase whitefly populations. Water sprays may also be useful in dislodging adults. Watering can also reduce the hot, dry dusty conditions that favour whiteflies and inhibit their natural enemies (Grafton-Cardwell *et al.*, 2015).

ORGANIC INSECTICIDES. Whiteflies can be controlled by insecticidal soaps, neem oil or petroleum-based oils (Grafton-Cardwell *et al.*, 2015). In Turkey, one application of mineral oils in winter is recommended to control citrus whitefly. If winter application is missed, two applications will be needed in late spring (end of May to beginning of June) and 25 days later (Davarci, 1996). A low concentration of petroleum spray oils (0.25–0.5%) has been found to provide acceptable control in China on sweet orange and pummelo (Rae *et al.*, 2000).

IV Thysanoptera

IV-1 Thripidae

Several species are economically significant pests. For citrus, two species are considered as major pests, ***Scirtothrips dorsalis*** Hood and ***Pezothrips kellyanus*** (Bagnall), while two others are treated as minor pests, ***Scirtothrips citri*** (Moulton) and ***Scirtothrips aurantii*** Faure.

Scirtothrips dorsalis Hood, yellow tea thrips or chilli thrips

Distribution

Yellow tea thrips is widely distributed in Asia, the Americas and Australia, with a restricted distribution in Africa (Uganda, South Africa) (Grafton-Cardwell and Ouyang, 1995; Hodges *et al.*, 2001; Kumar *et al.*, 2013; Minaei *et al.*, 2015).

Bio-ecology

Eggs are deposited in plant tissue. Hatching occurs over about 5 days. The immature stages take around 5–7 days to complete their development and it takes 3–3.5 days to complete the pupal stages (Grafton-Cardwell, 1995). Citrus thrips, particularly immatures, prefer to feed on young plant tissue, such as shoots, blossoms and young fruit (Niu *et al.*, 2014). They can grow at minimal temperature as low as 10°C and maximal temperature as high as 33.0°C. Their thermal requirement from egg to adult is 265 degree days (Kumar *et al.*, 2013). Populations are multivoltine with up to eight generations/year in temperate regions and 18 generations/year in warm regions.

Damage

Yellow tea thrips cause two kinds of symptom according to the citrus fruit stage. At the young fruit stage, it makes a silver-white-coloured circular scar injury symptom at the apex of the fruit (Hwang *et al.*, 2016), while it makes a rust-coloured symptom when the fruit is affected at the mature stage. Although infestation with yellow tea thrips does not reduce the yields of mature trees, it will often delay the maturity of young trees, which are not yet bearing fruit.

Pezothrips kellyanus (Bagnall), Kelly's citrus thrips (KCT)

Distribution

KCT is found in Australia, New Zealand, Hawaii and Chile and has recently been recorded as the main thrips in the Mediterranean Basin, Italy, Spain, Greece, Turkey and Tunisia (Conti *et al.*, 2003; Baker *et al.*, 2011; Navarro-Campos *et al.*, 2013a, b).

Bio-ecology

P. kellyanus is found frequently in citrus flowers of lemon (*Citrus lemon*), followed by orange (*Citrus sinensis*), bergamot (*Citrus bergamia*) and grapefruit (*Citrus paradisi*) (Conti *et al.*, 2003; Nas and Atakan, 2008). Adults and larvae are found under the calyx and between touching fruit later in season; they stay year round in citrus canopies and do not live in cover crops or weeds of citrus orchards (Niu *et al.*, 2014). Thrips required 204.5 degree days to complete development from egg to adult stage, above a minimum threshold of 10.2°C and the maximum percentage of larval stages of KCT was observed at about 500 degree days (Varikou *et al.*, 2009). Temperature probably affects the synchronization between the peak in abundance of KCT larvae, and the period when fruitlets are susceptible to thrips damage.

Nevertheless, damage by *P. kellyanus* is highly variable from one geographical area to another and from year to year (Navarro-Campos *et al.*, 2013b).

Damage

Thrips can damage the fruit from petal fall up to 6 weeks later (Planes *et al.*, 2015). The feeding of KCT on young and mature fruit causes scurfing (or halo), marking and rind bleaching. These blemishes reduce fruit quality, thereby reducing the packout of export-quality fruit and cause its downgrading (Siscaro *et al.*, 2008).

Scirtothrips citri (Moulton), California citrus thrips

Distribution

S. citri is reported in the USA, India, Iran and China (EPPO, 2015).

Bio-ecology

During spring and summer, females lay about 25 eggs in new tissues of leaves, young fruits or green twigs; in the autumn, overwintering eggs are laid mostly in the last growth flush of the season. Overwintered eggs hatch in March about the time of the new spring growth. They feed actively on tender leaves and fruit, especially under the sepals of young fruit (Kerns *et al.*, 2001c). Third and fourth instars (propupa and pupa) do not feed and complete development on the ground or in the crevices of trees. When adults emerge, they move actively around the tree foliage. They can produce up to eight generations during the year if the weather is favourable.

Damage

The nymphs and adults suck the sap from fully developed flowers, leaf buds, young and mature fruits and also from leaves which become discoloured and deformed (Kerns *et al.*, 2001c). Two white lines parallel to the leaf midrib and a whitish silvery ring around the fruit neck are characteristics of thrips infestation (Prakash, 2012). Citrus thrips is of greatest economic importance on navel oranges and lemons.

Scirtothrips aurantii Faure, South African citrus thrips (SACT)

Distribution

SACT is widespread from South Africa to Egypt and also occurs in West Africa and the Cape Verde Islands (EPPO, 2005).

Bio-ecology

SACT feeds on soft new leaves and it does not feed on mature leaves. When leaves harden off, SACT attacks young fruit. Eggs are inserted into soft, young tissues of leaves, stems and fruits (EPPO, 2005). Adults, first and second larval stages are found on green plant parts where they are actively feeding. Pupation occurs in sheltered places such as among leaf litter on the ground, in the crevices of bark or occasionally beneath the calyx of fruit. The life cycle can be completed in less than 30 days; development is slowed in winter and breeding is almost continuous (Gilbert, 1990).

Damage

Feeding damage by SACT shows silvering of the leaf surface and thickening of young leaves and a silver-grey ring around the calyx on young fruit (Kamburov, 1991). A heavy infestation can result in early death of leaves and distorted fruit. Damage reduces the proportion of export-quality fruits. If flushes of young leaves are severely attacked late in the season, then the crop of the following season may be reduced (Kamburov, 1991).

Organic controls

MONITORING. To monitor adults of *S. dorsalis* and *S. aurantii*, sticky yellow traps can be used; for *P. kellyanus*, white sticky traps are the most attractive followed by blue traps. Sticky cards can be a useful supplement to the more accurate means of scouting, which is visual inspection of leaves and fruits from petal fall to calyx closure for the occurrence of thrips (Grafton-Cardwell and Ouyang, 1995). Sticky cards should be replaced every 7–10 days and could show the potential time to start thrips control (Conti *et al.*, 2003; Navarro-Campos *et al.*, 2013a). Monitoring for citrus thrips begins at the initiation of bloom, but becomes critical at petal fall (90% blossom drop). On mature trees, sampling should continue until 70–90% of the fruit reach a minimum size of 2.5 cm (1 inch) in diameter. Once the fruit reaches this size it is no longer susceptible to scarring by citrus thrips. On immature trees, sampling should continue until the autumn flush ceases (Kerns *et al.*, 2001c).

CULTURAL CONTROL. Weed destruction in the field and surrounding margins can help to reduce thrips populations, since these areas serve as overwintering and reinfestation sites (Kumar *et al.*, 2013). Growers should take care, however, that vegetation management does not conflict with strategies designed to reduce soil loss (through maintaining soil cover), to increase biodiversity, and to make the farm system more sustainable (Kuepper, 2004). Surround® crop protectant is labelled for suppression of thrips and several other insects for different crops including citrus. Surround® is a kaolin clay film sprayed on to the crop as a barrier to insect pests and it can be used with supplemental control measures (Kuepper, 2004). In Japan, synthetic reflective (vinyl) film has been used to protect citrus crops from *S. dorsalis* infestations (Kumar *et al.*, 2013).

MICROBIOLOGICAL CONTROL. Entomopathogenic fungi, *Beauveria bassiana* and *Metarhizium anisopliae*, are useful for thrips control and may be used in rotation with other allowed insecticides or in combination with spray oils (Kumar *et al.*, 2013). Spinosad is highly effective towards citrus thrips and works well under hot and cool conditions. It provides residual control for 10–14 days. Spinosad works best when used with a non-ionic surfactant or with a narrow-range (NR) oil (i.e. a lightweight oil) (Kerns *et al.*, 2001c; Kuepper, 2004; Knapp *et al.*, 2013). All these products are most effective when used early in spring, before large thrips populations have built up (Planes *et al.*, 2015).

ALTERNATIVE PESTICIDES. Several alternative pesticides are available for controlling thrips. Sulfur, insecticidal soap, horticultural oils and diatomaceous earth have all demonstrated efficacy in suppressing thrips (Kerns *et al.*, 2001c; Kuepper, 2004); but only in very frequent spray schedules (at least once a week). Three applications of superfine sulfur are recommended at monthly intervals in fruit crops for spring thrips control; lime sulfur has also been suggested as an alternative (Planes *et al.*, 2015). Petroleum spray oils can be used to control thrips; thorough coverage is very important.

BOTANICAL CONTROL. Neem, rotenone, ryania, pyrethrum, nicotine, NR oils (castor oil derivatives) and organic-compatible compounds such as garlic have been suggested for thrips control (Kuepper, 2004; Navarro-Campos *et al.*, 2013b), but organic growers should be aware that nicotine, tobacco and many other botanical products can be prohibited in their countries.

BIOLOGICAL CONTROL. Many beneficial organisms work to suppress thrips. These include ladybird beetles, minute pirate bugs, ground beetles, big-eyed bugs, lacewings, hoverflies, predatory mites and spiders. Predatory mites, *Iphiseiuss degenerans* Berlese, *Amblyseius swirskii*, *Neoseiulus cucumeris* Oudemans and *Neoseiulus barkeri* Hughes (Acari: Phytoseiidae) can be important in reducing citrus thrips; they are commercially available and can be used as classical biological control (by releases) (Elekcioglu and Uygun, 2006; Karamouna *et al.*, 2010; Kumar *et al.*, 2013). They are naturally present in the Mediterranean region (Israel, Italy, Cyprus, Spain, Tunisia and Egypt) as well as *Euseius tularensis* and *Euseius hibisci* Chant present in California and *Euseius sojaensis* Ehara in Japan. *Amblydromalus limonicus* Garman & McGregor (Phytoseiidae), naturally present in the Americas, Australia and New Zealand, is also commercially available and can be an excellent biological control agent against thrips.

These predatory mites that occur on fruits and leaves inside the tree are definitely capable of assisting in suppressing thrips, particularly late in the season as a means of preventing late thrips scribbling (Grafton-Cardwell *et al.*, 1999; Conti *et al.*, 2003). *Stratiolaelaps scimitus* Womersley, *Hypoaspis miles* Berlese and *Hypoaspis aculeifer* Canestrini (Acari: Laelapidae) are also predatory mites; but typical soil mites, feeding on soil organisms such as small mites and insects; they, also, eat thrips pupae. They are naturally present in Europe, North America and Japan and are commercially available and used in biological crop protection. Predatory thrips such as *Haplothrips* spp. attack KCT in Australia. Adults of *Orius insidiosus* Say have been observed to feed on all the developmental stages of thrips, and since it is a generalist predator which feeds on aphids, mites, moth eggs and pollen, its population does not decline when there are periodic drops in the thrips population (Kuepper, 2004).

V Coleoptera

V-1 Cerambicidae

The longhorn beetles are a cosmopolitan family, and several are serious pests. The larvae, called roundheaded borers, bore into wood, where they can cause extensive damage to living trees. In citrus, ***Anoplophora chinenis*** is considered as a major pest, while ***Anoplophora malasiaca*** (Thomson) occurs as a minor pest.

Anoplophora chinenis (Forster), citrus long-horned beetle (CLHB)

Distribution

China, Japan, Korea, Indonesia, Malaysia, Vietnam and Italy (Herard *et al.*, 2008; Noma *et al.*, 2010).

Bio-ecology

A. chinensis is a polyphagous xylophile. Each female can produce up to 200 eggs after mating, and each egg is separately deposited under the tree bark. Upon hatching, the larva bores into the stem and later enters the heartwood, tunnelling up and down (Adachi and Korenaga, 1989; EPPO, 2013). That is then used as a place for beetle pupation (the process of growing from larva to adult). From egg to adult, emergence can take 12–18 months. Adults are active during the daytime feeding on leaves, petioles and the young bark of host trees (Noma *et al.*, 2010). In tropical and subtropical regions there is one generation/year although further north there may be one generation every 2 years (EPPO, 2013).

Damage

The larva bores into the stem and destroys the pith and vascular system of the host (Adachi and Korenaga, 1989). Frass and woodpulp extruding from holes give a good indication of infestation. Damage to young trees is most serious. The adults can also cause damage by feeding on leaves, petioles and fruiting shoots resulting in economic losses (Noma *et al.*, 2010).

Organic controls

CULTURAL CONTROL. Infestations are eradicated by removing and destroying infested trees. In Japan, children are often paid by growers to collect and kill adult beetles in orchards in an attempt to reduce populations (Noma *et al.*, 2010).

MICROBIOLOGICAL CONTROL. Biological control has been used in Japan with the nematode *Steinernema feltiae*, and with the pathogenic fungi *Beauveria bassiana* and *Beauveria brongniartii*, which can decrease the emergence of the pest in a citrus orchard (Tsutsumi *et al.*, 1990).

BIOLOGICAL CONTROL. The egg parasitoid *Aprostocetus anoplophorae* Delvare is specific to CLHB. Most of the early larval ectoparasitoids, *Spathius erythrocephalus* Wesmael, *Eurytoma melanoneura* Walker, *Calosota vernalis* Curtis, *Cleonymus brevis* Boucek, *Trigonoderus princeps* (Westwood) and

Sclerodermus sp. attacked *A. chinensis* (Herard *et al.*, 2008).

Anoplophora malasiaca (Thomson), white-spotted longicorn beetle

Distribution

A. malasiaca is widely distributed in China, Japan, Korea and Taiwan.

Bio-ecology

Adult females live 14–66 days. They lay about 200 eggs in their lifespans, mainly in cracks at the base of the citrus tree (Adachi and Korenaga, 1989). The larvae bore immediately into the wood. Adults emerge in the early summer through small holes in the bark. The beetle can overwinter as an egg, a larva or a pupa. The time required to complete the life cycle of *A. malasiaca* is 1 or 2 years (Adachi, 1994).

Damage

The larvae, inside the wood, destroy the cambium in their early development stages and the xylem in late stages which causes tree deterioration or death (Adachi and Korenaga, 1989). The holes and tunnels made in the wood cause tree decline and eventually wilting. Weakened branches may break off. Infested trees are thus prone to secondary attack by other insects or diseases (Lee and Lo, 1996).

Organic controls

CULTURAL CONTROL. Cultural control consists of continually protecting oviposition sites in citrus trees, mainly on the lower parts of the trunk. To prevent *A. malasiaca* oviposition, two methods can be utilized: (i) repelling adult females from citrus trees; and (ii) eliminating female oviposition sites. For the latter method, using wire net or fishing net was very effective (Adachi and Korenaga, 1989).

MICROBIOLOGICAL CONTROL. *Beauveria brongniartii* has been commercially used against *A. malasiaca* on citrus trees in Japan (Tsutsumi *et al.*, 1990).

VI Acari

Phytophagous mites are important pests of citrus worldwide (Jeppson *et al.*, 1975). Vacante (2010) reported on 104 mite species associated with citrus, most of these species do not cause any problems in citrus groves. Few species are considered as major pests causing important economic losses. The economically important species attacking citrus are found in four families: Tetranychidae, Tenuipalpidae, Eriophyidae and Tarsonemidae (Gerson, 2003).

VI-1 Tetranychidae

According to Gerson (2003), four tetranychid species are considered as major pests to citrus: ***Eotetranychus sexmaculatus*** (Riley), ***Panonychus citri*** (Mc Gregor), ***Eutetranychus orientalis*** (Klein) and ***Eutetranychus banksi*** (McGregor). Other species occur on citrus trees but are considered as occasional or locally important species such as ***Tetranychus urticae*** (Koch), ***Eotetranychus lewisi*** (McGregor) and ***Eutetranychus africanus*** (Tucker) (Jeppson *et al.*, 1975; Gerson, 2003; Vacante, 2010).

Species and their distribution

E. sexmaculatus is found in Australia, Hawaii, New Zealand, the USA, Peru, China, India, Japan, Taiwan, Iraq and Republic of Korea (South) (Migeon and Dorkeld, 2006). Citrus red mite, *P. citri*, has a worldwide distribution (Migeon and Dorkeld, 2006). Commonly known as oriental red mite or citrus brown mite, *E. orientalis* is a cosmopolitan species (Migeon and Dorkeld, 2006). The Texas citrus mite, *E. banksi*, occurs on citrus in Hawaii, North, Central and South America, Spain, Portugal, India and Egypt (Migeon and Dorkeld, 2006). The two-spotted spider mite, *T. urticae* (Koch), is a cosmopolitan species; but Gerson (2003) considered this species of minor importance on citrus.

Bio-ecology

The length of time from egg to adult varies greatly depending on species and climatic conditions, especially temperature and humidity (Vacante, 2010). Tetranychids commonly prefer young leaves, some species prefer the underside of the leaf (e.g. *T. urticae* and *E. sexmaculatus*) and others the upper side (e.g. *P. citri* and *E. orientalis*). However, at greater densities, mites invade all plant surfaces and older leaves (Jeppson *et al.*, 1975). For *E. sexmaculatus* and during the 10–20 days that the females live, they may deposit 25–40 eggs. The total life cycle may occur in 8–12 days. This mite is adversely influenced by dry winds. The female *P. citri* lays an average of 30 eggs deposited at the rate of two or three/day, and it may live as long as 18 days in summer. The life cycle is completed in 14 days and 12 to 15 generations occur per year (Jeppson *et al.*, 1975; Childers and Abou-Setta, 1999). *E. banksi* is found throughout the year in Texas. Low relative humidity and high temperature favour its development. Its life cycle is similar to that of *E. orientalis*. Females lay up to eight eggs/day. Adults live about 12 days in summer (Jeppson *et al.*, 1975). The life duration of *T. urticae* females is about 30 days and the number of eggs deposited per female averages 90–110 (Jeppson, 1989).

Damage

Adults and nymphs feeding on leaves inhibit photosynthesis and can lead to necrosis (Corpuz-Raros, 1986; Jeppson, 1989; Childers and AbouSetta, 1999). With increasing damage the leaves die and the whole plant can eventually die (Jeppson, 1989). *E. sexmaculatus* feeds on the lower surfaces of leaves producing yellow depressions covered by webbing; it does not feed on the fruit except during severe infestations. With the increase of the infestation, the leaves become entirely yellow, and drop prematurely (Jeppson *et al.*, 1975). *P. citri* feeds on leaves, fruit and green bark, the feeding lesions appear as small silvery spots. With severe infestation, the trees become silver-grey and the leaves stop development and drop, and the shoots die back. The fruits are blemished and lose their market value. *T. urticae* feeding on fruits and leaves produces chlorotic areas and with severe injury infested leaves may drop (Jeppson, 1989).

Organic controls

MONITORING. Damage is usually noticed as stippled or yellow leaves. For detection of tetranychid mites, a magnifying glass is used to examine the leaves closely for mites, cast skins and webbing.

CULTURAL CONTROL. Cultural practices can influence the population density of injurious mites (Vacante, 2010). Adequate irrigation is an important factor in reducing economic damage; heavy spider-mite infestations combined with water stress can cause severe defoliation (Jeppson *et al.*, 1975; Dreistadt, 2012). Fertilization, especially the excessive use of nitrogen, can influence the reproductive rate of mites and their population density. Studies have found a positive correlation between high nitrogen concentration and density of phytophagous mites (Puttaswamy and Channabasavanna, 1982; Jackson and Hunter, 1983; Wermelinger *et al.*, 1985; Wilson *et al.*, 1988; Chen *et al.*, 2007).

BIOLOGICAL CONTROL. Predators belonging to the Phytoseiidae family are very important in regulating phytophagous mite populations (McMurtry and Croft, 1997) and several species are commercially available.

OTHER MEANS. Numerous natural enemies are well known throughout the world but biological control alone is insufficient (Vacante, 2010). To limit the development of mite populations, certain petroleum oil sprays are organically acceptable methods (Grafton-Cardwell *et al.*, 1995).

VI-2 Eriophyidae

Eriophyidae is a family of more than 200 genera of mites, which live as plant parasites, commonly causing galls or other damage to the plant tissues. This family is represented by two major species ***Aceria sheldoni*** (Ewing) and ***Phyllocoptruta oleivora*** (Ashmead) and one minor species ***Aculops pelekassi*** (Keifer).

Aceria sheldoni (Ewing), citrus bud mite

Distribution

A. sheldoni has worldwide distribution.

Bio-ecology

Each female deposits about 50 eggs singly on tender tissue where the mites are feeding. The period from egg to adult is 10–15 days in summer and autumn and two to three times longer in winter. Many generations (about 15 in the Mediterranean region) develop during the year depending on various ecological factors. The citrus bud mite lives on different species of citrus, orange, lime, grapefruit, etc., but it prefers lemon (Beattie and Gellatley, 2003). Its population develops inside both wood and flower buds or shelters under the flowering buds and the fruit rosette as soon as fruit growth begins.

Damage

Citrus bud mite feeding on foliage and blossom buds causes distortion of shoot growth, deformed blossoms and sometimes deformed fruits (Beattie and Gellatley, 2003). Mature trees may be affected and nursery trees can be severely damaged. In heavy infestation (100 mites in a single bud), the bud scales blacken by the mites' feeding and the bud may die.

Organic controls

MONITORING. In spring, developing blossom is inspected, looking for signs of malformation. This is the most important time for inspection for bud mite. However, the scout can also inspect each new flush to look for shortened internodes, or any other signs of malformation on the leaves.

BIOLOGICAL CONTROL. Among the predatory mites, some stigmaeids and phytoseiids (*Typhlodormips swirskii* and *Typhlodromus pyri*) have been recorded in different regions of the world.

MICROBIOLOGICAL CONTROL. The fungus, *Hirsutella thompsonii* Fisher var. *synematosa* has been found to develop on the citrus bud mite.

OTHER MEANS. *A. sheldoni* can be kept below economic injurious levels by thorough coverage spraying with sulfur or petroleum spray oils.

Phyllocoptruta oleivora (Ashmead), citrus rust mite (CRM)

Distribution

CRM is a serious pest of citrus in many tropical and subtropical areas of the world; especially in humid areas (Beattie and Gellatley, 2003).

Bio-ecology

The female lays eggs as soon as she reaches maturity with an average of 20 eggs on fruit or leaf surfaces (Rogers and Stansly, 2016). The CRM can be found on all citrus varieties, but it particularly prefers lemons, limes and grapefruit (Beattie and Gellatley, 2003). Favourable conditions for population development include warm weather, high rainfall and high relative humidity (more than 70%) (French, 2002). In spring, CRMs are found on new flushes and new leaves. Later, mites move on fruits as they appear. This mite can develop throughout the year, its entire cycle (egg to egg) requiring about a week in summer, several weeks in winter, and it may raise 30 annual generations (South America).

Damage

The CRM infests twigs, leaves and fruits. Feeding turns the green twigs and mature leaves to reddish brown, known as 'russeting' (French, 2002; Beattie and Gellatley, 2003). It causes blackening on green fruit and produces blemishes on the peel of the mature fruit, which reduces the market value

(Rogers *et al.*, 2016). Most damage occurs in summer–autumn. Heavy mite feeding can add to the visible symptoms resulting in: (i) a slower growth rate of fruits, so that the average size at harvest may be smaller; (ii) increased leaf drop and fruit drop; and (iii) a general loss of tree vigour.

Organic controls

MONITORING. Monitoring procedures and action thresholds are the backbone of the biocontrol programme. Growers should begin monitoring immediately after post-bloom; the monitoring frequency is every 2 weeks. In early spring, the leaves should be checked for mites and the fruits checked after they reach 1.3 cm (1/2 inch) in size (Rogers and Stansly, 2016). It takes 2 weeks or more for ring damage to become visible. The canopy density has an effect on rust mite populations and their ability to increase over a short period.

MICROBIOLOGICAL CONTROL. The epiphytotic fungi, *Hirsutella thompsonii* (Fisher) and *Beauveria bassiana* can provide biological control and maintain CRM populations below damaging levels (French, 2002); *H. thompsonii* has been used in Cuba and the USA (Texas and Florida) but only in very humid environments.

BIOLOGICAL CONTROL. These mites have no specific predators, but some natural enemies. These include predaceous mites of the families Phytoseiidae (*Amblyseius victoriensis*, *Amblyseius swirskii*, *Amblyseius elinae*, *Amblyseius deleoni*, *Amblyseius lentiginosus* and *Iphiseius degenerans*) and Stigmaeidae (*Agistemus exsertus*), small beetles and other small insect predators that prey on the CRM in different parts of the world (Beattie and Gellatley, 2003; Gerson, 2003).

Aculops pelekassi (Keifer), pink citrus rust mite (PCRM)

Distribution

PCRM occurs in Greece, Italy, southern Asia (including Japan and Thailand), the USA (Florida), Brazil and Paraguay.

Bio-ecology

A. pelekassi may be easily confused with *P. oleivora*. The PCRM and the CRM can coexist on the same leaf or fruit; the CRM is usually the prevalent species (Rogers and Stansly, 2016). However, the PCRM develops to greater damaging populations early in the spring season on new foliage. Peak density can vary by several weeks depending on geographical location and weather. PCRM are more abundant in drier weather conditions and it feeds on the lower leaf surfaces. From overwintering sites, it moves on to buds in the early spring and later on to fruits (Rogers and Stansly, 2016).

Damage

Leaf injury caused by feeding of PCRM is dramatic at mite densities exceeding 200 or more per leaf. Both mature and developing leaves can be affected with varying degrees of leaf distortion, crinkling, burning and dieback (Rogers and Stansly, 2016).

Organic controls

CULTURAL CONTROL. For the cultural control, canopy density has an effect on CRM and the PCRM populations and their ability to increase over a short period of time (Rogers *et al.*, 2016).

BIOLOGICAL CONTROL. Kostiainen and Hoy (1996) listed about 60 species in the family Phytoseiidae that were recorded from *Citrus* spp. In fact, no phytoseiids appear to be specialist predators of acarine citrus pests. The Stigmaeidae, another family of acarine predators, may feed on citrus mites, but there is no clear evidence of specialization on this prey (Gerson, 2003). Many species of ladybird beetles occur in coastal citrus orchards, *Halmus chalybeus*, *Serangium bicolor* and *Stethorus nigripes*. Large numbers of adults and larvae of these three ladybird beetles have been observed feeding on heavy infestations of citrus red mite.

VI-3 Tenuipalpidae

Three tenuipalpid species were considered as major pests to citrus: ***Brevipalpus obovatus***

Donnadieu, ***Brevipalpus phoenicis*** (Geijskes) and ***Brevipalpus californicus*** (Banks). While ***Brevipalpus lewisi*** McGregor was listed as a minor pest (Childers *et al.*, 2001; Gerson, 2003; Vacante, 2010).

Species and their distribution

B. phoenicis, *B. californicus* and *B. obovatus* have a worldwide distribution and are commonly associated. *B. lewisi* has a restricted distribution in Africa (Egypt), Asia (India, Iran, Japan, Taiwan, Turkey), Europe (Greece, France, Portugal, Spain), North America (USA, Mexico) and Australia (Jeppson *et al.*, 1975; Childers, 1994; Childers *et al.*, 2003a, b).

Bio-ecology

Peak populations of *Brevipalpus* species occur during the warmest months because periods of high temperature and low humidity have no deleterious influence upon the mite populations (Childers *et al.*, 2003a, b). These mites usually live on the fruit and lower leaf surface and aggregate along the mid-vein and major lateral veins (Haramoto, 1969). *Brevipalpus* species overwinter in the adult stage, but may be active throughout the year on citrus.

Damage

While feeding, *Brevipalpus* mites remove cell substances and inject toxic saliva into all parts of the citrus tree. Feeding injury symptoms include: chlorosis, blistering, bronzing, or necrotic areas on leaves (Childers *et al.*, 2003b). *B. californicus* and *B. obovatus* are suspected of transmitting citrus leprosis virus (CiLV) (Childers *et al.*, 2001). On lemons, silver scars can occur during severe infestations of *B. californicus* (Attiah, 1956; Elmer and Jeppson, 1957). *Brevipalpus* species damage leaves, twigs, branches and fruits of various citrus species and varieties (Vacante, 2010).

Organic controls

Predators generally fail to provide economic control. Predation activity is noticeable after *Brevipalpus* mites achieve a very high population density and severe plant damage has already occurred (Vacante, 2010).

Sulfur has been a widely used pesticide against this pest over a long period in almost all crops affected (Jeppson *et al.*, 1975).

In order to limit tenuipalpid populations, horticultural practices are important: (i) using CiLV-resistant varieties and healthy plants; (ii) removing infested trees or parts; and (iii) avoiding heavily infested hedges and windbreaks (Haramoto, 1969; Maia and Oliveira, 2004).

VI-4 Tarsonemidae

Polyphagotarsonemus latus (Banks), broad mite

Distribution

The broad mite occurs in citrus as a major pest and is cosmopolitan, especially in tropical and subtropical areas.

Bio-ecology

The life cycle is an emerging larval stage lasting about 1 day and then moulting; young mites become adults within a further 4 days. Pharate females (developing nymphs) are picked up by the males and moved to newly developing flush and young citrus fruit. Mating occurs immediately after the female emerges. Females produce five eggs/day in an average life of 10 days (Rogers and Stansly, 2016). Optimal environmental conditions include warm temperatures, high humidity and low light intensity. The broad mite prefers lemons and limes.

Damage

The broad mite is only capable of feeding on very young, tender leaf or fruit tissues. The toxic saliva that is injected by this mite can result in significant damage (Beattie and Gellatley, 2003). New leaf growth that is fed upon becomes distorted and feathered (Rogers and Stansly, 2016). Often the mites cannot be found on damaged tissue. A delayed terminal dieback can occur on infested citrus seedlings. Small fruit become silvered from intense feeding by the mite with subsequent reduced fruit growth.

Organic controls

Fungi that are known to be used for control of broad mites include *Beauveria bassiana*, *Hirsutella thompsonii* and *Verticium lecanii*. Known predators of broad mites include some phytoseiids such as *Neoseiulus californicus* (McGregor), *Neoseiulus barkeri*, *Amblyseius swirskii* (Athias-Henriot), *Typhlodromus athiasae* and *Iphiseius degenerans* (Berlese).

VII Hymenoptera

VII-1 Formicidae

The family Formicidae comprises over 12,000 described species (Lentini and Verdinelli, 2012). Among these, many species are considered as agricultural pests, and most of them are honeydew feeders (Haney, 1988). The importance of these pests is due mostly to the symbiosis that they establish with many honeydew-producing citrus pests (hemipterans) (De Bach *et al.*, 1951; Haney *et al.*, 1987; Itioka and Inoue, 1996; Pekas *et al.*, 2010; Dao *et al.*, 2014). Ants provide protection by building refuges around hemipteran colonies or by disturbing or killing their natural enemies (Phillips and Sherk, 1991; Katayama and Suzuki, 2003; Campos *et al.*, 2006). Very few ant species directly damage the trees (Banks and Lofgren, 1991). Some ants found in citrus orchards are considered as beneficial predators that kill a wide range of insects, including citrus pests, and exert an interspecific competitive pressure on harmful ant species (Boyce, 1948; Urbaneja *et al.*, 2006; Xiao *et al.*, 2007).

Honeydew-feeder species indirectly damage the citrus crops by disturbing or killing the natural enemies. Campos *et al.* (2006) found that the presence of ants reduced the parasitism activities of two parasitoids of the citrus mealybug, *A. pseudococci* and *L. dactylopii*, by 35%. Other ant species (e.g. fire ants: *Solenopsis* sp. and leafcutting ants: *Atta* sp.) damage plants when feeding on the tender tissue, often killing young citrus trees by girdling their trunks or cutting their leaves (Boyce, 1948; Banks and Lofgren, 1991).

Organic controls

Monitor the orchard in spring when honeydew-producing insects appear and periodically inspect for ants in young trees. No effective natural enemies of ants are known. The ants can be prevented from climbing trees by skirt pruning and the use of sticky materials applied to the trunk. The persistence of sticky material can be increased by applying it higher above the ground to reduce dust and dirt contamination and to decrease irrigation wash-off (Grafton-Cardwell, 2015).

VIII Orthoptera

VIII-1 Acrididae

These insects, commonly known as grasshoppers or locusts, are general feeders and sporadic pests of citrus orchards. These species usually occur in small numbers but, although rare, mass infestations may cause serious damage (Jeppson, 1989).

A number of species have been recorded as damaging citrus trees. These include: (i) *Schistocerca gregaria* Farsk in the Mediterranean region; (ii) *Schistocerca vagu* (Scud.), *Melanoplus mexicanus* (Scud.), *Melanoplus devastator* (Scud.), *Oedoleonatus enigma* (Scd.) and *Camnula pellucid* (Scud.) in California; (iii) *Locustana pardalina* Walk. and *Nomadacris septemfasciata* Sert. in South Africa; (iv) *Melanoplus mexicanus* and *Melanoplus differentialis* (Thom.) in Arizona; (v) the Egyptian tree locust, *Anacridium aegyptium* (Linnaeus), widely distributed from the Mediterranean zone to the Middle East; (vi) *Anacridium melanorhodon* (Walker) and *Anacridium wernerellum* are found in Africa, the Sahel and Sudan; (vii) the eastern lubber grasshopper,

Romalea microptera (Beauvois) and the American grasshopper, *Schistocerca americana* (Drury) in Florida; (viii) *A. melanorhodon* subspecies *arabafrum* (Dirsh) in East Africa and Arabia; and (ix) *Anacridium moestum* (Serville) in South Africa (Jeppson, 1989; Fauna Europaea, 2013).

Bio-ecology

These pests establish one generation per year. The adult female undergoes a photoperiod-controlled reproductive diapause that lasts from September to January, and deposits eggs in the spring. Juveniles develop during the hot season, from April to August and new adults appear in autumn (Jeppson, 1989).

Damage

These pests are feeding on leaves, fruits and green bark of twigs. Mass infestations may cause total defoliation of the trees (Jeppson, 1989).

Organic controls

Control measures are usually not needed for the occasional adults encountered. It is mainly the migratory species that require control and in this case, a community action is necessary. Keeping the vegetation mowed is very helpful, as short vegetation does not often support grasshoppers (Jeppson, 1989).

The recent development of effective oil formulations of *Metarhizium anisopliae* spores in Africa, Australia and Brazil opens new possibilities for organic control strategies. *Metarhizium* biopesticide kills 70% and 90% of treated locusts within 14 and 20 days, respectively (Lomer *et al.*, 2001).

Conclusion and Perspectives

Interest in organic production is increasing and this growth has not been supported adequately by rigorous research to address challenges such as pest management. However, some research conducted in aspects of integrated pest management programmes, may have applicability in organic systems (Zehnder *et al.*, 2007).

The principles of pest management in organic systems involve the adoption of ecological practices specified by international and national organic production standards (e.g. IFOAM, 2014; USDA-NOP, 2016). Priority is given to preventative management strategies, followed by more direct measures if necessary (Wyss *et al.*, 2005). Emphasis is placed on the use of multiple and compatible strategies to prevent damaging levels of pests, thus minimizing the need for curative solutions.

Choice of soils, rootstocks and cultivars requires careful planning and decisions such as these should be made during the planning stage. Healthy plants are better able to resist pests and deseases. Therefore, it is best to start by choosing varieties that are insect/disease resistant (van Emden, 1991).

Before taking action, monitoring is the basis of successful pest control. Close scrutiny of plants is the best way to discover the pest presence and level of damage. A variety of traps and pheromones used as monitoring tools are available and adapted to many pest arthropods.

Appropriate horticultural management practices contribute to limiting pest development. Moderate doses of fertilizers limit the development of arthropod pests. Field sanitation is important to manage many pest species; keeping a clean field throughout the season helps to minimize feeding and sheltering sites for many insects. In addition, it is important to clear out residues to remove insect pests. Tillage can destroy pupae and remove overwintering sites for many injurious pests. Pruning encourages ventilation of the internal canopy parts and directly exposes the juvenile stages of scales and mealybugs to solar radiation and rain, which results in their high mortality rates.

A 'healthy', biologically active soil will help keep the populations of insects below their economic thresholds. Recent studies have shown that pest attacks are linked to optimal biological and physical properties of soil (Altieri and Nicholls, 2003). Lower numbers of pest insects were reported on crops grown with organic compared with synthetic sources of fertilizer (Culliney and Pimentel, 1986; Kajimura *et al.*, 1993).

Biocontrol agents have been studied and used for decades and their failures and successes have been extensively reviewed (DeBach, 1964; Huffaker and Messanger, 1976; Van Driesche and Bellows, 1996). The impact of natural biological control on pests may often occur after economic thresholds are surpassed. Thus, inundatively or inoculatively applied controls have been developed for a wide variety of arthropod pests and can provide effective control.

The development of mass-rearing and release techniques for predators and parasitoids has facilitated management programmes for various pest species. Several control agents are commercially available for inundation or inoculation biocontrol and are in most cases specific to their host species and native to the region (Zehnder *et al.*, 2007). To achieve adequate levels of natural enemy activity, it would be costly to rely upon the releases alone. Habitat management could be used in combination with inoculation and inundation methods to improve the success of the release strategies in an approach of integrated biological control (Gurr and Wratten, 1999).

Ground cover management is an important component of conservation biological control strategies, especially in perennial cropping systems. Providing beneficial organisms with alternative food sources (pollen, nectar) when their food supply becomes limited is the key of natural enemy conservation strategies (Altieri *et al.*, 1985; Altieri, 1999; Landis and Haas, 1992; Landis *et al.*, 2000). In this context, intercropping and the use of trees or shrubs as hedges and windbreaks encourages infestations by various pests and consequently also that of their natural enemies (Onillon, 1988). The choice of botanical species thus requires careful evaluation. It is important to define the species that promote the 'no-citrus' pests, which provide alternative hosts for the beneficial parasitoids or predators without becoming a problem for citrus (Rodrigues *et al.*, 2006).

In organic agriculture, the application of organic chemical controls is used as a last option. Botanical insecticides such plant oils are commonly used in organic farming. In tree organic production, mineral oils are applied during winter dormancy to kill the overwintering developmental stages of pests. Spinosad is used on a variety of crops to control a number of insect pests (Isman, 2006). Sex pheromones are also used to disrupt mating of various lepidopteran pests. Non-insecticidal repellents such as kaolin clay are effective in controlling various insects and in some cases can even replace insecticide treatments. Insecticidal soaps are extracted from plant or animal lipids and work by direct contact on soft-bodied insects. The criteria for approved substances differ between the national organic standard organizations (Zehnder *et al.*, 2007). For a list of chemical controls that are permitted under an organic system, farmers have to contact their national list of allowed substances.

The volume of pest management research conducted on organic systems is small compared with the far wider literature on integrated pest management crops. Accordingly, there is a need for more research to be conducted on certified organic orchards.

The limited number of commercially available agents points to the need for research to successfully combine inundation and inoculation biological control agents with other organic pest management practices (Vacas *et al.*, 2012).

The researcher's interest concerning ground cover management is increasing, aiming at obtaining a better and fuller understanding of the effect of cover crops and the surrounding vegetation on the pests and on their natural enemies. Yet, it is not totally clear how effectively conservation biocontrol can be combined with other forms of biocontrol and other organic agricultural practices.

The success of biological controls will need future efforts by: (i) trying with improved rearing methods to raise more natural enemies; (ii) making more direct releases for field evaluation; and (iii) complete studies on the biology of the natural enemies and their behaviour in the same habitat (intraguild predation or competitive displacement).

Microbial control agents (viruses, bacteria, fungi and nematodes) have been studied as alternative control methods for various citrus pests (Kaya and Lacey, 2007). Aspects that warrant further studies are: (i) conducting more field trials rather than emphasis on laboratory colonization; (ii) finding successful combinations of entomopathogens, predators and parasitoids; and (iii) improving formulation and storage.

In general, it was found that proper management of cultural practices hampers the development of arthropod pests in citrus. However, the bibliography lacks matter, which constitutes a call for research to obtain better understanding about the effects of horticultural practices on pests and their natural enemies.

Biodiversity of predatory insects and mites generally increases in organic farming, compared with conventional practice. However, for an organic farmer, being within a landscape dominated by conventional farms with high pesticide input may be disadvantageous. Future studies of the effects of organic agriculture on natural enemies and pests will need to cover several scales, including the landscape scale (Bengtsson *et al.*, 2005; Zehnder *et al.*, 2007).

Finally, the efforts of specialists can be directed towards the application of modern technologies: (i) mathematical models; and (ii) population genetics and molecular biology to study population dynamics, the behaviour of prey and predators and to determine the correct systematic status of the various species, both injurious and beneficial.

References

Abbassi, M., Besri, M. and Bouhida, M. (1996) Citrus pest problems and their control in the Near East, Morocco. In: Morse, J.G., Juck, R.F. and Gumpf, D.J. (eds) *FAO Plant Production and Protection Paper* 135(402). Food and Agriculture Organization of the United Nations (FAO), Rome, pp. 91–109.

Abd-Rabou, S. (2011) New records of the soft scale insects hosts associated with the promising parasitoid, *Scutellista caerulea* (Fonscolombe) (Hymenoptera: Pteromalidae). *Egyptian Journal of Agricultural Research* 89(4), 1295–1302.

Abo-Sheaesha, M.A. and Agamy, E.A. (2004) Biological control utilizing *Trichogramma evanescens* (West.) and Agerin (B.T.) in comparison to ethion for controlling *Prays citri* (Mill.) in lime orchards. *Egyptian Journal of Biological Pest Control* 14(1), 31–35.

Adachi, I. (1994) Development and life cycle of *Anoplophora malasiaca* (Thomson) (Coleoptera: Cerambycidae) on citrus trees under fluctuating and constant temperature regimes. *Applied Entomology and Zoology* 29(4), 485–497.

Adachi, I. and Korenaga, R. (1989) Control methods for *Anaplophora malasiaca* Thomson (Coeloptera: Cerambycidae) in citrus groves. I. Comparison of effects in several methods for preventing oviposition. *Applied Entomology and Zoology* 24(3), 315–318.

Agarwal, M.L. and Kapoor, V.C. (1986) New records of some hymenopterous parasites of fruit flies (Diptera: Tephritidae) from India. *Bulletin of Entomology – New Delhi* 27(2), 193.

Allen, D.C. (1995) The soft scales. NY Forest Owner-Info 1995; pp. 18–19. Available at: http://www.dec.ny.gov/docs/lands_forests_pdf/scales.pdf (accessed 24 November 2016).

Altieri, M.A. (1999) The ecological role of biodiversity in agro-ecosystems. *Agriculture, Ecosystem and Environment* 74, 19–31. DOI: 10.1016/S0167-8809(99)00028-6.

Altieri, M.A. and Nicholls, C. (2003) Soil fertility and insect pests: harmonizing soil and plant health in agroecosystems. *Soil and Tillage Research* 72, 203–211.

Altieri, M.A., Wilson, R.C. and Schmidt, L.L. (1985) The effects of living mulches and weed cover on the dynamics of foliage and soil arthropod communities in three crop systems. *Crop Protection* 4, 201–213. DOI: 10.1016/0261-2194(85)90018-3.

Aluja, M., Guillen, J., Liedo, P., Cabrera, M., Rios, E., De la Rosa, G. and Celedonio, H. (1990) Fruit infesting tephritids (Dipt.: Tephritidae) and associated parasitoids in Chiapas, Mexico. *Entomophaga* 35, 39–48.

Amiri-Besheli, B. (2007) Efficacy of *Bacillus thuringiensis* and mineral oil against *Phyllocnistis citrella* Stainton (Lepidoptera: Gracillariidae). *International Journal of Agriculture and Biology* 9(6), 893–896.

Argov, Y. and Gerson, U. (2012) Gracillaridae, Yponomeutidae and Pyralidae. In: Vacante, V. and Gerson, U. (eds) *Integrated Control of Citrus Pests in the Mediterranean Region*. Bentham Science Publishers, London, pp. 223–230.

Argov, Y. and Rossler, Y. (1993) Biological control of the Mediterranean black scale, *Saissetia oleae* (Hom.: Coccidae) in Israel. *Entomophaga* 38, 89–100.

Argov, Y. and Rossler, Y. (1996) Introduction, release and recovery of several exotic natural enemies for biological control of the citrus leafminer, *Phyllocnistis citrella*, in Israel. *Phytoparasitica* 24(1), 33–38.

Argov, Y., Uygun, N., Porcelli, F. and Baspinar, H. (2012) Aleyrodidae. In: Vacante, V. and Gerson, U. (eds) *Integrated Control of Citrus Pests in the Mediterranean Region*. Bentham Science Publishers, London, pp. 156–171.

Attiah, H.H. (1956) The genus *Brevipalpus* in Egypt (Acarina: Tenuipalpidae). *Bulletin of the Entomological Society of Egypt* 40, 433–448.

Aubert, B. (1987) *Trioza erytreae* Del Guercio and *Diaphorina citri* Kuwayama (Homoptera: Psylloidea), the two vectors of citrus greening disease: biological aspects and possible control strategies. *Fruits* 42(3), 149–162.

Ayers, E.L. (1957) The two medfly eradication programs in Florida. *Proceedings of the Florida State Horticultural Society* 70, 67–69.

Baker, G.J., Keller, M.A., Crisp, P., Jackman, D., Barbour, D. and Purvis, S. (2011) The biological control of Kelly's citrus thrips in Australian citrus orchards. *Integrated Control in Citrus Fruit Crops. IOBC (International Organisation for Biological and Integrated Control) WPRS Bulletin* 62, 267–274.

Baloch, U.K. (1996) Citrus pest problems and their control in the Near East, Pakistan. In: Morse J.G., Juck R.F. and Gumpf D.J. (eds) *FAO Plant Production and Protection Paper* 135(402). Food and Agriculture Organization of the United Nations (FAO), Rome, pp. 111–128.

Banks, W.A. and Lofgren, C.S. (1991) Damage to young citrus trees by the red imported fire ant, *Solenopsis invicta* (Hymenoptera: Formicidae). *Journal of Economic Entomology* 84(1), 241–246. Available at: http://ddr.nal.usda.gov/bitstream/10113/23162/1/IND92044417.pdf (accessed 30 November 2016).

Barbagallo, S. and Patti, I. (1986) The citrus aphids: behaviour, damages and integrated control. In: Cavalloro, R. and Di Martino, E. (eds) *Integrated Pest Control in Citrus Groves*. Proceedings of the Experts' Meeting, 26–29 March 1985, Acireale, Italy. A.A. Balkema, Rotterdam, The Netherlands, pp. 67–75.

Bateman, M.A. (1968) Determinants of abundance in a population of the Queensland fruit fly, *Dacus tryoni*. *Symposia of the Royal Entomological Society London* 4, 119–131.

Batra, R.C., Sharma, D.R. and Chanana, Y.R. (1992) Screening of citrus germplasm for their resistance against citrus leaf miner, *Phyllocnistis citrella* Stainton. *Journal of Insect Science* 5(2), 150–152.

Beardsley, J.W. and Gonzalez, R.H. (1975) The biology and ecology of armored scales. *Annual Review of Entomology* 20, 47–73.

Beattie, G.A.C. and Gellatley, J.G. (2003) Mite pests of citrus. AGFACTS, p. 6. Available at: http://mvcitrus.org.au (accessed 10 November 2016).

Beattie, G.A.C. and Hardy, S. (2005) Using petroleum-based spray oils in citrus. AGFACTS, p. 7. Available at: http://mvcitrus.org.au (accessed 10 November 2016).

Beers, E.H., Brunner, J.F., Willett, M.J. and Warner, G.M. (1993) *Orchard Pest Management*. Good Fruit Grower. Yakima, Washington, p. 276.

Bengtsson, J., Ahnstrom, J. and Weibull, A.C. (2005) The effects of organic agriculture on biodiversity and abundance: a meta-analysis. *Journal of Applied Ecology* 42, 261–269.

Ben Salah, H., Mamouchi, N. and Cheikh, M. (1996) Citrus pest problems and their control in the Near East, Tunisia. In: Morse, J.G., Juck, R.F. and Gumpf, D.J. (eds) *FAO Plant Production and Protection Paper* 135(402). Food and Agriculture Organization of the United Nations (FAO), Rome, pp. 159–174.

Beroza, M., Green, N., Geltler, S.I., Steiner, L.F. and Miyashita, D.H. (1961) Insect attractants: new attractants for the Mediterranean fruit fly. *Journal of Agricultural and Food Chemistry* 9, 361–365.

Blackman, R.L. and Eastop, V.F. (1994) *Aphids on the World's Trees*. CAB International, Wallingford, UK, p. 987.

Boyce, A.M. (1948) Insects and mites and their control. In: Batchelor, L.D. and Webber, H.J. (eds) *The Citrus Industry: Production of the Crop*, Vol. 2. University of California Press, Berkeley, California, pp. 765–769.

Browning, H.W. (1994) Classical biological control of citrus scale insects. In: Rosen, D., Bennett, F.D. and Capinera, J.L. (eds) *Pest Management in the Subtropics: Biological Control – A Florida Perspective*. Intercept, Andover, UK, 49–78.

Bruwer, I.J. (1998) Longmussel scale: *Lepidosaphes gloverii* (Packard). In: Bedford, E.C.G., Van den berg, M.A. and De Villiers, E.A. (eds) *Citrus Pests in the Republic of South Africa*. Institute for Tropical and Subtropical Crops, Nelspruit, South Africa, pp. 153–157.

Burckhardt, D. and Ouvrard, D. (2012) A revised classification of the jumping plant-lice (Hemiptera: Psylloidea). *Zootaxa* 3509, 1–34.

Buschman, L.L. (1980) Parasites of *Nezara viridula* (Hemiptera: Pentatomidae) and other Hemiptera in Florida. *Florida Entomologist* 63, 154–162.

Byrne, D.N. and Bellows, T.S. Jr (1991) Whitefly biology. *Annual Review of Entomology* 36, 431–457. DOI: 10.1146/annurev.en.36.010191.002243.

CABI/European and Mediterranean Plant Protection Organization (EPPO) (1999) *Planococcus citri*. Distribution Maps of Plant Pests, No. 43. CAB International, Wallingford, UK.

CABI/European and Mediterranean Plant Protection Organization (EPPO) (2006) *Triozaerytreae*. Distribution Maps of Plant Pests, No. 234. CAB International, Wallingford, UK.

Caleca, V., Lo Verde, G., Lo Verde, V., Palumbo Piccionello, M. and Rizzo, R. (2010) Control of *Bactrocera oleae* and *Ceratitis capitata* in organic orchards: use of clays and copper products. International Society for Horticultural Science (ISHS) Organic Fruit Conference. *Acta Horticulturae* 873, 227 –234. Available at: https://doi.org/10.17660/ActaHortic.2010.873.24 (accessed 10 November 2016).

Campos, J.M., Martínez-Ferrer, M.T. and Fores, V. (2006) Parasitism disruption by ants of *Anagyrus pseudococci* (Girault) and *Leptomastix dactylopii* Howard (Hymenoptera: Encyrtidae), two parasitoids of the citrus mealybug *Planococcus citri* (Risso) (Homoptera: Pseudococcidae). *Integrated Control in Citrus Fruit Crops. IOBC (International Organisation for Biological and Integrated Control) WPRS Bulletin* 29(3), 33–46. Available at: http://www.iobc-wprs.org/pub/bulletins/iobc-wprs_bulletin_2006_29_03.pdf (accessed 2 November 2016).

Carver, M. (1978) The black citrus aphids, *Toxoptera citricidus* (Kirkaldy) and *T. aurantii* (Boyer de Fonscolombe) (Homoptera: Aphididae). *Journal of the Australian Entomological Society* 17, 263–270.

Cavalloro, R. and Di Martino, E. (eds) (1986) *Integrated Pest Control Control in Citrus Groves*. Proceedings of the Experts' Meeting, 26–29 March 1985, Acireale, Italy. A.A. Balkema, Rotterdam, The Netherlands, 616 pp.

Cesnik, R. and Medina, C.L. (1995) High pressure water sprays to control chaff scale (Homoptera: Diaspididae) in citrus. *Journal of Entomological Science* 30, 93–94.

Chakroun, S., Skander, M. and Grissa-Lebdi, K. (2015) Sterile insect technique against date moth, *Ectomyelois ceratoniae* Zeller, in Tunisia: research of gamma radiation sterilising dose. *International Journal of Scientific Research & Engineering Technology* 3(2), 55–62.

Chen, R.T., Chen, Y.H. and Huang, M.D. (1989) Biology of the green lacewing, *Chrysopa boninensis* and its predation efficiency on the citrus leaf miner, *Phyllocnistis citrella*. In: Huang, M.D. (ed.) *Studies on the Integrated Management of Citrus Insect Pests*. Academic Book & Periodical Press, Guangzhou, Guangdong, China, pp. 96–105.

Chen, Y., Opit, G.P., Jonas, V.M., Williams, K.A., Nechols, J.R. and Margolies, D.C. (2007) Two spotted spider mite population level, distribution, and damage on ivy geranium in response to different nitrogen and phosphorus fertilization regimes. *Journal of Economic Entomology* 100(6), 1821 1830.

Childers, C.C. (1994) Feeding injury to 'Robinson' tangerine leaves by *Brevipalpus* mites (Acari: Tenuipalpidae) in Florida and evaluation of chemical control on citrus. *Florida Entomologist* 77(2), 265–271.

Childers, C.C. and Abou-Setta, M.M. (1999) Yield reduction in Tahiti lime from *Panonychus citri* feeding injury following different pesticide treatment regimes and impact on the associated predaceous mites. *Experimental and Applied Acarology* 10, 771–783.

Childers, C.C., Kitajima, E.W., Welbourn, W.C., Rivera, C. and Ochoa, R. (2001) *Brevipalpus* mites on citrus and their status as vectors of citrus leprosis. *Manejo Integrado de Plagas* 60, 66–70.

Childers, C.C., Rodrigues, J.C.V. and Welbourn, W.C. (2003a) Host plants of *Brevipalpus californicus, B. obovatus,* and *B. phoenicis* (Acari: Tenuipalpidae) and their potential involvement in the spread of viral diseases vectored by these mites. *Experimental and Applied Acarology* 30(1–3), 29–105.

Childers, C.C., French, J.V. and Rodrigues, J.C.V. (2003b) *Brevipalpus californicus, B. obovatus, B. phoenicis,* and *B. lewisi* (Acari: Tenuipalpidae): a review of their biology, feeding injury and economic importance. *Experimental and Applied Acarology* 30, 5–28.

Christenson, L.D. and Foote, R.H. (1960) Biology of fruit flies. *Annual Review of Entomology* 5, 171–192.

Chua, T.H. and Wood, B.J. (1990) Other tropical fruit trees and shrubs. In: Rosen, D. (ed.) *Armoured Scale Insects, Their Biology, Natural Enemies and Control*, Vol. 4B. *World Crop Pests*. Elsevier, Amsterdam, pp. 543–552.

Clausen, C.P. (1978) Tephritidae (Trypetidae, Trupaneidae). In: Clausen, C.P. (ed.) *Introduced Parasites and Predators of Arthropod Pests and Weeds: A World Review*. United States Department of Agriculture Agricultural Handbook 480. United States Department of Agriculture, Washington, DC, pp. 320–335.

Coll, M. and Abd-Rabou, S. (1998) Effect of oil emulsion sprays on parasitoids of the black parlatoria, *Parlatoria ziziphi*, in grapefruit. *BioControl* 43, 29–37.
Commeau, J. and Sola, E. (1964) Une nouvelle cochenille des agrumes sur la Cote d'Azur. *Phytoma* 16, 49–50.
Conti, F., Tumminelli, R., Fisicaro, R., Perrotta, G. and Marullo, R. (2003) An IPM system for new citrus thrips in Italy. *Bulletin OILB/SROP [Cette Organisation Internationale de Lutte Biologique (OILB)-Section régionale ouest-paléarctique (SROP)]* 26(6), 203–208.
Corpuz-Raros, LA. (1986) Mites. In: *Guide to Philippine Flora and Fauna*. *Phillippine Agricultural Science* 84. Natural Resources Management Centre, Ministry of Natural Resources and University of Philippines, Diliman, Quezon City, the Philippines, pp. 341–351.
Corrêa-Ferreira, B.S. and Moscardi, F. (1995) Seasonal occurrence and host spectrum of egg parasitoids associated with soybean stink bugs. *Biological Control* 5, 196–202.
Culliney, T. and Pimentel, D. (1986) Ecological effects of organic agricultural practices on insect populations. *Agriculture, Ecosystem and Environment* 15, 253–266.
Cunningham, R.T. (1989) Parapheromones. In: Robinson, A.S. and Hooper, G. (eds) *World Crop Pests, Fruit Flies, Their Biology, Natural Enemies and Control*. Elsevier Science Publishers B.V., Amsterdam (3A), pp. 221–230.
Danzig, E.M. and Pellizzari, G. (1998) Diaspididae. In: Kozár, F. (ed.) *Catalogue of Palaearctic Coccoidea*. *Hungarian Academy of Sciences*. Akaprint Nyomdaipari Kft, Budapest, pp. 172–370.
Dao, H.T., Meats, A., Beattie, G.A.C. and Spooner-Hart, R. (2014) Ant-coccid mutualism in citrus canopies and its effect on natural enemies of red scale, *Aonidiella aurantii* (Maskell) (Hemiptera: Diaspididae). *Bulletin of Entomological Research* 104(02), 137–142.
Davarci, T. (1996) Citrus pest problems and their control in the Near East, Turkey. In: Morse, J.G, Juck, R.F. and Gumpf, D.J. (eds) *FAO Plant Production and Protection Paper* 135(402). Food and Agriculture Organization of the United Nations (FAO), Rome, pp. 175–206.
Davis, C.J. (1967) Progress in the biological control of the southern green stink bug, *Nezara viridula* variety smaragdula (Fabr.) in Hawaii (Heteroptera: Pentatomidae). *Mushi* 39, 9–16.
Dean, H.A. (1982) Reduced pest status of the Florida red scale on Texas citrus associated with *Aphytis holoxanthus*. *Journal of Economic Entomology* 75, 147–149.
DeBach, P. (1964) *Biological Control of Insect Pests and Weeds*. Chapman & Hall, London, p. 844.
DeBach, P. and White, E.B. (1960) Commercial mass culture of the California red scale parasitoid, *Aphytis lingnanensis*. *California Agricultural Experiment Station Bulletin* 770, 58.
DeBach, P., Fleschner, C. and Dietrick, E. (1951) A biological check method for evaluating the effectiveness of entomophagous insects. *Journal of Economic Entomology* 44, 763–766.
Deguine, J.P., Toulassi, A.N. and Quilici, S. (2011) Net choice is key to the augmentorium technique of fruit fly sequestration and parasitoid release. *Crop Protection* 30, 198–202.
Dekle, G.W. (1976) *Florida Armored Scale Insects*. Florida Department of Agriculture, Division of Plant Industry Library, Gainesville, Florida, p. 345.
Delrio, G. (1986) Tephritid pests in citriculture. In: Cavalloro, R. and Di Martino, E. (eds) *Integrated Pest Control in Citrus Groves*. Proceedings of the Experts' Meeting, 26–29 March 1985, Acireale, Italy. A.A. Balkema, Rotterdam, The Netherlands, pp. 135–149.
De Meyer, M., Copeland, R.S., Lux, S.A., Mansell, M., Quilici, S., Wharton, R., White, I.M. and Zenz, N.J. (2002) Annotated check list of host plants for Afrotropical fruit flies (Diptera: Tephritidae) of the genus *Ceratitis*. Koninklijk Museum noor Midden-Afrika Tervuren Belge, *Zoölogische Documentatie* 27, 1–91.
DeWoskin, R. (1981) *Medfly Training Manual*. Joint Cooperative Mediterranean Fruit Fly Eradication Project. United States Department of Agriculture (USDA), California Department of Food and Agriculture, Santa Clara County Agriculture Commissioner, Alameda County Department of Agriculture. USDA, Washington, DC, p. 215.
Dhouibi, M.H., Hermi, N., Soudani, D. and Thlibi, H. (2016) Biocontrol of the carob moth *Ectomyelois ceratoniae* (Lepidoptera, Pyralidae) in pomegranate and citrus orchards in Tunisia. *International Journal of Agriculture Innovations and Research* 4(5), 849–856.
Drake, C.J. (1920) The southern green stink-bug in Florida. *Quarterly Bulletin of the Florida State Plant Board* 4(3), 41–94.
Drees, B.M., Reinert, J.A. and Williams, M.L. (2005) *Florida Wax Scales: A Major Pest of Hollies and Other Landscape Shrubs and Trees*. Texas Cooperative Extension, the Texas A&M University System, Publication EEE-00023, College Station, Texas.

Dreistadt, S.H. (2012) *Integrated Pest Management for Citrus. Statewide Integrated Pest Management Program*, 3rd edn. University of California Agriculture and Natural Resources (UCANR) publication 3303. UCANR, Oakland, California, p 275.

Duyck, P.F. and Quilici, S. (2002) Survival and development of different life stages of three *Ceratitis* spp. (Diptera: Tephritidae) reared at five constant temperatures. *Bulletin of Entomological Research* 92(6), 461–469.

Dyck, V.A., Hendrichs, J. and Robinson, A.S. (2005) *Sterile Insect Technique: Principles and Practice in Area-wide Integrated Pest Management*. Springer, Dordrecht, The Netherlands, p. 787.

Ebeling, W. (1959) *Subtropical Fruit Pests*. University of California Division of Agricultural Science, University of California Press, Berkeley, California, 436 pp.

Elekcioglu, N.Z. and Uygun, N. (2006) The parasitoid complex of the citrus leafminer, *Phyllocnistis citrella* Stainton (Lepidoptera: Gracillariidae) in the east Mediterranean region of Turkey and their role in biological control. *Turkish Journal of Zoology* 30, 155–160.

Elekcioglu, N.Z. and Uygun, N. (2013) Population fluctuation of citrus leafminer, *Phyllocnistis citrella* Stainton (Lepidoptera: Gracillariidae) and its parasitoids in the eastern Mediterranean region of Turkey. *Pakistan Journal of Zoology* 45(5), 1393–1403.

Elmer, H.S. and Jeppson, L.R. (1957) Biology and control of citrus flat mite. *Journal of Economic Entomology* 50(5), 566–570.

El-Metwally, M.M., Ghanim, N.M. and El-Kady, S.M.L. (2010) Local bacterial isolates as entomopathogenic agents against the citrus flower moth, *Prays citri* Miller (Lepidoptera, Hyponomeutidae) in lime orchards at north Delta region, Egypt. *Bulletin of the Entomology Society of Egypt* 36, 171–184.

Epsky, N.D., Hendrichs, J., Katsoyannos, B.I., Vasquez, L.A., Ros, J.P., Zmreo-lu, A., Pereira, R., Bakri, A., Seewooruthun, S.I. and Heath, R.R. (1999) Field evaluation of female-targeted trapping systems for *Ceratitis capitata* (Diptera: Tephritidae) in seven countries. *Journal of Economic Entomology* 92, 156–164.

Epsky, N.D., Kendra, P.E., Penaa J. and Heath, R.R. (2011) Comparison of synthetic food-based lures and liquid protein baits for capture of *Anastrepha suspensa* (Diptera: Tephritidae) adults. *Florida Entomologist* 94, 18–185.

European Food Safety Authority (EFSA) (2008) Pest risk assessment made by France on *Prays citri* considered by France as harmful in French overseas departments of French Guiana, Guadeloupe and Martinique. *EFSA Journal* 680, 2–22.

European Food Safety Authority (EFSA) (2015) Scientific opinion on pest categorisation of *Circulifer haematoceps* and *C. tenellus*. *EFSA Journal* 13(1), 32.

European and Mediterranean Plant Protection Organization (EPPO) (2005) *Scirtothrips aurantii, Scirtothrips citri, Scirtothrips dorsalis*. *EPPO Bulletin* 35(2), 353–356.

European and Mediterranean Plant Protection Organization (EPPO) (2013) *Anoplophora chinensis*: procedures for official control. *EPPO Bulletin* 43(3), 518–526.

European and Mediterranean Plant Protection Organization (EPPO) (2015) PQR Database. EPPO, Paris. Available at: http://www.eppo.int/DATABASES/pqr/pqr.htm (accessed 1 November 2016).

Fasulo, T.R. and Weems, H.V. (1999) Citrus Whitefly, *Dialeurodes citri* (Ashmead) (Insecta: Hemiptera: Aleyrodidae). EDIS. ENY-084. (Revised, January 2014). Available at: http://edis.ifas.ufl.edu/pdffiles/in/in24100.pdf (accessed 4 November 2016).

Fauna Europaea (2013) version 2.6.2, last update 29 August 2013. Available at: http://www.faunaeur.org/external_databases.php (accessed 10 November 2016).

Felix, C.S., Uchôa, M.A. and Faccenda, O. (2009) Capture of *Anastrepha sororcula* (Diptera: Tephritidae) in McPhail and Jackson traps with food attractant and virgin adults. *Brazilian Archives of Biology and Technology* 52(1), 99–104.

Flint, M.L. and Dreistadt, S.H. (1998) *Natural Enemies Handbook: The Illustrated Guide to Biological Pest Control*. University of California Agriculture and Natural Resources (UCANR), Oakland, California.

Foldi, I. (2001) Liste des cochenilles de France (Hemiptera, Coccoidea). *Bulletin de la Société entomologique de France* 106, 303–308.

Food and Agriculture Organization of the United Nations (FAO) (2015) *Citrus Fruit Statistics 2015*. FAO, Rome, p. 47.

Food and Agriculture Organization of the United Nations (FAO) - International Atomic Energy Agency (IAEA) (2003) Trapping Guidelines for Area-wide Fruit Fly Programs. IAEA, Vienna, Austria, p. 47. Available at: http://www-pub.iaea.org/ MTCD/publications /PDF /TG-FFP_web.pdf (accessed 4 November 2016).

French, J.V. (2002) Arthropod pest management. In: Anciso, J.R. (ed.) *IPM in Texas Citrus*. All of the Texas A&M University Press, College Station, Texas, pp. 13–25.

Furuhashi, K. and Nishino, M. (1983) Biological control of arrowhead scale, *Unaspis yanonensis*, by parasitic wasps introduced from the People's Republic of China. *Entomophaga* 28, 277–286.

Garnsey, S.M., Gottwald, T.R. and Yokomi, R.K. (1998) Control strategies for citrus tristeza virus. In: Hadidi, A., Khetarpal, R.K. and Koganezawa, H. (eds) *Plant Viral Disease Control: Principles and Practices*. APS Press, St Paul, Minnesota, pp. 639–658.

Garrido, A., Busto, T. and Tarancon, J. (1984) Assessment of adults of *Prays citri* Mill (Lep. Hyponomeutidae) with a synthetic pheromone and the relation to damage. *Anales del Instituto Nacional de Investigaciones Agrarias, Agricola* 25, 147–154.

Gausman, H.W. and Hart, W.G. (1974) Reflectance of four levels of sooty-mold deposits produced from the honedew of three insect species. *Journal of the Rio Grande Valley Horticultural Society* 28, 131–136.

Gerson, U. (2003) Acarine pests of citrus: overview and non-chemical control. *Systematic and Applied Acarology* 8, 3–12.

Gilbert, M.J. (1990) Relative population levels of citrus thrips *Scirtothrips aurantii* on commercial *Citrus* and adjacent bush. *South African Journal of Zoology* 25(1), 72–76.

Gill, H.K., Goyal, G. and Gillett-Kaufman, J. (2013) *Citrus Mealybug,* Planococcus citri *(Risso) (Insecta: Hemiptera: Pseudococcidae)*. EENY-537. (Published 2012, revised 2013). Entomology and Nematology, Florida Cooperative Extension Service. University of Florida Institute of Food and Agricultural Sciences (IFAS), Gainsville, Florida.

Gill, R.J. (1988) *The Scale Insects of California. Part 1. The Soft Scales (Homoptera: Coccoidea: Coccidae). Technical Series in Agricultural Biosystematics and Plant Pathology*. California Department of Food and Agriculture, Sacramento, California, pp. 1–132.

Glenn, D.M. (2012) The mechanisms of plant stress mitigation by kaolin-based particle films and applications in horticultural and agricultural crops. *Horticultural Science* 47(6), 710–711.

Grafton-Cardwell, E.E. and Ouyang, Y. (1995) Augmentation of *Euseius tularensis* (Acari: Phytoseiidae) in *Citrus*. *Environmental Entomology* 24(3), 738–747.

Grafton-Cardwell, E.E., Ouyang, Y. and Striggow, R.A. (1999) Predacious mites for control of citrus thrips, *Scirtothrips citri* (Thysanoptera: Thripidae) in nursery citrus. *Biological Control* 14(1), 29–36.

Grafton-Cardwell, E.E., Morse, J.G., O'Connell, N.V., Phillips, P.A., Kallsen, C.E. and Haviland, D.R. (2015) *Citrus: Insects, Mites, and Snails. University of California IPM Pest Management Guidelines: Citrus*. University of California, Agriculture and Natural Resources (UCANR) Publication 3441. Available at: http://ipm.ucanr.edu/PMG/selectnewpest.citrus.html (accessed 25 October 2016).

Griffiths, J.T. and Thompson, W.L. (1957) Insects and mites found on Florida citrus. *University of Florida Agricultural Experiment Station Bulletin* 591, 30–33.

Guizar-Guzman, L. and Sanchez-Pena, S.R. (2013) Infection by *Entomophthora sensu stricto* (Entomophthoromycota: Entomophthorales) in *Diaphorina citri* (Hemiptera: Liviidae) in Veracruz, Mexico. *Florida Entomologist* 96(2), 624–627.

Gullan, P.J. and Cook, L.G. (2007) Phylogeny and higher classification of the scale insects (Hemiptera: Sternorrhyncha: Coccoidea). In: Zhang, Z.Q. and Shear, W.A. (eds) *Linnaeus Tercentenary: Progress in Invertebrate Taxonomy*. Zootaxa 1–766. Available at: http://www.mapress.com/zootaxa/2007f/ zt01668 p425.pdf (accessed 24 November 2016).

Gurr, G.M. and Wratten, S.D. (1999) Integrated biological control: a proposal for enhancing success in biological control. *International Journal of Pest Management* 45, 81–84.

Halbert, S.E. and Brown, L.G. (1998) *Toxoptera citricida* (Kirkaldy), Brown citrus aphid – identification, biology, and management strategies. *Entomology Circular* 374. Available at: http://www.doacs.state.fl.us/pi/enpp/ento/entcirc/ent374.pdf (accessed 24 November 2016).

Hall, D.G. (2008) Biological control of *Diaphorina citri*. I Taller Internacional sobre Huanglongbing de los cítricos (*Candidatus Liberibacter* spp) y el psílidoasiático de los cítricos (*Diaphorina citri*). Hermosillo, Sonora, Mexico, p. 7.

Haney, P.B. (1988) Identification, ecology and control of the ants in citrus. A world survey. In: Goren, R., Mendel, K. and Goren, N. (eds) *Citriculture: Proceedings of the Sixth International Citrus Congress*. Balaban Publishers, Warsaw, pp. 1227–1251.

Haney, P.B., Luck, R.F. and Moreno, D.S. (1987) Increases in densities of the citrus red mite, *Panonychus citri* (Acarina: Tetranychidae), in association with the Argentine ant, *Iridomyrmex humilis* (Hymenoptera: Formicidae), in southern California citrus. *Entomophaga* 32, 49–57.

Haramoto, F.H. (1969) Biology and control of *Brevipalpus phoenicis* (Geijskes) (Acarina: Tenuipalpidae). Technical Bulletin, Hawaii Agricultural Experiment Station, 1–63. Available at: https://scholarspace.manoa.hawaii.edu/bitstream/10125/11442/1/uhm_phd_6613704_r.pdf (accessed 24 October 2016).

Hendrichs, J. (2000) Use of the sterile insect technique against key insect pests. *Sustainable Development International* 2, 75–79. Available at: http://infohouse.p2ric.org/ref/40/39706.pdf (accessed 2 December 2016).

Herard, F., Ciampitti, M., Maspero, M., Cocquempot, C., Delvare, G., Lopez, J., Ramualde, N., Jucker, C. and Colombo, M. (2008) Potential natural enemies of the citrus longhorned beetle, *Anoplophora chinensis* (Col.: Cerambycidae), an invasive Asian pest in Italy. *Integrated Control in Citrus Fruit Crops. IOBC (International Organisation for Biological and Integrated Control) WPRS Bulletin* 38, 238.

Hoddle, M.S. and Pandey, R. (2014) Host range testing of *Tamarixia radiata* (Hymenoptera: Eulophidae) sourced from the Punjab of Pakistan for classical biological control of *Diaphorina citri* (Hemiptera: Liviidae: Euphyllurinae: Diaphorinini) in California. *Journal of Ecomonic Entomology* 107(1), 125–136.

Hodges, G., Ruter, J.M. and Braman, S.K. (2001) Susceptibility of *Ilex* species, hybrids and cultivars to Florida wax scale (*Ceroplastes floridensis* Comstock). *Journal of Environmental Horticulture* 19, 32–36.

Hoy, M.A. and Nguyen, R.U. (1997) Classical biological control of the citrus leafminer *Phyllocnistis citrella* Stainton (Lepidoptera: Gracillariidae): theory, practice, art and science. *Tropical Lepidoptera* 8(1), 1–19.

Huffaker, C.B. and Messenger, P.S. (eds) (1976) *Theory and Practice of Biological Control*. Academic Press, New York, 788 pp.

Hwang, R.Y., Hyun, J.W., Chio, C.W. and Chio, Y.H. (2016) Distinguishing injury symptoms by *Scirtothrips dorsalis* Hood and other pest and disease in citrus. In: Mattos, D.J., Carlos, E.F., Novelli, V.M., de Azevedo, A.F., Filho, H.D.C. and Zaccheo, P.V.C. (eds) *Sustainable Citriculture: The Role of Applied Knowledge*. Abstract Book, Printed in Brazil, p. 68.

Ibrahim, S.S. and Shahateh, W.A. (1984) Biological studies on the citrus flower moth *Prays citri* Miller in Egypt. *Arab Journal of Plant Protection* 2(1), 4–9.

International Federation of Organic Agriculture Movements (IFOAM) (2014) The IFOAM NORMS for Organic Production and Processing. Germany. Available at: http://www.ifoam.bio/sites/default/files/ifoam_norms_july_2014_t.pdf (accessed 30 October 2016).

International Plant Protection Convention (IPPC) (2005) Phytosanitary Procedures for Fruit Fly (Tephritidae) Management, p. 8. Available at: http://www.maff.go.jp/j/syouan/syokubo/keneki/ pdf/ 3e. pdf (accessed 30 October 2016).

Isman, M.B. (2006) Botanical insecticides, deterrents, and repellents in modern agriculture and an increasingly regulated world. *Annual Review of Entomology* 51, 45–66.

Itioka, T. and Inoue, T. (1996) The consequences of ant-attendance to the biological control of the red wax scale insect *Ceroplastes rubens* by *Anicetus beneficus*. *Journal of Applied Ecology* 33, 609–618.

Jacas, J.A., Karamaouna, F., Vercher, R. and Zappala, L. (2010) Citrus pest management in the northern Mediterranean Basin (Spain, Italy and Greece). In: Ciancio, A. and Mukerji, K.G. (eds) *Integrated Management of Arthropod Pests and Insect Borne Diseases*. Springer, Dordrecht, The Netherlands, pp. 3–26.

Jackson, P.R. and Hunter, P.E. (1983) Effects of nitrogen fertilizer level on development and populations of the pecan leaf scorch mite (Acari: Tetranychidae). *Journal of Economic Entomology* 76, 432–435.

James, D.G. (1989) Population biology of *Biprorulus bibax* Breddin (Hemiptera: Pentatomidae) in a southern New South Wales citrus orchard. *Journal of the Australian Entomological Society* 28, 279–286.

Jeppson, L.R. (1989) Biology of citrus insects, mites and mollusks. In: Reuther, W., Calavan, E.C. and Carman, G.E. (eds) *The Citrus Industry*. University of California, Oakland, California, pp. 1–87.

Jeppson, L.R., Keifer, H.H. and Baker, E.W. (1975) *Mites Injurious to Economic Plants*. University of California Press, Berkeley, California, 614 pp.

Jones, D.R. (2003) Plant viruses transmitted by whiteflies. *European Journal of Plant Pathology* 109, 195–219.

Jones, W.A. (1988) World review of the parasitoids of the southem green stink bug, *Nezara viridula* (L.) (Heteroptera: Pentatornidae). *Annals of the Entomological Society of America* 81, 262–273.

Kairo, M.T.K., Paraiso, O., Gautam, R.D. and Peterkin, D.D. (2013) *Cryptolaemus montrouzieri* (Mulsant) (Coccinellidae: Scymninae): a review of biology, ecology, and use in biological control with particular reference to potential impact on non-target organisms. *CAB Reviews* 8, No. 005. CAB International, Wallingford, UK.

Kajimura, T., Maeoka, Y., Widiarta, I.N., Sudo, T., Hidaka, K. and Nakasuji, F. (1993) Effects of organic farming of rice plants on population density of leafhoppers and planthoppers: population density and reproductive rate. *Japanese Journal of Applied Entomology and Zoology* 37, 137–144.

Kalaitzaki, A.P., Tsagkarakis, A.E. and Lykouressis, D.P. (2011) Population fluctuation of *Phyllocnistis citrella* and its parasitoids in two citrus species in western Crete (Greece). *Entomologia Hellenica* 20, 31–44.

Kamburov, S.S. (1991) Damage to fruit and the impact on crop set from late infestations of citrus thrips (*Scirtothrips aurantii* Faure). *Citrus Journal* 1(1), 33–34.

Karamaouna, F., Pascual, S., Urbaneja, A. and Jacas, J.A. (2008) Citrus leafminer *Phyllocnistis citrella* (Lepidoptera: Gracilariidae) and its parasitoids. Ten years after the implementation of classical biological control in Spain. *Integrated Control in Citrus Fruit Crops. IOBC (International Organisation for Biological and Integrated Control) WPRS Bulletin* 38, 177.

Karamaouna, F., Mylonas, P., Papachristos, D., Kontodimas, D., Michaelakis, A. and Kapaxidi, E. (2010) Main arthropod pests of *Citrus* culture and pest management in Greece. In: Ciancio, A. and Mukerji, K.G. (eds) *Integrated Management of Arthropod Pests and Insect Borne Diseases*. Springer, Dordrecht, The Netherlands, pp. 29–53.

Katayama, N. and Suzuki, N. (2003) Bodyguard effects for aphids of *Aphis craccivora* Koch (Homoptera: Aphididae) as related to the activity of two ant species, *Tetramorium caespitum* Linnaeus (Hymenoptera: Formicidae) and *Lasius niger* L. (Hymenoptera: Formicidae). *Applied Entomology and Zoology* 38(3), 427–433.

Kaya, H.K. and Lacey, L.A. (2007) Introduction to microbial control. In: Lacey, L.A. and Kaya, H.K. (eds) *Field Manual of Techniques in Invertebrate Pathology: Application and Evaluation of Pathogens for Control of Insects and Other Invertebrate Pests*, 2nd edn. Springer Scientific Publishers, Dordrecht, The Netherlands, pp. 3–7.

Kennett, C.E., McMurtry, J.A. and Beardsley, J.W. (1999) Biological control in subtropical and tropical crops. In: Bellows, T.S. and Fisher, T.W. (eds) *Handbook of Biological Control: Principles and Applications*. Academic Press, New York, p. 1046.

Kerns, D., Wright, G. and Loghry, J. (2001a) Citrus leafminer (*Phyllocnistis citrella*). In: *Citrus Arthropod Pest Management in Arizona*. The University of Arizona, College of Agriculture. Available at: http://cals.arizona.edu/crops/citrus/insects/citrusinsect.html (accessed 10 November 2016).

Kerns, D., Wright, G. and Loghry, J. (2001b) Citrus mealybug (*Planococcus citri*). In: *Citrus Arthropod Pest Management in Arizona*. The University of Arizona, College of Agriculture. Available at: http://cals.arizona.edu/crops/citrus/insects/citrusinsect.html (accessed 12 November 2016).

Kerns, D., Wright, G. and Loghry, J. (2001c) Citrus thrips (*Scirtothrips citri*). In: *Citrus Arthropod Pest Management in Arizona*. The University of Arizona, College of Agriculture. Available at: http://cals.arizona.edu/crops/citrus/insects/citrusinsect.html (accessed 12 November 2016).

Klassen, K. and Curtis, C.F. (2005) History of the sterile insect technique. Principles and practice in area wide integrated pest management. In: Dyck, V.A., Hendrichs, J. and Robinson, A. (eds) *Sterile Insect Technique. Principles and Practice in Area Wide Integrated Pest Management*. Springer, Dordrecht, The Netherlands, pp. 3–36.

Klungness, L.M., Jang, E.B., Mau, R.F.L., Vargas, R.I., Sugano, J.S. and Fujitani, E. (2005) Sanitation techniques for controlling tephritid fruit flies (Diptera: Tephritidae) in Hawaii. *Journal of Applied Sciences and Environmental Management* 9(2), 5–14.

Knapp, M., Van Houten, Y., Hoogerbrugge, H. and Bolckmans, K. (2013) *Amblydromalus limonicus* (Acari: Phytoseiidae) as a biocontrol agent: literature review and new findings. In: Schausberger, P. (ed.) *Acari in a Changing World: Proceedings of the 7th Symposium of EURAAC*, Vienna, 2012. *Acarologia* 53(2), 191–202.

Knipling, E.F. (1955) Possibilities of insect control or eradication through the use of sexually sterile males. *Journal of Economic Entomology* 48(4), 459–466.

Komazaki, S. (1987) Growth and reproduction in the first two and summer generations of two citrus aphids, *Aphis citricola* van der Goot and *Toxoptera citricidus* (Kirkaldy) (Homoptera: Aphididae), under different thermal conditions. *Applied Entomology and Zoology* 23, 220–227.

Kostiainen, T.S. and Hoy, M.A. (1996) The Phytoseiidae as biological control agents of pest mites and insects. A bibliography (1960–1994). Monograph 17. University of Florida, Agricultural Experiment Station, Gainesville, Florida, p. 355.

Kuepper, G. (2004) Thrips management alternatives in the field. Pest management technical note. *ATTRA* 6. Available at: www.attra.ncat.org (accessed 2 November 2016).

Kumar, V., Kakkar, G., McKenzie, C.L., Seal, D.R. and Osborne, L.S. (2013) An Overview of Chilli Thrips, *Scirtothrips dorsalis* (Thysanoptera: Thripidae): Biology, Distribution and Management. Weed and Pest Control, Conventional and New Challenges. Available at: http://www.intechopen.com/books/weed-and-pest-control-conventional-and-new challenges (accessed 10 November 2016).

Kuwana, I. and Ishii, T. (1927) On *Prospaltella smithi* Silv. and *Crytognatha* sp., the enemies of *Aleurocanthus spiniferus* Quaintance, imported from Canton, China. *Review of Applied Entomology* 15, 463.

Landis, D.A. and Haas, M.J. (1992) Influence of landscape structure on abundance and within-field distribution of European corn borer (Lepidoptera, Pyralidae) larval parasitoids in Michigan. *Environmental Entomology* 21, 409–416.

Landis, D.A., Wratten, S.D. and Gurr, G.M. (2000) Habitat management to conserve natural enemies of arthropod pests in agriculture. *Annual Review of Entomology* 45, 175–201.

Langford, E.A., Nielsen Uffe, N., Johnson, S.N. and Riegler, M. (2014) Susceptibility of Queensland fruit fly, *Bactrocera tryoni* (Froggatt) (Diptera: Tephritidae), to entomopathogenic nematodes. *Biological Control* 69, 34–39.

Lapointe, S.L., Hall, D.G., Murata, Y., Parra-Pedrazzoli, A.L., Bento, J.M.S., Vilela, E.F. and Leal, W.S. (2006) Field evaluation of a synthetic female sex pheromone for the leafmining moth *Phyllocnistis citrella* (Lepidoptera: Gracillariidae) in Florida citrus. *Florida Entomologist* 89(2), 274–276.

Lapointe, S.L., Keathley, C.P., Stelinski, L.L., Urrutia, W.H. and Mafra-Neto, A. (2015) Disruption of the leafminer *Phyllocnistis citrella* (Lepidoptera: Gracillariidae) in citrus: effect of blend and placement height, longevity of disruption and emission profile of a new dispenser. *Florida Entomologist* 98(2), 742–748.

Leblanc, L. and Mararuai, A. (1999) Fruit bagging to control fruit flies. Technical Report, Pest Advisory Leaflet No. 36. Plant Protection Service Secreteriat of the Pacific Community, New Caledonia, p. 5.

Lee, C.Y. and Lo, K.C. (1996) Reproductive biology of the white-spotted longicorn beetle, *Anoplophora macularia* (Thomson) (Coleoptera: Cerambycidae), on sweet oranges. *Journal of Agricultural Research in China* 45(3), 297–304.

Lentini, A. and Verdinelli, M. (2012) Formicidae. In: Vacante, V. and Gerson, U. (eds) *Integrated Control of Citrus Pests in the Mediterranean Region*. Bentham Science Publishers, London, pp. 231–241.

Lernoud, J. and Willer, H. (2016) Current statistics on organic agriculture worldwide: area, producers, markets, and selected crops. In: Willer, H. and Lernoud, J. (eds) *The World of Organic Agriculture. Statistics and Emerging Trends*, pp. 34–114. Available at: http://orgprints.org/ (accessed 2 November 2016).

Lez, M.M., Juan, A., Capllonch, M. and Alemany, A. (2008) Female-biased mass trapping vs. bait application techniques against the Mediterranean fruit fly, *Ceratitis capitata* (Dipt., Tephritidae). *Journal of Applied Entomology* 132(9–10), 753–761.

Lindegren, J.E., Wong, T.T.Y. and McInnis, D.O. (1990) Response of Mediterranean fruit fly (Diptera: Tephritidae) to the entomogenous nematode *Steinerma feltiae* in field tests in Hawaii. *Environmental Entomology* 19, 383–386.

Liquido, N.J., Shinoda, L.A. and Cunningham, R.T. (1991) *Host Plants of the Mediterranean Fruit Fly (Diptera Tephritidae): An Annotated World Review*. Entomological Society of America Miscellaneous Publication 77. Entomological Society of America, Lanham, Maryland, 52 pp.

Liu, H.Y., Gumpf, D.J., Oldfield, G.N. and Calavan, E.C. (1983) The relationship of *Spiroplasma citri* and *C. tenellus*. *Phytopathology* 73, 585–590.

Lo Genco, A., Ciotta, C. and Lo Pinto, M. (2008) Evolution of *Phyllocnistis citrella* Stainton (Lepidoptera, Gracillariidae) and its parasitoids in the last five years in citrus orchards of western Sicily (Italy). *Integrated Control in Citrus Fruit Crops. IOBC (International Organisation for Biological and Integrated Control) WPRS Bulletin* 38, 178–182.

Lomer, C.J., Bateman, R.P., Johnson, D.L., Langewald, J. and Thomas, M. (2001) Biological control of locusts and grasshoppers. *Annual Review of Entomology* 46, 667–702.

LoVerde, G., Caleca, V. and LoVerde, V. (2011) The use of kaolin to control *Ceratitis capitata* in organic citrus groves. *Bulletin of Insectology* 64(1), 127–130.

Lowery, D.T., Isman, M.B. and Brard, N.L. (1993) Laboratory and field evaluation of neem for the control of aphids (Homoptera: Aphididae). *Journal of Economic Entomology* 86, 864–870.

Luck, R.F., Forster, L.D. and Morse, J.G. (1997) An ecologically based IPM program for citrus in California's San Joaquin Valley using augmentative biological control. In: *Proceedings of the International Society of Citriculture VIII International Citrus Congress*, 12–17 May 1996, Sun City, South Africa. International Society of Citriculture, South Africa, pp. 499–503.

Maddison, P.A. (1976) Interim report to the South Pacific Bureau of Economic Co-operation, on pests of a limited range of crops. In UNDP/FAO survey of agricultural pests and diseases. Interim report March 1976. Part 2 Nematology Entomology. *South Pacific Bureau of Economic Development* 4, 1–57.

Maia, O.M.A. and Oliveira, C.A.L. (2004) Colonization capacity of *Brevipalpus phoenicis* (Geijskes) (Acari: Tenuipalpidae) on plants used as hedge, windbreak and on weeds. *Neotropical Entomology* 33(5), 625–629.

Maltby, W.L., Jimenez, E. and DeBach, P. (1968) Biological control of armored scale insects in Mexico. *Journal of Economic Entomology* 61, 1086–1088.

Manjunath, K.L., Halbert, S.E., Ramadugu, C., Webb, S. and Lee, R.F. (2008) Detection of *'Candidatus Liberibacter asiaticus'* in *Diaphorina citri* and its importance in the management of citrus Huanglongbing in Florida. *Phytopathology* 98(4), 387–396.

Martin, J.H. and Mound, L.A. (2007) An annotated check list of the world's whiteflies (Insecta: Hemiptera: Aleyrodidae). *Zootaxa* 1492, 1–84.

Martínez-Ferrer, M., Campos, J.M. and Fibla, J.M. (2011) Field efficacy of *Ceratitis capitata* (Diptera: Tephritidae) mass trapping technique on clementine groves in Spain. *Journal of Applied Entomology* 136(3), 181–190.

McMurtry, J.A. and Croft, B.A. (1997) Life-styles of phytoseiid mites and their roles in biological control. *Annual Review of Entomology* 42, 291–321.

McPhail, M. (1939) Protein lures for fruit flies. *Journal of Economic Entomology* 32, 758–761.

McPherson, J.E. (1982) *The Pentatomoidea (Hemiptera) of Northeastern North America with Emphasis on the Fauna of Illinois*. Southern Illinois University Press, Carbondale, Illinois, p. 241.

Mellado, J.J.S. (2011) Biological control of California red scale, *Aonidiella aurantii* (Hemiptera: Diaspididae): spatial and temporal distribution of natural enemies, parasitism levels and climate effects. Doctoral thesis, Escuela Tecnica Superior De Ingenieros Agronomos, Valencia, Spain, 196 pp.

Mello, A.F.S., Yokomi, R.K., Payton, M.E. and Fletcher, J. (2010) Effect of citrus stubborn disease on navel orange production in a commercial orchard in California. *Journal of Plant Pathology* 92, 429–438.

Meyerdirk, D.E. and Moratorio, M.S. (1987) *C. tenellus* (Baker), the beet leafhopper (Homoptera: Cicadellidae): laboratory studies on fecundity and longevity. *Canadian Entomologist* 119, 443–447.

Michaud, J.P. (2002) Biological control of Asian citrus psyllid, *Diaphorina citri* (Hemiptera: Psyllidae) in Florida: a preliminary report. *Entomological News* 113(3), 216–222.

Michaud, J.P. (2004) Natural mortality of Asian citrus psyllid (Homoptera: Psyllidae) in central Florida. *Biological Control* 29, 260–269.

Migeon, A. and Dorkeld, F. (2006) Spider Mites Web. Available at: http://www.montpellier.inra.fr/CBGP/spm-web (accessed 22 October 2016).

Miller, D.R. and Davidson, J.A. (1990) A list of the armored scale insect pests. In: *Armored Scale Insects, Their Biology, Natural Enemies and Control, [Series title: World Crop Pests, Vol. 4B]*. Boletin del Museo de Entomologia de la Universidad del Valle, Elsevier, Amsterdam, The Netherlands, p. 688.

Minaei, K., Bagherian, S.A.A. and Aleosfoor, M. (2015) *Scirtothrips dorsalis* (Thysanoptera: Thripidae) as a pest of citrus in Fars province, Iran. *Iranian Journal of Plant Protection Science* 46(2), 219–225.

Minas, R.D.S., Souza, R.M., Dolinski, C., Cavalloro, R. and Burla, R. (2016) Potential of entomopathogenic nematodes (Rhabditida: Heterorhabditidae) to control Mediterranean fruit fly (Diptera: Tephritidae) soil stages. *Nematoda* 14. Available at: 1http://dx.doi.org/10.4322/nematoda.02016 (accessed 2 November 2016).

Mineo, G. (1986) Lepidoptera on citrus trees and particularly on those inhabiting the Mediterranean area. In: Cavalloro, R. and Di Martino, E. (eds) *Integrated Pest Control in Citrus Groves*. Proceedings of the Experts' Meeting, 26–29 March 1985, Acireale, Italy. A.A. Balkema, Rotterdam, The Netherlands, pp. 127–133.

Mineo, G. (1993) Effects of cultural techniques on the control of citrus flower moth (*Prays citri* Mill.) infestations (Lep. Yponomeutidae). *Frustula Entomologica* 16, 89–95.

Mitchell, P.L. (2004) Heteroptera as vectors of plant pathogens. *Neotropical Entomology* 33(5). Available at: http://www.scielo.br/scielo.php?pid=S1519-566X2004000500001&script=sci_arttext (accessed 31 October 2016).

Mnif, I., Elleuch, M., Ellouze-Chaabouni, S. and Ghribi, D. (2013) *Bacillus subtilis* SPB1 biosurfactant: production optimization and insecticidal activity against the carob moth, *Ectomyelois ceratoniae*. *Crop Protection* 50, 66–72.

Mundinger, F.G. and Chapman, P.J. (1932) Plant bugs as pests of pear and other fruits in the Hudson Valley. *Journal of Economic Entomology* 25, 655–658.

Muniappan, R., Shepard, B.M., Carner, G.R. and Ooi, P.A.-C. (2012) *Arthropod Pests of Horticultural Crops in Tropical Asia: Pests of Citrus* (*Citrus* spp., Rutaceae). CAB International, Wallingford, UK, p. 167.

Nakahara, S. (1982) *Checklist of the Armored Scales (Homoptera: Diaspididae) of the Conterminous United States*. Animal and Plant Health Inspection Service, Plant Protection and Quarantine, United States Department of Agriculture, Washington, DC, p. 110.

Nanta, P., Morakote, R., Lamanakkanee, B. and Boonyong, S. (1996) The role of hymenopterous parasitoids attacking citrus leafminer *Phyllocnistis citrella* Stainton in pummelo orchards in Pichit Province, Thailand. *Thammasat International Journal of Science and Technology* 1(1), 44–46.

Nas, S. and Atakan, E.A. (2008) Thrips species on citrus in the Eastern Mediterranean region of Turkey: *Pezothrips kellyanus* Bagnall (Thysanoptera: Thripidae). ZiraatFakultesiDergisi, Mustafa Kemal Universitesi 13(1/2), 65–72.

Navarro-Campos, C., Pekas, A., Aguilar, A. and Garcia-Marí, F. (2013a) Factors influencing citrus fruit scarring caused by *Pezothrips kellyanus*. *Journal of Pest Science* 86(3), 459–467.

Navarro-Campos, C., Pekas, A., Aguilar, A. and Garcia-Marí, F. (2013b) Factors explaining variation in citrus fruit scarring by *Pezothrips kellyanus* (Thysanoptera: Thripidae). *IOBC (International Organisation for Biological and Integrated Control) WPRS Bulletin* 95, 71–76.

Newell, I.R. and Haramoto, F.H. (1968) Biotic factors influencing populations of *Dacus dorsalis* in Hawaii. *Proceedings of the Hawaiian Entomological Society* 20(1), 81–139.

Nguyen, R. and Sailer, R.I. (1979) Colonization of a citrus whitefly parasite, *Prospaltella lahorensis*, in Gainesville, Florida. *Florida Entomologist* 62, 59–65.

Nguyen, R., Brasil, J.R. and Poucher, C. (1983) Population density of the citrus blackfly, *Aleurocanthus woglumi* Ashby (Homoptera: Aleyrodidae), and its parasites in urban Florida in 1979–81. *Environmental Entomology* 12, 878–884.

Nguyen, R., Hamon, A.B. and Fasulo, T.R. (1998) *Citrus Blackfly,* Aleurocanthus woglumi *Ashby (Insecta: Hemiptera: Aleyrodidae).* EENY-042 1998. (Revised March 2010 and April 2016). Entomology and Nematology, Florida Cooperative Extension Service. University of Florida Institute of Food and Agricultural Sciences (IFAS), Gainsville, Florida.

Niccoli, A., Sacchetti, P. and Lupi, E. (1991) Observations on the capture of *Ceratitis capitata* in peach orchards in Tuscany. *Redia* 74, 641–658.

Niu, J.Z., Hull-Sanders, H., Zhang, Y.X., Lin, J.Z., Dou, W. and Wang, J.J. (2014) Biological control of arthropod pests in citrus orchards in China. *Biological Control* 68, 15–22.

Noma, T., Colunga-Garcia, M., Brewer, M., Landis, J. and Gooch, A. (2010) Citrus long-horned beetle *Anoplophora chinensi.* Michigan State University IPM Program, Michigan State University's invasive species factsheets, p. 2.

Nouh, G.M. and Hussein, M.A. (2014) The role of entomopathogenic nematodes as biocontrol agents against some tephritid flies. *Advances in Biological Research* 8(6), 301–306.

Onillon, J.C. (1988) Lutte biologique et intégrée dans les vergers de citrus en zone méditerranéenne. *Entomophaga* 33(4), 481–494.

Orphanides, G.M., Ioannou, N., Kyriakou, A., Phillis, J. and Americanos, P. (1996) Citrus pest problems and their control in the Near East, Cyprus. In: Morse, J.G., Juck, R.F. and Gumpf, D.J. (eds) *FAO Plant Production and Protection Paper* 135(402). Food and Agriculture Organization of the United Nations (FAO), Rome, pp. 7–23.

Ortu, S., Cocco, A., Delrio, G. and Lentini, A. (2004) Integrated pest management to control California red scale *Aonidiella aurantii* (Maskell) in the Mediterranean area (Sardinia, Italy). Paper presented at the International Congress of Entomology XXII, 15–21 August, Brisbane, Australia.

Ovruski, S.M., Colin, C., Soria, A., Orono, L.E. and Schliserman, P. (2003) Introduccion y produccion en laboratorio de *Diachasmimorpha tryoni* y *Diachasmimorpha longicaudata* (Hymenoptera: Braconidae) para el control biologico de *Ceratitis capitata* (Diptera: Tephritidae) en la Argentina. *Revista Sociedad Entomologica Argentina* 62, 49–59.

Öztürk, N., Ölçülü, M. and Ulusoy, M.R. (2011) The adult population dynamics of the carob moth [*Ectomyeloisceratoniae* Zell. (Lepidoptera: Pyralidae)] in citrus orchards in Adana and Mersin provinces. *BitkiKorumaBulteni* 51(4), 347–358.

Padmanaban, B. (1994) Screening of citrus germplasm for controlling citrus leaf-miner (*Phyllocnistis citrella*) (Lepidoptera: Phyllocnistidae). *Indian Journal of Agricultural Sciences* 64(10), 723–726.

Pandey, S., Pandey, J.P. and Tiwari, R.K. (2011) Effect of some neem based insecticides on wing shape and pigmentation in lemon-butterfly, *Papilio demoleus* L. *World Applied Science Journal* 13(6), 1356–1360.

Panhwar, F. (2005) Mediterranean Fruit Fly (*Ceratitis capitata*) Attack on Fruits and its Control in Sindh Pakistan. Digitalverlag GmbH, Germany. Available at: www.ChemLin.com (accessed 31 October 2016).

Panis, A. (1977) Pseudococcids (Homoptera, Coccoidea, Pseudococcidae) within the context of integrated control in citrus groves round the Mediterranean. *Boletin del Servicio de Defensa contra Plagas e Inspeccion Fitopatologica* 3, 139–145.

Panis, A. (1982) Scale insects (Homoptera, Coccoidea, Coccidae) within the framework of integrated control in Mediterranean citrus culture. *Revue de Zoologie Agricole et de Pathologie Végétale* 79, 12–22.

Panizzi, A.R. (2008) Southern green stink bug, *Nezara viridula* (L.) (Hemiptera: Heteroptera: Pentatomidae). In: Capinera, J.L. (ed.) *Encyclopedia of Entomology*. Springer, Heidelberg, Germany, pp. 3471–3472.

Panizzi, A.R., McPherson, J.E., James, D.G., Javahery, M. and McPherson, R.M. (2000) *Stink Bugs (Pentatomidae).* CRC Press, Boca Raton, Florida, pp. 421–474 [Chapter 13].

Papadopoulos, N.T. (1999) Study on the biology and ecology of the Mediterranean fruit fly *Ceratitis capitata* (Diptera: Tephritidae) in northern Greece. PhD dissertation, Aristotle University of Thessaloniki, Greece.

Papadopoulos, N.T., Katsoyannos, B.I., Kouloussis, N.A., Hendrichs, J., Carey, J.R. and Heath, R.R. (2001) Early detection and population monitoring of *Ceratitis capitata* (Diptera: Tephritidae) in mixed fruit orchard in Northern Greece. *Journal of Economic Entomology* 94, 971–978.

Parra-Pedrazzoli, A.L., Leal, W.S., Vilela, E.F., Mendonça, M.C. and Bento, J.M.S. (2009) Synthetic sex pheromone of citrus leafminer in Brazilian citrus groves. *Pesquisa Agropecuária Brasileira* 44(7), 676–680.

Pekas, A., Tena, A., Aguilar, A. and Garcia-Mari, F. (2010) Effect of Mediterranean ants (Hymenoptera: Formicidae) on California red scale (Hemiptera: Diaspididae) populations in citrus orchards. *Environmental Entomology* 39(3), 827–834.

Perovic, T. and Hrnčić, S. (2008) The control of *Citrus* leaf miner *Phyllocnistis citrella* Stainton with bioinsecticides. *Integrated Control in Citrus Fruit Crops. IOBC (International Organisation for Biological and Integrated Control) WPRS Bulletin* 38, 191–194.

Peterson, G.D. (1955) Biological control of the orange spiny white fly. *Guam Journal of Economic Entomology* 48, 681–683.

Phillips, P. and Sherk, C. (1991) To control mealybugs, stop honeydew-seeking ants. *California Agriculture* 45(2), 26–28.

Planes, L., Catalán, J., Jaques, J.A., Urbaneja, A. and Tena, A. (2015) *Pezothrips kellyanus* (Thysanoptera: Thripidae) nymphs on orange fruit: importance of the second generation for its management. *Florida Entomologist* 98, 3.

Polek, M., Vidalakis, G. and Godfrey, K. (2007) Citrus bacterial canker disease and Huanglongbing (citrus greening). University of California Publication 8218. University of California, Agriculture and Natural Resources, Oakland, California, p. 12.

Poprawski, T.J., Parker, P.E. and Tsai, J.H. (1999) Laboratory and field evaluation of hyphomycete insect pathogenic fungi for control of brown citrus aphid (Homoptera: Aphididae). *Environmental Entomology* 28(2), 315–321.

Prakash, O.M. (2012) *IPM Schedule for Citrus Pests*. National Horticulture Mission, Extension Bulletin No.4, New Delhi, p. 36.

Prinsloo, G.L. (1984) An illustrated guide to the parasitic wasps associated with citrus pests in the Republic of South Africa. *Department of Agriculture Science Bulletin* 402, 119.

Puttaswamy, A. and Channabasavanna, G.P. (1982) Influence of nitrogen fertilization of the host plant on the population of *Tetranychus ludeni* (Acari, Tetranychidae). *Indian Journal of Acarology* 6, 64–71.

Qureshi, A.J., Khan, A.A., Jones, M. and Stansly, P.A. (2013) Management of Asian citrus psyllid in organic citrus groves. *Citrus Industry* February, 6–10. Available at: http://www.crec.ifas.ufl.edu/ (accessed 31 October 2016).

Rae, D.J., Watson, D.M., Huang, M.D., Cen, Y.J., Wang, B.Z., Beattie, G.A.C, Liang W.G., Tan, B.L. and Liu, D.G. (2000) Efficacy and phytotoxicity of multiple petroleum oil sprays on sweet orange (*Citrus sinensis* (L.)) and pummelo (*C. grandis* (L.)) in Southern China. *International Journal of Pest Management* 46(2), 125–140.

Rana, G.L., Kaloostian, G.H., Oldfield, G.N., Granett, A.L., Calavan, E.C., Pierce, H.D., Lee, I.M. and Gumpf, D.J. (1975) Acquisition of *Spiroplasma citri* through membranes by homopterous insects. *Phytopathology* 65, 1143–1145.

Rea, J.H., Wratten, S.D., Sedcole, R., Cameron, P.J., Davis, S.L. and Chapman, R.B. (2002) Trap cropping to manage green vegetable bug *Nezara viridula* (L.) (Heteroptera: Pentatomidae) in sweet corn in New Zealand. *Agricultural and Forest Entomology* 4, 101–107.

Riba, M., Marti, J. and Sans, A. (2003) Influence of azadirachtin on development and reproduction of, *Nezara viridula* L. (Het., Pentatomidae). *Journal of Applied Entomology* 127(1), 37–41. DOI: 10.1046/j.1439-0418.2003.00684.x.

Robacker, D.C. and Moreno, D.S. (1988) Responses of female Mexican fruit flies at various distances from male-produced pheromone. *Southwest Entomology* 13, 95–100.

Rodrigues, P., Ilharco, F.A., da Silva, E.B. and Franco, J.C. (2006) Interactions between ground cover management, hedges and aphids in lemon orchards. *Integrated Control in Citrus Fruit Crops. IOBC (International Organisation for Biological and Integrated Control) WPRS Bulletin* 29(3), 143–150.

Rogers, M. (2012) Citrus pest spotlight: citrus snow scale. *Citrus Industry*, 12. Available at: http://www.crec.ifas.ufl.edu/extension/trade_journals/2012/2012_May_snow_scale.pdf (accessed 16 November 2016).

Rogers, M.E. and Stansly, P.A. (2016) Rust mites, spider mites, and other phytophagous mites. In: Rogers, M.E. and Dewdney, M.M. (eds) *Florida Citrus Pest Management Guide*. University of Florida, Gainesville, Florida, pp. 39–43.

Rogers, M.E., Stansly, P.A. and Stelinski, L.L. (2016) Asian citrus psyllid and citrus leafminer. In: Rogers, M.E. and Dewdney, M.M. (eds) *Florida Citrus Pest Management Guide*. University of Florida, Gainesville, Florida, pp. 33–38.

Rombach, M.C., Aguda, R.M., Shepard, B.M. and Roberts, D.W. (1986) Entomopathogenic fungi (Deuteromycotina) in the control of the black bug of rice, *Scotinophara coarctata* (Hemiptera: Pentatomidae). *Journal of Invertebrate Pathology* 48, 174–179.

Rousse, P., Chiroleu, F., Veslot, J. and Quilici, S. (2007) The host- and microhabitat olfactory location by *Fopius arisanus* suggests a broad potential host range. *Physiological Entomology* 32(4), 313–321. Available at: http://www.blackwell-synergy.com/loi/pen (accessed 10 November 2016).

Russo, A. and Longo, S. (2004) Diagnostic protocols for regulated pests: *Unaspis citri*. *European and Mediterranean Plant Protection Organization Bulletin* 34, 299–301.

Sarada, G., Gopal, K., Venkata Ramana, K.T., Mukunda Lakshmi, L. and Nagalakshmi, T. (2014) Citrus butterfly (*Papilio demoleus*) biology and management: a review. *Journal of Agriculture and Allied Sciences* 3(1), 17–25.

Selhime, A.G., Muma, M.H., Simanton, W.A. and McCoy, W.C. (1969) Control of Florida red scale in Florida with the parasite *Aphytis holoxanthus*. *Journal of Economic Entomology* 62, 954–955.

Serghiou, C.S. (1983) The citrus mealybug, *Planococcus citri* Risso–carob moth, *Ectomyelois ceratoniae* Zeller, pest complex on grapefruit and its chemical control. *Technical Bulletin* 56, 17.

Seymour, J., Bowman, G. and Crouch, M. (1995) Effects of neem seed extract on feeding frequency of *Nezara viridula* L. (Hemiptera: Pentatomidae) on pecan nuts. *Journal of Australian Entomological Soceity* 34, 221–223.

Silva, F.D., Rafael, F., Meirelles, N., Redaelli, L.R. and Dal Soglio, F.K. (2006) Diversity of flies (Diptera: Tephritidae and Lonchaeidae) in organic citrus orchards in the Vale do Rio Caí, Rio Grande do Sul, Southern Brazil. *Neotropical Entomology* 35(5), 666–670.

Siscaro, G., Longo, S., Maugeri, V, Reina, P. and Zappala, L. (1999) Establishment in Sicily of *Ageniaspis citricola*. *Informatore Agrario* 55(16), 85–86.

Siscaro, G., Perrotta, G., Conti, F. and Zappa, L. (2008) Field evaluation on citrus fruit scars in Italy: the role of *Pezothrips kellyanus* (Bagnall) (Thysanoptera: Thripidae). *Integrated Control in Citrus Fruit Crops. IOBC (International Organisation for Biological and Integrated Control) WPRS Bulletin* 38, 200–203.

Sivinski, J., Jeronimo, F. and Holler, T. (2000) Development of aerial releases of *Diachasmimorpha tryoni* (Cameron) (Hymenoptera: Braconidae), a parasitoid that attacks the Mediterranean fruit fly, *Ceratitis capitata* (Weidemann) (Diptera: Tephritidae), in the Guatemalan Highlands. *Biocontrol Science and Technology* 10, 15–25.

Spreen, T.H. (2012) The world citrus industry. In: Verheye, W.H. (ed.) *Encyclopedia of Life Support System*. Vol III: *Soil, Plant Growth and Crop Production*. United Nations Educational, Scientific and Cultural Organization, London, p. 21.

Stansly, P.A. and Rogers, M.E. (2016) Soft-bodied insects attacking foliage and fruit. In: Rogers, M.E. and Dewdney, M.M. (eds) *Florida Citrus Pest Management Guide*. University of Florida, Gainesville, Florida, pp. 45–47.

Stelinski, L.L., Miller, J.R. and Rogers, M.E. (2008) Mating disruption of citrus leafminer mediated by a noncompetitive mechanism at a remarkably low pheromone release rate. *Journal of Chemical Ecology* 34(8), 1107–1113.

Stelinski, L.L., Lapointe, S.L. and Meyer, W.L. (2010) Season-long mating disruption of citrus leafminer, *Phyllocnistis citrella* Stainton, with an emulsified wax formulation of pheromone. *Journal of Applied Entomology* 134(6), 512–520.

Sternlicht, M., Barzakay, I. and Tamim, M. (1990) Management of *Prays citri* in lemon orchards by mass trapping of males. *Entomologia Experimentalis et Applicata* 55, 59–67.

Stiller, M. (2009) Leafhoppers associated with grasslands of South Africa – grassland biome endemics. *Grassroots* 9(4), 13–15.

Stoetzel, M.B. (1994) Aphids (Homoptera: Aphididae) of potential importance on *Citrus* in the United States with illustrated keys to species. *Proceedings of the Entomological Society of Washington* 96, 74–90.

Syed, R.A., Ghani, M.A. and Murtaza, M. (1970) Studies on the Trypetids and their natural enemies in West Pakistan. III. *Dacus* (Strumeta) *zonatus* (Saunders). *Technical Bulletin, Commonwealth Institute of Biological Control* 13, 1–16.

Takagi, M. (2003) Biological control of citrus scale pests in Japan. In: *Proceedings of the 1st International Symposium on Biological Control of Arthropods*, 14–18 January 2002, Hawaii, pp. 351–355. Available at: https://www.bugwood.org/arthropod/ (accessed 3 November 2016).

Talhouk, A.M.S. (1975) Citrus pests throughout the world. Ciba-Geigy Agrochemicals, Basel, Switzerland. *Technical Monograph* 4, 21.

Tan, K.H., Nishida, R., Jang, E.B. and Shelly, T.E. (2014) Pheromones, male lures, and trapping of tephritid fruit flies. In: Shelly, T., Epsky, N., Jang, E.B., Reyes-Flores, J. and Vargas, R.I. (eds) *Trapping and the Detection, Control, and Regulation of Tephritid Fruit Flies*. Springer Science+Business Media, Dordrecht, The Netherlands, pp. 15–74.

Tang, Y.Q., Weathersbee, A.A. and Mayer, R.T. (2002) Effect of neem extract on the brown citrus aphid (Homoptera: Aphididae) and its parasitoid *Lysiphlebus testaceipes* (Hymenoptera: Aphididae). *Environmental Entomology* 31, 172–176.

Tena, A., Soto, A. and Garcia-Marí, F. (2008) Parasitoid complex of black scale *Saissetia oleae* on citrus and olives: parasitoid species composition and seasonal trend. *BioControl* 53(3), 473–487.

Thomas, D.B. (1993) Survivorship of the pupal stages of the Mexican fruit fly *Anastrepha ludens* in an agricultural and a non-agricultural situation. *Entomological Science* 28(4), 350–362.

Thomas, D.B. (1995) Predation on the soil inhabiting stages of the Mexican fruit fly. *Southwestern Entomologist* 20(1), 61–71.

Toledo, J., Liedo, P., Flores, S., Campos, S.E., Villaseñor, A. and Montoya, P. (2006) Use of *Beauveria bassiana* and *Metarhizium anisopliae* for fruit fly control: a novel approach. In: *Fruit Flies of Economic Importance: From Basic to Applied Knowledge*. Proceedings of the 7th International Symposium on Fruit Flies of Economic Importance, 10–15 September, Salvador, Brazil, pp. 127–132.

Tsutsumi, T., Kashio, T., Hashimoto, S., Gyoutoku, Y., Kai, I. and Narahara, M. (1990) Studies on biological control of the white-spotted longicorn beetle, *Anoplophora malasiaca*, by an entomogenous fungus, *Beauveria brongniartii*. IV. Field evaluation of hanging polyurethane foam sheet containing *B. brongniartii* conidia, on citrus trees for the control of adult beetle. *Proceedings of the Association for Plant Protection of Kyushu* 36, 173–176.

Ulrichs, C.H., Mewis, I. and Schnitzler, W.H. (2001) Efficacy of neem and diatomaceous earth against cowpea aphids and their deleterious effect on predating Coccinelidae. *Journal of Applied Entomology* 125, 571–575.

United States Department of Agriculture (USDA) National Organic Program (NOP) (2016) National Organic Program Standards. Available at: http://www.ams.usda.gov/nop/indexNet.htm (accessed 14 November 2016).

Urbaneja, A., Mari, F.G., Tortosa, D., Navarro, C., Vanaclocha, P., Bargues, L. and Castanera, P. (2006) Influence of ground predators on the survival of the Mediterranean fruit fly pupae, *Ceratitis capitata*, in Spanish citrus orchards. *BioControl* 51, 611.

Uygun, N. and Satar, S. (2008) The current situation of citrus pests and their control methods in Turkey. *Integrated Control in Citrus Fruit Crops. IOBC (International Organisation for Biological and Integrated Control) WPRS Bulletin* 38, 2–9.

Uygun, N., de Mendoza, A.H. and Baspinar, H. (2012) Aphididae. In: Vacante, V. and Gerson, U. (eds) *Integrated Control of Citrus Pests in the Mediterranean Region*. Bentham Science Publishers, London, pp. 126–136.

Vacante, V. (2010) *Citrus Mites: Identification, Bionomy and Control*. CAB International, Wallingford, UK, 378 pp.

Vacas, S., Vanaclocha, P., Alfaro, C., Primo, J., Verdu, M.J., Urbaneja, A. and Navarro-llopis, V. (2012) Mating disruption for the control of *Aonidiella aurantii* Maskell (Hemiptera: Diaspididae) may contribute to increased effectiveness of natural enemies. *Pest Management Science* 68(1), 142–148. DOI: 10.1002/ps.2239.

Van den Berg, M.A. and Greenland, J. (1997) Classical biological control of *Aleurocanthus spiniferus* (Hemiptera: Aleyrodidae), on citrus in Southern Africa. *Entomophaga* 42, 459–465.

Van den Berg, M.A., Deacon, V.E. and Thomas, C.D. (1991) Ecology of the citrus psylla, *Trioza erytreae* (Hemiptera: Triozidae). Mating, fertility and oviposition. *Phytophylactica* 23(3), 195–200.

Van Driesche, R.G. and Bellows, T.S. (1996) *Biological Control*. Chapman & Hall, New York, p. 539.

Van Emden, H.F. (1991) The role of host plant resistance in insect pest mismanagement. *Bulletin of Entomological Research* 81, 123–126.

Van Emden, H.F., Eastop, V.F., Hughes, R.D. and Way, M.J. (1969) The ecology of *Myzus persicae*. *Annual Review of Entomology* 14, 197–270.

Vargas, R.I., Souder, S.K., Hoffman, K., Mercogliano, J., Smith, T.R., Hammond, J., Davis, B.J., Brodie, M. and Dripps, J.E. (2014) Attraction and mortality of *Bactrocera dorsalis* (Diptera: Tephritidae) to STATIC Spinosad ME weathered under operational conditions in California and Florida: a reduced-risk male annihilation treatment. *Journal of Economic Entomology* 107(4), 1362–1369. Available at: http://dx.doi.org/10.1603/EC14121 (accessed 3 November 2016).

Varikou, K., Tsitsipis, I., Alexandrakis, V. and Hoddle, M. (2009) Effect of temperature on the development and longevity of *Pezothrips kellyanus* (Thysanoptera: Thripidae). *Annals of the Entomological Society of America* 102(5), 835–841.

Vincent, H.R. and Carde, R.T. (2009) *Encyclopedia of Insects*, 2nd edn. Academic Press, New York, pp. 559–586.

Volkl, W., Mackauer, M., Pell, J.K. and Brodeur, J. (2007) Predators, parasitoids and pathogens. In: Van Emden, H.F. and Harrington, R. (eds) *Aphids as Crop Pests*. CAB International, Wallingford, UK, pp. 187–215.

Waterhouse, D.F. and Sands, P.A. (2001) *Classical Biological Control of Arthropods in Australia. CSIRO Entomology*. Australian Centre for International Agricultural Research, Canberra, p. 450.

Wermelinger, B., Oertli, J.J. and Delucchi, V. (1985) Effect of host plant nitrogen fertilization on the biology of the two-spotted spider mite, *Tetranychus urticae*. *Entomologia Experimentalis et Applicata* 38(1), 23–28.

White, I.M. and Elson-Harris, M.M. (1992) *Fruit Flies of Economic Significance: Their Identification and Bionomics*. CAB International, Wallingford, UK, p. 601.

Williams, D.J. (1985) *Australian Mealybugs*. British Museum (Natural History), London, p. 431.

Wilson, L.J., Smilanick, J.M., Hoffmann, M.P., Flaherty, D.L. and Ruiz, S.M. (1988) Leaf nitrogen and position in relation to population parameters of pacific spider mite, *Tetranychus pacificus* (Acari: Tetranychidae) on grapes. *Environmental Entomology* 17(6), 964–968.

Wong, T.T.Y. and Ramadan, M.M. (1987) Parasitization of the Mediterranean and oriental fruit flies (Diptera: Tephritidae) in the Kula area of Maui, Hawaii. *Journal of Economic Entomology* 80, 77–80.

Wong, T.T.Y., Mochizuki, N. and Nishimoto, J.I. (1984) Seasonal abundance of parasitoids of the Mediterranean and oriental fruit flies (Diptera: Tephritidae) in the Kula area of Maui, Hawaii. *Environmental Entomology* 13, 140–145.

Wyss, E., Luka, H., Pfiffner, L., Schlatter, C., Uehlinger, G. and Daniel, C. (2005) Approaches to pest management in organic agriculture: a case study in European apple orchards. *Organic Research*, 33–36.

Xiao, Y. and Fadamiro, H.Y. (2010) Exclusion experiments reveal relative contributions of natural enemies to mortality of citrus leafminer, *Phyllocnistis citrella* (Lepidoptera: Gracillariidae) in Alabama satsuma orchards. *Biological Control* 54(3), 189–196.

Xiao, Y.F., Qureshi, J.A. and Stansly, P.A. (2007) Contribution of predation and parasitism to mortality of citrus leafminer, *Phyllocnistis citrella* Stainton (Lepidoptera: Gracillariidae), populations in Florida. *Biological Control* 40(3), 396–404.

Yokomi, R.K. (1995) Why the concern about spread of brown citrus aphids into new citrus areas? In: Lee, R., Roca-PeAa, M., Niblett, C.L., Ocoa, F., Garnsey,S.M., Yokomi, R.K. and Lastra R (eds) *Citrus Tristeza Virus and the Brown Citrus Aphid in the Caribbean Basin Management Strategies*. Proceedings of the Third International Workshop, University of Florida, Lake Alfred, Florida, pp. 27–31.

Zehnder, G., Gurr, G.M., Kuhne, S., Wade, M.R., Wratten, S.D. and Wyss, E. (2007) Arthropod pest management in organic crops. *Annual Review of Entomolology* 52, 57–80. Available at: http://www.colostate.edu/programs/northernipm/pdf/organicIPM.pdf (accessed 3 November 2016).

Zhou, C.A., Zou, J.J. and Huang, S.D. (1994) The effects of *Ageratum interplanting* in hilly citrus orchards in Hunan Province on citrus mite and insect populations. *Acta Phytophylactica Sinica* 21(1), 57–61.

Zolfagharieh, H., VafaeiShoushtari, R., Farazmand, H., Ardakani, M.R., Babaii, M. and Mostafavi, H. (2009) Application of nuclear technique for determination controlling dose of pomegranate fruit moth, *Ectomyelois ceratoniae* Zeller (Lep.: Pyralidae). *Journal of Entomological Research* 1(1), 35–42.

5 Pest Management in Organic Apple, Pear and Stone Fruit

Claudia Daniel,* Silvia Matray, Sibylle Stoeckli and Urs Niggli
Research Institute of Organic Agriculture, Forschungsinstitut für biologischen Landbau (FiBL), Frick, Switzerland

Introduction

Holistic approaches to pest management that aim at maximizing self-regulation and resilience of orchards are a key goal in organic farming (see Chapter 2, this volume). Pest management starts even before planting an orchard by site selection, orchard layout, planting systems (tree densities and pruning system), choice of cultivar and rootstock as well as cultivation techniques. Cultivation techniques and measures applied for disease control can also influence the dynamics of pest insects within orchards and need to be included in a holistic system view. In addition, the use of flowering strips to enhance natural enemies is a field of intense research. Direct control methods using biocontrol organisms or bioinsecticides are available for many pest insects. However, these methods can have side effects on beneficial arthropods and thus destabilize the self-regulating system. Therefore, selective methods combined with specific prevention strategies should be preferred and use of non-selective biopesticides should be limited to a minimum. This chapter describes currently applied and possible future strategies and methods for pest control in organic apple, pear and stone fruit production. Strategies and methods for control of most important pests are summarized in Tables 5.1 and 5.2 at the end of the chapter.

Influence of Cultivar and Rootstock Choice on Pest Control

The choice of cultivar is an important element for pest and disease management in organic farming. However, cultivar choice is strongly driven by market demands, adaptation to the local soil and climatic environment and disease resistance. At present, genetically anchored insect resistance or tolerance is rarely a selection criterion in the worldwide fruit breeding programmes and consequently there are many gaps in knowledge as well as a lack of insect-resistant cultivars. Insect resistance of available apple cultivars was first described for woolly apple aphid, *Eriosoma lanigerum* (Hausmann) (Underhill and Cox, 1938) and was subsequently investigated in several field experiments for pests such as: (i) aphids (*Dysaphis* sp. and *Aphis pomi* De Geer) (Habekuß *et al.*, 2000; Qubbaj *et al.*, 2005; Stoeckli *et al.*, 2011); (ii) mites (Habekuß

* Corresponding author: claudia.daniel@fibl.org

et al., 2000); (iii) sawflies *Hoplocampa* sp. (Tamosiunas *et al.*, 2013); (iv) codling moth, *Cydia pomonella* (Linnaeus); (v) oriental fruit moth, *Grapholita molesta* Busck; (vi) apple maggot, *Rhagoletis pomonella* (Walsk) (Hogmire and Miller, 2005; Myers *et al.*, 2006); and (vii) apple blossom weevil, *Anthonomus pomorum* Linnaeus (Kalinová *et al.*, 2000). Recently, more emphasis was placed on the mechanisms of insect resistance: chemical cues from the plant surface (Lombarkia and Derridj, 2008), fruit firmness (Stoeckli *et al.*, 2011) as well as plant-surface characteristics influencing egg adhesion (Al Bitar *et al.*, 2014) were found to influence infestation with *C. pomonella*. Information on pest resistance in pear and stone fruit cultivars is scarce and limited to aphids, such as *Myzus persicae* (Sulzer) (Staudt *et al.*, 2010), *Myzus cerasi* (Fabricius) (Arnaudov and Kolev, 2009) and *Dysaphis piri* (Boyer de Fonscolombe) (Evans *et al.*, 2008), as well as *Cacopsylla pyri* Linnaeus (Civolani *et al.*, 2013; Emami *et al.*, 2014).

Cultivar choice may not only influence pest insects directly, but also indirectly by influencing parasitization and predation (Bottrell *et al.*, 1998; Cortesero *et al.*, 2000). Parasitization rates of *A. pomorum* have been shown to differ consistently among apple cultivars (Mody *et al.*, 2011). An impact of leaf structures on predatory mites has been shown by Schmidt (2014): leaf hairs can influence ease of prey capture, alter abiotic conditions or increase pollen capture for use as a food source. In addition, plants can play an active role by mediating many interactions between pests and parasitoids: odours released by attacked plants are important cues for parasitoids and predators to locate their host/prey (Degen *et al.*, 2012; Peñaflor and Bento, 2013). There is still a huge gap in knowledge concerning the impacts of these effects in orchard systems.

Only very few references are available on the influence of rootstocks on pest populations, such as *E. lanigerum* (Sandanayaka *et al.*, 2003), *Synanthedon myopaeformis* (Borkhausen) (Ateyyat, 2006) or root-knot nematodes (*Meloidogyne* sp.) (Ye *et al.*, 2009). Rootstock research mainly focuses on growth regulation. However, as rootstocks can influence plant physiology, mineral and water uptake, as well as hormonal control within the tree (Hrotkó, 2007), certain effects on pest and disease infestations can be expected and should be further investigated.

Influence of Cultivation Techniques on Pest Control

Cultural practices (e.g. pruning, thinning, fertilization and irrigation) need to be adapted to the cultivar, rootstock, soil properties and weed competition. **Pruning systems** such as centrifugal training or extinction pruning improve light penetration into the tree canopy (Lauri *et al.*, 2009). Only few studies on the effects of pruning on pest control have been conducted. Simon *et al.* (2006) observed a significant decrease of *Dysaphis plantaginea* (Passerini) and *Panonychus ulmi* (Koch) in centrifugal-**B27B41** trained trees. They assumed that the longer distances between shoots in centrifugal-trained trees or the reduced availability of suitable shoots were the main reasons for this observation (Simon *et al.*, 2012). Tree training and pruning might therefore be an additional element contributing to crop protection in organic orchards. However, decreased branching densities can also have negative effects on pest regulation: due to higher light penetration and consequently increased temperatures within the tree canopies, development of *C. pomonella* larvae can be favoured (Kührt *et al.*, 2006). Vool *et al.* (2014) observed that increased light penetration into old high-stem apple tree canopies led to an increased leaf chlorophyll content and correspondingly higher abundances of sucking pests, such as *Psylla mali* (Schmidberger) and aphids.

Tree-pruning strategies go hand in hand with **fertilization strategies**, which can also influence pest levels (Grechi *et al.*, 2010). Nitrogen concentration in host plants is known to have an impact on sucking insects (Douglas, 1993), such as aphids (Kytö *et al.*, 1996; Sauge *et al.*, 2010) and mites (*P. ulmi*) (Papp *et al.*, 2001). However, nitrogen content is influenced not only by the level of fertilization, but also by the type of fertilizer, mineralization within the soil, soil microorganisms and nitrogen uptake by plants. Haltrich *et al.*

(2000) observed that *A. pomi*, *D. plantaginea* and *Dysaphis devecta* (Walker) showed different reactions depending on the nitrogen source (urea or ammonium nitrate – applied at the same nitrogen levels). The use of organic fertilizers was shown to lower the number of pest insects on crops compared with synthetic sources of fertilizer (Eigenbrode and Pimentel, 1988).

Blossom thinning is necessary to increase fruit size in apple and stone fruit and to break biannual bearing (Weibel *et al.*, 2012). Mechanical rope-thinners used in apple and peach (Reighard and Henderson, 2012) induce physiological stress: shortage of assimilates leads to the abortion of fruitlets (Weibel *et al.*, 2008). No studies are available on the effects of thinning on pest insects. However, due to the impact on assimilate distribution, an effect on pest insects (e.g. aphids which form first colonies during the blossom period) is very likely. As manual thinning is very labour intensive, pest insects which selectively destroy flowers such as *A. pomorum* in apple or *Argyresthia pruniella* Clerck in cherry are welcome to a certain level and are only controlled if they reach very high population densities. Hand thinning of small fruitlets mainly aims at obtaining a suitable fruit size, but also at destroying infested and damaged fruit. If fruitlets infested with young larvae of *Hoplocampa testudinea* (Klug) are dropped to the soil, the larvae cannot complete their development and damage to neighbouring fruitlets is prevented. In addition, apples showing signs of damage by *D. plantaginea* or by *Operophtera brumata* Linnaeus are removed. Thus, a certain level of damage by these pests is tolerable, because apple trees usually produce an excess of fruit which needs to be removed at thinning.

In many European apple production areas, the use of **anti-hail nets** is a prerequisite to reduce the risk of hailstorms injuring the fruits. Despite the wide use of anti-hail nets (mesh size of 3 × 10 mm), knowledge on the effects of the nets on orchard insects is scarce. Graf *et al.* (1999) observed that anti-hail nets form a physical barrier for immigration of *C. pomonella* into orchards, whereas emigration was not influenced. In addition, anti-hail nets enhance the efficacy of mating disruption by pheromones due to keeping the odour cloud in a defined area and by minimizing border effects (Graf *et al.*, 1999). Besides this effect, Tasin *et al.* (2008) observed that anti-hail nets have a disruptive effect on *C. pomonella* mating due to a decreased number of males being able to locate a calling female. Consequently, fewer injured fruit were observed in net-covered versus uncovered plots. Moreover, anti-hail nets can influence prey–predator interactions: experiments in Frick in a pesticide-free apple orchard equipped with boxes for bats showed that bats use the rooftops of the anti-hail nets for hunting during dusk, the period of highest flight activity of *C. pomonella*.

Waterproof crop covering to prevent fruit cracking and splitting caused by rain is used in cherry production. In conventional farming, these covers are installed a few weeks before harvest, whereas in organic farming the covers are already in place during flowering in order to reduce damage by diseases (Børve and Stensvand, 2003). Combining these covers with protective nets against birds or pest insects was therefore an appealing idea. Nets with a mesh size below 1.3 mm were shown to provide good control against *Rhagoletis cerasi* Linnaeus (Balmer, 2005; Ughini *et al.*, 2010). However, observations in a protected cherry orchard in Frick indicate that installation of netting also excludes aphid antagonists which leads to increased densities and damage of *M. cerasi* during summer. Similar observations were made in apple trees covered with anti-insect nets against *C. pomonella*: nets had a direct negative impact on *D. plantaginea* control by excluding natural enemies, especially Coccinellidae and Syrphidae (Dib *et al.*, 2010).

Other cultural practices, such as **irrigation** or **weed management**, can also influence pest incidence. Overhead watering has been shown to reduce damage by *C. pomonella* by up to 90% (Knight, 1998), but it might increase the incidence of apple diseases. Weed management can be achieved by tilling, flame or hot-steam treatments, the use of mulches of organic material or mulch layers of synthetic fibres or plastic (Niggli *et al.*, 1990). Mulches (organic or synthetic) as well as coverage by weeds result in lower soil temperatures and higher soil moisture levels (Mathews *et al.*, 2002), which can influence organisms at different trophic levels. However, different taxonomic groups react differently to particular treatments: mulching led to a reduced abundance of carabid beetles, but increased staphylinid catches (Miñarro *et al.*, 2009). Brown and Tworkoski (2004) described that compost mulches reduced weed growth for a year after application and affected arthropod abundance during 2 years after application: Higher densities of predatory insects led to reduced incidence of *Phyllonorycter blancardella* (Fabricius) and migrating nymphs of *E. lanigerum* (Brown and Tworkoski, 2004). Higher levels of soil moisture can increase the efficacy of entomopathogenic nematodes (Shapiro-Ilan *et al.*, 2006) in the upper centimetres of soil: the efficacy of *Steinernema* sp. against *C. pomonella* has been found to be higher in mulched compared with unmulched plots (Lacey *et al.*, 2006). Entomopathogenic fungi are also affected by interactions between soil moisture and temperature (Jaronski, 2010): higher moisture has been shown to influence the efficacy of *Metarhizium anisopliae* (Metchnikoff) Sorokin against puparia of *Ceratitis capitata* Wiedemann (Ekesi *et al.*, 2003). In addition, many pests spend at least part of their life cycle in the soil. Thus, weed management can influence pest incidence. A dense rooting system of centipede grass under peach trees has been shown to have a negative impact on pupation and emergence of *Conotrachelus nenuphar* (Herbst) whereas bare soil was most favourable for its development (Akotsen-Mensah *et al.*, 2012). *Rhagoletis* species, such as the apple maggot *R. pomonella*, the cherry fruit flies *R. cerasi* and *Rhagoletis indifferens* Curran, spend about 10 months each year pupated in the soil directly under the tree canopy. Soil moisture can influence emergence of these pests (Yee, 2013). Moreover, soil covering was shown to prevent emergence of *R. cerasi* and reduce infestation by over 90% (Daniel and Baker, 2013). There is still a major gap in knowledge on how different weed management strategies influence orchard pests. However, not only pest insects, but also many beneficials spend part of their life cycle in the soil. The European earwig, *Forficula auricularia* Linnaeus – an important natural enemy of a wide range of insect pests, such as *E. lanigerum*, *C. pomonella*, *C. pyri* or *D. plantaginea* – hibernates and oviposits in the underground. Badly timed soil tillage which destroys underground nests of hibernating adult earwigs in autumn or early spring can reduce the number of earwig nymphs in spring and summer by 50% (Moerkens *et al.*, 2012).

Conservation Biocontrol: Influence of Landscape and Orchard Design

Conservation biocontrol aims at enhancing the efficacy of natural enemies in orchard production systems by providing alternative food sources and shelter habitats, as well as by reducing negative impacts of management practices. Many measures can be established within the orchard. However, the **surrounding landscape** also needs to be taken into account. Vicinity to forests can increase incidence of pests such as *A. pomorum* (Toepfer *et al.*, 1999). Parasitoids (*Scambus pomorum* (Ratzeburg)) of *A. pomorum* are also enhanced by forests and can stabilize pest populations at tolerable levels (Zijp and Blommers, 2002; Mody *et al.*, 2011). *Drosophila suzukii* Matsumura and *R. cerasi* are also observed in higher densities close to forest borders. For *D. suzukii* this is mainly due to alternative host plants within the forest, for *R. cerasi* it might be a combination of alternative hosts and reduced wind speed close to the forest. In diverse, mosaic landscapes, a migration of beneficials between orchards and the surrounding landscape can be observed, thus mitigating negative effects of cultivation measures on orchard diversity. The impact of ecosystem service and landscape diversity on pest control was shown by Stutz and Entling (2011): suppression of *M. cerasi* was reduced by habitat isolation. In a monitoring of cherry tree arthropods, Schüepp *et al.* (2014) observed that most predatory arthropods declined with habitat isolation, whereas herbivores either increased with isolation or showed no significant response.

Wind breaks planted around orchards in windy areas also influence pest control (Maalouly *et al.*, 2013), especially if multispecies hedgerows (see next paragraph) are used. Ricci *et al.* (2011) observed that the number of *C. pomonella* larvae was significantly lower in areas shaded and protected from the wind by hedgerows. In wind-protected areas, *C. pyri* was observed in lower densities due to higher presence of antagonists (Debras *et al.*, 2008).

Positive effects of landscape parameters can be further increased by **orchard habitat management** to foster conservation biocontrol (Palm *et al.*, 2014). Functional agrobiodiversity in orchards has been a topic of intense research in recent years. A lot of work has been done in apple orchards (Landis *et al.*, 2000; Simon *et al.*, 2010). Little knowledge is available for peach (Penvern *et al.*, 2010; Sautereau *et al.*, 2013) and no published data are available for pear, cherry and other stone fruit. With the establishment of natural and semi-natural habitats within orchards, beneficial organisms can be actively promoted. Several measures are suggested: (i) diverse hedges around and within orchards; (ii) flowering strips at the orchard borders as well as within the alleyways in combination with adapted mowing and mulching systems for these flower strips; (iii) flowering understory within tree rows; (iv) nesting boxes for birds, bats and wild bees; (v) a pile of stones as shelters for lizards and other small reptiles and mammals; and (vi) overwintering boxes for Chrysopidae (Simon *et al.*, 2010; Weibel *et al.*, 2010). The created habitats need to provide shelters for resting, diapausing or hibernating, as well as refuge areas to escape disruptive agricultural practices. In addition, food (e.g. alternate preys and hosts), nectar and pollen, as well as reproduction sites are needed. There is a vast amount of literature on the impacts of flowering plants and hedgerows on orchard biodiversity. Many authors showed an increase in predator and parasitoid abundance by adding alternate food resources to orchard systems (Wyss, 1995, 1996; Zijp and Blommers, 2002; Bribosia *et al.*, 2005; Debras *et al.*, 2008; Miñarro and Prida, 2013). However, clear causal relationships for pest control were rarely demonstrated. Several studies are summarized in a comprehensive literature review (Simon *et al.*, 2010): in most studies (16), biodiversity management had a positive effect on pest control, in some studies (9) effects were negligible and some studies (5) showed negative effects. In addition, in most experimental work only single interaction effects were tested. Data quantifying the additive effects of multiple measures on system biodiversity and finally demonstrating an impact on yield and on the production economy are lacking.

Only few studies are available investigating the interactions between different predator and parasitoid species. Dib *et al.* (2011) showed that the aphid antagonists *F. auricularia* and *Episyrphus balteatus* (De Geer) did not negatively interact but had an additive effect on control *of D. plantaginea.* Similar observations were made by Wyss *et al.* (1999) for the aphid predators *Adalia bipunctata* (Linnaeus) and *E. balteatus.* However, buckwheat flowers led to an increase in captures of *Anacharis* sp., a parasitoid of the brown lacewing (*Micromus tasmaniae*), itself a beneficial insect (Stephens *et al.*, 1998). The mix of positive, negative and neutral effects of enemy diversity is caused by niche complementarity, intraguild predation and functional redundancy (Straub *et al.*, 2008).

Multi-species hedgerows around orchards can not only act as windbreaks but also influence arthropod communities in orchards (Miñarro and Prida, 2013). Several factors need to be taken into account when designing a hedgerow around orchards (Simon *et al.*, 2010): species which can be hosts for important orchard pests or diseases need to be avoided. Flowers, nectar, alternative prey/hosts and shelter for antagonists need to be present in a year-long succession. Selected plant species also need to be adapted to local climates. Therefore, no general recommendation for plant species can be given. Positive effects of hedgerows have been observed on pollinators (Miñarro and Prida, 2013) and on natural enemies of aphids and psyllids (Bribosia *et al.*, 2005; Debras *et al.*, 2008; Miñarro and Prida, 2013). Predatory mites are enhanced by increased pollen availability (Duso *et al.*, 2004). Hymenopteran parasitoids of *A. pomorum* are enhanced by the availability of secondary hosts in the hedgerows (Zijp and Blommers, 2002). In addition, hedgerows can act as a buffer mitigating negative impacts of pesticide use within the orchards by fostering recolonization of apple trees with predatory mites (Tuovinen, 1994) or spiders.

Flowering strips in the alleyways of orchards were also shown to have a positive effect on pest control. The beneficial aphidiphagous complex (spiders, predaceous Heteroptera, Coccinellidae and Chrysopidae) of the apple orchard is favoured by flowering strips, to the benefit of aphid control (Wyss, 1995, 1996). Antagonists of mites (Yan *et al.*, 1997), *E. lanigerum* (Gontijo *et al.*, 2013) and leafrollers (Stephens *et al.*, 1998) led to more stable ecosystems and kept densities below economic levels. Flowering resources for orchard habitats need to be carefully selected (Gontijo *et al.*, 2013). In order to have a positive impact on pest control, flowers must provide resources which are lacking in the orchard environment (e.g. if aphid honeydew or nectar is already available, additional flowers might not have an additional impact). In addition, nectar needs to be accessible for natural enemies but not for pest insects (Géneau *et al.*, 2012) and natural enemies need to move between flowers and crops instead of simply aggregating in the flowers (Landis *et al.*, 2000). Again, no general recommendation for plant species can be given, because flower species need to be adapted to local soil and climatic conditions. Negative impacts of flowering strips were observed in some cases when spider mites migrated into the trees after mulching of flowering strips. In addition, temporary destruction of the vegetation by mulching or mowing also deprives beneficial invertebrates of their shelter and alternative food sources (Horton *et al.*, 2003). Nevertheless, mulching between rows is needed in order to keep the orchards accessible to machinery for harvesting and applying plant protection products, and also in order to avoid competition for water and nutrients between apple trees and flowering plants (Brown and Glenn, 1999). In order to limit the collateral damage, alternate mulching of, for example, every second row at a time should be used.

Adapted **understorey management** can create another niche for beneficial arthropods. In most orchards, mechanical tillage is still used beneath fruit trees. However, the 'Swiss sandwich system' (Stefanelli *et al.*, 2009) allows for flowering plants underneath the trees: tree strips are not tilled over the whole surface but are managed as a sandwich system. A central strip of 0.3–0.5 m width with flowering cover plants (e.g. *Hieracium pilosella* Linnaeus or *Potentilla reptans* Linnaeus) is established, while only the two outer adjacent strips of 0.4–0.5 m width each are tilled. This system allows both soft soil management and an additional surface with flowering plants providing habitat for beneficials. Another strategy to increase diversity in orchards is interplanting of different fruits: interplanting of peach into apple orchards has been shown to reduce apple fruit injury from San Jose scale, *Quadraspidiotus perniciosus* (Comstock) and stink bugs (Pentatomidae) (Brown *et al.*, 2010).

Diverse systems can suppress pests in many ways: (i) by altered microclimatic conditions; (ii) by making host localization more difficult for pest insects due to quantitative barriers, reduced visual or semiochemical appearance; and (iii) by maintaining higher diversity of natural enemies, as well as increasing their longevity and fecundity (Simon *et al.*, 2010). Thus, conservation biocontrol can contribute to pest control provided that natural enemies are not negatively influenced by cultural practices such as applying an excess of pesticides.

Direct Control of Pest Insects

The use of biopesticides in organic farming is debated passionately: in theory, the need to use direct control measures is a sign of an unstable agroecosystem and thus undesirable. However, often there is an obvious necessity of direct control in order to meet market demands in quality and quantity. However, the use of direct control measures should be limited to the minimum and, whenever possible, selective methods for pest control (see Tables 5.1 and 5.2) should be preferred in order not to further destabilize the ecosystem. Insecticides allowed in organic production are of natural origin. Nevertheless, they can have side effects on beneficial arthropods (Jansen *et al.*, 2010). Especially spinosad – a natural metabolite of an actinomycete – with its broad-spectrum activity negatively affects parasitic hymenoptera, predatory bugs, syrphids and many other beneficial insects (Biondi *et al.*, 2012). In apple orchards, an increase of *E. lanigerum* can be observed after treatments with spinosad. This observation is most likely due to the side effects of spinosad on the main parasitoid *Aphelinus mali* Haldeman (Heunis and Pringle, 2003). As *E. lanigerum* is very difficult to control with available organic pesticides, the use of spinosad might solve one problem while creating another problem. The use of spinosad should therefore only be a last option if no other means for control are available. Most other insecticides used in organic production, such as pyrethrin, quassia, rotenone or neem, are plant extracts. Their advantage lies in their lack of persistence and bioaccumulation in the environment, because they generally degrade faster in sunlight, air and moisture than synthetic products (Grdiša and Gršić, 2013). In addition, they are usually more selective to non-target insects than conventional products (Grdiša and Gršić, 2013). Side effects are described for pyrethrin (Jansen *et al.*, 2010) and rotenone, whereas quassia and neem are harmless to most beneficial insects (El-Wakeil *et al.*, 2006). Another product widely used for pest control in organic farming is the clay mineral kaolin, which we consider to be more selective than spinosad, pyrethrin and rotenone, but less selective than neem or quassia. Only very few reports on side effects of kaolin on beneficial arthropods are available (Markó *et al.*, 2008). Soaps used for pest control are considered to have only a very small impact on beneficial insects whereas oil products can be detrimental. However, most oil products are only used for dormant sprays in very early spring when most beneficial insects are still hibernating. Highly selective methods, such as mating disruption by pheromones, granulosis virus and *Bacillus thuringiensis* Berliner (*Bt*), are available for most major orchard pests. In addition, biocontrol agents such as entomopathogenic fungi (Zimmermann, 2007), entomopathogenic nematodes, parasitoids (e.g. *Trichogramma* and *A. mali*), and predators (*Anthocoris*) can be used with little risk of side effects.

Fruit infesting pests: most detrimental damage in orchards is caused by pest insects with larvae developing in the ripening fruit, such as the Lepidoptera (*C. pomonella*, *Grapholita* sp., *Anarsia lineatella* (Zeller)), flies (*Rhagoletis* sp., *C. capitata*, *D. suzukii*) or plum curculio (*C. nenuphar*). Tolerable thresholds for these pests are very low, as they directly damage the marketable product. Therefore parasitoids and predators are usually not sufficiently effective in controlling these pests. However, for most lepidopteran pests, highly selective methods for control are available.

The main pest, *C. pomonella* (Pajač *et al.*, 2011), can be controlled with good success by mating disruption (Judd *et al.*, 1997; Angeli *et al.*, 2007) and granulosis virus. However, both methods are only efficient at low-to-medium pest densities and mating disruption is only successful if applied to large, homogeneous areas. In most European countries a combination of mating disruption and granulosis virus has to be applied for successful codling moth control (with the exception of South Tyrol, with 75% mating disruption). Climate change will alter codling moth phenological events and adaptations of plant protection strategies will be required (Stoeckli *et al.*, 2012): a combination of different management methods will be necessary due to higher pest densities under future climatic conditions. In addition, the shapes and sizes of organic apple orchards will have to be adapted to promote mating disruption. In areas of high codling moth pressure, mating disruption and granulosis virus may be combined with the application of entomopathogenic nematodes (Lacey and Shapiro-Ilan, 2008) or *Trichogramma* egg parasitoids (Zimmermann, 2004). In the south of France, exclusion nets are widely used to prevent damage in organic apple orchards (Dib *et al.*, 2010). Bagging of individual fruit (Grasswitz and Fimbres, 2013) seems only an option for small-scale production because of the high labour input needed. Thus, there is usually no need to apply non-selective products for codling moth control. Similar methods as for codling moth control can be applied to control most other fruit-infesting Lepidoptera (see Tables 5.1 and 5.2).

The release of sterile males (sterile insect technique, SIT) was developed for

C. pomonella: mass-reared males are sterilized by gamma radiation and released in high numbers. After the pairing with a sterile male, wild females are unable to produce viable offspring which leads to a reduction in pest population (Vreysen *et al.*, 2010). Apart from biological and technical challenges (e.g. mass rearing at reasonable costs; fitness of released males; sexing of released insects), there are also social challenges thwarting such programmes: SIT needs area-wide collaboration among farmers in order to deliver good results. A few uncooperative farmers can ruin the whole project. In addition, SIT is a fairly cost-intensive technology and funding arrangements acceptable to all stakeholders (small-scale and large-scale farmers, home gardeners) need to be found. After more than 30 years of research, SIT was brought into practice in a very ambitious, multimillion dollar programme in Canada (the OKSIR programme – Okanagan-Kootenay Sterile Insect Release Program): it started in 1992, first releases of sterile males were made in 1994. Currently, 8000 ha are covered (Vreysen *et al.*, 2010). The use of SIT has also been developed for *C. capitata* and has been applied in several countries (Hendrichs *et al.*, 2003; Enkerlin, 2005). For many highly mobile pest insect species (e.g. *D. suzukii*) area-wide approaches might be the only option to achieve sustainable control without broad-spectrum insecticides. As *D. suzukii* was only recently introduced to Europe and America (Hauser, 2011; Cini *et al.*, 2014; Deprá *et al.*, 2014), efficient control measures are still lacking. Crop sanitation (e.g. early and complete harvest, removal of infested fruit) and baited traps are currently the only control options (Landolt *et al.*, 2012; Basoalto *et al.*, 2013; Iglesias *et al.*, 2014). In order to achieve good performance of trapping strategies, landscape parameters (wild hosts, neighbouring crops) need to be taken into account.

Tephritid fruit flies are also among the most damaging orchard pests. Polyvoltine species (*C. capitata*) occur mainly in warmer climates; whereas univoltine species (*R. pomonella, R. cerasi, Rhagoletis cingulata* (Loew), *R. indifferens, Rhagoletis fausta* (Osten Sacken)) can cause damage in colder regions. They are most efficiently controlled by bait-formulated spinosad products (Pelz-Stelinski *et al.*, 2006; Yee, 2007; Reekie *et al.*, 2010): combining spinosad with a protein-based food bait attracts the flies and stimulates feeding which ensures insecticide uptake. At the same time, negative side effects are reduced, because application rates can be considerably reduced and whole tree canopies do not need to be covered in insecticide. Another approach is the use of physical barriers on the fruit surface to prevent oviposition: kaolin is used in apple production (Caleca *et al.*, 2010; Leskey *et al.*, 2010), but because of white residues on the fruit surface, it is not an option for fresh market production of stone fruits. Oil products might have a similar effect (Daniel, 2014); however, products at the optimal formulation are not yet available for this purpose. Entomopathogenic fungi have been shown to have good efficacy against various tephritid species (Castillo *et al.*, 2000; Daniel and Wyss, 2009; Daniel and Wyss, 2010; Garrido-Jurado *et al.*, 2011; Imoulan and Elmeziane, 2014); a commercial product based on *Beauveria bassiana* (Balsamo) Vuill. is registered for cherry fruit fly control in several European countries (Daniel and Grunder, 2012).

Leaf-damaging insects can usually be tolerated in higher numbers, because they do not directly damage the fruit. Applications of *Bt* provide sufficient control for most leaf-damaging lepidopteran larvae (Pinnock and Milstead, 1978; Smirle *et al.*, 2003). Aphid species, *E. lanigerum* and mites are often sufficiently controlled by naturally occurring antagonists as long as the antagonists are not impaired by the use of broad-spectrum pesticides. For *E. lanigerum* a release of the parasitoid *A. mali* can provide efficient control (Ateyyat *et al.*, 2001). However, applications of neem against *D. plantaginea* in early spring are standard in most organic apple orchards. The main driver for these applications is the low economic threshold of one to two aphids per 100 flower clusters (Blommers, 1994) which suggests that *D. plantaginea* is extremely harmful. Thus farmers are afraid to leave out an application. Surprisingly, this low threshold is not based on sound experimental data: the relationship between aphid densities and yield reduction is unknown (Whalon and Croft, 1984). The same is true for *E. lanigerum* (Gontijo *et al.*, 2013). Ten years ago, Hemptinne *et al.* (2003) already recognized the low threshold as a factor hampering new strategies of aphid control, but currently this threshold is still widely used. In addition, the presence of antagonists is not taken into account in this economic threshold. With high numbers of predators present in early spring, tenfold higher aphid densities could be tolerated without substantial yield loss. Thus, adapted thresholds, which take the presence of antagonists into account, are needed to encourage the farmers to reduce insecticide applications. If necessary, neem products provide sufficient control against most aphid species. Care should be taken in pear production as some varieties show phytotoxic reactions after neem applications. *Aphis* species which are not sensitive to neem applications (*A. pomi*) can be controlled by applications of pyrethrin. Most applications against aphids are conducted in early spring, focusing on aphid fundatrices. For species such as *M. cerasi*, for which spring applications do not provide sufficient control, autumn application against migrating forms might be worth testing. So far, autumn applications have only been tested against *D. plantaginea* (Wyss and Daniel, 2004; Bürgel *et al.*, 2005). Good effects of autumn applications were also shown against pear leaf blister mite (*Eriophyes pyri* (Pagenstecher)) (Daniel *et al.*, 2007) – a pest which is nearly impossible to control in spring.

Wood-infesting pest species, such as *Synanthedon* sp., *Zeuzera* sp. or *Xyleborus dispar* (Fabricius), are also threatening to farmers, because their damage might kill the whole tree. Damage can be reduced by preventive measures aimed at avoiding bark damage. *Synanthedon* sp. can be controlled by mating disruption or by mass trapping using juice baits (Audemard, 1988; Teixeira *et al.*, 2010; Marius Aurelian *et al.*, 2012). Ethanol-baited traps are suitable for mass trapping of *X. dispar* (Speranza *et al.*, 2009).

In conclusion, many direct control measures are available for organic fruit production. Recent literature shows that the use of biocontrol agents, such as nematodes or entomopathogenic fungi, is still a field of intense research. However, the availability to farmers of different products also depends on regional commercialization of products as well as on national registration: hurdles to registration of novel crop protection products make it difficult for companies to develop and commercialize products for minor crops in the relatively small organic sector (Tamm *et al.*, 2004).

Future Prospects and Research Needs

In order to reduce the need for direct control of pest insects, a redesign of orchard systems is vital. We need to rethink the whole cropping system in order to optimize ecosystem services. In a true holistic approach, the impacts of cultural practices, effects of surrounding landscape and of within-orchard biodiversity as well as side effects of pesticide applications all need to be taken into account. In addition, plants can play an active role by mediating many of the interactions between pests and parasitoids: odours released by attacked plants are important cues for parasitoids and predators to locate their host/prey. However, integrating single-factor laboratory research (e.g. plants 'calling' for parasitoids) into a multi-factorial agricultural system, and understanding interactions between different factors, is challenging. There are still huge gaps in knowledge on the efficacy of different measures as well as on the economic benefits of such interventions. Farmers, advisers and researchers need to take a holistic view with a system perspective rather than simple reductionist approaches. In particular, the economic thresholds for treatment decisions need to be adjusted: the presence of antagonists should be taken into account in order to encourage farmers to reduce insecticide applications.

Acknowledgements

Many thanks to Franco Weibel, David Granatstein, Hans-Jakob Schärer, Carolin Möller, Lucius Tamm, Monika Messmer, Andi Häseli, Ignazio Giordano and Beatrice Steinemann for help and comments on the manuscript.

Table 5.1. Main pests of apple and pear and strategies for control.

Pest	Host	Prevention	Antagonists	Selective control	Non-selective control
Codling moth (*Cydia pomonella*)	Apple, pear, quince, apricot	Conservation biocontrol	Birds, bats, parasitoids, predators (ants, carabids, spiders)	Granulosis virus,[a] mating disruption,[b] sterile insect technique,[c] nematodes,[d] *Trichogramma*,[e] fruit bagging[f]	Spinosad,[g] kaolin,[h] exclusion nets[i]
Other fruit-infesting Lepidoptera (*Grapholita lobarzewskii*, *Pammene rhediella*, *Adoxophyes orana*, *Pandemis* sp.)	Apple (pear)	Conservation biocontrol	Birds, bats, parasitoids, predators (ants, carabids, spiders)	*Bt*,[j] *G. lobarzewskii*: mating disruption,[k] *A. orana*: granulosis virus	Spinosad,[l] kaolin,[h] exclusion nets
Apple maggot (*Rhagoletis pomonella*)	Apple	Cultivar choice,[m] removal of unmanaged host trees, destruction of infested fruit	Carabids, ants, spiders	Mass trapping/perimeter trapping (dark-coloured spherical sticky-traps),[n] fruit bagging[f]	Spinosad bait application,[o] kaolin,[p] exclusion nets
European apple sawfly (*Hoplocampa testudinea*)	Apple	Conservation biocontrol, cultivar choice, removal of infested fruit, chicken keeping	Birds, spiders, carabids, ants, parasitoids[q]	Quassia,[r] nematodes[s]	Kaolin[h]
Apple blossom weevil (*Anthonomus pomorum*)	Apple	Site selection, conservation biocontrol, crop management	Birds, parasitoids		Spinosad,[t] kaolin[h]
Leaf-damaging Lepidoptera (*Choristoneura rosaceana*, *Operophtera brumata*, *Spilonota ocellana*, *Yponomeuta malinellus*)	Apple, cherry, apricot, nectarine, peach, pear, plum/prune	Conservation biocontrol	Birds, predators (ants, carabids, spiders) and parasitoids	*Bt*,[l] neem, *O. brumata*: sticky-bands, *S. ocellana*: mating disruption	Spinosad,[l] pyrethrin, rotenone
Rosy apple aphid (*Dysaphis plantaginea*), *Dysaphis* sp., pear-bedstraw aphid (*Dysaphis pyri*)	Apple (pear)	Cultivar choice, N fertilization, tree training, crop management, conservation biocontrol	Predators (coccinellids, syrphids, lacewings, spiders, earwigs, gall midges)	Neem[u]	Pyrethrin, rotenone, horticultural oils and soaps, kaolin[v]

Continued

Table 5.1. Continued.

Pest	Host	Prevention	Antagonists	Selective control	Non-selective control
Woolly apple aphid (*Eriosoma lanigerum*)	Apple	Broad-spectrum insecticide avoidance, cultivar and rootstock selection, conservation biocontrol, remove infested sprouts or use a strong water jet	Parasitoids (*Aphelinus mali*), earwigs, coccinellids, syrphids, lacewings	Release of *Aphelinus mali*[w]	Pyrethrin, sticky-bands, winter oil treatments
Apple clearwing moth (*Synanthedon myopaeformis*)	Apple, pear, plum, apricot	Crop management, wound prevention/ treatment	Parasitoids	Mating disruption,[x,y] oil treatments on trunk[z]	Mass trapping with apple-juice baits[y]
European shothole borer (*Xyleborus dispar*)	Apple, pear, others	Crop management (healthy trees, undamaged bark), remove infested branches or trees		Mass trapping with ethanol traps,[a*] *Bt*[b*]	
Pear psylla (*Cacopsylla pyri*)	Pear	Remove infested sprouts, conservation biocon- trol, crop management, cultivar choice	Predatory bugs (*Anthocoris* spp.), lacewings, earwigs	Release of anthocorids,[c*] entomopathogenic fungi[d*]	Kaolin,[e*] rotenone, sprinkling, oil, soap
Pear-leaf blister mite (*Eriophyes pyri*)	Pear	Cultivar choice	Predatory mites (*Typhlodromus pyri*)	Release of *T. pyri*[f*]	Autumn sulfur applications,[g*] winter oil treatments

[a](Arthurs *et al.*, 2007; Lacey *et al.*, 2008); [b](Judd *et al.*, 1997; Angeli *et al.*, 2007); [c](Judd and Gardiner, 2005; Vreysen *et al.*, 2010); [d](Lacey and Shapiro-Ilan, 2008); [e](Zimmermann, 2004); [f](Grasswitz and Fimbres, 2013); [g](Arthurs *et al.*, 2007; Delate *et al.*, 2008); [h](Markó *et al.*, 2008); [i](Dib *et al.*, 2010); [j](Smirle *et al.*, 2003), [k](Gambon *et al.*, 2009); [l](Smirle *et al.*, 2003); [m](Hogmire and Miller, 2005); [n](Yee and Goughnour, 2011; Wright *et al.*, 2012); [o](Yee, 2007; Reekie *et al.*, 2010; Wright *et al.*, 2012); [p](Leskey *et al.*, 2010); [q](Babendreier, 1996); [r](Sjöberg *et al.*, 2015); [s](Vincent and Bélair, 1992; Bélair *et al.*, 1998); [t](Daniel *et al.*, 2005b); [u](Weibel *et al.*, 2007); [v](Bürgel *et al.*, 2005); [w](Ateyyat *et al.*, 2001); [x](Audemard, 1988); [y](Marius Aurelian *et al.*, 2012); [z](Erler, 2010); [a*](Speranza *et al.*, 2009); [b*](Sezen *et al.*, 2008); [c*](Sigsgaard *et al.*, 2006; Daugherty *et al.*, 2007); [d*](Erler *et al.*, 2014); [e*](Daniel *et al.*, 2005a); [f*](Prasli\u010dka *et al.*, 2011); [g*](Daniel *et al.*, 2007).

Table 5.2. Main pests of stone fruit and strategies for control.

Pest	Host	Prevention	Antagonists	Selective control	Non-selective control
Spotted wing drosophila (*Drosophila suzukii*)	Stone and soft fruit	Site selection, removal of infested fruit, early and complete harvest	Parasitoids[a]	Mass trapping[b]	Spinosad bait application, exclusion nets, kaolin
Mediteranean fruit fly (*Ceratitis capitata*)	Stone fruit, citrus and many others	Cultivar choice[c]	Parasitoids, predators (ants, carabids, spiders)	Mass trapping/perimeter trapping, attract and kill, entomopathogenic fungi,[d] fruit bagging,[e] sterile insect technique[f]	Spinosad bait application,[g] kaolin[h]
Cherry fruit flies (*Rhagoletis cerasi*, *Rhagoletis cingulata*, *Rhagoletis indifferens*, *Rhagoletis fausta*)	Cherry	Cultivar choice;[i] complete harvest, remove infested fruit, remove wild cherry trees and honeysuckle in the proximity, keep chickens	Parasitoids, ants, carabids, spiders	*Beauveria bassiana*[j]	Nets,[k] sticky-traps[l] combined with baits, oil products,[m] bait application of spinosad[n]/neem,[o] kaolin[p]
Plum curculio (*Conotrachelus nenuphar*)	Apple and stone fruit	Remove fruit dropped in June/July, shake branches, soil/weed management,[q] cultivar choice	Ants, parasitoids	Nematodes,[r] baited traps,[s] *B. bassiana*[t]	Pyrethrin, kaolin[u]
Brown marmorated stinkbug (*Halyomorpha halys*) and other damaging bugs (Pentatomidae, Miridae)	Apple, pear, stone fruit	Ground cover management	Parasitoids, predators (coccinellids, earwigs, spiders)	Fruit bagging[v]	Pyrethrin, kaolin
Fruit-infesting Lepidoptera (*Grapholita funebrana*, *Grapholita molesta*, *Anarsia lineatella*)	Stone fruit	Cultivar choice, conservation biocontrol, *A. lineatella*: remove infested sprouts	Birds, predators (ants, carabids, spiders) and parasitoids	*G. molesta*: mating disruption,[w] entomopathogenic nematodes;[x] *G. funebrana*: *Trichogramma dendrolimi*;[y] *A. lineatella*: *Bt*, fruit bagging,[v] mating disruption[z]	Spinosad, nets, kaolin[a*]

Continued

Table 5.2. Continued.

Pest	Host	Prevention	Antagonists	Selective control	Non-selective control
Leaf-damaging Lepidoptera (*Hyphantria cunea*, *Archips argyrospila*, *Orthosia hibisci*, *Amphipyra pyramidoides*, *Orgyia vetusta*, *Platynota stultana*, *Paleacrita vernata*, *Alsophila pometaria*)	Apple, cherry, apricot, nectarine, peach, pear, plum/prune	Conservation biocontrol, *P. stultana*: remove and destroy fruit mummies; *H. cunea*: cut out infested twigs	Birds, predators (ants, carabids, spiders) and parasitoids	*Bt*,[b*] for Geomantriidae: sticky bands	Spinosad
Aphids (*Myzus cerasi*, *Myzus persicae*, *Brachycaudus helichrysi*, *Brachycaudus persicae*)	Stone fruit	Reduced N fertilization, crop management, cultivar choice, conservation biocontrol	Predators (coccinellids, syrphids, lacewings, spiders, earwigs)	Neem	Pyrethrin, rotenone, soap, oil
Spider mites (*Panonychus ulmi*, *Tetranychus urticae*)	Apple, cherry, apricot, nectarine, peach, pear, plum/prune	Conservation biocontrol, cultivar choice, avoid broad-spectrum insecticides and sulfur, reduced N fertilization	Predatory mites,[c*] and bugs	*B. bassiana*[e*]	Oils, soap
Peach tree borer (*Synanthedon exitiosa*)	Stone fruit	Crop management, wound prevention/treatment	Parasitoids	Mating disruption,[f*] nematodes[g*]	

[a](Chabert *et al.*, 2012); [b](Landolt *et al.*, 2012; Basoalto *et al.*, 2013; Iglesias *et al.*, 2014); [c](Tabilio *et al.*, 2013); [d](Castillo *et al.*, 2000; Garrido-Jurado *et al.*, 2011; Imoulan and Elmeziane, 2014); [e](Pastopoulos *et al.*, 2012); [f](Hendrichs *et al.*, 2003); [g](Braham *et al.*, 2007); [h](Braham *et al.*, 2007; Caleca *et al.*, 2010; Pastopoulos *et al.*, 2012); [i](Stamenković *et al.*, 1996); [j](Daniel and Wyss, 2009, 2010); [k](Daniel and Baker, 2013); [l](Daniel *et al.*, 2014); [m](Daniel, 2014); [n](Pelz-Stelinski *et al.*, 2006; Yee and Alston, 2006); [o](Böckmann *et al.*, 2014); [p](Yee, 2008); [q](Akotsen-Mensah *et al.*, 2012); [r](Shapiro-Ilan *et al.*, 2013); [s](Leskey *et al.*, 2008); [t](Pereault *et al.*, 2009); [u](Lalancette *et al.*, 2005); [v](Grasswitz and Fimbres, 2013); [w](Tollerup *et al.*, 2012; Kong *et al.*, 2014); [x](De Carvalho Barbosa Negrisoli *et al.*, 2013); [y](Zimmermann, 2004); [z](Audemard, 1988; Kutinkova *et al.*, 2012); [a*](Lalancette *et al.*, 2005); [b*](Pinnock and Milstead, 1978); [c*](Jaastad *et al.*, 2007); [e*](Duso *et al.*, 2008); [f*](Teixeira *et al.*, 2010; Grasswitz and Yao, 2013); [g*](Grasswitz and Yao, 2013).

References

Akotsen-Mensah, C., Boozer, R.T. and Fadamiro, H.Y. (2012) Influence of orchard weed management practices on soil dwelling stages of plum curculio, *Conotrachelus nenuphar* (Coleoptera: Curculionidae). *Florida Entomologist* 95, 882–889.

Al Bitar, L., Gorb, S.N., Zebitz, C.P.W. and Voigt, D. (2014) Egg adhesion of the codling moth *Cydia pomonella* L. (Lepidoptera, Tortricidae) to various substrates: II. Fruit surfaces of different apple cultivars. *Arthropod–Plant Interactions* 8, 57–77.

Angeli, G., Anfora, G., Baldessari, M., Germinara, G.S., Rama, F., De Cristofaro, A. and Ioriatti, C. (2007) Mating disruption of codling moth *Cydia pomonella* with high densities of Ecodian sex pheromone dispensers. *Journal of Applied Entomology* 131, 311–318.

Arnaudov, V.A. and Kolev, K.K. (2009) Susceptibility of some introduced sweet cherry cultivars to the attacks of black cherry aphids *Myzus cerasi* Fab. (Homoptera: Aphididae). *Acta Horticulturae* 825, 401–406.

Arthurs, S.P., Lacey, L.A. and Miliczky, E.R. (2007) Evaluation of the codling moth granulovirus and spinosad for codling moth control and impact on non-target species in pear orchards. *Biological Control* 41, 99–109.

Ateyyat, M.A. (2006) Effect of three apple rootstocks on the population of the small red-belted clearwing borer, *Synanthedon myopaeformis*. *Journal of Insect Science* 6, 1–5.

Ateyyat, M.A., Al-Awamleh, M. and El-Osofi, H. (2001) Rearing and release of *Aphelinus mali* (Hald) (Hymenoptera: Aphelinidae), the sole parasitoid of woolly apple aphid *Eriosoma lanigerum* (Hausmann) (Homoptera: Eriosomatidae) on apple orchards in Ash-Shoubak. *Academic Journal of Entomology* 4, 108–113.

Audemard, H. (1988) Confusion sexuelle avec des phéromones en Europe de l'Ouest. *Agriculture, Ecosystems and Environment* 21, 101–110.

Babendreier, D. (1996) Studies on two ichneumonid parasitoids as potential biological control agents of the European apple sawfly *Hoplocampa testudinea* Klug (Hymenoptera: Tenthredinidae). *Acta Horticulturae* 422, 236–240.

Balmer, M. (2005) Kulturschutznetze zur Kontrolle der Kirschfruchtfliege im überdachten Anbau. *Obstbau* 1, 14–16.

Basoalto, E., Hilton, R. and Knight, A. (2013) Factors affecting the efficacy of a vinegar trap for *Drosophila suzukii* (Diptera: Drosophilidae). *Journal of Applied Entomology* 137, 561–570.

Bélair, G., Vincent, C. and Chouinard, G. (1998) Foliar sprays with *Steinernema carpocapse* against early-season apple pests. *Journal of Nematology* 30, 599–606.

Biondi, A., Mommaerts, V., Smagghe, G., Viñuela, E., Zappalà, L. and Desneux, N. (2012) The non-target impact of spinosyns on beneficial arthropods. *Pest Management Science* 68, 1523–1536.

Blommers, L.H.M. (1994) Integrated pest management in European apple orchards. *Annual Review of Entomology* 39, 213–241.

Böckmann, E., Köppler, K., Hummel, E. and Vogt, H. (2014) Bait spray for control of European cherry fruit fly: an appraisal based on semi-field and field studies. *Pest Management Science* 70, 502–509.

Børve, J. and Stensvand, A. (2003) Use of a plastic rain shield reduces fruit decay and need for fungicides in sweet cherry. *Plant Disease* 87, 523–528.

Bottrell, D.G., Barbosa, P. and Gould, F. (1998) Manipulating natural enemies by plant variety selection and modification: a realistic strategy? *Annual Review of Entomology* 43, 347–367.

Braham, M., Pasqualini, E. and Ncira, N. (2007) Efficacy of kaolin, spinosad and malathion against *Ceratitis capitata* in citrus orchards. *Bulletin of Insectology* 60, 39–47.

Bribosia, E., Bylemans, D., Migon, M. and Van Impe, G. (2005) In-field production of parasitoids of *Dysaphis plantaginea* by using the rowan aphid *Dysaphis sorbi* as substitute host. *BioControl* 50, 601–610.

Brown, M.W. and Glenn, D.M. (1999) Ground cover plants and selective insecticides as pest management tools in apple orchards. *Journal of Economic Entomology* 92, 899–905.

Brown, M.W. and Tworkoski, T. (2004) Pest management benefits of compost mulch in apple orchards. *Agriculture, Ecosystems and Environment* 103, 465–472.

Brown, M.W., Mathews, C.R. and Krawczyk, G. (2010) Extrafloral nectar in an apple ecosystem to enhance biological control. *Journal of Economic Entomology* 103, 1657–1664.

Bürgel, K., Daniel, C. and Wyss, E. (2005) Effects of autumn kaolin treatments on the rosy apple aphid, *Dysaphis plantaginea* (Pass.) and possible modes of action. *Journal of Applied Entomology* 129, 311–314.

Caleca, V., Lo Verde, G., Lo Verde, V., Piccionello, M.P. and Rizzo, R. (2010) Control of *Bactrocera oleae* and *Ceratitis capitata* in organic orchards: use of clays and copper products. *Acta Horticulturae* 873, 227–234.

Castillo, M.A., Moya, P., Hernández, E. and Primo-Yúfera, E. (2000) Susceptibility of *Ceratitis capitata* Wiedemann (Diptera: Tephritidae) to entomopathogenic fungi and their extracts. *Biological Control* 19, 274–282.
Chabert, S., Allemand, R., Poyet, M., Eslin, P. and Gibert, P. (2012) Ability of European parasitoids (Hymenoptera) to control a new invasive Asiatic pest, *Drosophila suzukii*. *Biological Control* 63, 40–47.
Cini, A., Anfora, G., Escudero-Colomar, L.A., Grassi, A., Santosuosso, U., Seljak, G. and Papini, A. (2014) Tracking the invasion of the alien fruit pest *Drosophila suzukii* in Europe. *Journal of Pest Science* 87, 559–566. DOI: 10.1007/s10340-014-0617-z.
Civolani, S., Grandi, G., Chicca, M., Pasqualini, E., Fano, E.A. and Musacchi, S. (2013) Probing behaviour of *Cacopsylla pyri* on a resistant pear selection. *Journal of Applied Entomology* 137, 365–375.
Cortesero, A.M., Stapel, J.O. and Lewis, W.J. (2000) Understanding and manipulating plant attributes to enhance biological control. *Biological Control* 17, 35–49.
Daniel, C. (2014) *Rhagoletis cerasi*: oviposition reduction effects of oil products. *Insects* 5, 319–331.
Daniel, C. and Baker, B. (2013) Dispersal of *Rhagoletis cerasi* in commercial cherry orchards: efficacy of soil covering nets for cherry fruit fly control. *Insects* 4, 168–176.
Daniel, C. and Grunder, J. (2012) Integrated management of European cherry fruit fly *Rhagoletis cerasi* (L.): situation in Switzerland and Europe. *Insects* 3, 956–988.
Daniel, C. and Wyss, E. (2009) Susceptibility of different life stages of the European cherry fruit fly, *Rhagoletis cerasi*, to entomopathogenic fungi. *Journal of Applied Entomology* 133, 473–483.
Daniel, C. and Wyss, E. (2010) Field applications of *Beauveria bassiana* to control the European cherry fruit fly *Rhagoletis cerasi*. *Journal of Applied Entomology* 134, 675–681.
Daniel, C., Pfammatter, W., Kehrli, P. and Wyss, E. (2005a) Processed kaolin as an alternative insecticide against the European pear sucker, *Cacopsylla pyri* (L.). *Journal of Applied Entomology* 129, 363–367.
Daniel, C., Tschabold, J.L. and Wyss, E. (2005b) Bekämpfung des Apfelblütenstechers mit Spinosad im biologischen Anbau. *Schweizerische Zeitschrift für Obst- und Weinbau* 4, 9–12.
Daniel, C., Linder, C. and Wyss, E. (2007) Autumn acaricide applications as a new strategy to control the pear leaf blister mite *Eriophyes pyri*. *Crop Protection* 26, 1532–1537.
Daniel, C., Mathis, S. and Feichtinger, G. (2014) A new visual trap for *Rhagoletis cerasi* (L.) (Diptera: Tephritidae). *Insects* 5, 564–576.
Daugherty, M.P., Briggs, C.J. and Welter, S.C. (2007) Bottom-up and top-down control of pear psylla (*Cacopsylla pyricola*): fertilization, plant quality, and the efficacy of the predator *Anthocoris nemoralis*. *Biological Control* 43, 257–264.
Debras, J.F., Senoussi, R., Rieux, R., Buisson, E. and Dutoit, T. (2008) Spatial distribution of an arthropod community in a pear orchard (southern France). Identification of a hedge effect. *Agriculture, Ecosystems and Environment* 127, 166–176.
De Carvalho Barbosa Negrisoli, C.R., Negrisoli, A.S., Garcia, M.S., Dolinski, C. and Bernardi, D. (2013) Control of *Grapholita molesta* (Busck, 1916) (Lepidoptera: Tortricidae) with entomopathogenic nematodes (Rhabditida: Heterorhabditidae, Steinernematidae) in peach orchards. *Experimental Parasitology* 135, 466–470.
Degen, T., Bakalovic, N., Bergvinson, D. and Turlings, T.C.J. (2012) Differential performance and parasitism of caterpillars on maize inbred lines with distinctly different herbivore-induced volatile emissions. *PLoS One* 7, e47589.
Delate, K., McKern, A., Turnbull, R., Walker, J.T.S., Volz, R., White, A., Bus, V., Rogers, D., Cole, L., How, N., Guernsey, S. and Johnston, J. (2008) Organic apple production in two humid regions: comparing progress in pest management strategies in Iowa and New Zealand. *HortScience* 43, 12–21.
Deprá, M., Poppe, J.L., Schmitz, H.J., De Toni, D.C. and Valente, V.L.S. (2014) The first records of the invasive pest *Drosophila suzukii* in the South American continent. *Journal of Pest Science* 87, 379–383.
Dib, H., Sauphanor, B. and Capowiez, Y. (2010) Effect of codling moth exclusion nets on the rosy apple aphid, *Dysaphis plantaginea*, and its control by natural enemies. *Crop Protection* 29, 1502–1513.
Dib, H., Jamont, M., Sauphanor, B. and Capowiez, Y. (2011) Predation potency and intraguild interactions between generalist (*Forficula auricularia*) and specialist (*Episyrphus balteatus*) predators of the rosy apple aphid (*Dysaphis plantaginea*). *Biological Control* 59, 90–97.
Douglas, A.E. (1993) The nutritional quality of phloem sap utilized by natural aphid populations. *Ecological Entomology* 18, 31–38.
Duso, C., Malagnini, V., Paganelli, A., Aldegheri, L., Bottini, M. and Otto, S. (2004) Pollen availability and abundance of predatory phytoseiid mites on natural and secondary hedgerows. *BioControl* 49, 397–415.

Duso, C., Malagnini, V., Pozzebon, A., Castagnoli, M., Liguori, M. and Simoni, S. (2008) Comparative toxicity of botanical and reduced-risk insecticides to Mediterranean populations of *Tetranychus urticae* and *Phytoseiulus persimilis* (Acari Tetranychidae, Phytoseiidae). *Biological Control* 47, 16–21.

Eigenbrode, S.D. and Pimentel, D. (1988) Effects of manure and chemical fertilizers on insect pest populations on collards. *Agriculture, Ecosystems and Environment* 20, 109–125.

Ekesi, S., Maniania, N.K. and Lux, S.A. (2003) Effect of soil temperature and moisture on survival and infectivity of *Metarhizium anisopliae* to four tephritid fruit fly puparia. *Journal of Invertebrate Pathology* 83, 157–167.

El-Wakeil, N.E., Gaafar, N.M. and Vidal, S. (2006) Side effect of some neem products on natural enemies of *Helicoverpa* (*Trichogramma* spp.) and *Chrysoperla carnea*. *Archives of Phytopathology and Plant Protection* 39, 445–455.

Emami, M.S., Shishehbor, P. and Karimzadeh, J. (2014) The influences of plant resistance on predation rate of *Anthocoris nemoralis* (Fabricius) on *Cacopsylla pyricola* (Förster). *Archives of Phytopathology and Plant Protection* 47, 2043–2050.

Enkerlin, W.R. (2005) Impact of fruit fly control programmes using the sterile insect technique. In: Dyck, V.A., Hendrichs, J. and Robinson, A.S. (eds) *Sterile Insect Technique – Principles and Practice in Area-wide Integrated Pest Management*. Springer, Dordrecht, The Netherlands, pp. 651–676.

Erler, F. (2010) Efficacy of tree trunk coating materials in the control of the apple clearwing, *Synanthedon myopaeformis*. *Journal of Insect Science* 10, 63. DOI: 10.1673/031.010.6301. Available at: insectscience.org/10.63 (accessed 26 June 2017).

Erler, F., Pradier, T. and Aciloglu, B. (2014) Field evaluation of an entomopathogenic fungus, *Metarhizium brunneum* strain F52, against pear psylla, *Cacopsylla pyri*. *Pest Management Science* 70, 496–501.

Evans, K.M., Govan, C.L. and Fernández-Fernández, F. (2008) A new gene for resistance to *Dysaphis pyri* in pear and identification of flanking microsatellite markers. *Genome* 51, 1026–1031.

Gambon, N., Barro, P., Pavan, F. and Zandigiacomo, P. (2009) Mating disruption of the small fruit tortrix (*Grapholita lobarzewskii*) in organic apple orchards of northeastern Italy. *Bulletin of Insectology* 62, 125–129.

Garrido-Jurado, I., Torrent, J., Barrón, V., Corpas, A. and Quesada-Moraga, E. (2011) Soil properties affect the availability, movement, and virulence of entomopathogenic fungi conidia against puparia of *Ceratitis capitata* (Diptera: Tephritidae). *Biological Control* 58, 277–285.

Géneau, C.E., Wäckers, F.L., Luka, H., Daniel, C. and Balmer, O. (2012) Selective flowers to enhance biological control of cabbage pests by parasitoids. *Basic and Applied Ecology* 13, 85–93.

Gontijo, L.M., Beers, E.H. and Snyder, W.E. (2013) Flowers promote aphid suppression in apple orchards. *Biological Control* 66, 8–15.

Graf, B., Höpli, H., Rauscher, S. and Höhn, H. (1999) Hagelnetze beeinflussen das Migrationsverhalten von Apfel und Schalenwickler. *Schweizerische Zeitschrift für Obst- und Weinbau* 135, 289–292.

Grasswitz, T.R. and Fimbres, O. (2013) Efficacy of a physical method for control of direct pests of apples and peaches. *Journal of Applied Entomology* 137, 790–800.

Grasswitz, T.R. and Yao, S. (2013) Control of the greater peach tree borer (*Synanthedon exitiosa*) (Lepidoptera: Sesiidae) in small-scale organic orchards. *Acta Horticulturae* 1001, 111–114.

Grdiša, M. and Gršić, K. (2013) Botanical insecticides in plant protection. *Agriculturae Conspectus Scientificus* 78, 85–93.

Grechi, I., Hilgert, N., Sauphanor, B., Senoussi, R. and Lescourret, F. (2010) Modelling coupled peach tree-aphid population dynamics and their control by winter pruning and nitrogen fertilization. *Ecological Modelling* 221, 2363–2373.

Habekuß, A., Proeseler, G., Schliephake, E. and Fischer, C. (2000) Resistance of apple to spider mites and aphids. *Acta Horticulturae* 538, 271–276.

Haltrich, A., Papp, J., Fail, J. and Kis, L. (2000) Effect of nitrogen-fertilizers and apple cultivars on aphids under IPM treatment conditions. *Acta Horticulturae* 525, 209–216.

Hauser, M. (2011) A historic account of the invasion of *Drosophila suzukii* (Matsumura) (Diptera: Drosophilidae) in the continental United States, with remarks on their identification. *Pest Management Science* 67, 1352–1357.

Hemptinne, J.L., Dixon, A.F.G. and Wyss, E. (2003) Biological control of the rosy apple aphid, *Dysaphis plantaginea* (Passerini) (Homoptera: Aphididae): learning from the ecology of ladybird beetles. In: Soares, A.O., Ventura, M.A., Garcia, V. and Hemptinne, J.L. (eds) *8th International Symposium on Ecology of Aphidophaga: Biology, Ecology and Behaviour of Aphidophagous Insects*. *Arquipélago – Life and Marine Science*, Supplement 5, pp. 33–41.

Hendrichs, J., Robinson, A.S., Cayol, J.P. and Enkerlin, W. (2003) Medfly areawide sterile insect technique programmes for prevention, suppression or eradication: the importance of mating behavior studies. *Florida Entomologist* 85, 1–13.

Heunis, J.M. and Pringle, K.L. (2003) The susceptibility of *Aphelinus mali* (Haldeman), a parasitoid of *Eriosoma lanigerum* (Hausmann), to pesticides used in apple orchards in the Elgin area, Western Cape Province, South Africa. *African Entomology* 11, 91–95.

Hogmire, H.W. and Miller, S.S. (2005) Relative susceptibility of new apple cultivars to arthropod pests. *HortScience* 40, 2071–2075.

Horton, D.R., Broers, D.A., Lewis, R.R., Granatstein, D., Zack, R.S., Unruh, T.R., Moldenke, A.R. and Brown, J.J. (2003) Effects of mowing frequency on densities of natural enemies in three Pacific Northwest pear orchards. *Entomologia Experimentalis et Applicata* 106, 135–145.

Hrotkó, K. (2007) Advances and challenges in fruit rootstock research. *Acta Horticulturae* 732, 33–42.

Iglesias, L.E., Nyoike, T.W. and Liburd, O.E. (2014) Effect of trap design, bait type, and age on captures of *Drosophila suzukii* (Diptera: Drosophilidae) in berry crops. *Journal of Economic Entomology* 107, 1508–1512.

Imoulan, A. and Elmeziane, A. (2014) Pathogenicity of *Beauveria bassiana* isolated from Moroccan Argan forests soil against larvae of *Ceratitis capitata* (Diptera: Tephritidae) in laboratory conditions. *World Journal of Microbiology and Biotechnology* 30, 959–965.

Jaastad, G., Røen, D., Bjotveit, E. and Mogan, S. (2007) Pest management in organic plum production in Norway. *Acta Horticulturae* 734, 193–199.

Jansen, J.P., Defrance, T. and Warnier, A.M. (2010) Effects of organic-farming-compatible insecticides on four aphid natural enemy species. *Pest Management Science* 66, 650–656.

Jaronski, S.T. (2010) Ecological factors in the inundative use of fungal entomopathogens. *BioControl* 55, 159–185.

Judd, G.J.R. and Gardiner, M.G.T. (2005) Towards eradication of codling moth in British Columbia by complimentary actions of mating disruption, tree banding and sterile insect technique: five-year study in organic orchards. *Crop Protection* 24, 718–733.

Judd, G.J.R., Gardiner, M.G.T. and Thomson, D.R. (1997) Control of codling moth in organically-managed apple orchards by combining pheromone-mediated mating disruption, post-harvest fruit removal and tree banding. *Entomologia Experimentalis et Applicata* 83, 137–146.

Kalinová, B., Stránský, K., Harmatha, J., Čtvrtečka, R. and Žd'árek, J. (2000) Can chemical cues from blossom buds influence cultivar preference in the apple blossom weevil (*Anthonomus pomorum*)? *Entomologia Experimentalis et Applicata* 95, 47–52.

Knight, A.L. (1998) Management of codling moth (Lepidoptera: Tortricidae) in apple with overhead watering. *Journal of Economic Entomology* 91, 209–216.

Kong, W.N., Li, J., Fan, R.J., Li, S.C. and Ma, R.Y. (2014) Sex-pheromone-mediated mating disruption technology for the oriental fruit moth, *Grapholita molesta* (Busck) (Lepidoptera: Tortricidae): overview and prospects. *Psyche (New York)*. DOI: 10.1155/2014/253924. Available at: https://www.hindawi.com/journals/psyche/2014/253924/ (accessed 26 June 2017).

Kührt, U., Samietz, J. and Dorn, S. (2006) Plant architecture, hail nets and thermal behaviour influencing developmental rate and modelling of the codling moth. *Acta Horticulturae* 707, 197–203.

Kutinkova, H., Dzhuvinov, V. and Samietz, J. (2012) Control of peach twig borer and oriental fruit moth by mating disruption in an apricot orchard. *Acta Horticulturae* 966, 169–174.

Kytö, M., Niemelä, P. and Larsson, S. (1996) Insects on trees: population and individual response to fertilization. *Oikos* 75, 148–159.

Lacey, L.A. and Shapiro-Ilan, D.I. (2008) Microbial control of insect pests in temperate orchard systems: potential for incorporation into IPM. *Annual Review of Entomology* 53, 121–144.

Lacey, L.A., Granatstein, D., Arthurs, S.P., Headrick, H. and Fritts Jr, R. (2006) Use of entomopathogenic nematodes (Steinernematidae) in conjunction with mulches for control of overwintering codling moth (Lepidoptera: Tortricidae). *Journal of Entomological Science* 41, 107–119.

Lacey, L.A., Thomson, D., Vincent, C. and Arthurs, S.P. (2008) Codling moth granulovirus: a comprehensive review. *Biocontrol Science and Technology* 18, 639–663.

Lalancette, N., Belding, R.D., Shearer, P.W., Frecon, J.L. and Tietjen, W.H. (2005) Evaluation of hydrophobic and hydrophilic kaolin particle films for peach crop, arthropod and disease management. *Pest Management Science* 61, 25–39.

Landis, D.A., Wratten, S.D. and Gurr, G.M. (2000) Habitat management to conserve natural enemies of arthropod pests in agriculture. *Annual Review of Entomology* 45, 175–201.

Landolt, P.J., Adams, T. and Rogg, H. (2012) Trapping spotted wing drosophila, *Drosophila suzukii* (Matsumura) (Diptera: Drosophilidae), with combinations of vinegar and wine, and acetic acid and ethanol. *Journal of Applied Entomology* 136, 148–154.

Lauri, P.-É., Costes, E., Regnard, J.-L., Brun, L., Simon, S., Monney, P. and Sinoquet, H. (2009) Does knowledge on fruit tree architecture and its implications for orchard management improve horticultural sustainability? An overview of recent advances in the apple. *Acta Horticulturae* 817, 243–250.

Leskey, T.C., Piñero, J.C. and Prokopy, R.J. (2008) Odor-baited trap trees: a novel management tool for plum curculio (Coleoptera: Curculionidae). *Journal of Economic Entomology* 101, 1302–1309.

Leskey, T.C., Wright, S.E., Glenn, D.M. and Puterka, G.J. (2010) Effect of Surround WP on behavior and mortality of apple maggot (Diptera: Tephritidae). *Journal of Economic Entomology* 103, 394–401.

Lombarkia, N. and Derridj, S. (2008) Resistance of apple trees to *Cydia pomonella* egg-laying due to leaf surface metabolites. *Entomologia Experimentalis et Applicata* 128, 57–65.

Maalouly, M., Franck, P., Bouvier, J.C., Toubon, J.F. and Lavigne, C. (2013) Codling moth parasitism is affected by semi-natural habitats and agricultural practices at orchard and landscape levels. *Agriculture, Ecosystems and Environment* 169, 33–42.

Marius Aurelian, V., Evenden, M.L. and Judd, G.J. (2012) Small-plot studies comparing pheromone and juice baits for mass-trapping invasive *Synanthedon myopaeformis* in Canada. *Entomologia Experimentalis et Applicata* 145, 102–114.

Markó, V., Blommers, L.H.M., Bogya, S. and Helsen, H. (2008) Kaolin particle films suppress many apple pests, disrupt natural enemies and promote woolly apple aphid. *Journal of Applied Entomology* 132, 26–35.

Mathews, C.R., Bottrell, D.G. and Brown, M.W. (2002) A comparison of conventional and alternative understory management practices for apple production: multi-trophic effects. *Applied Soil Ecology* 21, 221–231.

Miñarro, M. and Prida, E. (2013) Hedgerows surrounding organic apple orchards in north-west Spain: potential to conserve beneficial insects. *Agricultural and Forest Entomology* 15, 382–390.

Miñarro, M., Espadaler, X., Melero, V.X. and Suárez-Álvarez, V. (2009) Organic versus conventional management in an apple orchard: effects of fertilization and tree-row management on ground-dwelling predaceous arthropods. *Agricultural and Forest Entomology* 11, 133–142.

Mody, K., Spoerndli, C. and Dorn, S. (2011) Within-orchard variability of the ecosystem service 'parasitism': effects of cultivars, ants and tree location. *Basic and Applied Ecology* 12, 456–465.

Moerkens, R., Leirs, H., Peusens, G., Beliën, T. and Gobin, B. (2012) Natural and human causes of earwig mortality during winter: temperature, parasitoids and soil tillage. *Journal of Applied Entomology* 136, 490–500.

Myers, C.T., Hull, L.A. and Krawczyk, G. (2006) Seasonal and cultivar-associated variation in oviposition preference of oriental fruit moth (Lepidoptera: Tortricidae) adults and feeding behavior of neonate larvae in apples. *Journal of Economic Entomology* 99, 349–358.

Niggli, U., Weibel, F.P. and Gut, W. (1990) Weed control with organic mulch materials in orchards, results from 8 year field experiments. *Acta Horticulturae* 285, 97–102.

Pajač, I., Pejić, I. and Barić, B. (2011) Codling moth, *Cydia pomonella* (Lepidoptera: Tortricidae) – major pest in apple production: an overview of its biology, resistance, genetic structure and control strategies. *Agriculturae Conspectus Scientificus* 76, 87–92.

Palm, C., Blanco-Canqui, H., DeClerck, F., Gatere, L. and Grace, P. (2014) Conservation agriculture and ecosystem services: an overview. *Agriculture, Ecosystems and Environment* 187, 87–105.

Papp, J., Jenser, G. and Haltrich, A. (2001) Effect of nitrogen supply on the population of European red spider mite and green apple aphid in an IPM apple orchard. *Acta Horticulturae* 564, 407–412.

Pastopoulos, S., Pliakoni, E.D. and Nanos, G.D. (2012) Kaolin sprays and individual fruit bagging effects on quince fruit quality. *Acta Horticulturae* 940, 381–386.

Pelz-Stelinski, K.S., Gut, L.J. and Isaacs, R. (2006) Behavioral responses of *Rhagoletis cingulata* (Diptera: Tephritidae) to GF-120 insecticidal bait enhanced with ammonium acetate. *Journal of Economic Entomology* 99, 1316–1320.

Peñaflor, M.F.G.V. and Bento, J.M.S. (2013) Herbivore-induced plant volatiles to enhance biological control in agriculture. *Neotropical Entomology* 42, 331–343.

Penvern, S., Bellon, S., Fauriel, J. and Sauphanor, B. (2010) Peach orchard protection strategies and aphid communities: towards an integrated agroecosystem approach. *Crop Protection* 29, 1148–1156.

Pereault, R.J., Whalon, M.E. and Alston, D.G. (2009) Field efficacy of entomopathogenic fungi and nematodes targeting caged last-instar plum curculio (Coleoptera: Curculionidae) in Michigan cherry and apple orchards. *Environmental Entomology* 38, 1126–1134.

Pinnock, D.E. and Milstead, J.E. (1978) Microbial control of the fruit tree leafroller, *Archips argyrospila* [Lep.: Tortricidae] in California. *Entomophaga* 23, 203–206.

Praslička, J., Schlarmannová, J., Matejovičová, B. and Tancík, J. (2011) The predatory mite *Typhlodromus pyri* (Acari: Phytoseiidae) as a biocontrol agent of *Eriophyes pyri* (Acari: Eriophyidae) on pear. *Biologia* 66, 146–148.

Qubbaj, T., Reineke, A. and Zebitz, C.P.W. (2005) Molecular interactions between rosy apple aphids, *Dysaphis plantaginea,* and resistant and susceptible cultivars of its primary host *Malus domestica. Entomologia Experimentalis et Applicata* 115, 145–152.

Reekie, J., Specht, E., Appleby, M., Wilson, K., Carter, K., Fraser, H. and Pink, L. (2010) The control of apple maggots in organic orchards with GF-120 NF Naturalyte* fruit fly bait. *Acta Horticulturae* 873, 283–288.

Reighard, G.L. and Henderson, W.G. (2012) Mechanical blossom thinning in South Carolina peach orchards. *Acta Horticulturae* 965, 117–122.

Ricci, B., Franck, P., Bouvier, J.C., Casado, D. and Lavigne, C. (2011) Effects of hedgerow characteristics on intra-orchard distribution of larval codling moth. *Agriculture, Ecosystems and Environment* 140, 395–400.

Sandanayaka, W.R.M., Bus, V.G.M., Connolly, P. and Newcomb, R. (2003) Characteristics associated with woolly apple aphid *Eriosoma lanigerum*, resistance of three apple rootstocks. *Entomologia Experimentalis et Applicata* 109, 63–72.

Sauge, M.H., Grechi, I. and Poëssel, J.L. (2010) Nitrogen fertilization effects on *Myzus persicae* aphid dynamics on peach: vegetative growth allocation or chemical defence? *Entomologia Experimentalis et Applicata* 136, 123–133.

Sautereau, N., Penvern, S., Fauriel, J., Petitgenet, M. and Bellon, S. (2013) Combining multiple performances for sustainable agriculture: is organic fruit farming a prototype? A comparison of performances with conventional fruit farming. *Acta Horticulturae* 1001, 79–90.

Schmidt, R.A. (2014) Leaf structures affect predatory mites (Acari: Phytoseiidae) and biological control: a review. *Experimental and Applied Acarology* 62, 1–17.

Schüepp, C., Uzman, D., Herzog, F. and Entling, M.H. (2014) Habitat isolation affects plant–herbivore–enemy interactions on cherry trees. *Biological Control* 71, 56–64.

Sezen, K., Muratoglu, H., Nalcacioglu, R., Mert, D., Demirbag, Z. and Kati, H. (2008) Highly pathogenic *Bacillus thuringiensis* subsp. *tenebrionis* from European shot-hole borer, *Xyleborus dispar* (Coleoptera: Scolytidae). *New Zealand Journal of Crop and Horticultural Science* 36, 77–84.

Shapiro-Ilan, D.I., Gouge, D.H., Piggott, S.J. and Fife, J.P. (2006) Application technology and environmental considerations for use of entomopathogenic nematodes in biological control. *Biological Control* 38, 124–133.

Shapiro-Ilan, D.I., Wright, S.E., Tuttle, A.F., Cooley, D.R. and Leskey, T.C. (2013) Using entomopathogenic nematodes for biological control of plum curculio, *Conotrachelus nenuphar:* effects of irrigation and species in apple orchards. *Biological Control* 67, 123–129.

Sigsgaard, L., Esbjerg, P. and Philipsen, H. (2006) Experimental releases of *Anthocoris nemoralis* F. and *Anthocoris nemorum* (L.) (Heteroptera: Anthocoridae) against the pear psyllid *Cacopsylla pyri* L. (Homoptera: Psyllidae) in pear. *Biological Control* 39, 87–95.

Simon, S., Lauri, P.-E., Brun, L., Defrance, H. and Sauphanor, B. (2006) Does manipulation of fruit-tree architecture affect the development of pests and pathogens? A case study in an organic apple orchard. *Journal of Horticultural Science and Biotechnology* 81, 765–773.

Simon, S., Bouvier, J.-C., Debras, J.-F. and Sauphanor, B. (2010) Biodiversity and pest management in orchard systems. A review. *Agronomy for Sustainable Development* 30, 139–152.

Simon, S., Morel, K., Durand, E., Brevalle, G., Girard, T. and Lauri, P.-E. (2012) Aphids at crossroads: when branch architecture alters aphid infestation patterns in the apple tree. *Trees – Structure and Function* 26, 273–282.

Sjöberg, P., Swiergiel, W., Neupane, D., Lennartsson, E., Thierfelder, T., Tasin, M. and Rämert, B. (2015) Evaluation of temperature sum models and timing of *Quassia amara* (Simaroubaceae) wood-chip extract to control apple sawfly (*Hoplocampa testudinea* Klug) in Sweden. *Journal of Pest Science* 88, 301–310. DOI: 10.1007/s10340-10014-10616-10340.

Smirle, M.J., Lowery, D.T. and Zurowski, C.L. (2003) Susceptibility of leafrollers (Lepidoptera: Tortricidae) from organic and conventional orchards to azinphosmethyl, spinosad, and *Bacillus thuringiensis*. *Journal of Economic Entomology* 96, 879–884.

Speranza, S., Bucini, D. and Paparatti, B. (2009) European shot-hole borer [*Xyleborus dispar* (F.)]: comparison between capture with chemio-chromotropic Rebell® rosso traps and modified Mastrap®L traps. *Acta Horticulturae* 845, 535–538.

Stamenković, S., Garić, R., Milenković, S., Nikolić, M. and Stamenkovic, T. (1996) Susceptibility of some sweet cherry cultivars to *Rhagoletis cerasi* (Diptera, Tephritidae). *Acta Horticulturae* 410, 555–560.

Staudt, M., Jackson, B., El-Aouni, H., Buatois, B., Lacroze, J.P., Poëssel, J.L. and Sauge, M.H. (2010) Volatile organic compound emissions induced by the aphid *Myzus persicae* differ among resistant and susceptible peach cultivars and a wild relative. *Tree Physiology* 30, 1320–1334.

Stefanelli, D., Zoppolo, R.J., Perry, R.L. and Weibel, F. (2009) Organic orchard floor management systems for apple effect on rootstock performance in the midwestern United States. *HortScience* 44, 263–267.

Stephens, M.J., France, C.M., Wratten, S.D. and Frampton, C. (1998) Enhancing biological control of leafrollers (Lepidoptera: Tortricidae) by sowing buckwheat (*Fagopyrum esculentum*) in an orchard. *Biocontrol Science and Technology* 8, 547–558.

Stoeckli, S., Mody, K., Dorn, S. and Kellerhals, M. (2011) Association between herbivore resistance and fruit quality in apple. *HortScience* 46, 12–15.

Stoeckli, S., Hirschi, M., Spirig, C., Calanca, P., Rotach, M.W. and Samietz, J. (2012) Impact of climate change on voltinism and prospective diapause induction of a global pest insect – *Cydia pomonella* (L.). *PLoS One* 7, e435723.

Straub, C.S., Finke, D.L. and Snyder, W.E. (2008) Are the conservation of natural enemy biodiversity and biological control compatible goals? *Biological Control* 45, 225–237.

Stutz, S. and Entling, M.H. (2011) Effects of the landscape context on aphid–ant–predator interactions on cherry trees. *Biological Control* 57, 37–43.

Tabilio, M.R., Fiorini, D., Marcantoni, E., Materazzi, S., Delfini, M., De Salvador, F.R. and Musmeci, S. (2013) Impact of the Mediterranean fruit fly (Medfly) *Ceratitis capitata* on different peach cultivars: the possible role of peach volatile compounds. *Food Chemistry* 140, 375–381.

Tamm, L., Häseli, A., Fuchs, J.G., Weibel, F.P. and Wyss, E. (2004) Organic fruit production in humid climates of Europe: bottlenecks and new approaches in disease and pest control. *Acta Horticulturae* 638, 333–339.

Tamosiunas, R., Duchovskiene, L. and Valiuskaite, A. (2013) Monitoring of sawfly populations (Hymenoptera, Symphyta: *Hoplocampa* spp.) in plum and apple orchards using visual traps. *Proceedings of the Latvian Academy of Sciences, Section B: Natural, Exact, and Applied Sciences* 67, 130–135.

Tasin, M., Demaria, D., Ryne, C., Cesano, A., Galliano, A., Anfora, G., Ioriatti, C. and Alma, A. (2008) Effect of anti-hail nets on *Cydia pomonella* behavior in apple orchards. *Entomologia Experimentalis et Applicata* 129, 32–36.

Teixeira, L.A.F., Miller, J.R., Epstein, D.L. and Gut, L.J. (2010) Comparison of mating disruption and mass trapping with Pyralidae and Sesiidae moths. *Entomologia Experimentalis et Applicata* 137, 176–183.

Toepfer, S., Gu, H. and Dorn, S. (1999) Spring colonisation of orchards by *Anthonomus pomorum* from adjacent forest borders. *Entomologia Experimentalis et Applicata* 93, 131–139.

Tollerup, K.E., Rucker, A. and Shearer, P.W. (2012) Whole-farm mating disruption to manage *Grapholita molesta* (Lepidoptera: Tortricidae) in diversified New Jersey orchards. *Journal of Economic Entomology* 105, 1712–1718.

Tuovinen, T. (1994) Influence of surrounding trees and bushes on the phytoseiid mite fauna on apple orchard trees in Finland. *Agriculture, Ecosystems and Environment* 50, 39–47.

Ughini, V., Malvicini, G.L., Pisaroni, F., Caruso, S. and Plessi, C. (2010) Trials on the use of nets in the Vignola cherry district against cherry fruit fly (*Rhagoletis cerasi* L.). *Acta Horticulturae* 873, 337–342.

Underhill, G.W. and Cox, J.A. (1938) Studies on the resistance of apple to the woolly apple aphid [*Eriosoma lanigerum* (Hausm.)]. *Journal of Economic Entomology* 31, 622–625.

Vincent, C. and Bélair, G. (1992) Biocontrol of the apple sawfly, *Hoplocampa testudinea*, with entomogenous nematodes. *Entomophaga* 37, 575–582.

Vool, E., Rätsep, R., Karp, K., Kruus, M., Luik, A., Veromann, E. and Mänd, M. (2014) Does thinning of old apple trees improve fruit quality and decrease pest incidence? *International Journal of Fruit Science* 14, 362–375.

Vreysen, M.J.B., Carpenter, J.E. and Marec, F. (2010) Improvement of the sterile insect technique for codling moth *Cydia pomonella* (Linnaeus) (Lepidoptera Tortricidae) to facilitate expansion of field application. *Journal of Applied Entomology* 134, 165–181.

Weibel, F.P., Tamm, L., Wyss, E., Daniel, C., Häseli, A. and Suter, F. (2007) Organic fruit production in Europe: successes in production and marketing in the last decade, perspectives and challenges for the future development. *Acta Horticulturae* 737, 163–172.

Weibel, F.P., Chevillat, V.S., Rios, E., Tschabold, J.L. and Stadler, W. (2008) Fruit thinning in organic apple growing with optimised strategies including natural spray products and rope-device. *European Journal of Horticultural Science* 73, 145–154.

Weibel, F.P., Daniel, C., Hammelehle, A., Pfiffner, L. and Wyss, E. (2010) Potential and limits of pesticide free apple growing by a self-regulating orchard set-up: project presentation and first experiences. In: *14th International Conference on Cultivation Technique and Phytopathological Problems in Organic Fruit-Growing*, 22–24 February, Weinsberg, Germany. Fördergemeinschaft Oekologischer Obstbau e.V. (FOEKO), Weinsberg, Germany, pp. 292–296.

Weibel, F.P., Lemcke, B., Monzelio, U., Giordano, I. and Kloss, B. (2012) Successful blossom thinning and crop load regulation for organic apple growing with potassium-bi-carbonate (Armicarb®): results of field experiments over 3 years with 11 cultivars. *European Journal of Horticultural Science* 77, 145–153.

Whalon, M.E. and Croft, B.A. (1984) Apple IPM implementation in North America. *Annual Review of Entomology* 29, 435–470.

Wright, S.E., Leskey, T.C., Jacome, I., Piero, J.C. and Prokopy, R.J. (2012) Integration of insecticidal, phagostimulatory, and visual elements of an attract and kill system for apple maggot fly (Diptera: Tephritidae). *Journal of Economic Entomology* 105, 1548–1556.

Wyss, E. (1995) The effects of weed strips on aphids and aphidophagous predators in an apple orchard. *Entomologia Experimentalis et Applicata* 75, 43–49.

Wyss, E. (1996) The effects of artificial weed strips on diversity and abundance of the arthropod fauna in a Swiss experimental apple orchard. *Agriculture, Ecosystems and Environment* 60, 47–59.

Wyss, E. and Daniel, C. (2004) Effects of autumn kaolin and pyrethrin treatments on the spring population of *Dysaphis plantaginea* in apple orchards. *Journal of Applied Entomology* 128, 147–149.

Wyss, E., Villiger, M. and Müller-Schärer, H. (1999) The potential of three native insect predators to control the rosy apple aphid, *Dysaphis plantaginea*. *BioControl* 44, 171–182.

Yan, Y.H., Yu, Y., Du, X.G. and Zhao, B.G. (1997) Conservation and augmentation of natural enemies in pest management of Chinese apple orchards. *Agriculture, Ecosystems and Environment* 63, 253–260.

Ye, H., Wang, W.J., Liu, G.J., Zhu, L.X. and Jia, K.G. (2009) Resistance mechanisms of *Prunus* rootstocks to root-knot nematode, *Meloidogyne incognita*. *Fruits* 64, 295–303.

Yee, W.L. (2007) Attraction, feeding, and control of *Rhagoletis pomonella* (Diptera: Tephritidae) with GF-120 and added ammonia in Washington state. *Florida Entomologist* 90, 665–673.

Yee, W.L. (2008) Effects of several newer insecticides and kaolin on oviposition and adult mortality in western cherry fruit fly (Diptera: Tephritidae). *Journal of Entomological Science* 43, 177–190.

Yee, W.L. (2013) Soil moisture and relative humidity effects during postdiapause on the emergence of western cherry fruit fly (Diptera: Tephritidae). *Canadian Entomologist* 145, 317–326.

Yee, W.L. and Alston, D.G. (2006) Effects of spinosad, spinosad bait, and chloronicotinyl insecticides on mortality and control of adult and larval western cherry fruit fly (Diptera: Tephritidae). *Journal of Economic Entomology* 99, 1722–1732.

Yee, W.L. and Goughnour, R.B. (2011) Differential captures of *Rhagoletis pomonella* (Diptera: Tephritidae) on four fluorescent yellow rectangle traps. *Florida Entomologist* 94, 998–1009.

Zijp, J.P. and Blommers, L.H.M. (2002) Survival mode between the yearly reproduction periods, and reproductive biology of *Scambus pomorum* (Hymenoptera: Ichneumonidae: Pimplinae), a parasitoid of the apple blossom weevil *Anthonomus pomorum* (Coleoptera: Curculionidae). *Entomologia Generalis* 26, 29–46.

Zimmermann, O. (2004) Use of *Trichogramma* wasps in Germany: present status of research and commercial application of egg parasitoids against lepidopterous pests for crop and storage protection. *Gesunde Pflanzen* 56, 157–166.

Zimmermann, G. (2007) Review on safety of the entomopathogenic fungi *Beauveria bassiana* and *Beauveria brongniartii*. *Biocontrol Science and Technology* 17, 553–596.

6 Organic Integrated Pest Management of Tropical Fruit Crops

Jorge E. Peña,[1]* Daniel Carrillo[1] and Ben Faber[2]
[1]*University of Florida, Homestead, Florida, USA;* [2]*University of California, Ventura, California, USA*

Introduction

Due to the current trend when consumers are looking for produce that is free of toxins and synthetic products, there is a significant opportunity for tropical fruit growers and marketers to capitalize on organic production of these crops (Zehnder *et al.*, 2007; Pritts, 2012). The same authors also emphasize that the fast growth of organic agriculture has not been adequately supported by vigorous research in order to address challenges such as arthropod pest management. This could be due to the fact that organic production is more technically difficult than traditional agriculture. Organic production is fraught with problems of lower yield, variable effectiveness of pest, disease and weed management tools, and market access (Suckling and Butcher, 2001; van Mele *et al.*, 2001). Production of tropical fruit ranges from sophisticated plantation-type crops to rudimentary backyard production coupled with economic solvency and production purpose, internal versus export production. Reviews of research for tropical fruit integrated pest management (IPM), for banana, avocado, annona, passion fruit, litchi, longan, guava, papaya and pineapple (Aguiar-Menezes *et al.*, 2002; Gould and Raga, 2002; Ooi *et al.*, 2002; Petty *et al.*, 2002; Waite and Hwang, 2002; Wysoki *et al.*, 2002) show that each crop has a good amount of information on technologies that can be incorporated into organic IPM programmes. There are wide gaps, however, mostly on sampling and monitoring tactics for some pests, which are the backbone of any IPM programme. To address the needs for organic IPM, Zehnder *et al.* (2007) used different phases of the Wyss's model to review plans for an organic arthropod pest management. There are four phases in this system (Wyss *et al.*, 2005). The first phase strategy consists of utilizing cultural practices such as crop rotation, soil management, non-transgenic plant resistance and farm/field location. The second phase strategy focuses on vegetation management to enhance natural enemy impact on pest populations. The third phase involves inundative and inoculative releases of biological control agents, and the fourth phase is based on the use of approved insecticides of biological and mineral origin, and the use of mating disruption.

The objective of this chapter is to review the tactics that could be used or are currently used for organic tropical fruit production. Here we will evaluate four tropical-subtropical

* Corresponding author: jepena@ufl.edu

crops: avocado, mango, papaya and banana. For each one of these crops, a few key pests will be selected and the phases of the IPM programmes (mentioned above) reviewed and discussed.

First Phase

Crop rotation

Many tropical fruit crops (mango, avocado, litchi, guava, longan, durian, jackfruit, etc.) provide a relatively long-term and stable environment, offering at least a 30-year cycle of continuing habitats for both pests and natural enemies. For these types of crops, crop rotation is seldom considered as a solution to a pest problem. However, other crops are less stable because their productivity is reduced within 5–12 years (i.e. passion fruit, banana, papaya and dragon fruit). Therefore, crop rotation is only considered here for both banana and papaya, as these crops will last a maximum of 10–12 years in the field. Crop rotation works well when the pest that we need to avoid is oligophagous. For instance, the host range of *Cosmopolites sordidus* (Germar) (Coleoptera: Curculionidae), a key pest of banana, is generally considered to be restricted to the *Musaceae*. Because of this, Price (1994) observed that residual weevil populations from previous crops can cause heavy damage to new plantings. Gold *et al.* (2002) considers that rotating crops with other hosts besides banana or other *Musaceae* can reduce weevil levels. Crop rotation is also a tactic considered in papaya production to reduce attack by aphids, whiteflies and other sucking-type pests (Pantoja *et al.*, 2002).

Sanitation

Banana

In banana, Gold *et al.* (2002) considered that use of clean planting material can reduce initial banana weevil infestations, retard pest build-up for several crop cycles, and protect new banana stands against nematodes and some diseases. Suckers used as planting propagules often contain weevil eggs, larvae and occasionally, adults. Removing these weevils from planting material eliminates the most important source of infestation in new plantations. Banana weevils readily oviposit on residues for extended periods after harvest (Abera, 1997). For some clones, attack on residues may be more extensive than that on growing plants, for example Gros Michel in Ecuador, Cavendish in Australia and Latin America, and Ney Poovan in Uganda (Vilardebo, 1960; Treverrow and Bedding, 1993; Gold and Bagabe, 1997).

Crop sanitation in banana has been widely recommended to eliminate weevil refuges and breeding by other authors cited by Gold *et al.* (2002), among them Waterhouse and Norris (1987). Methods include cutting residues at or below the soil surface and chopping or splitting old rhizomes and pseudostems. Peasley and Treverrow (1986) suggest that crop hygiene (i.e. sanitation) is the long-term key to weevil control and that without it all other control measures are pointless.

The use of tissue culture plantlets for banana weevil control has been recommended by Peasley and Treverrow (1986). Unlike other methods, tissue culture plants are likely to be 100% free of banana weevils and nematodes at the time of planting. However, tissue culture plantlets are not universally available or affordable (Seshu Reddy *et al.*, 1998).

Paring, or removal of the outer surface of the rhizome, has also been widely recommended (Seshu Reddy *et al.*, 1998). Paring can expose weevil galleries and allow the farmer to reject heavily damaged suckers. Removal of all of the leaf sheaths and paring of the entire rhizome will eliminate most weevil eggs and first instar larvae. Many later instar larvae are likely to be deeper within the rhizome and not removed by paring.

Hot-water treatment to kill weevil eggs and larvae continues to be promoted (Seshu Reddy *et al.*, 1993, 1998). Ordinarily, the rhizomes are pared and then completely submerged in hot water (e.g. 52°C for 27 min or 54°C for 20 min) which is also a highly effective control against banana nematodes (Seshu Reddy *et al.*, 1993; Speijer *et al.*, 1993). However, Arroyave (1985) reported that hot-water baths are not effective at killing larvae deep within the rhizome.

Taylor (1991) reported that Tanzanian farmers viewed the recommendation of rhizome paring with 'extreme disbelief'. In addition, implementation of hot-water baths for control of banana weevils and nematodes requires investment in a hot-water tank and a heating source (e.g. electricity, gas burner, wood). As a result, adoption by resource-poor farmers may be limited.

Papaya

Sanitation is used in papaya to manage four common species of root-knot nematodes: (i) *Meloidogyne incognita* (Kofoid and White) Chitwood; (ii) *Meloidogyne javanica* (Treub) Chitwood; (iii) *Meloidogyne arenaria* (Neal) Chitwood; and (iv) *Meloidogyne hapla* Chitwood. Cultural tactics such as rotation with non-hosts, sanitation and avoidance and destruction of residual crop root is considered by Poormina (2005) as a good solution to nematode infestation. Pantoja *et al.* (2002) reported that suppression of fruit flies in papaya fields can be achieved by several methods and sanitation should be the first step; fruits should be removed as they ripen, and all fallen or infested fruit should be destroyed (Pantoja *et al.*, 2002). In Hawaii, sanitation of papaya fruits is usually insufficient but early harvesting is an effective means to avoid fruit fly damage. Papayas are usually fruit fly free when picked less than one-quarter ripe (Liquido, 1990, 1991). However, harvesting too early to avoid fruit fly infestation can result in diminished fruit flavour as fruit will not ripen fully. Sanitation to remove papaya plants infected with viruses is one of the most common practices recommended to avoid dissemination of viruses carried by aphids.

Avocado

In avocado, Hernandez *et al.* (2000) recommended collection of fallen fruits, weed removal and fallowing the soil to expose pupae to control the weevil *Conotrachelus aguacatae* (Barber).

Mango

In mango, Jirón (1995) reported that *Anastrepha obliqua* (Macquart) populations could be reduced by increasing planting distances in order to reduce humidity and increase solar radiation within orchards, eliminating fruit fly host plants in the hedges of mango orchards, and removing early or late fruit from the trees. In India cultural control practices include removal of fallen fruit and inter-tree ploughing and raking. The mango seed weevil, *Sternochetus mangifereae* (Fabricius) and the mango pulp weevil, *Sternochetus frigidus* (Fabricius) (Coleoptera: Curculionidae), are important pests of mango. The weevils overseason under bark and stone walls, where they remain dormant until the next flowering season (Shukla and Tandom, 1985). Removal and disposal of all fallen fruits and seeds is very labour intensive and has been inconsistent in demonstrating pest control (De and Pande, 1987; Hansen and Amstrong, 1990).

Cropping systems and crop management

Banana

Mixed cropping systems often result in lower insect pressure by reducing immigration rates, interfering with host plant location and increasing emigration rates. Kehe (1985) found much lower incidence of weevil attack (confidence interval (CI) = 6%) in plantains mixed with older coffee stands (i.e. > 5 year) than in those mixed with younger coffee plants (CI = 91%), with cacao (CI = 88%) or with annual crops (CI = 79%). By contrast, Uronu (1992) cited by Gold *et al.* (2002), tested a series of intercrops and failed to find a viable crop mixture that would both reduce weevil numbers and produce satisfactory banana yields. Recent work has demonstrated that grass mulches may increase weevil damage by creating a more favourable environment (i.e. cool, moist conditions) for adult weevils (Price, 1994; Rukazambuga, 1996; Braimah, 1997; all cited by Gold *et al.*, 2002). Deep planting and earthing up have been recommended to render the rhizome inaccessible to ovipositing females and prevent the establishment of a good root system. Seshu Reddy *et al.* (1993) planted cooking bananas at depths of 15 cm, 30 cm, 45 cm and 60 cm in drums and reported that shallow-planted suckers were more prone to attack, although some weevils were able to find the deepest planted suckers. However, Abera (1997) showed that weevils freely oviposit in leaf sheaths. Aguilar *et al.* (2014) recommended planting aromatic plants (i.e. *Calendula officinalis* Linnaeus, *Cynbopogon* sp.) along the borders of banana fields to repel destructive insects in the Philippines.

Papaya

Barrier plants may act as natural sinks for non-persistent aphid-transmitted viruses and have proved to be an effective crop management strategy to protect papaya crops against virus infection. Use of the cover crop, *Wedelia trilobata* (L.) Hitchc. did not affect aphid density but resulted in lower aphid visits to papaya plants (Robles *et al.*, 2006). Villanueva and Ortega (1993) reported that aphid populations were reduced when cover mulch was used. Hernandez *et al.* (2000) maintained that the use of vegetable cover crops, such as *Zea mays* Linnaeus and *Hibiscus sabdariffa* Linnaeus, combined with citronella and neem sprays, using healthy plants during transplant, and the removal of infected plants, resulted in a longer period of fruiting.

Integrated crop management strategies for papaya aphids have been developed in Veracruz, Mexico and the Philippines (Pantoja *et al.*, 2002) to manage papaya ringspot virus and its vectors. Barrier crops such as sorrel (*H. sabdariffa*) planted around the fields 2 months before transplanting papaya may reduce the virus incidence by affecting the behaviour of aphids. The deep red to purple coloration of *H. sabdariffa* is believed to repel aphids from landing on papaya fields. Intercropping barriers of maize or sorghum are used as intermediate landing crops in the Philippines (Andrade *et al.*, 1994; cited by Pantoja *et al.*, 2002). Aphids carrying non-persistent virus clean their stylets after feeding on the companion crops, reducing the viral infection (Andrade *et al.*, 1994). Destruction of plants with viral symptoms reduces the source of primary inoculum and delays development of the infection in the field (Pantoja *et al.*, 2002). Protecting seedlings under polypropylene or anti-aphid covers is recommended to reduce rapid field infestations (Andrade *et al.*, 1994). Mora Aguiliera *et al.* (1996) suggested transplanting papaya plants early in February and June as additional alternatives to the traditional April–May transplanting dates, in order to delay the epidemic onset of the papaya ringspot virus in central Veracruz, Mexico.

The combination of orchard design, trap crops and border trapping has been proposed as a means to reduce *Toxotrypana curvicauda* Gerstaecker (Diptera: Tephritidae) damage in commercial orchards in Mexico (Aluja *et al.*, 1997a, b; cited by Pantoja *et al.*, 2002). A trap crop of 10 m surrounding the main block of papaya trees can reduce the incidence of *T. curvicauda* (Aluja *et al.*, 1997a, b).

Avocado and mango

To our knowledge there are no available studies to report on the advantage of crop rotation and intercropping for avocado and mango.

Soil management

Banana

McIntyre *et al.* (2003) evaluated the effects of mulch on banana weevil, soil and plant nutrient status, soil water, and banana growth and development. After 3 years, the fully mulched plots had significantly more soil Ca and Mg, and foliar K, than plots that did not receive mulch. The mulched plots had greater water retention. However, the mulched plots also exhibited significantly higher banana weevil densities and greater plant damage than the control plots. Despite greater weevil damage, the treatments that were mulched yielded significantly heavier bunches. Hence the effects of the mulch on soil water infiltration and banana foliar nutrient status outweighed the detrimental effects of banana weevil damage.

Thrips (i.e. *Hercinothrips bicinctus* (Bagnall), *Ceratothripoides brunneus* (Bagnall), *Frankliniella occidentalis* (Pergande), *Scirtothrips kenyensis* Mound) feed on the lower surface of banana buds and flowers, and their damage might result in premature fruit shedding (Gold *et al.*, 2002). Ploughing, harrowing and solarization are suggested practices that can kill thrips pupae in the soil from previously infested crops.

Avocado

Cultural control efforts at managing *Scirtothrips perseae* Nakahara in avocado orchards have focused primarily on controlling the pest when late second instars, prepupae and pupae are in the soil (Peña *et al.*, 2013). One strategy for increasing thrips pupal mortality rates beneath trees is to use composted organic yard waste applied to a depth of approximately 30 cm and spread to the edge of the canopy. The exact mechanism for thrips suppression is unknown but could be due to an abundance of generalist arthropod predators that colonize the mulch, the release of secondary plant compounds from decaying mulch, along with entomopathogenic nematodes (e.g. *Steinernema* spp.) or fungi (e.g. *Beauveria bassiana* (Balsamo) Vuill.) that have been isolated from mulch. In addition to *S. perseae* control, addition of mulch also assists with the control of the plant pathogen *Phytopthora cinnamomi* Rands, improved soil quality, water retention and weed suppression (Peña *et al.*, 2013).

Papaya

Hornworms (i.e. *Erinnys ello* (Linnaeus), *Erinnyis* sp., Lepidoptera: Sphingidae) are considered important papaya defoliators in the neotropics (Pantoja *et al.*, 2002). In North, Central and South America the larvae can be found year round, but are more abundant during the summer months. Weed control practices and soil preparation can reduce adult and pupal populations (Abreu, 1994).

Non-transgenic plant resistance

There is a vast amount of information on insect and mite plant resistance on avocado, mango, papaya and banana.

Avocado

Kerguelen and Hoddle (2000) compared the susceptibility of seven cultivars of avocado to feeding by the mite *Oligonychus perseae* Tuttle, Baker and Abbatiello. Based on the percentage of leaf area damaged by mites in February 1998, cultivars were categorized into three groups: (i) 'Hass' and 'Gwen' were *susceptible*; (ii) 'Fuerte', 'Lamb', 'Hass' and 'Reed' were *resistant*; and (iii) 'Esther' and 'Pinkerton' were of *intermediate* susceptibility. These authors suggest that seasonal changes in the nutritional quality of leaves may be the major factor determining susceptibility of avocado cultivars to *O. perseae*.

In Mexico, different avocado cultivars show varying susceptibilities to the mite *Oligonychus punicae* infestations. Reyes and Salgado (1994) demonstrated that the cultivars '175PLS', '54PLS', '131PLS', '120PLS', '18PLS', '137PLS' and '30PLS' show some tolerance against *O. punicae*. The cultivars 'Rincon' and 'Fuerte' have been considered as tolerant to the mite species.

In Israel, 'Nabal' is the most susceptible cultivar to the pyriform scale, *Protopulvinaria pyriformis* (Cockerell), followed by 'Ein Vered', 'Reed', 'Hass' and 'Fuerte'; whereas 'Ettinger' trees are sometimes attacked when located close to infested trees of the susceptible cultivars (De Meijer *et al.*, 1989; cited by Peña *et al.*, 2013). In Florida, Peña *et al.* (2013) reported the possible plants resistant among avocado cultivars to the avocado lace bug, *Pseudacysta perseae* (Heidemann). The cultivars 'Simmonds', 'Waldin', 'Booth 8' and 'Loretta' had the highest natural infestation levels for the lace bug and leaf photosynthesis was reduced by 50% when the leaves sustained 40% damage. Plants of cultivar 'Simmonds' with 100% of their leaves infested exhibited an early leaf drop and no fruit set. In contrast a West Indies × Guatemala hybrid was scarcely affected by the pest. Reyes and Salgado (1994) demonstrated that leaves of the avocado cultivars, '30PLS', '54PLS', 'Rincoatl', '18PLS' and '158PLS' were tolerant to infestations of *Scirtothrips* spp., as were flowers of the cultivars '18PLS', '44PLC', 'ColinV-101', '175PLS', '158PLS' and 'PV2'. Ebeling (1959) reported that 'Fuerte' and

'Dickinson' are moderately resistant to attack by *Scirtothrips* spp.

Mango

Yee (1987) reported that the most susceptible mango cultivars in Hawaii to the fruit fly *Bactrocera dorsalis* (Hendel) are 'Hawaiian', 'Pirie' and 'Sandersha'. Singh (1991) indicated that the frequency of *Bactrocera* injury in physiologically mature fruit of 'Dashehari' ranged from 3.6% to 10%, while in fully ripe fruit the frequency of injured fruit ranged from 10% to 25.9%. Highest damage was reported in fully ripe fruit of 'Mallika', followed by 'Totapari'.

Susceptibility of different mango cultivars to the attack of another fruit fly, *A. obliqua*, were measured by Carvalho *et al.* (1996) who observed that 'Espada' showed no infestation by *A. obliqua*, whereas 'Carlota' was highly infested. Furthermore, 'Espada' had an adverse effect on the longevity of *A. obliqua* females, possibly due to the presence of toxic susbtances or absence of essential nutrients. Other studies have shown that immature mango fruit showed less susceptibility to *Anastrepha suspensa* (Loew) than mature mangoes (Hennesey and Schnell, 2001).

Mango cultivars resistant to the mango weevil would be beneficial if they also have good agronomic characteristics. Potential mechanisms of resistance are cultivars that produce no seeds, those that form seeds early, or those that fruit off-season. Most cultivars grown in Hawaii and India are equally susceptible (Hansen *et al.*, 1989), although others, such as 'Itamaraca', have shown some resistance (Balock and Kozuma, 1964).

In Florida, cultivars 'Keenan', an unknown cultivar, 'cv. 9819', 'Brander', and 'Bombay Green' had significantly more mango bud mite, *Aceria mangiferae* Sayed, than cultivars 'Joellen', 'Duncan', 'Red Itamaraca', 'Smith', 'Wally' and 'Hindi' (Peña *et al.*, 2005).

Papaya

Pantoja *et al.* (2002) indicated that differences in papaya varietal susceptibility to fruit flies have been documented for the Hawaiian and Cera varieties. Virus-resistant papaya varieties are available, but further work is needed on aphid vectors sampling, host finding and colonization, and insect–pathogen relationships. Papaya bunchy top virus is still a limiting factor for papaya production in the Caribbean, but little work has been conducted on the vectors. Only poor to modest relationships have been shown between aphids and leafhopper vectors and the number of affected plants in a field.

Banana

Ortiz *et al.* (1995) found all plantains equally susceptible to the banana weevil, while wild banana accessions and some cooking and dessert banana cultivars showed high levels of resistance. Later, Kiggundu *et al.* (2003) evaluated 45 *Musa* clones, including endemic and introduced cultivars plus hybrids, for resistance to banana weevil in a field trial in Uganda. The East African Highland bananas (*Musa* spp. AAA) and plantains (*Musa* spp. AAB), as well as plantain-derived hybrids (AAB × AA), showed highest levels of susceptibility to this pest. For more information on plant resistance to the banana weevil, see Gold *et al.* (2002).

Work with resistance for other pests (i.e. mites, viruses and nematodes) can be found in Wardlaw (1961), Dale (1987), Drew *et al.* (1992) and Rodrigues and Irish (2012).

Farm/field location

Mango

A good example of the importance of field location is showed in the case of *Thaumatotibia leucotreta* (Meyrick) (Lepidoptera: Tortricidae) affecting mango in South Africa. *T. leucotreta* is an important pest of cultivated crops in sub-Saharan Africa and also occurs in a variety of wild and cultivated host plants. Grove *et al.* (2012) suggested that in some areas of South Africa, the proximity of mango to citrus increased *T. leucotreta* infestation in mango. Therefore, while difficult sometimes, location of a mango orchard at considerable distance from a citrus orchard might help with the pest management.

Banana

Rodrigues and Irish (2012) showed the importance of planting banana at a good distance from coconut plantations, as these were the source of *Raoiella indica* Hirst (Acari: Tenuipalpidae) infestations in banana.

Pruning

Avocado

Pruning is commonly practised to manage branch borers in avocado and mango. For instance, to control the weevil, *Copturus aguacatae* Kissinger, Hernandez *et al.* (2000) recommend removal of infested branches. Ambrosia beetles vector a lethal disease that is affecting avocados in Florida (Carrillo *et al.*, 2014). Removal and complete destruction (chipping) of infected trees is recommended to eliminate beetle breeding sites. Since pruning wounds are believed to attract ambrosia beetles, pruning should not be practised during periods of high ambrosia beetles activity.

Physical control

Papaya

According to Vincent *et al.* (2003) physical control methods can be classified as passive (e.g. trenches, fences, organic mulch, particle films, inert dusts and oils), active (e.g. mechanical, polishing, pneumatic, impact and thermal) and miscellaneous (e.g. cold storage, heated air, flaming and hot water immersion). We also consider as physical control the use of bags or covers to protect from insect attack. Krishna Kumar *et al.* (2010) stated that neither raising papaya plants inside a net house covered on all sides nor frequent applications of insecticides or plant nutrients were efficacious on decreasing papaya ringspot virus (PRSV) incidence in papaya in India.

Mango

Helopeltis sp. (Miridae) are minor pests of mango fruit in the Philippines and in northern Australia, where it feeds on the fruit and causes superficial corky blemishes that detract from the fruit's appearance. In the Philippines, bagging carried out to protect the fruit from fruit flies is also effective at controlling *Helopeltis* sp.

Banana

Fruit bagging using newspapers and old leaves is suggested as a control measure against banana silvering thrips, *Hercinothrips bicinctus* Bagnall and the fruit fly, *Ceratitis rosa* Karsch.

Second Phase

Vegetation management

A cover crop changes temperature and humidity regimes, making the plants less susceptible to cold damage in the winter and cooler in the summer. This changes the reproductive regime of arthropods. Plant diversification in agroecosystems can result in increased environmental opportunities for natural enemies and, consequently, improved biological control (Altieri and Letourneau, 1982).

Papaya

Habitat management or the concept of plant trapping was advocated against the papaya fruit fly. For instance, the fruit fly *Toxotrypana curvicauda* Gerstaecker mostly attack fruits in plants located within the first 10 m from the edge of a plantation. In Colima, Mexico, growers plant sorghum and sunflower rows before they plant papaya, as a way to maintain sources of pollen for mite predators (J.E. Peña, 1997, personal observation).

Third Phase

Biological control agents

The search for potential biological control agents of tropical fruit pests has received a vast input. The natural enemies of many tropical fruit pests are known but in few instances these are used in comprehensive biological control programmes. Biological control ranges from mass releases to inoculative releases of predators and parasitoids. Examples of the status of biological control agents against some of the most important groups of pests of each crop will be discussed here.

Avocado

The biological control of hemipterans in avocado will be used as an example of the effectiveness of this method of control. There are also other examples for mites, thrips and loopers. For more information, see Peña *et al.* (2013).

HEMIPTERANS. The success of biological control agents against hemipterans affecting tropical fruit is perhaps one activity that has achieved much success. For instance, in Cuba, *Aleurodicus cordini* (Back.) is a pest of avocados; the biocontrol agents of this pest include *Baccha clavata* (Sci.), *Baccha parvicornis* Loew (Syrphidae), *Chrysopa* sp. (Chrysopidae), *Isodromus iceryae* (Howard) and *Carthasis distinctus* (Harris) (Nabidae). In Israel, the Japanese bayberry whitefly, *Parabemisia myricae* (Kuwana), was discovered in 1978, causing heavy damage to avocado (Sternlicht, 1979). Local predaceous mites in the family Phytoseiidae, lacewings (Neuroptera), ladybird beetles (Coccinellidae), predaceous bugs of the family Anthocoridae and parasitic wasps did not effectively control the pest. Thus, the parasitoid *Eretmocerus debachi* Rose and Rosen (Aphelinidae) was imported from California, successfully established and eventually controlled *P. myricae* (Rose and DeBach, 1992). Other exotic natural enemies such as *Eretmocerus* sp. and *Encarsia* sp. (Aphelinidae) (from Japan), the ladybird beetles (*Nephaspis amnicola* Wingo) and *Delphastus pusillus* LeConte (from Hawaii), the beetle *Cybocephalus binotatus* Grouvelle (Cybocephalidae), and the fungus *Aschersonia aleyrodis* Webber were released in numerous avocado and citrus orchards but probably did not become established (Swirski *et al.*, 1987). The red-banded whitefly (*Tetraleurodes perseae* Nakahara) infests avocados in California (Hoddle, 2006). In California, the aphelinid *Cales noacki* Howard causes 30–100% parasitism (Hoddle, 2006). In Mexico, the parasitoid *E. perseae* Rose and Zolnerowich was reared from *T. perseae* (Rose and Zolnerowich, 2004). *Cryptochetum iceryae* (Williston) and *Rodolia iceryae* Janson (Coccinellidae) were introduced to Israel for the control of *Icerya purchasi* (Maskell) (Mendel and Blumberg, 1991). Another coccinellid, the Australian *Cryptolaemus montrouzieri* Mulsant, introduced to South Africa from California for the control of *Planococcus citri* (Risso) also feeds on *I. purchasi* in Australia (Smith *et al.*, 1997).

To enhance biological control of hemipteran pests, the control of ants is recommended. Mealybugs are usually kept at low numbers by their natural enemies and most outbreaks are associated with chemical disruption of their natural enemies. However, ants are attracted to and feed on the honeydew that mealybugs excrete and they protect the mealybugs against their natural enemies, hence ants can also help to cause a mealybug outbreak.

The soft brown scale (*Coccus hesperidum* Linnaeus) is a cosmopolitan pest of many tropical and subtropical plants including avocado (Ebeling, 1950). Fruit may be stained by sooty mould deposits causing the fruit to be downgraded. Ants of various species are invariably associated with infestations of this pest. The brown house ant, *Pheidole megacephala* (Fabricius), the pugnacious ant, *Anoplolepis custodiens* Smith, and related species are commonly encountered (Annecke and Georgala, 1978).

Peña *et al.* (2013) discussed that although ant control is usually not a standard recommendation in orchards of non-citrus tropical and subtropical crops, it is regarded as essential to suppress soft brown scale in orchards where it occurs.

The heart-shaped scale, *Protopulvinaria pyriformis* (Cockerell), feeds on avocado (mainly on cultivar 'Hass'), guava and on the ornamental plant *Aphanamixis polystachya* (Wall.) R.N. Parker (= *Aphanamixis moora rohituka*) (De Villiers, 1981). The following natural enemies of the heart-shaped scale have been identified in South Africa (Robertson and De Villiers, 1986): *Metaphycus galbus* Annecke, *Metaphycus helvolus*

(Compère), *Metaphycus stanleyi* (Compère) (Encyrtidae), *Coccophagus basalis* Compère, *Coccophagus pulvinariae* Compère (Aphelididae) and *Tetrastichus* sp. (Eulophidae). Two hyperparasitoids, *Cheiloneurus cyononotus* Waterston (Encyrtidae) and *Marietta javensis* (Howard) (Aphelinidae) were also found. Predators included *Chilocorus angolensis* Crotch and *Hyperaspis senegalensis hottentota* Mulsant. Du Toit *et al.* (1991) found that on average 70% parasitism occurred in the adult brown heart-shaped scale female stage, 4% in the second instar and 14% in the third instar. In Israel, the parasitic wasps, *Microterys flavus* (Howard) (Encyrtidae) and *Coccophagus lycimnia* (Walker) (Aphelinidae), and predators, *Chilocorus bipustulatus* Linnaeus, *Oenopia (Synharmonia) conglobata* Linnaeus, *Scymnus flavicollis* Redtenbacher (ladybird beetles, Coccinellidae), *Anisochrysa carnea* Stephens (green lacewings, Chrysopidae), spiders and the fungus *Verticillium lecanii* Zimmerman (Hadar, 1993) were unable to keep populations of pyriform scale below the economic threshold. Thus, efforts were exerted to import various natural enemies. *Metaphycus swirskii* Annecke & Mynhardt (imported from Kenya), initially the most abundant parasite, was soon replaced by *M. stanleyi* (imported from the USA, South Africa and Spain), which is today the dominant natural enemy of the pyriform scale in Israel.

Fruit spotting bugs, *Amblypelta nitida* Stal and *Amblypelta lutescens lutescens* Distant (Coreidae) are major and often devastating pests of avocados in Queensland and northern New South Wales. Egg parasitoids *Gryon* sp. (Scelionidae), *Ooencyrtus caurus* Huang and Noyes (Encyrtidae) and *Anastatus* sp. (Eupelmidae) may infest up to 90% of eggs but the immigrant adults which result from unparasitized eggs cause severe damage.

The major biological control agents of *Pseudacysta perseae* (Heidemann) in Florida are two egg parasitoids, an undetermined species of a Trichogrammatidae and the mymarid *Eyrthmelus klopomor* Triapitsyn (Peña *et al.*, 2009). Predators include *Stethoconus prafectus* (Distant) (Heteroptera: Miridae) (Henry *et al.*, 2009; Holguin *et al.*, 2009) and *Tingidoletes praelonga* Gagné (Diptera: Cecidomyiidae) (Gagné *et al.*, 2008).

Mango

Analysis of the current situation of biological control agents in mango will focus on controlling fruit flies.

FRUIT FLIES. Both classical biological control and repeated augmentative releases of mass-reared parasitoids have been used to suppress *Anastrepha*, *Ceratitis* and *Bactrocera* populations (Sivinski, 1996; Sivinski *et al.*, 1996, 1997; Montoya *et al.*, 2000). In Florida and Latin America (e.g. Mexico, Costa Rica, Brazil, Colombia and Peru), parasitoid species such as *Diachasmimorpha longicaudata* (Ashmead), *Biosteres vandenboschi* (Fullaway) and *Aceratoneuromyia indica* (Silvestri) have been imported and released in the USA, Mexico, Costa Rica, Brazil and Peru for the control of *A. suspensa*, *Anastrepha ludens* (Loew) and *Anastrepha fraterculus* (Wiedemann) (Ovruski *et al.*, 2000). We note that despite the widespread use of exotic parasitoids over the past 80–100 years, the current trend is to resort to native species which pose less of an environmental threat to local fauna (García-Medel *et al.*, 2007).

Use of parasitoids in mango is much hindered by the fact that the fruit are very large and therefore provide larvae with a refuge from parasitism (e.g. see López *et al.*, 1999). As a consequence Aluja (1993) and Montoya *et al.* (2000) recommend that parasitoids be released outside the mango orchards to attack fly larvae in their much-smaller native hosts and therefore significantly reduce the size of populations entering mango orchards.

Several parasitoids, for example *Opius fullawayi* (= *Diachasmimorpha fullawayi*) (Silvestri), *Opius kraussii* (= *Diachasmimorpha kraussii*) Fullaway, *Opius tryoni* (Cameron), *Opius bellus* Gahan, *Biosteres longicaudatus* Ashmead (= *Diachasmimorpha longicaudata*), *Bactrocera tryoni* (Couron) (= *Diachasmimorpha tryoni* (Cameron)) and *Opius oophilus* (Fullaway) (= *Fopius arisanus* (Sonan)), have been reported parasitizing *Ceratitis capitata* (Wiedemann) (Beardsley, 1961; Wharton and Marsh, 1978). Bess *et al.* (1961) reported that the most important parasitoids collected from *C. capitata* in Hawaii were *B. vandenboschi* (Fullaway) (= *F. arisanus*), *O. oophilus* (= *F. arisanus*) and *B. longicaudatus* (= *D. longicaudata*). In Brazil mainly *Doryctobracon areolatus* (Szépligeti) (97%) and *D. longicaudata* (3%) are found parasitizing fruit fly larvae attacking mango (Carvalho and de Queiroz, 2002). In Africa (Kenya, Ghana, Tanzania, Uganda and Cote D'Ivoire) the most important parasitoids obtained from *Ceratitis* spp. stemming from mangos were *D. fullawayi*, *Fopius caudatus* (Szépligeti), *Psyttalia cosyrae* (Wilkinson) and *Tetrastichus giffardianus* Silvestri (Lux *et al.*, 2003). In Mexico and other parts of Latin America, the most common parasitoids attacking fruit flies that infest mangos (e.g. *A. obliqua*, *A. ludens*, *Anastrepha pseudoparallela* (Loew), *Anastrepha turpiniae*) are *Doryctobracon areolatus* (Szépligeti), *Doryctobracon brasiliensis* (Szépligeti), *Doryctobracon crawfordi* (Viereck), *Doryctobracon fluminensis* (Lima) and *Utetes anastrephae* (Viereck) (López *et al.*, 1999; Ovruski *et al.*, 2000; Zucchi, 2000). Finally, in Pakistan the following parasitoids have been reported attacking parasitoids of *Bactrocera zonata* (Saunders): *Opius longicaudatus* (= *D. longicaudata*), *Dirhinus giffardii* Silvestri, and *Bracon* sp. In the same country, *O. longicaudatus* (= *D. longicaudata*), *D. giffardii* and *Spalangia grotiusi* Girault were commonly reported parasitizing *B. dorsalis*; however, their incidence is extremely low. *Opius* spp. (= *Diachasmimorpha* spp.) introduced from Malaysia into Hawaii became established against *B. dorsalis* (Clancy *et al.*, 1952); however, fruit flies directly damage produce that is to be marketed and a small fruit fly population can cause economic damage reducing success of classical biological control programmes.

Pathogens, nematodes and predators have also been used to control fruit flies in mango orchards. For instance, Peng and Christian (2006) used the weaver ant, *Oecophylla smaragdina* (Fabricius) for control of the Jarvis fruit fly, *Bactrocera jarvisi* (Tryon) in mango orchards in Australia. The most recent example is the work by van Mele *et al.* (2007) (and references therein) in Benin using an African weaver ant, *Oecophylla longinoda* (Latreille). Aluja and colleagues (Aluja *et al.*, 2005) also investigated the potential of ants as possible biological control agents in various tropical orchards and backyard gardens in which mangos were growing next to other fruit trees.

Papaya

The biological control of hemipterans in papaya will be used as an example of the effectiveness of this method of control.

SCALES. The white peach scale, *Pseudaulacaspis pentagona* (Targioni-Torzetti) (Homoptera: Diaspididae), has a cosmopolitan distribution and is one of the most economically important scale insects in the southeastern USA where it is a serious pest of peaches and other fruit and ornamental crops (Nakahara, 1982).

Philephedra tuberculosa Nakahara and Gill. is a scale insect pest of papaya, sugar apple, *Annona squamosa* Linnaeus, soursop, *Annona muricata* Linnaeus and several species of ornamentals (Peña *et al.*, 1987; Abreu, 1994). The life history, natural enemies and behaviour of *P. tuberculosa* have been studied by Peña *et al.* (1987). Important natural enemies include ten arthropods and *Verticillium lecanii* (Zimmermann).

MEALYBUGS. The papaya mealybug, *Paracoccus marginatus* Williams and Granara de Willink is a pest of papaya, cassava (*Manihot esculenta* Crantz), *Hibiscus* spp., aubergine (*Solanum melongena* Linnaeus), avocado (*Persea americana* Mill.), annona (*Annona* spp.) and sweetpotato (*Ipomoea batatas* (L.) Lam.). The papaya mealybug occurs in tropical and subtropical climates, principally in the coastal states of Mexico, and in several Caribbean islands (Williams and Granara, 1992). Biological control appears to be the main factor keeping the species under control in Mexico, where the most important natural enemies are *Anagyrus* sp., *Acerophagus* sp., ca. *texanus* Howard, and *Apoanagyrus* sp. (González *et al.*, 1999). Common predators are *Chrysopa* sp. and *Chilocorus cacti* Linnaeus but these are usually found in low densities.

Banana

The examples of the effectiveness of biological control agents in banana will be for weevils.

WEEVILS. According to Gold *et al.* (2001) some natural enemies of the banana weevil *C. sordidus* are reported, for example *Plaseius javanus* Erichson (Coleoptera: Histeridae) whose larvae and adults both attack banana weevil inmatures. Koppenhofer *et al.* (1992) and Koppenhofer (1994) identified 12 predators of the banana weevil in Kenya, the most important being the predator *Thyreocephalus intercularis* (Eppelsheim) (Coleoptera: Staphylinidae) and Castineiras and Ponce (1991) showed the efficacy of the ant *Pheidole megacephala* (Fabricius) in Cuba, and reported that the ant caused 60–70% weevil control. However, the biological control efforts against this weevil have been minimal. Some success has been obtained with the entomopathogen, *B. bassiana*, causing around 4–34% mortality of the weevil in Florida. Also several trials have been carried out, both in Australia and in other countries, using entomopathogenic nematodes (i.e. *Heterorhabditis* spp. and *Steinernema* spp.) and Gold *et al.* (2001) consider that the nematodes have been quite ineffective.

Fourth Phase

Biological and botanical insecticides

Avocado

Several pesticides including M-Pede (soap), citrus oil and Mycotrol (*B. bassiana*) were applied to an avocado orchard with an average of 15–28 avocado lace bugs per leaf. Seven days after spray application, avocado lace bug densities were significantly reduced by Mycotrol, and by M-Pede compared with the untreated control. Mycotrol significantly reduced avocado lace bug densities for 29 days compared with the untreated control.

Several insecticides have been rigorously evaluated under field conditions for efficacy against *S. perseae* and to determine application rates and frequencies. Aerial applications by helicopter of sabadilla, a botanical pesticide made from the ground seeds of *Schoenocaulon officinale* (Schltdl. and Cham.) A. Gray, were evaluated for control of *S. perseae* on avocado trees. Sabadilla (27–57 g active ingredient (AI)/ha and 11.1–23.7 kg/ha sugar bait) was applied in different volumes of water. On large trees, these volumes resulted in only 1–11% coverage; but on small trees, sabadilla in 374 l/ha significantly reduced numbers of larvae 3 days and 6 days after treatment. The residual activity of abamectin is increased when combined with horticultural oil or a surfactant because these products assist with translaminar movement that protects the AI from photodegradation that occurs on the leaf surface. Similar results are obtained with spinosad when it is applied with horticultural oil or surfactants to avocado trees (~ 4.5 m tall trees) for *S. perseae* control (Tollerup and Morse, 2006).

In California, thrips control methods for organic avocado production includes application of materials such as spinosad (Entrust) mixed with 1% narrow range 415 oil (NR415) once a year in March beginning in establishment year 3 when trees reach bearing age (Takele *et al.*, 2011).

In Israel, the giant looper, *Boarmia selenaria* Denis and Schiffermüller (Geometridae) is controlled by preparations containing *Bacillus thuringiensis* (*Bt*) var. *kurstaki* Berliner.

Mango

FRUIT FLIES. In the past 10 years, there has been a concerted effort to find environmentally friendly alternatives to the use of malathion to control fruit flies (e.g. Peck and McQuate, 2000). For example, cyromazine, imidacloprid (organochlorinated compound), spinosad (bacteria-derived insecticide) and phototoxic dyes (e.g. Phlocxine B) have been successfully tested against various fruit fly species (Díaz-Fleischer *et al.*, 1996; King and Hennessey, 1996; Peck and McQuate, 2000; Liburd *et al.*, 2004; McQuate *et al.*, 2005). But despite their success, and as is typical with insecticides intensively applied on a large scale, resistance has already been documented in the case of spinosad (Hsu and Feng, 2006) or collateral damage such as negative impact on natural enemies.

Use of pathogens/disease agents (e.g. fungi, bacteria, nematodes) has been attempted with varying degrees of success to control fruit flies. For example, the use of *Metarhizium anisopliae* (Metchnikoff) Sorokin has been evaluated in small-scale mango orchards in Ngruruman, Kenya using bait stations laced with the pathogen. Results do not show differences between use of pathogens and use of insecticides (malathion) (Lux *et al.*, 2003). Lezama-Gutierrez *et al.* (2000) also evaluated isolates of *M. anisopliae* against larvae of *A. ludens.* Their findings suggest that *M. anisopliae* can reduce 22–43% in adult emergence, depending on the soil where the larvae pupate. De la Rosa *et al.* (2002) evaluated the fungus *B. bassiana* under laboratory conditions and concluded that highest mortality was achieved at the adult stage. Poinar and Hislop (1981), Lindegren and Vail (1986) and more recently Toledo *et al.* (2006) have investigated the use of various nematodes against flies within the genera *Anastrepha, Bactrocera* and *Ceratitis*, among them *Heterorhabditis bacteriophora* Poinar, *Heterorhabditis heliothidis* (Khan, Brooks and Hirschmann) and *Steinernema feltiae* Filipjev. Finally, Robacker *et al.* (1996) and Toledo *et al.* (1999) have tested various strains/isolates of *Bt* against larvae of *A. ludens, A. obliqua* and *Anastrepha serpentina* (Wiedemann), all flies able to attack mango. For additional details on microbial control of pestiferous fruit flies, we recommend the recent review by Dolinski and Lacey (2007).

In Africa, the persistence and infectivity of aqueous, oil/aqueous (50:50) and granular formulations of *M. anisopliae* were evaluated against pupariating larvae of three species of fruit flies (*Ceratitis capitata*, *Ceratitis fasciventris* (Bezzi) and *Ceratitis cosyra* (Walker)) and their associated endoparasitoids (*Psyttalia concolor* (Szépligeti) and *P. cosyrae* (Wilkinson)). Granular fungal formulations were more effective than the aqueous and oil/aqueous formulations of conidia achieving 37%, 42% and 54% reduction in emergence in *C. capitata, C. fasciventris* and *C. cosyra*, respectively. The use of fungal pathogens could therefore be an important IPM component for the management of fruit fly species in African orchards (Ekesi *et al.*, 2005).

Banana

Peña *et al.* (1995) reported the efficacy of *B. bassiana* against the banana weevil infesting banana in Florida. Aguilar *et al.* (2014) recommended for insect control in banana a mix of natural products consisting of ginger, muscovado and sugar gin (4:1:5), or ginger, turmeric and gin (0.5:0.5:1).

Mineral insecticides

In Chile, mineral oil is considered effective against eggs and motile stages (Vargas and Rodriguez, 2008) of *Oligonychus yothersi* (MacGregor). Joubert *et al.* (2004) recommended the use of kaolin (Surround) as well as sulfur and lime sulfur for control of thrips, mango weevil and mango bug, and emphasized that use of these products caused few problems with resurgence of mango scale and mealybugs. Use of oils to reduce aphid probing on papaya plants has been recommended (Pantoja *et al.*, 2002).

Mating disruption

Avocado

The sex pheromone of *Stenoma catenifer* (Walsingham) is a highly unsaturated aldehyde, (9Z)-9,13-tetradecadien-11-ynal, and is a new class of natural product (Millar *et al.*, 2008). Synthesis instructions for the pheromone are available (Hoddle *et al.*, 2009; Zou and Millar, 2010). The best dispensers for releasing the pheromone in wing traps are 11 × 5 mm grey rubber septa and 11 × 5 mm white rubber septa and the longevity of the pheromone is at least 4 weeks (Hoddle *et al.*, 2009). The pheromone has been demonstrated to be attractive to male *S. catenifer* in Guatemala (source of the pest population for extractions), Mexico and Peru.

In Israel, timing of the control measures against the giant looper, *Boarmia selenaria* Schiffermiller is based on traps baited with virgin females (attracting males) and scouting the orchards for young caterpillars. Since mass production of giant looper virgin females for monitoring purposes is laborious and expensive, efforts were made to replace the virgin females by synthetic pheromone. The chemical compounds (Z, Z)-6,9-cis- 3S,4R epoxynonadecadiene and (Z, Z, Z)-3,6,9-nonadecatriene were identified as sex pheromone components (Becker *et al.*, 1990). Bioassays performed by electroantennograph (EAG) in a wind tunnel gave positive results, but in field tests males were not sufficiently attracted to these two compounds (Becker *et al.*, 1990). Following experiments involving decapitation of the giant looper and subsequent pheromone biosynthesis activating neuropeptide (PBAN) injections, a third compound was revealed.

Mango

Monitoring fruit flies attacking mango serves different purposes: (i) to apply a control or management tactic once the presence of the fruit fly is noticed; and (ii) to verify if fruit fly species in question will attack mango under natural conditions. In general, thresholds for adult fruit flies are quarantine mediated (Beers *et al.*, 1993). These 'thresholds' vary from location to location, but depending on the fruit fly species they are typically based on the capture of a single fruit fly. In other fruit crops, a threshold of five flies per trap is suggested allowing a reduction of four chemical sprays to 1.5 sprays per season (Beers *et al.*, 1993). Sampling for fruit flies in mango is mostly performed using adult traps. The McPhail trap has provided different results in mango orchards and is considered by Aluja (1999) as ineffective. Earlier, Balock and Lopez (1969) reported that high concentrations of McPhail traps reduced the build up of fly populations and protected mangoes from severe injury during certain periods of the year. Aluja *et al.* (1989), working in a mixed mango orchard in Chiapas, Mexico, found that only 31.1% of the flies (*Anastrepha* spp.) landing on the trap *were caught* by the McPhail trap, with many flies entering the trap but then escaping.

In Palau, Pacific Islands two lures are used to attract mango flies, *Bactrocera frauenfeldi* (Schiner) is attracted to cue-lure, and *Bactrocera occipitalis* (Bezzi) and *Bactrocera philippinensis* (Drew and Hancock) to methyl eugenol (Secretariat of the Pacific Community, 2005). Fruit flies *B. dorsalis* and *Bactrocera umbrosa* (Fabricius) were monitored and controlled by mass trapping of males with methyl eugenol and infestations were brought to sub-economic levels in Pakistan (Mohyuddin and Mahmood, 1993). Trimedlure is still considered one of the most important para-pheromones for the Mediterranean fruit fly, with exception of *C. cosyra* adults which are attracted to terpinyl acetate and not to trimedlure (Steck, 2003). In addition, the attractiveness of mango compounds is currently being investigated. For example, some of the volatiles emitted by 'Tommy Atkins' mangoes such as terpenes (e.g. *p*-cymene and limonene) are attractive to *C. capitata* adults (Hernández-Sánchez *et al.*, 2001).

Conclusions

Plant protection is agreed to be one of the major issues in organic farming (Deguine and Penvern, 2014). Arthropod pest management involves the adoption of ecologically sound practices specified by international and national organic standards (Zehnder *et al.*, 2007). Today, regardless of their geographical area of origin, tropical fruits are grown wherever there are tropical and subtropical climates in the world. Then, the composition of insects and mites for each tropical crop depend on the climatic and geographical characteristics of each production system. A good number of practices are available that could be included in a cohesive organic production system. However, organic tropical fruit protection currently relies on a limited number of methods that provide only partial control of pests, which produces lower yields and suboptimal economic performance. As a result, farmers hesitate to adopt these strategies (Deguine and Penvera, 2014). With few exceptions, tropical fruit IPM is dominated by reactive approaches such as application of curative fast-acting methods, rather than preventative methods, which are the basic aims of IPM. These preventative methods are: (i) management of the entire landscape not just the tropical fruit crop; (ii) identifying components in the landscape that attract or repel pests; (iii) improving soil health; (iv) introduction and/or conservation of beneficial organisms, plant resistance and cultural activities to maintain pest populations below the economic injury level; and (v) disruption of pest populations without toxins or use of pheromones (Peña, 2002; Pritts, 2012; Deguine and Penvera, 2014). Prevention is the key for organic IPM and it is necessary to emphasize the need to know and understand the biology and ecology of pestiferous arthropods and their relationship with the tropical fruits growing in each area.

One of the best examples for prevention is the work conducted by Verghese *et al.* (1988) determining the distribution and sampling techniques of *Thrips palmi* Karny affecting mango and their recommendation for the number of sample units needed. Another example is the work of Aluja *et al.* (1996) determining the population densities of *Anastrepha* species in Mexican mangoes. These authors determined that most flies (62.3%) were captured in traps placed in the periphery of orchards; the practical implication of this would be to expect that mass trapping concentrated in the periphery of the orchard would result in a reduction of fruit infestation.

The economic thresholds used in conventional production systems might need to be adjusted for organic production that relies more on preventative methods rather than on curative ones. For instance, Pattison *et al.* (2010) adjusted the economic thresholds for *Pratylenchus goodeyi* Sher and Allen (lesion nematode) for subtropical and tropical banana plantations. While the threshold in the subtropics was 20.5–35.5 root disease index, in the tropics the threshold fluctuated between 9.2 and 15.6 root disease index. Similarly, economic thresholds could be adjusted for organic production systems that are based on preventative strategies.

Zehnder *et al.* (2007) proposed a series of strategies for organic IPM. The first strategy is to make the crop unavailable to pests in space and time. This practice applies mostly for tropical–temporal fruit crops and involves activities such as crop rotation and planting time. Sanitation (i.e. removing any infested plant material) is recommended for fruit flies and weevils in banana, papaya and mango while sanitation is now one of the major strategies to remove trees infected with laurel wilt of avocado.

Adequate fertilization or using manure and other soil amendments may result in a system with fewer pest outbreaks and plants that are less suitable as a host for pests. Examples of these are shown for banana, but the other crop systems need more research to demonstrate the effect of fertilization on pest attacks. On the other hand, development of pest-resistant or -tolerant cultivars has several examples for both temporal as well as for perennial tropical fruit crops. In tropical crops it is common to find cultivars with partial insect or mite resistance. This type of resistance can be utilized in those areas where key pests are known and where the cultivar resistance programme is focused on a particular key pest, such as weevils, mites, fruit flies, leafhoppers or gall midges. Other tactics to make the crop unacceptable to pests include interfering with oviposition preferences, host plant discrimination, or host location involving the use of intercropping, as pointed our earlier for banana and papaya.

The richness of information on the biological control agents available for almost each crop or pest is presented in this chapter. Reduction of pest survival by enhancing natural enemies using polycultures, while practised in Eastern agriculture for commercial tropical crops, is seldom found in Western countries, as the norm is towards monoculture and seldom towards polyculture. Many pests can be managed by enhancing the efficacy and local abundance of the existing community of natural enemies through modification of the environment. Organic perennial tropical fruit crops are a perfect target because in this type of system, there is minimal use of disruptive broad-spectrum pesticides that will affect natural enemies. Increasing plant diversity can benefit natural enemies by providing them with a favourable microclimate (shelter), source of alternative hosts or prey, or a supply of plant-based foods (nectar, pollen).

Another example is the use of biological control agents that can control different pests in one crop. Peng and Christian (2005) used weaver ants, *Oecophyla smaragdina* (Hymenoptera: Formicidae) in an IPM programme in Australia, that attempted to manage: (i) the leafhopper, *Idisocopus nitidulus* (Walker) (Hemiptera: Cicadellidae); (ii) the red-banded thrips, *Selenothrips rubrocinctus* (Giard) (Thysanoptera: Thripidae); (iii) the mango tip borer *Penicillaria jocosatrix* (Guenee) (Lepidoptera: Noctuidae); (iv) the fruit spotting bug, *Amblypelta lutescens lutescens* (Distant) (Hemiptera: Coreidae); (v) the seed weevil, *Sternochetus mangiferae* (Coleoptera: Curculionidae); and (vi) the fruit fly, *B. jarvisi* (Diptera: Tephritidae). The use of weaver ants combined with the use of 'soft chemicals' (i.e. oil, potassium soap and Applaud) were considered key elements for organic mango production. Another basic element that contributed to the success of this programme was frequent monitoring of the main pests as well as monitoring of the ant density per tree. However, there were problems utilizing ants in these systems. One of them was their aggressive behaviour, annoying workers during harvest. This factor was solved by spraying water on to the trees 20 min before harvest.

Biological and cultural control tactics on papaya-based systems need further attention. Culturally based practices can provide a first line of defence against secondary pests. Research and extension protocols should emphasize integrating cultural and biologically based practices in order to develop IPM and integrated crop management programmes.

In the case of organic IPM the curative method is either the use of a botanical, mineral insecticide or a biological insecticide, which regularly will not act as fast as a synthetic organic insecticide. Then economic thresholds to be used in organic IPM systems might be much lower than those used in conventional production systems. The advantage of mass trapping using pheromones or attractants become more important in organic agriculture, because it reduces more and more the reliance on synthetic insecticides.

In order to succeed in the incorporation of multiple tactics into an organic IPM system, farmer participation is crucial. For instance, in Vietnam, van Mele *et al.* (2001) found that during a study with mango growers, only 10% of 93 participating farmers knew about natural enemies. Aguilar *et al.* (2014) emphasized that even the most studied theoretically drafted strategies to control pest populations are bound to fail unless farmers, who are the major stakeholders, are involved every step of the way. By engaging tropical fruit growers and enlisting their help, an organic pest management strategy that is sustainable, safe and cost-effective can be designed and implemented.

References

Abera, A.M.K. (1997) Oviposition preferences and timing of attach by the banana weevil (*Cosmopolites sordidus* Germar) in East African highland banana (*Musa* spp.). MSc thesis, Makerere University, Kampala, Uganda, p. 120.

Abreu, E. (1994) Insectos de la papaya y practicas de control. In: *Memorias Foro: Cultivo, Produccion y Manejo.* Estacion Experimental Agricola de Puerto Rico, Recinto de Mayaguez, University of Puerto Rico, Aguadilla, Puerto Rico, pp. 58–62.

Aguiar-Menezes, E., Menezes, E., Cassino, P. and Soares, M. (2002) Passion fruit pests. In: Peña, J.E., Sharp, J. and Wysoki, M. (eds) *Tropical Fruit Pests and Pollinators: Biology, Economic Importance, Natural Enemies and Control.* CAB International, Wallingford, UK, pp. 361–390.

Aguilar, C.H., Lasalita-Zapico, F., Namocatct, J., Fortich, A. and Bojadores, R.M. (2014) Farmers' perceptions about bananas insect pests and integrated pest management (IPM) systems in Soccsksargen, Mindanao,

Philippines. In: 2014 International Conference on Intelligent Agriculture. *International Proceedings of Chemical, Biological and Environmental Engineering* (*IPCBEE*) 63. IACSIT Press, Singapore, pp. 22–27.

Altieri, M.A. and Letourneau, D.K.(1982) Vegetation management and biological control in agroecosystems. *Crop Protection* 1, 405–430.

Aluja, M. (1993) *Manejo Integrado de la Mosca de la Fruta*. Editorial Trillas, Mexico D.F., Mexico.

Aluja, M. (1999) Fruit fly (Diptera: Tephritidae) research in Latin America: myths, realities and dreams. *Anais da Sociedade Entomologica do Brasil* 28, 565–594.

Aluja, M., Cabrera, M., Guillen, J., Celedonio, H. and Ayora, F. (1989) Behavior of *Anastrepha ludens, A. obliqua* and *A. serpentina* (Diptera: Tephritidae) on a wild mango tree (*Mangifera indica*) harbouring three McPhail traps. *Insect Science and its Application* 10, 309–318.

Aluja, M., Celedonio-Hurtado, H., Liedo, P., Cabrera, M., Castillo, F., Guillen, J. and Rios, E. (1996) Seasonal fluctuations and ecological implications for management of *Anastrepha* fruit flies (Diptera: Tephritidae) in commercial mango orchards in southern Mexico. *Journal of Economic Entomology* 89, 654–667.

Aluja, M., Jimenez, A., Pinero, J., Camino, M., Aldana, L., Valdes, M.E., Castrejon, V., Jacome, I., Davila, A.B. and Figueroa, R. (1997a) Daily activity patterns of papaya fruit fly (Diptera: Tephritidae) in Morelos and Veracruz, Mexico. *Journal of Economic Entomology* 90, 505–519.

Aluja, M., Jimenez, A., Camino, M., Pinero, J., Aldana, L., Castrejon, V. and Valdes, M.E. (1997b) Habitat manipulation to reduce papaya fruit fly (Diptera: Tephritidae) damage: orchard design, use of trap crops and border trapping. *Journal of Economic Entomology* 90, 1567–1575.

Aluja, M., Sivinski, J., Rull, J. and Hodgson, P.J. (2005) Behavior and predation of fruit fly larvae (*Anastrepha* spp.) (Diptera: Tephritidae) after exiting fruit in four types of habitats in tropical Veracruz, Mexico. *Environmental Entomology* 34, 1507–1516.

Andrade, H., Avila, C., Gracia, E., Mora, A., Nieto, D., Teliz, D. and Villanueva, J.A. (1994) La mancha anular del papaya en Veracruz, mexico y su manejo integrado. In: *Septima reunion cientifica del sector agropecuario y forestall del estado de Veracruz*. Grupo Interdisciplinario del Papayo (GIP), Veracruz, Mexico, pp. 87–92.

Annecke, D.P. and Georgala, M.B. (1978) Soft brown scale *Coccus hesperidum* L. In: Bedford, E.C.G. (ed.) *Citrus Pests in the Republic of South Africa. Science Bulletin, Department of Agricultural Technical Services, Republic of South Africa* 391, 99–102.

Arroyave, F.P. (1985) Control del picudo negro *Cosmopolites sordidus* Germar en semilla vegatativa de platano (*Musa* AAB Simmonds). Tesis Ing. Agr., Universidad de Caldas, Manizales, Colombia, p. 112.

Balock, J.W. and Kozuma, T. (1964) Notes on the biology and economic importance of the mango weevil, *Sternochetus mangiferae* (Fabricius), in Hawaii (Coleoptera: Curculionidae). *Proceedings of the Hawaiian Entomological Society* 8, 353–364.

Balock, J.W. and Lopez, D.F. (1969) Trapping for control of the Mexican fruit fly in mango and citrus groves. *Journal of Economic Entomology* 62, 53–56.

Beardsley, J.W. (1961) A new review of the Hawaiian Braconidae (Hymenoptera). *Proceedings of the Hawaiian Entomological Society* 27, 333–366.

Becker, D., Cyjon, R., Cosse, A., Moore, I., Kimmel, T. and Wysoki, M. (1990) Identification and enantioselective synthesis of (Z,Z)-6,9-cis-3S,4R- epoxynonadecadiene, the major sex pheromone component of *Boarmia selenaria*. *Tetrahedron Letters* 31, 4923–4926.

Beers, E., Hull, L. and Jones, V. (1993) Sampling pests and beneficial arthropods of apples. In: Pedigo, L. and Buntin, D. (eds) *Handbook of Sampling Methods for Arthropods in Agricultyure*. CRC Press, Boca Raton, Florida, pp. 383–416.

Bess, H.A., van den Bosch, R. and Haramoto, F.H. (1961) Fruit fly parasites and their activities in Hawaii. *Proceedings of the Hawaiian Entomological Society* 27, 367–378.

Braimah, H. (1997) Laboratory studies on the host plant searching behaviour and chemical ecology of the banana weevil, *Cosmopolites sordidus* (Germar 1824) (Coleoptera: Curculionidae). PhD thesis, University of Reading, Reading, UK, p. 311.

Carrillo, D., Duncan, R.E., Ploetz, J. Campbell, A., Ploetz, R. and Peña, J.E. (2014) Lateral transfer of a phytopathogenic symbiont among native and exotic ambrosia beetles. *Plant Pathology* 63, 54–62.

Carvalho, P.J. and de Queiroz, A.C. (2002) *A Cultura da Mangueira*. CIP-Brasil, Brasilia, Brasil, p. 162.

Carvalho, R. da S., Nascimento, A., Morgante, J. and Fonseca, N. (1996) Susceptibility of different mango varieties to the attack of the fruit fly, *Anastrepha obliqua*. In: McPheron, B. and Steck, G. (eds) *Fruit Fly Pests: A World Assessment of Their Biology and Management*. St Lucie Press, Delray Beach, Florida, pp. 325–331.

Castineiras, A. and Ponce, E. (1991) Efectividad de la utilizacion de *Pheidole megacephala* (Hymenoptera: Formicidae) en la lucha biologica contra *Cosmopolites sordidus* (Coleoptera: Curculionidae). *Proteccion de Plantas* 1, 15–21.

Clancy, D., Marucci, P. and Dresner, E. (1952) Importation of natural enemies to control oriental fruit fly in Hawaii. *Journal of Economic Entomology* 45, 85–90.

Dale, J.L. (1987) Banana bunchy top virus: evaluation of a continuing threat. In: Persley, G.L. and De Langhe, E.A. (eds) *Banana and Plantain Breeding Strategies*. Proceedings of an international workshop, 13–17 October 1986, Cairns, Australia. *ACIAR Proceedings* 21, 187.

De, K. and Pande, Y.D. (1987) Evaluation of certain non insecticidal methods of reducting infestation of the mango nuit weevil, *Sternochetus gravis* (F.) in India. *Tropical Pest Management* 33, 27–28.

Deguine, J.P. and Penvera, S. (2014) Agroecologic crop protection in organic farming: relevance and limits. In: Bellon, S. and Penvera, S. (eds) *Organic Farming Prototype for Sustainable Agriculture*. Springer, Dordrecht, New York, pp. 107–130.

De la Rosa, W., López, F.L. and Liedo, P. (2002) *Beauveria bassiana* as a pathogen of the Mexican fruit fly (Diptera: Tephritidae) under laboratory conditions. *Journal of Economic Entomology* 95, 36–43.

De Meijer, A.H., Wysoki, M., Swirski, E., Blumberg, D. and Izhar, Y. (1989) Susceptibility of avocado cultivars to the pyriform scale, *Protopulvinaria pyriformis* Cockerell (Homoptera: Coccidae). *Agriculture, Ecosystems and Environment* 25, 75–82.

De Villiers, E.A. (1981) Heart-shaped scale in avocados. *Farming in South Africa*, Leaflet H.4. 2 pp. Department of Agricultural Technical Services, Pretoria.

Diaz-Fleischer, F., Toledo, J., Enkerlin, W. and Fernandez, J. (1996) Cyromazine: effects on three species of *Anastrepha*. In: McPheron, B. and Steck, G. (eds) *Fruit Fly Pests: A World Assessment of Their Biology and Management*. St Lucie Press, Delray Beach, Florida, pp. 336–337.

Dolinski, C. and Lacey, L.A. (2007) Microbial control of arthropod pests of tropical tree fruits. *Neotropical Entomology* 36, 161–179.

Drew, R.A., Smith, M.K. and Anderson, D. (1992) Field evaluation of micropropagated bananas derived from plants containing banana bunchy top virus. *Plant Cell, Tissue and Organ Culture* 28, 203–205.

Du Toit, W.J., Schutte, M.S. and Steyn, W.P. (1991) Parasitoids of the heart-shaped scale, *Protopulvinaria pyriformis* (Cockerell) (Hemiptera: Coccidae) on avocados in South Africa. *South African Avocado Growers' Association Yearbook* 14, 74–77.

Ebeling, E. (1959) *Subtropical Fruit Pests*. University of California Press, Riverside, p. 436.

Ebeling, W. (1950) *Subtropical Entomology*. Lithotype Process Co., San Francisco, p. 747.

Ekesi, S., Maniania, N.K., Mohamed, S. and Lux, S.A. (2005) Effect of soil application of different formulations of *Metarhizium anisopliae* on African tephritid fruit flies and their associated endoparasitoids. *Biological Control* 35, 83–91.

Gagné, R.J., Peña, J.E. and Acevedo, F.E. (2008) A new lestodiplosine (Diptera: Cecidomyiidae) preying on the avocado lace bug, *Pseudacysta perseae* (Heteroptera: Tingidae) in Southern Florida. *Florida Entomologist* 91, 43–48.

García-Medel, D., Sivinski, J., Díaz-Fleischer, F., Ramirez-Romero, R. and Aluja, M. (2007) Foraging behavior by six fruit fly parasitoids (Hymenoptera: Braconidae) released as single- or multiple-species cohorts in field cages: influence of fruit location and host density. *Biological Control* 43, 12–22.

Gold, C.S. and Bagabe, M.I. (1997) Banana weevil, *Cosmopolites sordidus* Germar (Coleoptera, Curculionidae), infestation of cooking and beer bananas in adjacent stands in Uganda. *African Entomology* 5, 103–108.

Gold, C.S., Peña, J.E. and Karamura, E.B. (2001) Biology and integrated pest management for the banana weevil *Cosmopolites sordidus* (Germar) (Coleoptera: Curculionidae). *Integrated Pest Management Reviews* 6, 79–155.

Gold, C.S., Pinese, B. and Peña, J.E. (2002) Pests of banana. In: Peña, J.E., Sharp, J. and Wysoki, M. (eds) *Tropical Fruit Pests and Pollinators: Biology, Economic Importance, Natural Enemies and Control*. CAB International, Wallingford, UK, pp. 13–56.

González, H., Villanueva, J.A. and Miller, D.R. (1999) Parasitoides del piojo harinoso del papayo, *Paracoccus maerginatus* Williams y Granara de Willink (Homoptera: Pseudococcdae) en mexico. Memorias del XXII Congreso nacional de Control Biologico, October 1999, Montecillo, Mexico.

Gould, W.P. and Raga, A. (2002) Pest of guava. In: Peña, J.E., Sharp, J. and Wysoki, M. (eds) *Tropical Fruit Pests and Pollinators: Biology, Economic Importance, Natural Enemies and Control*. CAB International, Wallingford, UK, pp. 295–314.

Grove, T., De Beer, M. and Joubert, P. (2012) Host status of commercial mango cultivars to *Thaumatotibia leucotreta* (Lepidoptera: Tortricidae) in South Africa. *Journal of Economic Entomology* 105, 1954–1962.

Hadar, D. (1993) Population dynamics of the pyriform scale, *Protopulvinaria pyriformis* Cockerell, and its natural enemies in avocado groves. PhD thesis, The Hebrew University of Jerusalem, Israel, p. 144. (in Hebrew, with English summary)

Hansen, J.D. and Amstrong, J.W. (1990) The failure of field sanitation to reduce infestation by the mango nut weevil, *Cryptorhynchus mangiferae* (F.) (Coleoptera: Curculionidae). *Tropical Pest Management* 36, 359–361.

Hansen, J.D., Armstrong, J.W. and Brown, S. (1989) The distribution of the mango weevil, *Cryptorhynchus mangiferae* (Coleoptera: Curculionidae), in Hawaii. *Proceedings of the Hawaii Entomological Society* 29, 31–39.

Hennesey, M. and Schnell, R.J. (2001) Resistance of immature mango fruits to Caribbean fruit fly (Diptera: Tephritidae). *Florida Entomologist* 84, 318–319.

Henry, T.J., Peña, J.E., Long, D. and Acevedo, F. (2009) *Stethoconus praefectus* (Hemiptera: Miridae): first North American records of an Old World predatory plant bug preying on avocado lace bug, *Pseudacysta perseae* (Hemiptera: Tingidae) in Florida. *Proceedings of the Entomological Society of Washington* 111, 98–105.

Hernandez, H.G., Johansen, R., Gazca, L., Equihua, A., Salinas, A., Estrada, E., Duran, F. and Yalle, A. (2000) Plagas del aguacate. In: Teliz, D. (ed.) *El Aguacate y su Manejo Integrado*. Ediciones Mundi-Prensa, Mexico DF, pp. 117–136.

Hernández-Sánchez, G., Sanz-Barbosa, I., Casaña-Giner, V. and Primo-Yúfara, E. (2001) Attractiveness for *Ceratitis capitata* (Wiedemann) (Dipt., Tephritidae) of mango (*Mangifera indica*, cv. Tommy Atkins) airborne terpenes. *Journal of Applied Entomology* 125, 189–192.

Hoddle, M.S. (2006) Phenology, life tables, and reproductive biology of *Tetraleurodes perseae* (Hemiptera: Aleyrodidae) on California avocados. *Annals of the Entomological Society of America* 99, 553–559.

Hoddle, M.S., Millar, J.G., Hoddle, C.D., Zou, Y. and McElfresh, J.S. (2009) Synthesis and field evaluation of the sex pheromone of *Stenoma catenifer* (Lepidoptera: Elachistidae). *Journal of Economic Entomology* 102, 1460–1467.

Holguin, C.M., Peña, J.E., Henry, T.J. and Acevedo, F. (2009) Biology of *Stethoconus praefectus* (Distant) (Heteroptera: Miridae), a newly established predator of the avocado lace bug, *Pseudacysta perseae* (Heteroptera: Tingidae), in Florida. *Florida Entomologist* 92, 54–57.

Hsu, J.-C. and Feng, H.-T. (2006) Development of resistance to spinosad in oriental fruit fly (Diptera: Tephritidae) in laboratory selection and cross-resistance. *Journal of Economic Entomology* 99, 931–936.

Jirón, L.F. (1995) Opciones al uso de insecticidas en mango. In: Garcia, J., Fuentes, G. and Monge, J. (eds) *Opciones al uso unilateral de plaguicidas en Costa Rica*. Editorial Euned, San José, pp. 129–155.

Joubert, P.H., Grove, T., de Beer, M.S. and Steyn, W.P. (2004) Evaluation of kaolin (Surround WP) in an IPM program on mangoes in South Africa. *Acta Horticulturae* 645, 493–499.

Kehe, M. (1985) Les principaux insectes depredateurs du plantain en Cote D'Ivoire: importance des infestations et incidence agro-economique. Paper presented at 94-101 in IARPB meeting, 27–31 May 1985, Abidjan, Cote D'Ivoire.

Kerguelen, V. and Hoddle, M.S. (2000) Comparison of the susceptibility of several cultivars of avocado to the persea mite, *Oligonychus perseae* (Acari: Tetranychidae). *Scientia Horticulturae* 84, 101–114.

Kiggundu, A., Gold, C.S., Labuschagne, M., Vuylsteke, D. and Louw, S. (2003) Levels of host plant resistance to banana weevil, *Cosmopolites sordidus* (Germar) (Coleoptera: Curculionidae) in African Musa germplasm. *Euphytica* 133, 267–277.

King, J.R. and Hennessey, M.K. (1996) Spinosad bait for the Caribbean fruit fly (Diptera: Tephritidae). *Florida Entomologist* 79, 526–531.

Koppenhofer, A.M. (1994) Observations on the bionomics of *Thyreocephalus interocularis* (Eppelsheim) (Col., Staphylinidae), a predator of the banana weevil. *Journal of Applied Entomology* 117, 382–394.

Koppenhofer, A.M., Seshu Reddy, K.V., Madel, G. and Lubega, M.C. (1992) Predators of the banana weevil, *Cosmopolites sordidus* (Germar) (Col., Curculionidae) in Western Kenya. *Journal of Applied Entomology* 114, 530–533.

Krishna Kumar, N.K., Singh, H. and Kalleshwaraswamy, C.M. (2010) Aphid (Aphididae: Homoptera) vectors of papaya ringspot virus (PRSV), bionomics, transmission efficiency and factors contributing to epidemiology. In: Kumar, N.K. (ed.) Proceedings of an International Sympsium on Papayas. *Acta Horticulturae* 851, 431–442.

Lezama-Gutierrez, R., Trujillo, A., Molina-Ochoa, J., Rebolledo-Dominguez, O., Pescador, A., Lopez-Edwards, M. and Aluja, M. (2000) Virulence of *Metarrizium anisopliae* (Deuteromycotina: Hypomycetes) on *Anastrepha ludens* (Diptera: Tephritidae): laboratory and field trials. *Journal of Economic Entomology* 93, 1080–1084.

Liburd, O.E., Holler, T.C. and Moses, A.L. (2004) Toxicity of imidacloprid-treated speheres to Caribbean fruit fly, *Anastrepha suspensa* (Diptera: Tephritidae) and its parasitoid *Diachasmimoprha longicaudata* (Hymenoptera: Braconidae) in the laboratory. *Journal of Economic Entomology* 97, 525–529.

Lindegren, J. and Vail, P. (1986) Susceptibility of Mediterranean fruit fly, melon fly and oriental frui fly (Diptera: Tephritidae) to the entopmogenous nematode *Steinernema feltiae* in laboratory tests. *Environmental Entomology* 4, 465–468.

Liquido, N.J. (1990) Morphological defects in the blossom end of papaya fruits in relation to infestation rates by oriental fruit fly and melon fly (Diptera: Tephritidae). *Journal of Economic Entomology* 83, 1883–1887.

Liquido, N.J. (1991) Effect of ripeness and location of papaya fruits on the parasitization rates of oriental fruit fly and melon fly (Diptera: Tephritidae) by braconid (Hymenoptera) parasitoids. *Environmental Entomology* 20, 1732–1736.

López, M., Aluja, M. and Sivinski, J. (1999) Hymenopterous larval-pupal and pupal parasitoids of *Anastrepha* flies (Diptera: Tephritidae) in Mexico. *Biological Control* 15, 119–129.

Lux, S.A., Ekesi, S., Dimbi, S., Mohamed, S. and Billah, M. (2003) Mango-infesting fruit flies in Africa: perspectives and limitations of biological approaches to their management. In: Neuenschwander, P., Borgemeister, C. and Langewald, J. (eds) *Biological Control in IPM Systems in Africa*. CAB International, Wallingford, UK, pp. 277–293.

McIntyre, B.D., Gold, C.S., Ssali, H. and Riha, S.J. (2003) Effects of mulch location on banana weevil, soil and plant nutrients, soil water and biomass in banana fields. *Biology and Fertility of Soils* 39, 74–79.

McQuate, G.T., Peck, S.L., Barr, P.G. and Sylva, C.D. (2005) Comparative evaluation of spinosad and phloxine B as toxicants in protein baits for suppression of three fruit fly (Diptera: Tephritidae) species. *Journal of Economic Entomology* 98, 1170–1178.

Mendel, Z. and Blumberg, D. (1991) Colonization trials with *Cryptochetum iceryae* and *Rodolia iceryae* for improvement biological control of *Icerya purchasi* in Israel. *Biological Control* 1, 68–74.

Millar, J.G., Hoddle, M.S., McElfresh, J.S., Zou, Y. and Hoddle, C.D. (2008) (9Z)-9,13-tetradecadiene-11-ynal, the sex pheromone of the avocado seed moth, *Stenoma catenifer*. *Tetrahedron Letters* 49, 4820–4823.

Mohyuddin, A.I. and Mahmood, R. (1993) Integrated control of mango pests in Pakistan. *Acta Horticulturae* 341, 467–483.

Montoya, P., Liedo, P., Benrey, B., Cancino, J., Barrera, J.F., Sivinski, J. and Aluja, M. (2000) Biological control of *Anastrepha* spp. (Diptera: Tephritidae) in mango orchards through augmentative releases of *Diachasmimorpha longicaudata* (Ashmead) (Hymenoptera: Braconidae). *Biological Control* 18, 216–224.

Mora Aguliera, G., Nieto-Angel, D., Campbell, C.L., Teliz, D. and Garcia, E. (1996) Multivariate comparison of papaya ringspot epidemics. *Phytopathology* 86, 71–78.

Nakahara, S. (1982) *Checklist of the Armored Scales (Homoptera: Diaspididae) of the Conterminous United States*. United States Department of Agriculture, Animal and Plant Health Inspection Service, Plant Protection and Quarantine, Beltsville, Maryland, 110 pp.

Ooi, P., Winotai, A. and Peña, J.E. (2002) Pests of minor tropical fruits. In: Peña, J.E., Sharp, J. and Wysoki, M. (eds) *Tropical Fruit Pests and Pollinators: Biology, Economic Importance, Natural Enemies and Control*. CAB International, Wallingford, UK, pp. 315–330.

Ortiz, R., Vuylseke, D., Dumpe, B. and Ferris, R.S. (1995) Banana weevil resistance and corm hardness in *Musa* germplasm. *Euphytica* 86, 95–102.

Ovruski, S., Aluja, M., Sivinski, J. and Wharton, R. (2000) Hymenopteran parasitoids on fruit infesting Tephritidae (Diptera) in Latin America and the southern United Status: diversity, distribution, taxonomic status and their use in fruit fly biological control. *Integrated Pest Management Reviews* 5, 81–107.

Pantoja, A., Follett, P.A. and Villanueva-Jimenez, J. (2002) Pests of papaya. In: Peña, J.E., Sharp, J. and Wysoki, M. (eds) *Tropical Fruit Pests and Pollinators: Biology, Economic Importance, Natural Enemies and Control*. CAB International, Wallingford, UK, pp. 131–156.

Pattison, A.B., Stanton, J.M., Cobon, J.A. and Doogan, V.J. (2010) Population dynamics and economic threshold of the nematodes *Radopholus similis* and *Pratylenchus goodeyi* on banana in Australia. *International Journal of Pest Management* 48, 107–111.

Peasley, D.L. and Treverrow, N. (1986) Count, cut and dry: a banana weevil borer management program. Information from Department of Agriculture, New South Wales, Australia, 4 pp.

Peck, S.L. and McQuate, G.T. (2000) Field tests of environmentally friendly malathion replacements to suppress wild Mediterranean fruit fly (Diptera: Tephritidae) populations. *Journal of Economic Entomology* 93, 280–289.

Peña, J.E. (2002) Introduction. In: Peña, J.E., Sharp, J. and Wysoki, M. (eds) *Tropical Fruit Pests and Pollinators: Biology, Economic Importance, Natural Enemies and Control*. CAB International, Wallingford, UK, pp. 1–11.

Peña, J.E., Baranowski, R.M. and Litz, R.E. (1987) Life history, behavior and natural enemies of *Philephedra tuberculosa* n. sp. (Homoptera: Coccidae). *Florida Entomologist* 70, 423–427.

Peña, J.E., Giblin-Davis, R.M. and Duncan, R. (1995) Impact of indigenous *Beauveria bassiana* (Balsamo) Vuillemin on banana weevil and rotten sugarcane weevil (Coleoptera: Curculionidae) populations in banana in Florida. *Journal Agricultural Entomology* 12, 163–167.

Peña, J.E., Palevsky, E., Otero-Colina, G., Ochoa, R. and Meister, C.W. (2005) Mango bud mite, *Aceria mangiferae* bionomics and control under Florida conditions. *Proceedings of the Florida State Horticultural Society* 119, 21–24.

Peña, J.E., Triapitsyn, S.V., Long, D., Evans, G. and Roltsch, W. (2009) First record of *Erythmelus klopomor* (Hymenoptera: Mymaridae) as a parasitoid of the avocado lace bug, *Pseudacysta perseae* (Heteroptera: Tingidae). *Florida Entomologist* 92, 394–395.

Peña, J.E., Hoddle, M., Aluja, M., Palevsky, A., Ripa, R. and Wysoki, M. (2013) Insect and mite pests. In: Schaffer, B., Wolstenholme, N. and Whiley, A.W. (eds) *The Avocado: Botany, Production and Uses*, 2nd edn. CABI, Wallingford, UK, pp. 423–488.

Peng, R.K. and Christian, K. (2005) Integrated pest management in mango orchards in the northern territory Australia, using the weaver ant, *Oecophylla smaragdina* (Hymenoptera: Formicidae) as a key element. *International Journal of Pest Management* 51, 149–155.

Peng, R.K. and Christian, K. (2006) Effective control of Jarvis's fruit fly *Bactrocera jarvisi* (Diptera: Tephritidae), by weaver ant, *Oecophyla smaragdina* (Hymenoptera: Formicidae), in mango orchards in the Northern Territory of Australia. *International Journal of Pest Management* 52, 275–282.

Petty, G.J., Stirling, G.R. and Bartholomew, D.P. (2002) Pests of pineapple. In: Peña, J.E., Sharp, J. and Wysoki, M. (eds) *Tropical Fruit Pests and Pollinators: Biology, Economic Importance, Natural Enemies and Control.* CAB International, Wallingford, UK, pp. 157–196.

Poinar, G.O. Jr and Hislop, R.G. (1981) Mortality of Mediterranean fruit fly adults (*Ceratitis capitata*) from parasitic nematodes (*Neoaplectana* and *Heterorhabditis* spp.). IRCS *Medical Science, Microbiology Parasitology and Infectous Disease* 9, 641.

Poormina, K. (2005) Nematode management in banana and papaya. In: Balamohan, T.N., Yijayakumar, R., Soorianathasundaram, K., Poormina, K. and Vadivel, E. (eds) *Management of Production Problems in Tropical Fruit Crops.* Department of Fruit Crops, Tamil Nadu Agricultural University, PSri Sahthi Promotional Litho Process, Coinbatore, India, pp. 152–155.

Price, N.S. (1994) Alternate cropping systems in the management of *Radopholus similis* and *Cosmopolites sordidus*, two important pests of bananas and plantain. *International Journal of Pest Management* 40, 237–244.

Pritts, M. (2012) Managing farming systems, landscapes, pests and pathogens to improve consumer perception of berries. *Journal of Berry Research* 2, 1–6.

Reyes, J.C. and Salgado, M. (1994) *Tolerancia de Algunas Selecciones de Aguacate al Attaque de Araña Roja (*O. punicae*) y Thrips (*Scirtothrips *sp).* Memorias, Centro de Investigaciones Cientificas y Tecnologicas del Aguacate en el Estado de Mexico, Fundacion Salvador Sanchez Colin, Coatepec Harinas, Mexico, pp. 175–178.

Robacker, D.C., Martinez, A.J., Gracia, J.A., Diaz, M. and Romero, C. (1996) Toxicity of *Bacillus thuringiensis* to Mexican fruit fly (Diptera: Tephritidae). *Journal of Economic Entomology* 89, 104–110.

Robertson, C.M. and De Villiers, E.A. (1986) Parasites of avocado pest bite the dust. *Citrus and Subtropical Fruit Research Institute, Information Bulletin* 168, 9.

Robles, W., Pantoja, A., Abreu, E., Peña, J.E., Ortiz, J., Lugo, M., Cortes, M. and Macchiavelli, R. (2006) Effects of cultural practices on the incidence of alate aphids and virosis on *Carica papaya. Manejo integrado de plagas y agroecología* (Costa Rica) 77, 38–43.

Rodrigues, J.C.V. and Irish, B.M. (2012) Effect of coconut palm proximities and *Musa* spp., germplasm resistance to colonization by *Raoiella indica* (Acari: Tenuipalpidae). *Experimental and Applied Acarology* 57, 309–316.

Rose, M. and DeBach, P. (1992) Biological control of *Parabemisia myricae* (Kuwana) (Homoptera: Aleyrodidae) in California. *Israel Journal of Entomology* 25/26, 73–95.

Rose, M. and Zolnerowich, G. (2004) *Eretmocerus persea* n.sp., (Hymenoptera: Chalcidoidea: Aphelinidae) reared from *Tetraleourodes perseae* Nakahara (Homoptera: Aleyrodidae: Aleyrodinae) living on avocado in Mexico. *Vedalia* 11, 45–52.

Rukazambuga, N.D.T.M. (1996) The effects of banana weevil (*Cosmopolites sordidus* Germar) on the growth and productivity of bananas (*Musa* AAA EA) and the influence of host vigour on attack. PhD thesis, University of Reading. Reading, UK, p. 249.

Secretariat of the Pacific Community (2005) *Fruit Flies in Palau.* Pest Advisory Leaflet No. 44, Palau, Pacific Islands.

Seshu Reddy, K.V., Koppenhofer, A.M. and Uronu, B. (1993) Cultural practices for the control of the banana weevil. In: Gold, C.S. and Gemmill, B. (eds) *Biological and Integrated Control of Highland Banana and Plantain Pests and Diseases. Proceedings of a Research Coordination Meeting.* International Institute of Tropical Agriculture (IITA), Cotonou, Benin, pp. 140–147.

Seshu Reddy, K.V., Prasad, J.S. and Sikora, R.A. (1998) Biointensive management of crop borers of banana. In: Saini, S.K. (ed.) *Proceedings of a Symposium on Biological Control in Tropical Crop Habitats: Third International Conference on Tropical Entomology*, 30 October–4 November 1994. International Institute of Tropical Agriculture (IITA), Cotonou, Benin, pp. 261–287.

Shukla, R.P. and Tandom, P.L. (1985) Bio-ecology and management of the mango weevil, *Sternochetus mangiferae* (Fabricius) (Coleoptera: Curculionidae). *International Journal of Tropical Agriculture* 3, 293–303.

Singh, G. (1991) Loss assessment, ecology and management of mango fruit fly, *Dacus* sp. *Acta Horticulturae* 291, 425–436.

Sivinski, J.M. (1996) The past and potential of biological control of fruit flies. In: McPheron, B.A. and Steck, G.J. (eds) *Fruit Fly Pests. A World Assessment of Their Biology and Management.* St Lucie Press, Delray Beach, Florida, pp. 365–375

Sivinski, J.M., Calkins, C.O., Baranowski, R.M., Harris, D., Brambila, J., Diaz, J., Bums, R.E., Holler, T. and Dodson, D. (1996) Suppression of Caribbean fruit fly (*Anastrepha suspensa* (Loew) Diptera: Tephritidae) population through releases of the parasitoid *Diachasmimorpha longicaudata* (Ashmead) (Hymenoptera: Braconidae). *Biological Control* 6, 177–185.

Sivinski, J., Aluja, M. and López, M. (1997) Spatial and temporal distributions of parasitoids of Mexican *Anastrepha* species (Diptera: Tephritidae) within the canopies of fruit trees. *Annals of the Entomological Society of America* 90, 604–618.

Smith, D., Beattie, G.A.C. and Broadley, R. (eds) (1997) *Citrus Pests and Their Natural Enemies: Integrated Pest Management in Australia.* Queensland Department of Primary Industries Information Series Q 197030. Queensland Department of Primary Industries, Brisbane, Australia, p. 272.

Speijer, P.R., Budenberg, B. and Sikora, R.A. (1993) Relationships between nematodes, weevils, banana and plantain cultivars and damage. *Annals of Applied Biology* 123, 517–525.

Steck, G.J. (2003) Mango Fruit Fly, Marula Fruit Fly, *Ceratitis cosyra* (Walker) (Insecta: Diptera: Tephritidae). Available at: http://edis.ifas.ufl.edu/IN563 (accessed 25 June 2017).

Sternlicht, M. (1979) *Parabemisia myricae* (Homoptera: Aleyrodidae) a new pest in Israel. *Hassadeh* 59, 1830–1831. (in Hebrew)

Suckling, D.M. and Butcher, R. (2001) Plant protection challenges in organic production. In: *Proceedings of a New Zealand Plant Protection Society Symposium*, August 2000, Christchurch, New Zealand. New Zealand Plant Protection Society Inc., Lincoln University, Canterbury, New Zealand, p. 88.

Swirski, E., Blumberg, D., Wysoki, M. and Izhar, Y. (1987) Biological control of the Japanese bayberry whitefly, *Parabemisia myricae* (Kuwana) (Homoptera Aleyrodidae), in Israel. *Israel Journal of Entomology* 21, 11–18.

Takele, E., Faber, B. and Vue, M. (2011) *Avocaco Sample Establishment and Production Costs and Profitability Analysis for Ventura, Santa Barbara, and San Luis Obispo Counties, 2011*. Organic Production Practices. University of California, Agriculture and Natural Resources, Ventura, 30 pp. Available at: http://www.californiaavocadogrowers.com/sites/default/files/documents/Organic%20Production%20for%20Ventura,%20Santa%20Barbara%20and%20San%20Luis%20Obispo.pdf (accessed 7 August 2015).

Taylor, B. (1991) Research field work on uplant bananas, *Musa* spp., principally acuminate triploid AAA types in the Kagera region of Tanzania. *Rivista di Agricultura Subtropicalee e Tropicale* 85, 349–392.

Toledo, J., Liedo, P., Williams, T. and Ibarra, J. (1999) Toxicity of *Bacillus thuringiensis* β-exotoxin to three species of fruit flies (Diptera: Tephritidae). *Journal of Economic Entomology* 92, 1052–1056.

Toledo, J., Rasgado, M.A., Ibarra, J.E., Gómez, A., Liedo, P. and Williams, T. (2006) Infection of *Anastrepha ludens* following soil applications of *Heterorhabditis bacteriophora* in a mango orchard. *Entomologia Experimentalis et Applicata* 119, 155–162.

Tollerup, K.E. and Morse, J.G. (2006) The effect of horticultural spray oil and surfactants on the residual efficacy of spinosad against avocado thrips, *Scirtothrips perseae* (Thysanoptera: Thripidae). *Journal of Agriculture and Urban Entomology* 22, 127–131.

Treverrow, N.L. and Bedding, R.A. (1993) Devlopment of a system for the control of the banana weevil borer, *Cosmopolites sordidus* with entompathogenic nematodes. In: Bedding, R., Akhurst, R. and Kaya, H. (eds) *Nematodes and the Biological Control of Insect Pests.* Commonwealth Scientific and Industrial Research Organisation (CSIRO), Melbourne, Victoria, pp. 41–47.

Uronu, B.E.M.A. (1992) The effect of plant resistance and cultural practices on the population densities of banana weevil *Cosmopolites sordidus* (Germar) and on banana yield. PhD thesis, Kenyatta University, Nairobi, p. 216.

van Mele, P., Nguyen, T.T.C. and van Huis, A. (2001) Farmer's knowledge, perceptions and practices in mango pest management in the Mekong Delta, Vietnam. *International Journal of Pest Management* 47, 7–16.

van Mele, P., Vayssières, J.-F., van Tellingen, E. and Vrolijks, J. (2007) Effects of an African weaver ant, *Oecophylla longinoda,* in controlling mango fruit flies (Diptera: Tephritidae) in Benin. *Journal of Economic Entomology* 100, 695–701.

Vargas, R. and Rodriguez, S. (2008) Avocado red mite, *Oligonychus yothersi* (McGregor). In: Ripa, R. and Larral, P. (eds) *Manejo de Plagas en Paltos y Cítricos. Colleccion de Libros INIA* 23. Instituto de Investigaciones Agropecuarias, Ministerio de Agricultura, La Cruz, Chile, pp. 239–246.

Verghese, A., Tandon, P.L. and Prasada Rao, G.S. (1988) Ecological studies relevant to the management of *Thrips palmi* Karny on mango in India. *Tropical Pest Management* 34, 55–58.

Vilardebo, A. (1960) *Los Insectos y Nematodos de las Bananeras del Ecuador*. Institut Franco-Équatorien de Recherches Fruitières, Paris, p. 78.

Villanueva, J.A. and Ortega, L.D. (1993) *Cintas Reflejantes Contra Áfidos Alados Vectores del Virus de la Mancha Anular, en Manejointegrado de Papayo*. Congreso Nacional de Horticultura (5, Veracruz, 1993) Programa Científico y Memoria. Veracruz, Mexico, p. 104.

Vincent, C., Hallman, G., Panneton, B. and Fleurat-Lessard, F. (2003) Management of agricultural insects with physical control methods. *Annual Review of Entomology* 48, 261–281.

Waite, G.K. and Hwang, J.S. (2002) Pests of litchi and longan. In: Peña, J.E., Sharp, J. and Wysoki, M. (eds) *Tropical Fruit Pests and Pollinators: Biology, Economic Importance, Natural Enemies and Control.* CAB International, Wallingford, UK, pp. 331–360.

Wardlaw, C.W. (1961) *Banana Diseases Including Plantains and Abaca*. Longmans, London.

Waterhouse, D.F. and Norris, K.R. (1987) *Biological Control: Pacific Prospects*. Inkata Press, Melbourne, Victoria, p. 454.

Wharton, R.A. and Marsh, P.M. (1978) New World Opiinae (Hymenoptera: Braconidae) parasitic on Tephritidae (Diptera). *Journal of the Washington Academy of Science* 68, 147–165.

Williams, D.J. and Granara, M.C. (1992) *Mealybugs of Central and South America*. CAB International, Wallingford, UK, p. 635.

Wysoki, M., van den Berg, M., Ish-Am, G., Gazit, S., Peña, J.E. and Waite, G.K. (2002) Pests and pollinators of avocado. In: Peña, J.E., Sharp, J. and Wysoki, M. (eds) *Tropical Fruit Pests and Pollinators: Biology, Economic Importance, Natural Enemies and Control.* CAB International, Wallingford, UK, pp. 223–294.

Wyss, E., Luka, H., Pfeiffner, L., Schlatter, C., Uehlnger, G. and Daniel, C. (2005) Approaches to pest management in organic agriculture: a case study in European apple orchards. *Organic Research* (May), 33–36.

Yee, W. (1987) The mango in Hawaii. *Cooperative Extension Service, University of Hawaii Circular* 388, 19–22.

Zehnder, G., Gurr, G.M., Kuhne, S., Wade, M.R., Wratten, S.D. and Wyss, E. (2007) Arthropod pest management in organic crops. *Annual Review of Entomology* 52, 57–80.

Zou, Y. and Millar, J.G. (2010) Improved synthesis of (9Z)-9,13-tetradecadien-11-ynal, the sex pheromone of the avocado seed moth, *Stenoma catenifer*. *Tetrahedron Letters* 51, 1336–1337.

Zucchi, R.A. (2000) Espécies de *Anastrepha,* sinonímias, plantas hospedeiras e parasitóides. In: Malavasi, A. and Zucchi R.A. (eds) *Moscas-Das-Frutas de Importância Econômica no Brasil: Conhecimento Básico e Aplicado*. Holos Editora, Ribeirao Preto, Brasil, p. 327.

7 Pest Management in Organic Grape Production

Serge Kreiter
Montpellier SuPAgro UMR CBGP, Montferrier-sur-Lez cedex, France

Introduction

Grapes are grown in a great variety of climates and agricultural situations, ranging from extreme hot and dry conditions (e.g. Israel, Greece, south of Spain, Arizona) to cool-climate conditions (e.g. Canada, New Zealand, Moldavia) that confer a strong specificity to the final product (Dominé, 2010). In each of these situations, appropriate and relevant information must be acquired in order to develop sustainable pest management programmes adapted to the given wine- or table-grape-producing areas, and the local complex of diseases and arthropod pests.

From a crop protection point of view, fungal diseases (powdery mildew, bunch rot, downy mildew) are the more serious concerns and major drivers of pest management programmes in the major areas of production. In several regions of the world, a suite of diseases require several fungicide sprayings per season to achieve optimal vine health.

However, arthropod pests also pose serious threats that must be addressed. As stated in Bentley *et al.* (2008), the absolute and relative importance of nematodes, insects and mites in grape production depends on the crop market (fresh versus processed grapes) and environmental conditions. The literature concerning arthropods associated with vineyards consists of more than 5000 scientific papers and book chapters from 1973 to 2016. Historically, research in vineyards focused on arthropod pests and consequently, relatively little is known about the biodiversity of arthropods in unmanaged or lightly managed vineyards. A notable exception are the biodiversity studies conducted systematically in the vineyards of Quebec, in which a high level of arthropod biodiversity was found and this demonstrates the potentially rich array of natural enemies that may occur in this agrosystem for biological control of pests (Bostonian *et al.*, 2012).

Such information should prove to be useful for the development of strategies to manage vineyards' arthropod pests with relatively little use of insecticides, such as in organic or biodynamic vineyards.

The More Important Pests of Grapes

In his review of grape insects, Bournier (1976) produced a long list of insects related to the parts of grapevines attacked (roots, woody trunk, very young shoots, shoots, buds, berries and leaves) worldwide. He also included miscellaneous classes, such as gall makers,

* serge.kreiter@supagro.fr

honeydew producers, aerial polyphagous insects and soil insects. That list which also reported the main geographical regions where the pests are found has been updated in Bentley *et al.* (2008) who stated that approximately 150 arthropod species are considered as pests of grapes worldwide. Actually, since the review made by these two authors, several new entomological problems have arisen in vineyards, because of the considerable increase of commercial exchanges worldwide and the increasing list of invasive species (Germain *et al.*, 2014). Several cicadellid species are, for example, now considered as economically important pests because they are vectors of phytoplasma diseases (Foissac and Wilson, 2010). Mani *et al.* (2014) reported recently 113 nematodes, 41 mites, 459 insects and 40 vertebrates, totalling 653 pests, known to damage the crop in different grape-growing regions of the world, providing information regarding the distribution of these pests and with references. Despite this great number of pests listed, Mani *et al.*'s (2014) list is incomplete, there are some redundancies, especially for Europe, and they do not mention the economic importance of all these animals. It thus represents more a list of phytophagous organisms potentially occurring on grapes, among which only few are real pests that are economically damaging.

A more realistic list of pests is given by Kreiter (2008) for France with a hierarchy among the list of pests. The number of really damaging pests is low compared with the long list provided (Kreiter, 2008).

Damage caused by arthropods falls into two categories: (i) irreversible; and (ii) reversible. Fortunately most types of damage are reversible, pest damage varying from very minor effects to complete loss of crops.

It is very difficult to classify the pests in relation to the damage they cause. Some species are very dangerous in some areas while they are less dangerous or not dangerous at all in others. Not all the species found and collected from vines are mentioned here in this chapter; only the dangerous species or the most common are briefly described. The damage level accepted on grapes varies between grapes for wine and table grapes. It is also a subjective parameter, depending on the social level, the quality requirement and human dietary customs.

The major pests of vineyards belong to the grape berry moths, leafhoppers, mealybugs, flies, Coleoptera, phytophagous mites and nematodes. A list of potential pests is given in Table 7.1 along with the region of the world where they cause problems and the plant part that is damaged and some of these species are discussed in more detail in the next section.

The Grape Moths

Most of the grape moths (see Table 7.1) are grape berry moths but some species also affect buds, shoots and leaves and one important pest is a root pest. Grape berry moths affect production in terms of yield and in terms of quality (sugar content and the amount of sugar converted into alcohol) and facilitate contamination by *Botrytis* causing Botrytis bunch rot. The grape moths are thus considered as key pests for integrated management. Botrytis bunch rot is caused by *Botrytis cinerea* Persoon and results in great losses. It is always present on the fruit set. However, it requires a wound to start a bunch rot infection. Wounds can come from insects, wind and accidental damage, etc. To control Botrytis bunch rot, a number of efficient fungicides are available on the market but they are very expensive, their efficiency is variable, resistance is quickly acquired and thus, preventive methods of contamination (and especially grape berry moth management) are a major concern. All efficient available fungicides are of synthetic origin and thus cannot be used in organic production.

The two European grape berry moths, *Eudemis* or *Lobesia botrana* (Denis & Schiffermüller) and *Cochylis* or *Eupoecilia ambiguella* (Hübner)

Diagnosis and biology

The adult moth of *Lobesia botrana* is 6–8 mm long. The forewings are a light, creamy

Table 7.1. Main pests (more damaging species) found on grapes in the various grape regions of the world.

Order	Family	Genus	Species	Region	Plant part damaged[a]
Dorylaimida[b]	Longidoridae	*Xiphinema*	*diversicaudatum* (Micoletzky)	Europe	Ro
	Longidoridae	*Xiphinema*	*index* (Thorne & Allen)	World	Ro
Rhabditida[c]	Heteroderidae	*Meloidogyne*	*arenaria* (Neal)	World	Ro
	Pratylenchidae	*Pratylenchus*	spp.	World	Ro
Thromidiforma[d]	Eriophyidae	*Calepitrimerus*	*vitis* (Nalepa)	World	Bu, Le, Fr
	Eriophyidae	*Colomerus*	*vitis* (Pagenstecher)	World	Bu, Le, Fr
	Tenuipalpidae	*Brevipalpus*	*chilensis* Baker	Chile	Bu, Le, Fr
	Tenuipalpidae	*Tenuipalpus*	*granati* Sayed	Mediterranean area	Bu, Le, Fr
	Tetranychidae	*Eotetranychus*	*carpini* (Oudemans)	South Europe	Bu, Le, Fr
	Tetranychidae	*Eotetranychus*	*willamettei* McGregor	USA	Bu, Le, Fr
	Tetranychidae	*Olignonychus*	*mangiferae* (Rahman & Sapra)	India	Bu, Le, Fr
	Tetranychidae	*Panonychus*	*ulmi* (Koch)	Nearctic, Palearctic	Bu, Le, Fr
	Tetranychidae	*Tetranychus*	*mcdanieli* McGregor	West USA, Champagne, France	Bu, Le, Fr
	Tetranychidae	*Tetranychus*	*pacificus* (McGregor)	West USA	Bu, Le, Fr
	Tetranychidae	*Tetranychus*	*urticae* Koch	World	Bu, Le, Fr
Coleoptera	Cerambycidae	*Acalolepta*	*vastator* (Newman)	Australia	Ro, Wo
	Chrysomelidae	*Fidia*	*viticida* (Walsh)	Eastern USA	Ro
			longipes (Melsheimer)	Eastern USA	Ro
	Scarabeidae	*Costelytra*	*zealandica* (White)	New Zealand	Sh, Le, Fr
	Scarabeidae	*Popilia*	*japonica* Newman	Japan, eastern USA, Canada, the Azores, Italy	Le, Fr
Diptera	Drosophilidae	*Drosophila*	*melanogaster* Meigen	Europe	Fr
	Drosophilidae	*Drosophila*	*suzukii* Matsumura	North America, Europe, Asia	Fr
Hemiptera	Cicadellidae	*Draeculacephala*	*minerva* Ball	Western USA	Le, Sh, Bu
	Cicadellidae	*Erythroneura*	spp.	North America, China	Le, Sh, Bu
	Cidadellidae	*Graphocephala*	*atropunctata* (Signoret)	Western USA	Le, Sh, Bu
	Cicadellidae	*Homalodisca*	*vitripennis* (Germar)	USA	Le, Sh, Bu
	Cicadellidae	*Oncopsis*	*alni* (Schrank)	Europe	Le, Sh, Bu
	Cicadellidae	*Scaphoideus*	*titanus* Ball	Southern Europe	Le, Sh, Bu
	Cicadellidae	*Xyphon*	*fulgida* (Nottingham)	Western USA	Le, Sh, Bu
	Cixiidae	*Hyalesthes*	*obsoletus* Signoret	Europe	Le
	Coccidae	*Parthenolecanium*	*corni* (Bouché)	Argentina, Brazil, Chile, Europe	Le, Sh, Bu, Fr
	Coccidae	*Parthenolecanium*	*persicae* (Fabricius)	Australia, Chile, Europe	Le, Sh, Bu, Fr
	Coccidae	*Neopulvinaria*	*innumerabilis* (Rathvon)	Europe	Le, Sh, Bu, Fr
	Pseudococcidae	*Dismicoccus*	*brevipes* (Cockerell)	Brazil	Le, Sh, Bu, Fr

Continued

Table 7.1. Continued.

Order	Family	Genus	Species	Region	Plant part damaged[a]
	Pseudococcidae	*Ferrisia*	*gilli* (Gullan)	Western USA	Le, Sh, Bu, Fr
	Pseudococcidae	*Heliococcus*	*bohemicus* Sulc	Europe	Le, Sh, Bu, Fr
	Pseudococcidae	*Maconellicoccus*	*hirsutus* Green	India	Le, Sh, Bu, Fr
	Pseudococcidae	*Phenacoccus*	*aceris* (Signoret)	Europe	Le, Sh, Bu, Fr
	Pseudococcidae	*Planococcus*	*citri* Risso	Brazil, Europe, South Africa	Le, Sh, Bu, Fr
	Pseudococcidae	*Planococcus*	*ficus* Signoret	Western USA, Argentina, Europe, Middle East, South Africa	Le, Sh, Bu, Fr
	Pseudococcidae	*Pseudococcus*	*calceolariae* (Maskell)	New Zealand	Le, Sh, Bu, Fr
	Pseudococcidae	*Pseudococcus*	*longispinus* (Targioni Tozzetti)	Western USA, New Zealand	Le, Sh, Bu, Fr
	Pseudococcidae	*Pseudococcus*	*maritimus* (Ehrhorn)	North America, Chile, South Africa, Asia	Le, Sh, Bu, Fr
	Pseudococcidae	*Pseudococcus*	*viburni* (Signoret)	Western USA, Chile, South Africa	Le, Sh, Bu, Fr
Lepidoptera	Tortricidae	*Argyrotaenia*	*franciscana* (Walsingham)	Western USA	Le, Fr
	Tortricidae	*Argyrotaenia*	*ljungiana* (Thunberg)	Europe	Le, Fr
	Tortricidae	*Argyrotaenia*	*velutinana* (Walker)	Eastern USA	Le, Fr
	Tortricidae	*Epiphyas*	*postvittana* Walker	Australia, New Zealand	Le, Fr
	Tortricidae	*Eupoecilia*	*ambiguella* (Hübner)	Europe	Fr
	Tortricidae	*Lobesia*	*botrana* (Denis & Schiffermüller)	Europe, USA	Fr
	Tortricidae	*Paralobesia*	*viteana* (Clemens)	Eastern USA	Fr
	Tortricidae	*Platynota*	*stultana* Walsingham	Western USA	Bu, Le, Fr
	Tortricidae	*Sparganothis*	*pilleriana* (Denis & Schiffermüller)	Europe	Bu, Le, Fr
	Sesiidae	*Vitacea*	*polistiformis* (Harris)	Eastern USA	Ro

[a]Bu, Buds; Fr, fruits or berries; Le, leaves; Ro, roots; Sh, shoots; Wo, wood.
[b]Order Dorylaimida in sub-class Dorylaimia, of class Enoplea, of phylum Nematoda.
[c]Order Rhabditida in sub-class Chromadoria, of class Chromadorea, of phylum Nematoda.
[d]Order Thromidiforma in sub-class Acari, of class Arachnida.

white to tannish colour, with black, brown and grey mottling. Their hindwings are a greyish colour. Females tend to be larger than males. The pupae are a dark brown colour and usually 4–9 mm long. Larvae are about 1 mm when hatched, and can grow to 10–12 mm long for late instar larvae. The larvae are a pale, yellowish/whitish colour when newly hatched and usually become light green to light brown. The eggs of *Eudemis* are typically laid singly on the host plant, and have a rounded, flat shape (CABI, 2014; Gilligan and Epstein, 2014). The forewing of *Eudemis ambiguella* is yellow or yellowish orange with a well-defined dark-brown to black median fascia. Males and females exhibit no sexual dimorphism in wing pattern although females may be slightly larger than males. Males lack a forewing costal fold. The head, prothoracic shield and legs are dark brown to black. Body colour varies from brown to yellow and green.

Distribution

Lobesia botrana is native to southern Italy and is thought to have originated from Austria. It has been introduced to all Europe, North and West Africa, the Middle East, eastern Russia and Japan (Table 7.1). It was introduced recently (October, 2009) into the western USA in Napa County, California (Zalom *et al.*, 2011) and in Argentina and Chile but seems to occur in very restrictive area and in low populations in those countries, thus is not such as an invasive pest (Gilligan and Epstein, 2014).

Economic importance

These two grape moths give the most trouble to European growers. They are often confused because: (i) they occur often in the same place and at the same time; (ii) they have fairly similar biologies and synchronous flight periods; and (iii) they cause nearly the same type of damage. *L. botrana* has three generations (four in southern Europe) and lives principally in hot and dry situations, whereas *E. ambiguella* has only two generations (three in the hottest places) and lives in fresh and humid situations. Their biologies have been the object of many papers published in a lot of different countries. The caterpillars penetrate into the grapes and the entry holes favour the establishment of the fungus causing grey mould, *B. cinerea*. Studies on the interaction between *L. botrana* and *B. cinerea* have demonstrated a mutualistic relationship between these two organisms (Mondy *et al.*, 1998a, b). The larvae act as vectors of *B. cinerea* inoculum (Fermaud and Le Menn, 1992). In addition, tunnelling larvae facilitate rapid penetration and development of the mycelium on the grape berries (Fermaud and Le Menn, 1989). The preferred host is *Vitis vinifera* (wine grape) and *L. botrana* is one of the most important pests of this plant in the Palearctic region. However, larvae are polyphagous and have been recorded feeding on more than 40 species belonging to approximately 20 families (Table 7.2). Although the grape is its most economically important host, *E. ambiguella* has also been recorded on other plants (Table 7.3) (Thiéry, 2008; Gilligan and Epstein, 2014).

The American grape berry moth, *Paralobesia viteana* (Clemens)

Diagnosis and biology

Adults of *Paralobesia viteana* and *L. botrana* cannot be separated by wing pattern; a genitalia dissection is necessary to confirm identity. Thus *L. botrana* which was discovered in California in 2008–2009 can be confused with the American species. Late instar larvae are 10–15 mm long with a yellowish green to pale brown abdomen. The head and prothoracic shield are yellowish brown and the shield is variably shaded with dark brown to black on the posterior and lateral margins. No morphological characters have been identified to reliably separate the larvae of *Paralobesia* and *Lobesia*. Should *P. viteana* be introduced to the West Coast of the USA or Europe, or *L. botrana* expand out of California, molecular diagnostics would be required to identify larvae of *Paralobesia*/*Lobesia* found on grape (Gilligan and Epstein, 2014). *P. viteana* has three to four generations in Virginia with a possible fifth generation appearing for some hot years (Gilligan and Epstein, 2014).

Distribution

Paralobesia viteana is distributed in North America (Gilligan and Epstein, 2014) in particular the eastern USA (Table 7.1).

Economic importance

This species is the primary lepidopteran pest of grapes in eastern North America and has been considered one of the more severe grape pests in several states. It has the same

Table 7.2. Main host plants of *Lobesia botrana*. (From Gilligan and Epstein, 2014.)

Family	Genus/species	Common name
Actinidiaceae	*Actinidia chinensis* Planch.	Chinese gooseberry
Araliaceae	*Hedera helix* L.	English ivy
Asteraceae	*Tanacetum vulgare* L.	Common tansy
Berberidaceae	*Berberis vulgaris* L.	Common barberry
Caprifoliaceae	*Lonicera tatarica* L.	Tatarian honeysuckle
Caprifoliaceae	*Viburnum lantana* L.	Wayfaring tree
Caryophyllaceae	*Dianthus* sp.	Carnation
Cornaceae	*Cornus mas* L.	Cornelian cherry
Cornaceae	*Cornus sanguinea* L.	Bloodtwig dogwood
Cornaceae	*Cornus* L.	Dogwood
Ebenaceae	*Diospyros kaki* L. f.	Japanese persimmon
Ebenaceae	*Diospyros virginiana* L.	Common persimmon
Ericaceae	*Arbutus unedo* L.	Strawberry tree
Grossulariaceae	*Ribes nigrum* L.	European blackcurrant
Grossulariaceae	*R. rubrum* L.	Cultivated redcurrant
Grossulariaceae	*Ribes uva-crispa* L.	European gooseberry
Lamiaceae	*Rosmarinus officinalis* L.	Rosemary
Liliaceae	*Urginea maritima* (L.) Baker	Red squill
Menispermaceae	*Menispermum canadense* L.	Common moonseed
Oleaceae	*Ligustrum vulgare* L.	European privet
Oleaceae	*Olea europaea* L.	Olive
Oleaceae	*Syringa vulgaris* L.	Common lilac
Punicaceae	*Punica granatum* L.	Pomegranate
Ranunculaceae	*Clematis vitalba* L.	Old man's beard
Rhamnaceae	*Ziziphus jujuba* (L.) Karst.	Common jujube
Rosaceae	*Malus pumila* Mill.	Apple
Rosaceae	*Prunus amygdalus*	Sweet almond
Rosaceae	*Prunus avium* (L.) L.	Sweet cherry
Rosaceae	*Prunus domestica* L.	European plum
Rosaceae	*Prunus dulcis* (Mill.) D.A. Webb	Sweet almond
Rosaceae	*Prunus persica* (L.) Batsch var. *nucipersica* (Suckow) Schneider	Nectarine
Rosaceae	*Prunus salicina* Lindl.	Japanese plum
Rosaceae	*Prunus spinosa* L.	Blackthorn
Rosaceae	*Pyrus communis* L.	Common pear
Rosaceae	*Rubus caesius* L.	European dewberry
Rosaceae	*Rubus fruticosus* L.	Shrubby blackberry
Rosaceae	*Rubus* sp.	Raspberry
Thymeleaceae	*Daphne laureola* L.	Spurge- or dahne-laurel
Thymeleaceae	*Daphne gnidium* L.	Flax-leaved daphne
Thymeleaceae	*Thymelaea hirsuta* (L.) Endl.	Thymelaea
Vitaceae	*Parthenocissus quinquefolia* (L.) Planch.	Virginia creeper
Vitaceae	*Vitis vinifera* L.	Grapevine

biology and causes the same damage as the French *Eudemis* (*L. botrana*) but pupation occurs not under the bark of the vine stock but in a rolled-up leaf on the ground. Grape is the unique larval host of *P. viteana*, although several secondary hosts have been documented (T.M. Gilligan, Animal and Plant Health Inspection Service, Fort Collins, North Carolina, 2017, personal communication).

The Australian grape berry moth or light brown apple moth, *Epiphyas postvittana* (Walker)

Diagnosis and biology

Both sexes have forewings (6–10 mm in males, 9–13 mm in females) that are light brown to pale yellow with brown to dark-

Table 7.3. Main host plants of *Eupoecilia ambiguella*. (From Gilligan and Epstein, 2014.)

Family	Genus/species	Common name
Aceraceae	*Acer campestre* L.	Hedge maple
Araliaceae	*Eleutherococcus* Maxim.	Ginseng
Araliaceae	*Hedera helix* L.	English ivy
Araliaceae	*Hedera* L.	Ivy
Caprifoliaceae	*Lonicera* L.	Honeysuckle
Caprifoliaceae	*Lonicera periclymenum* L.	European honeysuckle
Caprifoliaceae	*Lonicera ramosissima* Franch. & Sav. ex Maxim.	
Caprifoliaceae	*Symphoricarpos* Dunham.	Snowberry
Caprifoliaceae	*Viburnum* L.	Viburnum
Cornaceae	*Cornus* L.	Dogwood
Cornaceae	*Cornus mas* L.	Cornelian cherry
Cuscutaceae	*Cuscuta* L.	Dodder
Cuscutaceae	*Cuscuta reflexa* Roxb.	Giant dodder
Grossulariaceae	*Ribes* L.	Currant
Oleaceae	*Ligustrum* L.	Privet
Oleaceae	*Syringa* X *persica* L.	Persian lilac
Rhamnaceae	*Frangula alnus* Mill.	Glossy buckthorn
Rhamnaceae	*Rhamnus* L.	Buckthorn
Rosaceae	*Prunus* L.	
Vitaceae	*Parthenocissus quinquefolia* (L.) Planch.	Virginia creeper
Vitaceae	*Vitis vinifera* L.	

brown markings (females with a dark mark on the dorsum of each forewing and two dark spots on the posterior of the thorax). Males are more variable than females, with three different forewing phenotypes in California. The hindwing in both males and females is mottled with dark scales (Gilligan and Epstein, 2014). Male genitalia are specific and distinctive, and examination of these structures is essential for reliable identification.

Larvae are generally yellowish green but early and mid-instar larvae range from translucent to opaque reddish brown. The head of all instars is pale brown and the prothoracic shield is approximately the same colour as the rest of the body. The head, prothoracic shield and legs of mid- to late instars are not dark and do not have contrasting markings. The anal shield is pale brownish green. Dimensions vary from 1.6 mm (first instar) to 10–20 mm (last instar) long. Molecular diagnostics are necessary for a reliable positive identification of *E. postvittana* larvae, especially early instars. *E. postvittana* has two to four generations/year, depending on temperature and latitude.

Distribution

Epiphyas postvittana is native to Australia but has been introduced into Tasmania, New Zealand, Hawaii and in California (Gilligan and Epstein, 2014) (Table 7.1).

Economic Importance

Epiphyas postvittana is a pest in various crops: pome and stone fruits and other horticultural crops. It has more than 500 host plants distributed in 121 families and 363 genera. Larvae prefer herbaceous plants over woody ones and they can feed on the leaves, buds and flowers of their hosts. However, the majority of economic damage is caused by fruit injury. They feed on the surface of fruits under webbed leaves, causing scarring as well as providing a site for rot or infection. Larval damage to fruit crops in Australia and New Zealand during years with outbreaks can reach 70%.

The small grape moths – *Argyrotaenia* spp.

There are several small grape moths of the genus *Argyrotaenia* on grapes in the world including the orange tortrix or apple skinworm, *Argyrotaenia franciscana* (Walsingham), the grape tortrix or eulia, *Argyrotaenia ljungiana* (Thunberg) and the red-banded leafroller, *Argyrotaenia velutinana* (Walker). *A. ljungiana* was mentioned for the first time in France 60 years ago (1954, cited in Bournier, 1976), coming from orchards or confused with *Eudemis* to which it resembles.

Diagnosis and biology

- *Argyroteania franciscana* has grey to brownish grey forewings with a dark, variably defined median fascia and an outer spot on the costa. Hindwings are primarily grey. However, wing pattern and size can be variable. Males lack a forewing costal fold. Head and prothoracic shield of larvae are light brown and unmarked. Larvae are usually pale to dark green even if abdomen colour varies with host plant. *A. franciscana* larvae are similar to those of many other Archipini so molecular diagnostics are necessary for a reliable positive identification. It is a bivoltine or multivoltine species, depending on temperature and thus on location. In warmer areas of California, larvae aestivate during summer with only two generations. In cooler coastal areas, there may be up to five continuous overlapping generations with adults present year-round (Gilligan and Epstein, 2014).
- *Argyroteania ljungiana* has pale brown to silvery white forewings with dark reddish brown to grey markings (a well-defined median fascia and outer costal spot). Hindwings are greyish brown. Males lack a forewing costal fold. Adults may appear similar to other species of *Argyrotaenia* and thus genitalia dissection is necessary for a reliable identification. Late instar larvae are pale green with a yellowish brown head, a yellowish green prothoracic shield with a black posterolateral mark and black shading on the posterior margin (Gilligan and Epstein, 2014).
- *Argyroteania velutinana* has pale brown to golden brown forewings with a reddish brown median fascia. Generally there is a dark mark or partial fascia at the base of the wing, a reddish brown outer costal spot, and a row of near-white scales along the termen that may extend to the median fascia in some individuals. Hindwings are greyish brown. Males lack a forewing costal fold. Adults can appear similar to other species of *Argyrotaenia* and thus genitalia dissection is necessary for a reliable identification. Late instar larvae are 13–18 mm long with a green to yellowish green abdomen; the head, prothoracic shield and thoracic legs are yellowish green and unmarked. The green unmarked larva can be confused with the larva of many other tortricids (e.g. other species of *Argyrotaenia*, *E. postvittana* and *Choristoneura rosaceana* (Harris)) and molecular diagnostics may be necessary for a reliable positive identification. *A. velutinana* has two to three generations depending on latitude, two with a partial third possible in the north and a possible fourth generation in the hotter south. Overwintering occurs at the pupal stage (Gilligan and Epstein, 2014).

Distribution

A. velutinana occurs in the eastern part of the USA while *A. franciscana* occurs in the western part (Gilligan and Epstein, 2014) and *A. ljungiana* in Europe (Bournier, 1976) (Table 7.1).

Economic importance

The damages inflicted by the three species are the same. *A. franciscana* has been mentioned as one of the most polyphagous tortricid species in North America. Its host list includes plants in more than 40 families, many of which being important crops (Gilligan and Epstein, 2014). Larvae of *A. velutinana* are highly polyphagous and have been described by Freeman (1958) as feeding 'on almost any plant'. This includes several conifers (Gilligan and Epstein, 2014). Larvae of these species can cause economic damage by directly feeding on developing fruit in citrus, apple and grape. Larvae may also feed on stems, causing fruit to drop.

The omnivorous leafroller, *Platynota stultana* Walsingham

Diagnosis and biology

The male moth has dark brown (basal half) to golden brown (distal half) forewings although the female has more uniform golden brown to dark brown forewings with markings that are usually less distinct. Both sexes have labial palpi that are extremely elongated. Males have a forewing costal fold. This species is very similar to other *Platynota* species and genitalia dissection is necessary to confirm identity.

Late instar larvae are 12–15 mm long, with a cream-coloured and translucent abdomen, and a yellowish brown head and prothoracic shield, with the posterolateral margins on the prothoracic shield shaded with dark brown in some individuals (Gilligan and Epstein, 2014).

Platynota stultana has four to six generations/year with adults that may be present year round. Newly hatched larvae move towards the top of the plant and feed within a bud or between two leaves. They disperse to other hosts by ballooning in the wind on a silk thread. Late instar larvae feed within a shelter constructed of rolled or folded leaves.

Distribution

This species is recorded from Arizona, California, Florida, Hawaii, Mexico and Texas (Gilligan and Epstein, 2014) (Table 7.1).

Economic importance

Larvae of *P. stultana* feed on a lot of plant species belonging to more than 20 families (Gilligan and Epstein, 2014). It is a serious pest in greenhouses, vineyards and other commercially important hosts (alfalfa, *Citrus* sp., maize, cotton, grape, peach, pear and pepper).

The grape tortricid, *Sparganothis pilleriana* (Denis & Schiffermüller)

Diagnosis and biology

This species is a nocturnal moth 20–25 mm long, that has yellow forewings with copper highlights marked by three brown bands, and hindwings that are purplish brown with a lighter fringe. The caterpillars have a greenish body and black head and can reach up to 20–30 mm long. The first instar hibernates under the bark and they devour buds in spring before attacking the leaves and fruits later. They also surround leaves and bunches of grapes in countless silk-like threads, disturbing their normal development.

Distribution

This species is found only in Europe (Gilligan and Epstein, 2014) (Table 7.1).

Economic importance

This species in univoltine and not too polyphagous. Larvae feed on *Salix repens*, *Fragaria* spp. and grape.

The grape root borer, *Vitacea polistiformis* (Harris)

Diagnosis and biology

This moth belongs to the family Sesiidae and not to the family Tortricidae, which is the family of all the other moths already mentioned. The moths are wasp-like in appearance with a brown body, an orange head with orange antennae with brown-black markings, a dark brown abdomen with reddish brown markings, a very narrow yellow band on the posterior edge of segments two, four and sometimes six and orange legs with brown-black markings. The forewings are dark and mostly opaque but the hindwings are more transparent.

The eggs hatch on the soil surface and the larvae tunnel into the root system. Overwintering of *V. polistiformis* intervenes at two larval stages. The life cycle takes 2 years to complete (but some studies indicate a 3-year life cycle). Almost all the life cycle is spent as larvae feeding on grape roots. They bore into the roots and crown below the soil surface, and the damage caused by the borer results in reduced vine growth, small leaves and reduced berry size, thus there is a great reduction in the productivity of the vine. The first sign of the presence of this pest is the lack of plant vigour. The presence of cast pupal skins protruding from the soil near the base of the trunk in late July and August is another clear sign. Larvae leave the roots, pupate in cocoons near the soil surface in June and adults emerge 35–40 days later in July, with greatest numbers present in the last 2 weeks of July.

Distribution

It is the most serious threat to grapes in eastern USA, especially in Florida (Table 7.1).

Economic importance

Because damage is restricted to below ground, problems go unnoticed until the vines start to die by complete root destruction.

The Grape Hemiptera Leaf- and Planthoppers

A number of cicadellid species are vectors of phytopathogenic agents (see Table 7.1 for their distribution). They feed in the phloem or xylem tissues with their piercing-sucking mouthparts. They lay eggs in stems or leaves or under the bark. Sharpshooters and most species of grape leafhoppers feed on many types of green vegetation during winter. They enter vineyards soon after bud break to feed and reproduce. The cicadellids *Scaphoideus titanus* Ball and *Oncopsis alni* (Schrank) overwinter as eggs under the bark of their host plants, whereas the cixiid *Hyalesthes obsoletus* Signoret hibernates as nymphs on plant roots (Sforza *et al.*, 1999). *S. titanus* is univoltine whereas other species have two to four generations and sometimes five to seven in hot places. Bentley *et al.* (2008) listed 17 species of leafhoppers as grape pests worldwide because of their feeding habits. Within the group of phloem-sap feeders, two leafhopper species are recognized to be vectors of phytoplasma diseases in grape: (i) *S. titanus*, the vector of flavescence dorée or golden yellow (Bournier, 1976); and (ii) *O. alni*, the vector of Palatinate grapevine yellows. The planthopper *H. obsoletus* is demonstrated to be the vector of bois noir or black wood (Sforza and Boudon-Padieu, 1998). *S. titanus* and *H. obsoletus* are considered to be of great economic importance.

Among the xylem-sap feeders, numerous species of sharpshooters are known to be vectors of *Luella fastidious* Wells, the bacterial causal agent of Pierce's disease recently introduced in Italy and Corsica. The blue-green sharpshooter, *Graphocephala atropunctata* (Signoret), the red-headed sharpshooter, *Carneocephala fulgida* Nottingham, the green sharpshooter, *Draeculacephala minerva* Ball, and the glassy-winged sharpshooter, *Homalodisca vitripennis* (Germar) (formerly known as *Homalodisca coagulata* (Say)), are among those most frequently mentioned (Olivier *et al.*, 2012).

The invasion risk posed by the xylem-feeding hemipteran, *H. vitripennis*, and the xylem-dwelling phytopathogenic bacterium, *Xylella fastidiosa*, was examined using the computer climate modelling program CLIMEX by Hoddle (2004). Model predictions indicated that suitable climatic conditions for *H. vitripennis* and strains of *X. fastidiosa* exist in almost all grape production areas of the world (Hoddle, 2004). CLIMEX predicted that cold stress accumulation would exclude Pierce's disease causing strains of *X. fastidiosa* from France and northern and central grape-producing areas of Spain and Italy. This result is incongruous with Pierce's disease reports from Kosovo in the Balkans (Hoddle, 2004) and with recent introductions in Italy and France. *Xylella fastidiosa* subsp. *pauca* was recorded in 2013 in Puglia (Italy) causing serious damage to olive groves and it was detected in numerous other host plants, mainly ornamentals (EPPO, 2016). *Xylella fastidiosa* subsp. *multiplex* was discovered in France on the island of Corsica (France) on *Polygala myrtifolia* L. (ornamentals) 2 years later in October 2015. It was then found on the French mainland, first in Nice and Mandelieu-la-Napoule (Alpes-Maritimes department), and then in other places in Alpes-Maritimes and Var departments (EPPO, 2016). Most infected plants were *P. myrtifolia* and the subspecies which occurs is *X. fastidiosa* subsp. *multiplex* in

France, thus differing from the subspecies found in Italy. Eradication measures have been implemented immediately in all infected areas in Italy and France (EPPO, 2016).

The Typhlocybinae cause direct damage to grapes by emptying the mesophyll cell contents or by cell rupture feeding. This subfamily comprises many serious grape pests such as *Erythroneura comes* (Say), *Erythroneura elegantula* Osborn, *Erythroneura variabilis* Beamer, *Erythroneura vitis* (Harris), *Erythroneura vulnerata* Fitch and *Erythroneura ziczac* Walsh, as well as *Empoasca fabae* (Harris) in North America, *Arboridia adanae* (Dlabola) in Mediterranean regions, *Jacobiasca lybica* (Bergevin and Zanon), *Empoasca vitis* (Göthe) and *Zygina rhamni* Ferrari in Europe.

The number of recognized leafhopper pests is expected to increase a lot due to global warming which causes changes in areas of distribution and the need for development of new techniques to sample, detect and identify pathogens and their vectors (Foissac and Wilson, 2010).

The grape mealybugs and soft scales

Economic losses due to vineyard mealybugs and soft scales (Table 7.1) have increased dramatically during the past decade, due to major changes in the use of insecticides, many of the efficient ones now being forbidden.

The grape mealybugs

DIAGNOSIS AND BIOLOGY. These insects are named mealybugs because of the powdery secretions covering their bodies. Species mentioned in Table 7.1 are referred to here as grape mealybugs although they are polyphagous. Outwardly, the vineyard mealybugs look similar. Mealybug females are wingless with an elongate oval-shaped body 3–5 mm long that can be covered with wax secretions forming distinct spine-like filaments. Each species has distinct biological characteristics that result in different geographic ranges, host plant preferences and economic injuries. Complications in their reliable identification are due to great similarities for some species, with the need of a great expertise in identification and sometimes molecular tools for few species, and to the fact that these pests have been often moved from their geographic origin and that many are now found in many parts of the world in mixed populations (Daane *et al.*, 2012). All species have generally three larval instars for the female and four for the male. The unsettled first instar or crawler, 0.6 mm long, moves quickly to find a feeding spot and this is considered to be the dispersal stage. The successive stages increase in size and amount of wax secretion. Females are unwinged and mature males are winged. To attract adult males, females emit a sex pheromone. Most vineyard mealybugs place their eggs in cotton-like ovisacs. *Pseudococcus longispinus* Targioni Tozzetti, *Ferrisia gilli* (Gullan), *Dismicoccus brevipes* (Cockerell) and *Heliococcus bohemicus* Sulc are ovoviviparous. The number of offspring ranges from 50 to 800 depending on the species, environmental conditions and food supply. The number of generations is very variable even in one region for the same species, depending on temperature. Except for *Planococcus ficus* Signoret and *H. bohemicus*, there is no diapause. Mealybug populations overwinter under the bark. From spring to summer, mealybug populations follow the movement of plant resources from roots to shoots to leaves to berry clusters, when berries ripen and sugars increase (Daane *et al.*, 2012).

DISTRIBUTION. *Ps. longispinus* is present in western USA and New Zealand, *P. ficus* in western USA, Argentina, Europe, the Middle East and South Africa, *F. gilli* in western USA, *D. brevipes* in Brazil, and *H. bohemicus* only in Europe (Table 7.1).

ECONOMIC IMPORTANCE. Mealybugs are phloem-feeders that use long mouthparts to suck plant fluids. Most of them can feed on vine roots, the trunk, buds, shoots, canes, leaves and berries. However, the extent of damage caused varies for each species. They excrete honeydew that serves as a substrate for the development of sooty mould fungi that can result in further vine damage, especially on table grapes. Feeding damage can result in defoliation and in the case of repeated annual infestations, cause vine death (Daane *et al.*, 2012). In most of the world's wine-grape-growing regions, the transmission of viruses rather than mealybug feeding or contamination is the primary concern. Grapevine leaf-roll disease (GLD) is caused by a complex of several viruses (closterovirus), collectively known as grapevine leaf-roll-associated viruses (GLRaVs). In cool climate conditions, the pathogens can damage vines, the crop and wine quality. The most obvious GLD symptoms become apparent when red-grape cultivars display leaf reddening with green venation. Symptoms are not so apparent in white-grape cultivars, other than a slight leaf chlorosis. Both red- and white-grape cultivars develop the classic downward rolling of leaf margins and phloem disruption. GLRaV infection impacts berry development and growth by delaying bud break, flowering and berry maturation, including changes in colour, reduced sugar content and increases in acidity of the juice. *P. ficus* was first demonstrated to be a vector but then, several species of soft scales have been shown to transmit GLRaVs: (i) *Pseudococcus calceolariae* (Maskell); (ii) *Ps. longispinus*; (iii) *Pseudococcus maritimus* (Ehrhorn); (iv) *Pseudococcus viburni* (Signoret); (v) *Planococcus citri* Risso; and (vi) *H. bohemicus*. Additionally, GLRaVs can be transmitted by *Parthenolecanium corni* (Bouché) (see below). Although all mealybug and scale life stages may be capable of transmitting GLRaVs, the smaller stages appear to be more efficient. Virus and mealybugs can both survive many years on the vine roots after the vine above ground has been removed (Daane *et al.*, 2012). The rugose wood complex is caused by several viruses, including rupestris stem pitting associated virus (RSPaV), grapevine virus A (GVA) and grapevine virus B (GVB). Spread is mainly through propagation. GVA and GVB can be transmitted by some species of mealybugs.

The grapevine soft scales

DIAGNOSIS AND BIOLOGY. The group of soft scales includes several species of the family Coccidae, the most important ones being the grape soft scales, *Parthenolecanium corni* (Bouché) and *Parthenolecanium persicae* (Fabricius) and the cottony maple scale *Neopulvinaria innumerabilis* (Rathvon). Grapevine soft scales are small oval-shaped sucking insects up to 6 mm long that live beneath a protective dark brown wax cover. During spring and summer, mature scales deposit hundreds of eggs under their bodies and then die. Crawlers hatch and move to the leaves to feed. Later, they move back to the canes to feed and remain on during winter (Dunn and Zurbo, 2014).

Neopulvinaria innumerabilis is a small, flattened, brown scale insect about 3 mm long. In early summer mature females begin to secrete white, waxy, cottony-appearing egg sacs in which they lay as many as 1500 eggs. Severely infested trees look like they are covered with strings of popcorn.

DISTRIBUTION. *Parthenolecanium corni* is present in Europe and South America while *P. persicae* is present in Europe, Chile and Australia and *N. innumerabilis* only in Europe (Table 7.1).

ECONOMIC IMPORTANCE. The insects feed mainly on the stems or canes and if large populations occur, vine growth and grape production can be reduced. The major problem with grapevine scale is that they excrete honeydew which falls on to grapevine leaves and bunches leading to sooty mould development and photosynthesis reduction, with subsequent growth and productivity reduction. *P. persicae* has one annual generation whereas *P. corni* has two. Immature scales overwinter on the previous season's wood and begin maturing in spring. *N. innumerabilis* can be found on all species of maples (*Acer* spp.) but have a strong preference for silver maple. It is also known to be able to survive on honey and black locust, white ash, *Euonymus*, oak, boxelder, dogwood, hackberry, sycamore, beech, elm, willow, basswood, poplar and vines. While conspicuous, *N. innumerabilis* infestations usually have little impact on established trees and vines. However, in large numbers they can cause premature leaf drop and twig dieback. Heavy infestations can cause leaves to turn yellow to light green and may cause stunted leaf growth. The rugose wood complex is caused by several viruses, including RSPaV, GVA and GVB. Spread is mainly through propagation. *N. innumerabilis* can transmit GVA.

Other Insects

The fig longicorn borer *Acalolepta vastator* (Newman)

Diagnosis and biology

The adult beetle is about 3 cm long and has antennae longer than the body. Adult emergence is protracted over the months between spring and autumn. Females lay eggs in fissures or cracks in the grapevine bark or near the base of canes. Larvae hatch and bore into the vine wood and can tunnel throughout the trunk and into roots. They are cream with a brown head and grow to 4 cm in length. Pupation occurs in the tunnel and the adult emerges from the trunk by chewing a hole. Larval excrement and sawdust are often visible in tunnels and around the vine trunk indicating an infestation.

Distribution

This species has become a major pest in Australia, limited at the moment to a small area of New South Wales in Australia (Table 7.1).

Economic importance

The fig longicorn borer can cause extensive damage to the vine trunk which can lead to dieback and significant crop losses. Borers are difficult to control because boring stages are not accessible (Dunn and Zurbo, 2014).

The grape rootworms *Fidia viticida* (Walsh) and *Fidia longipes* (Melsheimer)

Diagnosis and biology

Grape rootworm larvae overwinter in the soil around grapevine roots. The entire life cycle takes 1 or 2 years to complete. The feeding period is in spring. Beetle larvae pupate near the surface in late May and June when the grapevines bloom. Pupation takes about 2 weeks. Adult beetles are 6 mm long, brown and have yellowish hairs. Adults feed on leaves for about 1 month, leaving chain-like holes in the leaves. This feeding damage is similar to that made by *Altica* spp. Female beetles lay clusters of 20–30 eggs under the loose bark on canes. After 1–2 weeks, larvae hatch, drop to the ground, burrow into the soil and begin to feed on grapevine roots.

Distribution

The two species are common in eastern USA (Table 7.1), of different importance in each state and very difficult to separate (SRIC, 2012).

Economic importance

Smaller roots may be totally consumed. Larger ones become pitted from feeding activity. Damage to the root system can be very severe: (i) vines become unthrifty; (ii) fewer grapes are produced; and (iii) death occurs after several years.

The grass grub beetle, *Costelytra zealandica* (White)

Diagnosis and biology

Between 10 and 15 days after female adults have been mated, they lay their 100 small, white and oval-shaped eggs into wet or loose soil. The adult beetle can grow up to a mature length of 13 mm. It has a shiny brown colouring. These beetles can be seen flying in the evenings during the summer months. The eggs are around 1.5 mm in diameter. About 3–4 days after being laid, the eggs turn smooth, spherical and start to swell by absorbing moisture. Once the eggs are mature they turn a dark brown/black colour. Once the mature eggs have hatched, larvae have a whitish/grey C-shaped segmented body, with a darker coloured head and dark brown jaws. The size of the larvae ranges from 5 mm to 25 mm in length. The abdomen of the larva is very swollen and looks bigger than any other part of the body. There are many thin fine hairs along both sides of the body, especially around the abdomen. The pupae start off the same creamy white colour as the larvae, but slowly turn to a dark brown during pupation. They have a soft body and are generally shorter and thicker than the grass grub larvae, they grow to between 10 and 30 mm in length.

The New Zealand grass grub beetle has an annual life cycle, taking typically 1 year under normal climatic conditions but can take up to 2 years in adverse weather such as drought or significant cold periods (this also includes periods where the soil moisture is very high), or due to a food shortage. Over that 1-year period the adult beetles live for 4–6 weeks emerging in spring (October–December in New Zealand) when the air temperature is above 10°C, and undertake two main flights, the first flight is to mate and the second flight is to feed and lay clutches of eggs. Females are the last to emerge from the soil, this leads to localized infestations due to them being unable to reach a significant distance before being mated. Egg cluster size can range from five to 80 eggs. *C. zealandica* larvae hatch between November and March.

Distribution

This species is endemic to New Zealand and one of the most common insects. It is found throughout except in high altitudes above 1200 m (Table 7.1).

Economic importance

Insects start to feed on roots of clover and pasture plants from the time they hatch until September but they can cause damage to vineyards in late spring, when they defoliate leaves and damage shoots and inflorescences. Depending on population densities, beetles will first damage leaf edges which may be consumed right down to the main leaf veins (NZW, 2013).

The Japanese beetle, *Popilia japonica* Newman

Diagnosis and biology

Larvae can be distinguished from other common scarabaeid larvae by the V-shaped arrangement of the last two rows of spines on the ventral surface of the last abdominal segment. Adult beetles are 12 mm long, metallic, with a shiny golden green thorax, lateral tufts of white hair on the abdomen, and two patches of white hair on the pygidium. It overwinters as a larva (usually the third instar) in a cell, about 15–30 cm deep in the soil. In the spring, when the soil temperature exceeds 10°C, the larvae resume feeding on plant roots at about 5 cm depth. Pupation usually occurs after a few weeks' feeding and the beetles emerge in late May to early July, depending on latitude. The average life of the adults is 30–45 days and the eggs are laid in the soil. After hatching from the eggs, the larvae feed in the soil. Normally, there is one generation/year but, at the northern edge of its range a few individuals may need 2 years to complete the life cycle. The adult beetles feed on foliage, flowers and fruits. They are gregarious and many beetles can be collected together on a single plant.

Distribution

Popilia japonica is found in Japan, eastern USA, Canada, the Azores and Italy (Table 7.1).

Economic importance

In the USA, *P. japonica* has been recorded feeding on at least 295 species of plants, economic damage being recorded on 106. The food preference of the beetles changes during the year but preferred hosts include species of: *Acer*, *Aesculus*, *Betula*, *Castanea*, *Glycine*, *Juglans*, *Malus*, *Platanus*, *Populus*, *Prunus*, *Rosa*, *Rubus*, *Salix*, *Tilia*, *Ulmus* and *Vitis*. In Japan, the host range appears to be smaller than in North America. Within the EPPO (European and Mediterranean Plant Protection Organization) region the host range of *P. japonica* would be similar: *Malus*, *Prunus*, *Rubus* and *Vitis*, with their wide distribution and intensive cultivation especially providing favourable food sources for the Japanese beetle. Individual plants may be completely defoliated, while adjacent ones remain virtually undamaged. The beetles prefer the young leaves and they eat out holes between the leaf veins, skeletonizing the leaves. Such injury becomes very severe in young vines, very vulnerable to complete defoliation. Feeding on flowers and fruits is not unusual. Pastureland is the preferred habitat of larvae while grape is the preferred adult food source (SRIC, 2012).

The drosophila or vinegar flies, *Drosophila melanogaster* Meigen and *Drosophila suzukii* Matsumura

Drosophila flies are not usually serious pests of grapes. Sometimes, outbreaks of *Drosophila melanogaster* and *Drosophila simulans* can cause bacteria development which causes acetic sour rot in the wine produced.

Diagnosis and biology

Drosophila suzukii belongs to the *melanogaster* species group. *D. suzukii* adults are 2–3 mm long with red eyes, a pale brown or yellowish brown thorax and black transverse stripes on the abdomen. The antennae are short and stubby with branched arista. Sexual dimorphism is evident: males display a dark spot on the leading top edge of each wing and females are larger than males and possess a large serrated ovipositor. The eggs are oval 0.2 mm, milky white, with two filaments at one end, 0.4–0.6 mm long. The maggot-like larvae are white with visible internal organs and black mouthparts. They grow throughout three larval stages and when fully grown can reach 5.5 mm long and 0.8 mm wide. The pupae are spindle-shaped, reddish brown and bear two stalks with small finger-like projections, 3.5 mm long and 1.2 mm wide. This species has a high reproductive rate and short generation time: *D. suzukii* can theoretically have up to 13 generations/year, which may contribute towards rapid spread, given available suitable hosts (SRIC, 2012).

Distribution

The spotted wing drosophila (SWD) fly, *D. suzukii* is of major concern in North America and in Europe (Bordeaux) (Table 7.1). This species is thought to be native to Eastern and South-east Asia, including China, Japan and Korea (Walsh *et al.*, 2011).

Economic importance

Drosophila suzukii is a fruit-crop pest and is a serious economic threat to soft summer fruit. As a polyphagous pest, it infests a wide range of fruit crops, including grape, as well as an increasing number of wild fruits. It is an economically damaging pest because the females are able to infest thin-skinned fruits before harvest and the larvae destroy the fruit pulp by feeding. It was first recorded as invasive in Hawaii in 1980 and then simultaneously in California and in Europe in 2008. Since 2008 it has spread rapidly throughout the temperate regions of North America and Europe, due to global trade and the initial lack of regulation over the spread of any *Drosophila* spp. Larvae cause damage by feeding on the pulp inside fruit and berries. The infested fruit begins to collapse around the feeding site causing a depression or visible blemish on the fruit. The oviposition scar exposes the fruit to secondary attack by pathogens and other insects, which may cause rotting (Hauser *et al.*, 2009; Walton *et al.*, 2010). Fruits become susceptible to *D. suzukii* as they start to change colour, which coincides with softening skins and high sugar levels (Burrack *et al.*, 2013). Costs of losses due to this species are estimated to be several hundred millions euros.

The Grape Mites

The first infestations of spider mites in European vineyards were detected in the second half of the 19th century after the invasion of powdery mildew, downy mildew and phylloxera from North America. At that time, problems with mites injurious to grapes were negligible but the situation suddenly changed after World War II. Most of these problems were immediately associated with the extensive use of chlorinated hydrocarbon insecticides and later, by ethylene-*bis*-dithiocarbamate (EBDC) fungicides and organophosphate insecticides. Actually, even if mites are very often very minor pests, they are an excellent indicator of problems induced by intensive chemical sprayings. Mites are in the subclass Acari within the class Arachnida and are therefore closely related to spiders and scorpions. They have two body regions, no antennae, no visible eyes and usually four pairs of legs. The easiest way to distinguish between different mite pests of grapevines is, when possible, by the damage they cause. However, accurate identification of mite specimens is often required and a microscopic magnification of at least 40× is necessary.

The erineum mite, *Colomerus vitis* (Pagenstecher)

Diagnosis and biology

The grape leaf erineum mite *Colomerus vitis* is 0.15–0.18 mm long, creamy white in colour, wormlike, with two pairs of anterior legs (the two posterior pairs have disappeared in immature and adult stages). Adult females lay eggs during spring inside the swelling bud. Immature mites move to the lower surface of leaves and cause erinea, which are very obvious blisters on the upper surface of leaves and the lower leaf surface of each blister has white or brown hairy growths within the raised blister. Erineum mites overwinter inside buds as females but after budburst they move on to leaves to feed and complete their life cycle within the hairy blister.

Distribution

This species has a worldwide distribution (Table 7.1).

Economic importance

Damage can be widespread but does not usually have any economic importance, except in some regions where blisters appear in spring on the peduncles of young berries, causing berry drop. Australians distinguish a bud strain, possibly another species (but still called *Co. vitis*). This 'strain' can lead to malformed leaves, aborted or damaged bunches, tip and even bud death. Many of these symptoms are caused by *Calepitrimerus vitis* (Nalepa) (discussed next) but if this 'bud mite' exists and is a distinct species, it is far more dangerous than the erineum mite (Duso *et al.*, 2012). *Co. vitis* was recently demonstrated to transmit the grapevine Pinot Gris virus (Malagnini *et al.*, 2015).

The grape-leaf rust mite, *Calepitrimerus vitis* (Nalepa)

Diagnosis and biology

The grape-leaf rust mite is 0.12 mm long, and cream to pink in colour. Other characteristics are just like those of *Co. vitis.* Rust mites overwinter under the bark or inside buds. About 2 weeks after bud burst, most of the mites have migrated to the developing shoots and leaves. During the growing season, rust mites can disperse by active movement across overlapping parts of the canopy. Between three and 12 generations a year are likely. Mites start to migrate to their winter shelters very early, from early August to the end of September. Early season rust mite damage can be confused with bud mite or cold injury or with *Eutypa* spp. and esca symptoms (esca is a fungal disease of mature grapevines) as the leaf distortion or crinkling symptoms and poor shoot growth can be similar. The damage is most obvious from bud burst to when five to eight leaves have emerged. The damage then becomes much less visible as the shoots recover and grow out. Severe early spring damage can still be detected in mature leaves through the growing season. Symptoms of restricted spring growth have been recently attributed to feeding by rust mite. The most visible and easily recognizable symptoms of rust mites occur from late July to early September in the northern hemisphere. The leaves start to darken and take on a bronzed appearance because of the presence of rust mites feeding on and damaging the surface cells of the leaf.

Distribution

This species has a worldwide distribution (Table 7.1).

Economic importance

Early spring damage can be severe for vines growing at the extremes of their normal growing area where conditions are not suitable for spring growth (i.e. at altitude, on north-facing slopes, or humid areas). The damage caused can be very harmful for young plants. Despite the high number of rust mites, no impact on wine has been found during summer symptoms but there may be a possible effect on reserves and flowering the next year (Duso *et al.*, 2012).

The flat mites, *Brevipalpus chilensis* Baker and *Tenuipalpus granati* Sayed

Diagnosis and biology

The flat mite adults are 0.3 mm long, flat, shield shaped and yellowish to reddish brown in colour. Eggs are oval and bright red for some species and deposited throughout the vine. The six-legged larvae, more lightly coloured than the adults, moult to become eight-legged nymphs which moult into adults. In spring, flat mites feed on developing canes and after that on the under surface of leaves.

Distribution

Brevipalpus chilensis is found in Chile and *Tenuipalpus granati* is found all over the Mediterranean area. It has been mentioned on grapes in Egypt (Hassan *et al.*, 1986), Greece (Papaioannou-Souliotis *et al.*, 1994), Sicily (Vacante and Tropea-Garzia, 1987), Tunisia (Kreiter *et al.*, 2002) and Morocco (Serge Kreiter, unpublished data) (Table 7.1).

Economic importance

Early season damage is confined to small dark spots or scars around the base of canes. The mites then move to the bunch stalks, berry pedicels and berries. These mites are sap suckers and causes chlorosis or yellowing of leaves. Damage to the bunch stalks and pedicels can partly starve the berries, preventing sugar accumulation. Dark spots may occur and cause aesthetic damage to the table grape. The adults spend the winter under the outer bud scales and under the rough bark at the base of the canes.

Tenuipalpus granati is not a major pest but is very often present on grapes in the Mediterranean.

Some other species occur on vines around the world, for example *Brevipalpus californicus* (Banks) and *Brevipalpus lewisi* McGregor in Australia and Europe. These polyphagous species are pests in other crops and have occasionally induced outbreaks on grapes for disturbance reasons. They are not considered as real pests.

Mites of the genus *Tetranychus*

The two-spotted spider mite, *Tetranychus urticae* Koch, the Pacific spider mite, *Tetranychus pacificus* McGregor and the McDaniel spider mite, *Tetranychus mcdanieli* McGregor, are 0.5 mm long. They are pale coloured and have two distinct dark spots on their body, especially *T. urticae*. Development is similar to the flat mites, is as short as 1 week and a lot of generations can be completed in a season. The three species cannot be distinguished by comparing females and accurate identification needs male genitalia dissection or at least microscopic observation.

Tetranychus urticae

Tetranychus urticae is a polyphagous species (more than 1100 host plants) that colonizes grapes as well as several weeds occurring in vineyards (Boller *et al.*, 1985). *T. urticae* has, however, lower performance on grape than on apple leaves. The potential of this mite to increase its population level increases with high temperatures (at 33–38°C, the egg to egg period is about 5.5 days) associated with low rates of relative humidity (Van de Vrie *et al.*, 1972). Females are usually the overwintering stage. However, in dry, warm climatic conditions (e.g. south-western Spain, Morocco) juveniles can also be found overwintering on weeds at protected sites and some non-diapausing strains may be found (Allam, 2000). In spring, females can resume their activity before bud swell, and mites have to colonize weeds where the first generation appears. In central Europe (e.g. Switzerland) females come out of diapause in March and April and most of them move to weeds (Baillod *et al.*, 1989). In Spain mites disperse from weeds to grapevine leaves in spring (Arias and Nieto, 1991). In Switzerland and France, dispersal from weeds to grapevine occurs in early summer (Kreiter *et al.*, 1991). In Spain, *T. urticae* can develop up to 15 generations/year (Arias and Nieto, 1991). In late summer, the spider mites enter into diapause, depending on temperature and photoperiod, and move to overwintering sites.

The use of some herbicides favours the migration of spider mites to vines (Kreiter *et al.*, 1991).

Tetranychus urticae feeds predominantly on the spongy mesophyll and palisade cells on the leaf under-surface, causing a loss of chlorophyll, leaf discoloration and leaf drop (Candolfi, 1991). On defoliated vines, *T. urticae* can feed on any green tissue including shoots and berries (Arias and Nieto, 1981). In semi-dry climatic conditions, serious damage can be observed from July onwards, and the level of grapevine defoliation is related to early intense feeding (Arias and Nieto, 1978, 1981). Feeding intensity of *T. urticae* dramatically increases from 10°C to 35°C (Candolfi *et al.*, 1991) which explains why spider mite populations cause more damage in hot climate areas. Additional factors affecting the degree of damage are: (i) the time and duration of infestation; (ii) the cultivar; and (iii) soil moisture. Increasing mite densities result in significant reduction of the net photosynthesis, transpiration, as well as stomatal and mesophyll conductance (Candolfi *et al.*, 1992). These cumulative population densities caused adverse effects to 1–2-year-old wood and roots but not to yield and berry quality and to sugar contents (Candolfi, 1991).

Tetranychus pacificus

Tetranychus pacificus has the same characteristics as *T. urticae* but is less polyphagous (around 50 host plants). It feeds on both sides of the leaves by puncturing the cell walls and extracting the contents. Initial colonies cause stippling on the leaves, which increases with increasing population densities, leading to a general browning or scorching of the leaves and premature leaf fall. When *T. pacificus* is abundant, the mites cover the shoot terminal with webbing. In severe attacks, especially in hot weather, the shoots are distorted. Damage occurs from June to August. Like other tetranychid mites which infest plant leaves, the mites remove photosynthetic tissue and therefore reduce the plant's photosynthetic capacity (on grapes, Welter *et al.*, 1989) and can, depending on infestation levels, reduce vegetative growth and crop yield. The degree of damage depends on infestation levels, the crop and other biotic and abiotic stresses. *T. pacificus* can occur with *Eotetranychus willamettei* McGregor on grapes.

The McDaniel mite, T. mcdanieli

Tetranychus mcdanieli is a serious pest of deciduous fruit trees and grapes in North America (15 host plants mentioned). In Europe, it was first detected in 1981 in the Champagne region of France, where it seemed to be localized. Females overwinter under the bark, and they feed in spring first on vine buds and then on leaves. In North America the development from egg to adult requires about 8 days and seven to nine generations can be completed within a year. In France, depending on climatic conditions, populations peak in July and August. High temperature, low humidity and cultivar attributes are key factors promoting the growth of mite populations. Mites spin profuse webbing, and at high infestation levels, leaves can mat together. The first adverse effect of feeding is chlorophyll loss, followed by leaf discoloration and drop. Leaf discoloration and drop are more pronounced in hot and dry seasons. The impact of this spider mite on grapevine physiology and yield parameters is unknown (Duso *et al.*, 2012).

Mites of the genus *Panonychus*

The European red spider mite, *Panonychus ulmi* Koch is considered the most important spider mite in European vineyards with

T. urticae and *Eotetranychus carpini*. It is widespread in temperate regions and less damaging in hot regions. It is a polyphagous tetranychid mite (with around 140 host plants mentioned). It overwinters as eggs laid on 1- or 2-year-old branches (usually at the insertion of shoots) from late summer onwards. Eggs hatch from April to May and the first generation develops at sprouting. The first generation can be completed in about 20 days. The number of generations per year is four to six in central Europe to six or seven in southern Europe. Demographic parameters of *P. ulmi* have been estimated using apple leaves as a plant substrate. The spatio-temporal distribution of *P. ulmi* has been thoroughly studied. The head of the trunk (double Guyot system) is the preferred oviposition site, nevertheless a significant percentage of winter eggs are also laid in the basal part of branches. At the beginning of the growing season, the spider mites prefer basal leaves and later, leaves located at the middle of shoots. Spider mite densities are higher on main leaves than on lateral leaves. *P. ulmi* seasonal dynamics seem to follow a general pattern despite differences between regions. After winter egg hatch, the spider mite population remains at low levels until June and begins to increase in early summer, peaking in mid- to late summer. Infestation by *P. ulmi* can vary significantly among grapevine cultivars, and leaf hairiness promotes spider mite population increases. *P. ulmi* feeds on the spongy mesophyll and palisade cells causing leaf discoloration (Duso *et al.*, 2012).

Mites of the genus *Eotetranychus*

The yellow spider mite, Eotetranychus carpini *(Oudemans)*

Eotetranychus carpini is an important mite pest of grapes in southern Europe including southern Switzerland, even if it is a polyphagous species mentioned on 34 host plants including other cultivated plants, especially fruit trees. The life cycle of *E. carpini* is completed on grapevine, where females overwinter under bark crevices. In April, they move to the new vegetation. Populations persist on basal leaves in spring and spread along the shoots after bloom. The mite can complete from four to six generations in Switzerland and from seven to eight generations in Italy and France. In late summer, females migrate from the leaves to overwintering sites. The lower thermal threshold for development is about 7 ± 1°C and the optimal temperature for development and reproduction is 26 ± 1°C (Bonato *et al.*, 1990). Little is known about the spatial distribution of overwintering females. In spring *E. carpini* colonizes leaf under-surfaces, in particular leaf portions located along the main veins. Therefore, most eggs are laid at the conjunction of the midrib and veins or along the veins. During the growing season, the mites are more concentrated on leaves located in mid-shoots. In southern France, generally populations peak between mid-July and the beginning of August, with a second peak in September. This species feeds on spongy mesophyll and palisade cells. Yellow (on some cultivars reddish) spots appear on leaves, mainly along the main veins. As these symptoms spread to the entire surface, leaves progressively dry and abscise. This symptom is easily observed in mid- and late summer and involves first basal leaves, and then leaves located in the middle part of the shoots. The influence of *E. carpini* on vine physiology, plant growth, yield and berry quality has not been studied (Duso *et al.*, 2012).

Willamette spider mite, Eotetranychus willamettei *McGregor*

Eotetranychus willamettei is a more specialist species, feeding on grape and a few other host plants and especially fruit trees. It is a significant pest of cultivated grapes in viticultural regions in the western USA, including Oregon, Washington and California. Willamette spider mites can cause leaf injuries and outbreaks over long periods cause crop damages. Compared with the Pacific spider mite (*T. pacificus*) and the McDaniel spider mite (*T. mcdanieli*) present in Washington state, two species that can be found on grapes in the west in warmer climates, the Willamette spider mite has lower optimal developmental temperatures. It is often found in cooler climates. In warmer climates it begins feeding earlier in the growing season.

Eotetranychus willamettei has been used as an 'inoculation' agent. Its feeding induces plant resistance to the more injurious Pacific spider mite (Karban *et al.*, 1997). The Willamette spider mite responds to host plant conditions including: (i) plant water status (Karban and English-Loeb, 1990); and (ii) plant nutrient status, especially nitrogen (Wood and Reilly, 2000). Cultural practices such as irrigation and fertilization can consequently influence feeding and reproductive rate, and so, population density (Chen *et al.*, 2007).

The Grape Nematodes

These roundworms attack the roots causing a decline in vigour and yield but they rarely kill vines. Nematode damage is usually not uniform throughout the vineyard but is localized in certain areas, often associated with soil type. Nematodes that have the potential to cause problems include the root knot nematodes (*Meloidogyne* spp.), the dagger nematodes (*Xiphinema* spp.) and the lesion nematodes (*Pratylenchus* spp.) (see Table 7.1 for distribution). It is necessary to sample the soil prior to planting in order to determine if there are damaging populations of nematodes present because there are only very few management options once vines are planted (Dunn and Zurbo, 2014).

Root knot nematodes, *Meloidogyne* spp.

These nematodes are more damaging in sandy soils. They are thus not likely to be a problem in most grape-growing regions in the piedmont and mountains of the world. These nematodes feed on the inside of roots. Feeding sites are characterized by swellings (galls) on young feeder roots and large galls on older roots. Large populations result in reduced vine vigour and yield. Symptoms are more pronounced under water stress or where grapes have nutritional deficiencies.

Dagger nematodes, *Xiphinema* spp.

These nematodes are common in many soils in various regions of the world. They feed on the tips of the fine feeder roots, which become necrotic and stop growing, resulting in small galls or a 'witch's broom' appearance as new roots appear and are damaged. Large populations can result in a significant reduction in vine vigour. In addition to causing damage to the roots, dagger nematodes can transmit several virus diseases including grape fan-leaf virus (GFLV), tomato ringspot virus (TomRSV) and tobacco ringspot virus (TRSV) (Dunn and Zurbo, 2014).

Lesion nematodes, *Pratylenchus* spp.

These nematodes are widespread in many growing regions of the world. They feed in the finer roots causing lesions, which result in poor root development and reduced vine vigour (Dunn and Zurbo, 2014).

Strategies, Means and Tools in Organic Production

The first step in pest control in organic production is to identify exactly what pests are present. An understanding of their biology, life cycle and susceptible stages further assists in making the optimal selection for successful prevention and management. Thorough knowledge of the weather in the region, cultivars, planting density, vine vigour, canopy characteristics, pest complex, and past problems is important for optimizing control decisions. Despite the growth of organic viticulture, there has been a lack of research-based information to address the need for a greater understanding of the mechanisms operating in organic farming systems, including plant–pest interactions. However, there has been increasing interest among the scientific research community in organic systems research. The European Union (European Commission, 2013) discusses the growth of organic research in Europe and Willer and Kilcher (2009) do this at the world level for each region. The principles of arthropod pest management in organic production imply the adoption of alternative practices specified by international and national organic production standards. The use of multiple and varied tactics to prevent damaging levels of pests, thus minimizing the need for curative solutions and costs, is emphasized. Kogan (1998) drew attention to early integrated pest management (IPM) proponents that emphasized ecological approaches to establish more permanent solutions to pest problems. Reactive approaches have, however, continued to dominate pest management decision making in conventional agriculture, because of the low cost of inorganic/synthetic pesticides. None the less, IPM has provided a framework for the development of pest management programmes in organic systems, including preventative tactics such as: (i) enhancement of natural enemies of pests; (ii) cultural methods; and (iii) plant resistance. A conceptual model for pest management in organic farming (see Fig. 1.1 in Chapter 1, this volume) was proposed by Wyss *et al.* (2005), refined by Zehnder *et al.* (2007) and complemented by Luka (2012; cited in Forster *et al.*, 2013) (see Chapter 2 and 6, this volume). The present chapter uses the five phases of Wyss's model to review arthropod pest management strategies suitable for organic viticulture. While the focus of this chapter is arthropod pest management, pest management normally considers all pest taxa including pathogens, weeds, nematodes and vertebrates. A better understanding of the progress and prospects for arthropod pest management research pertinent to organic farming worldwide is necessary.

Concrete Management Strategies in Grape Organic Production

Management of grape berry moths

Monitoring

The threshold for corrective measures is fixed at 5% of clusters injured. One hundred clusters should be randomly inspected inside the vine plot in order to determine the percentage of clusters damaged. This method will tell if treatment of the plot is necessary. A local treatment of perimeter rows may be all that is necessary to control this pest in some years. The most important thing is to have good control of the maturing larvae in mid- to late July (in the northern hemisphere).

Pheromone traps should be used in vineyards. Trapping of adult male moths indicates the beginning of flight activity. Three traps for monitoring 4–5 ha is the minimum recommended. These traps should be checked weekly for male moths, and pheromone caps should be changed every 1 or 2 months maximum to obtain accurate flight information. Normally, growers should not rely on pheromone trap data from other vine regions. Mating and egg laying will occur over a 2–3 week period following the first detection of flight activity. Theoretically, early season controls of this pest prevent it from becoming well established within the vineyard, and eliminate the need for control later in the season. It should be noted that the second and third flight activity periods occurring in July and in August–September are the most important because of the proximity of the harvest. Actually, these summer adult moths produce the larvae capable of causing major damage to the maturing and then mature fruits. Scouting should be implemented each week after bloom. A protective cover spray should be applied if berry cluster damage reaches 6% for grapes used for processing or 3% for those grown for the fresh market.

Cultural methods

In light infestations, injured berries can be removed by hand. Several cultural methods have been used in the past to reduce the overwintering grape berry moth population but this is not an easy task for moths that overwinter on the trunk. Cultural controls can be used to kill the overwintering pupae in leaves for *P. viteana*. Leaves can be gathered and destroyed in the autumn, or leaves can be buried within the soil in the spring, 2 weeks before bloom, by rototilling (also called rotovating) or cultivating. Some direct (pest-killing) and indirect (microclimate-modifying) practices to reduce *L. botrana* infestation levels can be used, including: (i) pruning the vine canopy; (ii) leaf stripping; (iii) irrigation; (iv) earthing up; (v) weeding; and especially (vi) moving the harvesting date. However, cultural methods have a limited efficiency by themselves, and are often inapplicable in major vineyards where the possibilities of changing cultural schedules are restricted. For example, a systematic advance of the harvesting date to reduce larval damage in the third generation is often incompatible with high quality wine production.

Pesticide sprayings

Some organic insecticides (neem oil, pyrethrums, spinosad) typically have relatively short residual activity and hence may work better if applied twice per generation, spaced about 7 days apart. They are broad-spectrum insecticides and thus may have side effects on beneficials. *Bacillus thuringiensis* Berliner, or *Bt*, is an alternative useful in controlling larvae of grape berry moths. However, careful timing must be observed to effectively kill the larvae before they have a chance to feed internally on grape clusters. Two applications of *Bt* should be applied during the egg-laying period of each harmful generation.

Biological control

Numerous predators and parasitoids are reported in the literature. Among the parasitoids are several species of tachinid flies and several hundred species of parasitic wasp in the ichneumonoid and chalicidoid superfamilies. The parasites that are reported to cause the greatest impact are those that attack the overwintering pupa. In Spain for *L. botrana*, these include the pteromalids *Dibrachys affinis* and *Dibrachys cavus*, which are reported to cause up to 70% pupal mortality, whereas in Italy the ichneumonoids *Dicaelotus inflexus* and *Campoplex capitator* are the most important (Moreau *et al.*, 2010). Numerous attempts to release egg parasitoids have been made in different American and European countries using different *Trichogramma* species. The results obtained varied a lot at first and could not yet clearly promote the use of this technique in vineyards. However, recent promising results induce great hope in the possibility to use *Trichogramma* spp. in the future. Numerous species of predators (insects and arachnids) also feed on the juvenile stages of grape berry moths and even pathogens (fungi, viruses and microsporidia) on some occasions seem to be very important.

Mating disruption

An alternate method of control using pheromone rope ties to disrupt the males of the grape berry moth was approved. This method confuses the males so they are unable to find the females, thus preventing mating, reducing the number of fertile grape berry moth females in a treated vineyard. This method is most effective in vineyards of at least 2 ha for *P. viteana*, 5 ha for *E. ambiguella* and 10 ha for *L. botrana*. Ties are dispensed manually at a rate of 400–1000 ties/ha depending on the species involved and the system and the commercial model used. Vineyards utilizing these ties should continue to scout their plantings.

Management of the grape root borer

Monitoring

In order to easily monitor this pest, pheromone traps are the only reliable means, as the response by male root borers to the female pheromone is quite high. A minimum of three pheromone traps should be placed in a diagonal manner within the plot by late June and checked each week thereafter. A single pheromone cap within a trap is enough for the entire season.

Cultural methods

Weed management is critical for the control of this pest. Good under-row weed control is important in limiting the number of sites available for ovipositioning and in providing an area under the trellis suitable for spraying an insecticide. Good control of root borers with polyethylene mulch has also been achieved in North Carolina (OSU-OARDC,

2009), which can be accomplished at planting. It works well and for a long time but the mulch must be maintained over the years in order to be successful. It is costly to install and discard after the season. Several companies are now offering rather expensive biodegradable versions (OSU-OARDC, 2009).

Biological control

The use of beneficial nematodes has been a proven success in controlling the grape root borer (Williams and Grewal, 2002).

Mating disruption and massive trapping

The use of mating disruption of the grape root borer is presently being tested in the USA (Ohio State University). This method confuses the male root borers that become unable to locate the female borers. Matings become very rare, reducing fertile root borer females and the number of eggs laid in a treated vineyard. The method needs the use of 250 ties/ha dispensed every six to seven vines at the top of the plant. It needs several years to bring the borer population to acceptable levels. External vegetation with wild grapes is a very good source of grape root borers and should be considered for managing this pest. Elimination of wild grapes from these areas helps to reduce the root borer population. Sticky traps are placed around the vineyard in external vegetation in late June at 10–15 m intervals. Traps should be checked weekly. If infestations are high, many borers will be trapped. This method requires a continued effort over several years to reduce populations step by step.

Management of leafhoppers

Monitoring

Vigorous vines are preferred by leafhoppers. The heaviest populations are normally found on the end vines and on outside rows. This is partly because these vines are usually the most vigorous and therefore the most attractive. It is also partly because of the border or boundary effect. Vigorous vines fortunately can tolerate the highest populations.

Cultural methods

Several studies showed vine vigour could be lowered to reduce leafhopper densities, fecundity and adult immigration, without reducing crop yield (Daane *et al.*, 2005). Some cover crop species, especially grasses, can reduce the available water and nutrients, resulting in lower vine vigour. Therefore, a portion of the observed leafhopper reduction found in vineyards with cover crops may result from lowered vine vigour rather than an increase in natural enemies. Cover crop species selections should thus be considered first for their impact on soil health and vine growth, rather than as a primary tool for pest management (Daane *et al.*, 2005). Cultivation and clean-up of adjacent weedy land in the autumn will eliminate favourable overwintering sites in and near a vineyard. Weeds and trash in and around a vineyard is a source of leafhoppers. If this material is cultivated before spring the adults lose their protection and feeding sites, although in areas with extensive agriculture this practice has less value as the adults will just move to an adjacent crop or weedy area. Certain cultivars are likely to suffer higher leafhopper populations than others. Wine and table grape varieties fit this criteria. Moreover, late-producing cultivars are more likely to favour leafhoppers than early maturing cultivars. However, the level of reduction experienced was too small to be economically important. Further, the mechanism(s) leading to this reduction remains unclear because the addition of cover crops did not consistently lead to higher predator densities. One off-shoot of cover cropping that has been utilized in a few north coast vineyards of California is the establishment of a refuge corridor or strip of annual and perennial plants that provide pollen, nectar and alternative prey throughout the season. Only one study has looked at the impact of such refuge corridors and the authors report a reduction in leafhopper densities (Nicholls *et al.*, 2001).

Pesticide sprayings

When high populations of leafhoppers are encountered an application of an insecticidal soap may be required. The use of insecticidal soaps with pyrethrum, neem oil, petroleum or mineral or paraffinic oils, or garlic juice is most effective when the application covers the leafhoppers with little or no residue on the leaves. Kaolin clay repellent is efficient. Surround® has been found to be effective in repelling leafhoppers, leafrollers and the glassy-winged sharpshooter.

Flavescence dorée and the mandatory control of its vector *Scaphoideus titanus* Ball are considered as important issues in organic vineyards. In France, reducing the three mandatory insecticide applications against *S. titanus* was tested from 2000 to 2003 and since 2004 has been increasingly applied. Rules determining the number of treatments depend on regional settings. In the Saint-Emilion vineyard (east of Bordeaux) plots situated 2–4 km from a flavescence dorée focus receive no larvicide treatments, while one or two treatments against nymphs are applied in those that are closer. Treatment against adults is made if more than three adults are caught on yellow sticky traps during a week. In Bordeaux, 90% of the vineyards receive only two applications. However, this insecticide reduction is costly (€26/ha) because of increased monitoring (Chuche and Thiéry, 2014). An efficient alternative insect pest control strategy conforming to the guidelines of organic production is pyrethrin + sesame oil, the only organic product showing an efficacy higher than 90% against the immature stages of *S. titanus*. However, the product has no effect on adult leafhoppers and repeated applications of this product are proved to be toxic against the predatory mite species *Amblyseius andersoni* (Chant). Despite this toxicity and this lack of effect on adults this is the only efficient and recommended control strategy in organic vineyards. It should be applied three times at an interval of 10 days after the first appearance of individuals of the third nymphal stage (Gusberti *et al.*, 2008). However, a single application of pyrethrin demonstrated only 45–61% efficacy in Slovenia (Žežlina *et al.*, 2013).

Biological control

Leafhoppers have few natural enemies. Mymarid egg parasitoids, *Anagrus* species, are the most important natural enemies and are present in all leafhopper-infested vineyards. These tiny parasitoids typically control leafhoppers, with egg parasitism levels often reaching > 90% mid-way through the growing season (Daane *et al.*, 2005). All of the predators found feeding on leafhoppers are generalist predators. Spiders form the most abundant and diverse group, with more than 50 species identified in vineyard collections. Other leafhopper predators found include the whirligig mite, *Anystis agilis* (Banks), and green lacewings. Five different green lacewing species have been collected in vineyards but the most common seems to be *Chrysoperla carnea* (Stephens). While spiders are often quite visible and abundant, lacewing larvae are more difficult to find and are far lower in density. Their low larval abundance relative to the number of lacewing eggs found suggest either the larvae leave the vine, fall prey to other predators or cannibalism, or have high natural mortality.

Scaphoideus titanus population levels observed on *V. vinifera* in France are significantly higher than those in the USA (Maixner *et al.*, 1993), which suggests that the native natural enemies, which are more efficient in the original area, were not introduced with the vector. However, natural enemy pressure against *S. titanus* is lower in both continents (Malausa and Sentenac, 2011) than for other grape pests. Biological control of *S. titanus* has been attempted, either by releasing natural enemies from their native habitat or by increasing local natural enemy populations; however, these two strategies have yielded poor results so far (Malausa and Sentenac, 2011). In France, the release of *Gonatopus flavipes* Olmi (Hymenoptera: Dryinidae) was performed for 3 successive years in two experimental sites (Burgundy and Côtes-du-Rhône vineyards). In total, 368 adults were released and 46,000 *S. titanus* were collected and examined. The best parasitism rate was very low, at < 0.4% (Malausa and Sentenac, 2011). The apparently very low efficiency of this parasitoid is magnified by the difficulty of rearing it.

Management of grape mealybugs and soft scales

These insects have received attention recently due to their ability to vector GLRaVs (grapevine leaf-roll-associated viruses). These pests rarely reach population levels which require treatment and treatment will not stop virus transmission.

Monitoring

There are no simple and effective methods to visually monitor vineyard mealybugs. The number of samples needed for an accurate count is often high because most mealybugs have an aggregative distribution pattern (Geiger and Daane, 2001). It is unfortunately impossible to have a single sampling method for all vineyard mealybugs as they have a lot of biological differences (e.g. number of generations, preferred feeding locations).

The signals of an infested vine are:

- ants that are closely associated with mealybugs (Addison and Samways, 2000);
- honeydew on the leaves or on the trunk (Daane *et al.*, 2011);
- leaves turning yellow or brown and dropping from the vine (Daane *et al.*, 2011); and
- berry clusters in direct contact with the spurs or trunk are more likely to be infested (Geiger and Daane, 2001).

The use of sticky traps with sex pheromone could be a faster sampling method than visual monitoring (Bierl-Leonhardt *et al.*, 1981) as the pheromone lures in and traps adult winged males. Numerous sex pheromones have recently been identified (Daane *et al.*, 2012) and they are being tested as management tools to detect mealybug populations. Trap counts can be used to predict berry damage (Walton *et al.*, 2004) and these sex pheromones are commercially available. However, both conventional sampling and pheromone trapping have advantages and disadvantages and both methods should be used in combination.

Cultural methods

A number of cultural controls are used but only very few have been sufficiently evaluated and many practices are restricted to the table grape market. They include:

- When vines are thinned for berry size increase, grape clusters that are more susceptible to mealybug infestation (e.g. those that come in direct contact with the trunk or cordon) can also be removed (Geiger and Daane, 2001).
- Trellising systems for cane-pruned cultivars result in grape clusters that hang away from the trunk and cordons, and this significantly reduces cluster infestation.
- Harvest date also impacts mealybug infestations which can be higher in cultivars harvested later in the season because of a greater exposure time (Daane *et al.*, 2011).
- Mealybugs are found under the bark of the trunk, cordon, spurs and canes, where they are protected from insecticides, natural enemies and environmental conditions. Stripping the bark exposes more mealybugs to these mortality factors. The infested bark should be carefully destroyed rather than left in the row as the mealybugs can move back on to the vine. After bark stripping, possible treatments are pesticides or flaming or banding the trunk with sticky bands to reduce movement of mealybugs and ants from the trunk to the clusters. This is efficient but labour intensive and generally very costly in most grape markets worldwide.
- Cover crops with non-host plants for mealybugs can be used to improve soil health and lower pest densities by increasing numbers or diversity of beneficial organisms. Parasitoids attacking mealybugs in vineyards could use floral nectaries found on some species as a food source. Generalist predators (lacewings, ladybird beetles, phytoseiid mites) might also use these extrafloral resources and phytophagous organisms in the cover crop as alternate prey. More work on the effectiveness of ground covers on mealybugs and their natural enemies is, however, needed.
- Vines with high vigour can increase mealybug populations in two ways: (i) excess nitrogen has been shown to increase the size of mealybug females

and the number of eggs in each ovisac; and (ii) the increased foliage associated with vigorous vines provides better shelter for the mealybugs by reducing temperatures inside the vine leaf canopy and reducing the amount of applied foliar insecticide that reaches the mealybug. Controlling vine vigour is therefore a practice that can help to improve mealybug control.

Pesticide sprayings

In the event that moderate to high populations develop, dormant oils can be applied prior to bud break and may provide some control. Insecticides applied during the season should be timed to coincide with production of crawlers. Dabbing individual pests with an alcohol-soaked cotton swab or neem-based leaf shine will work when infestations are light. Natural pesticides, such as insecticidal soap and d-Limonene and neem-oil products or garlic juice can also be used to kill the larvae. However, these products have very little persistence in the environment, so several applications during egg hatching will be required for effective control. Horticultural oils work by smothering insects and will control all pest stages, including adults which are protected from most other insecticides by their armour coverings.

Biological control

Hundreds of natural enemies can attack mealybugs. García *et al.* (2015) offer ScaleNet, a literature-based model of scale insect biology and systematics, which is a good reference source for a list of organisms that can be used for biological control. Numerous predators contribute to mealybug control; only a few are specialized whereas most are generalists preying on the small, soft-bodied arthropods. There are no studies of the impact on mealybug populations for most of these natural enemies. *Cryptolaemus montrouzieri* Mulsant, a ladybird beetle native to Australia but exported throughout the world, is the most well-known predator. Both adults and larvae of this coccinellid kill mealybugs. The larvae are mealybug mimics and this 'camouflage' allows beetle larvae to forage without disturbance from mealybug-tending ants (Daane *et al.*, 2007). There have only been a very few studies on the impact of *C. montrouzieri* on mealybug densities (see, for example, Mani and Thontadarya, 1989). Many coccinelled beetles also attack mealybugs in the subfamily Scymninae (*Hyperaspis*, *Nephus* (= *Scymnobius*) and *Scymnus*). Because taxonomic keys for these beetles poorly differentiate among species, many of the observed beetles are misidentified. Migratory ladybird beetles in the subfamily Coccinellinae are often attracted to large mealybug outbreaks and their honeydew. However, more studies are needed to document the effectiveness of the native ladybird beetles found as mealybug predators.

Lacewings have long been associated with mealybugs. Their larvae are effective predators of smaller mealybugs but eggs in the mealybug ovisac have waxy secretions which provide protection from the predator. Larger mealybugs also excrete an ostiolar fluid that can act as a defensive mechanism.

Cecidomyiid flies (i.e. predaceous midges) are other common mealybug predators (Abbas, 1999), but very little is known about their impact on mealybug population densities. Species include *Dicrodiplosis californica* Felt in California (Geiger and Daane, 2001), *Diadiplosis koebelei* in New Zealand (Charles, 1985) and *Triommata coccidivora* Felt in India (Mani *et al.*, 1987).

Most successful biocontrol programmes concern parasitoids of the family Encyrtidae that are mealybug specialists attacking only a few species (Noyes and Hayat, 1994; le Vieux and Malan, 2013). They are typical internal endobionts that may be solitary or gregarious, attacking various host stages.

Some parasitoid species are attracted to the mealybug's sex pheromones (Walton *et al.*, 2006), which may act as a kairomone (Franco *et al.*, 2008) and studies are presently screening the attractiveness of different parasitoids to different mealybug sex pheromones. Some studies are testing the hypothesis that some parasitoids spend more time searching in vineyards where a mating disruption programme is used, significantly increasing parasitism rates.

Ants have long been associated with outbreaks of hemipterans producing honeydew.

The mutualistic association has clear benefits for the ants (carbohydrate food sources) and in return ants are protecting hemipterans from natural enemies. Ants have been actually shown to disrupt mealybug biological control in vineyards in North America (Daane *et al.*, 2007) and in South Africa (Mgocheki and Addison, 2009). Ant species vary in vineyard regions but the Argentine ant, *Linepithema humile* (Mayr), is one of the world's most damaging invasive insects and it is now common in many vineyards of the world in association with mealybugs and soft scale pests (Daane *et al.*, 2012).

The potential of entomopathogenic nematodes as biocontrol agents of *P. ficus* was studied (le Vieux and Malan, 2013). The two indigenous species with the most promising results were *Heterorhabditis zealandica* and *Steinernema yirgalemense*, which were responsible for 96% and 65% mortality, respectively. The capability of both *H. zealandica* and *S. yirgalemense* to complete their life cycles in the host and to produce a new cohort of infective juveniles was demonstrated. Both *H. zealandica* and *S. yirgalemense* were able to move 15 cm vertically downwards and infect *P. ficus* with a respective mortality of 82% and 95%. This study showed *P. ficus* to be a suitable host for *H. zealandica* and *S. yirgalemense*, with both nematode species showing considerable potential for future use in the field control of *P. ficus* (le Vieux and Malan, 2013).

Mating disruption

Mating disruption was first attempted against *P. ficus* in North America (Walton *et al.*, 2006) and is currently gaining in popularity. It is likely that future mealybug control programmes will rely more heavily on novel control strategies using semiochemicals, especially if the price of synthetic sex pheromones for mealybugs can be reduced.

Managing other insects

The fig longicorn borer

Borers are difficult to control because the boring stage is usually not accessible to insecticides or biocontrol agents. Careful pruning and removal of prunings may remove many of the larvae. Retraining of vines may be necessary following pruning of vines with serious infestations.

The rootworms Fidia *spp.*

Visual monitoring consists of scouting for adults feeding on leaves. Pupae can be destroyed by intensive, shallow cultivation of soil until adult emergence occurs in late June. Control of the grape rootworms is most easily accomplished through treatments directed at the adult stage. The key to effective control of adults is proper timing of treatments. Treatments applied too early may not persist long enough to kill rootworm adults during the 3–4-week period when most emerge from the soil. Treatments applied too late will allow some eggs to hatch and the larvae to enter the soil unharmed. Treatments should be made when the first beetles are observed in vineyards. This period will vary from late May to early to mid-July, depending upon location. Growers should carefully check their vineyards each week following application of treatments. A second application should be made if any adult rootworms are detected.

The grass grub beetle

MONITORING. This takes place from late October to the end of November. Methods using a sex pheromone are still in progress. Grass grub beetles have two distinct flight periods:

- Beetles first emerge and then start a flight within 1 m of the ground for mating, almost not feeding on foliage at this stage.
- After 2–3 weeks, the flight becomes situated above 1 m from the ground and at this time, beetles are feeding intensively on foliage.

It is possible to use buckets of water placed under the vines at the ends of rows and to check for beetles each day from late October to the end of November. Such traps are, however, not attractive at all for beetles. An alternative method is to place a torch on the ground and monitor beetle activity every evening during the same period. Alimentary traps are not available.

PESTICIDE SPRAYINGS. If vineyards are sprayed too early, a new growth period may start later during a period with no protection but if sprayings are too late, damage may already have occurred. Spraying while beetles are flying will be the most effective, and help to determine which areas of the vineyard require control. It is thus possible to determine when to spray in space and time. If warm evenings are combined with high beetle activity, this is the best period to start spraying and decide where to make applications. Control in organic vineyards is based on the use of pyrethrum insecticides. In grazed pastures, grass grub beetles have been controlled by the use of a biocontrol product with the bacteria *Serratia entomophila* introduced into the soil. Once in the soil, the bacteria cause a disease in the grass grub beetle larvae that first stop them feeding and eventually kill them.

The Japanese beetle

Control measures should be applied at beetle emergence and thereafter if both susceptible cultivars are grown and high populations of Japanese beetles have occurred. Infestations can be controlled by hand-picking in small vineyards. Several natural enemies including bird predators, parasitic Hymenoptera and diseases may occur but in an unreliable way.

MONITORING. There is no economic threshold on the number of beetles or amount of damage that requires treatment.

CULTURAL METHODS. Preventative, cultural, mechanical and physical methods must be the first choice for pest control, and conditions for use of cultural methods of pest control must be documented in the organic system plan. Clean harvesting, which prevents an accumulation of overripe fruit, helps to prevent beetles from being attracted to the crop. Ploughing or cultivation can destroy pupae in the soil.

PESTICIDE SPRAYINGS. One solution to controlling adult Japanese beetles is rotenone but this insecticide is forbidden in several countries of the world. Under most circumstances, if rotenone is allowed, this product should not be used for the organic control of Japanese beetle adults due to its toxicity to beneficials but it may be considered as a 'last resort' control measure.

BIOLOGICAL CONTROL. The nematode *Heterorhabditis bacteriophora* Poinar can also be applied to control Japanese beetle populations. Nematodes burrow themselves as far as 15 cm and feed on the grubs. Since nematodes are living organisms and are pest specific, they are widely accepted as organic pest controls. The key to controlling Japanese beetles this way requires proper timing of applications to kill the Japanese beetle grubs. Milky spore disease (*Paenibacillus popilliae* Dutky) is available for the control of Japanese beetles, but it is only effective in protecting grassy areas from large larval populations; winged adults will continue to enter vineyards from untreated areas. Milky spore disease, while not being toxic to humans or animals, is highly effective in killing Japanese beetles. It is available in both powder and granular form. Milky spore is a bacterium, *P. popilliae* that attracts the Japanese beetle grub. Parasitic wasps and flies have been used to control Japanese beetles. Two species of tiphiid wasps (*Tiphia vernalis* Rohwer and *Tiphia popilliavora* Rohwer) have been introduced to control Japanese beetle larvae. The tachinid fly, *Istocheta aldrichi* (Mesnil) is known to parasitize adult beetles. Another option is to trap and release native pests where control is desired.

The drosophilid flies

MONITORING. Traps using bait of either yeast or apple cider vinegar may be used for monitoring drosophilid flies. Traps using apple cider vinegar alone are attractive and less odorous to work with than those with yeast added; traps with added yeast may be somewhat more sensitive, but the fluid should be replaced with each service of the traps. Instructions for construction of simple plastic-cup traps are presented by Walsh *et al.* (2011). Traps should be checked at least weekly. Most of the *Drosophila* spp. flies collected will not be spotted wing or vinegar drosophilids, so the flies collected must be checked carefully. Although multi-component bait blends may provide a more selective lure to increase the reliability of risk assessments for SWD (= *Drosophila suzukii* Matsumura), additional research is much needed to quantify relationships between adult trap catch and egg/larval infestations in susceptible fruit. This information is critical to the development of a formal economic threshold for the pest, as there are currently no reliable metrics linking adult presence with pest damage (Asplen *et al.*, 2015).

CULTURAL CONTROL. Fruit should be harvested promptly to eliminate breeding sites. This issue should be kept in mind once SWD are established in an area, since at times grape growers may leave berries on the vine to allow greater development of some harvest parameters. Any overripe or rotten fruit nearby should be destroyed. In vineyards, pomace produced during the crushing process should not be dumped near the producing vineyard block as this can become a source for many SWD. Harvest timing and sanitation are critical to control fruit fly populations because they will reinfest fallen fruit. Proper waste disposal is very important for wineries located near fields. These prophylactic methods are the only available methods for *D. suzukii* in Western Europe as this species is not attacking the fruits for the moment.

PESTICIDE SPRAYINGS. Control measures are directed against the adults of *D. melanogaster* and other vinegar drosophilids; there are no effective controls for larvae in the fruit. As vulnerable fruit approach ripeness, weekly spray applications should be made. When using organic materials, shorter spray intervals will be needed because of the shorter residual life of botanical insecticides. Azadirachtin (neem oil) has been shown to provide about 25% control of SWD. Pyrethrins (often used in a mix with other ingredients) provide about 80% control of SWD but have no residual activity. Azadirachtin and pyrethrins (either alone or in combination) thus show poor efficacy. Spinosyns offer an additional mode of action class, with spinetoram having greater efficacy than spinosad. In 2011, apparent tolerance to pyrethrin developed after repeated application in Oregon. This highlights the challenge for organic growers who must rotate their use of spinosyns with another chemical class to manage insecticide resistance. These spinosyns generally provide a relatively high level of adult control, and 5–7 days residual activity, depending on conditions. Under rainy conditions, this residual period may be shorter.

Application just before berries close in clusters must be done if SWD needs to be controlled as oviposition sites are actually in the protected inner surfaces of the cluster. Mode of action classes and the maximum number of applications per season of each insecticide must be taken into consideration. Rotation of insecticides of differing modes of action in controlling this insect is necessary, even in organic production systems. In these production systems, the number of organic tools remains quite limited. Some tools exist but management in organic viticulture remains a challenge. The main organic products are spinosad, pyrethrin and kaolin. The use of cultural methods (pruning to improve air circulation, prompt harvesting) for minimizing insecticide sprayings is a good option.

BIOLOGICAL CONTROL. In European vineyards SWD is not a problem on fruits and biological control methods do not concern this species in Europe. Evidence for successful levels of natural biological control of SWD in the USA is lacking. Native parasitoid wasps appear to have limited population level impacts in Pacific production regions (Rossi Stacconi *et al.*, 2013, 2015). Potential native predators of SWD include several species of *Orius* (Hemiptera: Anthocoridae), which were observed feeding on SWD larvae (Walsh *et al.*, 2011). Preliminary laboratory studies with *Orius insidiosus* (Walsh *et al.*, 2011) indicated that they could feed on SWD larvae infesting blueberries, although the effects of this predation on population levels are currently unknown. Despite the lack of successful biological control programmes against SWD, the importance of this control tactic within the IPM framework is well recognized by researchers. Approximately 50 parasitoid wasp species, belonging to four families and at least 16 genera, are known to develop on

Drosophila spp. (Carton *et al.*, 1986). Parasitoids that attack frugivorous *Drosophila* spp. are diverse, but the most important genera are the larval parasitoids *Leptopilina* (Figitidae) and *Asobara* (Braconidae), and the pupal parasitoids *Pachycrepoideus* and *Trichopria* (Diapriidae) (Wertheim *et al.*, 2006). Under laboratory conditions, French and Spanish populations of two generalist pupal parasitoids have shown effectiveness against SWD (Chabert *et al.*, 2012; Kacsoh and Schlenke, 2012; Gabarra *et al.*, 2015). Of these, *Pachycrepoideus vindemmiae* Rondani has the widest host range, having been reported to attack over 60 fly species worldwide (Carton *et al.*, 1986; Wang and Messing, 2004). *P. vindemmiae* was recently collected using SWD-baited sentinel traps in commercial soft fruits and natural habitats of northern Italy and Spain. In the laboratory, its parasitization efficacy was confirmed with up to 80% parasitism on infested raspberries (Chabert *et al.*, 2012; Rossi Stacconi *et al.*, 2013; Gabarra *et al.*, 2015). The most promising development with respect to this putative biological control agent occurred recently with the demonstration of its development in SWD under standard laboratory conditions (Rossi Stacconi *et al.*, 2015). *Trichopria* c.f. *drosophilae* is a more specialized species on frugivorous *Drosophila* spp., occupying a worldwide geographic range including Europe, Africa, North America and Australia (Carton *et al.*, 1986). Despite the more attractive (from a biological control perspective) feature of a narrower host range, very little is known about the capacity of these pupal parasitoids to control natural populations of *Drosophila*. Its basic biology was preliminarily described in the laboratory (Gabarra *et al.*, 2015). As with *P. vindemmiae*, Rossi Stacconi *et al.* (2015) confirmed the ability of this species to develop in SWD under laboratory conditions. A greater understanding of the host–parasitoid interactions between these two generalist parasitoids and SWD is now both warranted and needed.

Classical biological control is a potentially useful management strategy for an invasive pest species whenever effective indigenous or resident natural enemies are lacking in the new distribution range. While there is a large literature on the biology and ecology of *Drosophila* parasitoids, little published literature is available on the natural enemies of SWD and their impact on populations of this species. It is especially concerning that virtually no information on this topic is available from China or Korea, despite SWD being widespread in eastern China, Korea and Japan. Many *Drosophila* species in Japan, including SWD, are attacked by several larval (*Asobara* spp., *Ganaspis* spp. and *Leptopilina* spp.) and pupal parasitoids (*Trichopria* spp.) (Mitsui *et al.*, 2007; Kasuya *et al.*, 2013). Larval *Drosophila* parasitoids include species that are host generalists and others that are apparently quite species specific (Kasuya *et al.*, 2013; Nomano *et al.*, 2014), whereas pupal parasitoids tend to be host generalists. Increasingly, government regulatory agencies that issue permits for new biological control agents require that potential agents exhibit a high degree of host specificity. Thus, the search for candidate classical biocontrol agents of SWD is likely to focus on those larval parasitoid species with higher degrees of host specificity and is still an ongoing issue.

No current biological control methods are recommended.

Managing grape mites

Although the broad management principles for the control of rust, bud and blister mites are similar, recommended control strategies differ for each species. Several predatory insects and spiders feed on mites but the most efficient natural predators of mite pests are predatory mites.

Cultural methods

DUST CONTROL. Along with dry conditions, there has long been an association between mite outbreaks and dusty roads. It is a common cultural practice to oil roads and requires crews to drive slowly in order to reduce dusty conditions (Daane *et al.*, 2005).

VINE STRESS. There is a standing recommendation that to reduce mite outbreaks vineyard managers should maintain vine vigour as spider mite outbreaks are often associated with dry conditions and vine stress. In fact, it is not uncommon to observe no mite damage in vineyards which are irrigated throughout the season. However, the impact of water-stressed vines on spider mite densities or the mechanisms behind any observed changes in mite density are not well understood. For example, the influence of water stress on the two-spotted spider mite may be negative or non-linear (Daane *et al.*, 2005).

SULFUR TREATMENTS. Sulfur can be a miticide for spider mites (Guichou, 2003) but sulfur applications can change mite species composition as sulfur applications can reduce densities of predatory mites (Daane *et al.*, 2005).

GRAPE CULTIVAR. The characteristics of the leaf surface may impact mite abundance, with cultivars having pubescent leaf under-surfaces supporting higher populations of predaceous mites. For example, Duso (1992) and Kreiter *et al.* (2002) reported that *Kampimodromus aberrans* (Oudemans) and *Typhlodromus* (*Typhlodromus*) *pyri* Scheuten were more abundant on cultivars with hairy leaf under-surfaces and concluded that predaceous mite abundance was largely independent of prey density, but rather was more closely associated with host plant suitability. In another study of 20 grape (*Vitis*) species, 25% of the variability in abundance of the phytoseiid *Typhlodromus* (*Anthoseius*) *caudiglans* Schuster was determined by leaf characteristics, such as the presence of leaf domatia (tiny tufts of hair on the underside of the leaves), rather than spider mite abundance (Karban *et al.*, 1995).

Pesticide sprayings

Since biological control is a successful strategy to manage mites that are injurious to grapes, chemical control would be necessary only when predatory mites have been eradicated and their re-colonization is difficult. Chemical control may be required even in young vineyards to control *Cal. vitis* populations. The following technical aspects should be considered when selecting and applying pesticides for the control of phytophagous mites: (i) action thresholds based on economic injury levels; (ii) pesticide spectrum (ovicide, larvicide, nymphicide and/or adulticide); (iii) modes of action taking into account resistance management strategies; and (iv) side effects on beneficial arthropods. For organically managed vineyards, soaps, oils, neem oil and botanicals all are popularly reported to have some impact on mite abundance, although no scientific studies that document their effectiveness could be found in the literature. As mentioned previously, the botanically based pyrethrums are broad-spectrum materials. These organically approved pesticides should be handled in a manner similar to that used for synthetic pesticides with respect to their negative non-target impacts. Before the expense of developing new materials for mite control, these materials should be tested and the results published to provide clear guidelines (Daane *et al.*, 2005). An acaricide application schedule can be followed according to grapevine phenological stages. The use of acaricides to control overwintering forms was frequent in the past but is no longer in recent protocols. From bud swell to bud burst or later, *Cal. vitis* can be reduced by using mineral oils or sulfur. From bud burst to two or three unfolded leaves, acaricides may be applied to control eggs or juveniles of *P. ulmi*. When inflorescences are clearly visible different developmental stages of mites occur and compounds active against motile stages and mite growth inhibitors can be applied. Similar strategies are followed in early or mid-summer. The development of resistant strains is an important issue for chemical control. This is of particular importance for spider mites (EU, 2015). Spider mites rapidly develop resistant strains and the principal genetic and ecological factors involved are: (i) arrhenotokous reproduction; (ii) high reproductive rate; (iii) inbreeding; (iv) a very short life cycle; and (v) lack of dispersal of the phytophagous mites from treated areas and a low level of immigration from untreated areas (Duso *et al.*, 2012). In Europe, resistance in *P. ulmi* and *T. urticae* has been reported in fruit orchards and other crops. The hypothesis of the occurrence of resistant strains of spider mites in vineyards has been drawn from field observations, but no characterization of resistance has been carried out.

Predatory mites are susceptible to several insecticides and fungicides, so chemicals that are friendly to predatory mites should be selected to ensure high numbers of predatory mites in the vineyard.

Biological control

Spider mites and rust mites are classic examples of pesticide-induced pests. Their occurrence is negligible in minimally disturbed vineyards and markedly reduced in organic vineyards. This is because of the presence of several macro-predators and predatory mites, especially those belonging to the family Phytoseiidae. The most important macro-predators in European and American vineyards belong to the Thysanoptera (Aelothripidae), Heteroptera (Anthocoridae and Miridae), Coleoptera (Coccinellidae and Staphilinidae) and Neuroptera (Chrysopidae). Macro-predators have relatively long developmental times, high reproductive potential and voracity, their impact is often significant at high mite infestation levels, but they do not persist when prey are scarce (Duso *et al.*, 2012). Predatory mites can build stable populations in vineyards representing the most important component of biocontrol resources (Duso *et al.*, 2012). The Stigmaeidae have a potential in controlling eriophyids (Duso *et al.*, 2012) but their response to spider mite build up is slower when compared with that exhibited by phytoseiids. Therefore, the latter have attracted most interest of researchers. Surveys have revealed a great diversity of phytoseiids in European vineyards (Tixier *et al.*, 2013). As an example, more than 20 phytoseiid species have been recorded in Italian and French vineyards (Kreiter *et al.*, 2000; Duso *et al.*, 2012). Among them, *T. pyri* dominates in central Europe, *K. aberrans* in southern Europe where *Amblyseius andersoni* (Chant), *Typhlodromus exhilaratus* and *Phytoseius finitimus* Ribaga are also present (Kreiter *et al.*, 2000; Duso *et al.*, 2012). It should be stressed that these species are generalist predators, type III, after McMurtry *et al.* (2013). They can persist when prey densities decline by surviving on alternative foods. Some of them show a narrow association with the host plant, a capacity to regulate their densities and to compete with other predators (McMurtry *et al.*, 2013). Knowledge of their feeding habits and relationships with plants is fundamental for conservation biological control tactics.

The above-mentioned phytoseiid species develop and reproduce on spider mites (e.g. *P. ulmi* and *E. carpini*) and show comparable demographic parameters when fed with eriophyoids (e.g. *Col. vitis*) and pollen (Duso *et al.*, 2012). *A. andersoni* exhibits shorter developmental times and higher oviposition rates than *T. pyri* and *K. aberrans* (Duso *et al.*, 2012). However, these two species respond better than *A. andersoni* to spider mite population increases (Duso *et al.*, 2012). *T. pyri* and *K. aberrans* are favoured on grape cultivars with hairy leaf under-surfaces. In contrast, *A. andersoni* shows an opposite trend (Duso *et al.*, 2012). Leaf morphology strongly affects colonization patterns when these species co-occur and can mediate interspecific competition (Duso *et al.*, 2012). *T. pyri*, *K. aberrans* and *A. andersoni* have been proven to be effective in preventing rust mite infestations (Duso *et al.*, 2012). *T. exhilaratus* exhibited a higher intrinsic rate of increase on *E. carpini* and pollen than on *P. ulmi* (Duso *et al.*, 2012). The adaptation of *T. exhilaratus* to low relative humidity is a fundamental requirement to colonize vineyards of southern Europe (Duso *et al.*, 2012). This species also proved to have potential to manage *Cal. vitis* (Duso *et al.*, 2012). The economic importance of *P. finitimus* is controversial. *P. finitimus* may have some potential to manage *P. ulmi* (Duso *et al.*, 2012) but its ability has not been compared with that of other phytoseiids. Its populations reach high densities on grape cultivars with hairy leaf under-surfaces, where it competes successfully with other predatory mites (Duso *et al.*, 2012). Generalist phytoseiids occurring in vineyards can consume tenuipalpids, eriophyids, tydeids, winterschmidtiids, and young stages of thrips or coccids as alternative prey. The presence of these prey can enhance phytoseiid performance, and the role of eriophyids in this context is crucial (Duso *et al.*, 2012). However, it should be stressed that some alternative prey (e.g. *Cal. vitis*) is difficult to manage.

Pollen is a fundamental food source for generalist predatory mites. Grape leaves are excellent pollen traps, and their analysis shows definite trends in pollen fluctuations. In France and Italy, pollen densities are relatively high at sprouting, after bloom and in late summer (Kreiter *et al.*, 2002; Duso *et al.*, 2012). When pollen is abundant on grape leaves, population sizes of *T. pyri*, *K. aberrans* and *A. andersoni* increase (Kreiter *et al.*, 2002; Duso *et al.*, 2012). Plant pathogenic fungi can constitute additional food resources for generalist phytoseiids. Grape downy mildew (GDM) and grape powdery mildew (GPM) are the most important worldwide grape pathogens. The spread of GDM foliar symptoms can promote population increases of *A. andersoni* and *T. pyri* (Duso *et al.*, 2012) as these species can develop and reproduce on GDM in the laboratory (Duso *et al.*, 2012). GPM is a supplementary food for *A. andersoni* and *T. pyri* (Duso *et al.*, 2012). Interactions between powdery mildew and phytoseiids have been poorly documented, despite the economic importance of this pathogen in the Mediterranean region. The management of non-prey foods for generalist phytoseiids is fundamental for conservation biocontrol tactics. Hedgerows can provide pollen (and phytoseiids) for contiguous vineyards. In an experimental farm comprising a hedgerow and a contiguous vineyard, elders produced large amounts of pollen allowing for population increases of the predatory mite *Euseius finlandicus* (Oudemans). However, the importance of this strategy for grapes is unclear (Kreiter *et al.*, 2002; Duso *et al.*, 2012). Pollen produced by species belonging to the *Poaceae* family is a major component of wind-borne pollen in vineyards of various regions, and *Poaceae* are widely used as cover crops. Experiments conducted in northern Italy showed that a reduction in grass mowing increased pollen densities on grape canopy and consequently phytoseiid densities (Duso *et al.*, 2012). Plant diversity in uncultivated areas surrounding crops and inside crops is assumed to increase natural enemy density and diversity. These uncultivated areas outside or inside crops are supposed to provide alternative, stable and durable food and habitat resources for natural enemies (Kreiter *et al.*, 2000; Liguori *et al.*, 2011; Duso *et al.*, 2012; Tixier *et al.*, 2015).

Managing grape nematodes

Monitoring

Preferences for host plants differ between parasitic nematodes. Root knot and root lesion nematodes can infect and reproduce in roots of many crops, fruit trees, ornamentals and weeds while the citrus nematode can infect grapevines, olive, citrus and pear. Many species can survive for 2 years or even more in soil without host plants. Before establishing a new planted vineyard, a soil count of nematode numbers and species is needed for rootstock and cultivar choices, particularly if the site has been used previously for horticultural crops.

In order to make nematode management decisions, it is important to know exactly what species are present and to estimate their population size, especially in the case of a previous orchard or vineyard that had problems caused by nematodes. In this case, population levels may be high enough to cause severe damage to the newly planted young vines. If species have not previously been identified, it is necessary to take soil samples (sampling time depends on the region, type of nematode and variety of grapes) and send them for identification. Bloom and harvest times influence a lot of nematode populations and differ according to region. It is necessary to irrigate 3 days before sampling (a sample of soil of five vines/block, mixed for a composite sample of 1 l/block, each block sample separated and identified with a label and put in a freezer for later analysis) or to wait for 3 consecutive days of rain for a good reliable indication of nematode populations. Soil samples must be taken at 0–40 cm depth with roots encountered between the vine trunks. The vineyard must be divided into sampling blocks representative of important factors: (i) cropping history; (ii) crop injury; and (iii) soil texture. The farm adviser should be contacted for more details about sampling, to help find a laboratory for extracting and identifying nematodes, and for help in interpreting sample results. Care should be taken to look for nematode symptoms in the vineyard late in the growing season to prepare for future management.

Cultural methods

Nematode-tolerant rootstocks can provide some protection from nematodes and other management benefits. Nematode-free planting material should be used that has been treated with hot water to eliminate any possible introduction of nematodes from nursery to vineyards.

To naturally rid an old vineyard site of the effects of *Xiphinema index* Thorne and Allen and grapevine fan-leaf virus, it is necessary to forgo planting grapes for more than 10 years. This period of time is required to allow old roots to decompose and nematode numbers to decrease. This will increase the length of time before a new vineyard exhibits virus symptoms, but will not prevent reinfestation. Crop rotation with crops or plant cover crops that are hosts to nematodes should not be carried out on vineyard sites. No single rootstock is resistant to all root knot nematodes and there are numerous other nematodes of concern. Selection of a rootstock is a risky endeavour because of their excessive or inadequate growth in certain situations and their limited breadth of resistance. Manures and other soil amendments can improve vine vigour and frequently reduce the effect of nematode infestation. To reduce stress on vines, measures should be taken to prevent soil compaction and stratification, to improve soil tilth and drainage, and to control other pests. Proper irrigation and fertilizer application also reduce stress on vines and help lessen the effect of nematodes such as root knot nematodes.

Cover crops may provide effective control by planting *Brassica* species. *Brassica* species suppress nematodes through the release of a chemical known as isothiocyanate as the plant material breaks down in the soil. Some mustard cultivars are commercially available. The best reduction of nematodes is achieved if the mustard is grown close to the vine row, slashed and covered with soil under the vine rows.

Resistant rootstocks can be a good tool for managing vineyard nematode populations. Selection of an appropriate rootstock requires accurate identification of the nematode genera present and availability of a corresponding resistant rootstock. Nematode-resistant rootstocks are derived from native grapes such as: (i) *Vitis aestivalis*, *Muscadinia rotundifolia*, *Vitis rufotomentosa* and *Vitis rupestris* × *M. rotundifolia* selections (*Meloidogyne incognita* [Kofoid and White], *X. index*); (ii) *Vitis arizonica* and *Vitis candicans* (*X. index*); and (iii) *Vitis champini*, *Vitis cinerea* and *V. rupestris* (*M. incognita*).

Pesticide sprayings

Vineyards planted in fumigated ground are known to have generally improved growth and yields compared with those planted on non-fumigated ground. For established vineyards, biofumigation is possible. Certain plants are able to kill or repel pests, disrupt their life cycle, or discourage them from feeding. Some of these (marigolds, sesame, castor bean and various brassicas) have been tested as nematode-suppressive cover crops. But extracts or essential oils of those plants can be applied as nematicides.

Biological control

A wide range of fungi, bacteria and invertebrates parasitize or prey on nematodes. Potential biological control agents include the bacterium *Pasteuria penetrans* (Thorne) which occurs naturally in many soils and which might provide some natural suppression in long-established vineyards.

New Trends for the Future

General

In general, improving plant diversity both in the crop (e.g. by planting associated cultivated plants, agroforestry, weed planting or weed management) and out of the crop (e.g. with hedges and/or fallows) at farm scale: (i) will improve the diversity of arthropod natural enemies (species richness and abundance) (see for example Letourneau *et al.*, 2011); (ii) is not favourable to arthropod pests (Tonhasca and Byrne, 1994; Letourneau *et al.*, 2011); and (iii) reduces the damage caused by arthropod pests (Letourneau *et al.*, 2011).

Semi-natural habitats (SNH) are very important for hibernation (Häni *et al.*, 1998; Sarthou *et al.*, 2014). Landscape heterogeneity improves the diversity of arthropod natural enemies (Bianchi *et al.*, 2006; Chaplin-Kramer *et al.*, 2011; Winqvist *et al.*, 2011; Veres *et al.*, 2013), improves biological control (Veres *et al.*, 2013) and reduces the damage caused by arthropod pests (Bianchi *et al.*, 2006; Winqvist *et al.*, 2011; Veres *et al.*, 2013). The biodiversity of natural enemies has positive effects on arthropod pest regulations (Halaj and Wise, 2001; Letourneau *et al.*, 2009). Management of SNH around crops improves the natural enemy diversity (Landis *et al.*, 2000; Collins *et al.*, 2003; Gurr *et al.*, 2012). In conclusion, improving natural or creating SNH diversity and quality improves the diversity of natural enemies which improves biological control and reduces the level of damage caused.

Gurr *et al.* (2012) have introduced an acronym, 'SNAP', with S for shelter or habitats for beneficials to live in, N for nectar since many parasitoids and some predators use nectar as an alternative food source, A as alternative prey for beneficials and P for pollen, which is a very important food source for beneficials, especially predators and particularly phytoseiid mites. This concept is, for example, illustrated by maintaining cover crops in vineyards which can enhance arthropod natural enemy populations by providing a habitat, shelter, nectar and alternative food for predators and parasitoids (Irvin, 2013). These data have to be taken into account by organic farming as conservative biological control is cheap and makes sense for environmentally compatible methods for arthropod pest management. However, how to manage and what to plant in and around crops are still open to questions and data produced so far is fragmentary and research is still in progress. Also organic farming has large positive effects on biodiversity compared with conventional farming (Tuck *et al.*, 2014).

Grape berry moths

Infestations of grape berry moth often recur on the same sites year after year. Low-risk areas will not develop economic infestations in most years but for high-risk areas, two alternatives are available: (i) pheromone mating disruption; and (ii) *Bacillus thuringiensis* (*Bt*).

Pheromone mating disruption is an efficient control method that is highly specific to grape berry moth and non-toxic for health and the environment. Dispensers of pheromone are placed on the top of vines in early May. The new pheromone dispensers diffuse pheromone over a period of nearly 6 months. Male moths are unable to locate and mate with females over the growing season and this suppresses oviposition and reduces damage caused by larvae.

Bt is an insecticide of biological origin (a toxin of the bacteria) that is effective in controlling berry moth larvae. The only critical point is that the larvae need to ingest the toxin *before* burrowing into the grape cluster to prevent damage and thus, careful timing of spraying is very important. But pheromone traps give a good indication of when flight begins and thus, allow a predictition of when mating might occur, when eggs are laid and when eggs hatch. Two applications during the extended egg-laying period of each generation are required, because this material persists for < 3 days in the field.

Besides mating disruption techniques and progress in monitoring techniques (e.g. the use of food traps against females), biological control may reveal itself to be very efficient at controlling grape moth populations. Parasitoids that are active at controlling grape moths in vineyards have been known for a long time; a few of them were already described in the middle of the 19th century in French vineyards and their efficiency was already recognized, especially against the diapausing and the first spring generations of the moths. Numerous attempts to release egg parasitoids have been made in different European countries using different *Trichogramma* species. The results obtained have varied a lot and could not yet clearly promote the use of this technique in vineyards. Recently, many researchers have investigated use of these beneficial insects, including *Trichogramma minutum* and *Trichogramma pretiosum* wasps as biological control agents to manage the populations of grape berry moth, *P. viteana* in the USA (Seaman *et al.*, 1990; Nagarkatti *et al.*, 2002, 2003). It has been reported that both *T. minutum* and *T. pretiosum* wasps in nature can parasitize over 40% eggs of grape berry moth (Seaman *et al.*, 1990; Nagarkatti *et al.*, 2002, 2003). Some other species are being tested presently in France and results so far are promising.

A biological control based on larval parasitoids could probably be efficiently developed as a valuable alternative to chemical control. The focus has been on larval parasitoids among which ichneumonids and chalcidoids (Hymenoptera) dominate, and results obtained in different French vineyards (Bordeaux vineyard, Perpignan and Montpellier area, Côtes du Rhône and Alsace) are sometimes promising. Many factors may favour or reduce their efficiency as biocontrol agents.

Leafhoppers

Leafhoppers are the most common vineyard pests. Are there organic farming techniques for leafhoppers that are as effective and inexpensive as the new synthetic pesticides? First, the more damaging variegated leafhopper (VLH), *Erythroneura variabilis* is not found in all vineyard regions and, when present, leafhopper densities (and damage) can vary depending on vine vigour, regional temperature and management practices. Secondly, natural enemies provide excellent control of western grape leafhopper (WGLH), *Erythroneura elegantula*, which is the more widespread species. Thirdly, wine, raisin and juice grape commodities have a higher tolerance for leafhopper damage. Leafhopper populations encountered in the north coast and central coast regions of California, USA can be effectively managed without synthetic pesticides. In regions where VLH predominates and pesticides are needed, there are organically approved botanicals, oils and soaps. All of these materials are reported, in grower testimonials or industry advertisements, as providing leafhopper control. There are few scientific studies that verify their impact, although Bentley *et al.* (2008) showed that horticultural mineral oils, applied prior to bloom, suppressed both WGLH and VLH populations. Another problem is the inconsistency reported with different formulations of botanical materials. This also needs to be addressed. If any of these organic materials are to be used, the following suggestions are provided to maximize pesticide effectiveness and minimize impact on biological controls:

1. For all arthropod pests the farm managers should utilize effective monitoring programmes and record pest populations in each block and from year to year. Many pesticide applications are made before leafhopper nymphs reach damaging thresholds.

2. Use of botanicals for the overwintering adult generation should be avoided. Only the first generation nymphs should be treated when the population numbers are very high as little damage is caused during this time of the season, and delaying applications until the later generations allows time for natural enemies to establish. These botanicals are broad spectrum and can have a negative impact on natural enemies.

3. The application should be timed so that it affects the most appropriate leafhopper stage. Most of the conventional pesticides kill both by contact and by a systemic or fumigation action and, therefore, can be effective against adult and nymph stages. Soaps and oils kill by contact only, and may be most effective against the smaller nymphal development stages.

Mealybugs

Pesticides

A recent online discussion of mealybug control materials, by members of the Association of Applied Insect Ecologists, suggests that mealybugs could be controlled by a number of organically approved materials. However, there appear to be no scientific studies on the use of oils, lime sulfur or soaps, which might be more conducive to natural enemies, and the botanicals have not yet been tested. Trials should be conducted with organically approved pesticide materials to determine their effectiveness.

Biological controls

There is effective biological control for the grape mealybug (*Pseudococcus maritimus* (Ehrhorn)) and there is no need for renewed importation efforts of potential natural enemies. Instead, the vineyard must be properly managed to reduce ants and pesticide treatments that can disrupt grape mealybug biological control. In contrast, biological control of the obscure mealybug (*Pseudococcus viburni*), long-tailed mealybug (*Pseudococcus longispinus*) and vine mealybug (*Planococcus ficus* (Signoret)) is incomplete. Furthermore, there are numerous parasitoid species that have been identified as potential obscure or vine mealybug natural enemies which have never been released in Californian vineyards. Renewed foreign exploration efforts should be a primary goal for these pests. Also, there have been no studies on the biological controls or population dynamics of the long-tailed mealybug. This work should be conducted to assess needed biological controls for the long-tailed mealybug. While the mealybug destroyer (*Cryptolaemus montrouzieri* Mulsant.) and green lacewings are used in commercial augmentation programmes, there are no studies that have evaluated the impact of these programmes. For example, lacewing larvae were observed to be effective predators of immature mealybugs, although they have had a more difficult time feeding on eggs in the mealybug ovisac or on mealybug adults, suggesting that synchronizing release to mealybug development stage may be critical. In contrast, experimental studies found that releases of *Pseudaphycus angelicus* (Howard) and *Anagyrus pseudococci* (Girault) suppressed the grape and vine mealybugs, respectively, but at this time there are no commercial insectaries for these parasitoids. Recently, there has also been grower-generated interest in testing augmentative releases of predaceous mites and cecidomyiid flies, and yet the biologies of these natural enemies, as mealybug predators, are relatively unknown and there is no information on their use in an augmentative release programme. It appears that there is still much to be accomplished in the development of mealybug biological controls.

Monitoring and control decisions

Early detection of mealybug infestations, when the population is small and isolated on a few vines, would improve efficacy of control treatments (Daane *et al.*, 2012). However, visual sampling of vineyard mealybugs, especially at low densities, is labour intensive. The use of sex pheromone-baited traps, for the winged adult male mealybugs, offers a more effective sampling tool. Already, the identification and synthesis of vine mealybug sex pheromone has resulted in a highly successful commercial monitoring programme (Millar *et al.*, 2002; Walton *et al.*, 2004). Still not yet determined is the relationship between pheromone trap counts and mealybug damage. Future control measures will focus on novel methods to monitor mealybugs, using synthetic sex pheromones that may even find commercial use in mating disruption programmes (Daane *et al.*, 2012).

Mating disruption

The synthetic vine mealybug sex pheromone proved so effective that it is being tested for use in mating disruption programmes. If this proves to be a viable option, mating disruption may be the primary alternative to pesticide treatments for control in organic vineyards. Some parasitoids species seem to spend more time searching for mealybugs in vineyards using a disruption programme, thereby increasing parasitism rates (Daane *et al.*, 2008).

Ant controls

As mentioned previously, ants can exacerbate mealybug pest problems by disrupting natural enemy activity in vineyards (Phillips and Sherk, 1991; Daane *et al.*, 2004). Unfortunately, pesticide controls for ants are often more disruptive than those materials applied for the mealybugs. Therefore, if biological control is to be developed, ants must also be controlled with pesticide materials that fit into the IPM and/or organic programmes. Currently, researchers are working with different protein and sugar ant baits to deliver small amounts of pesticides. This work will be a crucial development for the implementation of mealybug biological control.

Mealybugs as vectors

While laboratory studies have shown that mealybugs can transmit GLRaVs, there is no information on the natural infectivity level of mealybugs collected in the field, or their transmission efficiency. There is a need to better understand GLD and the role of mealybugs and other scale insects in the dispersion of the viruses that cause this important plant disease.

Mites

Pesticides

For organically managed vineyards, soaps, oils, neem oil and botanicals are all popularly reported to have some impact on mite abundance, although no scientific studies that document their effectiveness could be found in the literature. As mentioned previously, the botanically based pyrethrums are broad-spectrum materials. These organically approved pesticides should be handled in the same way as synthetic pesticides with respect to their negative non-target impacts. Before the expense of developing new materials for mite control, these materials should be tested and the results published to provide clear guidelines.

Biological controls

There are many effective biological control agents of spider mites and further foreign exploration for new natural enemies is not warranted unless new exotic phytophagous mites are found. To improve their presence in the vineyard, a systems approach needs to be considered to balance vine vigour, pesticide sprays and cultural practices. What is surprising is that augmentation of phytoseiids has not become a more popular practice. Research has clearly demonstrated the effectiveness of this programme. It would appear that most vineyard managers choose miticides for their immediate impact on pest populations and their suitability for 'timed' applications. In contrast, most augmentation programmes require more labour – especially in sampling pest and natural enemy populations – to know when and what to release. Predatory mite releases will become cost-effective if targeted to augment naturally occurring predation, with the number of predators released dependent on the abundance of the naturally occurring predators rather than dependent on prohibitively costly inundative releases. To improve pest control decisions, such as augmentation, binomial sampling techniques using early season ratios of predator:spider mites have been developed. However, practical use of this technique needs better adoption by vineyard managers. There also needs to be a better understanding of the importance of the 'secondary' mite pests – the Willamette mite and the two-spotted spider mite, as well as their associated natural enemy complex and the impact of vineyard management practices on their densities. As discussed with leafhopper controls, proper sampling and treatment decisions will greatly reduce unnecessary pesticide applications.

Cultural controls

The presence of leaf domatia can increase the abundance of fungal feeding mites, leading to a discussion of engineered or selected grape cultivars with leaf domatia. However, there has not been a similar interest in breeding cultivars that increase the presence of phytoseiid mites.

Conclusions

Further studies of the effects of organic agriculture practices on pests and natural enemies will need to cover field to landscape scales. Organic agriculture is a corpus of good agricultural practices but the results may depend on several factors at various scales. It is not a panacea neither a prerequisite for achieving ecological pest management.

The presence of riparian habitats has been shown to enhance predator colonization and abundance on adjacent vineyards, although this influence was limited by the distance to which natural enemies dispersed into the vineyard (Nicholls *et al.*, 2001). However, the corridor amplified this influence by enhancing timely circulation and dispersal movement of predators into the centre of the field.

Finally, the number of pest management studies conducted on organic systems is very few compared with the huge literature on IPM in conventional crops. Consequently, there is an urgent need for more studies to be conducted on certified organic land on IPM in organic production. Use of modern synthetic pesticides that are more friendly for the environment and human health are not allowed in organic agriculture, but many genetic, biological and cultural methods developed for conventional crops may be very useful.

References

Abbas, M.S.T. (1999) Studies on *Dicrodiplosis manihoti* Harris (Diptera, Cecidomyiidae), a common predator of mealybugs. *Journal of Pest Sciences* 72, 133–134.

Addison, P. and Samways, M.J. (2000) A survey of ants (Hymenoptera: Formicidae) that forage in vineyards in the Western Cape Province, South Africa. *African Entomology* 8(2), 251–260.

Allam, L. (2000) Dynamique comparée des populations de *Tetranychus urticae* Koch (Acarina, Tetranychidae) et rôle des contraintes du milieu dans l'acquisition des stratégies démographiques. Thèse d'Etat, Université Moulay Ismaïl, Faculté des Sciences, Meknès, Soutenue le-25.11.2000.

Arias, A. and Nieto, J. (1978) Observaciones sobre la biología de la Araña amarilla (*Tetranychus urticae*) en las viñas de Tierra de Barros (Badajoz) durante 1976 y 1977. Servicio de Defensa contra Plagas e Inspección Fitopatológica, Comunicaciones, Serie Estudios y experiencias, enero n 31/78, 46 pp.

Arias, A. and Nieto, J. (1981) Observaciones sobre la biologia de la Arana amarilla (*Tetranychus urticae*) y correlación entre síntomas y pérdidas en una viña de Tierra de Barros (Badajoz), durante 1980. Servicio de Defensa contra Plagas e Inspección Fitopatológica, Comunicaciones, Serie Estudios y experiencias, mayo ns 9/81, 41 pp.

Arias, A. and Nieto, J. (1991) La arana amarilla común (*Tetranychus urticae* Koch) en Tierra de Barros. I: Bioecología. *Vitivinicultura* II, 38–41.

Asplen, M.K., Anfora, G., Biondi, A., Choi, D.S., Chu, D., Daane, K.M., Gibert, P., Gutierrez, A.P., Hoelmer, K.A., Hutchinson, W.D., Isaacs, R., Jiang, Z.L., Kárpáti, Z., Kimura, M.T., Pascual, M., Philips, C.R., Plantamp, C., Ponti, L., Vétek, G., Vogt, H., Walton, V.M., Yu, Y., Zappalà, L. and Desneux, N. (2015) Invasion biology of spotted wing drosophila (*Drosophila suzukii*): a global perspective and future priorities. *Journal of Pest Science* 88, 469–494.

Baillod, M., Antonin, P. and Mittaz, C. (1989) Migrations, estimation des populations et nuisibilité de l'acarien jaune commun *Tetranychus urticae* dans la viticulture valaisanne. *Revue Suisse de Viticulture, d'Arboriculture et d'Horticulture* 21, 179–183.

Bentley, W.J., Varela, L.G., Zalom, F.G., Smith, R.J., Purcell, A.H., Phillips, P.A., Haviland, D.R., Daane, K.M. and Battany, M.C. (2008) *University of California Pest Management Guidelines: Grape Leafhopper.* University of California Integrated Pest Management (IPM). Available at: http://www.ipm.ucdavis.edu/PMG/r302300111.html (accessed 14 July 2016).

Bianchi, F.J.J.A., Booij, C.J.H. and Tscharntke, T. (2006) Sustainable pest regulation in agricultural landscapes: a review on landscape composition, biodiversity and natural pest control. *Proceedings of the Royal Society of London B: Biological Sciences* 273, 1715–1727.

Bierl-Leonhardt, B.A., Moreno, D.S., Schwarz, M., Fargerlund, J. and Plimmer, J.R. (1981) Isolation, identification and synthesis of the sex pheromone of the citrus mealybug, *Planococcus citri* (Risso). *Tetrahedron Letters* 22, 389–392.

Boller, E.F., Janser, E., Zahner, S. and Potter, C. (1985) Kann Herbizideinsatz im Weinbau Spinnmilbenprobleme verursachen? *Schweizer Zeitschrift für Obst- und Weinbau* 121, 527–531.

Bonato, O., Cotton, D., Kreiter, S. and Gutierrez, J. (1990) Influence of temperature on the life-history parameters of the yellow grapevine mite *Eotetranychus carpini* (Acari: Tetranychidae). *International Journal of Acarology* 16, 241–245.

Bostonian, N.J., Vincent, C. and Isaacs, R. (eds) (2012) *Arthropod Management in Vineyards: Pests, Approaches, and Future Directions*. Springer, New York, 505 pp.

Bournier, A. (1976) Grape insects. *Annual Review of Entomology* 22, 355–276.

CABI (2014) *Drosophila suzukii*. In: Ioriatti, C., Rossi Stacconi, M.V., Anfora, G. (eds) *Invasive Species Compendium*. CAB International, Wallingford, UK. Available at: http://www.cabi.org/isc (accessed 14 July 2016).

Candolfi, M.P. (1991) Einfluss von *Tetranychus urticae* Koch und *Panonychus ulmi* Koch (Acari) auf Gaswechsel. PhD dissertation, Wachstum, Ertrag und Traubenqualität der Weinrebe.

Candolfi, M.P., Keller, M. and Boller, E.F. (1991) Mite-load function improves precision of feeding damage estimation in *Tetranychus urticae*. *Entomologia Experimentalis et Applicata* 58, 289–293.

Candolfi, M.P., Boller, E.F. and Wermelinger, B. (1992) Influence of the two-spotted spider mite, *Tetranychus urticae*, on the gas exchange of Pinot noir grapevine leaves. *Vitis* 31, 205–212.

Carton, Y., Boulétreau, M., Van Alphen, J.J. and Van Lenteren, J.C. (1986) The drosophila parasitic wasps. In: Ashburner, M., Carson, H.L. and Thompson, D.J. (eds) *The Genetics and Biology of Drosophila*. Academic Press, London, pp. 347–393.

Chabert, S., Allemand, R., Poyet, M., Eslin, P. and Gibert, P. (2012) Ability of European parasitoids to control a new invasive Asiatic pest, *Drosophila suzukii*. *Biological Control* 63(1), 40–47.

Chaplin-Kramer, R., O'Rourke, M.E., Blitzer, E.J. and Kremen, C. (2011) A meta-analysis of crop pest and natural enemy response to landscape complexity. *Ecology Letters* 14, 922–932.

Charles, J.G. (1985) *Diadiplosis koebelei* new record (Diptera: Cecidomyiidae) a predator of *Pseudococcus longispinus* (Homoptera: Pseudococcidae) from New Zealand. *New Zealand Journal of Zoology* 12, 331–334.

Chen, Y., Opit, G.P., Jonas, V.M., Williams, K.A., Nechols, J.R. and Margolies, D.C. (2007) Two spotted spider mite population level, distribution, and damage on ivy geranium in response to different nitrogen and phosphorus fertilization regimes. *Journal of Economic Entomology* 100(6), 1821–1830.

Chuche, J. and Thiéry, D. (2014) Biologie et écologie de *Scaphoideus titanus*, cicadelle vectrice de la flavescence dorée. *Phytoma* 679, 25–29.

Collins, K.L., Boatman, N.D., Wilcox, A. and Holland, J.M. (2003) A 5-year comparison of overwintering polyphagous predator densities within a beetle bank and two conventional hedgebanks. *Annals of Applied Biology* 143, 63–71.

Daane, K.M., Sime, K.R., Cooper, M.L. and Battany, M.C. (2004) Ants in your vineyard? *University of California Plant Protection Quarterly* 14(2) 6–10.

Daane, K.M., Smith, R.J., Klonsky, K.M. and Bentley, W.J. (2005) Organic vineyard management in California. Paper presented at symposium 'IPM in Organic Systems', XXII International Congress of Entomology, 16 August 2004, Brisbane, Australia. *Organic Research* May, 37–55.

Daane, K.M., Sime, K.R., Fallon, J. and Cooper, M.L. (2007) Impacts of Argentine ants on mealybugs and their natural enemies in California's coastal vineyards. *Ecological Entomology* 32, 583–596.

Daane, K.M., Cooper, M.L., Triapitsyn, S.V., Walton, V.M., Yokota, G.Y., Haviland, D.R., Bentley, W.J., Godfrey, K.E. and Wunderlich, L.R. (2008) Vineyard managers and researchers seek sustainable solutions for mealybugs, a changing pest complex. *California Agriculture* 62, 167–176.

Daane, K.M., Bentley, W.J., Smith, R.J., Haviland, D.R., Weber, E. and Gispert, C. (2011) Vine mealybug. In: Bettiga, L. and Bentley, W.J. (eds) *University of California Grape Pest Management Manual*. University of California Press, Oakland, California, pp. 125–135.

Daane, K.M., Almeida, R.P.P., Bell, V.A., Walker, J.T.S., Botton, M., Fallahzadeh, M., Mani, M., Miano, J.L., Sforza, R., Walton, V.M. and Zaviezo, T. (2012) Biology and management of mealybugs in vineyards. In: Bostonian, N.J., Vincent, C. and Isaacs, R. (eds) *Arthropod Management in Vineyards: Pests, Approaches, and Future Directions*. Springer, New York, pp. 271–307.

Dominé, A. (2010) *Wine*. H.F. Ullmann, Postdam, Germany.

Dunn, G. and Zurbo, B. (2014) *Grape Vine Pests and their Management. Primefact* 511, 2nd edn. New South Wales Department of Primary Industries, Orange, New South Wales, Australia. Available at: www.dpi.nsw.gov.au/factsheets (accessed 14 July 2016).

Duso, C. (1992) Role of *Amblyseius aberrans* (Oud.), *Typhlodromus pyri* Scheuten and *Amblyseius andersoni* (Chant) (Acari, Phytoseiidae) in vineyards. III: Influence of a variety characteristics on the success of *A. aberrans* and *T. pyri* releases. *Journal of Applied Entomology* 114, 455–462.

Duso, C., Pozzebon, A., Kreiter, S., Tixier, M.-S. and Candolfi, M.P. (2012) Management of phytophagous mites in European vineyards. In: Bostonian, N.J., Vincent, C. and Isaacs, R. (eds) *Arthropod Management in Vineyards: Pests, Approaches, and Future Directions*. Springer, New York, pp. 191–217.

EPPO (European and Mediterranean Plant Protection Organization) (2016) First reports of *Xylella fastidiosa* in the EPPO region. EPPO, Special Alert. Available at: https://www.eppo.int/QUARANTINE/special_topics/Xylella_fastidiosa/Xylella_fastidiosa.htm (accessed 14 July 2016).

European Commission (2013) *Facts and Figures on Organic Agriculture in the European Union*. Agriculture and Rural Development. European Commission, Brussels, 44 pp.

European Union (EU) (2015) EU – Pesticides Database. Available at: http://ec.europa.eu/food/plant/pesticides/eu-pesticides-database/public/?event=homepage&language=EN (accessed 14 July 2016).

Fermaud, M. and Le Menn, R. (1989) Association of *Botrytis cinerea* with grape berry moth larvae. *Phytopathology* 79, 651–656.

Fermaud, M. and Le Menn, R. (1992) Transmission of *Botrytis cinerea* to grapes by grape berry moth larvae. *Phytopathology* 82, 1393–1398.

Foissac, X. and Wilson, M.R. (2010) Current and possible future distributions of phytoplasma diseases and their vectors. In: Weintraub, P.G. and Jones, P. (eds) *Phytoplasmas: Genomes, Plant Hosts, and Vectors*. CAB International, Wallingford, UK, pp. 309–324.

Franco, J.C., Silva, E.B., Cortegano, E., Campos, L., Branco, M., Zada, A. and Mendel, Z. (2008) Kairomonal response of the parasitoid *Anagyrus* spec. nov. near *pseudococci* to the sex pheromone of the vine mealybug. *Entomologia Experimentalis et Applicata* 126, 122–130.

Freeman, T.N. (1958) The Archipinae of North America. *The Canadian Entomologist* 90(Supplement 7), 1–89.

Gabarra, R., Riudavets, J., Rodríguez, G.A., Pujade-Villar, J. and Arnó, J. (2015) Prospects for the biological control of *Drosophila suzukii*. *BioControl* 60(3), 331–339.

García, M., Denno, B., Miller, D.R., Miller, G.L., Ben-Dov, Y. and Hardy, N.B. (2015) ScaleNet: A Literature-based Model of Scale Insect Biology and Systematics. Available at: http://scalenet.info DOI: 10.1653/0015-4040(2005)88[482:ssighc]2.0.co;2 (accessed 2 July 2015).

Geiger, C.A. and Daane, K.M. (2001) Seasonal movement and sampling of the grape mealybug, *Pseudococcus maritimus*, in San Joaquin Valley vineyards. *Journal of Economic Entomology* 94, 291–301.

Germain, J.-F., Balmès, V., Mouttet, R., Ramel, J.-M., Fried, G. and Reynaud, P. (2014) *Espèces Invasives: 3 Années d'alertes et de Signalements l'ANSES, Unité, Entomologie et Plantes Invasives*. Association Française de Protection des Plantes (AFPP)-Colloque Ravageurs et Insectes Invasifs Emergents, Montpellier, France.

Gilligan, T.M. and Epstein, M.E. (2014) Tortricids of Agricultural Importance by Interactive Keys developed in Lucid 3.5. Last updated August. Colorado State University. Available at: http://idtools.org/id/leps/tortai/ (accessed 1 March 2017).

Guichou, S. (2003) Activité acaricide du soufre sur *Tetranychus urticae* (Acari: Tetranychidae): action des facteurs du milieu et variabilité des réponses. Thèse de doctorat, Ecole Nationale Supérieure Agronomique de Montpellier (ENSA), Montpellier, France, 126 pp. + annexes.

Gurr, G.M., Wratten, S.D. and Snyder, W.E. (2012) *Biodiversity and Insect Pests: Key Issues for Sustainable Management*. Wiley-Blackwell, Oxford, 360 pp.

Gusberti, M., Jermini, M., Wyss, E. and Linder, C. (2008) Efficacy of insecticides against *Scaphoideus titanus* in organic vineyards and their side effects. *Revue Suisse de Viticulture, d'Arboriculture et d'Horticulture* 40, 173–177.

Halaj, J. and Wise, D.H. (2001) Terrestrial trophic cascades: how much do they trickle? *The American Naturalist* 157, 262–281.

Häni, F.J., Boller, E.F. and Keller, S. (1998) Natural regulation at the farm level. In: Pickett, C.H. and Bugg, R.L. (eds) *Enhancing Biological Control – Habitat Management to Promote Natural Enemies of Agricultural Pests*. University of California Press, Berkeley, California, pp. 161–210.

Hassan, N.M.F., Afifi, A.M. and Nawar, M.S. (1986) Mites inhabiting plants and soil in Sinai: and newly reclaimed lands. *Bulletin of the Entomological Society of Egypt* 66, 211–225.

Hauser, M., Gaimari, S. and Damus, M. (2009) *Drosophila suzukii* new to North America. *Fly Times* 43, 12–15.

Hoddle, M.S. (2004) The potential adventive geographic range of glassy-winged sharpshooter, *Homalodisca coagulate* and the grape pathogen *Xylella fastidiosa*: implications for California and other grape growing regions of the world. *Crop Protection* 23, 691–699.

Irvin, N.A. (2013) Assessing the use of cover crops for sustainable pest control in California (funded by Western SARE). University of California Riverside, Department of Entomology. Available at: http://biocontrol.ucr.edu/irvin/research/wsare.html#Top%20of%20Page (accessed 14 July 2016).

Kacsoh, B.Z. and Schlenke, T.A. (2012) High hemocyte load is associated with increased resistance against parasitoids in *Drosophila suzukii*, a relative of *D. melanogaster*. *PLoS One* 7(4), e34721. DOI: 10.1371/journal.pone.0034721.

Karban, R. and English-Loeb, G.M. (1990) A vaccination of Willamette mites (Arachnida: Tetranychidae) to protect grape plants from high populations of Pacific mites. *Journal of Economic Entomology* 83, 2252–2257.

Karban, R., English-Loeb, G., Walker, M.A. and Thaler, J. (1995) Abundance of phytoseiid mites on *Vitis* species: effects of leaf hairs, domatia, prey abundance and plant phylogeny. *Experimental and Applied Acarology* 19, 189–197.

Karban, R., English-Loeb, G. and Hougen-Eitzman, D. (1997) Mite vaccinations for sustainable management of spider mites in vineyards. *Ecological Applications* 7(1), 183–193.

Kasuya, N., Mitsui, H., Ideo, S., Watada, M. and Kimura, M. (2013) Ecological, morphological and molecular studies on *Ganaspis* individuals (Hymenoptera: Figitidae) attacking *Drosophila suzukii* (Diptera: Drosophilidae). *Applied Entomology and Zoology* 48, 87–92.

Kogan, M. (1998) Integrated pest management: historical perspectives and contemporary development. *Annual Review of Entomology* 43, 243–270.

Kreiter, S. (ed.) (2008) *Les Ravageurs de la Vigne*, 2nd edn. Féret Editor, Bordeaux, France, 389 pp.

Kreiter, S., Brian, F., Magnien, C., Sentenac, G. and Valentin, G. (1991) Spider mites and chemical control of weeds: interactions. In: Dusbabek, F. and Bukva, V. (eds) *Modern Acarology*, Vol. 2. Academia, Prague and SPB Academic Publishing bv, The Hague, The Netherlands, pp. 725–736.

Kreiter, S., Tixier, M.-S., Auger, P., Muckensturm, N., Sentenac, G., Doublet, B. and Weber, M. (2000) Phytoseiid mites of vineyards in France (Acari: Phytoseiidae). *Acarologia* 61, 77–96.

Kreiter, S., Tixier, M.-S., Croft, B.A., Auger, P. and Barret, D. (2002) Plants and leaf characteristics influencing the predaceous mite *Kampimodromus aberrans* in habitats surrounding vineyards. *Environmental Entomology* 31, 648–660.

Landis, D.A., Wratten, S.D. and Gurr, G.M. (2000) Habitat management to conserve natural enemies of arthropod pests in agriculture. *Annual Review of Entomology* 45, 175–201.

Letourneau, D.K., Jedlicka, J.A., Bothwell, S.G. and Moreno, C.R. (2009) Effects of natural enemy biodiversity on the suppression of arthropod herbivores in terrestrial ecosystems. *Annual Review of Ecology, Evolution and Systematics* 40, 573–592.

Letourneau, D.K., Armbrecht, I., Rivera, B.S., Lerma, J.M., Carmona, E.J., Daza, M.C., Escobar, S., Galindo, V., Gutiérrez, C., López, S.D., Mejía, J.L., Rangel, A.M., Rangel, J.H., Rivera, L., Saavedra, C.A., Torres, A.M. and Trujillo, A.R. (2011) Does plant diversity benefit agroecosystems? A synthetic review. *Ecological Applications* 21, 9–21.

Le Vieux, P. and Malan, A.P. (2013) An overview of the vine mealybug (*Planococcus ficus*) in South African vineyards and the use of entomopathogenic nematodes as potential biocontrol agent. *South African Journal of Enology and Viticulture* 34, 108–118.

Maixner, M., Pearson, R.C., Boudon-Padieu, E. and Caudwell, A. (1993) *Scaphoideus titanus*, a possible vector of grapevine yellows in New York. *Plant Disease* 77, 408–413.

Malagnini, V., de Lillo, E., Saldarelli, P., Beber, R., Duso, C., Raiola, A., Zanotelli, L., Valenzano, D., Giampetruzzi, A., Morelli, M., Ratti, C., Causin, R. and Gualandri, V. (2015) Preliminary data on the transmission of grapevine Pinot Gris virus by *Colomerus vitis*. In: *Proceedings of the 18th Meeting of International Council for the Study of Virus and Virus-like diseases of the Grapevine (ICVG)*, 7–11 September, Ankara, Turkey. ICVG, Geneva, New York, pp. 217–218.

Malausa, J.-C. and Sentenac, G. (2011) Parasitoïdes de *Scaphoideus titanus*. In: Sentenac, G. (ed.) *La Faune Auxiliaire des Vignobles de France*. La France Agricole, Paris, pp. 143–146.

Mani, M. and Thontadarya, T.S. (1989) Development of the encyrtid parasitoid *Anagyrus dactylopii* How. on the grape mealybug *Maconellicoccus hirsutus* Green. *Entomon* 14(1–2), 49–52.

Mani, M., Thontadarya, T.S. and Singh, S.P. (1987) Record of natural enemies of the grape mealybug *Maconnellicoccus hirsutus* Green. *Current Science* 56, 624–635.

Mani, M., Shivaraju, C. and Srinivasa Rao, M. (2014) Pests of grapevine: a worldwide list. *Pest Management in Horticultural Ecosystems* 20(2), 170–216.

McMurtry, J.A., De Moraes, G.J. and Sourasso, N.F. (2013) Revision of the lifestyles of phytoseiid mites and implications for biological control strategies. *Systematic and Applied Acarology* 18(4), 297–320.

Mgocheki, N. and Addison, P. (2009) Interference of ants (Hymenoptera: Formicidae) with biological control of the vine mealybug *Planococcus ficus*. *Biological Control* 49, 180–185.

Millar, J.G., Daane, K.M., McElfresh, J.S., Moreira, J., Malakar-Kuenen, R., Guillen, M. and Bentley, W.J. (2002) Development and optimization of methods for using sex pheromone for monitoring the mealybug

Planococcus ficus (Homoptera: Pseudococcidae) in California vineyards. *Journal of Economic Entomology* 95(4) 706–714.
Mitsui, H., Van Achterberg, K., Nordlander, G. and Kimura, M.T. (2007) Geographical distributions and host associations of larval parasitoids of frugivorous Drosophilidae in Japan. *Japan Natural History* 41, 1731–1738.
Mondy, N., Charrier, B., Fermaud, M., Pracros, P. and Corio-Costet, M.-F. (1998a) A mutualism between a phytopathogenic fungus (*Botrytis cinerea*) and a vineyard pest (*Lobesia botrana*): positive effects on insect development and oviposition behaviour. *Comptes Rendus De l'Académie des Sciences, Life Sciences* 321, 665–671.
Mondy, N., Pracros, P., Fermaud, M. and Corio-Costet, M.-F. (1998b) Olfactory and gustatory behaviour by larvae of *Lobesia botrana* in response to *Botrytis cinerea*. *Entomologia Experimentalis et Applicata* 88, 1–7.
Moreau, J., Villemant, C., Benrey, B. and Thiery, D. (2010) Species diversity of larval parasitoids of the European grapevine moth (*Lobesia botrana*, Lepidoptera: Tortricidae): the influence of region and cultivar. *Biological Control* 54(3), 300–306. DOI: 10.1016/j.biocontrol.2010.05.019.
Nagarkatti, S., Muza, A.J., Saunders, M.C. and Tobin, P.C. (2002) Role of the egg parasitoid *Trichogramma minutum* in biological control of the grape berry moth, *Endopiza viteana*. *BioControl* 47, 373–385.
Nagarkatti, S., Tobin, P.C., Saunders, M.C. and Muza, A.J. (2003) Release of native *Trichogramma minutum* to control grape berry moth. *The Canadian Entomologist* 135, 589–598.
Nicholls, C.I., Parrella, M. and Altieri, M.A. (2001) The effects of a vegetational corridor on the abundance and dispersal of insect biodiversity within a northern California organic vineyard. *Landscape Ecology* 16(2), 133–146.
Nomano, F.Y., Mitsui, H. and Kimura, M.T. (2014) Capacity of Japanese *Asobara* species (Hymenoptera: Braconidae) to parasitize a fruit pest *Drosophila suzukii*. *Journal of Applied Entomology* 139, 105–113.
Noyes, J.S. and Hayat, M. (1994) *Oriental Mealybug Parasitoids of the Anagyrini (Hymenoptera: Encyrtidae)*. CAB International, Wallingford, UK, 554 pp.
Ohio State University, Ohio Agricultural Research and Development Center (OSU-OARDC) (2009) Available at: http://www.oardc.ohio-state.edu/organicviticulture/root_borer.html (accessed 19 June 2017).
Olivier, C., Vincent, C., Saguez, J., Galka, B., Weintraub, P.G. and Maixner, M. (2012) Leafhoppers and planthoppers: their bionomics, pathogen transmission and management in vineyards. In: Bostonian, N.J., Vincent, C. and Isaacs, R. (eds) *Arthropod Management in Vineyards: Pests, Approaches, and Future Directions*. Springer, New York, pp. 253–270.
Papaioannou-Souliotis, P., Ragusa di Chira, S. and Tsolakis, H. (1994) Phytophagous mites and their predators observed on cultivated plants in Greece during 1975–1990. *Annales de l'Insitut Phytopathologique Benali* (Numéro special) 17, 35–87.
Phillips, P.A. and Sherk, C.J. (1991) To control mealybugs, stop honeydew-seeking ants. *California Agriculture* 45(2) 26–28.
Rossi Stacconi, M.V., Grassi, A., Dalton, D.T., Miller, B., Ouantar, M., Loni, A., Ioriatti, C., Walton, V.M. and Anfora, G. (2013) First field records of *Pachycrepoideus vindemiae* as a parasitoid of *Drosophila suzukii* in European and Oregon small fruit production areas. *Entomologia* 1, e3.
Rossi Stacconi, M.V., Buffington, M., Daane, K.M., Dalton, D.T., Grassi, A., Kaçar, G., Miller, B., Miller, J.C., Braser, N., Ioriatti, C., Walton, V.M., Wiman, N.G., Wang, X. and Anfora, G. (2015) Host stage preference, efficacy and fecundity of parasitoids attacking *Drosophila suzukii* in newly invaded areas. *Biological Control* 84, 28–35.
Sarthou, J.-P., Badoz, A., Vaissière, B., Chevallier, A. and Rusch, A. (2014) Local more than landscape parameters structure natural enemy communities during their overwintering in semi-natural habitats. *Agriculture, Ecosystems and Environment* 194, 17–28.
Seaman, A.J., Nyrop, J.P. and Dennehy, T.J. (1990) Egg and larval parasitism of the grape berry moth (Lepidoptera: Tortricidae) in three grape habitats in New York. *Environmental Entomology* 19, 764–770.
Sforza, R. and Boudon-Padieu, E. (1998) Le principal vecteur de la maladie du Bois noir: Faisons connaissance avec cet insecte fulgoromorphe, *Hyalesthes obsoletus*, depuis le vignoble jusqu'au laboratoire. *Phytoma – La Défense des Végétaux* 510, 33–37.
Sforza, R., Bourgoin, T., Wilson, S.W. and Boudon-Padieu, E. (1999) Field observations, laboratory rearing and descriptions of immatures of the planthopper *Hyalesthes obsoletus*. *European Journal of Entomology* 96(4), 409–418.
Southern Region IPM Center (SRIC) (2012) *Pest Management Strategic Plan for Wine Grapes in Virginia and North Carolina*. Virginia Tech, Blacksberg, Virginia and North Carolina State University, Raleigh, North Carolina, 42 pp.

Thiéry, D. (2008) Les tordeuses nuisibles au vignoble. In: Kreiter, S. (ed.) *Les Ravageurs de la Vigne*, 2nd edn. Féret Editor, Bordeaux, France, pp. 214–246.

Tixier, M.-S., Baldasar, A., Duso, C. and Kreiter, S. (2013) Phytoseiidae in European grape (*Vitis vinifera* L.): bio-ecological aspects and keys to species (Acari: Mesostigmata). *Zootaxa* 3721(2), 101–142.

Tixier, M.-S., Arnaud, A., Douin, M. and Kreiter, S. (2015) Effects of agroforestry on Phytoseiidae communities (Acari: Mesostigmata) in vineyards. A synthesis of a 10-year period of observations. *Acarologia* 55(4), 361–375.

Tonhasca, A. Jr and Byrne, D.N. (1994) The effects of crop diversification on herbivorous insects: a meta-analysis approach. *Ecological Entomology* 19, 239–244.

Tuck, S.L., Winqvist, C., Mota, F., Ahnström, J., Turnbull, L.A. and Bengtsson, J. (2014) Land-use intensity and the effects of organic farming on biodiversity: a hierarchical meta-analysis. *Journal of Applied Ecology* 51(3), 746–755.

Vacante, V. and Tropea Garzia, G. (1987) Grape mites in Sicily. Contribution I. In: Cavalloro, R. (ed.) *Integrated Pest Control in Viticulture*. Proceedings of Meeting Commission of the European Communities (CEC) 'Experts' Group, 26–28 September 1985, Portoferraio, Italy. Balkema A.A., Rotterdam, The Netherlands, pp. 207–215.

van de Vrie, M., McMurty, J.A. and Huffaker, C.B. (1972) Ecology of tetranychid mites and their natural enemies: a review III. Biology, ecology and pest status and host-plant relations of tetranychids. *Hilgardia* 41, 343–432.

Veres, A., Petit, S., Conord, C. and Lavigne, C. (2013) Does landscape composition affect pest abundance and their control by natural enemies? A review. *Agriculture, Ecosystems & Environment* 166, 110–117.

Walsh, D.B., Bolda, M.P., Goodhue, R.E., Dreves, A.J., Lee, J., Bruck, D.J., Walton, V.M., O'Neal, S.D. and Zalom, F.G. (2011) *Drosophila suzukii*: invasive pest of ripening soft fruit expanding its geographic range and damage potential. *Journal of Integrated Pest Management* 2, G1–G7.

Walton, V.M., Daane, K.M. and Pringle, K.L. (2004) Monitoring *Planococcus ficus* in South African vineyards with sex pheromone-baited traps. *Crop Protection* 23, 1089–1096.

Walton, V.M., Daane, K.M., Walter, J., Bentley, W.J., Millar, J.G., Larsen, T.E. and Malakar-Kuenen, R. (2006) Pheromone-based mating disruption of *Planococcus ficus* (Hemiptera: Pseudococcidae) in California vineyards. *Journal of Economic Entomology* 99, 1280–1290.

Walton, V., Lee, J., Bruck, D., Shearer, P., Parent, E., Whitney, T. and Dreves, A.J. (2010) Recognizing fruit damaged by spotted wing drosophila, *Drosophila suzukii*. Oregon State University, Extension Service Publication, EM 9021, p. 4. Available at: http://ir.library.oregonstate.edu/xmlui/bitstream/handle/1957/19525/em9021.pdf (accessed 14 July 2016).

Wang, X. and Messing, R. (2004) Fitness consequences of body-size-dependent host species selection in a generalist cleptoparasitoid. *Behavioral Ecology and Sociobiology* 56, 513–522.

Welter, S., Farnham, D., McNally, P. and Freeman, R. (1989) Effect of Willamette mite and Pacific spider mite on grape photosynthesis and stomatal conductance. *Environmental Entomology* 18(6), 953–957.

Wertheim, B., Allemand, R., Vet, L.E.M. and Dicke, M. (2006) Effects of aggregation pheromone on individual behaviour and food web interactions: a field study on *Drosophila*. *Ecological Entomology* 31, 216–226.

Willer, H. and Kilcher, L. (2009) *The World of Organic Agriculture – Statistics and Emerging Trends 2009*. International Federation of Organic Agriculture Movements (IFOAM), Bonn, Germany and Research Institute of Organic Agriculture (FiBL), Geneva, Switzerland.

Williams, R.N. and Grewal, P.S. (2002) Assessing the potential of entomopathogenic nematodes to control the grape root borer through laboratory and greenhouse bioassays. *Biocontrol Science and Technology* 12(1), 35–42.

Winqvist, C., Bengtsson Jan Aavik, T., Berendse, F., Clement, L.W., Eggers, S., Fischer, C., Flohre, A., Geiger, F., Liira, J., Pärt, T., Thies, C., Tscharntke, T., Weisser, W.W. and Bommarco, R. (2011) Mixed effects of organic farming and landscape complexity on farmland biodiversity and biological control potential across Europe. *Journal of Applied Ecology* 48, 570–579.

Wood, B.W. and Reilly, C.C. (2000) Pest damage to pecan is affected by irrigation, nitrogen application and fruit load. *Horticultural Science* 35, 669–672.

Wyss, E., Luka, H., Pfiffner, L., Schlatter, C., Uehlinger, G. and Daniel, C. (2005) Approaches to pest management in organic agriculture: a case study in European apple orchards. Paper presented at symposium 'IPM in Organic Systems', XXII International Congress of Entomology, 16 August 2004, Brisbane, Australia. *Organic Research* May, 33–36.

Zalom, F.G., Varela, L.G. and Cooper, M. (2011) European Grapevine Moth, *Lobesia botrana*. Cooperative Extension and Statewide Intergrated Pest Management (IPM) Program. University of California, Davis. Available at: http://ipm.ucdavis.edu/EXOTIC/eurograpevinemoth.html (accessed 14 July 2016).
Zehnder, G., Gurr, G.M., Kühne, S., Wade, M.R., Wratten, S.D. and Wyss, E. (2007) Arthropod pest management in organic crops. *Annual Review of Entomology* 52, 57–80.
Žežlina, I., Škvarč, A., Bohinc, T. and Trdan, S. (2013) Testing the efficacy of single applications of five insecticides against *Scaphoideus titanus* on common grapevines. *International Journal of Pest Management* 59(1), 1–9.

8 Pest Management in Organic Olive

Gavino Delrio and Andrea Lentini*
University of Sassari, Sassari, Italy

Introduction

Olive production occupies an estimated area of some 10 million ha worldwide, although more than 98% of the *c.*800 million cultivated olive trees are located around the Mediterranean Basin. Olive trees are very long lived, and their productivity fluctuates considerably from year to year. Most traditional groves are sited in hilly, rocky and arid areas, in which other crops cannot be cultivated; as a result, they represent a characteristic component of the Mediterranean landscape. In more recent times, a combination of planting dense stands of early fruiting cultivars, using irrigation and artificial fertilizers and spraying with pesticides, has wrought many changes to traditional practice (Lòpez-Villalta, 1996).

More than 200 insect pests, fungi and weeds are potentially harmful to the olive tree, but the number of species responsible for economically significant levels of damage is only around ten. The most prominent bacterial disease is caused by *Pseudomonas savastanoi* pv. *savastanoi* (olive knot), while the four most important fungal pathogens are: (i) *Spilocaea oleagina* (Cast.) Hugh. (olive leaf spot); (ii) *Verticillium dahliae* Kleb. (verticillium wilt); (iii) *Colletotrichum gloeosporioides* (Penz.) Penz. and Sacc. (anthracnose); and (iv) *Mycocentrospora cladosporioides* (Sacc.) U. Braun (cercosporiosis). These diseases are controlled by a combination of cultural measures and spraying with various copper-based compounds (Roca *et al.*, 2007; López-Escudero and Mercado-Blanco, 2011). An outbreak in Apulia (Italy) of the bacterial pathogen *Xylella fastidiosa*, which is transmitted by xylem-fluid feeding insects, has been recently documented; the disease caused leaf scorching and dieback over an area of some 8000 ha (Cariddi *et al.*, 2014).

The olive grove is a relatively stable agroecosystem, in which the incidence of exotic insect pests is rare and most of the pest species do not cause crop yield losses. Approximately 30 insect and mite species are commonly found in Mediterranean Basin olive groves, but the only ones which consistently cause economic damage are the olive fruit fly, the olive moth and the black scale. A number of thrips, psyllid, scale, moth, cecidomyiid and coleopteran species cause occasional damage, as their populations are limited by abiotic factors and natural enemies. Overall, pests and pathogens are responsible for yield losses of around 30%, half of which are due to insects. The annual

* Corresponding author: lentini@uniss.it

cost of control measures in the Mediterranean Basin exceeds €100 million (Haniotakis, 2005).

Until the 1950s, the management of diseases and pests was based on cultural methods, along with the use of a few chemical compounds of uncertain effectiveness. The development of inexpensive synthetic pesticides was associated with a major improvement in control. Although olive insect pests have not so far developed any practical resistance to traditional insecticides, spraying has been associated with environmental pollution, reductions in natural enemy population sizes, some severe outbreaks of black scale and other secondary pests, and the presence of toxic residues in the olive oil. Excessive insecticide applications have been declining as a result of the development of area-wide integrated pest management (IPM) protocols, but spraying is still widely used to maintain yield and quality (Delrio, 1992).

A rising demand for certified organic products has resulted in the spread of organic production methods to some 570,000 ha in the Mediterranean Basin by 2012 (Willer *et al.*, 2014). The olive crop is generally tolerant of pest attacks and can be effectively protected through biological control methods. In fact, economic thresholds of the main pests are quite high in olives for oil production, allowing the employment of control strategies that are less effective but also environmentally safer than insecticide applications.

Major Pests

Olive fruit fly

Bactrocera oleae (Rossi) (Diptera Tephritidae) is the most damaging olive pest in the Mediterranean Basin and California. This insect passes through several generations per year and develops within the fruit of both the cultivated olive (*Olea europaea* subsp. *europaea*) and its wild relatives *Olea oleaster* and, in Africa, *O. europaea* subsp. *cuspidata*.

The adult insect can live for up to several months, feeding on scale honeydew, pollen, bird faeces and phylloplane bacteria. Their gut harbours the bacterial symbiont '*Candidatus* Erwinia dacicola', which is housed both in an oesophageal bulb and along the whole gut, and is transmitted to the offspring during oviposition (Capuzzo *et al.*, 2005). The symbiotic relationship implies that the bacteria are required for larval development and as a food source for the adult insect (Tzanakakis, 1984). An adult female can fly up to 1 or 2 km during the day searching for fruits suitable for oviposition. Sexual maturity is affected by a combination of the fly's nutritional status and prevailing weather conditions, and is usually reached a few days after adult emergence. Mating lasts some hours and is preceded by a complicated courtship behaviour within the olive tree canopy at dusk. Females produce a four-component sex pheromone which attracts males over long distances, while the males produce a pheromone of unknown function. The major component of the blend is 1,7-dioxaspiro[5.5]undecane, of which the S-(+) enantiomere acts as the long-range male attractant and the R-(-) enantiomere as the aggregation, arrestant and aphrodisiac pheromone (Haniotakis and Pitarra, 1994). Each female is generally monogamous, whereas a male can mate once a day several times during its lifespan (Cavalloro and Delrio, 1974). After a pre-oviposition period of approximately 1 week, the female lays up to 10–20 eggs/day, producing in all several hundred eggs during her lifespan. The most favourable oviposition temperature range is 20–30°C, while the Mediterranean Basin winter temperatures are generally too low for eggs to mature successfully. Females tend to be barren in June–July because, despite enjoying favourable temperature conditions, no fruit is available for oviposition (Delrio and Prota, 1976). Fruits are chosen by the ovipositing female based on their size, shape and colour, and the emission of certain volatile compounds which stimulate oviposition. Since fruits are not chosen until they have reached a minimum size, summer infestations tend to begin in the most precocious cultivars. The female pierces the fruit with her ovipositor and lays an egg in a sub-epidermal chamber, visible from the outside as a brown-coloured triangular lesion. The larvae feed within the fruit mesocarp, tunnelling their way through the fruit. Mature larvae pupate inside unripe fruits in summer preparing an exit hole for the adult, whereas pupation occurs in the soil at 2–3 cm depth in autumn–winter. The duration of the pupal stage is approximately 10 days in summer and up to 4 months during the winter. Thus, the overall generation time is less than a month during the summer, but 3 or 4 months over the winter. Depending on the prevailing temperature conditions and the supply of suitable fruit, the fly can pass through up to six generations/year. In certain coastal areas where large trees are cultivated and where fruit is available almost year round, two generations can be achieved during the spring and a further three or four between late June and November. Overwintering occurs largely at the pupal stage, even though adult flies have been observed in some areas during particularly mild winters (Delrio and Prota, 1976).

The level of infestation is governed by a number of both abiotic and biotic factors: chief among the former is the ambient temperature, while the latter include the extent of the host cultivar's resistance, the number of fruits produced by a tree and the size of the natural enemy population (Delrio and Prota, 1990). Very high summer temperatures can reduce the survival rate of adults, eggs and larvae, while unusually low temperatures (experienced in more northerly or high altitude sites) reduce the progress of an infestation. Cultivars are known to differ with respect to their susceptibility to the pest, due to both physical and chemical factors (Delrio *et al.*, 2010; Rizzo *et al.*, 2012). In low-yielding years, infestation can be severe enough to cause 100% fruit drop, but damage levels are much less severe in high-yielding years.

Known predators of the olive fly (olive fruit fly) are generalists, mainly spiders and soil-dwelling insects which feed on pupae in the soil. The Mediterranean Basin parasitoid complex comprises a small number of polyphagous chalcidoids and one braconid (Neuenschwander *et al.*, 1983). The eulophid *Pnigalio mediterraneus* Ferrière & Delucchi and the eurytomid *Eurytoma martellii* Domenichini are the most important parasitoids of the larvae, while the eupelmid *Eupelmus urozonus* Dalman feeds mainly on pupae and has evolved a hyperparasitism to other chalcidids. The braconid *Psyttalia concolor* (Szépligeti) is an endophagous parasitoid of tephritid pupae, including those of the olive fly. Parasitism rates can vary markedly from site to site, and are thought to be encouraged by a stable agroecosystem. The parasitoid complex can reduce olive fly populations by as much as 70%, but has only a limited effect on infestation intensity, because the olive fly has a high biotic potential (Delrio and Prota, 1990).

Olive fly infestation is especially damaging for table olives, where the economic threshold is extremely low and oviposition stings cause aesthetic damage that strongly reduce the fruit marketability. Quantitative damage on oil olives results from fruit drop and reduction in oil production caused by feeding larvae. The pest also has a negative effect on oil quality, as the oil derived from infested pulp has an increased content of oleic acid, a higher peroxide number and a larger K232 and K270 spectrophotometric value. Oils produced from heavily infested olives show a marked reduction in polyphenol content and spectrum, while changes in their fatty acid or sterol composition are less evident (Delrio *et al.*, 1995).

Olive moth

Prays oleae (Bernard) (Lepidoptera Praydidae) develops on a range of host species (*Olea*, *Phillyrea*, *Ligustrum*, *Jasminum*), however, it completes three generations/year only on the olive tree. Newly hatched anthophagous generation larvae feed initially on the anthers and ovaries, eventually damaging between ten and 20 flowers by fusing them with silk. In the carpophagous generation, the eggs are laid on the calyx of a young fruit, and the emerging larvae drill a hole into the pea-sized fruit and enter the endocarp. Fruit damaged in this way often become dehydrated and drop. Those which remain on the tree provide a source of nutrition to the developing larva which feeds on the endocarp; finally, the mature larva emerges causing the fruit to drop. The phyllophagous generation develops over the period from November until the following spring. During this time, the larvae burrow into the leaf, later emerging to feed on the outside of the leaf as well as to mine newly emerging shoots (Arambourg and Pralavorio, 1986).

Populations of the insect can fluctuate significantly, governed by a combination of weather conditions, cultural practices, cultivar tolerance and activity of natural enemies. Ambient air temperature exerts a major influence over the insects' reproductive potential, particularly through its effect on the number of eggs laid. Temperatures above 30°C in conjunction with a low (< 50%) relative humidity can kill both eggs and newly hatched larvae of the carpophagous generation.

The parasitoid complex, which includes approximately 40 species parasitizing mainly larvae and pupae of the anthophagous generation, can cause up to 60% mortality (Arambourg and Pralavorio, 1986). The most important species are the egg parasitoids belonging to the genus *Trichogramma* (Trichogrammatidae), and the egg-larval parasitoids *Ageniaspis fuscicollis* (Dalman) (Encyrtidae) and *Chelonus eleaphilus* Silvestri (Braconidae). Carpophagous generation eggs are predated by a number of anthocorid, mirid and chrysopid species, most notably *Chrysoperla carnea* (Stephens). More recently it has been shown that a number of ant and spider species highly abundant in olive groves prey upon the pre-imaginal stages of the olive moth. Although predation as a whole can cut the pest population size by more than 90%, in some cases it does not impose a sufficiently strong brake on olive moth infestation.

Of the three generations, only those developing on flowers and fruits can cause economic losses. Generally, the anthophagous generation is not harmful to the olive cultivars that produce high numbers of flowers and have low fruit set. In Italy, infestations exceeding 40% of flowers in olive oil cultivars and 10% in table olive cultivars are considered dangerous; in Portugal, the economic thresholds for small-fruit cultivars are 4–6% in high-yielding years and 8–11% in low-yielding ones.

Olive cultivars differ markedly in their susceptibility to infestation by the carpophagous generation, due to host preference and resistance. Table cultivars with bigger fruit size (> 4 g when ripe) are more susceptible to attack. The higher resistance observed in small-fruit cultivars may be related to the extent of the fruit drop after the fruit set stage, which affects mainly the infested drupes, the rapidity with which the endocarp hardens, and the food quality of the endocarp. In Italy, the economic threshold was tentatively set at 40–50% of infested olives in small-drupe cultivars and at 5–10% in big-fruit cultivars.

Black scale

Saissetia oleae (Olivier) (Hemiptera Coccidae) is the most frequently occurring scale pest of olive trees. The insect is highly polyphagous, but it concentrates on olive, citrus and oleander in the Mediterranean Basin. The black scale reproduces by thelytokous parthenogenesis and typically completes just one generation/year, although occasionally two are achieved. Second and third stage nymphs are the major overwintering forms, although some young females have also been observed. Ovipositing females are present during the spring, whereas active crawlers are abundant in summer when they spread through the canopy and settle mainly on the abaxial leaf surface. The fecundity can vary from 150 to 2500 eggs, depending on the insect's nutritional status, which is enhanced if the crop receives either nitrogenous fertilizer and/or irrigation (Pellizzari, 1997). When dispersing, active crawlers are challenged by heavy rainfall, direct sunlight and/or high temperature and the mortality rate can reach 90–95%.

Black scale infestations are generally controlled by a natural enemy complex, involving about 50 species all over the world. In the Mediterranean Basin, the most common predator species are the chrysopid *C. carnea* and the coccinellids *Exochomus quadripustulatus* (Linnaeus), *Chilocorus bipustulatus* (Linnaeus), *Rhyzobius lophanthae* (Blaisdell) and other species. The main parasitoids are the aphelinid *Coccophagus lycimnia* (Walker) and the encyrtids *Metaphycus helvolus* (Compere) feeding on second and third instars, *Metaphycus hageni* Daane & Caltagirone parasitizing third instars and young females, and *Metaphycus lounsburyi* (Howard) (= *M. bartletti*) feeding on third instars and young and ovipositing females. Scale eggs are eaten by larvae of the pteromalids *Scutellista caerulea* (Fonscolombe) and *Moranila californica* (Howard) (Delrio and Foxi, 2010).

The damage caused by this pest results from sap sucking and consequent production of honeydew. The latter promotes the growth of sooty mould, which compromises photosynthesis and transpiration, and can induce premature leaf drop. A particularly heavy infestation can affect yield for several years.

Minor Pests

Olive thrips

Liothrips oleae (Costa) (Thysanoptera Phlaeothripidae) is a widespread pest in the Mediterranean Basin (Arroyo Varela and Lacasa Plasencia, 1986). It overwinters as an adult, sheltering in trees damaged by olive knot disease or bark beetles, or beneath the bodies of dead black scales presenting the exit hole of the parasitoid *S. caerulea*. Three or four generations can be completed in a year; the first starts in spring, infesting buds, flowers, leaves and growing fruits. The feeding punctures damage leaves and fruits, which develop black spots on their surface. Heavy infestations induce leaf dieback with negative consequences on crop yield. The thrips' most important natural enemies are the eulophid parasitoid *Thripastichus gentilei* Del Guercio and the anthocorid *Anthocoris nemoralis* (Fabricius).

Meadow froghopper

Philaenus spumarius (Linnaeus) (Hemiptera Aphrophoridae) feeds on the xylem fluid of a number of herbaceous and tree species. The insect completes a single generation/year and the nymphs develop in a foam nest. In olive trees, nymphs can develop on root suckers, whereas adults feed on the canopy, not causing direct damage. However, the pest can serve as a vector of *Xylella fastidiosa*, a xylem-limited bacterium involved in a severe decline syndrome of the olive tree, as recently reported in Italy (Saponari *et al.*, 2014).

Olive psyllids

Four *Euphyllura* species (Hemiptera Psyllidae) are economically important pests of olive. *Euphyllura olivina* (Costa) is found in the western end of the Mediterranean Basin and in California, *Euphyllura phillyreae* Foester throughout the Mediterranean Basin, particularly in Greece and Turkey, *Euphyllura straminea* Loginova has been reported in West Asia and *Euphyllura pakistanica* Loginova is widespread in the Indian subcontinent as well as in Syria (Asadi *et al.*, 2011).

E. olivina can complete up to three generations/year, even though usually it develops in a single spring generation. Adults overwinter in the plant canopy, and oviposition begins in the spring on young floral buds and leaves. Nymphs produce a waxy covering over the new shoots and flower clusters, causing flower abortion and fruit drop at high population density. The species is predated by various syrphids and lacewings and by the encyrtid endoparasitoid *Psyllaephagus euphyllurae* (Masi).

Cerococcid scale

Pollinia pollini (Costa) (Hemiptera Cerococcidae) is a specific pest of olive and is widely distributed across the whole olive production area. It completes one or two generations/year, colonizing twigs damaged by bark beetles or crevices in the bark, often passing unnoticed (Alexandrakis, 1980). Its feeding punctures at the insertion of leaves and inhibits shoot growth, leading to shoot dieback. The scale is predated by the ladybird beetles

C. bipustulatus and *Exochomus quadripustulatus* (L.).

Viburnum cushion scale

Lichtensia viburni Signoret (Hemiptera Coccidae) colonizes olive, but also viburnum, ivy, myrtle and mastic, infesting leaves, twigs and fruits. Damage is caused by sap depletion, excretion of abundant honeydew and discoloration of fruit, which reduces table olive marketability. The scale passes through two generations/year, and is usually well controlled by a number of predatory ladybird beetles and parasitoid wasps (Longo, 1986).

Filippia follicularis

The coccid *Filippia follicularis* (Targioni-Tozzetti) is commonly found on the abaxial leaf surface, where it completes one generation/year without causing any economic damage (Longo, 1986).

Armoured scales

At least 12 species of polyphagous Diaspididae species can colonize olive, but only the widely distributed *Parlatoria oleae* (Colvée) (olive scale) and *Aspidiotus nerii* Bouché (oleander scale) have an economic importance (Benassy, 1986).

The olive scale is a polyphagous insect, favouring *Rosaceae* species hosts, along with olive and a few ornamentals. It completes two or three generations/year and overwinters as an adult female. It settles on twigs, leaves and fruits, causing twig dieback and premature leaf and fruit drop. In addition, salivary excretions induce the development of purple spots on immature fruit and dark spots on ripe ones, reducing the marketability of table olives. The most important natural enemies of the olive scale are the coccinellid predator *C. bipustulatus* and the aphelinid parasitoids *Aphytis maculicornis* (Masi) and *Coccophagoides utilis* Doutt.

The oleander scale includes multiple cryptic parthenogenetic and sexual species. It attacks more than 100 families of woody plants, including olive, citrus, palms and oleander, passing through three to five generations/year, and overwintering predominantly as adult females. The scale colonizes all parts of the host, causing leaf drop, shoot dieback, green spots on purple fruits and fruit deformation possibly representing a major problem in table fruit cultivars. On the other hand, scale infestations are less important for olive oil producers, even though severe attacks can significantly reduce oil yield. Large populations can develop on trees either covered with dust or treated repeatedly with insecticide, as both of these situations inhibit biological control normally exerted by coccinellid beetles and the aphelinid parasitoids *Encarsia citrina* (Crawford) and several species of *Aphytis*.

Leopard moth

Zeuzera pyrina Linnaeus (Lepidoptera Cossidae) attacks many tree and shrub species and has been associated with damage to some olive cultivars. This moth is a problem in the eastern Mediterranean Basin but only rarely in the west. The insect completes one generation/year, with adults emerging from June to August in Italy and from late April to October in Egypt (Hegazi *et al.*, 2014). Eggs are laid in clumps in bark crevices, and the larvae first penetrate into young shoots, with an entrance hole marked by sap, sawdust and frass, moving later to bigger branches and the trunk. The degree of damage depends on the age of the tree, as a single larva can kill a young tree, whereas a large tree can tolerate colonization by at least ten larvae. A number of parasitoids target the larva, but parasitism is not effective in limiting infestation.

Small olive leafminer

Metriochroa latifoliella (Millière) (Lepidoptera Gracillariidae) completes two generations/year on both olive and *Phyllirea* spp., with larvae mining the leaves not causing economic damage. On the contrary, the small olive leafminer seems beneficial since it serves as an alternative host for *P. mediterraneus*, the most effective parasitoid of the olive fly.

Jasmine moth

Palpita vitrealis (Rossi) (= *P. unionalis*) (Lepidoptera Crambidae) is distributed widely over Europe, North Africa and West Asia, developing predominantly on *Oleaceae* species (olive, privet, ash and jasmine). *P. vitrealis* is a non-diapausing species able to complete four or five overlapping generations/year in Italy and Spain, six in Israel, and up to ten in Egypt (Hegazi *et al.*, 2012). The larvae develop on young shoots and suckers and feed on young leaves, affecting the growth of young trees. However, at high population density larvae can pierce the fruits, being of economic importance also in olive-producing trees.

Olive pyralid moth

Euzophera pinguis (Haworth) (Lepidoptera Pyralidae) develops on olive and ash trees completing two overlapping generations/year. The pest is widely distributed in the Mediterranean Basin, but economic damage has been so far reported in Spain. Adults can be actively flying for over 10 months of the year, while the larvae develop throughout the year. Eggs are laid in bark crevices and on lesions caused by pruning or mechanical harvesting. The larvae tunnel under the bark at the base of branches or at the bottom of the trunk, obstructing the flow of sap, weakening the branch and leading to leaf drop in mature trees and even the death of young plants (Quesada-Moraga *et al.*, 2013). In some rare cases, significant damage caused by the related quince moth *Euzophera bigella* (Zeller) has been reported (Simoglou *et al.*, 2012).

Olive weevil

Otiorhynchus cribricollis Gyllenhal (Coleoptera Curculionidae) is the most common weevil pest of olive in the Mediterranean Basin, and can also be harmful in California and Australia (Arambourg, 1986). The pest completes one generation/year and its larvae develop on the roots of herbaceous wild plants. Adults are polyphagous and cause damage to olive trees mainly in summer through feeding on the leaves, leaving characteristic semicircular incisions. In nurseries and young olive groves, the damage to leaves and stems can be severe enough to kill the plant.

Rhodocyrtus cribripennis

Rhodocyrtus cribripennis (Desbrochers) (Coleoptera Rhynchitidae) is a weevil found in association with both cultivated and wild olive trees, and completes one generation every 2 years. In July, females oviposit on the surface of the seed inside a pierced fruit. Larvae penetrate the kernel and consume the endosperm, emerging in autumn and overwintering in the soil. Pupation occurs the following summer and the new adults emerge during the second year. Feeding wounds can cause fruit drop and, when severe infestations occur, significant yield losses are possible (Perdikis *et al.*, 2009).

Olive bark beetle

Phloeotribus scarabeoides (Bernard) (Coleoptera Curculionidae Scolytinae) attacks olive and some other *Oleaceae* species throughout the Mediterranean Basin and West Asia. The number of generations completed per year is temperature dependent, varying from two or three in southern Europe to four in North Africa. Adults overwinter in short tunnels excavated at the base of branches. In March, the females excavate a subcortical breeding tunnel in the pruning wood of medium to large branches. The spring generation is completed in 4–6 weeks. Adults emerge in May–June and feed by excavating a short tunnel in the shoot of a vigorous tree, causing shoot weakening, dieback or wind breakage. Other generations occur in the following months. The natural enemy complex includes more than ten predator species (mites, beetles and ants) and more than 20 hymenopteran parasitoids, the most frequent of which is the pteromalid *Dinotiscus colon* (Linnaeus) (Jarraya, 1986).

Olive borer

Hylesinus toranio (Danthoine) (= *H. oleiperda*) (Coleoptera Curculionidae Scolytinae) colonizes a number of *Oleaceae* species (*Olea*, *Fraxinus*, *Siringa* and *Ligustrum*). The species is usually univoltine, but can complete a second generation per year in warmer climates (North Africa). This insect develops in vigorous trees and is therefore a primary pest. In summer, females excavate deep brood galleries reaching the sapwood of young trees. The larvae produce small branched annular tunnels which can induce dieback, while the maternal tunnels form reddish cracking plates which weaken the branches (Jarraya, 1986).

Olive fruit midge

Lasioptera berlesiana Paoli (Diptera Cecidomyiidae) completes four or five generations/year on olive fruit, as well on leaf galls produced by mites on the mastic tree (*Pistacia lentiscus*). The females oviposit into punctures made by the olive fly and vector the fungus *Botryosphaeria dothidea* (Moug. Fr.) Ces. & De Not. (= *Camarosporium dalmaticum*), which produces brown spots on the fruit and induces premature fruit drop. Egg development is faster than that of the olive fly, and the emerging larvae feed both on olive fly eggs and on the fungal mycelium (Solinas, 1967). Its value as a biological control agent of the olive fly is limited by the damage caused by its association with *B. dothidea* that can be of economic importance in table olive cultivars.

Olive leaf gall midge

Dasineura oleae (F. Low) (Diptera Cecidomyiidae) is present throughout the Mediterranean Basin, but is a serious pest only in Greece, Syria and Turkey (Simoglou *et al.*, 2012). The species completes one or two generations/year and the larvae produce leaf galls and deform the flower peduncle. The olive leaf gall midge is rarely of economic importance, being usually controlled by a number of parasitoids, but severe infestations can cause leaf and flower drop.

Olive bark midge

Resseliella oleisuga (Targioni-Tozzetti) (Diptera Cecidomyiidae) completes three or four generations/year on olive, ash and *Phyllirea* spp. The females lay their eggs in bark crevices and the larvae develop beneath the bark, causing cortical necrosis and twig drying. The most serious infestations occur in the nursery, originating from graft wounds, and in new growth induced by rejuvenation pruning (Coutin and Katlabi, 1986).

Pest Control in Organic Oliviculture

The principles of pest management applied by organic growers reflect those developed for IPM strategies and consist of adopting ecologically sound practices recommended by international and national organizations (Malavolta and Perdikis, 2012). Emphasis is given to cultural practices which avoid the build-up of pest populations. Preventative measures include the use of resistant or tolerant cultivars, appropriate planting and training systems, adequate cultural practices, and a management regime for the canopy and vegetation designed to increase the abiotic mortality factors and to support populations of predators and parasitoids. Only direct control measures authorized by certified organizations can be used in organic groves, and insecticide applications should be based on economic thresholds set for each specific growing region. Populations of major pests should be regularly monitored in order to assess the risk of crop damage.

Preventative measures and cultural control

When establishing a new olive grove, pest and diseases control can be eased by providing an environment in which plant and animal diversity is encouraged. A high level of biodiversity should promote natural enemy populations, as these species often require alternative hosts, food and shelter. Organic olive groves have been shown to represent more suitable environments than conventionally managed ones for maintaining predators and parasitoids (Santos *et al.*, 2007).

The choice of a cultivar with both known resistance (or at least tolerance) to the major diseases and pests, and good adaptation to the local soil and climatic conditions is critical. In fact, in different olive-growing areas, cultivars tolerant to diseases and pests are already cultivated (Bellini *et al.*, 2008). In areas where the risk of olive fly damage is high, the grower should avoid interplanting susceptible large-fruited cultivars among more tolerant olive oil cultivars. Nursery material should always be certified as being free of viruses, diseases and pests (particularly scales). Trees should be shaped to maximize air circulation and the penetration of direct sunlight, as this avoids creating a micro-environment favourable for the development of diseases and scale colonization.

The soil management should avoid deep tillage and spray of herbicides in order to preserve biodiversity and increase ecological stability. Organic fertilization, allowed in organic agriculture, reduces the negative effects of inorganic nitrogen fertilization that promotes the development of sap-feeding pests such as psyllids and scales. Excessive irrigation should also be avoided, as this increases the risk of olive fly and black scale infestation, by reducing the killing effect of high summer temperatures on pest eggs and larvae.

Regular pruning can produce unfavourable microclimatic conditions for pest development inside the canopy. In fact, the excessive shading due to the canopy density favours the increase of humidity and reduces the importance of mortality by abiotic factors in summer (direct sunlight exposure and high temperatures), which are the most important natural limiting factors of populations of scales and other pests. The ventilation of the canopy is also an important preventative measure against some diseases, such as olive leaf spot. In addition regular pruning prevents the development of some pests by removing branches severely infested by diseases (olive knot) and insects such as scales, bark beetles and the leopard moth. Moreover, olive thrips populations can be reduced by removing bark beetle- and olive knot-infested branches where the pest overwinters. However, care should be taken to seal pruning cuts of large branches with pruning paint to avoid colonization of the wound by the olive pyralid moth. The traditional control technique used to reduce the olive bark beetle populations consists of pruning branches in early spring as a bait to attract ovipositing females; this material is later removed or destroyed before the emergence of adults.

The harvest should be carried out at commercial maturity of fruits to obtain the highest quantity of oil per plant. The main considerations surrounding the timing of harvest are: (i) the fruit removal force; (ii) the fruits' oil content; (iii) the quality of the oil; and (iv) the incidence of olive fly. Damage caused by the olive fly can be reduced by anticipating the harvest, because the increase of oil accumulation in olives does not offset the risk of qualitative and quantitative losses due to heavy infestation during the autumn. Early harvest was for a long time the only control means to significantly reduce olive fruit fly attacks. In Italy, the optimal harvest period for most cultivars cannot be delayed beyond October, with a few late-ripening cultivars able to be harvested in November (Delrio *et al.*, 1995). Harvesting all the fruit helps to reduce the olive fly population as it minimizes the size of the spring generations.

Biological control

Sustainability of the olive grove agroecosystem can be achieved by preserving the functional biodiversity and the natural equilibrium between pests and natural enemies. Olive groves harbour a high number of arthropod predator species and about 300–400 parasitoid species, which greatly contribute to the ecosystem stability. Conservation and maximum exploitation of naturally occurring biological control agents are traditionally considered very important for control of olive pests.

Conservation and increase of natural enemies can be promoted by: (i) a combination of maintaining a diversity of vegetation in and around the olive groves and using selective insecticides and/or other non-disruptive pest control techniques (conservation biological control); (ii) introducing exotic natural enemies (classical biological control); and (iii) rearing and subsequently releasing beneficial insects (augmentative biological control).

Conservation biological control implies the manipulation of the environment with a view to increasing the abundance of natural enemies. Maintaining herbaceous and woody vegetation inside and around the olive grove should favour predator and parasitoid organisms, and so bear down on the populations of some pests, such as olive psyllids and the olive moth (Paredes *et al.*, 2013). Also the populations of *Pnigalio mediterraneus* (the main parasitoid of the olive fly) increase when oak, apple or *Anagyris foetida* trees are present in the vicinity of the olive grove, because these tree species all support alternative hosts of the parasitoid. *Dittrichia viscosa* was considered for 50 years the most important plant species to enhance natural enemies, as it is infested by a gall-forming tephritid, *Myopites stylatus* (Fabricius), an alternative host for two olive fruit fly parasitoids, *Eupelmus urozonus* and *Eurytoma martellii* (Isaakides, 1957). However, recently *E. urozonus* has been found to be mainly a hyperparasitoid, and the *Eurytoma* species developing on *M. stylatus* was described as *Eurytoma inulae* Domenichini, an organism unable to parasitize the olive fruit fly (Delrio *et al.*, 2007; Delrio, 2010). This case points out the need to identify plants harbouring alternative hosts for the most important natural enemies of olive pests. As a general point, the management of the non-crop vegetation needs to be arranged so that it does not conflict with predator food supply, the presence of alternative hosts and preys, and shelter. Zealous weed control, as practised by many olive producers, is likely to be harmful in the context of conserving beneficial arthropod populations.

One effect of spraying with broad-spectrum insecticides is to decrease the size of predator and parasitoid populations, resulting in a higher incidence of minor pest outbreaks. Repeated applications of dimethoate to control the olive fly have led to more frequent and more severe infestations of black scale (Delrio, 1992). A particularly striking example is represented by the heavy infestation of oleander scale which affected olive production in north-western Crete in the late 1970s, following repeated use of chemical control to suppress the three major pests. The oleander scale flourished as the population of its major parasitoid *Aphytis chilensis* Howard had crashed (Alexandrakis and Benassy, 1979). Bait sprays against the olive fly proved to be much less detrimental to *Aspidiotus nerii* parasitoids than cover sprays, and eventually restored the natural equilibrium. The efficacy and environmental impact of organic-certified insecticides on the olive biocoenosis are poorly researched; even the application of processed kaolin and the use of bait traps to control the olive fly are known to have a negative effect on some natural enemies (Pascual *et al.*, 2010).

Although parasitoids contribute to the control of pest populations, their overall effect is not strong enough to provide an economic level of protection against the major olive pests (Paraskakis *et al.*, 1980; Bigler *et al.*, 1986; Delrio and Prota, 1990; Bento *et al.*, 1998). Over the years, three exotic predator and 35 exotic parasitoid species have been introduced in the Mediterranean Basin in attempts to provide adequate biological control over the three main olive pests (Jervis *et al.*, 1992). Most of these introductions involved the release of rather limited numbers of individuals, and few species have resulted in their successful establishment. Nevertheless, this strategy should be persevered with, as it can produce major long-term benefits where it is successful.

In the Mediterranean area, the olive fly is parasitized by some polyphagous chalcidoids and a braconid endoparasitoid that are unable to achieve an effective control (Delrio and Prota, 1990). None the less, biological control has a great potential, as many parasitoid species are able to effectively parasitize the olive fruit fly in Africa, including at least six additional braconid wasps (Silvestri, 1914; Hoelmer *et al.*, 2004). Early in the 1900s, Silvestri attempted, although without success, to introduce into Italy nine African olive fly parasitoids and six other tephritid predators collected from various parts of the world. By the 1980s, improvements in the capacity to mass-rear natural enemies allowed the braconid larval–pupal parasitoid *Diachasmimorpha longicaudata* (Ashmead) to be released in Spain, Sardinia and Greece, although it too failed to establish any sustainable population. A further introduction target was the egg–pupal parasitoid *Fopius arisanus* (Sonan), which was introduced in central Italy, Sardinia and Israel, with some success in the latter (Delrio, 2010; Argov *et al.*, 2012). Current projects are focusing on the introduction of *Psyttalia lounsburyi* (Silvestri) and *Psyttalia humilis* (Silvestri) in the Mediterranean Basin but to date poor establishment has been obtained (Argov *et al.*, 2012). However, *P. lounsburyi* was introduced in California and it is now established in the coastal regions (Daane *et al.*, 2015).

The potential of predators and parasitoids to control black scale is well recognized. In recent years, 17 hymenopteran parasitoid species and the coccinellid predator *Rhyzobius forestieri* (Mulsant) have been introduced in the Mediterranean Basin (Katsoyannos, 1984; Argov and Rossler, 1993). Among the former, *Metaphycus helvolus*, *M. lounsburyi* and *M. hageni* have all become widely established and are thought to contribute significantly to control the scale infestations (Tena *et al.*, 2008; Delrio and Foxi, 2010).

The augmentative biological control strategy has been tested in olive groves to control olive fly, olive moth and black scale. The success of this strategy depends on: (i) having a firm understanding of the agro-ecosystem ecology; (ii) the ability to economically mass rear the natural enemy; and (iii) the existence of an already established complex of predators and parasitoids in addition to those being released. Two strategies of augmentative release have been considered: inundative release and inoculative release. Inundative release attempts to mimic a pesticide application by the release of massive numbers of biological agents as a means of rapidly reducing the target pest population. By contrast, inoculative release uses relatively small numbers of beneficials in the hope that their numbers will increase naturally so that the progeny will ultimately provide pest suppression. Cost is usually the decisive factor determining the choice of approach. In some Mediterranean Basin regions, the olive fly is attacked by the braconid wasp *Psyttalia concolor*, although the level of biological control exerted is weak, probably because of asynchrony between the parasitoid and pest. In fact, olive fly larvae are typically unavailable for parasitism when the wasp females emerge in the spring. A substantial effort has been made to develop an effective level of biological control over the olive fly by inundative releases of *P. concolor*. Since 1960, a mass-rearing method was developed using the Mediterranean fruit fly as a host, since this insect can be reared very inexpensively. Large-scale experiments have demonstrated the effectiveness of inundatively releasing *P. concolor* (Genduso, 1981). Some promising results have been obtained recently from the use of the inoculative strategy, in which a small number of parasitoid adults per tree was released early in the season. In this way, the olive fly control is achieved by new generations of parasitoids emerging from the first olive fly generation (Delrio *et al.*, 2005). However, biological control against *B. oleae* has been mainly limited by: (i) high production costs; (ii) lack of mass-rearing laboratories; and (iii) reduced quality of mass-reared insects.

The effect of an inundative release of mass-reared egg parasitoid *Trichogramma* spp. against the olive moth and the jasmine moth has been evaluated in several countries.

These parasitoids can be mass produced on alternative hosts at a low cost. The best results have been obtained from releases of large numbers (up to 3,000,000/ha) of locally collected wasp strains, achieving an egg parasitism rate ranging from 35% to 91% in different trials (Herz *et al.*, 2005). These data indicate this strategy is promising since high mortality rates were observed in some studies, even though further optimization is needed.

Scales could also be profitably controlled through the augmentative release of parasitoids and predators. *Metaphycus helvolus* is the most commonly mass-reared parasitoid for control of the black scale, while the coccinellid predators *C. bipustulatus* and *R. lophanthae* could be released to reduce the population density of various diaspidid pests (Daane *et al.*, 1991; Stathas *et al.*, 2005).

Microbiological control

A number of entomopathogens have been identified as potential biological control agents of olive pests. Sprays based on *Bacillus thuringiensis* (*Bt*) have been widely used to give effective protection against larvae of both the olive moth and the jasmine moth. In order to reduce the population size of the olive moth carpophagous generation, *Bt* spraying needs to be directed to larvae of the anthophagous generation (Yamvrias and Young, 1977). *Bt* strains effective to both adults and larvae of the olive fly have been identified and tested using the bait spray technique. However, achieving biological control via this approach is likely to be rather impractical since repeated applications are required (Alberola *et al.*, 1999; Navrozidis *et al.*, 2000; Delrio, 2010).

The entomopathogenic fungi *Beauveria bassiana* and *Metarhizium anisopliae* were isolated in the soil of olive groves and in the tree canopy and are natural biocontrol agents of the olive moth and the olive pyralid moth (Quesada-Moraga *et al.*, 2007; Oliveira *et al.*, 2012). These entomopathogens were also tested in the laboratory against the olive fly, showing high effectiveness against pupariating third instar larvae and pupae. *B. bassiana* is also effective against olive fly adults and is registered for use in olive groves (Konstantopoulou and Mazomenos, 2005). However, weekly applications of the *B. bassiana* formulation are required to effectively protect olives, dramatically increasing the seasonal costs of the control programme. This entomopathogenic fungus applied to pruning wounds is an effective tool for the microbial control of the olive pyralid moth (Quesada-Moraga *et al.*, 2013).

Behaviour modifying chemicals

With the aim to control the major olive pests, a number of visual and food attractants, pheromones and repellents are widely used (Delrio *et al.*, 1983; Delrio, 1984).

As olive flies are attracted by yellow (frequency range of 500–580 nm) high reflectance surfaces, yellow sticky traps are sometimes used for monitoring and control purposes. However, these traps are not selective to beneficials so they are not used in mass trapping control programmes. Food odour attractants have long been used for monitoring adult olive fly populations and bait-spray control measures. McPhail traps have also been employed for population monitoring and mass trapping; these devices consist of a bell-shaped glass container with an invaginated opening at the bottom and a water reservoir baited with attractants. At one time the usual attractant was molasses, but nowadays this has been replaced by protein hydrolysate and ammonium salts, as ammonia is the essential volatile attractant of protein hydrolysates. McPhail traps represent an effective means of capturing adults (especially females) flying within a 20 m radius from the trap and are partially selective to beneficials. In recent times, traditional McPhail traps have been replaced by plastic bottles, which are much cheaper. The most common design involves a 1.5 l plastic bottle with two to four holes in the upper portion baited with three to four *Torula* yeast tablets/l of water.

Food attractants have been used against olive fly adults, both for mass trapping and for lure-and-kill purposes. Mass trapping involves capturing large numbers of adults, while the lure-and-kill control strategy aims to attract the flies into a trap containing insecticide. The two commonest versions of lure-and-kill traps consist of either a light green paper envelope containing ammonium bicarbonate, or a piece of cardboard to which an ammonium bicarbonate dispenser is attached. In both cases, the trap is impregnated with a pyrethroid insecticide, which has a highly knockdown effect and is stable for about 4 months (Lentini *et al.*, 2010). Sex pheromone dispensers can be added to improve the trap effectiveness. Traps hung in the tree canopy at a density of 100–150/ha before the pea-size fruit stage provide season-long protection against the olive fly. Lure-and-kill strategies are effective provided that the pest population is not too high, and particularly if the olive trees are either isolated from other groves, or if neighbouring groves are also applying effective control measures (Broumas *et al.*, 2002).

Sex pheromones are commonly used to monitor lepidopteran pests, although they have also been tested for mass trapping and mating disruption. Mass trapping based on ultraviolet-light sticky traps baited with sexual pheromones has proved to be highly effective against the leopard moth (Hegazi *et al.*, 2009a). Mating disruption experiments targeting the olive moth, jasmine moth and olive pyralid moth have all given promising results (Ortiz *et al.*, 2004; Hegazi *et al.*, 2007; Hegazi *et al.*, 2009b).

Chemical control

When cultural and biological protection measures are insufficient to reduce pest populations below the economic thresholds and chemical control techniques are required, the preferred products should be those with minimum impact on human health, non-target species and the environment. European Union regulations (EC 889/2008) covering organic agriculture only permit the use of plant protection products of natural origin (kaolin clay, copper compounds, mineral oils, azadirachtin, hydrolysed proteins, pheromones, pyrethrins, spinosad) and pyrethroids only in lure-and-kill traps. The control decision has to be based on economic thresholds, a risk assessment and a prognosis provided by an official forecasting service. For this purpose, appropriate pest monitoring programmes are needed in each region.

Conventionally, scales and psyllids have been controlled by sprayings with mineral oil, which provide high efficacy and low impact on natural enemies. For effective control of black scale, a high-volume spray should be directed against newly hatched crawlers and not be continued once the scales have entered the adult stage.

Cover spraying with azadirachtin is effective against the olive moth, jasmine moth, olive psyllids and black scale, but not against the olive fly (Lentini *et al.*, 2005; Ibraheem *et al.*, 2012; Mahmoud, 2014). Azadirachtin seems to be harmless to lacewings and other predators and parasitoids (i.e. natural enemies of olive fly) in olive groves (Medina *et al.*, 2004; Iannotta *et al.*, 2007).

Sprays containing the insecticidal compound spinosad, a fermentation product of the bacterium *Saccharopolispora spinosa*, provide effective control over the anthophagous generation of olive moth, jasmine moth, psyllids and olive thrips (Mandour *et al.*, 2008). Bait sprays poisoned with spinosad represent a useful preventative treatment against olive fly adults, but because of the short persistence of the compound, repeated applications are generally required (Gonçalves *et al.*, 2012). Spinosyns have a minimum impact on beneficial arthropods, but over-use of spinosad has already led to the development of resistance in the olive fly in California (Kakani *et al.*, 2010).

A number of natural compounds have demonstrated repellent and/or oviposition-deterrent properties against the olive fly. Chief among these are various copper salts, which are also widely used as fungicides. The effectiveness of these compounds against the olive fly is thought to lie in their deterrence to ovipositing females, but they may also act to inhibit gut bacterial symbiosis (Prophetou-Athanasiadou *et al.*, 1991; Belcari *et al.*, 2005). Certain clay-based products have been used in the past as repellents; a renewed interest in them has followed the recognition that kaolin particle films appear to deter oviposition by olive fly and olive moth (De la Roca, 2003; Saour and Makee, 2004). Kaolin applications should be repeated as the size of fruits increases thereby reducing the cover protection and after heavy rainfall. Kaolin cover sprays have no documented negative effects on the chemical or organoleptic properties of the olive oil (Benincasa *et al.*, 2008).

Conclusions

The area cultivated with organic olive trees has greatly increased as a consequence of the increasing demand for organic products. Olive cultivation is suitable for organic farming due to the tolerance of oil varieties to most pests, and to the high biodiversity of the olive agroecosystem that allows the development of many natural enemies. In fact, predators and parasitoids are usually able to naturally control most of olive pests, and successful introduction in the Mediterranean Basin of some exotic natural enemies effectively contributed to control the black scale, that caused high crop loss in the past. In addition, cultural techniques used in organic agriculture can greatly reduce pest infestation.

Direct control measures are usually needed only against the olive fruit fly and the olive moth. Some control techniques allowed in organic agriculture are highly effective and are also widely used in conventionally managed groves. Copper salts, usually used to control pathogens, contribute to reduce olive fly attacks, whereas applications of kaolin particle film were as effective as synthetic insecticides, even though the collateral effects to natural enemies need to be further evaluated. Mass trapping and lure and kill show a high effectiveness against low populations of the olive fly and will be further improved by developing more effective attractants and traps. Also spinosad bait sprays are effective against the olive fly but repeated applications are required, which increase the cost of intervention.

Other control methods against major and secondary pests are developing rapidly, even though their use in olive groves require additional scientific and technological improvements and cost reduction.

Inoculative and inundative biological control has been so far limited by lack of mass-rearing facilities and high production costs. Similarly, the wide use of mating disruption against lepidopteran pests is constrained by high costs. Microbiological control with *Bt* is effectively used to reduce infestation of the olive moth while further studies are needed for its use against the olive fruit fly. Other microbiological agents such as entomopathogenic bacteria, fungi and nematodes have provided promising results and some products have been registered.

The sterile insect technique (SIT), widely used against the Mediterranean fruit fly, was tested during the 1970s and 1980s and has been recently recovered in order to develop a commercial SIT-based strategy to control the olive fruit fly.

Effective control measures in organic oliviculture are already available and can be further spread with adequate information and demonstration. Current research should be focused on: (i) improving pest control efficiency; (ii) reducing the cost of the control tools; and (iii) developing new alternative techniques allowed in organic agriculture.

References

Alberola, T.M., Aptsoglou, S., Arsenakis, M., Bel, Y., Delrio, G., Ellar, D.J., Ferré, J., Granero, F., Guttmann, D.M., Koliais, S., Martinez-Sebastian, M.J., Prota, R., Rubino, S., Satta, A., Scarpellini, G., Sivropoulou, A. and Vasara, E. (1999) Insecticidal activity of strains of *Bacillus thuringiensis* on larvae and adults of *Bactrocera oleae* Gmelin (Dipt. Tephritidae). *Journal of Invertebrate Pathology* 74, 127–136.

Alexandrakis, V. (1980) Bio-ecological data on *Pollinia pollini* (Hom. Coccoidea, Asterolecanidae) on olive in Crete. *Annales de la Société Entomologique de France* 16(1), 9–17.

Alexandrakis, V. and Benassy, C. (1979) Le problème *Aspidiotus nerii* Bouché (Homoptera, Diaspididae) en Crète: exemple récent de pullulation provoquée. *Fruit* 34(9), 535–541.

Arambourg, Y. (1986) Curculionidae. In: Arambourg, Y. (ed.) *Traité d'Entomologie Oléicole*. Conseil Oléicole International, Madrid, pp. 11–19.

Arambourg, Y. and Pralavorio, R. (1986) Yponomeutidae. In: Arambourg, Y. (ed.) *Traité d'Entomologie Oléicole*. Conseil Oléicole International, Madrid, pp. 47–70.

Argov, Y. and Rossler, Y. (1993) Biological control of the Mediterranean black scale, *Saissetia oleae* (Hom.: Coccidae) in Israel. *Entomophaga* 38, 89–100.

Argov, Y., Kuslitzky, W. and Hoelmer, K. (2012) Biological control of olive fruit fly, *Bactrocera oleae*, in Israel. *IOBC (International Organisation for Biological and Integrated Control) WPRS Bulletin* 79, 79–85.

Arroyo Varela, M. and Lacasa Plasencia, A. (1986) Phloeothripidae. In: Arambourg, Y. (ed.) *Traité d'Entomologie Oléicole*. Conseil Oléicole International, Madrid, pp. 289–300.

Asadi, R., Talebi, A.A., Khalghani, J., Fathipour, Y., Moharramipour, S. and Burckhardt, D. (2011) Comparative development and demographic parameters of *Euphyllura pakistanica* on four olive cultivars. *Bulletin of Insectology* 64(2), 159–165.

Belcari, A., Sacchetti, P., Rosi, M.C. and Del Pianta, R. (2005) The use of copper products to control the olive fly (*Bactrocera oleae*) in central Italy. *IOBC (International Organisation for Biological and Integrated Control) WPRS Bulletin* 28(9), 45–48.

Bellini, E., Giordani, E. and Rosati, A. (2008) Genetic improvement of olive from clonal selection to cross-breeding programs. *Advances in Horticultural Science* 22(2), 73–86.

Benassy, C. (1986) Diaspididae. In: Arambourg, Y. (ed.) *Traité d'Entomologie Oléicole*. Conseil Oléicole International, Madrid, pp. 206–286.

Benincasa, C., Briccoli Bati, C., Iannotta, N., Pellegrino, M., Pennino, G., Rizzuti, B. and Romano, E. (2008) Efficacy of kaolin and copper based products on olive-fruit fly (*B. oleae* Gmelin) and effects on nutritional and sensory parameters of olive oils. *La Rivista Italiana di Scienze dell'Alimentazione* 37(4), 21–31.

Bento, A., Ilideo, J., Campos, M. and Torres, L. (1998) Parasitismo associado a traca da oliveira *Prays oleae* Bern. em Tras-os-Montes (Nordeste de Portugal). *Boletin de Sanidad Vegetal Plagas* 24, 949–954.

Bigler, F., Neuenschwander, P., Delucchi, V. and Michelakis, S. (1986) Natural enemies of preimaginal stages of *Dacus oleae* Gmel. (Dipt., Tephritidae) in Western Crete II: impact on olive fly populations. *Bollettino del Laboratorio di Entomologia Agraria Filippo Silvestri* 43, 79–96.

Broumas, T., Haniotakis, G., Liaropoulos, C., Tomazou, T. and Ragoussis, N. (2002) The efficacy of an improved form of the mass-trapping method, for the control of the olive fruit fly, *Bactrocera oleae* (Gmelin) (Dipt., Tephritidae): pilotscale feasibility studies. *Journal of Applied Entomology* 126, 217–223.

Capuzzo, C., Firrao, G., Mazzon, L., Squartini, A. and Girolami, V. (2005) '*Candidatus* Erwinia dacicola', a coevolved symbiotic bacterium of the olive fly *Bactrocera oleae* (Gmelin). *International Journal of Systematic and Evolutionary Microbiology* 55, 1641–1647.

Cariddi, C., Saponari, M., Boscia, D., De Stradis, A., Loconsole, G., Nigro, F., Porcelli, F., Potere, O. and Martelli, G.P. (2014) Isolation of a *Xylella fastidiosa* strain infecting olive and oleander in Apulia, Italy. *Journal of Plant Pathology* 96(3), 1–5.

Cavalloro, R. and Delrio, G. (1974) Mating behavior and competitiveness of gamma-irradiated olive fruit flies. *Journal of Economic Entomology* 67, 253–255.

Coutin, R. and Katlabi, H. (1986) Cecidomyiidae. In: Arambourg, Y. (ed.) *Traité d'Entomologie Oléicole*. Conseil Oléicole International, Madrid, pp. 95–113.

Daane, K.M., Barzman, M.S. and Caltagirone, L.E. (1991) Augmentative release of *Metaphycus helvolus* for control of black scale, *Saissetia oleae*, in olives. *Plant Protection Quarterly* 2, 6–9.

Daane, K.M.,Wang, X., Nieto, D.J., Pickett, C.H., Hoelmer, K.A., Blanchet, A. and Johnson, M.W. (2015) Classic biological control of olive fruit fly in California, USA: release and recovery of introduced parasitoids. *BioControl* 60(3), 317–330.

De la Roca, M. (2003) Surround® crop protectant: la capa protectora natural para cultivos como el olivar. *Phytoma España* 148, 82–85.

Delrio, G. (1984) Biotechnical methods for olive pest control. In: Cavalloro, R. and Crovetti, A. (eds) *Integrated Pest Control in Olive-Groves*. Proceedings of the Commission of European Communities (CEC)/ Food and Agriculture Organization of the United Nations (FAO)/International Organisation for Biological and Integrated Control (IOBC) International Joint Meeting, 3–6 April 1984, Pisa, Italy. Balkema A.A., Rotterdam, The Netherlands, pp. 394–410.

Delrio, G. (1992) Integrated control in olive groves. In: Van Lenteren, J.C., Minks, A.K. and de Ponti, O.M.B. (eds) *Towards Environmentally Safer Agriculture*. Proceedings of the IOBC (International Organisation for Biological and Integrated Control)/WPRS Biological Control and Integrated Crop Protection International Conference. Pudoc Scientific Publisher, Wageningen, The Netherlands, pp. 67–76.

Delrio, G. (2010) Biological control of olive pests in the Mediterranean region. *IOBC (International Organisation for Biological and Integrated Control) WPRS Bulletin* 53, 85–92.

Delrio, G. and Foxi, C. (2010) Current status of *Saissetia oleae* biological control in Sardinia (Italy). *IOBC (International Organisation for Biological and Integrated Control) WPRS Bulletin* 59, 89–92.

Delrio, G. and Prota, R. (1976) Osservazioni eco-etologiche sul *Dacus oleae* Gmelin nella Sardegna nord-occidentale. *Bollettino di Zoologia Agraria e di Bachicoltura* 13, 49–118.

Delrio, G. and Prota, R. (1990) Determinants of abundance in a population of the olive-fruit fly. *Frustula Entomologica* 11, 47–55.

Delrio, G., Economopoulos, A.P., Economopoulos, P.V., Haniotakis, G.H. and Prota, R. (1983) Comparative field studies on food, sex, and visual attractants for the olive fruit fly. In: Cavalloro, R. (ed.) *Fruit Fruit Flies of Economic Importance*. Proceedings of the Commission of European Communities (CEC)/IOBC (International Organisation for Biological and Integrated Control) International Symposium, Athens, November 1982. A.A. Balkema, Rotterdam, The Netherlands, pp. 465–472.

Delrio, G., Lentini, A., Vacca, V. and Serra, G. (1995) Influenza dell'infestazione di *Bactrocera oleae* (Gmel.) sulla produzione e sulle caratteristiche qualitative dell'olio di oliva. *Rivista Italiana Sostanze Grasse* 72, 5–9.

Delrio, G., Lentini, A. and Satta, A. (2005) Biological control of olive fruit fly through inoculative releases of *Opius concolor* Szepl. *IOBC (International Organisation for Biological and Integrated Control) WPRS Bulletin* 28(9), 53–58.

Delrio, G., Lentini, A. and Satta, A. (2007) Augmentative releases of *Eupelmus urozonus* Dalm. against the olive fruit fly and observations on its facultative hyperparasitism. *IOBC (International Organisation for Biological and Integrated Control) WPRS Bulletin* 30(9), 7.

Delrio, G., Deliperi, S. and Lentini, A. (2010) Susceptibility of Sardinian olive cultivars to *Bactrocera oleae* infestations. *IOBC (International Organisation for Biological and Integrated Control) WPRS Bulletin* 53, 17–21.

Genduso, P. (1981) Attuali conoscenze sulla lotta biologica contro la mosca delle olive a mezzo di entomofagi. *Informatore Fitopatologico* 31, 57–59.

Gonçalves, M.F., Santos, S.A.P. and Torres, L.M. (2012) Efficacy of spinosad bait sprays to control *Bactrocera oleae* and impact on non-target arthropods. *Phytoparasitica* 40, 17–28.

Haniotakis, G.E. (2005) Olive pest control: present status and prospects. *IOBC (International Organisation for Biological and Integrated Control) WPRS Bulletin* 28(9), 1–9.

Haniotakis, G.E. and Pitarra, I.S. (1994) Response of *Bactrocera* (*Dacus*) *oleae* males (Diptera: Tephritidae) to pheromones as affected by concentration, insect age, time of day, and previous exposure. *Environmental Entomology* 23(3), 726–731.

Hegazi, E.M., Konstantopoulou, M.A., Milonas, P., Herz, A., Mazomenos, B.E., Khafagi, W.E., Zaitun, A., Abdel-Rahman, S.M., Helal, I. and El-Kemny, S. (2007) Mating disruption of the jasmine moth *Palpita unionalis* (Lepidoptera: Pyralidae) using a two pheromone component blend: a case study over three consecutive olive growing seasons in Egypt. *Crop Protection* 26, 837–844.

Hegazi, E., Khafagi, W.E., Konstantopoulou, M., Raptopoulos, D., Tawfik, H., Abd El-Aziz, G.M., Abd El-Rahman, S.M., Atwa, A., Aggamy, E. and Showeil, S. (2009a) Efficient mass-trapping method as an alternative tactic for suppressing populations of leopard moth (Lepidoptera: Cossidae). *Annals of the Entomological Society of America* 102(5), 809–818.

Hegazi, E.M., Kostantopoulou, M.A., Herz, A., Mazomenos, B.E., Khafagi, W.E., Agamy, E., Zaitun, A., Abd El-Aziz, G.M., Showiel, S. and Abdel-Rahman, S.M. (2009b) Is mating disruption effective in controlling the olive moth, *Prays oleae*? *Crop Protection* 28, 181–189.

Hegazi, E.M., Konstantopoulou, M.A., Khafagi, W.E., Schlyter, F., Herz, A., Raptopoulos, D.G., Hassan, S. and Atwa, A. (2012) The population trend of *Palpita unionalis* in different olive varieties in Egypt. *Phytoparasitica* 40, 451–459.

Hegazi, E., Schlyter, F., Khafagi, W., Atwa, A., Agamy, E. and Konstantopoulou, M. (2014) Population dynamics and economic losses caused by *Zeuzera pyrina*, a cryptic wood-borer moth, in an olive orchard in Egypt. *Agricultural and Forest Entomology* 17, 9–19. DOI: 10.1111/afe.12075.

Herz, A., Hassan, S.A., Hegazi, E., Nasr, F.N., Youssef, A.A., Khafagi, W.E., Agamy, E., Ksantini, M., Jardak, T., Mazomenos, B.E., Konstantopoulou, M.A., Torres, L., Gonçalves, F., Bento, A. and Pereira, J.A. (2005) Entwicklung nachhaltiger pflanzenschutzstrategien zur bekämpfung von schadschmetterlingen im olivenanbau. *Gesunde Pflanzen Pflanzenschutz-Verbraucherschutz-Umweltschutz* 57, 117–128.

Hoelmer, K.A., Kirk, A., Wharton, R.A. and Pickett, C.H. (2004) Foreign exploration for parasitoids of the olive fruit fly, *Bactrocera oleae*. In: Woods, D. (ed.) *Biological Control Program Annual Summary, 2003*. California Department of Food and Agriculture, Sacramento, California, pp. 12–14.

Iannotta, N., Belfiore, T., Brandmayr, P., Noce, M.E. and Scalercio, S. (2007) Evaluation of the impact on entomocoenosis of active agents allowed in organic olive farming against *Bactrocera oleae* (Gmelin, 1790). *Journal of Environmental Science and Health, Part B* 42, 783–788.

Ibraheem, A.M., Al-Arnaouty, S.A.G., Moussa, S.F.M. and Helmy, S.M.Y. (2012) Efficiency of certain biopesticides against the olive black scale insect, *Saissetia oleae* (Olivier) on olive trees at Giza Governorate, Egypt. *Egyptian Academic Journal of Biological Sciences* 5(2), 87–93.

Isaakides, C. (1957) Sur la lutte biologique contre *Dacus oleae* Gmel. *Entomophaga* 2, 245–249.

Jarraya, A. (1986) Scolytidae. In: Arambourg, Y. (ed.) *Traité d'Entomologie Oléicole*. Conseil Oléicole International, Madrid, pp. 21–36.

Jervis, M.A., Kidd, N.A.C., McEwen, P., Campos, M. and Lozano, C. (1992) Biological control strategies in olive pest management. In: Haskell, P.T. (ed.) *Research Collaboration in European IPM Systems. Monograph No 52*. Lavenham Press Limited, Lavenham, Suffolk, UK, pp. 31–39.

Kakani, E.G., Zygouridis, N.E., Tsoumani, K.T., Seraphides, N., Zalom, F.G. and Mathiopoulos, K.D. (2010) Spinosad resistance development in wild olive fruit fly *Bactrocera oleae* (Diptera: Tephritidae) populations in California. *Pest Management Science* 66, 447–453.

Katsoyannos, P. (1984) The establishment of *Rhyzobius forestieri* (Col., Coccinellidae) in Greece and its efficiency as an auxiliary control agent against a heavy infestation of *Saissetia oleae* (Hom. Coccidae). *Entomophaga* 29, 387–397.

Konstantopoulou, M.A. and Mazomenos, B.E. (2005) Evaluation of *Beauveria bassiana* and *B. brongniartii* strains and four wild-type fungal species against adults of *Bactrocera oleae* and *Ceratitis capitata*. *BioControl* 50, 293–305.

Lentini, A., Delrio, G. and Foxi, C. (2005) Experiments for the control of olive fly in organic agriculture. *IOBC (International Organisation for Biological and Integrated Control) WPRS Bulletin* 28(9), 73–76.

Lentini, A., Deliperi, S. and Delrio, G. (2010) Seasonal changes in efficiency of traps used in the lure and kill method for the control of *Bactrocera oleae*. *IOBC (International Organisation for Biological and Integrated Control) WPRS Bulletin* 53, 69–72.

Longo, S. (1986) Notes on the behaviour of *Filippia follicularis* (Targ.-Tozz.) and *Lichtensia viburni* Sign. (Homoptera, Coccidae) in Sicily. *Bollettino del Laboratorio di Entomologia Agraria Filippo Silvestri* 43, 173–177.

López-Escudero, F.J. and Mercado-Blanco, J. (2011) Verticillium wilt of olive: a case study to implement an integrated strategy to control a soil-borne pathogen. *Plant Soil* 344, 1–50.

Lòpez-Villalta, L.C. (1996) Production techniques. In: International Olive Oil Council (ed.) *World Olive Encyclopedia*. Egedsa, Sabadell, Spain, pp. 145–190.

Mahmoud, F.M. (2014) Efficacy of eco-smart insecticides against certain biological stages of jasmine moth, *Palpita unionalis* Hb. (Lepidoptera: Pyralidae). *Pesticides and Phytomedicine* 29(1), 55–65.

Malavolta, C. and Perdikis, D. (2012) Guidelines for integrated production of olives. *IOBC (International Organisation for Biological and Integrated Control) WPRS Bulletin* 77, 1–19.

Mandour, N.S., Osman, M.A.M., Mahmoud, M.F. and Mosleh, Y.Y. (2008) Evaluation of spinosad as a biopesticide for controlling the jasmine moth, *Palpita unionalis* Hb. (Lepidoptera: Pyralidae). *Egyptian Journal of Biological Pest Control* 18(1), 207–213.

Medina, P., Budia, F., Del Estal, P. and Viñuela, E. (2004) Influence of azadirachtin, a botanical insecticide, on *Chrysoperla carnea* (Stephens) reproduction: toxicity and ultrastructural approach. *Journal of Economic Entomology* 7(1), 43–50.

Navrozidis, E.L., Vasara, E., Karamanlidou, G., Salpiggidis, G.K. and Koliais, S.L. (2000) Biological control of *Bactrocera oleae* (Diptera: Tephritidae) using a Greek *Bacillus thuringiensis* isolate. *Journal of Economic Entomology* 93, 1657–1661.

Neuenschwander, P., Bigler, F., Delucchi, V. and Michelakis, S. (1983) Natural enemies of preimaginal stages of *Dacus oleae* in Western Crete. I. Bionomics and phenologies. *Bollettino del Laboratorio di Entomologia Agraria Filippo Silvestri* 40, 3–32.

Oliveira, I., Pereira, J.A., Lino-Neto, T., Bento, A. and Baptista, P. (2012) Fungal diversity associated to the olive moth, *Prays oleae* Bernard: a survey for potential entomopathogenic fungi. *Microbial Ecology* 63(4), 964–974.

Ortiz, A., Quesada, A. and Sanchez, A. (2004) Potential for use of synthetic sex pheromone for mating disruption of the olive pyralid moth, *Euzophera pinguis*. *Journal of Chemical Ecology* 30, 991–1000.

Paraskakis, M., Neuenschwander, P. and Michelakis, S. (1980) *Saissetia oleae* (Oliv.) (Hom., Coccidae) and its parasites on olive trees in Crete, Greece. *Zeitschrift für Angewandte Entomologie* 90, 450–464.

Paredes, D., Cayuela, L., Gurr, G.M. and Campos, M. (2013) Effect of non-crop vegetation types on conservation biological control of pests in olive groves. *PeerJ* 1, e116. DOI: 10.7717/peerj.116.

Pascual, S., Cobos, G., Seris, E. and Gonzàlez-Nùñez, M. (2010) Effects of processed kaolin on pests and non-target arthropods in a Spanish olive grove. *Journal of Pest Sciences* 83, 121–133.

Pellizzari, G. (1997) Coccid pests of important crops: olive. In: Ben-Dov, Y. and Hodgson, C.J. (eds) *Soft Scale Insects. Their Biology, Natural Enemies and Control*. Elsevier, Amsterdam, pp. 217–229.

Perdikis, D., Garantonakis, N., Giatropoulos, A., Paraskevopoulos, A., Lykouressis, D. and Kitsis, P. (2009) Damage evaluation of *Rhynchites cribripennis* (Col., Attelabidae) in olive fruits. *Journal of Applied Entomology* 133, 512–517.

Prophetou-Athanasiadou, D.A., Tzanakakis, M.E., Myroyannis, D. and Sakas, G. (1991) Deterrence of oviposition in *Dacus oleae* by copper hydroxide. *Entomologia Experimentalis et Applicata* 61, 1–5.

Quesada-Moraga, E., Navas-Cortés, J.N., Maranhao, E.A.A., Ortiz-Urquiza, A. and Santiago-Álvarez, C. (2007) Factors affecting the occurrence and distribution of entomopathogenic fungi in natural and cultivated soils. *Mycological Research* 111(8), 947–966.

Quesada-Moraga, E., Yousef, M., Ortiz, A., Rui, M., Torres, Z., Garrido-Jurado, I. and Estévez, A. (2013) *Beauveria bassiana* (Ascomycota: Hypocreales) wound dressing for the control of *Euzophera pinguis* (Lepidoptera: Pyralidae). *Journal of Economic Entomology* 106, 1602–1607.

Rizzo, R., Caleca, V. and Lombardo, A. (2012) Relation of fruit color, elongation, hardness, and volume to the infestation of olive cultivars by the olive fruit fly, *Bactrocera oleae*. *Entomologia Experimentalis et Applicata* 145, 15–22.

Roca, L.F., Moral, J., Viruega, J.R., Avila, A., Oliveira, R. and Trapero, A. (2007) Copper fungicides in the control of olive diseases. *Olea* 26, 48–50.

Santos, S.A., Pereira, J.A., Torres, L.M. and Nogueira, A.J. (2007) Evaluation of the effects, on canopy arthropods, of two agricultural management systems to control pests in olive groves from north-east of Portugal. *Chemosphere* 67(1), 131–139.

Saour, G. and Makee, H. (2004) A kaolin-based particle film for suppression of the olive fruit fly *Bactrocera oleae* Gmelin (Dip., Tephritidae) in olive groves. *Journal of Applied Entomology* 128, 28–31.

Saponari, M., Loconsole, G., Cornara, D., Yokomi, R.K., De Stradis, A., Boscia, D., Bosco, D., Martelli, G.P., Krugner, R. and Porcelli, F. (2014) Infectivity and transmission of *Xylella fastidiosa* by *Philaenus spumarius* (Hemiptera: Aphrophoridae) in Apulia, Italy. *Journal of Economic Entomology* 107, 1316–1319.

Silvestri, F. (1914) Viaggio in Eritrea per cercare i parassiti della mosca delle olive. *Bollettino del Laboratorio di Zoologia Generale e Agraria di Portici* 9, 186–226.

Simoglou, K.B., Karataraki, A., Roditakis, N.E. and Roditakis, E. (2012) *Euzophera bigella* (Zeller) (Lepidoptera: Pyralidae) and *Dasineura oleae* (F. Low) (Diptera: Cecidomyiidae): emerging olive crop pests in the Mediterranean? *Journal of Pest Science* 85, 169–176.

Solinas, M. (1967) Osservazioni biologiche condotte in Puglia sulla *Prolasioptera berlesiana* Paoli, con particolare riferimento ai rapporti simbiotici col *Dacus oleae* Gmel. e con la *Sphaeropsis dalmatica* (Thum.) Gigante. *Entomologica* 3, 129–176.

Stathas, G.J., Bouras, S.L., Eliopoulos, P.A. and Emmanouel, N.G. (2005) Control of diaspidid scales on olive trees by releasing coccinellid predators. *IOBC (International Organisation for Biological and Integrated Control) WPRS Bulletin* 28(9), 157–166.

Tena, A., Soto, A. and Garcia-Mari, F. (2008) Parasitoid complex of black scale *Saissetia oleae* on citrus and olives: parasitoid species composition and seasonal trend. *BioControl* 53, 473–487.

Tzanakakis, M.E. (1984) Consideration on the possible usefulness of olive fruit fly symbioticides in integrated control in olive groves. In: Cavalloro, R. and Crovetti, A. (eds) *Integrated Pest Control in Olive-Groves*. Proceedings of the Commission of European Communities (CEC)/Food and Agriculture Organization of the United Nations (FAO)/International Organisation for Biological and Integrated Control (IOBC) International Joint Meeting, 3–6 April 1984, Pisa, Italy. Balkema A.A., Rotterdam, The Netherlands, pp. 386–393.

Willer, H., Lernoud, J. and Schlatter, B. (2014) Current statistics on organic agriculture worldwide: organic area, producers and market. In: Willer, H. and Lernoud, J. (eds) *The World of Organic Agriculture. Statistics and Emerging Trends 2014. FiBL-IFOAM Report, revised version of 24 February* 2014. Research Institute of Organic Agriculture (FiBL), Frick, Switzerland and International Federation of Organic Agriculture Movements (IFOAM), Bonn, Germany, pp. 34–124.

Yamvrias, C. and Young, E.C. (1977) Trials using *Bacillus thuringiensis* to control the olive moth, *Prays oleae* in Greece in 1976. *Zeitschrift für Angewandte Entomologie* 84, 436–440.

9 Control of Pests in Soybean in Organic Farming

Marie-Stéphane Tixier*
Montpellier SupAgro UMR CBGP, Montferrier-sur-Lez cedex, France

Soybean Origins

The soybean (*Glycine max* (L.) Merr.) belongs to the family *Leguminosae* and the subfamily *Papilionideae*. This species originated from East Asia. It was first cultivated in China about 5000 years ago (Hymowitz, 2004) and then was introduced all over the world. Some pest species have thus also been introduced in these new production areas as well. The 'native' species are usually oligophagous, whereas other pests native to the areas where soybean has been introduced have adapted to this plant and are usually polyphagous (Table 9.1).

World Production of Soybean

Soybean is the most economically important member of the family *Leguminosae* in the world, providing vegetable protein for millions of people and ingredients for hundreds of chemical products. It is the largest oilseed crop in the world accounting for more than 50% of world oilseed production. The USA, Brazil and Argentina are the three biggest producers with 32%, 30% and 18% of the production, respectively. China, India and Paraguay produce approximatively 4–5% of the world production and Canada 2%. The other 86 countries producing soybean account for 6% of the total production (FAOSTAT, 2012). Seven countries account for 90% of soybean organic production: China (58%), the USA (15%), Canada (4%), India (3%), Austria (3%), Argentina (3%) and Italy (3%). However, organic soybean production has increased especially in Brazil to meet market requirements. The total production of organic soybean represents 0.2% of the world production for a total harvested area of 278,000 ha.

General Consideration on Crop Protection in Organic Farming

Organic farming is constrained by the non-use of chemical pesticides and genetically modified organisms. Thus, farmers have to use organic pesticides, biological pesticide formulations and alternative strategies. The authorization of organic and biological pesticides depends not only on organic farming rules but also on the legislation of each country or production area. The main organic insecticides are pyrethrin, neem

* marie-stephane.tixier@supagro.fr

Table 9.1. Pest species reported on soybean in different production areas, organized according to damage caused.

Main damage caused	Order	Family	Species	Other damage	Status	Distribution
Leaf defoliation	Coleoptera	Chrysomelidae	*Cerotoma trifurcata*	Damages on roots by larvae	Polyphagous	North and South America
			Diabrotica speciosa		Polyphagous	South America
			Maecolaspis calcarifera		Polyphagous	South America
			Phaedonia inclusa		Polyphagous	Indonesia
			Colaspis brunnea		Polyphagous	USA
			Colaspis crinicornis		Polyphagous	USA
		Curculionidae	*Calomycterus setarius*		Polyphagous	USA, Asia
			Aracanthus mourei		Polyphagous	Brazil
			Sternechus subsignatus		Oligophagous (*Fabaceae*)	South America
		Scarabaeideae	*Popillia japonica*		Polyphagous	Europe, Japan, North America
			Popillia quadriguttata		Polyphagous	China
	Lepidoptera	Noctuidae	*Anticarsia gemmatalis*	Damage to pods, seedlings and sometimes grains	Polyphagous	North and South America
			Chrysodeixis includes		Polyphagous	North and South America
			Chrysodeixis acuta		Polyphagous	Asia, Japan, Australia and Pacific Islands
			Chrysodeixis chalcites		Polyphagous	Asia, Japan, Australia and Pacific Islands
			Chrysodeixis eriosoma		Polyphagous	Australia, Africa, Asia, Brazil and Euorpe
			Helicoverpa zea		Polyphagous	America
			Helicoverpa armigera		Polyphagous	Australia, Africa, Asia, Brazil and Euorpe
			Helicoverpa punctigera		Polyphagous	Asia, Japan, Australia and Pacific Islands
			Spodoptera litura		Polyphagous	Mainly in Asia
			Spodoptera frugiperda		Polyphagous	Asia
			Spodoptera eridania		Polyphagous	Brazil
			Spodoptera cosmioides		Polyphagous	Brazil
			Spodoptera albula		Polyphagous	Brazil
			Spodoptera exigua		Polyphagous	Asia
			Spodoptera ornithogalli		Polyphagous	North America
			Heliothis virescens		Polyphagous	North and South America
			Rachiplusia nu		Polyphagous	North and South America
			Trichoplusia ni		Polyphagous	North and South America
			Trichoplusia orichalcea		Polyphagous	Australia
			Mocis alterna		Polyphagous	Australia
			Mocis frugalis		Polyphagous	Australia
			Mocis trifasciata		Polyphagous	Australia
		Erebidae	*Hypena scabra*		Polyphagous	North America
		Crambidae	*Omiodes indicatus*		Polyphagous	Brazil
			Lamprosema indicata		Polyphagous	Asia
		Nymphalidae	*Vanessa cardui*		Polyphagous	Worldwide distributed (origin: USA)
		Arctiidae	*Spilosoma obliqua*		Polyphagous	Asia
	Orthoptera	Acrididae	*Melanoplus differentialis*		Polyphagous	North America
			Melanoplus femurrubrum		Polyphagous	North America
			Rhammatocerus schistocercoides		Polyphagous	Brazil
		Gryllidae	*Anurogryllus muticus*		Polyphagous	Brazil

Continued

Table 9.1. Continued.

Main damage caused	Order	Family	Species	Other damage	Status	Distribution
Seed, and pod feeders	Hemiptera	Pentatomidae	*Euschistus heros*		Polyphagous	Brazil
			Nezarta viridula		Polyphagous	Brazil, Australia
			Piezodorus guildinii		Polyphagous	Brazil
			Dichelops furcatus		Polyphagous	Brazil
			Dichelops melacanthus		Polyphagous	Brazil
			Edessa meditabunda		Polyphagous	Brazil
			Acrosternum hilare		Polyphagous	USA
			Euschistus quadrotor		Polyphagous	USA
			Piezodorus oceanicus		Polyphagous	Australia
			Piezodorus hybneri		Polyphagous	Asia and Africa
			Piezodorus punctiventris		Polyphagous	Asia and Africa
		Alydidae	*Riptortus clavatus*		Polyphagous	Asia and Africa
			Riptortus dentipes		Polyphagous	Asia and Africa
			Neomegalotomus parvus		Polyphagous	Brazil
			Melanacanthus scutellaris		Polyphagous	Australia
			Riptortus serripes		Polyphagous	Australia
	Diptera	Cecidomyidae	*Asphondylia yushimai*		Oligophagous (*Fabaceae*)	Japan, Indonesia and China
	Lepidoptera	Pyralidae	*Etiella zinckenella*		Polyphagous	Worldwide distributed
			Elasmopalpus lignosellus		Polyphagous	North and South America
		Tortricidae	*Laspeyresia glycinivorella*		Polyphagous	North East of Asia
Leaf discoloration	Heteroptera	Miridae	*Halticus bractatus*		Polyphagous	North, Central and South America, Hawaii, the West Indies and the Galápagos Islands
	Hemiptera	Cicadellidae	*Empoasca fabae*		Polyphagous	North, South America and Europe
			Empoasca terminalis		Polyphagous	Indonesia
		Aphididae	*Aphis glycines*		Polyphagous	North America, Australia, China, Japan, Malaysia, Philippines, Taiwan and Thailand
	Thysanoptera	Thripidae	*Frankliniella schultzei*		Polyphagous	Brazil
			Caliothrips brasiliensis		Polyphagous	Brazil
			Frankliniella fusca		Polyphagous	North America
			Sericothrips variablilis		Polyphagous	North America
			Frankliniella intonsa		Polyphagous	Europe and Asia
			Scirtothrips dorsalis		Polyphagous	America and Asia
	Hemiptera	Aleyrodidae	*Bemisia tabaci*		Polyphagous	Worldwide distributed
	Heteroptera	Piataspidae	*Megacopta cribraria*		Oligophagous (*Fabaceae*)	Asia, North America
	Hemiptera	Membracidae	*Spissi stilus festinus*		Polyphagous	USA, Mexico
	Acari	Tetranychidae	*Tetranychus urticae*		Polyphagous	Worldwide distributed
			Mononychellus planki		Polyphagous	North and South America
			Tetranychus ludeni		Polyphagous	Worldwide distributed
		Tarsonemidae	*Polyphagotarsonemus latus*		Polyphagous	Worldwide distributed

Continued

Table 9.1. Continued.

Main damage caused	Order	Family	Species	Other damage	Status	Distribution
Stem borers	Coleoptera	Cerambycidae	*Dectes texanus*		Polyphagous	North America
			Obereopsis brevis		Polyphagous	India
Seed development and germination	Diptera	Anthonomiidae	*Delia platura*		Polyphagous	Worldwide distributed
Leaf miners	Coleoptera	Chrysomelidae	*Odontota horni*		Polyphagous	North America
			Odontota dorsalis		Polyphagous	North America
	Diptera	Agromyzidae	*Melanagromyza sojae*		Oligophagous (*Fabaceae*)	Asia, Africa, Australia
			Ophiomyia phaseoli		Oligophagous (*Fabaceae*)	Asia, Australia and Africa
			Melanagromyza dolichostigma		Oligophagous (*Fabaceae*)	Asia
			Ophiomyia centrosomatis		Oligophagous (*Fabaceae*)	Asia
	Lepidoptera	Gelechiidae	*Aproaerema modicella*		Oligophagous (*Fabaceae*)	Asia and introduced in Africa
Root feeders	Coleoptera	Scarabaeidae	*Phyllophaga implicita*		Polyphagous	North America
			Phyllophaga rugosa		Polyphagous	North America
			Phyllophaga congrua		Polyphagous	North America
			Phyllophaga cuyabana		Polyphagous	Brazil
			Phyllophaga capillata		Polyphagous	Brazil
			Cyclocephala flurida		Polyphagous	USA
			Lyogenis suturalis		Polyphagous	Brazil
			Phyllophaga serrata		Polyphagous	India
			Holotricha sp.		Polyphagous	China
			Anomala sp.		Polyphagous	China
			Maladera sp.		Polyphagous	China
	Hemiptera	Cydnidae	*Atarsocoris brachiariae*		Polyphagous	South America
			Scaptocoris castanea		Polyphagous	South America

oil (azadirechtin) and spinosad. The main biological insecticide is *Bacillus thuringiensis* (*Bt*). The decision to spray these pesticides is based on an integrated crop protection approach; applying pesticide when pest densities surpass a level of economic damage. However, as organic pesticides are not always safe from an environmental point of view, especially for natural enemies, alternative strategies such as biological control, cultural practices and resistant cultivars should be adopted (Bahlai *et al.*, 2010). Details of when and how these strategies are applied and/or are in development will be presented in this chapter. Table 9.2 provides details of the known economic thresholds to control pests in soybean, Table 9.3 outlines the alternative agronomic practices that can be used and Table 9.4 lists botanical preparations reported in the literature to control pests in soybean.

Pests in Soybean

Soybean is attacked by many pests damaging all parts of the plant (Kogan and Turnipseed, 1987). Many websites and reports in the literature list soybean pests especially in the United States Department of Agriculture (USDA) state services agency. Some keys to soybean pests as well as risk period synthesis are also available online. Many pests are polyphagous; they have adapted to the soybean in the areas where this plant has been introduced. Some species are oligophagous; usually they originated from Asia, the centre of origin of the soybean (Table 9.1). In this chapter pests will be presented according to the plant parts they attack. However, this structure is not always easy to apply as: (i) adults and larvae of the same species can affect different plant organs; and (ii) the same pest stage can cause damage to different

Table 9.2. Economic thresholds reported to control pests in soybean.

Pest species	Thresholds proposed in the literature
Noctuid species	30% defoliation throughout the plant 2 weeks prior to blooming (R1)
	15% defoliation 2 weeks prior to blooming until the pods have filled (R7–R8)
Cerotoma trifurcata	30% defoliation 2 weeks prior to blooming (R1)
	15% defoliation 2 weeks prior to blooming until the pods have filled (R7–R8)
Popillia japonica	30% defoliation before R1
	20% after R1
Vanessa cardui	30% defoliation is reached prior to bloom (before R1)
	20% after bloom or pod set (after R1)
Grasshoppers	18 or more specimens/m^2 within the field, 49 or more specimens/m^2 in field borders
	Or 20–25% defoliation of soybean leaves
Sternechus subsignatus	0.4 adults/m on the rows at stages between two and five trifoliate leaves (V2–V5)
Pentatomidae in Brazil	Four specimens/2 m of soybean row between the development of pods to fill of grains
Pentatomidae in USA	40 specimens are found on 100 sweeps
Elasmopalpus lignosellus	Two to three larvae/m
Empoasca fabae	For early vegetative stages: two leafhoppers per plant
	For flowering fields: one leafhopper per trifoliate leaf
	When pods are developing: two leafhoppers per trifoliate leaf
Aphis glycines	One hundred aphids per plant from R1 through mid-seed set (R5.5)
Thripidae species	75% of damaged trifoliates
	Or more than eight thrips per leaf are observed
Megacopta cribraria	One nymph-stage bug per sweep
Spissistilus festinus	At least 10% plants less than 25–30 cm (10–12 inches) tall infested
	For plants setting pods, one hopper per sweep (100 sweeps sampled)
Mites	10–15% leaf discoloration from R1 (beginning bloom) through R5 (beginning seed)
Obereopsis brevis	Two thresholds (different authors – see text): (i) 0.3–1.0 bud per row; and (ii) between 0.8 and 3.2 buds per row
Odontota horni	At seedlings stage: nine adults/row-ft (29 adults/row-m)
	At development stage V6 to maturity , 125 adults/row-ft (410 adults/row-m)
Melanagromyza sojae	Four to ten insects collected on 100 plants for the third generation and eight to 12 for the fourth one
Aproaerema modicella	One larva per five plants 30–45 days after sowing

plant organs (Table 9.1). Information on distribution, damage and control strategies in organic farming systems will be provided. Although this chapter tends to provide a description of a wide range of pests, the most complete information is provided for pests reported in the main soybean production in the world (the USA and Brazil) because information is more accessible.

Leaf defoliation

Several species belonging to different orders and families (Coleoptera: Chrysomelidae, Curculionidae, Scarabaeidae; Lepidoptera: Noctuidae, Nymphalidae; Orthoptera: Acrididae) feed on soybean leaves reducing photosynthesis yield and agronomic performance. Soybeans are usually able to compensate for such damage by producing large amounts of foliage; thus little effect on yield is usually observed (Haile *et al.*, 1998). However, some stages are critical especially from bloom (R1) to pod fill (R4), when seed development is highly dependent on photosynthesis.

Species of the family Chrysomelidae

DISTRIBUTION. Mainly three Chrysomelidae species cause damage on soybean: (i) *Cerotoma trifurcata* (Forster); (ii) *Diabrotica speciosa*

Table 9.3. Current agricultural practices proposed to control pests in soybean crops.

Agricultural practice	Pest	References
Trap plant		
Cayaponia martiana	*Diabrotica speciosa*	Hohmann and Carvalho (1989), Hoffmann-Campo *et al.* (2003)
Sorghum	*Helicoverpa armigera*	Virk *et al.* (2004)
Pigeon pea and okra	*H. armigera*	Regupathy (2005)
Sunflower, buckwheat, sorghum, millet, *Fagopyrum sagittatum*, *Triticale hexaploide*	Pentatomidae	Mizell *et al.* (2008), Kamminga *et al.* (2012), Russell and Mizell (2015)
Planting a perimeter of earlier maturing soybean	Pentatomidae	McPherson and Newsom (1984), Panizzi and Alves (1993), Kamminga *et al.* (2012)
Tagetes patula	Thripidae	Peres *et al.* (2009)
Sunflower	*Dectes texanus*	Michaud *et al.* (2007)
Soil fertilization		
Good N fertilization for ensuring plant compensation	*D. speciosa*	Hammack *et al.* (2010)
	Popillia japonica	Potter and Held (2002)
Soil acidification	*P. japonica*	Hammond and Stinner (1987)
Autumn fertilization	*P. japonica*	Potter and Held (2002)
Potassium deficiency	Favours aphids	Myers *et al.* (2005a), Myers and Gratton (2006), Walter and DiFonzo (2007)
Soil management		
Winter ploughing	Noctuidae	Fathipour and Sedariatan (2013)
Autumn ploughing	*Laspeyresia glycinivorella*	Kuwayama (1928)
Autumn tillage and stalks buried to a depth of at least 10–15 cm (4–6 inches)	*Dectes texanus*	Johnson (2011)
Disc ploughing	*Pollyphaga* sp.	Oliveira (1997), Oliveira *et al.* (2000)
Spring tillage	*Pollyphaga* sp.	Oliveira (1997), Oliveira *et al.* (2000)
Digging the crop 2–3 weeks after soybean germination	*Ophiomyia phaseoli*	Van der Groot (1930)
No-till fields	Acrididae	Hein and Campbell (2008)
	Delia platura	Hammond (1990), Pope (1998)
Mulch	*Elasmopalpus lignosellus*	Harsimran *et al.* (2010)
Transparent and black plastic mulchings	*D. platura*	Tae Heung (1992)
Rice straw mulch	*O. phaseoli*	Van der Groot (1930)
	Aproaerema modicella	Logiswaran and Mohanasundaram (1985)
Used coffee grounds (UCG) when applied as a 20% UCG mixture	*Megacopta cribraria*	Whitehouse and Zehnder (2015)
High rates of organic matter	*D. platura*	Tae Heung (1992)
Planting		
Direct sowing	*Sternechus subsignatus*	Hoffmann-Campo *et al.* (1999)
	E. lignosellus	Hoffmann-Campo *et al.* (2000), Gill *et al.* (2010)
Early planting	Noctuidae	Fathipour and Sedaratian (2013)
	S. subsignatus	Hoffmann-Campo *et al.* (1999)
Late planting	*Dectes texanus*	Johnson (2011)
	M. cribraria	Seiter *et al.* (2013)
	D. platura	Bessin (2004)
High planting density	*E. lignosellus*	Hoffmann-Campo *et al.* (2000)
	Spissistilus festinus	Higley and Boethel (1994)
	D. platura	Bessin (2004)
Avoid planting near lucerne fields	*Empoasca fabae*	Poston and Pedigo (1975)

Continued

Table 9.3. Continued.

Agricultural practice	Pest	References
Crop rotation		
With maize or wheat	*Diabrotica speciosa*	OEPP/EPPO (2005)
With maize, sorghum, millet and sunflowers	*S. subsignatus*	Hoffmann-Campo *et al.* (1999)
With cotton, *Crotalaria spectabilis* and *Crotalaria juncea* to a lesser extent	*Pollyphaga cuyabana*	Oliveira (1997), Oliveira *et al.* (2007)
Weed management and plant management around the field		
Avoid grassy weeds within soybean fields and in its border	Noctuidae	Rogers and Brier (2010)
	Acrididae	Hodgson (2014)
	S. festinus	Higley and Boethel (1994)
	Mites	Suekane *et al.* (2012)
Avoid grassy weeds especially *Leguminosae*	*Halticus bractatus*	Iowa State University (2012)
Reducing thistles near field	*Vanessa cardui*	NDSU (no date)
Removal of nearby food sources like flowering weeds and sweet-smelling substances	*D. platura*	Kessing and Mau (1991)
Avoid *Salix* spp. in field border and prefer evergreens	*Pollyphaga* sp.	Glogoza *et al.* (1998), Garcia *et al.* (2003)
Avoid planting sunflower near soybean fields	*D. texanus*	Wright and Hunt (2011)
Maintain weeds	*P. japonica*	Homes and Barret (1997)
	Empoasca fabae	Altieri *et al.* (1981), Hammond and Stinner (1987), Hammond and Jeffers (1990), Lamp *et al.* (1984), Lam and Pedigo (1998), Smith *et al.* (1988), Buckelew *et al.* (2000), Miklasiewicz and Hammond (2001)
	Aphis glycines	Gardiner *et al.* (2009a, b), Noma *et al.* (2010)
Irrigation management		
Limit soil moisture during flight	*P. japonica*	Potter *et al.* (1996)
Good irrigation management, limiting drought	*Aproaerema modicella*	Logiswaran and Mohanasundaram (1985)
Natural enemy release		
Entomopathogenic nematodes	*P. japonica*	Gaugler and Klein (1998)
Milky disease	*P. japonica*	Potter and Held (2002)
Baculovirus anticarsia	*Anticarsia gemmatalis*	Moscardi and Carvalho (1993), Sujii *et al.* (2002), Szewczyk *et al.* (2006)
Baculovirus (*Heliothis zea* single nuclear polyhedrosis virus, HzSNPV)	*Heliothis* sp.	Ignoffo and Couch (1981)
	Helicoverpa sp.	Ignoffo and Couch (1981)
	Spodoptera frugiperda	Valicente and Cruz (1991), Moscardi (1999)
Bacillus thuringiensis subspecies *kurstaki*	*Heliothis* sp.	Alexandre (2010)
	Spodoptera sp.	Alexandre (2010)
Bacillus sphaericus	*Spilosoma obliqua*	Singh *et al.* (2015)
B. thuringiensis	*Aproaerema modicella*	Shirale *et al.* (2010)
Beauveria bassiana	*A. modicella*	Joshi and Patel (2011)

Continued

Table 9.3. Continued.

Agricultural practice	Pest	References
Nosema locustae, Entomophaga grylli, Metarhizium anisopliae var. *acridum*	Acrididae	Magalhães *et al.* (2000), Schmidt *et al.* (2007a)
Trichogramma pretiosum	*A. gemmatalis, Chrysodeixis includens, Trichoplusia ni*	Parra and Zucchi (2004), Massaroli *et al.* (2014), Carvalho *et al.* (2014)
Trissolcus basalis, Telenamus podisi	*Nezara viridula, Piezodorus guildinii, Euschistus heros*	Corrêa-Ferreira (2003)
Plant resistance/tolerance		
	Noctuidae	Fathipour and Naseri (2011), Souza *et al.* (2014)
	Pentatomidae	Gilman *et al.* (1982), Jones and Sullivan (1982), Kamminga *et al.* (2012)
	Riptortus clavatus	Lopes *et al.* (1997)
	Aphis glycines	Zhang *et al.* (2009), Tilmon *et al.* (2011), Hodgson *et al.* (2012)
	Bemisia tabaci	Do Valle and Lourençao (2002), Vieira *et al.* (2011), Da Silva *et al.* (2012), Zhang *et al.* (2013)
	Tetranychus urticae	Razmjou *et al.* (2009), Dehghan *et al.* (2009)
Cultivars with many short trichomes are susceptible	*Etiella zinckenella*	Talekar and Lin (1994), Permana *et al.* (2012)
Cultivars with sparse or very short leaf pubescence are susceptible	*Empoasca fabae*	Yeargan *et al.* (1994)
Glabrous soybeans are highly resistant	*Laspeyresia glycinivorella*	Morse and Carter (1937)
Cultivars with very thin stems, high trichome density and high concentrations of polyphenols are resistant	*Melanagromyza sojae*	Chiang and Talekar (1980), Chiang (1984), Chiang and Norris (1985)
Genotypes having thick, dark green leaves	*Aproaerema modicella*	Mundhe (1980), Shetgar and Thombre (1984), Taware *et al.* (2001), Ambenagare *et al.* (2011)
Cultivars with rapid root growth	*Pollyphaga* sp.	Oliveira *et al.* (1997)
Intercropping		
With dwarf sorghum, tall sorghum and rye	*P. japonica*	Homes and Barret (1997)
With carrot	*P. japonica*	Altieri and Letourneau (1982)
With *Paeonia* sp.	*P. japonica*	Altieri and Letourneau (1982)
With pigeon pea variety, maize or sorghum	*Spilosoma obliqua*	Prasad and Sharma (1989)
With buckwheat	Pentatomidae	Youn and Jung (2008)
With cereal (barley, wheat or 'Indian corn')	*Laspeyresia glycinivorella*	Kuwayama (1928)
With wheat and rye	*Empoasca fabae*	Miklasiewicz and Hammond (2001), Koch *et al.* (2012)
Soybean with maize and/or adjacent bands of maize	*A. glycines*	Wang and Ba (1998), Wu *et al.* (2004), Hasibuan and Lumbanraja (2012), Koch *et al.* (2012)
With *Zanthoxylum zanthoxyloides* and *Lippi adoensis*	*Riptortus dentipes*	Olufolaji (2011)
With *Fagopyrum esculentum*	*A. glycines*	Woltz *et al.* (2012)
Lucerne cover crops co-planted	*A. glycines*	Schmidt *et al.* (2007b)
With aubergine	*Ophiomyia phaseoli*	Van der Groot (1930)

Table 9.4. Organic and botanical insecticides effects on soybean pests.

Botanical insecticides	Targeted pest species and known effects (if different than mortality)	References
Spinosad	Noctuid species	Valles and Capinera (1993)
	Pentatomidae species	Overall (2008), Kamminga *et al.* (2009, 2012)
	Thrips species	Bethke *et al.* (2014)
	Aproaerema modicella	Joshi and Patel (2011)
Neem oil	Noctuid species	Valles and Capinera (1993)
	Pentatomidae species	Overall (2008), Kamminga *et al.* (2009, 2012)
	Aphis glycines	Pecinovsky and Lang (2011)
	Bemisia tabaci	Carvalho *et al.* (2012)
	Obereopsis brevis	Singh *et al.* (2014)
	Ophiomyia phaseoli	Schmutterer (1984)
	Melanagromyza dolichostigma	Schmutterer (1984)
	Aproaerema modicella	Joshi and Patel (2011)
Pyrethrin	Noctuid species	Valles and Capinera (1993)
	Pentatomidae species	Overall (2008), Kamminga *et al.* (2009, 2012)
	Halticus bractatus	Capinera (2005)
	Megacopta cribraria	Seiter *et al.* (2015)
Extracts of *Cassia fistula* seeds	*Cerotoma trifurcata* fertility	Shilpi and Tomar (2014)
Mixture of 85% sesame oil, 2% garlic, 2% clove and 1% rosemary oil	*Popillia japonica*	Ranger *et al.* (2009, 2013)
Allyl Isothiocyanate essential oil		
Red thyme essential oil		
Garlic essential oil		
Eugenol essential oil		
Clove essential oil		
Wintergreen, peppermint, and combinations of wintergreen with ginger essential oils	*P. japonica* reduced attraction	Youssef *et al.* (2009)
Anise, bergamontmint, cedarleaf, dalmation sage, tarragon, and wormwood essential oils		
Eucalyptus urograndis oil	*Euschistus heros*	Ferreira De Souza and Favero (2015)
Soja oil	*Bemisia tabaci*	Costa *et al.* (2010)
Garlic oil	Decrease oviposition of *Tetranychus urticae*	Ismail *et al.* (2011)
Fresh *Zingiber cassumunar* Roxb. and *Curcuma longa*	*Lamprosema* sp.	Leksawasdi (1986)
Neem seed, sugar apple leaf and sugar apple seeds extracts	Suppress *Lamprosema indica* egg laying	Indiati (2014)
Ryania and *Sabadilla* extracts	*Spodoptera eridania*	Valles and Capinera (1993)
Sabadilla extracts	*Halticus bractatus*	Capinera (2005)
Spraying aqueous leaf extracts of *Euphorbia royleana* or *Lantana camara*	*Spilosoma obliqua*	Sharma *et al.* (1982)
Leaf extracts of *Linodenbergia grandifolia, Macuna cochinensis*, *Passiflora mollissima*, *Nyctanthes arbortritis*, *Swertia chirayita, Ailanthum excelsa* and *Schima khasiana* and seed oil fraction of *Pongamia glabra* and seed oil of *Psoralea corylifolia*	Antifeedant activities on *S. obliqua*	Tripathi *et al.* (1987), Mohanty *et al.* (1988), Premchanda (1989)

Continued

Table 9.4. Continued.

Botanical insecticides	Targeted pest species and known effects (if different than mortality)	References
Extracts of *Chamomilla recutita*	*Tetranychus urticae*	El Moneim *et al.* (2012)
Extracts of *Marjorana hortensis*	*T. urticae*	El Moneim *et al.* (2012)
Extracts of and *Eucalyptus*	*T. urticae*	El Moneim *et al.* (2012)
Kochia scoparia extract	*T. urticae*	Shi *et al.* (2006)
Onion herbal extract	*T. urticae*	Abramishvili and Chkhaidze (2013)
Compounds extracted from seeds of *Abrus precatorius* seed and *Trigonella foenumgraceum*	*Obereopsis brevis* 4th larval stage	Tomar (2009)
Spray of garlic and clay	*Halticus bractatus*	Atthowe *et al.* (2010)
HIPVs (methyl salicylate, cis-3-hexen-1-ol, and phenylethyl alcohol)[a]	Repellent for *H. bractatus*	Braasch *et al.* (2012)
Milstop® (85% potassium bicarbonate)	*Aphis glycines*	Rozeboom *et al.* (2014), Hesler (2013)
Sugar + Peroxide treatment	*A. glycines*	Pecinovsky and Lang (2011)
Wood vinegar	Repellent effect on *Bemisia tabaci*	Pangnakorn *et al.* (2010)
Silicon applications	*B. tabaci*	Ferreira *et al.* (2011)
10% cow urine, and cow dung ash	*Ophiomyia phaseoli*	Schmutterer (1984)
Soap solutions or organic oil	Reduce the larval hatching of *Melanagromyza dolichostigma*	Schmutterer (1984)

[a]HIPVs, herbivore-induced plant volatiles.

Germar; and (iii) *Maecolaspis calcarifera* Bechyné. *C. trifurcata* is the most frequent and widespread one. It is present in North and South America whereas *D. speciosa* and *M. calcarifera* cause damage mainly in South America. In Indonesia, *Phaedonia inclusa* (Stal) is one of the major pests of soybean leaves and pods. In Brazil, another species of Chrysomelidae is reported to damage soybean: *Myochrous armatus* (Baly). Some species of the genus *Colaspis* are also sometimes reported, such as *Colaspis brunnea* (Fabricius) and *Colaspis crinicornis* Schaeffer in the USA but damage is usually secondary. All these species are polyphagous.

DAMAGE

- **Direct damage** – Adults cause defoliation whereas larvae feed on roots. An adult of *C. trifurcata* consumes an average of 0.38 cm^2 of leaf/day. High densities of larval infestation can reduce soybean root node formation by up to 45%, impairing plant nitrogen (N) fixation. *M. armatus* adults are reported to attack seedlings and to cause plant tipping. Attacks of *P. inclusa* in Indonesia are reported to cause 80% of yield losses.
- **Indirect damage** – *C. trifurcata* can transmit the bean pod mottle virus (BPMV) that delays maturity of the soybean stem. Soybean infected by BPMV produces 3–52% less yield depending on the time of infection. *Colaspis* species are also reported to transmit this virus. Furthermore, the feeding injury on the pod surface facilitates bacterial and fungal infection (e.g. *Phomopsis* spp. infections of seed).

HOW TO CONTROL?

Thresholds and use of organic pesticides. For *C. trifurcata* the economic threshold is 30% defoliation throughout the plant 2 weeks prior to blooming (R1) and 15% defoliation 2 weeks prior to blooming until the pods have filled (R7–R8). Shilpi and Tomar (2014) showed that extracts of *Cassia fistula* seeds affect *C. trifurcata* fertility and concluded that this is an extremely promising bioinsecticide.

Trap plants. Cabrera Walsh (2005) studied the attractivity of several plant species. Ventura *et al.* (2000) found that traps baited with 1,4-dimethoxybenzene caught 29.4 times more beetles than control fields. In Brazil, trap plants such as *Cayaponia martiana* (*Curcurbitaceae*) are used for controlling *D. speciosa* (Hohmann and Carvalho, 1989). These plants attract adults that can then be killed by localized pesticide applications. In organic farming systems, parts of roots or stems of *C. martiana* can be mixed with an organic pesticide; the traps containing this mixture are placed preferentially in field borders when the first adults appear (30–40 traps/ha) (Hoffmann-Campo *et al.*, 2003).

Agricultural practices

- **Delaying the planting date** – Planting later may allow soybean seedlings to escape the highest *C. trifurcata* populations occurring earlier in the spring (first generation). However, some studies show that late-maturing fields may be attractive to second generation adults and create more pod injury. Altering the planting date is thus controversial and seems not to be a general solution.
- **Control of N fertilization and soil management** – Hammack *et al.* (2010) showed that the numbers of beetle were lower in N-treated plots (because of plant compensation) and in ridge-tilled compared with chisel-tilled plots.
- **Crop rotation** – Some studies recommend crop rotation with maize or wheat to control *D. speciosa* combined with no-till agriculture (OEPP/EPPO, 2005).
- **Intercropping** – Koch *et al.* (2012) showed that *C. trifurcata* densities were significantly reduced in plots with rye (*Secale cereale*) compared with those without but only on one of two plots studied. By contrast, Smith *et al.* (1988) found greater densities of *C. trifurcata* in plots with a rye cover crop than without and attributed this effect to increased soil moisture, which may have favoured oviposition of *C. trifurcata* (Marrone and Stinner, 1983). Finally, House and Stinner (1983) suggested that abundance of carabids in no-till rye plots can reduce foliar feeding on soybean by *C. trifurcata*. The response of *C. trifurcata* to grassy weeds or cover crops was variable or non-significant (Troxclair and Boethel, 1984; Buntin *et al.*, 1995; Lam and Pedigo, 1998; Buckelew *et al.*, 2000; Jackson and Pitre, 2004).

Biological control. No strategy based on natural enemy release has been developed. Some studies report the natural occurrence of biological control agents in soybean fields. However, the number of parasitoid species reported is low. The parasitoid species *Centistes gasseni* Shaw (Braconidae) and *Celatoria bosqi* Blanchard (Diptera: Tachinidae) have been found in South America parasitizing *D. speciosa* but with low parasitism rates (around 2%) (Heineck-Leonel and Salles, 1997; Cabrera Walsh *et al.*, 2003; Cabrera Walsh, 2004). Micheli (2005) observed a high parasitism rate of *D. speciosa* (60%) by *Celatoria* sp. in Brazilian soybean fields. *Celatoria diabroticae* (Shimer) was recorded in North America parasitizing *C. trifurcata* with parasitism rates ranging from 1% to 20% (in Cabrera Walsh, 2005). The most effective natural enemy of *C. trifurcata* is a tachinid species, *Medina* sp. in the USA; parasitism rates ranging from 40% to 90% (overwintering populations) were observed in soybean fields near lucerne crops. However, the average parasitism rate was much lower (3%) in a soybean field that was not next to lucerne (Loughran and Ragsdale, 1986). In Indonesia, an attempt to import a pteromalid wasp, *Schizonotus latus* (Walker) to control *P. inclusa* did not result in establishment (Shepard *et al.*, 1999). Some laboratory tests have shown the efficacy of entomopathogenic nematodes and entomopathogens (fungi) (Consolo *et al.*, 2003; Santos *et al.*, 2011). Natural parasitism of *D. speciosa* with *Beauveria bassiana* was observed (23%) in Brazil (Bastos Dequech *et al.*, 2006). Natural parasitism has been also reported in South America for *D. speciosa* with the following nematode species: *Hexamermis* sp. (parasitism rates ranging from 2% to 90%) and *Micoletzkya vidalae* (in Cabrera Walsh, 2005).

Species of the family Curculionidae

DISTRIBUTION. *Calomycterus setarius* Roelofs was introduced into the USA in 1932 from Japan (Johnson, 1944). It is a polyphagous species (more than 100 host plants). In Brazil, another species of this family, *Aracanthus mourei* (Rosado Neto), is considered as a secondary pest. In South America, another species occurs, *Sternechus subsignatus* (Boheman). This latter species is oligophagous and only consumes plants that are members of the family *Fabaceae* (Hoffmann-Campo *et al.*, 1999).

DAMAGE. Adults eat leaf edges with mean day consumption rates for *C. setarius* ranging from 0.16 cm^2 to 0.21 cm^2 (Hunt *et al.*, 2003); larvae are stem borers. Usually very little economic damage is caused by this species. The damage is much more important at the beginning of soybean cultivation when leaf compensation is not possible.

HOW TO CONTROL? *C. setarius* usually does not require specific control. By contrast, control of *S. subsignatus* is necessary.

Thresholds. The recommended threshold for *S. subsignatus* is 0.4 adults/m on the rows at stages between two and five trifoliate leaves (V2–V5) (Silva, 2000). However, chemical control is difficult and somewhat inefficient because of the occurrence of adults on lower leaves.

Agricultural practices

- **Crop rotation** – To break the life cycle of *S. subsignatus* crop rotation with plants not belonging to the family *Fabaceae* is proposed. Hoffmann-Campo *et al.* (1999) provide a list of interesting plants to alternate with soybean such as maize, sorghum, millet and sunflowers.
- **Planting dates** – Some studies show that when soybean was sown in October in Brazil there was much less damage than when sowing was implemented from November to December (Hoffmann-Campo *et al.*, 1999).
- **Soil management** – It seems that soil management does not increase *S. subsignatus* mortality. However, in some cases direct sowing is reported to limit damage caused by this species (Hoffmann-Campo *et al.*, 1999).

Species of the family Scarabaeidae

DISTRIBUTION. Two species of this family cause soybean defoliation: (i) *Popillia japonica* Newman, a species that is widespread across the world; and (ii) *Popillia quadriguttata* (F.) present in China and Korea. The Japanese beetle, *P. japonica* was introduced into the USA in 1916 from Japan. This species is not a pest in Japan but rapidly expanded in North America where it was free from natural enemies. It is also recorded from the Azores (Portugal), where it escaped from a US air base in the early 1970s. Reports of its occurrence in China, Korea, India and Russia are questionable and some authors suspect misidentifications (Potter and Held, 2002).

DAMAGE. *P. japonica* is polyphagous; it has been recorded on more than 300 plants belonging to 79 plant families (e.g. trees, maize, grass). Adults feed on leaves and larvae on roots. Adults cause the main damage.

HOW TO CONTROL?

Thresholds and organic insecticides. Pesticide use is recommended when 30% of defoliation is reached before R1 and 20% after R1. Ranger *et al.* (2009) studied the toxic effect of eight botanical formulations on larvae of *P. japonica*. The mixture that was composed of 85% sesame oil, 2% garlic, 2% clove and 1% rosemary oil was the most effective product (LC_{50} (the concentration of a chemical that will kill 50% of the sample population being tested) 0.42 ml/l). In 2013, Ranger *et al.* studied 24 plant-based essential oils. After topic applications on larvae (third instar), the most toxic compounds were allyl isothiocyanate, red thyme and to a lesser extent garlic, eugenol and clove.

Use of pheromone traps. Pheromone traps are used for monitoring adult flight. The attractive substances in traps are Japonilure (the synthetic sex pheromone) combined with a mixture of volatile oils such as phenethyl propionate, eugenol and geraniol (Potter and Held, 2002). In China, a study showed that mass trapping with Japonilure allowed 72% and 75% adult and larval reduction of *P. quadriguttata*, respectively, with a density of 30 traps/ha (Chen *et al.*, 2014). Addition of a floral lure increased this efficiency, and a reduction of 90% of overwintering larvae was obtained. However, mass trapping for controlling *P. japonica* is not recommended as large numbers of adults can be attracted to fields.

Agricultural practices

- **Soil management** – Practices that acidify soil (application of aluminum sulfate) are reported to reduce larval densities of *P. japonica*. By contrast, application of dolomitic limestone that elevates soil pH did not affect larvae populations (Hammond and Stinner, 1987). Nor was the application of a heavy roller to compact the soil (to prevent beetle flight) effective. Finally, fertilizing in autumn to promote root growth can enhance plant tolerance of pest attack (Potter and Held, 2002).
- **Crop rotation** – Some studies have shown that crop rotation could increase damage especially when maize, rather than soybeans, was the previous crop.
- **Intercropping systems** – Many studies show the positive effects of the presence of other plants in the soybean crops such as weeds, dwarf sorghum and tall sorghum and rye. These plants can provide alternative food and/or discourage pest dispersal (by acting as a physical barrier). A list of favourable and unfavourable plant species to be co-planted with soybean is provided in Homes and Barret (1997).
- **Irrigation management** – During dry summers, female beetles seek irrigated areas for oviposition as soil moisture is essential for egg survival and larval development. Irrigation management during peak beetle flight activity might thus reduce the larval population (Potter *et al.*, 1996).

Use of plant extracts as repellents and attractants. Studies on repellent plants are not new as in the 1930s some authors studied the impact of essential oils on *P. japonica* behaviour (Metzger and Grant, 1932). Youssef *et al.* (2009) tested the ability of 41 plant essential oils to reduce the attraction of *P. japonica* adults, under field conditions. Eight extracts significantly reduced *P. japonica* attraction. Wintergreen, peppermint, and combinations of wintergreen with ginger, produced the greatest reductions; the highest effect being observed for wintergreen and ginger oil combined. The other products with a significant effect were anise, bergamot mint, cedarleaf, dalmation sage, tarragon and wormwood oils. By contrast, six essential oils increased Japanese beetle attraction and among them camphor was the most attractive (Youssef *et al.*, 2009).

Biological control

- **Natural occurrence of biological control agents** – Many surveys have been carried out to search for natural enemies (e.g. Petty *et al.*, 2012). Endemic generalist predators, especially ants, staphylinids and carabids cause substantial mortality of eggs and young larvae. Some birds feed on larvae or adults and in apple trees it has been shown that chickens fed on several potential crop pests, including *P. japonica*. Hlivko and Rypstra (2003) showed that spiders affect feeding behaviour of *P. japonica*. It seems that the cues left by the spiders are unfavourable to *P. japonica*, suggesting a negative effect of spiders that is not based on a direct predation effect.

  Forty-nine natural enemy species were imported into the USA from Asia and Australia from 1920 to 1933 but only a few became established. The most widely distributed are *Tiphia vernalis* Rohwer that parasitizes overwintered larvae in spring, *Tiphia popilliavora* Rohwer which attacks young grubs in late summer and *Istocheta aldrichi* (Mesnil) a tachinid fly that parasitizes the newly emerged adults. Some authors report that they are not efficient enough whereas others argue that they are important biological control agents (Potter and Held, 2002). Altieri and Letourneau (1982) report that association between carrot and soybean negatively affects the population of *P. japonica* by favouring the occurrence of *T. papilliavora*. Planting *Paeonia* sp. is also sometimes recommended to increase the populations of *T. vernalis*.

- **Release of natural enemies** – Entomopathogenic nematodes of the families Steinernematidae and Heterorhabditidae have been found to naturally parasitize *P. japonica* larvae. Some species such as *Steinernema glaseri*, *Heterorhabditis bacteriophora* and *Steinernema feltiae* are also commercially available (Gaugler and Klein, 1998). Autodissemination of the fungus *Metarrhizium anisopliae* has also been proposed to suppress *P. japonica* populations in the Azores and the USA (Klein and Lacey, 1999; Vega *et al.*, 2007).

  Milky disease is the common name for spores of the bacteria *Paenibacillus popillae* and *Paenibacillus lentimorbus*. It leads to larval death after ingestion. Only one application is required as the disease continues to multiply on its own. It has been widely spread and some good results have been obtained. However, some loss of virulence of *P. popilliae* in fields has been observed. It should thus be considered as one of many factors that contribute to population suppression of *P. japonica*, rather than as an effective microbial insecticide. *Bt* serovar. *japonensis* strain Buibui has also been found to be efficient against larvae, but this product is not commercialized (Potter and Held, 2002).
- **Mass sterilization trials** – Some trials have been carried out releasing sterilized males to reduce mating and thus egg and larva populations. More than 235,000 sterilized males were released weekly in the USA in two successive summers. Ratios of sterile to normal males increased progressively and fertility of field-collected females declined, but these effects did not occur early enough to suppress the population. Release of sterile males has not been further explored as a management strategy (Potter and Held, 2002).

Species of the families Noctuidae and Erebidae (super-family Noctuoidea)

Many Lepidoptera species cause soybean defoliation. These species mainly belong to the families Noctuidae, Erebidae and Crambidae. They are polyphagous pests and some of them are characterized by long-distance migration within the same year, especially because of necessity to disperse to a cool climate during winter. Among these numerous species, the most important on soybean are *Anticarsia gemmatalis* Hubner, *Chrysodeixis includens* Walker, *Helicoverpa zea* (Boddie), *Helicoverpa armigera* (Hubner), *Spodoptera litura* (Fabricius), *Spodoptera frugiperda* Smith and *Lamprosema indicata* Fabricius, depending on the region.

DISTRIBUTION. *A. gemmatalis* and *C. includens* are present in North and South America. The two morphologically close species *H. zea* and *H. armigera* have disjunct distribution. *H. armigera* occurs in Australia, Africa, Asia and Europe; it was recently recorded in Brazil (Tay *et al.*, 2013). *H. zea* is mainly found in North and Central America, with some records in South America. Species of the genus *Spodopotera* also damage soybean, including: (i) *Spodoptera eridania* (Cramer), *Spodoptera cosmioides* (Walker) and *Spodoptera albula* (Walker) in Brazil; and (ii) *S. frugiperda*, *Spodoptera exigua* (Hubner) (originated from Asia and causes damage there) and *Spodoptera ornithogalli* (Guenée) in North America. Other Noctuidae species such as *Heliothis virescens* (Fabricius), *Rachiplusia nu* (Guenée) and *Trichoplusia ni* (Hübner) are also recorded from North and South America in soybean crops. In Australia, *Trichoplusia orichalcea* (Fabricius), *Mocis alterna* (Walker), *Mocis frugalis* (Fabricius) and *Mocis trifasciata* (Stephens) are reported as defoliator Noctuidae species. Other species reported from Asia, Japan, Australia and the Pacific islands include: (i) *Helicoverpa punctigera* Wallengren; (ii) *Chrysodeixis acuta* Walker; (iii) *Chrysodeixis chalcites* (Esper); and (iv) *Chrysodeixis eriosoma* (Doubleday). The two latter species are particularly difficult to distinguish and some authors question their synonymy (Murillo *et al.*, 2013). *Hypena scabra* (Erebidae) occurs in North America. Larvae of *Omiodes indicata* (Fabricius) (Crambidae) in Brazil and *L. indicata* (Crambidae) in Asia are also reported as soybean leaf defoliators, the latter being a key soybean pest in China. Some of these species were considered as secondary pests until the 1990s but the use of pesticides has reduced natural enemies leading to important outbreaks (Sosa-Gómez *et al.*, 2003; Specht *et al.*, 2015).

DAMAGE. Damage is caused by larvae that consume leaves, seedlings and sometimes grains and pods. The caterpillars of *A. gemmatalis* consume from 85 cm^2 to 150 cm^2 of leaf area and can even consume all the leaves in cases of high population densities. It is the most damaging foliage-feeding pest of soybean in the southern USA and Brazil. It may also attack tender stems, buds and small bean pods. *H. zea and H. armigera* larvae can feed on foliage, flowers and fruit depending on the moth flight period. Soybeans that bloom early, before the moth flight, usually escape infestation. Leaf area loss of plants averaged between 841 cm^2 and 1025 cm^2 per larva of *H. armigera* in Australia (Rogers and Brier, 2010). The most serious yield losses occur on soybean seed when the caterpillar stage coincides with this plant stage. *S. frugiperda* is certainly the most damaging species of this genus in America. The other species (*S. exigua*, *S. ornithogalli*, *S. eridania* and *S. albula*) are considered as occasional pests in soybean; however, *S. cosmioides* seems to cause more and more problems in South America.

HOW TO CONTROL?

Thresholds and organic insecticides. For most Noctuidae species all over the world, the threshold is 30% defoliation throughout the plant 2 weeks prior to blooming (R1) and 15% defoliation 2 weeks prior to blooming until the pods have filled (R7–R8). In Brazil, these thresholds correspond to 20 or more caterpillars/m (Bueno *et al.*, 2010a; Pomari-Fernandes *et al.*, 2015). In Brazil, *S. cosmioides* defoliates twice as much leaf area as the other species, thus more attention should be paid to this species, as the threshold will be reached more rapidly. For *H. zea*, treatment is usually recommended when defoliation reaches 40% before R1. An online calculator for determining the threshold level in soybeans can be found using the Corn Earworm Threshold Calculator for Soybeans (http://webipm.ento.vt.edu/cew/) (National Integrated Pest Management Network, no date). Thresholds for *H. armigera* have been defined as eight larvae/m^2 (Fathipour and Sedariatan, 2013) and as three larvae/m^2 for *S. litura*.

Pest flights can be monitored using light and/or pheromone traps. However, as for many other pests, trap catch data do not provide a quantitative threshold for the decision to spray, because there is not a clear relationship between catch number and subsequent crop damage.

The organic insecticides available according to national rules are spinosad, neem oil and pyrethrin (Valles and Capinera, 1993). Leksawasdi (1986) showed the negative effect of extracts of fresh *Zingiber cassumunar* Roxb. and *Curcuma longa* L. on several species of *Lamprosema*, the former extract providing the most interesting results. Indiati (2014) showed that extracts of neem seed, sugar apple leaf and sugar apple seeds were effective to suppress *L. indicata* egg laying. Abdullah *et al.* (2001) also reported the toxicity of neem extract and *Bt* on this latter species. Finally, Valles and Capinera (1993) also showed some efficiency of *Ryania* and sabadilla extracts on *S. eridania* in laboratory tests.

Biological control

- **Natural occurrence of natural enemies –** Many studies report the occurrence of parasitoids of noctuid species in soybean crops. Beach and Todd (1985) observed 12 parasitoid species of *C. includens* on soybean in the USA. The most common were *Copidosoma truncatellum* (Dalman) (Encyrtidae), *Meteorus autographae* Muesebeck and *Cotesia marginiventris* (Cresson) (Braconidae). These three species are polyphagous and have been reported on other Noctuidae soybean pests: (i) *M. autographae* on *A. gemmatalis*, *H. zea*, *S. eridania*, *S. frugiperda*, *S. ornithogalli* and *T. ni* (Grant and Shepard, 1984, 1986; Sourakov and Mitchell, 2014); (ii) *C. truncatellum* on *T. ni* (Jones *et al.*, 1982); and (iii) *C. marginiventris* on *Agrotis ipsilon*, *H. zea*, *H. virescens*, *P. scabra*, *S. eridania*, *S. exigua*, *S. frugiperda*, *S. ornithogalli* and *T. ni* (Sourakov and Mitchell, 2014). Archer and Bynum (1994) report low natural parasitism rates on *H. zea* except for eggs parasitized by *Trichogramma* spp. and *Telenomus* spp. (Hymenoptera: Scelionidae). Other *Trichogramma* spp. are naturally reported as *Trichogramma bruni* parasitizing *A. gemmatalis* and *Rachiplusia nu* in soybean crops in Argentina (Valverde *et al.*, 2009, 2014). Some larval parasitoid species of *H. zea* have also been observed (*Cotesia* spp., *Microplitis croceipes* (Cresson) (Hymenoptera: Braconidae), *Campoletis* spp. (Hymenoptera: Ichneumonidae), *Eucelatoria armigera* (Coquillett) and *Archytas marmoratus* (Townsend) (Diptera: Tachinidae). In Brazil, Foerster and Avanci (1999) and Polaszek and Foerster (1997) reported parasitism of *A. gemmatalis* by *Trichogramma pretiosum* Riley, *Trichogramma rojasi* Nagaraja & Nagarkatti, *Trichogramma atopovirilia* Oatman & Platner (Trichogrammatidae), *Trichogramma acacioi* Brun, Moraes & Soares, *Trichogramma lasallei* Pinto, *Telenomus cyamophylax* Polaszek (Scelionidae) and *Encarsia porteri* (Mercet) (Aphelinidae). Among the biological control agents of species of *Spodoptera*, the egg parasitoid *Telenomus remus* Nixon (Hymenoptera: Platygastridae) seems to be highly efficient in Brazil (Bueno *et al.*, 2010b; Pomari *et al.*, 2012). *Xanthopimpla punctata* (Fabricius) is commonly found in soybean crops in Asia. Dung *et al.* (2001) reported many parasitoid species of this latter species but also of *S. litura* and *S. armigera* in Vietnam soybean crops. Fathipour and Sedariatan (2013) provide a table listing parasitoids (and also predators and diseases) reported for *H. armigera*. In Australia, *C. eriosoma* is frequently parasitized by braconids (*Apantales* sp.) as well as by tachinid flies and ichneumonid wasps. In Hawai, several parasitoids contribute to controlling populations of this latter pest in soybean crops. These parasites include *Chaetogaedia monticola* (Bigot), *Copidosoma truncatellum* (Dalman) (85% parasitism), and *Eucelatoria armigera* (Coquillett) (Zimmerman, 1958). Wu and Jiang (2011) report four species of parasitoids, *Dolichogenidea* sp. (Hymenoptera: Braconidae), *X. punctata* (Hymenoptera: Ichneumonidae), *Brachymeria lasus* Walker and *Exorista sorbillans* Wiedemann; *Dolichogenidea* sp. being dominant with the highest larval parasitism at 80%. Braconid parasitoids of the genus *Orgilus* were observed to attack the first instar larvae of *Lamprosema diemenalis*. In the USA, southern armyworm *S. eridania* is attacked by naturally occuring natural enemies such as *Cotesia marginiventris* (Cresson), *Chelonus insularis* Cresson, *Meteorus autographae* Muesebeck and *Meteorus laphygmae* Viereck (Hymenoptera: Braconidae) (Tingle *et al.*, 1978).

Generalist predators have also been observed feeding on noctuid species in soybean fields. Common predators are Coccinellidae such as *Hippodamia convergens* Guerin-Meneville and *Coleomegilla maculata* DeGeer, *Collops* spp. (Coleoptera: Melyridae), *Chrysopa* sp. (e.g. *Chrysopa rufilabris* (Burmeister)), *Chrysoperla* spp. (Neuroptera: Chrysopidae), *Orius tristicolor* (White) (Hemiptera: Anthocoridae), *Geocoris punctipes* Say., *Geocoris uliginosus* (Say) (Hemiptera: Lygaeidae), *Reduviolus roseipennis* (Reuter), *Tropiconabis capsiformis* (Germar), *Hoplistoscelis deceptivus* (Harris) (Hemiptera : Nabiidae) and carabid species such as *Calleida decora* (F.) and *Chlaenius viridis* Menetries. Birds can also feed on earworms, but rarely are adequately abundant to be effective (Richman *et al.*, 1980; Pfannenstiel and Yeargan, 2002). The pentatomid *Stiretrus anchorago* (F.) and the reduviid *Arilus cristatus* (L.) consumed important quantities of larvae of *C. includens* in the USA. The pentatomid species of the genus *Podisus* is also reported to predate *A. gemmatalis* (Vinha Zanuncio *et al.*, 2012).

Besides parasitoids and predators, noctuid populations can also be affected by pathogens. The most common are *Entomophthora gamma*, *Nomuraea rileyi* and polyhedrosis virus (Roberts *et al.*, 1977; Carner, 1980; Boucias *et al.*, 1984). *N. releyi* is reported as an important biocontrol agent of *A. gemmatalis* in soybean crops in Brazil and the USA. It is also reported from *L. indicata*. Natural infestation by *Metarhizium*, *Beauveria bassiana* and *Paecilomyces* were also observed in Brazil (Sosa-Gómez *et al.*, 2001; Rossoni *et al.*, 2013). Kalkar and Craner (2005) reported natural infestation of *Lathypena scabra* by *Entomophaga aulicae* in the USA. Finally, some authors report parasitism of *S. frugiperda* in Brazil by entomopathogenic nematodes (*Steinernematidae*, *Heterorhabibditidae*) with mortality rates ranging from 28% to 56% (De Marco Salvadori *et al.*, 2012).

- **Release and spread of biological control agents**
  - **Baculovirus *Anticarsia gemmatalis* nucleopolyhedrovirus (*Ag*MNPV)** – A formulation of *Ag*MNPV was first developed in Brazil to control *A. gemmatalis*. This virus is specific to *A. gemmatalis*. To be active, the larvae must eat the virus, and the mortality occurs 4–7 days after ingestion. In the USA, a single spray application is recommended when 15–20 caterpillars are observed after beating, as the larvae should be old enough (1 cm in length) to ingest a lethal concentration of virus. In Brazil, the current recommendation for spraying is when 40 small caterpillars (less than 1.5 cm) or 30 small carterpillars and 10 larger caterpillars are trapped after beating (Moscardi and Carvalho, 1993; Sujii *et al.*, 2002; Szewczyk *et al.*, 2006). Soybean producers can buy this product but can also make the preparation by themselves. Other baculoviruses have been developed for other noctuid species in the world. A nuclear polyhedrosis virus (NPV) of *H. zea* was developed in the 1960s in the USA (Ignoffo and Couch, 1981) and sold all over the world. This virus (*H. zea* NPV, *Hz*SNPV) has a relatively broad range, infecting other species belonging to the genera *Helicoverpa* and *Heliothis*. Locally obtained isolates of *H. armigera* SNPV have also been produced and applied to many crops in China, India and Australia (Zhang *et al.*, 1995; Mettenmeyer, 2002; Sun and Peng, 2007; Erlandson, 2008; Srinivasa *et al.*, 2008). An *S. exigua* NPV has been used to control this species on vegetable crops in the USA, Europe, China and Thailand. Also, an NPV of *S. lituralis* is used in China, India and Thailand. In Brazil, an indigenous isolate of *S. frugiperda* NPV (*Sf*NPV) was used to control the insect in maize (Valicente and Cruz, 1991; Moscardi, 1999). Owing to difficulties and the high cost of *Sf*NPV production by the Brazilian Agricultural Research Corporation (EMBRAPA), this programme has been temporarily discontinued (Sun and Peng, 2007; Erlandson, 2008; Kumari and Singh, 2009; Szewczyk *et al.*, 2009; Moscardi *et al.*, 2011).
  - ***Bacillus thuringiensis (Bt)* subspecies *kurstaki*** – *Bt* is used to control *Heliothis* sp. and *Spodoptera* sp. The commercialized biotype for controlling Lepidoptera soybean pests is *Bt* subspecies *kurstaki* (Alexandre, 2010). To be active this biopesticide has to be eaten by the larvae. The caterpillar dies within 3–4 days afterwards but stops feeding between 1 and 2 days after the ingestion. Thus, *Bt* is more rapid and efficient than baculovirus. However, it persists in fields at low levels and repeated applications

are required (Sujii *et al.*, 2002). Sometimes, *Bt* can be mixed with baculovirus to ensure a more efficient control. Application is recommended when 30% of defoliation is reached (Moscardi *et al.*, 2011).

- **Release of *Trichogramma* sp.** – Parra and Zucchi (2004), Massaroli *et al.* (2014) and Carvalho *et al.* (2014) report natural parasitism of *A. gemmatalis*, *C. includens* and *Trichlopusia ni* in Brazil by *Trichogramma pretiosum* with parasitism rates of 90% in case of the two former species. *T. pretiosum* is also known to parasitize eggs of *H. armigera* with 70–90% parasitism reported. Various studies have aimed to characterize the best conditions for ensuring optimal results after mass release of *Trichogramma* spp. in Brazil (e.g. Bueno *et al.*, 2009, 2012). Despite this interest, there are not many commercialized examples of mass releasing (Hassan, 1993; Van Lenteren and Bueno, 2003). In Australia, mass releases of *Trichogramma* sp. are used to control *Heliothis* eggs in organic soybean.The species *Sycanus indagator* (Stal) was imported from India to the USA to control *C. includens* and *T. ni*, but did not establish (Greene and Shepard, 1974). In another programme, the reduviid species *Pristhesancus papuensis* Stal was introduced from Australia to the USA to control *H. zea* (Fathipour and Sedariatan, 2013).

- **Cultural practices**
  - **Ploughing** – Ploughing in late maturing crops in winter increases the mortality of pupae by exposing them to heat and predation. Residue on the soil surface may be attractive to egg-laying females (Fathipour and Sedariatan, 2013).
  - **Early planting** – This can avoid the seasonal peaks of population and limit larval infestations (Fathipour and Sedaratian, 2013).
  - **Weed management** – In the USA, armyworm populations can be reduced by managing grassy weeds within soybean fields and in their borders (Rogers and Brier, 2010).
  - **Trap cropping** – A sorghum trap crop is reported to be efficient for managing *H. armigera*, as it increases rates of parasitism by *Trichogramma chilonis* Ishii (Virk *et al.*, 2004). However, the trap crop should be destroyed prior to the pupation of the first large *H. armigera* larvae. In India, the pigeon pea or okra trap crop is also reported to limit *H. armigera* populations. Then this trap plant is treated with neem solutions (Regupathy, 2005).
- **Plant resistance** – Various soybean cultivars have partial resistance to noctuid species; usually this resistance is not sufficient to ensure complete control. Fathipour and Naseri (2011) presented detailed information regarding evaluation of soybean resistance to *H. armigera* in a book chapter entitled 'Soybean cultivars affecting performance of *Helicoverpa armigera* (Lepidoptera: Noctuidae)'. Souza *et al.* (2014) reported on the effect of plant defence stimulators (calcium and magnesium silicate on soil and sodium silicate and acibenzolar-S-methyl (ASM) sprayed on leaves) on *C. includens*. These authors showed that the resistant cultivars 'IAC 100' and 'IAC 17' associated with the calcium and magnesium silicate inducers and sodium silicate lengthened larval stage and induced high mortality in the *C. includens* adults. In 1985, Rabin and Pacovsky showed the effect of mycorrhizal fungus on soybean resistance to two Noctuidae species (*H. zea* and *S. frugiperda*). Hoffmann-Campo *et al.* (2006) showed negative effects of rutin on *A. gemmatalis* development. This compound is one of the flavonol glycosides identified in leaves of the wild soybean PI 227687.

Species of the family Nymphalidae

DISTRIBUTION. One species belonging to this family causes damage to soybean: *Vanessa cardui* (L.). This species originated from North America and is now distributed worldwide. It is very polyphagous and reported from more than 300 plant species (Poston *et al.*, 1977).

DAMAGE. Larvae damage soybean by consuming leaves in the upper plant canopy and produce silk to web the leaves together. Even if the extent of defoliation is spectacular, the economic impact is usually low (Poston *et al.*, 1977).

HOW TO CONTROL?

Insecticide treatment and thresholds. Treatments are required when 25–30% defoliation is reached prior to bloom (before R1) and 20% after bloom or pod set (after R1). This usually corresponds to about four to eight caterpillars/row-ft (equivalent to four to eight caterpillars/0.3 m row).

Weed control. Adult females prefer to lay eggs on thistles, so reducing thistles near field edges may reduce the number of larvae that migrate to the soybean crop (NDSU, no date). There are no data on natural enemies.

Species of the family Arctiidae

DISTRIBUTION. One species belonging to this family causes damage to soybean: *Spilosoma obliqua* (Walker). It is reported on soybean from Asia and is a polyphagous species (Singh and Varatharajan, 1999; Shrivastava *et al.*, 1999).

DAMAGE. This species eats leaves that take on the appearance of a net or web.

HOW TO CONTROL?

Biological control. Some natural enemies have been recorded: parasitoid species such as *Apanteles obliqua* Wlk. (Broconidae Hymenoptera) (18% of parasitism rates on larvae), *Stratiomyid* sp. (Stratiomyidae, Diptera) (17% of parasitism on pupae) and entomopathogen *N. rileyi* in India (Singh *et al.*, 2015). Singh *et al.* (2015) showed the high larvicidal activity of a strain of *Bacillus sphaericus* (82% mortality). In India, Pawar and Thombre (1992) also showed the natural occurence of NPV.

Botanical insecticides. Many studies have investigated the effects of botanical compounds on this pest (Agrawal and Mall, 1988; Ahmed and Bhattacharya, 1991; Hariprasad and Kanaujia, 1992; Tripathi and Singh, 1994; Tarkeshwar and Singh, 2014). Deshmukh *et al.* (1979) showed high mortality in the larvae due to a diet of cannabis. Losses could be decreased by spraying aqueous leaf extracts of *Euphorbia royleana* or *Lantana camara* (Sharma *et al.*, 1982). Leaf extracts of *Lindenbergia grandiflora* (a perennial diffuse herb), velvet bean (*Macuna cochinensis*), *Passiflora mollissima*, *Nyctanthes arbortristis* (an ornamental shrub from the Himalayas), *Swertia chirayita* (a perennial herb in north India), *Ailanthum excelsa* as well as needle wood, *Schima khasiana* (a perennial tree found in east Himalaya) and a seed oil fraction of *Pongamia glabra* and seed oil of babchi (*Psoralea corylifolia*) showed antifeedant activities (Sharma *et al.*, 1982; Tripathi *et al.*, 1987; Mohanty *et al.*, 1988; Premchanda, 1989).

Intercropping. Intercropping soybean with a variety of pigeon pea, maize or sorghum is reported to decrease the *S. obliqua* population (Prasad and Sharma, 1989).

Species of grasshoppers, families Acrididae and Gryllidae

DISTRIBUTION. Several species belonging to the families Acrididae and Gryllidae can damage soybean crops. In the USA the most frequent species belong to the family Acrididae: *Melanoplus differentialis* (Thomas) and *Melanoplus femurrubrum* (De Geer). In Brazil the most frequent species, *Rhammatocerus schistocercoides* (Rehn), also belongs to this latter family but an additional species belonging to the family Gryllidae, *Anurogryllus muticus* (De Geer), can damage the soybean crop especially during the initial plant growth stage. All these species are polyphagous; they can damage other crops and can also feed on uncultivated plants of several families (Cunningham and Sampson, 2000; Hoffmann-Campo *et al.*, 2000).

DAMAGE. Grasshoppers on soybean fields are sporadic but can cause severe damage during periodic outbreaks. Both immature stages and adults feed on leaves and the pod wall (and then the seeds). They cause irregular holes in the leaf. A yield reduction of 3–7% may occur if 40% defoliation occurs during any vegetative growth stage or if 20% leaf defoliation occurs during the pod forming and fill stages. Hot, dry weather conditions contribute to the development of large populations of grasshoppers (Cunningham and Sampson, 2000; Hoffmann-Campo *et al.*, 2000).

HOW TO CONTROL?

Thresholds. As grasshopper eggs hatch in the borders of soybean fields, these areas should be first observed especially in spring and early summer (Browde *et al.*, 1992). Then, 20–25% defoliation of soybean leaves would lead to insecticide application. The most efficient period to control grasshoppers is at the third and fourth instars. In some cases, thresholds are based on the number of specimens counted per surface. Treatments are recommended when densities are higher than 15 specimens in field borders and higher than 40 specimens in the field.

Biological control. Many natural enemies have been reported, including insects, mites, fungi and nematodes. Larvae of Meloidae and Carabidae (feeding on grasshopper eggs), larvae of Bombylidae (parasitizing eggs), Asilidae, Scelionidae wasps, Sarcophagidae flies and Nemestrinidae flies have been reported in soybean fields. The mite *Trombidium locustarum* (Walsh) is also an important natural enemy that feeds on eggs, nymphs and adults (University of Nebraska, Lincoln Extension, Institute of Agriculture and Natural Resources, 2008).

The fungus *Entomophthora grylli* often causes locally high mortality in grasshopper populations under warm and humid conditions. Fungal spores are airborne and infect other grasshoppers. Some nematodes can also infect grasshoppers, reducing oviposition.

Several biological products, such as the single-celled protozoan *Nosema locustae* and the fungus *E. grylli*, are commercially available to manage grasshoppers. The grasshopper larvae eat the baited product and then die, but slowly. The bait could be made up of flaky wheat bran treated with *N. locustae* (Wright *et al.*, 2015).

In Brazil, *Metarhizium anisopliae* var. *acridum* has been studied and seems to be a very good candidate for controlling *Rhammatocerus schistocercoides* (Magalhães *et al.*, 2000). Some practical advice is provided in Schmidt *et al.* (2007a).

Agricultural practices

- **Tillage** – As female grasshoppers prefer to lay eggs in undisturbed soil, this practice can be used to discourage adults from laying eggs in fields (Hein and Campbell, 2008).
- **Weed control** – Reducing grasses and other weeds within and around fields usually discourages adults from feeding and mating in that area. However, this can negatively affect natural enemy occurrence (Hodgson, 2014).

Insects that feed on seeds and pods

Species of the family Pentatomidae

DISTRIBUTION. Species belonging to the Pentatomidae family are important seed feeders. In Brazil the main species are *Euschistus heros* (F.), *Nezara viridula* (L.) and *Piezodorus guildinii* (Westwood) even if other species are cited in the literature (e.g. *Dichelops furcatus* (Fabricius), *Dichelops melacanthus* (Dallas), *Edessa meditabunda* (Fabricius)). In the USA, the main species are *Acrosternum hilare* (Say) and *Euschistus quadrotor* (Rolston). In Australia, the species *N. viridula* and *Piezodorus oceanicus* (Montrouzier) are reported as the main Pentatomidae pests of soybean. In Asia and Africa other species of this family also damage soybean pods, such as *Piezodorus hybneri* (Gmelin) and *Piezodorus punctiventris* (Dallas) (Panizzi, 1997; Hoffmann-Campo *et al.*, 2000; Kamminga *et al.*, 2012).

DAMAGE. These species feed on pods causing hardened, knotty spots on pods or pod drop from plants. Foliage, flowers and stems can also be attacked. During feeding, insects inject toxic saliva that provokes leaf deformation. The species *A. hilare* is also able to transmit the bean pod mottle virus. The susceptible soybean stages are between R4 and R5 (Panizzi, 1997; Hoffmann-Campo *et al.*, 2000; Kamminga *et al.*, 2012).

HOW TO CONTROL?

Thresholds. In Brazil, economic damage is observed when densities reach four specimens/2 m of soybean row between the development of pods to the fill of grains (Correa-Ferreira and Panizzi, 1999). In the USA, the control of *A. hilare* is recommended when 40 specimens are found on 100 sweeps. Another source proposes a lower threshold of 20 specimens for 100 sweeps always when the pods are still green. In some publications a threshold of one stink bug/row-ft (equivalent to one stink bug/0.3 m row) in the reproductive stage of soybean is reported to be a common threshold (Todd and Herzog, 1980). In many southern states of the USA, a beat sheet sample containing above one stink bug/0.3 m of row may require an insecticide application (Parker, 2012).

Stink bugs can be monitored using black light traps (Kaminga *et al.*, 2012). Some studies have been carried out to determine the efficiency of several pheromones for attracting stink bugs (i.e. Aldrich *et al.*, 1991, 2007; Tillman *et al.*, 2010). Other studies show that yellow pyramid traps constitute an attractive visual stimulus for *Euschistus* spp. (Mizell and Tedders, 1995; Leskey and Hogmire, 2005). Finally, tube traps made from clear plastic tubes with wire mesh cones on the ends were used in several studies with varying results, and are still commercially available for monitoring stink bugs.

Few studies concern organic insecticides. Some efficiency has been observed with the application of azdirectin, pyrethrin and spinosad (Overall, 2008; Kamminga *et al.*, 2009, 2012). Ferreira De Souza and Favero (2015) showed lethal effects of eucalyptus oil (*Eucalyptus urograndis*) on *E. heros.*

Plant resistance. Resistant varieties of soybean have been shown to affect pentatomid populations (Gilman *et al.*, 1982; Jones and Sullivan, 1982; Kamminga *et al.*, 2012).

Biological control. Important impact of natural enemies has been reported in organic soybean in Brazil (Perez and Correa-Ferreira, 2004). The most important natural enemies are the Hymenoptera: *Trissolcus basalis* (Wollaston) and *Telenamus podisi*, that both parasitize eggs. *Trichopodes penipes* (F.) (Diptera: Tachinidae) has been reported to parasitize adults of *N. viridula* with parasitism rates of 95% in winter. This species is also reported to naturally parasitize *A. hilare* adults and nymphs in the USA with parasitism rates of 93% (Mahr, 1999). *Hexacladia smithii* (Encyrtidae) is reported as the most important parasitoid of adults of *E. heros* in Brazil (Corrêa-Ferreira *et al.*, 1998). This species is present in high density populations between December and January (Corrêa-Ferreira and Panizzi, 1999) reaching sometimes parasitism rates of 98%. In Brazil, biological control of Pentatomidae is carried out with the egg parasitoids *T. basalis* and *T. podisi*, released at the end of flowering stages in the border of the plots. They parasitized eggs of *N. viridula*, *P. guildinii* and *E. heros* (Corrêa-Ferreira, 2003). *T. basalis* is also reported as an efficient parasitoid of *N. virudula* in Australia, as well as another introduced tachinid species *Trichopoda giacomellii* (Blancard) that parasitizes adults and the fifth immature stage. These two species can also parasitize *Piezodorus oceanicus* in Australia (Coombs and Khan, 1998).

Besides parasitoids, generalist predators have been reported to feed on Pentatomidae species, such as lacewing larvae, spined soldier bugs (*Podisus maculiventris* (Say)) and birds (Underhill, 1934; McPherson, 1982; McPherson and McPherson, 2000). Also some entomopathogenic pathogens are reported (Fuxa *et al.*, 2000; Ribeiro and Castiglioni, 2008).

Trap crops. Trap cropping is well documented in the literature to control Pentatomidae species in soybean plots all around the world (Newsom and Herzog, 1977; McPherson and Newsom, 1984; Todd and Schumann, 1988; Panizzi, 1997; Corrêa-Ferreira and Panizzi, 1999). During the summer to autumn period, sunflower, buckwheat, sorghum and millet are recommended as trap crops (Kamminga *et al.*, 2012). Additional trap crops include *Fagopyrum sagittatum* Gilib and *Triticale hexaploide* Lart. It seems that ratooning a trap crop of pearl millet, *Pennisetum glaucum* (L.) R. Br., buckwheat or sorghum may extend the efficacy of the trap crop (Mizell *et al.*, 2008; Russell and Mizell, 2015). Finally, some studies report the impact of planting a perimeter of earlier maturing soybean (McPherson and Newsom, 1984; Panizzi and Alves, 1993; Kamminga *et al.*, 2012).

Species of the family Alydidae

DISTRIBUTION. Two main species cause damage in soybean in Africa and Asia: *Riptortus clavatus* (Thunberg) and *Riptortus dentipes* (Fabricius). In Brazil, *Neomegalotomus parvus* (Westwood) is reported in soybean crops (Santos and Panizzi, 1998). In Australia, *Riptortus serripes* (Fabricius) and *Melanacanthus scutellaris* (Dallas) are also reported on soybean.

DAMAGE. The damage caused is similar to that caused by members of the Pentatomidae. These species are polyphagous. Kimura *et al.* (2008) showed *R. clavatus* can transmit yeast to pods (*Eremothecium ashbyi*). Santos and Panizzi (1998) report that eight and 16 specimens of *N. parvus* per plant cause 13.8% and 22% of pod losses, respectively.

HOW TO CONTROL?
Biological control. In Japan, natural egg parasitism of *R. clavatus* was observed with more than 70% parasitism rates with the following species: *Gryon japonicum* (Ashmead), *Ooencyrtus nezarae* Ishii and *Anastatus japonicus* Ashmead (Schaefer and Panizzi, 2000). Natural parasitism of *R. dentipes* was also observed (Schaeffer and Panizzi, 2000). Non-viable eggs of *Riptortus* sp. added to aggregation pheromone traps were found to enhance field parasitism by *O. nezarae* (Takasu *et al.*, 2004; Alim and Lim, 2011).

Intercropping. Youn and Jung (2008) showed that association of buckwheat with soybean enhances egg parasitism of *R. clavatus* with parasitism levels ranging between 70% and 83%. One of the most frequent parasitoids was *O. nezarae*. No effect on another parasitoid species *G. japonicum* was observed. In Africa, one reference reports the negative effect of two plants on *R. dentipes* (*Zanthoxylum zanthoxyloides* and *Lippi adoensis*) (Olufolaji, 2011).

Resistant varieties. Some studies deal with soybean genotypes resistant to *R. clavatus* (Lopes *et al.*, 1997). When planted late, the small-seeded soybean cultivar, 'Kyushu-143', suffered much less damage than large-seeded cultivars in Japan (Wada *et al.*, 2006).

Species of the family Cecidomyidae

DISTRIBUTION. *Asphondylia yushimai* Yukawa & Uechi is one of the major pests of soybean in Japan, Indonesia and China (Uechi *et al.*, 2005). It seems that *Prunus zippeliana* Miquel is a winter host of this species in Japan (Yukawa *et al.*, 2003, 2016). During summer, this species develops essentially on members of the family *Fabaceae*.

DAMAGE. Females lay eggs inside young pods; malformation of pods is caused by larval infestations. Furthermore when adults lay eggs in soybeans they inject a symbiotic fungus that proliferates and forms a whitish mat over the bean wall, and larvae grow feeding on the fungus (Uechi *et al.*, 2005).

HOW TO CONTROL? Colour traps can be used for monitoring soybean pod gall midge (RADA, 2010). There is no economic threshold. Many parasitic wasps attack the soybean pod gall midge. Controlling weeds beneath or around the plants may reduce the infestation of gall midge (RADA, 2010).

Species of the family Pyralidae

DISTRIBUTION. Two Pyralidae species are reported as pests on soybean, *Etiella zinckenella* (Treitschke) and *Elasmopalpus lignosellus* (Zeller). The former species is widespread all over the world and the latter occurs in North and South America. Both are polyphagous.

DAMAGE. Larvae of *E. zinckenella* consume flowers and young pods. It is the main pest of soybean in Indonesia (Van den Berg *et al.*, 1998a). *E. lignosellus* tunnels into the stem at the soil surface. Mean seed loss can reach 12%.

HOW TO CONTROL? No accurate threshold is reported in the literature. However, it seems that two to three larvae of *E. lignosellus*/m seem sufficient to apply a protection method (Molinari and Gamundi, 2010).

Plant resistance. Several authors have studied the tolerance of soybean cultivars. It seems that *E. zinckenella* prefers to deposit eggs on cultivars with many short trichomes. The preference test based on soybean phenology showed that the adult prefers the R4 stage (fully filled pod) for oviposition (Talekar and Lin, 1994; Permana *et al.*, 2012).

Biological control. Taghizadeh *et al.* (2012) and Reflinaldon and Ganeshi (2014) provide a review of parasitoids and entomopathogens naturally occuring in Indonesia. Van den Berg *et al.* (1995) report that *E. zinckenella* is parasitized by six parasitoids with a mean parasitism rate of 15%. In Argentina, some natural enemies of larvae of *E. lignosellus* were registered (one pathogen and a parasitoid *Horimenus* sp.) (Molinari and Gamundi, 2010). In the USA, several species of parasitoids are widespread and account for more than 98% of larval parasitism (*Orgilus elasmopalpi* Muesebeck, *Pristomerus spinator* (F.) and *Chelonus elasmopalpi* McComb). Two other species are also recorded but less often (*Stomatomyia floridensis* Townsend and *Cotesia* sp.) (Heatherly and Hodges, 1998). Predominant, widespread pathogens were a granulosis virus and *Beauveria* sp. A microsporidium was occasionally found infecting larvae of *E. lignosellus*. Mortality from parasites and pathogens frequently ranges between 20% and 30% for small and medium larvae (Funderburk *et al.*, 1984).

Cultural practices. In Brazil, it seems that direct seeding limits the occurrence of *E. lignosellus* (Hoffmann-Campo *et al.*, 2000; Gill *et al.*, 2010). Furthermore, mulching constitutes an unfavourable practice to the pest as it limits soil warming and increases humidity (Harsimran *et al.*, 2010). An increase in planting density could also limit this pest by increasing soil humidity (Hoffmann-Campo *et al.*, 2000).

Species of the family Tortricidae

DISTRIBUTION. One species of this family is reported on soybean. This species is *Laspeyresia glycinivorella*, synonym of *Leguminivora glycinivorella*. It is reported from North East of Asia.

DAMAGE. The larvae destroy seeds and can often consume as much as half the seed. It is one of the most destructive soybean pests in North China. Kuwayama (1928) provides a complete study of the biology of this species.

HOW TO CONTROL?

Biological control. Ichneumonidae species as *Epicurus hakonensis* and *Epicurus glycinivorellae* are reported to parasitize larvae with some high parasitism rates (Kuwayama, 1928). Besides parasitoids, some predators of the family Asilidae (*Asilus atripes* Loew) and the ant *Lasius* (*Chthonolasius*) *umbratus* Nylander subspecies *mixtus* are reported. It is also probable that birds, spiders and ground beetles feed on this pest. Kuwayama (1928) also reports two entomopathogen fungi of the genera *Entomophthora* and *Isaria*. More recently, Li-Wei Song *et al.* (2015) studied egg parasitism by *Trichogramma leucaniae* Pang & Chen and concluded that this species could be a valuable natural enemy.

Hu *et al.* (2013) studied the volatile attraction of different pheromones and plant extracts (i.e. linalool, (Z)-3-hexenyl acetate, geraniol, (E)-2-hexenal, benzaldehyde, and phenylacetaldehyde). Pheromone-baited traps are more attractive when combined with (E)-2-hexenal, benzaldehyde and phenylacetaldehyde and less when combined with linalool, (Z)-3-hexenyl acetate and geraniol.

Agronomic practices. This pest can be controlled by autumn ploughing. Intercropping with cereal (barley, wheat or 'Indian corn' a variant of maize) with a delayed soybean sowing is reported to limit the pest population. This measure seems to protect the plant as: (i) no pod occurs when the young larvae are present; and (ii) the height of 'Indian corn' obstructs the flight of the moth to a certain degree.

Plant resistance. Damage caused differs according to the soybean cultivars. Pod time production could be one of the factors that determine this resistance. Furthermore, it seems that glabrous soybeans are highly resistant whereas pubescent varieties are highly susceptible (Morse and Carter, 1937). Matsukawa (1928) reports two cultivars in Japan ('Hadaka-Daidzu' and 'Kenashi-Daidzu') little injured by this pest.

Leaf discoloration

Species of the family Miridae

DISTRIBUTION. One species of this family damages soybean: *Halticus bractatus* (Say). It is reported from North, Central and South America and also from Hawaii, the West Indies and the Galápagos Islands. It is polyphagous and a major pest of lucerne and can also cause some damage on soybean (Day and Saunders, 1990; Henry and Wilson, 2004; Capinera, 2005).

DAMAGE. Nymphs and adults both cause damage by sucking the cell sap. They cause a whitish or yellowish speckling on the foliage (Capinera, 2005).

HOW TO CONTROL?

Organic and botanical insecticides. The best time to scout for this pest is from VE through to V5. The economic threshold is not reported. Suppression is easily accomplished with insecticides, including botanical products such as rotenone, pyrethrin and sabadilla extracts (Capinera, 2005). Sprays of garlic and clay are also cited to be efficient (Atthowe *et al.*, 2010). Recently, Braasch *et al.* (2012) showed that three common herbivore-induced plant volatiles (HIPVs) (methyl salicylate, *cis*-3-hexen-1-ol and phenylethyl alcohol) were repellent for *H. bractatus*. The highest effect was observed for a mixture of the three compounds, opening new insights for controlling this pest.

Agrosystem management. High densities of fleahoppers are common on weeds, thus the destruction of such plants can limit the population of this pest. Also, this species is commonly found on leguminous crops such as lucerne and clover, so these nearby vegetables may be a risk for neighbouring soybean crops (Iowa State University, 2012).

Biological control. Leiophron uniformis (Gahan) (Hymenoptera: Braconidae) was observed to lead to 50% of mortality of nymphal stages in the USA (Capinera, 2005). Beyer (1921) reported several egg parasitoids such as *Anaphes perdubius* Girault, *Anagrus* sp. (Hymenoptera: Mymaridae), *Oligosita americana* Girault and *Paracentrobia subflava* Howard (Hymenoptera: Trichogrammatidae). In Canada, *Peristenus clematidis* Loan (Hymenoptera: Braconidae) is known to parasitize *H. bractatus*. An unidentified nematode and predatory mite also have been detected in the USA (Beyer, 1921; Day and Saunders, 1990).

Species of the family Cicadellidae

DISTRIBUTION. *Empoasca fabae* (Harris) is present in North and South America and Europe. In Indonesia, another *Empoasca* species, *Empoasca terminalis* (Distant), is reported from soybean (Nasruddin *et al.*, 2014). The former species is reported from more than 200 host plants. Leguminous hosts are preferred by *E. fabae*, followed by potatoes and other species of *Solanum* with 61.8% of the host species belonging to the family *Fabaceae* (Lamp *et al.*, 1994).

DAMAGE. Adults and immatures suck plant sap and inject toxic saliva that causes localized stippling, yellowish to reddish yellow discoloration of the leaf (especially at the tip), leaf vein distortion and leaf curling. These symptoms are commonly referred to as hopperburn (Nielsen *et al.*, 1990). Dwarfing may occur in leaves, stems and/or petioles, and floral development may be reduced. Young soybean leaves tend to have softer hairiness which favours movement, feeding and oviposition of leafhoppers. By contrast, the short leaf hairs found on older soybean leaves deter feeding and oviposition.

HOW TO CONTROL?

Threshold. The economic threshold depends on the plant age. For early vegetative stages, a treatment is required when there are two leafhoppers per plant; flowering fields can be treated when there is one leafhopper per trifoliate leaf. While pods are developing, treatments are recommended when there are two leafhoppers per trifoliate leaf (Ogunlana and Pedigo, 1974; Hunt and Pedigo, 2000; Krupke *et al.*, 2013).

Plant resistance. Soybean varieties with sparse or very short leaf pubescence are most susceptible to *E. fabae*. Also, plants under moisture stress appear to be more vulnerable to damage (Yeargan *et al.*, 1994). The normal dense pubescence commonly found in most soybean cultivars provides a high level of resistance to this pest, and significantly reduces feeding damage and oviposition with increased adult mortality (Ogunlana and Pedigo, 1974; Elden and Lambert, 1992).

Intercropping and weed management. The presence of weeds, particularly grasses, reduces densities of *E. fabae* in soybeans because of low suitability of grasses for *E. fabae* development and some repellent effects (i.e. Altieri *et al.*, 1981; Lamp *et al.*, 1984; Hammond and Stinner, 1987; Smith *et al.*, 1988; Hammond and Jeffers, 1990; Lam and Pedigo, 1998; Buckelew *et al.*, 2000; Miklasiewicz and Hammond, 2001). In a soybean–wheat cropping system as well as in soybean planted with rye, lower densities of *E. fabae* were observed (Miklasiewicz and Hammond, 2001; Koch *et al.*, 2012). However, heavy infestations could occur when soybean fields are nearby a lucerne crop (Poston and Pedigo, 1975).

Biological control. Some trials showed that chrysopids and coccinellids consumed 23 and ten nymphs, respectively, over a period of 5 days (Fenton and Hartzell, 1923). Parasitism of eggs by *Anagrus armatus* (Ashmead) (Hymenoptera: Mymaridae) is also reported (parasitism rates of 40%) but is usually too low to control this pest under field conditions (Appleton *et al.*, 2003). Natural parasitism by the fungus *Zoophthora radicans* (Brefeld) Batko (Zygomycetes: Entomophthorales) has been observed (Magalhães *et al.*, 1991) but because specific conditions are required (hygrometry and temperature), natural infestation does not occur often. Rodehouse *et al.* (1992) showed the negative effect of grassy corridors within soybean fields on *E. fabae* and natural enemies.

Species of the family Aphididae

DISTRIBUTION. The soybean aphid (*Aphis glycines* Matsumara), a pest that originated from Asia, was first observed in North America in 2000. It is reported from China, Japan, Malaysia, the Philippines, Taiwan and Thailand (Wang *et al.*, 1962; Wu *et al.*, 2004). It has also invaded Australian soybean fields. This insect's life cycle requires two host plants. The aphid survives as eggs during winter on the twigs of buckthorn species (*Rhamnus* sp.). In spring, aphids undergo two generations as wingless females on the buckthorn and the third generation migrates to soybean plants. Up to 18 generations a year are observed on soybeans (Wang *et al.*, 1962).

DAMAGE. This species is a very serious pest especially in organically farmed soybean. The estimated yield loss can reach more than 40% (sometimes 58%) when high densities of aphids are observed. This sap-sucking species provokes: (i) plant stunting; (ii) distorted foliage; (iii) premature defoliation; (iv) stunted stems and leaves; (v) reduced branch, pod and seed numbers; (vi) lower seed weight; and (vii) underdevelopment of root tissue. Furthermore, detrimental sooty moulds develop on honeydew produced by this aphid. Finally, this species can be a vector of alfalfa (lucerne) mosaic virus, soybean mosaic virus and bean yellow mosaic virus

(Li and Pu, 1991; Venette and Ragsdale, 2004; Tilmon *et al.*, 2011).

HOW TO CONTROL?

Thresholds and control. The threshold is 250 aphids per plant from R1 through to mid-seed set (R5.5). However, in organically farmed soybean some authors recommend applying an organic insecticide, such as pyrethrin for a density of 100 aphids per plant. This species has developed resistance to insecticides (Ragsdale *et al.*, 2007; Hodgson *et al.*, 2012). Application of botanical insecticides such as neem, insecticidal soap, PyGanic and tea compost does not show great efficiency (Coulter *et al.*, 2010). Rozeboom *et al.* (2014) and Hesler (2013) showed that an application of Milstop® (85% potassium bicarbonate), a broad-spectrum foliar fungicide, reduced aphid populations from 24% to 46% and increased soybean yield by 48% compared with water controls. Pecinovsky and Lang (2011) showed that sugar + peroxide treatment reduced aphid populations by approximately 28% and neem oil reduced aphid populations by about 58%.

A sex pheromone produced by females was identified (nepetalactol); it attracts males in the autumn. This pheromone might be used for attracting aphids when they are dispersing to the winter host plant (*Rhamnus* sp.) (Zhu *et al.*, 2006).

Biological control. In Asia, where *A. glycines* is originally from, natural enemies have a much greater impact on this pest than in areas where it has been introduced. However, some naturally occurring generalist biological control agents can affect aphid densities. In Australia, ladybird beetles and hoverfly larvae are reported to significantly impact this pest. In the USA, the important predators of *A. glycines* are: (i) ladybird beetle larvae (*Harmonia axyridis* Pallas, *Coccinella septempunctata* (L.)); (ii) lacewing larvae (*Chrysoperla* spp. and *Hemerobius* spp.); (iii) predatory bugs (*Orius insidious* (Say), *Nabis* spp., *Geocoris* spp., *Podisus maculiventris*); (iv) hoverflies (Syrphidae spp.); and (v) the aphid midge (*Aphidoletes aphidimyza* (Rondani)). Predators belonging to the Carabidae family (*Elaphropus anceps* (Le Conte), *Clavina impressefrons* Le Conte, *Bembidion quadrimaculatum* Say) and spiders (Salticidae and Lycosidae) are also reported (Fox *et al.*, 2004; Rutledge *et al.*, 2004; Tilmon *et al.*, 2011). Seven species of parasitoid have been recorded in North America but with low impact. However, over the last few years, an augmentation of the Asian native parasitoid *Aphelinus certus* Yasnosh (accidently introduced in North America) was observed; it is reported to be the main parasitoid species in soybean (Brodeur, 2013). Entomopathogens have also been recorded: infestation rates of 84% by *Pandora neoaphidis* (R. & H) Humber (Zygomycetes: Entomophtorales) were observed on high aphid densities (Nielsen and Hajek, 2005). Some authors question the indirect impact of fungicides on these natural enemies explaining the pullulations of aphids in some cases (Ragsdale *et al.*, 2006).

Natural enemy releases have also been carried out. The parasitoid *Aphelinus atriplicis* Kurdjumov was introduced in the USA from Europe in 2002 but has not been recovered since then (Heimpel *et al.*, 2004). A parasitoid from China, *Binodoxys communis* (Gahan), was introduced in 2007 in the USA but this introduction did not lead to successful results either (Ragsdale *et al.*, 2011). In an ongoing programme another parasitoid that originated from China was released in 2014 in the USA, *Aphelinus glycinis* Hopper and Woolley. Another species, *Aphelinus rhamni* Hopper & Woolley, also originated from Asia and is presently being studied (USDA ARS, 2016).

Some studies focus on the effect of landscape on populations of *A. glycines* and its natural enemies. Greater landscape diversification, especially perennial vegetation, favours the occurrence of natural enemies such as ladybird beetles (Gardiner *et al.*, 2009a, b; Noma *et al.*, 2010). By contrast, landscape dominated by soybean and maize is particularly favourable to *A. glycines*.

Finally, methyl salicylate lures were used within organic soybean fields to attract natural enemies, mimicking plants wounded by phytophagous species. Attraction of syrphid flies (Diptera: Syrphidae) and green lacewings (Neuroptera: Chrysopidae) was observed and an abundance of soybean aphids was significantly reduced (Mallinger *et al.*, 2011).

Plant resistance. Some resistant soybean cultivars exist and have been commercialized since 2009 in the USA. Resistant plants reduce survival and offspring numbers of the aphid pest and are also repellent. Several genes have been identified. The tolerant plant varieties contain the gene *Rag1* (resistance *Aphis glycines*). The resistant plants are not always aphid free and some economic injuries can occur. Several other resistant genes have been identified and implementation of multigenic resistance would reduce the possibility that aphids adapt to the resistant plants (Zhang *et al.*, 2009; Tilmon *et al.*, 2011; Hodgson *et al.*, 2012).

Intercropping. Intercropping with flowering crops such as buckwheat (*Fagopyrum esculentum*) can enhance biological control by attracting natural enemies (Woltz *et al.*, 2012). Schmidt *et al.* (2007b) showed that lucerne cover crops co-planted with soybean increase colonization by natural enemies. It is also recommended that natural grass or woodland areas are maintained to attract beneficial predators of soybean aphids (Landis *et al.*, 2008). One study provides evidence that a winter rye cover crop can impact pest populations in soybeans in the USA (Heidel, 2011). Similarly, in China, Indonesia and Canada, intercropping soybean with maize, *Zea mays* L., reduces populations of *A. glycines* (Wu *et al.*, 2004; Hasibuan and Lumbanraja, 2012; Koch *et al.*, 2012). It also seems that adjacent bands of maize would decrease the visual and olfactory attraction of soybean fields (Wang and Ba, 1998). However, Noma *et al.* (2010) showed that habitats dominated by maize and soybean were associated with greater soybean aphid populations.

Cultural practices. Aphids are affected by soybean age. Most studies show that a late planting date positively affects aphid populations as colonization will occur on young plants that are more susceptible to this pest (Myers *et al.*, 2005b; Hodgson *et al.*, 2012). However, other studies do not show the effect of planting date on aphid populations (Rutledge and O'Neil, 2006). Potassium deficiency is also reported to favour aphid pullulations (Myers *et al.*, 2005a; Myers and Gratton, 2006; Walter and DiFonzo, 2007). However, management of these practices cannot limit *A. glycines* populations alone.

Species of the family Thripidae

DISTRIBUTION. Several species of thrips cause damage in soybean crops all over the world. In Brazil, the two mains species are *Frankliniella schultzei* (Trybom) and *Caliothrips brasiliensis* (Morgan) whereas in the USA the most frequent species are *Frankliniella fusca* (Hinds) and *Sericothrips variabilis* Beach. Sometimes, the species *Frankliniella intonsa* (Trybom) (in Europe and Asia) and *Scirtothrips dorsalis* Hood (in America and Asia) are reported on soybean (Highley and Boethel, 1994; Hoffmann-Campo *et al.*, 2003).

DAMAGE. Thrips make tiny, linear, pale-coloured scars on soybean leaves. Soybean is susceptible early in the growing season from growth stages VE to V6. In Brazil the main damage caused is due to transmission of tobacco streak virus (Hoffmann-Campo *et al.*, 2003). In the USA, *S. variabilis* was shown to transmit soybean vein necrosis virus (Bloomingdale *et al.*, 2008).

HOW TO CONTROL? There is not much work on thrips management on soybean. Thresholds for applying organic pesticide (e.g. spinosad) correspond to 75% of damaged trifoliolates and when more than eight thrips per leaf are observed (Bethke *et al.*, 2014). Peres *et al.* (2009) suggest that *Tagetes patula* can be used as a trap crop as this plant attracts several thrips species including those damaging soybean. Natural predators reported are insects of the genus *Orius* and mites of the family Phytoseiidae (Casey and O'Neil, 2008). In Brazil, cards containing eggs of *Orius* sp. are released.

Species of the family Aleyrodidae

DISTRIBUTION. One species, *Bemisia tabaci* (Gennadius), of this family attacks soybean. This species is distributed worldwide and is considered to be one of the main crop pests in the world. Its damage to soybean has been more important since the introduction of the biotype B in Brazil in 1991 (Lourenção and Nagai, 1994; Hoffmann-Campo *et al.*, 2003; Lima and Lara, 2004). This species is highy polyphagous and is reported from more than 500 host plants.

DAMAGE. *B. tabaci* transmits carlavirus in Puerto Rico, which is a serious disease of soybean nurseries (Belay *et al.*, 2012). *B. tabaci* causes leaf discoloration and development of sooty mould on the honeydew produced that limits plant photosynthesis.

HOW TO CONTROL?

Plant resistance. Various studies have shown that some soybean cultivars are somewhat tolerant of *B. tabaci* (Do Valle and Lourençao, 2002; Vieira *et al.*, 2011; Da Silva *et al.*, 2012; Zhang *et al.*, 2013). In Brazil, the most resistant genotypes were 'Barreiras', 'IAC 17' and 'IAC 19' (Vieira *et al.*, 2011). Ferreira *et al.* (2011) showed that silicon applications combined with cultivar 'IAC-19' significantly decrease *B. tabaci* populations.

Organic insecticides. In Thailand, Pangnakorn *et al.* (2010) report a repellent effect of wood vinegar. Carvalho *et al.* (2012) showed a significant effect of neem oil. Finally, Costa *et al.* (2010) tested various vegetable oils and emphasized that Agrex (soya oil) had the highest toxicity.

Biological control. Natural enemies have been observed, especially parasitoid species of the Aphelinidae family. The most common parasitoids in Florida are *Encarsia pergandiella* (Howard), *Encarsia sophia* (Girault & Dodd) (synonym of *E. transvena* (Timberlake)) and *Encarsia nigricephala* Dozier. *Encarsia formosa* (Gahan), the most used parasitoid to control *B. tabaci* throughout the world, showed laboratory efficiency on *B. tabaci* reared on soybean (Takahashi *et al.*, 2008). In Brazil, natural parasitism was observed. Parasitism rates were low at the beginning of crop growth but then increased reaching 49%, suggesting that natural parasitism can contribute to the control of *B. tabaci* (Souza *et al.*, 2014).

Species of the family Plataspidae

DISTRIBUTION. One species, *Megacopta cribraria* (Fabricius), of this family attacks soybean. *M. cribraria* originated from Asia (Japan) and was first observed in the USA in 2009. It is present in Asia and Australia (Eger *et al.*, 2010). This species is mainly reported on plants of the family *Fabaceae* (Zhang *et al.*, 2012). This species is suspected to be a synonym of *Megacopta punctatissima* (Montandon) (Ruberson *et al.*, 2013).

DAMAGE. Adults and nymphs feed on the soybean stem, while small nymphs can feed on leaf veins. Yield losses can sometimes be significant (47% in Georgia, USA) and this species is considered as a serious threat. In its native area, this species has little economic impact.

HOW TO CONTROL?

Threshold and organic compounds. A threshold of one nymph-stage bug per sweep (i.e. 15 nymphs per 15-sweep sample) is recommended and pyrethrin can be used if the threshold is passed. A single insecticide application targeting nymphs is usually sufficient (Seiter *et al.*, 2015).

Biological control. Natural occurrence of various natural enemies is reported in the USA, but efficiency is limited. Predators of nymphs (*Geocoris uliginosus* (Say), *Zelus* sp., *Hippodamia convergens* Guerin-Meneville, *Chrysoperla rufilabris* (Burmeister)) and of adults (*Euthyrhynchus floridanus* (L.)) have been observed. Parasitoids have also been reported but again with low efficiency (i.e. tachinid *Phasia robertsonii* (Townsend)) (Ruberson *et al.*, 2013). Research is presently focusing on the introduction of a natural enemy that originated from Asia, *Paratelenomus saccharalis* (Dodd) (host specific, climate matching) that has been released in several of the US states (Gardner *et al.*, 2013).

Beauveria bassiana is naturally occurring and some laboratory tests showed a 60–80% mortality of adults and nymphs (Borah and Dutta, 2002; Borah and Sarma, 2009; Ruberson *et al.*, 2013).

Cultural practices. Whitehouse and Zehnder (2015) showed that used coffee grounds (UCG), when applied as a 20% UCG mixture, decreases *M. cribraria* preference for soybeans. However, UCG do not deter *M. cribraria* when applied as a mulch. Seiter *et al.* (2013) also showed that infestations usually occur near field edges and that late planting can limit *M. cribraria* attacks.

Species of the family Membracidae

DISTRIBUTION. One species, *Spissistilus festinus* (Say), of this family causes damage on soybean. It is reported from the USA and Mexico on soybean. It is a polyphagous species able to feed on lucerne, grasses, cowpeas, melon and tomatoes (Medal *et al.*, 1995).

DAMAGE. *S. festinus* is an occasional pest of soybean. Nymphs and adults girdle plant stems and petioles resulting in punctures that circle the stem or petiole. Feeding in later plant stages may cause aborted pods and reduction in seed weight (Highley and Boethel, 1994).

HOW TO CONTROL?

Thresholds. Treatment is recommended when at least 10% plants less than 25–30 cm (10–12 inches) tall are infested. In some US states treatment is recommended when 50% of seedling plants are girdled. For plants setting pods, a treatment threshold of one hopper per sweep (100 sweeps sampled) is recommended (Highley and Boethel, 1994). Ramsey *et al.* (2013) report that treatment should be applied when one or two *S. festinus* per sweep are observed.

Biological control. Medal *et al.* (1995) suggest that *S. festinus* nymphs can be predated by *Geocoris punctipes* (Say) and *Nabis roseipennis* (Reuter) in the USA. Roach (1987) also reported natural predation by the spider *Phidippus audax.* Nickerson *et al.* (1977) report that ants in soybean fields limit the predation of eggs of this pest.

Cultural practices. Infestation can be reduced by destroying weeds around fields and by seeding a little more heavily (to reduce the importance of the loss of a few plants) (Higley and Boethel, 1994).

Species of the mite families Tetranychidae and Tarsonemidae

DISTRIBUTION. Two main mite species are reported in soybean fields, the Tarsonemidae *Polyphagotarsonemus latus* (Banks) and the Tetranychidae *Tetranychus urticae* (Koch). Other Tetranychidae species such as *Mononychellus planki* McGregor and *Tetranychus ludeni* Zacher are also reported as soybean pests in Brazil (Poe, 1980).

DAMAGE. *P. latus* causes leaf discoloration (leaves turn maroon in colour). *T. urticae* causes white or yellow discoloration (stipples). Feeding by these pests causes damage to plant cells and results in yield loss through reduction in dry matter and grain production. In soybean, outbreaks are irregular and are usually associated with hot and dry conditions (Hoffmann-Campo *et al.*, 2000; De Freitas Bueno *et al.*, 2009).

HOW TO CONTROL?

Threshold and organic compounds. The recommended threshold for controlling *T. urticae* is 10–15% leaf discoloration from R1 (beginning to bloom) through to R5 (beginning to form seed) (Suekane *et al.*, 2012).

Many studies focus on the effect of some botanical compounds on *T. urticae*. For instance, El Moneim *et al.* (2012) showed that extracts of *Chamomilla recutita* had important effects on this species, followed by extracts of *Marjorana hortensis* and *Eucalyptus*. Ismail *et al.* (2011) showed that garlic oil caused a considerable reduction in oviposition. Shi *et al.* (2006) also reported the detrimental effects of *Kochia scoparia* extracts. Finally, Abramishvili and Chkhaidze (2013) reported a toxic effect (75%) of onion herbal extract.

Weed control. Spider mites refuge on weeds within and around field crops. Controlling weeds can thus limit mite pullulations (Suekane *et al.*, 2012).

Plant resistance. Many studies have focused on the evaluation of soybean varieties' resistance to *T. urticae* emphasizing that some of them showed some tolerance rates (Razmjou *et al.*, 2009; Dehghan *et al.*, 2009).

Biological control. The mite pathogenic fungus *Neozygites floridana* is the most effective natural enemy of *T. urticae* in soybean. It attacks all mite stages, and is specific to spider mites. However, its efficiency depends on temperatures and high relative humidity (90%) (Cullen and Schramm, 2009).

Several species of insects, mites and spiders prey on spider mites in soybean. Mites of the family Phytoseiidae are the most important pedators of *T. urticae* and *P. latus*. Even if augmentative releases of predatory mites are used in some crops, the high cost of this technique means that it is not used in soybean crops. However, these predators can naturally occur within soybean plots and in the neighbouring uncultivated environment. The most frequent species found in Brazil soybean agrosystems are: (i) *Neoseiulus idaeus* Denmark & Muma; (ii) *Neoseiulus californicus* (McGregor); (iii) *Typhlodromalus aripo* De Leon; (iv) *Neoseiulus anonymus* (Chant & Baker); (v) *Phytoseiulus fragariae* Denmark & Schicha; (vi) *P. macropilis* (Banks); (vii) *Proprioseiopsis cannaensis* (Muma); and (viii) *Galendromus annectens* (De Leon) (Guedes *et al.*, 2007; Roggia *et al.*, 2009; Rezende *et al.*, 2012; Reichert *et al.*, 2014). The fungus *N. floridana* was also quite frequently observed (Roggia *et al.*, 2009).

Stem borers

Only species of the family Cerambycidae are recorded as key stem borers.

Species of the family Cerambycidae

DISTRIBUTION. Two Cerambycidae species damage soybean. *Dectes texanus* LeConte is reported from North America. This species is polyphagous; it feeds on *Xanthium strumarium*, *Ambrosia artemisiifolia* and *Ambrosia trifida* (Michaud and Grant, 2005; Tindall *et al.*, 2010). *Obereopsis brevis* (Gahan) is also found in the literature under the name: *Oberea brevis*. It is only reported from India and is a polyphagous species (Gangrade and Singh, 2009).

DAMAGE. Adults of *D. texanus* deposit eggs in soybean petioles; the larvae then tunnel within the main stem. In soybean, yield losses due to *D. texanus* larvae are sporadic (estimated at 10%) (Richardson, 1975) although greater losses can result from the lodging of mature plants. However, this species is considered as a minor pest (Michaud and Grant, 2005). The larva of *Obereopsis brevis* feeds into the petiole and the stem causing the plant to fall (Kapoor *et al.*, 1972; Gangrade and Singh, 1975). It is a serious pest in soybean in India. It infests the plants both in the early and the late phases of growth (Gangrade and Singh, 2009).

HOW TO CONTROL? Chemical control of *D. texanus* is difficult and not efficient as the larvae are within the stem for a period of 10 months (Campbell and Duyn, 1977). For *O. brevis*, Ahirwar *et al.* (2015) and Netam *et al.* (2013) report density thresholds ranging between 0.3 and 1.0 buds per row and between 0.8 and 3.2 buds per row, respectively. Singh *et al.* (2014) report the use of a neem-leaf-based insecticide by soybean Indian producers. Tomar (2009) showed the toxicity of compounds extracted from seeds of *Abrus precatorius* seed and *Trigonella foenum-graceum* on the fourth larval stage of *Oberea brevis*.

Agronomic practices

- **Avoid favourable plants** – As sunflowers are more suitable hosts than soybean, planting soybean near sunflower may favour infestation of *D. texanus*. In the same way, crop rotation avoiding sunflower can reduce populations of this pest. Furthermore, good management of known weed hosts in and around the plot may reduce populations (Wright and Hunt, 2011).
- **Trap plants** – The suitability of sunflower can be used to develop trap

crops. Sunflower is strongly preferred over soybean for oviposition. Prevention of soybean infestation has been observed by planting rows of cultivated sunflowers around the field and a reduction of 65% of the population of *D. texanus* has been observed (Michaud *et al.*, 2007).

- **Tillage** – Autumn tillage is reported to significantly reduce overwintering larvae of *D. texanus*, but stalks must be buried to a depth of at least 10–15 cm (4–6 inches). Research in North Carolina has shown that burying stubble 5–6.5 cm (2–3 inches) deep in the autumn by discing or bedding can reduce winter survival by as much as 60–70% (Johnson, 2011).
- **Soil environment** – Wet soils seemed to be less favourable than well-drained soils; thus clay soils usually limit infestations. Stubble in low, wet areas led to 50–70% larval mortality, while well-drained areas experienced only 11–38% mortality (Johnson, 2011).
- **Cultural practices** – Several cultural practices are recommended to limit populations of *Obereopsis brevis*, such as: (i) deep summer ploughing; (ii) planting at the beginning of the monsoon; (iii) avoiding intercropping with maize and sorghum; and (iv) avoiding excess use of nitrogenous fertilizers (Chaudhary *et al.*, 2012). Thooyavathy *et al.* (2013) also report the use of healthy seeds, light traps and spraying cow's urine and cow dung solution.

Plant resistance. No soybean-resistant cultivars to *D. texanus* exist (Niide *et al.*, 2012). However, field observations suggest that early planted, short-season varieties might decrease lodging damage (Johnson, 2011).

Biological control. Several natural enemies of *D. texanus* larvae have been reported on soybean such as pteromalid parasitoids and a tachinid fly, *Zelia tricolor* (Coquillett) (Tindall and Fothergill, 2010). Tindall and Fothergill (2012) observed that *Dolichomitus irritator* (Ichneumonidae) parasitizes *D. texanus* larvae within soybeans, however, the parasitism rates were very low (1%).

Seed development and germination

Only one species in the family Anthomyiidae is recorded to affect seed development and germination in soybean.

Species of the family Anthomyiidae

DISTRIBUTION. *Delia platura* (Meigen) is a widespread species, introduced in the USA in 1865 (Weston and Miller, 1989; Griffiths, 1991). It is a polyphagous species feeding on more than 40 host plants (Ristich, 1950; Kessing and Mau, 1991). It is an important pest of germinating soybeans and maize (Funderburk *et al.*, 1983; Bessin, 2004).

DAMAGE. *D. platura* is an early season pest of soybean. Larvae consume seeds and cotyledons (Higley and Boethel, 1994).

HOW TO CONTROL?

Cultural practices. Some studies show that no-till fields are less attractive to females and thus less damage is observed (Pope, 1998). Hammond (1990) observed high densities of the pest when cover crops were incorporated in the soil (lucerne followed by *Secale cereale* (L.), soybean residue and maize). It seems that more *D. platura* females were more attracted when a live green cover was incorporated than with dead crop residue. Furthermore the author also showed that more adults were collected when a legume was incorporated compared with a grass. When no-tillage planting methods were used, no enhancement of populations was observed. Bessin (2004) reported preventative measures for controlling *D. platura* such as: (i) late planting; (ii) shallow planting; (iii) higher seeding rates; (iv) a well-prepared seedbed; and (v) turning the cover crops early. Female attractiveness can also be reduced by removal of nearby food sources like flowering weeds and sweet-smelling substances from surrounding areas (Kessing and Mau, 1991). Tae Heung (1992) showed that transparent and black plastic mulchings reduce attack by *D. platura* larvae. Finally, high organic matter content in the soil enhanced the speed of emergence resulting in less larval damage.

Biological control. The main part of the life cycle of *D. platura* occurs in the soil; thus the number of reported natural enemies is quite low. Isolated incidences of predation by spiders, ants and birds upon adults and of fungal diseases infecting larvae have been reported. However, none are significant in controlling the pest (Reid, 1940). Jaramillo *et al.* (2013) studied the effect of seven species of entomopathogenic nematodes in Colombia and showed a high susceptibility of *D. platura* to one species of *Steinernema* sp.

Leaf miners

Species of the family Chrysomelidae

DISTRIBUTION. *Odontota horni* Smith is only reported from North America. It is a polyphagous species.

DAMAGE. *O. horni* rarely causes economic damage. Larvae and adults feed on leaves. Larval mines are generally 1.4–4.8 cm^2 (Buntin, 1994). Adults eat through the leaves avoiding the main leaf veins, which results in a skeletonized appearance (Buntin, 1994). This species can transmit bean pod mottle virus. The species *Odontota dorsalis* (Thunberg) is also sometimes reported to feed on soybeans but the damage caused is usually minor (Werner *et al.*, 2003).

HOW TO CONTROL? There is not much literature on control management of this species. Some economic thresholds are reported at the seedling stage: (i) nine adults/row-ft (29 adults/row-m); and (ii) at the development stage V6 to maturity, 125 adults/row-ft (410 adults/row-m).

Species of the family Agromyzidae

DISTRIBUTION. The most important species of Agromyzidae causing damage in soybean are *Melanagromyza sojae* Zehntner, *Ophiomyia phaseoli* (Tryon) (in Asia, Australia and Africa), *Melanagromyza dolichostigma* De Meijere and *Ophiomyia centrosomatis* De Meijere (in Asia). The four species only feed on *Fabaceae* plants. Among them, the most damaging are *O. phaseoli* and *M. sojae* (Van den Berg *et al.*, 1998b).

DAMAGE. Larvae mine leaves and stems. Attacks on cotyledons are reported in Indonesia for *O. phaseoli*. Larval mines on leaves are initially silver white and then turn brown. The critical period is during 3–5 weeks after germination. Furthermore, damage is usually more serious in the dry season than in the rainy wet season (Van den Berg *et al.*, 1995, 1998b).

HOW TO CONTROL?

Threshold and application of biopesticides. There is not much information on thresholds for the control of these pests in soybean. One Chinese publication shows that pesticide should be used for *M. sojae* when four to ten insects are collected on 100 plants for the third generation and eight to 12 for the fourth one (Pan, 1996). Several studies report the efficiency of neem oil but also of 10% cow urine, and cow dung ash on *O. phaseoli*. Spraying young plants, at egg-laying time, with soap solutions or organic oil can reduce the larval hatching of *M. dolichostigma*. The effect of neem oil has been also studied (Schmutterer, 1984).

Plant resistance. Several publications report studies of soybean varieties that are tolerant of these pests. Chiang and Talekar (1980) found wild *Glycine soja* accessions to be highly resistant to *M. sojae*, these cultivars being characterized by very thin stems. Talekar and Tengkano (1993) also identified sources of resistance to *O. phaseoli* in soybean. Resistance seems to be positively correlated with trichome density and concentrations of polyphenols in the stem (Chiang, 1984; Chiang and Norris, 1985).

Biological control. Talekar and Chen (1985) provide a synthesis of the parasitoid species of *M. sojae*, *O. phaseoli*, *O. centrosematis* and *M. dolichostigma* in Indonesia, the Philippines, Taiwan, Hawaii, Madagascar and Mauritius. Although parasitoid species naturally occur in Asia, where *M. sojae* is originally from, parasitism rates rarely exceed 50%.

Furthermore, high parastism rates are observed late in the season, when damage has already occurred. In India, Jayappa *et al.* (2002) recorded many natural enemies of *M. sojae* (*Cynipoidea* sp., *Sphegigaster* sp., *Eurytoma melanagromyzae* Narendran, *Syntomopus carinatus* Sureshan & Narendran and *Aneuropria kairali* Rajmohana & Narendran) with low parasitism rates (3–20%). However, Van den Berg *et al.* (1995) observed in Indonesian soybean fields that seven hymenopterous parasitoids have a high impact on *M. sojae*. The eucoilid *Gronotoma* sp. was the prevalent species during the early and mid-season; and a complex of pteromalids became dominant towards the end of the season. In India, *Chlorocytus* sp. was recorded parasitizing *O. phaseoli* on soybean (Kundu, 1985). In Indonesia, Djuwarso *et al.* (1992) (cited in Shepard and Barrion, 1998) reported various natural enemies of *M. sojae* and *O. phaseoli* (*Cynipoide* sp., *Eurytomapoloni* (Hymenoptera: Braconidae), *Trigongastra agromyza* Dodd (Hymenoptera: Braconidae), *Eurytoma* sp. and *Secondella* sp.). In Africa Abate (1991) reports 17 species of parasitoids; among these the pteromalids *Sphegigaster stepicola* Bouc. and *Sphegigaster brunneicornis* (Ferr.) were the most common.

In Hawaii, where *O. phaseoli* was accidentally introduced in 1968, *Opius phaseoli* Fischer and *Opius importatus* Fischer were introduced from East Africa in 1969 (Davis, 1971). The installation of these two natural enemies is controversial (Greathead, 1975; Raros, 1975). These two parasitoids were also introduced into Brunei, but their impact on *O. phaseoli* populations has not been assessed.

Agronomic practices. Some studies show that agronomic practices ensuring good plant growth (e.g. soil fertility) will limit the damage caused by members of the family Agromyzidae. It seems that ridging the crop 2–3 weeks after soybean germination would reduce plant mortality caused by *O. phaseoli* (Van der Goot, 1930). Furthermore, some studies show that a rice straw mulch limits plant access to *O. phaseoli* females and thus egg laying in Indonesia (Van der Groot, 1930). However, this technique seems to be only efficient for the *O. phaseoli* strains able to attack cotyledons. For other species of Agromyzidae that cause damage on leaves, this technique would not be sufficient to hide the soybean from females. Van der Groot (1930) also showed negative effects of co-planting soybean with aubergine on *O. phaseoli*. He suggested that the shade caused by the companon plants limits the access of females to small soybean plants. However, intercropping with 60 crop plants belonging to 14 botanical families failed to protect soybean from *M. sojae* in Taiwan (AVRDC, 1981a, b).

Species of the family Gelechidae

DISTRIBUTION. *Aproaerema modicella* Deventer occurs in Asia and Africa (introduced in South Africa at the begining of the 2000s). This species feeds almost exclusively on *Fabaceae* plants (Shanower *et al.*, 1993a, b).

DAMAGE. *A. modicella* is a serious pest of soybean in Asia. Larvae cause small, blister-like mines on leaves near the midrib; affected leaves turn brown, rolled, shrivelled and dried. The feeding activity of a single larva can lead to 179.3 mm^2 to 34.8 cm^2 of leaf tissue destruction (Shanower *et al.*, 1993a, b).

HOW TO CONTROL?

Thresholds and bioinsecticides. There is no information on the threshold level of this species. A single study carried out in India reports that one larva per five plants would justify the application of pesticides 30–45 days after sowing (Kalayanasundaram and Murugesan, 1989). Joshi and Patel (2011) show the efficiency of neem oil, spinosad and *Beauveria bassiana* formulations. Shirale *et al.* (2010) also report the efficiciency of *Bt* and *B. bassiana*.

Plant resistance. Many studies focus on the evaluation of soybean cultivars that are tolerant of this pest and some of them analysed resistance mechanisms (Mundhe, 1980; Shetgar and Thombre, 1984; Taware *et al.*, 2001; Ambenagare *et al.*, 2011). The lowest incidence of *A. modicella* was observed in genotypes having thick, dark green leaves.

Several soybean cultivars are proposed for their resistance to this pest.

Biological control. Shanower and Ranga (2010) observed larvae of *Chlaenius* sp. eating larvae of *A. modicella* inside soybean leaf mines in India. Shetgar and Thombre (1984) observed that this pest can be parasitized by *Stenomesioideus ashmeadi* Subba Rao and Sharma and species of *Microchelonus*, *Apanteles* and *Goniozus*, with parasitism rates ranging from 4% to 84%. In Asia, more than 30 parasitoid species have been recorded with mean parasitism rates ranging between 20% and 50%, and some peak parasitism rates reaching 53–91%. Predators and pathogens have been poorly studied (Shanower *et al.*, 1992; Kenis and Cugala, 2006).

Agronomic practices. Mulching with rice straw had no effect on leafminer levels but had a positive effect on parasitism levels (Logiswaran and Mohanasundaram, 1985). The effect of sowing date is controversial even if poorly studied (Lewin *et al.*, 1979; Logiswaran *et al.*, 1982). Plants that are drought stressed are much more susceptible to leafminer attack than irrigated plants. Thus good irrigation management would limit damage caused by this pest.

Root feeders

Species of the family Scarabaeidae

DISTRIBUTION. Various species of the genus *Phyllophaga* are reported as soybean pests. In the USA, the most frequent are *Phyllophaga implicita* (Horn), *Phyllophaga rugosa* (Melsheimer) and *Phyllophaga congrua* (LeConte) whereas in Brazil the most frequent species are *Phyllophaga cuyabana* (Moser) and *Phyllophaga capillata* (Blanchard). All of them are very polyphagous; adults feed on foliage of trees and herbaceous plants, as well as on flowers and fruits. Larvae (white grubs) feed especially on the roots of fibrous-rooted plants (Hammond, 1944). The life cycle of these species spans 3–4 years with a long larval stage causing damage to roots (Turnipseed and Kogan, 1976). Sometimes, *Cyclocephala lurida* Bland (Coleoptera: Scarabaeidae) and *Lyogenis suturalis* are also reported in soybean in the USA and Brazil, respectively. However, both are occasional pests. In India, damage caused by *Phyllophaga serrata* (F.) is reported and in China species of the genera *Holotricha*, *Anomela* and *Maladera* have been observed.

DAMAGE. Newly hatched larvae first feed on organic matter, then on crop roots. The larvae feed on seedling roots, but rarely cause significant losses. Larval feeding can cause plant death especially during germination and the second and third larvae instars are the most harmful (Lentz, 1985). Infestations by *Phyllophaga* spp. are reported to be more common in light, sandy soils (that are well drained) than in poorly drained, heavy clay soils. Plant attacks can be detected because of yellow coloration and poor development (Oliveira *et al.*, 1996).

HOW TO CONTROL? Due to its subterranean habits, the management of the white grub is difficult. Some authors report that the presence of these species can even be beneficial for the soil because of the breeding galleries that favour water infiltaration, root development and straw incorporation (Oliveira, 1997).

Crop rotation. One important control strategy for this pest is crop rotation, using alternative host plants (Oliveira *et al.*, 2007) attractive to adults but unsuitable to larvae. Some plants such as cotton, *Crotalaria spectabilis* and to a lesser extent *Crotalaria juncea* are known to cause larval mortality of *P. cuyabana*, in its early developmental stages (Oliveira, 1997). By contrast, the rotation of soybean, maize and sunflower is not recommended.

Management of field borders. A tactic to prevent white grub injury is to remove the shelter belts where the adults feed and mate. Glogoza *et al.* (1998) observed that high infestation is due to the abundance of *Salix* spp. (preferred food plants of *Phyllophaga* sp. adults) around the field. Fields that are not bordered by trees do not contain *P. implicita*. It is also possible to plant trees that the beetles do not prefer such as evergreens (Garcia *et al.*, 2003).

Soil management. Disc ploughing can mechanically kill larvae. Furthermore, spring tillage can reduce larval populations by removing them near the surface, exposing them to radiation from the sun and natural enemies such as birds (Oliveira, 1997; Oliveira *et al.*, 2000). In Brazil, authors recommend that farmers avoid the formation of hardened soil layers and manage the soil acidity in order to favour root development (Oliveira *et al.*, 1997, 2000).

Tolerant cultivars. Cultivars with rapid root development are usually more tolerant of true white grubs. Inoculation of seeds with bacteria for N fixation (e.g. *Bradyrhizobium*) can favour root development and limit the harmfulness of these pest species (Oliveira *et al.*, 1997).

Biological control. There are few reports of natural enemies of these pests. Some specimens of *P. cuyabana* have been found in Brazil naturally infested by *Beauveria bassiana*, *Metarhizium anisopliae* and *Bacillus* sp. (Oliveira *et al.*, 2000).

Species of the family Cydnidae

DISTRIBUTION. Two main species are known to attack soybean especially in South America: *Atarsocoris brachiariae* (Becker) and *Scaptocoris castanea* (Perty). Both species are polyphagous (Hoffmann-Campo *et al.*, 2000).

DAMAGE. Adults and larvae feed by sucking the root sap, leading to poor plant development (Hoffmann-Campo *et al.*, 2000).

HOW TO CONTROL?

Agricultural practices. As for Scarabaeidae species, ploughing and soil disturbance limits populations of these pests.

Biological control. Several species of entomopathogens have been observed isolated in fields (species of the genera *Metarhizium*, *Beauveria* and *Paecilomyces*) and good results have been observed in laboratory experiments (Oliveira *et al.*, 2000). Amaral *et al.* (1996) and Oliveira *et al.* (2000) showed an interesting efficiency of *Metarhizium anisopliae* associated with organic matter. In laboratory conditions, Sartori *et al.* (2001) showed the entomopathogenic nematode *Steinernema carpocapsae* was highly efficient.

References

Abate, T. (1991) The bean fly, *Ophiomyia phaseoli* (Tryon) (Dipt., Agromyzidae), and its parasitoids in Ethiopia. *Journal of Applied Entomology* 111(1–5), 278–285.

Abdullah, M.D., Sarnthoy, O., Isichaikul, S. and Tantakom, S. (2001) Efficacy of cypermethrin, neem extract and *Bacillus thuringiensis* for controlling insect pests of vegetable soybean. *Kasetsart Journal [Natural Science]* 35, 14–22.

Abramishvili, T. and Chkhaidze, N. (2013) Efficacy of onion extract and two synthetic acaricides against *Tetranychus urticae* on soybean in Georgia. Working Group 'Integrated Control of Plant-feeding Mites'. In: Palevsky, E., Ridsdill-Smith, J., Weintraub, P., Gerson, U., Simoni, S., McMurtry, J. and Zemek, R. (eds) *Proceedings of the Third Working Group Meeting*, 13–16 September 2011, Český Krumlov, Czech Republic. International Organisation for Biological and Integrated Control (IOBC) West Palaearacti Regional Section (WPRS), Czech Republic, pp. 1–3. ISBN 978-92-9067-272-2[XIII + 142pp.

Agrawal, I.L. and Mall, S.B. (1988) Studies on the insecticidal and antifeedant activity of some plant extracts on Bihar hairy caterpillar, *Diacrisia obliqua* Walker (Lep.: Arctiidae). *Journal of Applied Entomology* 105, 529–532.

Ahirwar, R., Devi, P. and Gupta, R. (2015) Seasonal incidence of major insect-pests and their biocontrol agents of soybean crop (*Glycine max* L. Merrill). *Scientific Research and Essays* 12, 402–406.

Ahmed, S.R. and Bhattacharya, A.K. (1991) Growth inhibitory effect of some plants for *Spilosoma obliqua* Walker. *Indian Journal of Entomology*, 53(3), 453–474.

Aldrich, J.R., Hoffmann, M.P., Kochansky, J.P., Lusby, W.R., Eger, J.E. and Payne, J.A. (1991) Identification and attractiveness of a major pheromone component for Nearctic *Euschistus* spp. stink bugs (Heteroptera: Pentatomidae). *Environmental Entomology* 20, 477–483.

Aldrich, J.R., Khrimian, A. and Camp, M.J. (2007) Methyl 2,4,6-decatrienoates attract stink bugs and tachinid parasitoids. *Journal of Chemical Ecology* 33(4), 801–815.

Alexandre, T.M. (2010) Práticas de manejo de pragas utilizadas na soja e seu impacto sobre a cultura. Phd thesis, Universidade Federal do Paraná, Curitiba, Brazil, p. 104.

Alim, M.A. and Lim, U.T. (2011) Refrigerated eggs of *Riptortus pedestris* (Hemiptera: Alydidae) added to aggregation pheromone traps increase field parasitism in soybean. *Journal of Economic Entomology* 104(6), 1833–1839.

Altieri, M.A. and Letourneau, D.K. (1982) Vegetation management and biological control in agroecosystem. *Crop Protection* 1(4), 405–430.

Altieri, M.A., Todd, J.W., Hauser, E.W., Patterson, M., Buschman, G.A. and Walker, R.H. (1981) Some effects of weed management and row spacing on insect abundance in soybean fields. *Protection Ecology* 3(4), 339–343.

Amaral, J.L. do, Medeiros, M.O., Oliveira, C., Borges, V. and Souza, J.R. (1996) Efeito da Associação da matéria orgânica e do fungo *Metarhizium anisopliae* no controle do percevejo castanho das raízes '*Atarsocoris brachiariae* Becker. In: *Simpósio Sobre Recursos Naturais E Sócio-Econômicos Do Pantanal, 7.* Corumbá, 1996. EMBRAPA-SPI, Manejo e Conservação, Corumbá, Brazil, pp. 121–123.

Ambenagare, R.M., Shetgar, S.S. and Takankhar, V.G. (2011) Biochemical basis of resistance in soybean cultivars to leafminer *Aproaerema modicella* Deventer. *Indian Journal of Entomology* 73(1), 93–94.

Appleton, A.S.B., Gillard, C. and Schaafsma, A.W. (2003) Biology and management of the potato leafhopper, *Empoasca fabae* (Harris) (Homoptera: Cicadellidae) on field crops in Ontario. *Journal of the Entomological Society of Ontario* 134, 3–17.

Archer, T.L. and Bynum, E.D. Jr (1994) Corn earworm (Lepidoptera: Noctuidae) biology on food corn on the High Plains. *Environmental Entomology* 23, 343–348.

Asian Vegetable Research and Developement Center (AVRDC) (1981a) *AVRDC Progress Report 1979.* AVRDC, Shanhua, Taiwan, Republic of China, p. 107.

Asian Vegetable Research and Developement Center (AVRDC) (1981b) *AVRDC Progress Report 1980.* AVRDC, Shanhua, Taiwan, Republic of China, p. 110.

Atthowe, H., Gilkeson, L.A., Kite, P., Michalak, P., Pleasant, B., Reich, L. and Scheider, A.F. (2010) *The Organic Gardener's Handbook of Natural Pest and Disease Control: A Complete Guide to Maintaining a Healthy Garden and Yard the Earth-friendly Way.* Rodale Inc., Emmaus, Pennsylvania, 416 pp.

Bahlai, C.A., Xue, Y., McCreary, C.M., Schaasfma, A.W. and Hallet, R.H. (2010) Choosing organic pesticides over synthetic pesticides may not effectively mitigate environmental risk in soybeans. *PLoS One* 5(6), e1125.

Bastos Dequech, S.T., Sausen, C.D., Martins, J.D., Grendene Lima, C., Egewarth, R., Zumba, R.C. and Carús Guedes, J.V. (2006) Occurrence of parasitoids and pathogens on adults of Chrysomelidae (Coleoptera) in greenhouse, in Santa Maria, RS. *Ciência Rural* 36(6), 1912–1915.

Beach, R.M. and Todd, J.W. (1985) Parasitoids and pathogens of the soybean looper, *Pseudoplusia includens* (Walker), in South Georgia soybean. *Journal of Entomological Science* 20(3), 318–323.

Belay, D.K., Huckaba, R.M., Ramirez, A.M., Rodrigues, J.C.V. and Foster, J.E. (2012) Insecticidal control of *Bemisia tabaci* (Hemiptera: Aleyrodidae) transmitting carlavirus on soybeans and detection of the virus in alternate hosts. *Crop Protection* 35, 53–57.

Bessin, R. (2004) Corn: cool soils favor damage by some insect pests. *Kentucky Pest News* No.985.

Bethke, J.A., Dreistadt, S.H. and Varela, L.G. (2014) *Thrips: Integrated Pest Management for Home Gardeners and Landscape Professionals.* Pest notes. Statewide Integrated Pest Management Program, Publication 1429. University of California, Agriculture and Natural Resources, Davis, California.

Beyer, A.H. (1921) Garden flea-hopper in alfalfa and its control. *United States Department of Agriculture Bulletin* 964, 1–27.

Bloomingdale, C., Damon Smith, D. and Groves, R. (2008) Thrips dispersal and soybean vein necrosis virus (SVNV) in Wisconsin soybean. *Proceedings of the 2014 Wisconsin Crop Management Conference* 53, 95–98.

Borah, B.K. and Dutta, S.K. (2002) Entomogenous fungus, *Beauveria bassiana* (Balsamo) Vuillemin: a natural biocontrol agent against *Megacopta cribrarium* (Fab.). *Insect Environment* 8, 7–8.

Borah, B.K. and Sarma, K.K. (2009) Pathogenicity of entomopathogenous fungus, *Beauveria bassiana* (Balsamo) Vuillemin on *Megacopta cribrarium* (Fab.): a sucking pest of pigeonpea. *Insect Environment* 14, 159–160.

Boucias, D.G., Bradford, D.L. and Barfield, C.S. (1984) Susceptibility of the velvetbean caterpillar and soybean looper (Lepidoptera: Noctuidae) to *Nomuraea rileyi*: effects of pathotype, dosage, temperature and host age. *Environmental Entomology, Lanham* 77, 247–253.

Braasch, J., Wimp, G.M. and Kaplan, I. (2012) Testing for phytochemical synergism: arthropod community responses to induced plant volatile blends across crops. *Journal of Chemical Ecology* 38, 1264–1275.

Brodeur, J. (2013) *Aphis gycines* Matsumra, soybean aphid (Hemiptera: Aphididae). In: Mason, P.G. and Gillespie, D.R. (eds) *Biological Control Programmes in Canada 2001–2012*. CAB International, Wallingford, UK, pp. 93–98.

Browde, J.A., Pedigo, L.P., Degooyer, T.A., Higley, L.G., Wintersteen, W.K. and Zeiss, M.R. (1992) Comparison of sampling techniques for grasshoppers (Orthoptera, Acrididae) in soybean. *Journal of Economic Entomology* 85(6), 2270–2274.

Buckelew, L.D., Pedigo, L.P., Mero, H.M., Owen, M.D.K. and Tylka, G.L. (2000) Effects of weed management systems on canopy insects in herbicide-resistant soybeans. *Journal of Economic Entomology* 93, 1437–1443.

Bueno, R.C., Parra, J.R., Bueno, A.F. and Haddad, M.L. (2009) Desempenho de tricogramatídeos como potenciais agentes de controle de *Pseudoplusia includens* Walker (Lepidoptera, Noctuidae). *Neotropical Entomology* 38, 389–394.

Bueno, R.C., Bueno, A.F., Moscardi, F., Parra, J.R. and Hoffmann-Campo, C.B. (2010a) Lepidopteran larva consumption of soybean foliage: basis for developing multiple-species economic thresholds for pest management decisions. *Pest Management Science* 67(2), 170–174.

Bueno, R.C., Carneiro, T.R., Bueno, A.F., Pratissolli, D., Fernandes, O.A. and Vieira, S.S. (2010b) Parasitism capacity of *Telenomus remus* Nixon (Hymenoptera: Scelionidae) on *Spodoptera frugiperda* (Smith) (Lepidoptera: Noctuidae) eggs. *Brazilian Archives of Biology and Technology* 53, 133–139.

Bueno, R.C., Parra, J.R. and Bueno, A. (2012) *Trichogramma pretiosum* parasitism of *Pseudoplusia includens* and *Anticarsia gemmatalis* eggs at different temperatures. *Biological Control* 60, 154–162.

Buntin, G.D. (1994) Soybean leafminer. In: Higley, L.G. and Boethel, D.J. (eds) *Handbook of Soybean Insect Pests*. Entomological Society of America, Lanham, Maryland.

Buntin, G.D., Hargrove, W.L. and McCracken, D.V. (1995) Populations of foliage-inhabiting arthropods on soybean with reduced tillage and herbicide use. *Agronomy Journal* 87, 789–794.

Cabrera Walsh, G. (2004) Distribution, host specificity, and overwintering of *Celatoria bosqi* Blanchard (Diptera: Tachinidae), a South American parasitoid of *Diabrotica* spp. (Coleoptera: Chrysomelidae: Galerucinae). *Biological Control* 29, 427–434.

Cabrera Walsh, G. (2005) Diabroticina (Coleoptera: Chrysomelidae: Galerucinae) de la Argentina y el Cono Sur: una visión biogeográfica y evolutiva de su biología y la de sus enemigos naturales, en relación con la factibilidad del control biológico de las especies plagas. PhD acultad de Ciencias Exactas y Naturales Universidad de Buenos Aires, Argentina, 152 pp.

Cabrera Walsh, G., Athanas, M.M., Salles, L.A. and Schroder, R.F. (2003) Distribution, host range, and climatic constraints on *Centistes gasseni* (Hymenoptera: Braconidae), a South American parasitoid of cucumber beetles, *Diabrotica* spp. (Coleoptera: Chrysomelidae). *Bulletin of Entomological Research* 93(6), 561–567.

Campbell, W.B. and Van Duyn, J.W. (1977) Cultural and chemical control of *Dectes texanus texanus* on soybeans. *Journal of Economic Entomology* 70, 256–258.

Capinera, J.L. (2005) Garden fleahopper, *Halticus bractatus* (Say). Florida Cooperative Extension Services EENY-0078:1-4. Available at: http://edis.ifas.ufl.edu/pdffiles/IN/IN23500.pdf (accessed 25 June 2017).

Carner, G.R. (1980) Sampling pathogens of soybean insect pests. In: Kogan, M. and Herzog, D.C. (eds) *Sampling Methods In Soybean Entomology*. Springer, New York, pp. 559–574.

Carvalho, J.M., Pratissoli, D., Pin Dalvi, L., Silva, A.M., Oliveira de Freitas Bueno, R.C. and Freitas Bueno, A. (2014) Parasitism capacity of *Trichogramma pretiosum* on eggs of *Trichoplusia ni* at different temperatures. *Acta Scientiarum Agronomy* 36(4), 417–424.

Carvalho, S., Djair, V.J., Major, P.R. and Rossi, F.M. (2012) Efficiency of neem oil nanoformulations to *Bemisia tabaci* (GENN.) Biotype B (Hemiptera: Aleyrodidae). *Ciências Agrárias, Londrina* 33(1), 193–202.

Casey, D.B. and O'Neil, R.J. (2008) Voracity and prey preference of insidious flower bug (Hemiptera: Anthocoridae) for immature stages of soybean aphid (Hemiptera: Aphididae) and soybean thrips (Thysanoptera: Thripidae). *Environmental Entomology* 37(4), 964–972.

Chaudhary, H.R., Baldev, R., Meghwal, H.P. and Jadon, C. (2012) Evaluation of organic modules against insect pests of soybean. *Indian Journal of Entomology* 74(2), 163–166.

Chen, R.Z., Klein, M.G., Li, Q.Y. and Li, Y. (2014) Mass trapping *Popillia quadriguttata* using *Popillia japonica* (Coleoptera: Scarabaeidae) pheromone and floral lures in Northeastern China. *Environmental Entomology* 43(3), 774–781.

Chiang, H.S. (1984) Soybean resistance to stem-mining agromyzid beanflies (Diptera: Agromyzidae). *Dissertation Abstracts International, B (Sciences and Engineering)* 44(11), 3296B.
Chiang, H.S. and Norris, D.M. (1985) Expression and stability of soybean resistance to agromyzid beanflies. *International Journal of Tropical Insect Science* 6(3), 265–270.
Chiang, H.S. and Talekar, N.S. (1980) Identification of sources of resistance to the beanfly and two other agromyzid flies in soybean and mungbean. *Journal of Economic Entomology* 73, 197–199.
Consolo, V.F., Salerno, G.L. and Beron, C.M. (2003) Pathogenicity, formulation and storage of insect pathogenic Hyphomycetous fungi tested against *Diabrotica speciosa*. *BioControl* 48, 705–712.
Coombs, M. and Khan, S.A. (1998) Fecundity and longevity of green vegetable bug, *Nezara viridula*, following parasitism by *Trichopoda giacomellii*. *Biological Control* 12, 215–222.
Corrêa-Ferreira, B.S. (2003) *Soja Orgânica: Alternativas Para o Manejo dos Insetos-Pragas*. EMBRAPA-CNPSo, Londrina, Brazil, 83 p.
Corrêa-Ferreira, B.S. and Panizzi, A.R. (1999) Percevejos da soja e seu manejo. *Circular Técnica* 24. EMBRAPA-CNPSo, Londrina, Brazil, 45 pp.
Corrêa-Ferreira, B.S., Nunes, M.C. and Uguccioni, L.D. (1998) Ocorrência do parasitóide *Hexacladia smithii* Ashmead em adultos de *Euschistus heros* (F.) no Brasil. *An. Soc. Entomologia Brasileira* 27(3), 495–498.
Costa, G., Boiça Junior, A.L., Gonçalves Jesus, F. and Chagas Filho, N.R. (2010) Efeito do uso de óleos vegetais, associados ou não a inseticida, no controle de *Bemisia tabaci* (Gennadius, 1889) e *Thrips tabaci* (Lindeman, 1888), em feijoeiro, na época 'das águas'. *Bioscience Journal, Uberlândia* 26(1), 15–23.
Coulter, J., Moncada, K. and Sheaffer, C. (2010) Soybean production. In: Moncada, K.M. and Sheaffer, C. (eds) *Risk Management Guide for Organic Producers*. Arlene West Communications, St Paul, Minnesota, pp. 10.1–10.18.
Cullen, E. and Schramm, S. (2009) *Two-spotted Spider Mite Management in Soybean and Corn*. University of Wisconsin-Extension, Cooperative Extension, Madison, Wisconsin.
Cunningham, G.L. and Sampson, M.W. (2000) *Grasshopper Integrated Pest Management User Handbook. United States Department of Agriculture (USDA) Technical Bulletin* No. 1809. USDA, Washington, DC, 74 pp.
Da Silva, J.P., Lopes Baldin, E.L., Santana de Souza, E. and Lourenção, A.L. (2012) Assessing *Bemisia tabaci* (Genn.) biotype B resistance in soybean genotypes: antixenosis and antibiosis. *Chilean Journal of Agricultural Research* 72(4), 516–522.
Davis, C.J. (1971) Recent introductions for biological control in Hawaii. XVI. *Proceedings of the Hawaiian Entomological Society* 21, 59–62.
Day, W.H. and Saunders, L.B. (1990) Abundance of the garden leahopper (Hemiptera: Miridae) on alfalfa and parasitism by *Leiophron uniformis* (Gahan) (Hymenoptera: Braconidae). *Journal of Economic Entomology* 83, 101 106.
De Freitas Bueno, A., Oliveira de Freitas Bueno, R.C., Nabity, P.D., Higley, L.G. and Fernandes, O.A. (2009) Photosynthetic response of soybean to two-spotted spider mite (Acari: Tetranychydae) injury. *Brazilian Archives of Biology and Technology* 52(4), 825–834.
Dehghan, M.S., Allahyari, H., Saboori, S., Nowzari, J. and Naveh, V.H. (2009) Fitness of *Tetranychus urticae* Koch (Acari: Tetranychidae) on different soybean cultivars: biology and fertility life-tables. *International Journal of Acarology* 35(4), 341–347.
De Marco Salvadori, J., Schumacher Defferrari, M., Ligabue-Braun, R., Yamazaki Lau, E. and Salvadori, J.R. (2012) Characterization of entomopathogenic nematodes and symbiotic bacteria active against *Spodoptera frugiperda* (Lepidoptera: Noctuidae) and contribution of bacterial urease to the insecticidal effect. *Biological Control* 63(3), 253–263.
Deshmukh, P.D., Rathore, Y.S. and Bhattacharya, A.K. (1979) Larval survival of *Diacrisia obliqua* Walker on several plant species. *Indian Journal of Entomology* 41(1), 5–12.
Do Valle, G.E. and Lourençao, E.A.L. (2002) Resistência de genótipos de soja a *Bemisia tabaci* (Genn.) biótipo B (Hemiptera: Aleyrodidae). *Neotropical Entomology* 31(1), 285–295.
Dung, D.T., Hien Phuong, L.T. and Long, K.D. (2001) Insect parasitoid composition on soybean, some eco-biological characteristics of the parasitoid, *Xanthopimpla punctata* Fabricius on soybean leaffolder *Omiodes indicata* (Fabricius) in Hanoi, Vietnam. *Journal of International Society for Southeast Asian Agricultural Sciences* 17(2), 58–69.
Eger, J.E. Jr, Ames, L.M., Suiter, D.R., Jenkins, T.M., Rider, D.A. and Halbert, S.E. (2010) Occurrence of the Old World bug *Megacopta cribraria* (Fabricius) (Heteroptera: Plataspidae) in Georgia: a serious home invader and potential legume pest. *Insect Mundi* 0121, 1–11.

Elden, T.C. and Lambert, L. (1992) Mechanism of potato leafhopper resistance in soybean lines isogenic for pubescence type. *Crop Science* 32, 118–1191.

El Moneim, A.B., Fatma, S.A. and Turky, A.F. (2012) Control of *Tetranychus urticae* Koch by extracts of three essential oils of chamomile, marjoram and *Eucalyptus*. *Asian Pacific Journal of Tropical Biomedicine* 2(1), 24–30.

Erlandson, M. (2008) Insect pest control by viruses. In: Mahy B.W.J. and Van Regenmortel M.H.V (eds) *Encyclopedia of Virology*, 3rd edn. Elsevier, Oxford, pp. 125–133.

Fathipour, Y. and Naseri, B. (2011) Soybean cultivars affecting performance of *Helicoverpa armigera* (Lepidoptera: Noctuidae). In: Ng, T.B. (ed.) *Soybean – Biochemistry, Chemistry and Physiology*. ISBN: 978-953-307-219-7. Available at: http://www.intechopen.com/books/soybeanbiochemistry-chemistry-and-physiology/soybean-cultivars-affecting-performance-of-helicoverpa-armigeralepidoptera-Noctuidae (accessed 25 June 2017).

Fathipour, Y. and Sedariatan, A. (2013) Integrated management of *Helicoverpa armigera* in soybean cropping systems. In: El-Shemy, H.A. (ed.) *Soybean – Pest Resistance*, pp. 231–280. Available at: https://www.intechopen.com/books/soybean-pest-resistance/integrated-management-of-helicoverpa-armigera-in-soybean-cropping-systems (accessed 25 June 2017).

FAOSTAT (2012) Available at: http://faostat.fao.org/site/567/default.aspx#ancor (accessed 25 June 2017).

Fenton, F.A. and Hartzell, A. (1923) Bionomics and control of the potato leafhopper *Empoasca mali* Le Baron. *Iowa State Agriculture Experiment Station Research Bulletin* 78, 380–440.

Ferreira De Souza, T. and Favero, S. (2015) Avaliação de óleo essencial de *Eucalyptus urograndis* (Myrtaceae) no controle de Pentatomidae. *Revista Ciência Agronômica* 46(1), 216–222.

Ferreira, R.S., Moraes, J.C. and Antunes, C.S. (2011) Silicon influence on resistance induction against *Bemisia tabaci* biotype B (Genn.) (Hemiptera: Aleyrodidae) and on vegetative development in two soybean cultivars. *Neotropical Entomology* 40(4), 495–500.

Foerster, L.A. and Avanci, M.R.F. (1999) Egg parasitoids of *Anticarsia gemmatalis* Hübner (Lepidoptera: Noctuidae) in soybeans. *Anals da Societade de Entomologia Brasileira* 28(3), 545–548.

Fox, T.B., Landis, D.A., Cardoso, F.F. and DiFonzo, C.D. (2004) Predators suppress *A. glycines* Matsumura population growth in soybean. *Environmental Entomology* 33, 608–618.

Funderburk, J.E., Pedigo, L.P. and Berry, E.C. (1983) Seedcorn maggot (Diptera: Anthomyiidae) emergence in conventional and reduced-tillage soybean systems in Iowa. *Journal of Economic Entomology* 76, 131–134.

Funderburk, J.E., Boucias, D.G., Herzog, D.C., Sprenkel, R.K. and Lynch, R.E. (1984) Parasitoids and pathogens of larval lesser cornstalk borers (Lepidoptera: Pyralidae) in northern Florida. *Environmental Entomology* 13(5), 1319–1323.

Fuxa, J.E., Fuxa, J.R, Richter, A.R. and Weidner, E.H. (2000) Prevalence of a trypanosomatid in the southern green stink bug, *Nezara viridula*. *Journal of Eukaryotic Microbiology* 47, 388–394.

Gangrade, G.A. and Singh, O.P. (1975) Soybean plant response to the attack of *Oberea brevis* Swed. (Col., Cerambycidae). *Zeitschrift Fur Angewandte Entomologie* [*Journal of Applied Entomology*] 79, 285–290.

Gangrade, G.A. and Singh, O.P. (2009) Assessment of losses to soybean by *Oberea brevis* Swed. (Col., Cerambycidae). *Journal of Applied Entomology* 81, 26–30.

Garcia, M.A., Oliveira, L.J. and Oliveira, M.C.N. (2003) Aggregation behavior of *Phyllophaga cuyabana* (Moser) (Coleoptera: Melolonthidae): relationships between sites chosen for mating and offspring distribution. *Neotropical Entomology* 32(4), 537–542.

Gardiner, M.M., Landis, D.A., Gratton, C., DiFonzo, C.D., O'Neal, M.E., Chacon, J., Wayo, M.T., Schmidt, N.T., Mueller, E.E. and Heimpel, G.E. (2009a) Landscape diversity impacts biocontrol services in north-central US soybean. *Ecological Applications* 9, 143–154.

Gardiner, M.M., Landis, D.A., Gratton, C., Schmidt, N.P., O'Neal, M.E., Mueller, E., Chacon, J., Heimpel, G.E. and DiFonzo, S.D. (2009b) Landscape composition mediates coccinellid community structure. *Diversity and Distributions* 15, 554–564.

Gardner, W.A., Blount, J.L., Golec, J.R., Jones, W.A., Hu, X.P., Talamas, E.J., Evans, R.M., Dong, X., Ray, C.H. Jr, Buntin, G.D., Gerardo, N.M. and Couret, J. (2013) Discovery of *Paratelenomus saccharalis* (Dodd) (Hymenoptera: Platygastridae), an egg parasitoid of *Megacopta cribraria* F. (Hemiptera: Plataspidae) in its expanded North American range. *Journal of Entomological Science* 48, 355–359.

Gaugler, R. and Klein, M.G. (1998) Insect Parasitic Nematodes: Tools for Pest Management. Available at: http://www.cabi.org/isc/datasheet/43599 (accessed 25 June 2017).

Gill, H.K., McSorley, R., Goyal, G. and Webb, S.E. (2010) Mulch as a potential management strategy for lesser cornstalk borer, *Elasmopalpus lignosellus* (Insecta: Lepidoptera: Pyralidae), in bush bean (*Phaseolus vulgaris*). *Florida Entomologist* 93(2), 183–190.

Gilman, D.F., Mcpherson, R.M., Newsom, L.D., Herzog, D.C. and Williams, C. (1982) Resistance in soybeans to the southern green stink bug. *Crop Science* 22(3), 573–576.

Glogoza, P.A., Weiss, M.J. and Rao, M.B. (1998) Spatial distribution of *Phyllophaga implicita* (Horn) (Coleoptera: Scarabaeidae) larvae in relation to distance from the adult food source. *Journal of Economic Entomology* 91(2), 457–463.

Grant, J.F. and Shepard, M. (1984) Laboratory biology of *Meteorus autographae* (Hymenoptera: Braconidae), an indigenous parasitoid of soybean looper (Lepidoptera: Noctuidae) larvae. *Environmental Entomology* 13, 838–842.

Grant, J.F. and Shepard, M. (1986) Seasonal incidence of *Meteorus autographae* on soybean looper larvae on soybean in South Carolina, and the influence of the host density on parasitization. *Journal of Entomological Science* 21, 338–345.

Greathead, D.J. (1975) Biological control of the bean fly *Ophiomyia phaseoli* (Diptera: Agromyzidae) by *Opius* spp. (Hymenoptera: Braconidae) in the Hawaiian Islands. *Entomophaga* 20(3), 313–316.

Greene, G.L. and Shepard, M. (1974) Biological studies of a predator, *Sycanus indagator* II. Field survival and predation potential. *Florida Entomologist* 57(1), 33–38.

Griffiths, G.C.D. (1991) *Flies of the Nearctic Region*, Volume VIII, Part 2, Number 7. *Cyclorrhapha II (Schizophora: Calyptrate) Anthomyiidae*. E. Schweizerbart'sche Verlagsbuchhandlung (Nagele u. Obermiller), Stuttgart, Germany.

Guedes, J.V.C., Navia, D., Lofego, A.C. and Dequech, S.T.B. (2007) Ácaros associados à cultura da soja no Rio Grande do Sul, Brasil. *Neotropical Entomology* 36, 288–293.

Haile, F.J., Higley, L.G. and Specht, J.E. (1998) Soybean cultivars and insect defoliation: yield loss and economic injury levels. *Agronomy Journal* 90, 344–352.

Hammack, L., Pikul, J.L. and West, M.S. (2010) Phenology and abundance of bean leaf beetle (Coleoptera: Chrysomelidae) in eastern South Dakota on alfalfa and soybean relative to tillage, fertilization, and yield. *Environmental Entomology* 39, 727–737.

Hammond, G.H. (1944) The feeding habits of *Phyllophaga* larvae. *Annual Report of Entomological Society of Ontario* 71, 20.

Hammond, R.B. (1990) Influence of cover crops and tillage on seed corn maggot (Diptera: Anthomyiidae) populations in soybean. *Environmental Entomology* 19, 510–514.

Hammond, R.B. and Jeffers, D.L. (1990) Potato leafhopper (Homoptera: Cicadellidae) populations on soybean relay intercropped into winter wheat. *Environmental Entomology* 19, 1810–1819.

Hammond, R.B. and Stinner, B.R. (1987) Soybean foliage insects in conservation tillage systems: effects of tillage, previous cropping history and soil insecticide application. *Environmental Entomology* 16, 524–531.

Hariprasad, Y. and Kanaujia, K.R. (1992) Evidences of allelopathic chemicals in castor and maize against *Spilosoma oblique* Walk. *Proceedings of the First National Symposium on Allelopathy in Agro-ecosystems* 175–176.

Harsimran, K.G., McSorley, R., Goyal, G. and Webb, S.E. (2010) Mulch as a potential management strategy for lesser cornstalk borer, *Elasmopalpus lignosellus* (Insecta: Lepidoptera: Pyralidae), in bush bean (*Phaseolus vulgaris*). *Florida Entomologist* 93(2), 183–190.

Hasibuan, R. and Lumbanraja, J. (2012) The impact of soybean and corn intercropping system and soil fertility management on soybean aphid populations *Aphis glycines* (Hemiptera: Aphididae) and soybean growth performance. *Jurnal Hama dan Penyakit Tumbuhan Tropika* 12(1), 23–35.

Hassan, S. (1993) The mass rearing and utilization of *Trichogramma* to control lepidopterous pests: achievements and outlook. *Pesticide Science* 37, 387–391.

Heatherly, L.G. and Hodges, H.F. (1998) *Soybean Production in the Midsouth*. CRC Press, Boca Raton, Florida, 416 pp.

Heidel, T. (2011) Efficacy of intercropping in organic soybeans for management of the soybean aphid. MSc thesis, University of Minnesota, Saint Paul, Minnesota, p. 7.

Heimpel, G.E., Lee, J.C., Wu, Z.S., Weiser, L., Wackers, F. and Jervis, M.A. (2004) Gut sugar analysis in field-caught parasitoids: adapting methods originally developed for biting flies. *International Journal of Pest Management* 50, 193–198.

Hein, G.L. and Campbell, J.B. (2008) *A Guide to Grasshopper Control in Cropland*. Extension, a Division of the Institute of Agriculture and Natural Resources at the University of Nebraska–Lincoln. Available at: http://www.soybeanresearchinfo.com/pdf_docs/grasshoppers_NE_G1627.pdf (accessed 25 June 2017).

Heineck-Leonel, M.A. and Salles, L.A.B. (1997) Incidencia de parasitoides e patogenos em adultos de *Diabrotica speciosa* (Germ.) (Coleoptera: Chrysomelidae) na regiao de Pelotas, RS. *Annais da Societade Entomologica do Brasil* 26, 81–85.

Henry, T.J. and Wilson, M.R. (2004) First records of eleven true bugs (Hemiptera: Heteroptera) from the Galapágos Islands, with miscellaneous notes and corrections to published reports. *Journal of the New York Entomological Society* 112, 5–86.

Hesler, L. (2013) Efficacy of inorganic compounds against soybean aphid, laboraory tests 2012. *Arthropod Management Tests* 38.

Higley, L.G. and Boethel, D.J. (1994) *Handbook of Soybean Insect Pests*. Entomological Society of America, Lanham, Maryland, p.136.

Hlivko, J.T. and Rypstra, A.L. (2003) Spiders reduce herbivory: non-lethal effects of spiders on the consumption of soybean leaves by beetle pests. *Annals of the Entomological Society of America* 96(6), 914–919.

Hodgson, E. (2014) *Grasshopper Activity Observed*. Extension and Outreach, Integrated Crop Management, Iowa State University. Available at: http://crops.extension.iastate.edu/cropnews/2014/07/grasshopper-activity-observed (accessed 25 June 2017).

Hodgson, E., McCornack, B.P., Tilmon, K. and Knodel, J.J. (2012) Management recommendations for soybean aphid (Hemiptera: Aphididae) in the United States. *Journal of Integrated Pest Management* 3(1), 1–10.

Hoffmann-Campo, C.B., da Silva, M.T.B. and Oliveira, L.J. (1999) Aspectos biológicos e manejo integrado de *Sternechus subsignatus* na cultura da soja. *Circular Técnica* 22. EMBRAPA Soja/Fundacep, Londrina, Brazil, 32 pp.

Hoffmann-Campo, C.B., Moscardi, F., Corrêa-Ferreira, B.S., Oliveira, L.J., Sosa-Gomez, D.R., Panizzi, A.R., Corso, I.C., Gazzoni, D.L. and Oliveira, E.B. (2000) Pragas da soja no Brasil e seu manejo integrado. *Circular Técnica* 30. EMBRAPA Soja, Londrina, Brazil, 70 pp.

Hoffmann-Campo, C.B., Oliveira, L.J., Moscardi, F., Gazzoni, D.L., Corrêa-Ferreira, B.S., Lorini, I.A., Borges, M., Panizzi, A.R., Sosa-Gomez, D.R. and Corso, I.C. (2003) Integrated pest management in Brazil. In: Maredia, K.M., Dakouo, D. and Mota-Sanches, D. (eds) *Integrated Pest Management in the Global Arena*. CAB International, Wallingford, UK, pp. 285–299.

Hoffmann-Campo, C.B., Ramos Neto, J.A., Neves de Oliveira, M.C. and Lenita Oliveira, J. (2006) Efeito prejudicial de rutina em *Anticarsia gemmatalis*. *Pesquisa Agropecuaria Brasileira, Brasília* 41(10), 1453–1459.

Hohmann, C.L. and Carvalho, S.M. (1989) Pragas e seu controle. *IAPAR Circular* 63. Instituto Agronômico do Paraná (IAPAR), Londrina, Brazil, pp. 217–246.

Homes, D.M. and Barret, G.W. (1997) Japanese beetle *(Popillia japonica)* dispersal behavior in intercropped vs. monoculture soybean agroecosystems. *American Midland Naturalist* 137(2), 312.

House, G.J. and Stinner, B.R. (1983) Arthropods in no-tillage soybean agroecosystems: community composition and ecosystem interactions. *Environmental Management* 7, 23–28.

Hu, D., Feng, J., Wang, Z., Wu, H. and Zhang, X. (2013) Effect of nine plant volatiles in the field on the sex pheromones of *Leguminivora glycinivorella*. *Natural Product Communications* 8(3), 393–396.

Hunt, T.E., Higley, L.G. and Haile, F.J. (2003) Imported longhorned weevil (Coleoptera: Curculionidae) injury to soybean: physiological response and injury guild-level economic injury levels. *Journal of Economic Entomology*, 96(4), 1168–1173.

Hunt, T.L.H. and Pedigo, L. (2000) A re-examination of economic injury levels for potato leafhopper (Homoptera: Cicadellidae) on soybean. *Journal of Entomological Science* 35, 97–104.

Hymowitz, T. (2004) Speciation and cytogenetics. In: Boerma, H.R. and Specht, J.E. (eds) *Soybeans: Improvement, Production, and Uses*, 3rd edn. American Society of Agronomy, Inc., Madison, Wisconsin. pp. 97–136.

Ignoffo, C.M. and Couch, T.L. (1981) The nucleopolyhedrosis virus of *Heliothis* species as a microbial insecticide. In: Burges, H.D. (ed.) *Microbial Control of Pests and Plant Diseases*. Academic Press, London, pp. 329–62.

Indiati, S.W. (2014) The use of sugar apple and neem extract to control leaf eating pest on soybean. *Journal of Experimental Biology and Agricultural Sciences* 2(2), 208–214.

Iowa State University (2012) Field Crop Insects. Available at: http://soybeanresearchinfo.com/pdf_docs/FieldCropInsects.pdf (accessed 25 June 2017).

Ismail, M.S.M., Ghallab, M.M.A, Soliman, M.F.M. and AboGhalia, A.H. (2011) Acaricidal activities of some essential and fixed oils on the two-spotted spider mite, *Tetranychus urticae*. *Egyptian Academic Journal of Biological Sciences* 3, 41–48.

Jackson, R.E. and Pitre, H.N. (2004) Influence of Roundup Ready® soybean production systems and glyphosate application on pest and beneficial insects in wide-row soybean. *Journal of Agricultural Urban Entomology* 21, 61–70.

Jaramillo, C.M., Celeita, J.J. and Sáenz, A. (2013) Suceptibility of *Delia platura* to seven entomopathogenic nematode isolates from the Central Andes region of Colombia. *Universitas Scientiarum* 18(2), 165–172.

Jayappa, A.H., Reddy, K.M.S. and Kumar, N.G. (2002) Lepidopteran caterpillars of soyabean. *Insect Environment* 8(4), 186–188.

Johnson, D. (2011) *Dectes Stem Borer (aka Soybean Stem Borer) in Kentucky Grown Soybeans*. Extension of University of Kentucky, Lexington, Kentucky.

Johnson, J.P. (1944) The imported long-horned weevil, *Calomycterus setarius* Roelofs. *Connecticut Agricultural Experiment Station Bulletin* 479, 212–142.

Jones, D., Jones, G., Van Steenwyk, R.A. and Hammock, B.D. (1982) Effect of the parasite *Copidosoma truncatellum* on development of its host *Trichoplusia ni*. *Annals of the Entomological Society of America* 75, 7–11.

Jones, W.A. Jr and Sullivan, M.J. (1982) Role of host plants in population dynamics of stinkbug pests of soybean in South Carolina. *Environmental Entomology* 11, 867–875.

Joshi, M.D. and Patel, V.N. (2011) Effectiveness of *Beauveria bassiana* on soybean leaf miner, *Aproaerema modicella* Deventer. *Legume Research* 34(1), 73–74.

Kalayanasundaram, M. and Murugesan, S. (1989) Yield loss assessment and economic injury level for the leaf miner *Aproaerema modicella* (Deventer) (Lepidoptera: Gelechiidae) on soyabean in India. *International Journal of Tropical Insect Science* 10, 545–550.

Kalkar, O. and Craner, G.R. (2005) Characterization of the fungal pathogen, *Entomophaga aulicae* (Zygomycetes: Entomophthorales) in larval populations of the green cloverworm, *Plathypena scabra* (Lepidoptera: Noctuidae). *Turkish Journal of Biology* 29, 243–248.

Kamminga, K.L., Herbert, D.A. Jr, Kuhar, T.P., Malone, S. and Doughty, H. (2009) Toxicity, feeding preference, and repellency associated with selected organic insecticides against *Acrosternum hilare* and *Euschistus servus* (Hemiptera: Pentatomidae). *Journal of Economic Entomology* 102, 1915–1921.

Kamminga, K.L., Koppel, A.L., Herbert, D.A. and Kuhar, T.P. (2012) Biology and management of the green stink bug. *Journal of Integrated Pest Management* 3(3), 1–8.

Kapoor, K.N., Gujrati, J.P. and Gangrade, G.A. (1972) Parasites of *Oberea brevis* (Coleoptera: Lamiidae). *Annals of the Entomological Society of America* 65(3), 755.

Kenis, M. and Cugala, D. (2006) Prospects for the biological control of the groundnut leaf miner, *Aproaerema modicella*, in Africa. *CAB Reviews: Perspectives in Agriculture, Veterinary Science, Nutrition and Natural Resources* 1(031), p. 9.

Kessing, J.L.M. and Mau, R.F.L. (1991) Seed corn maggot, *Delia platura* (Meigen). Crop Knowledge Master, Department of Entomology, Honolulu, Hawaii.

Kimura, S., Tokumaru, S. and Kuge, K. (2008) *Eremothecium ashbyi* causes soybean yeast-spot and is associated with stink bug, *Riptortus clavatus*. *Journal of Genetic Pathology* 74, 275–280.

Klein, M.G. and Lacey, L.A. (1999) An attractant trap for the autodissemination of entomopathogenic fungi into populations of the Japanese beetle, *Popillia japonica* (Coleoptera: Scarabaeidae). *Biocontrol Science and Technology* 9, 151–158.

Koch, R.L., Porter, P.M., Harbur, M.M., Abrahamson, M.D., Wyckhuys, K.A.G., Ragsdale, D.W., Buckman, K., Sezen, Z. and Heimpel, G.E. (2012) Response of soybean insects to an autumn-seeded rye cover crop. *Environmental Entomology* 41(4), 750–760.

Kogan, M. and Turnipseed, M.G. (1987) Ecology and management of soybean arthropods. *Annual Review of Entomology* 32, 507–538.

Krupke, C.H., Obermeyer, J.L. and Bledsoe, L.W. (2013) *Soybean Insect Control Recommendations-2013*. Purdue Extension Publication E-77-W. Available at: http://extension.entm.purdue.edu/publications/e-77.pdf (accessed 25 June 2017).

Kumari, V. and Singh, N.P. (2009) *Spodoptera litura* nuclear polyhedrosis virus (NPV-S) as a component in integrated pest management (IPM) of *Spodoptera litura* (Fab.) on cabbage. *Journal of Biopesticides* 2, 84–86.

Kundu, G.G. (1985) *Melanagromyza phaseoli* (Tryon) (Agromyzidae: Diptera) – a new host of *Chlorocytus* sp. (Pteromalidae: Chalcidoidea) from India. *Journal of Entomological Research* 91, 110–111.

Kuwayama, S. (1928) Notes on *Laspeyresia glycinivorella* Matsumura, the soybean pod borer. *Journal of the College of Agriculture* (Hokkaido Imperial University, Sapporo, Japan) 19(5), 261–281.

Lam, W.K.F. and Pedigo, L.P. (1998) Response of soybean insect communities to row width under crop-residue management systems. *Environmental Entomology* 27, 1069–1079.

Lamp, W.O., Morris, M.J. and Armbrust, E.J. (1984) Suitability of common weed species as host plants for the potato leafhopper, *Empoasca fabae*. *Entomologia Experimentalis and Applicata* 36, 125–131.

Lamp, W.O., Nielsen, R.G. and Danielson, S.D. (1994) Patterns among host plants of potato leafhopper, *Empoasca fabae* (Homoptera: Cicadellidae). *Journal of the Kansas Entomology Society* 67(4), 354–368.

Landis, D.A., Gardiner, M.M., van der Werf, W. and Swinton, S.M. (2008) Increasing corn for biofuel production reduces biocontrol services in agricultural landscapes. *Proceedings of the National Academy of Sciences of the United States of America, Biology* 105, 20552–20557.

Leksawasdi, P. (1986) Control of the soybean leaf roller, *Lamprosema diemenalis* Guenee by using 'Phlai' turmeric and a braconid parasite. *Agris* 20(6), 451–460.

Lentz, G.L. (1985) Occurrence of *Phyllophaga congrua* (LeConte) and *Phyllophaga implicita* (Horn) (Coleoptera: Scarabaeidae) on soybeans. *Journal of the Kansas Entomological Society* 58(2), 202–206.

Leskey, T.C. and Hogmire, H.W. (2005) Monitoring stink bugs (Hemiptera: Pentatomidae) in mid-Atlantic apple and peach orchards. *Journal of Economic Entomology* 98(1), 143–153.

Lewin, H.D., Saroja, R.S., Sundaraju, D. and Padmanabhan, M.D. (1979) Influence of sowing time and weather on the incidence of groundnut leafminer. *Indian Journal of Agricultural Sciences* 49, 886–891.

Li, W.M. and Pu, Z.Q. (1991) Population dynamics of aphids and epidemics of soybean mosaic virus in summer sown soybean fields. *Acta Phytophylactica Sinica* 18(3), 123–126.

Lima, A.C.S. and Lara, F.M. (2004) Resistance of soybean genotypes to the silverleaf whitefly *Bemisia tabaci* (Genn.) biotype B (Hemiptera: Aleyrodidae). *Neotropical Entomology* 33, 71–75.

Li-Wei Song, L.W., Wen, X.Y., Zang, L.S., Ruan, C.C., Shi, S.S., Shao, W.W. and Zhang, F. (2015) Parasitism and suitability of different egg ages of the *Leguminivora glycinivorella* (Lepidoptera: Tortricidae) for three indigenous *Trichogramma* species. *Journal of Economic Entomology* 108(3), 933–939.

Logiswaran, E. and Mohanasundaram, M. (1985) Effect of intercropping, spacing and mulching in the control of groundnut leaf miner, *Aproaerema modicella* Deventer (Gelechiidae: Lepidoptera). *Madras Agriculture Journal* 72, 695–700.

Logiswaran, G., Madhava Rao, S., Vasudevan, G. and Kannan, V. (1982) Influence of time of sowing and weather factors on the infestation of leafminer *Aproaerema modicella* Deventer and yield in rainfed groundnut. *Madras Agriculture Journal* 69, 359–363.

Lopes, D., Destro, R., Montalvan, M.U. and Ventura, E.P. (1997) Genetic gain and correlations among traits for stink bug resistance in soybean. *Euphytica* 97, 161–166.

Loughran, J.C. and Ragsdale, D.W. (1986) *Medina* sp. (Diptera: Tachinidae): a new parasitoid of the bean leaf beetle: *Cerotoma trifurcata* (Coleoptera: Chrysomelidae). *Journal of Kansas Entomological Society* 59, 468–473.

Lourenção, A.L. and Nagai, H. (1994) Surtos populacionais de *Bemisia tabaci* no Estado de São Paulo. *Bragantia* 53, 53–59.

Magalhães, B.P., Humber, R.A., Shields, E.J. and Roberts, D.W. (1991) Effects of environment and nutrition on conidium germination and appressorium formation by *Zoophthora radicans* (Zygomycetes: Entomophthorales): a pathogen of the potato leafhopper (Homoptera: Cicadellidae). *Environmental Entomology* 20, 1460–1468.

Magalhães, B.P., Lecoq, M., Faria, M.R., Schmidt, F.G.V. and Guerra, W.D. (2000) Field trial with the entomopathogenic fungus *Metarhizium anisopliae* var. *acridum* against bands of the grasshopper *Rhammatocerus schistocercoides* in Brazil. *Biocontrol Science and Technology* 10, 427–441.

Mahr, M. (1999) *Trichopodes pennipes*, parasitoid of true bugs. *Midwest Biological Control News*, online vol. 6(5). Available at: http://www.entomology.wisc.edu/mbcn/kyf605.html (accessed 25 June 2017).

Mallinger, R.E., Hogg, D.B. and Gratton, C. (2011) Methyl salicylate attracts natural enemies and reduces populations of soybean aphids (Hemiptera: Aphididae) in soybean agroecosystems. *Journal of Economic Entomology* 104(1), 115–124.

Marrone, P.G. and Stinner, R.E. (1983) Effects of soil moisture and texture on ovipositional preferences of the bean leaf beetle, *Cerotoma trifurcata* (Forster) (Coleoptera: Coccinellidae). *Environmental Entomology* 12, 426–428.

Massaroli, A., Butnariu, A.R. and Doetzer, A.K. (2014) Occurrence of *Trichogramma* parasitoids in eggs of soybean Lepidopteran pests in Mato Grosso, Brazil. *International Journal of Biology* 6(2), 97–103.

McPherson, J.E. (1982) *The Pentatomoidea (Hemiptera) of Northeastern North America with Emphasis on the Fauna of Illinois*. Southern Illinois University Press, Carbondale and Edwardsville, Illinois.

McPherson, J.E. and McPherson, R.M. (2000) *Stink Bugs of Economic Importance in North America and Mexico*. CRC Press, Boca Raton, Florida.

McPherson, R. and Newsom, L. (1984) Trap crops for control of stink bugs in soybean. *Journal of the Georgia Entomological Society* 19, 470–480.

Medal, J.C., Mueller, A.J., Kring, T.J. and Gbur, E.E. (1995) Developmental stages of *Spissistilus festinus* (Homoptera: Membracidae) most susceptible to hemipteran predators. *Florida Entomologist* 78(4), 561–564.

Mettenmeyer, A. (2002) Viral insecticides hold promise for bio-control. *Farming Ahead* 124, 50–51.

Metzger, F.W. and Grant, D.H. (1932) *Repellency of the Japanese Beetle of Extracts Made From Plants Immune to Attack. United States Department of Agriculture (USDA) Technical Bulletin* No. 299. USDA, Beltsville, Maryland, p. 21.

Michaud, J.P. and Grant, A.K. (2005) The biology and behavior of the longhorned beetle, *Dectes texanus* on sunflower and soybean. *Journal of Insect Science* 5(25), 1–15.

Michaud, J.P., Qureshi, J.A. and Grant, A.K. (2007) Sunflowers as a trap crop for reducing soybean losses to the stalk borer *Dectes texanus* (Coleoptera: Cerambycidae). *Pest Management Science* 63(9), 903–909.

Micheli, A. (2005) Variabilidade intraespecífica, inimigos naturais e avaliação da mistura de fungos entomopatogênicos e inseticidas para o controle de *Diabrotica speciosa* (Germar, 1824) (Coleoptera: Chrysomelidae). Dissertação (Mestrado em Ciências Biológicas), Curso de Pós-graduação em Ciências Biológicas, Área de Concentração Entomologia, Universidade Federal do Paraná, Curitiba, Brazil.

Miklasiewicz, T.J. and Hammond, R.B. (2001) Density of potato leafhopper (Homoptera: Cicadellidae) in response to soybean–wheat cropping systems. *Environmental Entomology* 30, 204–214.

Mizell, R.F. and Tedders, W.L. (1995) A new monitoring method for detection of the stink bug complex in pecan orchards. *Proceedings of the Southeastern Pecan Growers Association* 88, 36–40.

Mizell, R.F., Riddle, T.C. and Blount, A.S. (2008) Trap cropping system to suppress stink bugs in the southern coastal plain. *Proceedings of the Florida State Horticulture Society* 121, 377–382.

Mohanty, K.K., Chakraborty, D.P. and Roy, S. (1988) Antifeedant activity of oil fractions of some leguminous plants against *Diacrisia oblique*. *Indian Journal of Agricultural Sciences* 58 (7), 579–580.

Molinari, M.M. and Gamundi, J.C. (2010) *Elasmopalpus lignosellus* (Zeller), un barrenador esporádico en soja. *Para Mejorar la Produccion – inta eea oliveros* 45, 121–122.

Morse, W.J. and Carter, J.L. (1937) Improvement in soybeans. In: *Yearbook of Agriculture 1937*. United States Department of Agriculture (USDA), Washington, DC, pp. 1154–1159.

Moscardi, F. (1999) Assessment of the application of baculoviruses for control of Lepidoptera. *Annual Review of Entomology* 44, 257–289.

Moscardi, F. and Carvalho, R.C.Z. (1993) Consumo e utiliza . cao de folhas de soja por *Anticarsia gemmatalis* Hub. (Lepidoptera: Noctuidae) infectada, em diferentes estadios larvais, por seu vırus de poliedrose nuclear. *Analas da Societade Entomologica Brasileira* 22, 267–280.

Moscardi, F., Lobo de Souza, M., Batista de Castro, M.E., Moscardi, M.L. and Szewczyk, B. (2011) Baculovirus pesticides: present state and future perspectives. In: Ahmad, I., Ahmad, F. and Pichtel, P. (eds) *Microbes and Microbial Technology: Agricultural and Environmental Applications*. Springer, New York, pp. 415–445.

Mundhe, D.R. (1980) Insect pest complex on soybean [*Glycine max* (L.)Merill] in Marathwada region. *Journal of Maharastra Agriculture University* 5, 259–260.

Murillo, H., Hunt, D.W.A. and VanLaerhoven, S.L. (2013) First records of *Chrysodeixis chalcites* (Lepidoptera: Noctuidae: Plusiinae) for east-central Canada. *The Canadian Entomologist* 145(3), 338–342.

Myers, S.W. and Gratton, C. (2006) Influence of potassium fertility on soybean aphid, *Aphis glycines* Matsumura (Hemiptera: Aphididae), population dynamics at a field and regional scale. *Environmental Entomology* 35, 219–227.

Myers, S.W., Gratton, C., Wolkowski, R.P., Hogg, D.B. and Wedberg, J.L. (2005a) Effect of soil potassium availability on soybean aphid (Hemiptera: Aphididae) population dynamics and soybean yield. *Journal of Economic Entomology* 98, 113–120.

Myers, S.W., Hogg, D.B. and Wedberg, J.L. (2005b) Determining the optimal timing of foliar insecticide applications for control of soybean aphid (Hemiptera: Aphididae) on soybean. *Journal of Economic Entomology* 98, 2006–2012.

Nasruddin, A., Fattah, A., Baco, M.S. and Said, A.E. (2014) Potential damages, seasonal abundance and distribution of *Empoasca terminalis* distant (Homoptera: Cicadellidae) on soybean in South Sulawesi. *Journal of Entomology Indonesia* 11, 93–102.

National Integrated Pest Management Network (no date) Corn Earworm Threshold Calculator for Soybeans. Copyright © 2000 Virginia Tech, Based on *Corn Earworm Management in Soybean* by Van Duyn, J.W. and Eckel, C.S., Cooperative Extension Service, North Carolina State University. Available at: http://webipm.ento.vt.edu/cew/ (accessed 25 June 2017).

Netam, A.K., Gupta, R. and Soni, S. (2013) Seasonal incidence of insect pests and their biocontrol agents on soybean. *Journal of Agriculture and Veterinary Science* 2(2), 2319–2372.

Newsom, L.D. and Herzog, D.C. (1977) Trap crops for control of soybean pests. *Louisiana Agriculture* 20, 14–15.

Nickerson, J.C., Kay, C.A.R., Buschman, L.L. and Whitcomb, W.H. (1977) The presence of *Spissistilus festinus* as a factor affecting egg predation by ants in soybeans. *Florida Entomologist* 60, 193–199.
Nielsen, C.J. and Hajek, A.E. (2005) Control of invasive soybean aphid, *Aphis glycines* (Hemiptera: Aphididae), populations by existing natural enemies in New York State, with emphasis on entomopathogenic fungi. *Environmental Entomology* 34, 1036–1047.
Nielsen, G.R., Lamp, O.W. and Stutte, G.W. (1990) Potato leafhopper (Homoptera: Cicadellidae) feeding disruption of phloem translocation in alfalfa. *Journal of Economic Entomology* 83, 807–813.
Niide, T., Higgins, R.A., Withworth, R.J. and Buschman, L.L. (2012) Antibiosis resistance in soybean plant introductions to *Dectes texanus* (Coleoptera: Cerambycidae). *Journal of Economic Entomology* 105(2), 598–607.
Noma, T., Gratton, C., Colunga-Garcia, M., Brewer, M.J., Mueller, E.E., Wyckhuys, K.A.G., Heimpel, G.E. and O'Neal, M.E. (2010) Relationship of soybean aphid (Hemiptera: Aphididae) to soybean plant nutrients, landscape structure, and natural enemies. *Environmental Entomology* 39, 31–41.
North Dakota State University (NDSU) (no date) Entomology Extension/Thistle Caterpillar/Painted Lady. Available at: https://www.ag.ndsu.edu/extensionentomology/field-crops-insect-pests/Documents/soybean/thistle-caterpillar (accessed 25 June 2017).
OEPP/EPPO (European and Mediterranean Plant Protection Organization) (2005) *Diabrotica speciosa*. Data sheets on quarantine pests. *Bulletin OEPP/EPPO Bulletin* 35, 374–376.
Ogunlana, M. and Pedigo, L. (1974) Pest status of the potato leafhopper. *Journal of Economic Entomology* 201–202.
Oliveira, L.J. (1997) Ecologia comportamental e de interações com plantas hospedeiras em *Phyllophaga cuyabana* (Moser) (Coleoptera: Melolonthidae, Melolonthinae) e implicações para o seu manejo em cultura de soja. 1997. Tese (Doutorado em Ciências) – Universidade Estadual de Campinas, Campinas, Brazil, 148 pp.
Oliveira, L.J., Santos, B., Parra, J.R.P., Amaral, M.L.B. do and Magri, D.C. (1996) Ciclo biologico de *Phyllophaga cuyabana* (Moser) (Scarabaeidae: Melolonthinae). *Anais da Sociedade Entomologica do Brasil* 25(3), 431–437.
Oliveira, L.J., Garcia, M.L., Hoffmann-Campo, C.B., Sosagomez, D.R., Farias, J.R.B. and Corso, I.C. (1997) Coró-da-soja *Phyllophaga cuyabana* (Moser 1918). *Circular Técnica* 20. EMBRAPA-CNPSo, Londrina, Brazil, 30 pp.
Oliveira, L.J., Hoffmann-Campo, C.B. and Garcia, M.A. (2000) Effect of soil management on the white grub population and damage in soyabean. *Pesquisa Agropecuaria Brasileira* 35(5), 887–894.
Oliveira, L.J., Garcia, M.A., Hoffmann-Campo, C.B. and Amaral, M.L.B. do (2007) Feeding and oviposition preference of *Phyllophaga cuyabana* (Moser) (Coleoptera: Melolonthidae) on several crops. *Neotropical Entomology* 36(5), 759–764.
Olufolaji, D.B. (2011) Prospects of large-scale use of natural products as alternatives to synthetic pesticides in developing countries. In: Dubey, N.K. (ed.) *Natural Products in Plant Pest Management*. CAB International, Wallingford, UK, pp. 191–205.
Overall, L.M. (2008) Evaluation of organic insecticides to control harlequin bug, *Murgantia histrionica* (Hahn), and yellow margined leaf beetle, *Microtheca ochrloma* Stål, on leafy greens. MSc thesis, Oklahoma State University, Stillwater, Oklahoma.
Pan, X. (1996) Study on the action threshold of *Melanagromyza sojae*. *Acta Zoonutrimenta Sinica* 22(1), 22–23.
Pangnakorn, U., Watanasorn, S., Kuntha, C. and Chuenchooklin, S. (2010) Effects of wood vinegar and fermented liquid organic fertilizer on soybean (Srisamrong 1) cultivated under drought conditions. *Journal of International Society for Southeast Asian Agricultural Sciences* 16(2), 67–73.
Panizzi, A.R. (1997) Wild host of pentatomids: ecological significance and role in their pest status on crops. *Annual Review of Entomology* 42, 99–122.
Panizzi, A.R. and Alves, R.M. (1993) Performance of nymphs and adults of the southern green stink bug (Heteroptera: Pentatomidae) exposed to soybean pods at different phenological stages of development. *Journal of Economic Entomology* 86, 1088–1093.
Parker, J.L. (2012) Assessment of stink bug feeding damage in Louisiana soybean: use of a no-choice feeding field protocol. MSc thesis, Louisiana State University, Baton Rouge, Louisiana.
Parra, J.R.P. and Zucchi, R.A. (2004) *Trichogramma* in Brazil: feasibility of use after twenty years of research. *Neotropical Entomology* 33, 271–281.
Pawar, V.M. and Thombre, U.T. (1992) Prospects of baculovirus in integrated pest management of pulses. In: Ananthkrishnan, T.N. (ed.) *Emerging Trends in Biocontrol of Phytophagous Insects*. Oxford and IBH Publishing Company Private Limited, New Delhi, pp. 253–258.

Pecinovsky, K.T. and Lang, B.J. (2011) Effect of certified organic products on soybean aphid. *Iowa State Research Farm Progress Reports* 87. Iowa State University, Ames, Iowa.

Peres, F., Fernandes, O., Silveira, L. and Silva, C. (2009) Cravo-de-defunto como planta atrativa para tripes em cultivo protegido de melão orgânico. *Bragantia* 68(4), 953–960.

Perez, W.A.A. and Correa-Ferreira, B.S. (2004) Methodology of mass multiplication of *Telenomus podisi* Ash. and *Trissolcus basalis* (Woll.) (Hymenoptera: Scelionidae) on eggs of *Euschistus heros* (Fab.) (Hemiptera: Pentatomidae). *Neotropical Entomology* 33(4), 457–462.

Permana, A.D., Asni, J.A., Putra, R.E., Sastrodihardjo, S. and Ahmad, I. (2012) The influence of trichome characters of soybean (*Glycine max* Merrill) on oviposition preference of soybean pod borer *Etiella zinckenella* Treitschke (Lepidoptera: Pyralidae) in Indonesia. *Journal of Entomology and Nematology* 4(3), 15–21.

Petty, B.M., Johnson, D.T. and Steinkraus, D.C. (2012) Survey of pathogens and parasitoids of *Popillia japonica* (Coleoptera: Scarabaeidae) in northwest Arkansas. *Journal of Invertebrate Pathology* 111, 56–59.

Pfannenstiel, R.S. and Yeargan, K.V. (2002) Identification and diel activity patterns of predators attacking *Helicoverpa zea* (Lepidoptera: Noctuidae) eggs in soybean and sweet corn. *Environmental Entomology* 31(2), 232–241.

Poe, S.L. (1980) Sampling mites on soybean. In: Kogan, M. and Hezrog, D.C. (eds) *Sampling Methods in Soybean Entomology*. Springer-Verlag, New York, pp. 312–323.

Polaszek, A. and Foerster, L.A. (1997) *Telenomus cyamophylax*, n. sp. (Hymenoptera: Scelionidae) attacking eggs of the velvetbean caterpillar, *Anticarsia gemmatalis* Hübner (Lepidoptera: Noctuidae). *Anais da Sociedade Entomológica do Brasil* 26, 177–181.

Pomari, A.F., Bueno, A.F., Bueno, R.C.O.F. and Menezes, A.O. Jr (2012) Biological characteristics and thermal requirements of the biological control agent *Telenomus remus* (Hymenoptera: Platygastridae) reared on eggs of different species of the genus *Spodoptera* (Lepidoptera: Noctuidae). *Annals of the Entomological Society of America* 105, 72–81.

Pomari-Fernandes, A., Freitas Bueno, A. and Sosa-Gómes, D.R. (2015) *Helicoverpa armigera*: current status and future perspectives in Brazil. *Current Agricultural Science and Technology* 21, 1–7.

Pope, R. (1998) *Pests of Germinating Corn and Soybean*. Iowa State University, Extension, IC-480. Iowa State University, Ames, Iowa, pp. 65–66.

Poston, F.L. and Pedigo, L.P. (1975) Migration of plant bugs and the potato leafhopper in a soybean–alfalfa complex. *Environmental Entomology* 4, 8–10.

Poston, F.L., Hammond, R.B. and Pedigo, L.P. (1977) Growth and development of the painted lady on soybeans (Lepidoptera: Nymphalidae). *Journal of Kansas Entomological Society* 50, 31–36.

Potter, D.A. and Held, D.W. (2002) Biology and management of the Japanese beetle. *Annual Review of Entomology* 47, 175–205.

Potter, D.A., Powell, A.J., Spicer, P.G. and Williams, D.W. (1996) Cultural practices affect root-feeding white grubs (Coleoptera: Scarabaeidae) in turfgrass. *Journal of Economic Entomology* 89, 156–164.

Prasad, D.P. and Sharma, V.N. (1989) Influence of planting pattern of intercropping and fertilizer levels on the incidence of Bihar hairy caterpillar *Spilosoma obliqua* Walker (Lepidoptera : Arctiidae) in soybean. *Trop-Grain Legume Bulletin* 36, 19–21.

Premchanda, X.X. (1989) Presence of feeding deterrent in the velvet bean, *Mucuna cochinensis* Roxb. *Indian Journal of Entomology* 51(2), 217.

Rabin, L.B. and Pacovsky, R.S. (1985) Reduced larva growth of two Lepidoptera (Noctuidae) on excised leaves of soybean infected with a mycorrhizal fungus. *Journal of Economic Entomology* 78, 1358–1363.

Ragsdale, D., Hodgson, E.W., McCornack, B.P., Koch, K.A., Venette, R.C. and Potter, B.D. (2006) Soybean aphid and the challenge of integrating recommendations within an IPM system. Paper presented at Illinois Crop Protection Technology Conference, University of Illinois at Urbana-Champaign, Champaign, Illinois.

Ragsdale, D.W., McCornack, B.P., Venette, R.C., Potter, B.D., Macrae, I.V., Hodgson, E.W., O'Neal, M.E., Johnson, K.D., O'Neil, R.J., Difonzo, C.D., Hunt, T.E., Glogoza, P.A. and Cullen, E.M. (2007) Economic threshold for soybean aphid (Hemiptera: Aphididae). *Journal of Economic Entomology* 100(4), 1258–1267.

Ragsdale, D.W., Landis, D.A., Brodeur, J., Heimpel, G.E. and Desneux, N. (2011) Ecology and management of soybean aphid in North America. *Annual Review of Entomology* 56, 375–399.

Ramsey, J., Catchot, A., Musser, F., Gore, J. and Cook, D. (2013) The pest status of three cornered alfalfa hopper, *Spissistilus festinus*, in Mississippi soybeans. *Annual Report Mississipi State University* 39. Mississipi State University, Oktibbeha County, Mississippi.

Ranger, C.M., Reding, M.E. and Oliver, J. (2009) Acute toxicity of essential oils to Japanese beetle larvae and their corresponding mass spectral analysis. Paper presented at the National Meeting of the Entomological Society of America, 13–16 December 2009, Indianapolis, Indiana.

Ranger, C.M., Reding, M.E., Oliver, J.B., Moyseenko, J.J., Yousef, N. and Krause, C.R. (2013) Acute toxicity of plant essential oils to scarab larvae (Coleoptera: Scarabaeidae) and their analysis by gas chromatography-mass spectrometry. *Journal of Economic Entomology* 106(1), 159–167.

Raros, E.S. (1975) Bionomics of bean fly, *Ophiomyia phaseoli* (Tryon) (Diptera: Agromyzidae) and its parasites in Hawaii. PhD thesis, Department of Entomology, University of Hawaii, Honolulu, Hawaii.

Razmjou, J., Tavakkoli, H. and Nemati, M. (2009) Life history traits of *Tetranychus urticae* on three legumes (Acari: Tetanychidae). *Munis Entomology and Zoology* 4(1), 204–211.

Reflinaldon, T.H. and Ganeshi, J. (2014) Pod borer of peanut and potential entomopathogenic fungi for its control in West Sumatra. *International Journal on Advanced Science, Engineering and Information Technology* 4, 59–63.

Regupathy, D.P. (2005) Push–pull strategy with trap crops, neem and nuclear polyhedrosis virus for insecticide resistance management in *Helicoverpa armigera* (Hübner) in cotton. *American Journal of Applied Science* 2, 1042–1048.

Reichert, M.B., Da Silva, G.L., Dos Santos Rocha, M., Johann, L. and Juarez Ferla, N. (2014) Mite fauna (Acari) in soybean agroecosystem in the northwestern region of Rio Grande do Sul State, Brazil. *Systematics and Applied Acarology* 19(2), 123–136.

Reid, W.J. Jr (1940) Biology of seed-corn maggot in the coastal plain of the South Atlantic states. *United States Department of Agriculture Technical Bulletin* 723, 43.

Rezende, J.M., Lofego, A.C., Navia, D. and Roggia, S. (2012) Mites (Acari: Mesostigmata, Sarcoptiformes and Trombidiformes) associated to soybean in Brazil, including new records from the Cerrado areas. *Florida Entomologist* 95(3), 683–693.

Ribeiro, A. and Castiglioni, E. (2008) Characterization of populations of natural enemies of *Piezodorus guildinii* (Westwood) (Hemiptera: Pentatomidae). *Agrociencia* 12, 48–56.

Richardson, L.G. (1975) Resistance of soybeans to a stem borer, *Dectes texanus texanus* LeConte. PhD thesis, Department of Entomology, North Carolina State University, Raleigh, North Carolina.

Richman, D.B.R., Hemenway, R.C. and Whitcomb, W.H. (1980) Field cage evaluation of predators of the soybean looper, *Pseudoplusia includens* (Lepidoptera: Noctuidae). *Environmental Entomology* 9, 315–317.

Ristich, S.S. (1950) Biology of seed-corn maggot, *Hylemya cilicrura* (Rondani), and preliminary studies on its control. PhD dissertation, Cornell University, Ithaca, New York.

Roach, S.H. (1987) Observations on feeding and prey selection by *Phidippus audax* (Hentz) (Araneae: Salticidae). *Environmental Entomology* 16, 1098–1102.

Roberts, S.J., Mellors, W.K. and Aemburst, E.J. (1977) Parasites of lepidopterous larvae in alfalfa and soybeans in central Illinois. *The Great Lakes Entomology* 10, 87–93.

Rodehouse, N.L., Barrett, G.W., Zimmerman, G.W. and Kemp, J.C. (1992) Effects of uncultivated corridors on arthropod abundances and crop yields in soybean agroecosystems. *Agriculture, Ecosystems and Environment* 38(3), 179–191.

Rogers, D.J. and Brier, H.B. (2010) Pest-damage relationships for *Helicoverpa armigera* (Hübner) (Lepidoptera: Noctuidae) on vegetative soybean. *Crop Protection* 29(1), 39–46.

Roggia, S., Guedes, J.V.C., Roppa Kuss-Roggia, R.C., Nascimento de Vasconcelos, G.J., Navia, D. and Delalibera, I. Jr (2009) Predatory mites and the fungus *Neozygites floridana* associated with spider mites on soybean, in Rio Grande do Sul, Brazil. *Pesquisa Agropecuaria Brasileira, Brasília* 44(1), 107–110.

Rossoni, C., Oliveira Kassab, S., Hidalgo Barbosa, R., Souza Silva, A., Alexandre Mota, T. and de Souza Loureiro, E. (2013) Natural occurrence of *Beauveria bassiana* (Balsamo) Vuillemin (Hypocreales: Clavicipitaceae) on *Anticarsia gemmatalis* Hübner (Lepidoptera: Erebidae) in the State of Mato Grosso do Sul. *EntomoBrasilis* 6(2), 162–164.

Rozeboom, P., Tilmon, K. and Hesler, L.S. (2014) Organic foliar Milstop shows efficacy against soybean aphid (*Aphis glycines*) on soybean (*Glycine max*). Abstract of paper presented at Entomology 2014, Entomological Society of America's 62nd Annual Meeting, 16–19 November 2014, Portland, Oregon.

Ruberson, J.R., Takasu, K., Buntin, G.D., Eger, J.E. Jr, Gardner, W.A., Greene, J.K., Jenkins, T.M., Jones, W.A., Olson, D.M., Roberts, P.M., Suiter, D.R. and Toews, M.D. (2013) From Asian curiosity to eruptive American pest: *Megacopta cribraria* (Hemiptera: Plataspidae) and prospects for its biological control. *Applied Entomology and Zoology* 48, 3–13.

Rural Agricultural Development Authority (RADA) (2010) *Biology, Behaviour and Management of the Gall Midge Complex on Hot Peppers*. Rural Agricultural Development Authority, Division of Technology, Training & Technical Information, RADA. Available at: http://www.rada.gov.jm/get_articles.php?ai=5 (accessed 13 August 2011).

Russell, F. and Mizell, R.F. III (2015) Stink Bug Management Using Trap Crops in Organic Farming. University of Florida Extension. Available at: http://articles.extension.org/pages/61596/stink-bug-management-using-trap-crops-in-organic-farming#.VWn34zjGPIV (accessed 25 June 2017).

Rutledge, C.E. and O'Neil, R.J. (2006) Soybean plant stage and population growth of soybean aphid. *Journal of Economic Entomology* 99, 60–66.

Rutledge, C.E., O'Neil, R.J., Fox, T.B. and Landis, D.A. (2004) Soybean aphid predators and their use in integrated pest management. *Annals of the Entomological Society of America* 97(2), 240–248.

Santos, C.H. and Panizzi, A.R. (1998) Danos qualitativos causados por *Neomegalotomus parvus* (Westwood) em sementes de soja. *Annais da Societade Entomologica do Brasil* 27, 387–393.

Santos, V., Alcides Moino, A. Jr, Andolo, V., Costa Moreira, C. and Alves de Olinda, R. (2011) Virulence of entomopathogenic nematodes (Rhabditida: Steinernematidae and Heterorhabditidae) for the control of *Diabrotica speciosa* germar (Coleoptera: Chrysomelidae). *Ciências e Agrotecnologia* 35(6), 1149–1156.

Sartori, J.E., Rosa, J.M.O., Wilcken, S.R.S., De Angelis, S. and Aguillera, M.M. (2001) Suscetibilidade de *Scaptocoris castanea* (Hemiptera: Cydinidae) a *Steinernema carpocapsae* (Rhabdtida: Steinermatidae) em condições de laboratório. In: *Reunião Sul Brasileira De Pragas De Solo* 8, 2001, Londrina, Anais. *EMBRAPA Soja. Documentos* 172. EMBRAPA Soja, Londrina, Brazil, pp. 231–232.

Schaefer, C.W. and Panizzi, A.R. (2000) *Heteroptera of Economic Importance*. CRC Press, Boca Raton, Florida.

Schmidt, F.G.V., Silva, J.B.T., Faria, M.R., Magalhães, B.P., Alves, R.T. and Lecoq, M. (2007a) Metodologia de aplicação do fungo *Metarhizium anisopliae* var. *acridum* para o controle do gafanhoto *Rhammatocerus schistocercoides* em campo. *EMBRAPA Recursos Genéticos e Biotecnologia Boletim de Pesquisa e Desenvolvimento* 208, 11 pp.

Schmidt, N.P., O'Neal, M.E. and Singer, J.W. (2007b) Alfalfa living mulch advances biological control of soybean aphid. *Environmental Entomology* 36, 416–424.

Schmutterer, H. (1984) Neem research in the Federal Republic of Germany since the First International Neem Conference. In: *Proceedings of the Second International Neem Conference*, 1983, Rauischholzhausen, Ebsdorfergrund, Germany. German Agency of Technical Cooperation, Eschborn, Germany, pp. 21–30.

Seiter, N.J., Reay-Jones, F.P.F. and Greene, J.K. (2013) Within-field spatial distribution of *Megacopta cribraria* (Hemiptera: Plataspidae) in soybean (*Fabales*: *Fabaceae*). *Environmental Entomology* 42, 1363–1374.

Seiter, N.J., Del Pozo-Valdivia, A.I., Greene, J.K., Reay-Jones, F.P.F, Roberts, P.M. and Reisig, D.R. (2015) Action thresholds for managing *Megacopta cribraria* (Hemiptera: Plataspidae) in soybean based on sweep-net sampling. *Journal of Economic Entomology* 108(4), 1818–1829.

Shanower, T.G. and Ranga, G.V.R. (2010) *Chlaenius* sp. (Col.: Carabidae): a predator of groundnut leaf miner larvae. *International Arachis Newsletter* 8, 19–20.

Shanower, T.G., Wightman, J.A., Gutierrez, A.P. and Rao, G.V.R. (1992) Larval parasitoids and pathogens of the groundnut leaf miner, *Aproaerema modicella* (Lep.: Gelechiidae), in India. *Entomophaga* 37, 419–427.

Shanower, T.G., Gutierrez, A.P. and Wightman, J.A. (1993a) Effect of temperature on development rates, fecundity and longevity of the groundnut leaf miner, *Aproaerema modicella* (Lepidoptera: Gelechiidae), in India. *Bulletin of Entomological Research* 83(3), 413–419.

Shanower, T.G., Wightman, J.A. and Gutierrez, A.P. (1993b) Biology and control of the groundnut leafminer, *Aproaerema modicella* (Deventer) (Lepidoptera: Gelechiidae). *Crop Protection* 12(1), 3–10.

Sharma, I.N.S., Singh, A.K. and Singh, S.P. (1982) Allelopathic potential of some plant substances as anti-feedants against insect pests of jute. In: *Proceedings of the First National Symposium on Allelopathy on Agroecosystems*. Indian Society of Allelopathy, Haryana Agricultural University, Hisar, Haryana, India, pp 157–176.

Shepard, B.M. and Barrion, A.T. (1998) Parasitoids of insects associated with soybean and vegetable crops in Indonesia. *Journal of Agricultural Entomology* 15(3), 239–272.

Shepard, B.M., Carner, G.R., Barrion, A.T., Ooi, P.A.C. and Van den Berg, H. (1999) *Insects and their Natural Enemies Associated with Vegetables and Soybean in Southeast Asia*. Quality Printing Company, Orangeburg, South Carolina, 108 pp.

Shetgar, S.S. and Thombre, U.T. (1984) Occurrence of natural enemies on soybean leaf miner and relative susceptibility of some soybean varieties to its attack. *Journal of Maharashtra Agricultural Universities* 9(2), 218–219.

Shi, G.L., Zhao, L.L., Liu, S.Q., Cao, H., Clarke, S.R. and Sun, J.H. (2006) Acaricidal activities of extracts of *Kochia scoparia* against *Tetranychus urticae*, *Tetranychus cinnabarinus*, and *Tetranychus viennensis* (Acari: Tetranychidae). *Journal of Economic Entomology* 99(3), 858–863.

Shilpi, K. and Tomar, D. (2014) Effect of *Cassia fistula* on *Ceratoma trifurcata*. *Biolife* 2(2), 672–678.

Shirale, D., Patil, M. and Bidgire, U. (2010) Field evaluation of biopesticides and synthetic insecticide against leaf miner, *Aproaerema modicella* population in soybean. *Journal of Biopesticides* 3(1), 358–360.

Shrivastava, S.K., Verma, R. and Sharma, S. (1999) Feeding potential of Bihar hairy caterpillar, *Spilosoma obliqua* Walker on soybean. *Journal of Insect Sciences* 12(2), 156–157.

Silva, M.T.B. (2000) Nível de controle e danos de *Sternechus subsignatus* (Boheman) (Coleoptera: Curculionidae) em soja, no sistema de plantio direto. *Anais da Societade de Entomologia do Brasil* 29(4), 809–816.

Singh, R.K., Dwivedi, B.S., Singh, A. and Tripathy, S. (2014) Farmers' knowledge and creativity in eco-friendly pest management: lessons in sustainable agriculture. *Indian Journal of Traditional Knowledge* 13(3), 574–581.

Singh, S.K., Mishra, P.K. and Tandon, S.M. (2015) Bioefficacy of *Bacillus sphaericus* R3 against *Spilarctia obliqua* wlk (Lepidoptera: Arctiidae). *Nature and Science* 13(3), 58–62.

Singh, Y.R. and Varatharajan, R. (1999) Host range of Bihar hairy caterpillar, *Spilosoma obliqua* (Walker) (Arctiidae: Lepidoptera). *Hexapoda* 11(2), 65–74.

Smith, A.W., Hammond, R.B. and Stinner, B.R. (1988) Influence of rye-cover crop management on soybean foliage arthropods. *Environmental Entomology* 17, 109–114.

Sosa-Gómez, D.R., Delpin, K.E., Mocardi, F. and Farias, J.R.B. (2001) Natural occurrence of the entomopathogenic fungi *Metarhizium, Beauveria* and *Paecilomyces* in soybean under till and no-till cultivation systems. *Neotropical Entomology* 30(3), 407–410.

Sosa-Gómez, D.R., Delpin, K.E., Moscardi, F. and Nozaki, M.H. (2003) The impact of fungicides on *Nomuraea rileyi* (Farlow) Samson epizootics and on populations of *Anticarsia gemmatalis* Hubner (Lepidoptera: Noctuidae), on soybean. *Neotropical Entomology* 32, 287–291.

Sourakov, A. and Mitchell, E. (2014) *Diadegma insulare* (Cresson) (Insecta: Hymenoptera: Ichneumonidae). University of Florida, publication EENY-124. Available at: http://entnemdept.ufl.edu/creatures/misc/wasps/diadegma_insulare.htm (accessed 25 June 2017).

Souza, P.V., Machado, B.R., De Freitas, M.M., Correa, F., Sousa Almeida, A.C. and Gonzalves de Jesus, F. (2014) *Chrysodeixis includens* (Lepidoptera: Noctuidae) on soybean treated with resistance inducers. *African Journal of Biotechnology* 13(50), 4562–4567.

Specht, A., Viera de Paula-Moraes, S. and Sosa-Gomez, D.R. (2015) Host plants of *Chrysodeixis includens* (Walker) (Lepidoptera, Noctuidae, Plusiinae). *Revista Brasileira de Entomologia* 59, 343–345.

Srinivasa, M., Jagadeesh Babu, C.S., Anitha, C.N. and Girish, G. (2008) Laboratory evaluation of available commercial formulations of HaNPV against *Helicoverpa armigera* (Hub.). *Journal of Biopesticides* 1, 138–139.

Suekane, R., Degrande, P.E., Pontes de Melo, E., Ferreira Bertoncello, T., dos Santos de Lima, I. and Kodama, C. (2012) Damage level of the two-spotted spider mite *Tetranychus urticae* Koch (Acari: Tetranychidae) in soybeans. *Revista Ceres* 59(1), 77–81.

Sujii, E.R., Tigano, M.S. and Sosa-Gomes, D. (2002) Simulação do impacto do fungo *Nomuraea rileyi* em populações da lagarta da soja, *Anticarsia gemmatalis*. *Pesquisa Agropecuaria Brasileira* 37(11), 1551–1558.

Sun, X.L. and Peng, H. (2007) Recent advances in biological pest insects by using viruses in China. *Virologica Sinica* 22, 158–162.

Szewczyk, B., Hoyos-Carvajal, L., Paluszek, M., Skrzecz, I. and Lobo de Souza, M. (2006) Baculoviruses re-emerging biopesticides. *Biotechnology Advances* 24, 143–160.

Szewczyk, B., Rabalski, L., Krol, E., Sihler, W. and Souza, M.L. (2009) Baculovirus biopesticides – a safe alternative to chemical protection of plants. *Journal of Biopesticides* 2, 209–216.

Tae Heung, K. (1992) *Delia platura* as a soybean insect pest. *Korean Journal of Applied Entomology* 31(2), 174–181.

Taghizadeh, R., Talebi, A.A., Fathipour, Y. and Khalghani, J. (2012) Effect of ten soybean cultivars on development and reproduction of lima bean pod borer, *Etiella zinckenella* (Lepidoptera: Pyralidae) under laboratory conditions. *Applied Entomology and Phytopathology* 79, 15–28.

Takahashi, K.M., Berti Filho, E. and Lourenção, A.L. (2008) Biology of *Bemisia tabaci* (Genn.) B-biotype and parasitism by *Encarsia formosa* (Gahan) on collard, soybean and tomato plants. *Scientia Agricola* 65(6), 639–642.

Takasu, K., Takano, S.I., Mizutani, N. and Wada, T. (2004) Flight orientation behavior of *Ooencyrtus nezarae* (Hymenoptera: Encyrtidae), an egg parasitoid of phytophagous bugs in soybean. *Entomological Science* 7, 201–206.

Talekar, N.S. and Chen, B.S. (1985) The beanfly pest complex of tropical soybean. In: Asian Vegetable Research and Development Center (AVRDC) *Soybean in Tropical and Subtropical Cropping Systems*. AVRDC, Shanhua, Taiwan, Republic of China, pp. 257–271.

Talekar, N.S. and Tengkano, W. (1993) Mechanism of resistance to bean fly (Diptera, Agromyzidae) in soybean. *Journal of Economic Entomology* 86(3), 981–985.

Talekar, S.N. and Lin, P.C. (1994) Characterization of resistance to limabean podborer (Lepidoptera: Pyralidae) in soybean. *Journal of Economic Entomology* 87, 821–824.

Tarkeshwar, K. and Singh, A.K. (2014) Insecticidal and antifeedant activity of *Melia azedarach* (L.) fruits, on *Spilosoma obliqua* (Walker) (Lepidoptera: Arctiidae) larvae. *Journal of Agriculture and Veterinary Science* 7(1), 15–20.

Taware, S.P., Raut, V.M., Halvankar, G.B. and Varghese, P. (2001) Field screening of elite soybean (*Glycine max*) lines for resistance to leaf miner (*Aproaerema modicella*) and stem fly (*Melanagromyza sojae*). *Indian Journal of Agricultural Sciences* 71(11), 740–741.

Tay, W.T., Soria, M.F., Thomazoni, D., Silvie, P., Behere, G.T., Anderson, C. and Downes, S. (2013) A brave New World for an Old World Pest: *Helicoverpa armigera* (Lepidoptera: Noctuidae) in Brazil. *PLoS One* 8(11), e80134.

Thooyavathy, R.A., Sridhar, S., Subramanian, K. and Vijayalakshmi, K. (2013) *Seed Production Techniques for Oilseeds and Pulses*. Revitalising Rainfed Agriculture Network, Centre for Indian Knowledge Systems (CIKS), Chennai, India, 40 pp.

Tillman, P.G., Aldrich, J.R., Khrimian, A. and Cottrell, T.E. (2010) Pheromone attraction and cross-attraction of *Nezara, Acrosternum*, and *Euschistus* spp. stink bugs (Heteroptera: Pentatomidae) in the field. *Environmental Entomology* 39(2), 610–617.

Tilmon, K.J., Hodgson, E.W., O'Neal, M.E. and Ragsdale, D.W. (2011) Biology of the soybean aphid, *Aphis glycines* (Hemiptera: Aphididae) in the United States. *Journal of Integrated Pest Management* 2(2), 1–7.

Tindall, K.V. and Fothergill, K. (2010) *Zelia tricolor* (Coquillett) (Diptera: Tachinidae): first record of an internal parasite of *Dectes texanus* Leconte (Coleoptera: Cermabycidae) in soybean. *Florida Entomologist* 93, 635–636.

Tindall, K.V. and Fothergill, K. (2012) *Dolichomitus irritator* (Hymenoptera: Ichneumonidae): a new parasite of *Dectes texanus* (Coleoptera: Cerambycidae) in soybeans. *Florida Entomologist* 95(1), 238–240.

Tindall, K.V., Steward, S., Musser, F., Lorenz, G., Bailey, W., House, J., Henry, R., Hastings, D., Wallace, M. and Fothergill, K. (2010) Distribution of the long-horned beetle, *Dectes texanus*, in soybeans of Missouri, Western Tennessee, Mississippi, and Arkansas. *Journal of Insect Sciences* 10, 178.

Tingle, F.C., Ashley, T.R. and Mitchell, E.R. (1978) Parasites of *Spodoptera exigua, S. eridania* (Lep.: Noctuidae) and *Herpetogramma bipunctalis* (Lep.: Pyralidae) collected from *Amaranthus hybridus* in field corn. *Entomophaga* 23, 343–347.

Todd, J.W. and Herzog, D.C. (1980) Sampling phytophagous Pentatomidae on soybean. In: Kogan, M. and Herzog, D.C. (eds) *Sampling Methods in Soybean Entomology*. Springer, New York, pp. 438–478.

Todd, J.W. and Schumann, F.W. (1988) Combination of insecticide applications with trap crops of early maturing soybean and Southern peas for population management of *Nezara viridula* in soybean (Hemiptera: Pentatomidae). *Journal of Entomological Science* 23, 192–199.

Tomar, D. (2009) Control of fertility of soybean insect pest of soybean by some plant glycoside as an insecticidal agent. PhD thesis, Dr Hari Singh Gour Central University, Sagar, Madhya Pradesh, India.

Tripathi, A.K. and Singh, D. (1994) Screening of natural products for insect antifeedant activity. Part-I: Plant extracts. *Indian Journal of Entomology* 56, 129–133.

Tripathi, A.K., Rao, S.M., Singh, D., Chakravarty, R.B. and Bhakuny, D.S. (1987) Antifeedant activity of plant extracts against *Spilosoma obliqua* Walker. *Current Science* 56(12), 607–608.

Troxclair, N.N. and Boethel, D.J. (1984) Influence of tillage practices and row spacing on soybean insect populations in Louisiana. *Journal of Economic Entomology* 77, 1571–1579.

Turnipseed, S.G. and Kogan, M. (1976) Soybean entomology. *Annual Review of Entomology* 21 247–282.

Uechi, M., Yukawa, J. and Usuba, S. (2005) Discovery of an additional winter host of the soybean pod gall midge, *Asphondylia yushimai* (Diptera: Cecidomyiidae) in Japan. *Applied Entomology and Zoology* 40(4), 597–607.

Underhill, G.W. (1934) The green stinkbug. *Virginia Agriculture Experiment Station Bulletin* 294, 1–26.

United States Department of Agriculture (USDA) Agricultural Research Service (ARS) (2016) Integrated Area-wide Management of Soybean Aphid with Biological Control Introductions and Host Plant Resistance – Ohio. Available at: https://www.ars.usda.gov/research/project/?accnNo=429131 (accessed 25 June 2016).

University of Nebraska, Lincoln Extension, Institute of Agriculture and Natural Resources (2008) A Guide to Grasshopper Control in Cropland. Available at: http://soybeanresearchinfo.com/pests/grasshoppers.html (accessed 25 June 2017).

Valicente, F.H. and Cruz, I. (1991) Controle biológico da lagarta do cartucho, *Spodoptera frugiperda*, com *Baculovirus spodoptera*. *Circular Técnica* 15. EMBRAPA CNPMS, Sete Lagoas, Brazil, 23 pp.

Valles, S.M. and Capinera, J.L. (1993) Response of larvae of the Southern armyworm, *Spodoptera eridania* (Cramer) (Lepidoptera: Noctuidae), to selected botanical insecticides and soap. *International Journal of Agricultural Entomology* 10(3), 145–153.

Valverde, L., Virla, E.G. and Querino, R. (2009) Primera cita de *Trichogramma bruni* Nagaraja (Hymenoptera: Trichogrammatidae) en el cultivo de soja del noroeste argentino (Tucumán), con mención de un nuevo hospedador. *Boletin Sanidad Vegetal Plagas* 35, 25–27.

Valverde, L., Querino, R. and Virla, E.G. (2014) First record of *Rachiplusia nu* (Lepidoptera: Noctuidae) as host of the egg parasitoid *Trichogramma bruni* (Hymenoptera: Trichogrammatidae). *Acta Zoológica Lilloana* 58(2), 250–252.

Van den Berg, H., Ankasah, K.H., Muhammad, A., Widayanto, H.A., Wirasto, H.B. and Yully, I. (1995) Soybean stem fly, *Melanagromyza sojae* (Diptera: Agromyzidae) on Sumatra: seasonal incidence and the role of parasitism. *International Journal of Pest Management* 41, 127–133.

Van den Berg, H., Shepard, B.M. and Nasikin (1998a) Damage incidence by *Etiella zinckenella* in soybean in East Java, Indonesia. *International Journal of Pest Management* 44(3), 153–159.

Van den Berg, H., Shepard, B.M. and Nasikin (1998b) Response of soybean to attack by stemfly *Melanagromyza sojae* in farmer's fields in Indonesia. *Journal of Applied Ecology* 35(4), 514–522.

Van der Goot, P. (1930) *De Agromyza-vliegjes der Inlandse katjang-gewassen op Java. Mededeelingen van het Instituut voor Plantenziekten* 78. [English translation: *Agromyzid Flies of Some Native Legume Crops in Java. Tropical Vegetable Information Service* 78.] Asian Vegetable Research and Development Center (AVRDC), Shanhua, Taiwan, Republic China.

Van Lenteren, J.C. and Bueno, V.H.P. (2003) Augmentative biological control of arthropods in Latin America. *BioControl* 48, 123–139.

Vega, F.E., Dowd, P.F., Lacey, L.A., Pell, J.K., Jackson, D.M. and Klein, M.G. (2007) Dissemination of beneficial microbial agents by insects. In: Lacey, L.A. and Kaya, H.K. (eds) *Field Manual of Techniques in Invertebrate Pathology*. Springer, Dordrecht, The Netherlands, pp. 127–148.

Venette, R.C. and Ragsdale, D.W. (2004) Assessing the invasion by soybean aphid (Homoptera: Aphididae): where will it end? *Annals of the Entomological Society of America* 97, 219–226.

Ventura, M.U., Martins, M.C. and Pasini, M.C. (2000) Responses of *Diabrotica speciosa* and *Cerotoma arcuata tingomariana* (Coleoptera: Chrysomelidae) to volatile attractants. *Florida Entomologist* 83(4), 403–410.

Vieira, S.S., Bueno, F.F., Boff, M.I.C., Bueno, R.C.O.F. and Hoffman-Campo, C.B. (2011) Resistance of soybean genotypes to *Bemisia tabaci* (Genn.) biotype B (Hemiptera: Aleyrodidae). *Neotropical Entomology* 40(1), 117–122.

Vinha Zanuncio, T., Azevedo de Freitas, F., de Souza Tavares, W., Guimarães de Menezes, C.W., Serrão, J.E. and Cola Zanuncio, J. (2012) *Podisus nigrispinus* (Heteroptera: Pentatomidae) and IAC-24 soybean cultivar are compatible. *Sustainable Agriculture Research* 1(2), 41–50.

Virk, J.S., Brar, K.S. and Sohi, A.S. (2004) Role of trap crops in increasing parasitation efficiency of *Trichogramma chilonis* Ishii in cotton. *Journal of Biological Control* 18, 61–64.

Wada, T., Endo, N. and Takahashi, M. (2006) Reducing seed damage by soybean bugs by growing small-seeded soybeans and delaying sowing time. *Crop Protection* 25(8), 726–731.

Walter, A.J. and DiFonzo, C.D. (2007) Soil potassium deficiency affects soybean phloem nitrogen and soybean aphid populations. *Environmental Entomology* 36, 26–33.

Wang, C.L., Siang, N.I., Chang, G.S. and Chu, H.F. (1962) Studies on the soybean aphid, *Aphis glycines* Matsumura. *Acta Entomologica Sinica* 11, 31–44.

Wang, Y. and Ba, F. (1998) Study on the optimum control of soybean aphid. *Acta Phytophylacica Sinica* 25, 151–155.

Werner, B.J., Krell, R.K., Pedigo, L.P. and Hill, J.H. (2003) The soybean leaf miner (Coleoptera: Chrysomelidae) as a vector of bean pod mottle virus. *Journal of the Kansas Entomological Society* 26, 125.

Weston, P.A. and Miller, J.R. (1989) Ovipositional responses of seedcorn maggot, *Delia platura* (Diptera: Anthomyiidae), to developmental stages of lima bean. *Annals of the Entomology Society of America* 82, 387–392.

Whitehouse, S. and Zehnder, C. (2015) Effects of used coffee grounds on *M. cribraria* preferences, soil characteristics, and soybean growth. *Metamorphosis*. Available at: http://kb.gcsu.edu/src/2015/friday/41/ (accessed 25 June 2017).

Woltz, J.M., Isaacs, I. and Landis, L. (2012) Landscape structure and habitat management differentially influence insect natural enemies in an agricultural landscape. *Agriculture, Ecosystems and Environment* 152, 40–49.

Wright, B., Bradshaw, J. and Peterson, J. (2015) Monitoring grasshopper levels in field borders. University of Nebraska. Available at: http://cropwatch.unl.edu/monitoring-grasshoppers (accessed 25 June 2017).

Wright, R.J. and Hunt, T.H. (2011) *Soybean Stem Borers in Nebraska*. University of Nebraska–Lincoln Extension Educational Programs. Available at: http://extensionpublications.unl.edu/assets/pdf/g2082.pdf (accessed 25 June 2017).

Wu, Z.S., Schenk-Hamlin, D., Zhan, W.Y., Ragsdale, D.W. and Heimpel, G.E. (2004) The soybean aphid in China: a historical review. *Annals of the Entomological Society of America* 97, 209–218.

Yeargan, K., Braman, S. and Barney, W. (1994) Effects of potato leafhoppers on soybean plant growth and yield. *Journal of Kansas Entomological Society* 67, 29–36.

Youn, H.S. and Jung, C. (2008) Effect of trap cropping on the egg parasitism of *Riptortus clavatus* (Thunberg) in soybean fields. *Journal of Asia-Pacific Entomology* 11, 73–76.

Youssef, N.N., Oliver, J.B., Ranger, C.M., Reding, M.E., Moyseenko, J.J., Klein, M.G. and Pappas, R.S. (2009) Field evaluation of essential oils for reducing attraction by the Japanese beetle (Coleoptera: Scarabaeidae). *Journal of Economic Entomology* 102, 1551–1558.

Yukawa, J., Uechi, N., Horikiri, M. and Tuda, M. (2003) Description of the soybean pod gall midge, *Asphondylia yushimai* sp. n. (Diptera: Cecidomyiidae), a major pest of soybean and findings of host alternation. *Bulletin of Entomological Research* 93, 73–86.

Yukawa, J., Ichinose, M., Kim, W., Uechi, N., Gyoutoku, N. and Fujii, T. (2016) Lower development threshold temperatures and thermal constants for four species of *Asphondylia* (Diptera: Cecidomyiidae) in Japan and their larval developmental delay caused by heat stress. *Applied Entomology and Zoology* 51(1), 71–80.

Zhang, G.Y., Sun, X.L., Zhang, Z.X., Zhang, Z.F. and Wan, F.F. (1995) Production and effectiveness of the new formulation of *Helicoverpa* virus pesticide-emulsifiable suspension. *Virologica Sinica* 10, 242–247.

Zhang, G., Gu, C. and Wang, D. (2009) Molecular mapping of soybean aphid resistance in PI 567541B. *Theoretical and Applied Genetics* 118, 473–482.

Zhang, J., Li, W., Zhang, L., Dai, H., Ci, D. and Xu, R. (2013) QTL mapping of soybean resistance to whitefly (*Bemisia tabaci* Gennadius) under multienvironment conditions. *Australian Journal of Crop Sciences* 7(8), 1212–1218.

Zhang, Y., Hanula, J.L. and Horn, S. (2012) The biology and preliminary host range of *Megacopta cribraria* (Heteroptera: Plataspidae) and its impact on kudzu growth. *Journal of Economic Entomology* 41, 40–50.

Zhu, J., Zhang, A., Park, K.C., Baker, T., Lang, B., Jurenka, R., Obrycki, J.J., Graves, W.R., Picket, J.A., Smiley, D., Chauhan, K.R. and Klun, J.A. (2006) Sex pheromone of the soybean aphid, *Aphis glycines* Matsumura, and its potential use in semiochemical-based control. *Environmental Entomology* 35(2), 249–257.

Zimmerman, E.C. (1958) *The Insects of Hawaii. Volume 8: Pyraloidea*. University of Hawaii Press, Honolulu, Hawaii, 456 pp.

10 Pest Management in Organic Chestnut

Hüseyin Baspinar* and Mehmet Karagöz
Adnan Menderes University, Aydın, Turkey

Introduction

Chestnut (*Castanea sativa* Miller) is an important broadleaved tree in many countries of the world. This tree is a member of the *Fagaceae* family (which also includes oak trees) and comprises four important species. These species include the American, Asiatic and European species, and the Asiatic one is further divided into Chinese and Japanese chestnut (Hageneder, 2005). Hence, the plants of this tree can be found in the continents of Asia and Europe as well as Australia and America. Natural hybridization of chestnut plants occurred where plants from two species were grown by growers in close proximity. Also, artificial hybridization of different species was accomplished for the sake of improving the yield potential of the original varieties. In recent times, the hybrids of chestnut can be found in some countries (Mudge and Brennan, 1999; Juma, 2014).

Chestnut trees shed their leaves in winter in order to tolerate the extreme cold, hence, they are categorized as deciduous trees. Harvesting is done before or during the onset of this process. A chestnut tree may have a height of a few metres (~ 5 m) to nearly 30 m (New Zealand Chestnut Council, 2000; Jacobs and Severeid, 2004).

Chestnut is grown for fruit and wood production in the mountainous areas of the world where it has very important economic value. In addition to carbohydrates, proteins, vitamins, minerals and oil, the fresh fruits of this tree may comprise water equal to half of their weight. The chestnut fruit needs to be peeled before it is consumed, whether it is fresh or processed. The chestnut fruit can be processed in several ways but cooked and roasted is the most important. The other ways of chestnut fruit processing include cooking, boiling and being made into sweets.

China is of primary importance in world chestnut production with 1,683,815 t being produced in 2014 followed by the other countries such as Bolivia, Turkey, Korea and Italy (Table 10.1) (FAOSTAT, 2014). In addition, many of the European countries (particularly the ones with a Mediterranean climate) have good production of chestnut.

Chestnut trees are attacked by a variety of pests and diseases. The pests include insect pests, mammals and birds. Insect pests are the most important. The most important birds which can damage chestnut trees are

* Corresponding author: hbaspinar@adu.edu.tr

Table 10.1. The chestnut production values (t) in the main chestnut-producing countries. (From FAOSTAT, 2014.)

	Year					
Countries	2009	2010	2011	2012	2013	2014
China	1,550,000	1,643,862	1,693,502	1,709,649	1,719,410	1,683,815
Bolivia	55,000	60,718	62,928	59,744	76,035	77,890
Turkey	61,697	59,171	60,270	57,881	60,019	63,762
Korea	75,911	68,630	64,586	62,345	64,184	56,551
Italy	56,755	56,628	56,853	59,764	55,086	51,959
Japan	21,700	23,500	19,100	20,900	21,000	21,400
Portugal	24,305	22,350	18,271	19,130	24,739	18,465
Others	60,798	70,061	67,041	76,561	77,593	77,722
Total	1,906,166	2,004,920	2,042,551	2,065,974	2,098,066	2,051,564

the sulfur crested cockatoo and rosella while the mammals which can damage these trees are grey squirrels, cattle, horses, rabbits and wallabies.

Both the insect pests and diseases are a serious constraint in chestnut production in several of the countries of the world. For example, *Phytophthora cambivora* (Petri) Buisman and *Ph. cinnamomi* Rands are reported to cause ink disease in the chestnut tree (Robin *et al.*, 2006). Similarly, chestnut blight is a fungal disease which is caused by the pathogen named *Cryphonectria parasitica* (Murrill) M.E. Barr (Anagnostakis, 1987). In addition, many insect pests also attack chestnut plants. The most important of these are the chestnut weevil, the goat moth, the chestnut tortrix, the yellow-legged clearwing and others.

The insect pests of chestnut can be controlled by employing several management techniques. These can comprise both chemical control and biological control. Chestnut is a fruiting tree whose seed is eaten directly as fresh or cooked. Hence, use of poisonous insecticides would be a dangerous move because of the hazardous effects on human health and the environment. Hence, this chapter includes a brief morphological description for the identification of the most important insect pests of chestnut plants, in addition to information on their bio-ecology (life cycle, damage, distribution) and control in organic production.

Major Pest Species and their Control in Organic Production

Curculio elephas (Gyllenhall) (Coleoptera: Curculionidae) (chestnut weevil)

Description

Adults are grey-yellow and sometimes reddish in colour, and are between 6.0 mm and 10.5 mm long. Adult females have a distinctive long snout on the front of their head, almost as long as their body length. The male's snout length is half of its body size. The name of the pest originated from its snout (Robert *et al.*, 1974; Anonymous, 2008a).

The egg is 0.5 × 0.3 mm long, oval and white. Larvae are legless, stout and bent into a 'C' shape, and 7–15 mm long. Pupae develop from fourth instar larvae (Robert *et al.*, 1974; Ploye and Menu, 2000; Anonymous, 2008a).

Life cycle

Curculio elephas overwinters in the soil in the larval stage, mostly 3–15 cm deep (Önuçar and Ulu, 1989; Anonymous, 2008a). The majority of larvae develop into pupae in the spring, but, some of them stay in the soil as pupae for 2–3 years, or even 4 years (USDA, 1983; Soula and Menu, 2003; Anonymous, 2008a). Overwintering larvae hibernate in a fragile chamber in the soil (USDA, 1983; Önuçar and Ulu, 1989; Manel and Debouzie,

1997; Menu *et al.*, 2000; Soula and Menu, 2003; Anonymous, 2008a).

The overwintering larva develops into a pupa in June (Anonymous, 2008a). Adults appear in August and they are active until the beginning of October (Soula and Menu, 2003; Anonymous, 2008a). Adults copulate in 8–10 days after emergence, and eggs are laid on the leaves near to the burrs; 20–50 eggs are laid by a single female (Ploye and Menu, 2000; Soula and Menu, 2003; Paparatti and Speranza, 2005; Anonymous, 2008a).

Damage

When eggs hatch the young larvae tunnel into the immature fruit, they cause damage to the young fruit by tunnelling and deposition of their frass. Larvae finish development in the nut (Önuçar and Ulu, 1989; Anonymous, 1995). More than one larva can feed and develop in one fruit (Hoffmann, 1963; Bovey *et al.*, 1975; Bürges and Gal, 1981).

The last larval stage leaves the fruit after harvesting during storage (USDA, 1983; Önuçar and Ulu, 1989; Anonymous, 2008a). This pest has one generation/year (Anonymous, 2008a).

Distribution and host plants

C. elephas is a common and widespread insect pest in all chestnut-growing areas in Turkey (Anonymous, 2008a). This weevil is generally distributed in the USA, Europe and parts of North Africa and the Middle East (USDA, 1983). Chestnut and oak are the known host plants.

Management

CULTURAL PRACTICES. Larvae usually leave the fruits after harvesting. In some countries, harvested fruits with burrs are left on the soil until they are transported to market. During this period, the larvae leave the fruit and move into the soil. The larvae can hibernate for 8–9 months in the soil. So the following cultural practices can help to control the population:

1. The infested fruits need to be destroyed in the packing house.

2. The floor of the packing house should be constructed from concrete, and the walls should be whitewashed to stop the larvae moving into the cervices in the packing house.

3. The nuts that are left on the soil surface after harvesting should be collected. Particularly in the month of June, the hatched larvae should be destroyed.

BIOLOGICAL CONTROL. The soil, where the nuts are stored, should be sprayed with entomopathogenic fungi and parasitic nematodes to control the emerging larval population. Entomopathogenics effectively control the larvae of the chestnut weevil (Kepenekci *et al.*, 2004; Paparatti and Speranza, 2005; Karagöz *et al.*, 2009; Avtzis and Cognato, 2013).

Cydia splendana (Hubner) (Lepidoptera: Totricidae) (chestnut tortrix)

Description

Adult moths are grey or dark grey-brown in colour and the wingspan is 17–19 mm. Forewings are ash coloured or brownish grey with light-coloured stripes. Eggs are oval, lentil shaped and ivory coloured, and 0.55 × 72 mm in size. Larvae are white or whitish pink, and 14 mm long. The larva changes its skin five times before it pupates. In the last larval stage the larva makes an oval-shaped cocoon in the soil and pupates in it (Seçkin, 1981; Anonymous, 2008a).

Life cycle

Cydia splendana overwinters in the last larval stage in a cocoon under the bark or in the soil until the next spring. The pupa develops from the larva in the same cocoon. Adults develop wings in June–July in central and northern Europe, and in August to the end of September in southern Europe (Seçkin, 1981; Anonymous, 2008a, b). Females deposit many single eggs on young fruit or on the leaves near the fruit. A single female can deposit between 60 and 300 eggs (Bonnemaison, 1962; Bovey *et al.*, 1975;

Seçkin, 1981; Anonymous, 1995). The larva emerges from the egg in 10–12 days, and the larva directly penetrates a burr and then into the fruit. The larva completes the whole of its development in the fruit (Bovey *et al.*, 1975; Seçkin, 1981; Anonymous, 2008a).

Damage

A single larva can develop in one fruit in 40–45 days (Anonymous, 2008a). The larva feeds in the tunnel inside the fruit and destroys it. *C. splendana* has one generation/ year. Larvae of *C. splendana* differ from *C. elephas* larvae with tunnelling irregular tunnels. The emergence hole of *C. elephas* is larger than *C. splendana*'s (Seçkin, 1981; Önuçar and Ulu, 1989; Anonymous, 2008a).

Distribution and host plants

C. splendana is distributed all over Europe (Anonymous, 2008b). The main host is chestnut for the larvae of *C. splendana* while *Quercus*, *Fagus* and *Juglans* are the other hosts (Anonymous, 2008a, b).

Management

CULTURAL PRACTICES. The cultural practices for control of *C. splendana* are the same as for *C. elephas*.

BIOLOGICAL CONTROL. Entomopathogenic nematodes are effective against *C. splendana* in soil where chestnuts are being stored (Karagöz *et al.*, 2009).

Cossus cossus (L.) (Lepidoptera: Cossidae) (goat moth)

Description

The adult wingspan is generally 70–100 mm and forewings are grey with a creamy colour at the base and marked with wavy cross lines. Young larvae of *Cossus cossus* are light pinky reddish and when they reach the late instar, the larvae are reddish on the dorsal side and yellow on the ventral side. Last instar larvae are 9–10 mm long. Pupae are 50–60 mm long and reddish brown in colour (AgroAtlas, 2008; Anonymous, 2008c).

Life cycle

Larvae overwinter in tunnels burrowed into the trunks or branches. The larva develops a cocoon during the spring (Anonymous, 2008c). The adults fly from July onwards. Females deposit 85% of their eggs on the base of the tree trunk where the soil touches the trunk, and the rest of the eggs are laid in a group of 10–15 eggs on the trunk or branches near to the ground level where the bark is damaged. A single female can deposit 700 eggs (Keith and Alexander, 2002; AgroAtlas, 2008; Anonymous, 2008c). After hatching from the eggs, the larvae move in different directions and they penetrate into the branch or the trunk of the tree through its bark. *C. cossus* hibernates in the larval stage in the tunnels under the bark of the branch or the trunk of the trees. The old caterpillar abandons the tree trunk and goes into the soil where it forms a cocoon. They feed on the cambium, phloem, xylem and parenchyma tissues in the trunk of the trees. The larvae are very destructive when they reach the wood of the trunk 10–15 cm under the soil. One generation is completed every 2–3 years (AgroAtlas, 2008; Anonymous, 2008c).

Damage

C. cossus damages the wood and can cause death of chestnut trees even if the trees are very large, if the pest is not controlled (Keith and Alexandar, 2002; Karagöz and Erincik, 2007; Anonymous, 2008c).

Distribution and host plants

This insect pest is found in Asian and European countries (AgroAtlas, 2008; Anonymous, 2008c). The hosts of this insect pest are pears, apple, plums, cherries, quince, apricot, walnut, chestnut, persimmon, olive, mulberries, sea buckthorn, and also willow, poplar, aspen, alder, ash trees, birches, beech, oak, maple, elm and oleaster (AgroAtlas, 2008; Anonymous, 2008c).

Management

CULTURAL PRACTICES. Whitewashing the trunks of trees and closing wounds on branches and trunk with grafting wax are among the control measures. Additionally, branches infested with the larvae must be pruned and removed from the plantations (AgroAtlas, 2008; Anonymous, 2008c).

PHYSICAL PRACTICES. Excavation of the ground surrounding the tree trunk (15–20 cm deep) is suggested in order to destroy the larvae in the trunk of the tree. Additionally, individual larvae can be destroyed with a wire by inserting it into the hole on the branch or the trunk and turning it (Anonymous, 2008c).

BIOLOGICAL CONTROL. There are many predators such as ants and birds, in particular woodpeckers, which feed on the larvae and eggs of *C. cossus*. It has been reported that use of pheromone traps could reduce *Cossus insularis* damage in pear orchards in Japan (Nakanishi *et al.*, 2013).

Synanthedon vespiformis (L.) (Lepidoptera: Sesiidae) (yellow-legged clearwing)

Description

The wingspan of adults is 18–20 mm and yellow bands on the abdomen are distinctive for adults. The mature larva is 16–22 mm long.

Life cycle

Adults are active during May and June in chestnut plantations in Turkey. The females of *Synanthedon vespiformis* deposit their eggs in a sequence on the bark or in a crevice of the trunk (Keith and Alexandar, 2002). The larvae feed in the gallery by tunnelling in the cambium between the bark and wood of chestnut trees.

Damage

The larvae also feed on the galls on the trunk caused by *Agrobacterium tumefaciens*, a bacterium causing gall development in the fruit trees. *S. vespiformis* larvae cause yellowing of the leaves and growth failure of the tree which cause the death of the tree (Önuçar and Ulu, 1989; Karagöz and Erincik, 2007). The pest prefers freshly cut trees, feeding and completing its life cycle in about 2 years (Blunck, 1953; Stovin, 1958; Real and Balachowsky, 1966).

Distribution and host plants

S. vespiformis is widespread in the Mediterranean and central Europe (Levi-Zada *et al.*, 2011). It is a polyphagous pest, and it bores into the woody parts of the plants. It feeds on forest and ornamental trees and shrubs as well as stone fruits (Halperin and Sauter, 1991–1992; Ben Yehuda *et al.*, 2007; Szántóné-Veszelka *et al.*, 2010).

Management

CULTURAL PRACTICES. Soil drainage is essential in wet areas for better growth of the trees. Larvae under the bark can be removed with a small knife. Pruning of branches infested with the larvae can help to control the pest (Anonymous, 2008c).

It was reported that mass trapping with lure traps (one trap for every five trees) was effective to control pest populations (Anonymous, 2008c).

Dryocosmus kuriphilus Yasumatsu (Hymenoptera: Cynipidae) (chestnut gall wasp)

Description

The adult female is black in colour and 2.5–3.0 mm long. The scapus and pedicels of the antennae, the apex of the clypeus and the mandibles are yellow brown in adults. The adult female looks like a European oak cynipid wasp, *Dryocosmus cerriphilus* Giraud. Eggs are milky white in colour and 0.1–0.2 mm long with a long stalk. The larvae are milky white in colour and 2.5 mm long when fully grown. The pupa is black or dark brown in colour and 2.5 mm long (Anonymous, 2005).

Life cycle

Dryocosmus kuriphilus is native to China, but has now spread to many countries in distant continents such as North America and Europe. It is classified as a quarantine pest by the European and Mediterranean Plant Protection Organization, EPPO (Anonymous, 2005). The adults of *D. kuriphilus* appear in summer, and they lay their eggs inside the bud. The first instar larvae overwinter inside the chestnut buds, and they continue their growth in the following spring when the buds start to develop (Anonymous, 2005). The gall formation is induced by feeding of the larvae inside the bud, and a heavy attack of *D. kuriphilus* can reduce the growth of the tree and the wood yield (Kato and Hijii, 1997). The chestnut gall wasp produces one generation/year and it reproduces asexually, by thelytokous parthenogenesis (Moriya *et al.*, 2003).

Damage

Damage caused by this pest on the fruits can lead to a 50–75% loss of yield (Payne *et al.*, 1983).

Distribution and host plants

This pest is found in countries such as China, Japan, Korea, Nepal, Canada, the USA, Croatia, the Czech Republic, France, Germany, Hungary, Italy, Slovenia, Switzerland (Anonymous, 2011, 2012, 2014) and Turkey (Çetin *et al.*, 2014). The list of host plants includes *Castanea crenata*, *Castanea dentata*, *Castanea mollissima*, *C. sativa* and *Castanea seguinii* (Anonymous, 2014).

Management

Resistant chestnut varieties could help to prevent the damage caused by the chestnut gall wasp in the beginning (Kajiura and Machida, 1961), but a biotype of the pest that attacks the resistant varieties was able to reproduce and increase its population, and it can become a dominant species (Shimura, 1972, 1973). As a result, resistance has been broken down in the resistant varieties of the chestnut.

MECHANICAL CONTROL. Infestations and populations can be controlled by pruning and destroying infested shoots in small chestnut orchards. However, this is not acceptable by growers in larger plantations because of the cost.

BIOLOGICAL CONTROL. Murakami (1981) reported that eight parasitoid species were reared from the galls in Japan, which were imported from China. Among these, *Torymus sinensis* Kamijo (Hymenoptera: Torymidae) has shown high host specificity and its life cycle has matched that of its host (Aebi *et al.*, 2006), and it has been imported to many other countries for rearing and release to control *D. kuriphilus*. As a result of biological control studies, many native parasitoids of *D. kuriphilus* from Europe were reported from the families Eurytomidae, Pteromalidae, Torymidae, Eupelmidae, Eulophidae and Ormyridae. As *D. kuriphilus* became established in a new habitat in Europe, it recruited parasitoids attacking local oak gall wasps (Aebi *et al.*, 2007; Matosevic and Melika, 2013; Quacchia *et al.*, 2013; Doganlar, 2014). It was also reported that many spider species were important predators, and the predation rate was estimated to be more than 20.2% in Japan (Nakamura and Nakamura, 1977).

References

Aebi, A., Schönrogge, K., Melika, G., Alma, A., Bosio, G., Quacchia, A., Picciau, L., Abe, Y., Moriya, S., Yara, K., Seljak, G. and Stone, G.N. (2006) Parasitoid recruitment to the globally invasive chestnut gall wasp *Dryocosmus kuriphilus*. In: Ozaki, K., Yukawa, J., Ohgushi, T. and Price, P.W. (eds) *Ecology and Evolution of Galling Arthropods and Their Associates*. Springer, Tokyo, pp. 103–121.

Aebi, A., Schönrogge, K., Melika, G., Quacchia, A., Alma, A. and Stone, G.N. (2007) Native and introduced parasitoids attacking the invasive chestnut gall wasp *Dryocosmus kuriphilus*. *EPPO Bulletin* 37, 166–171.

AgroAtlas (2008) AgroAtlas – Pests – *Cossus cossus* L. Available at: http://www.agroatlas.ru/en/content/pests/Cossus_cossus (accessed 2 November 2014).

Anagnostakis, S.L. (1987) Chestnut blight: the classical problem of an introduced pathogen. *Mycologia* 79, 23–37.

Anonymous (1995) *Technical Instruction for Plant Protection*, Volume 4. Republic of Turkey Ministry of Food, Agriculture and Livestock. Başak Press, Ankara.

Anonymous (2005) *Drycosmus kuriphilus*. *EPPO Bulletin* 35, 422–424.

Anonymous (2008a) *Technical Instruction for Plant Protection*, Volume 5. Republic of Turkey Ministry of Food, Agriculture and Livestock. Başak Press, Ankara.

Anonymous (2008b) Factsheet – *Cydia splendana*. Available at: http://idtools.org/id/leps/tortai/Cydia_splendana.htm (accessed 2 November 2014).

Anonymous (2008c) *Technical Instruction for Plant Protection*, Volume 4. Republic of Turkey Ministry of Food, Agriculture and Livestock. Başak Press, Ankara.

Anonymous (2011) *EPPO, 2011*. Archives of the European and Mediterranean Plant Protection Organization (EPPO) Reporting Service, EPPO, Paris. Available at: http://archives.eppo.org/EPPOReporting/Reporting_Archives.htm (accessed 2 November 2014).

Anonymous (2012) *Dryocosmus kuriphilus*, first revision (distribution map). In: *Distribution Maps of Plant Pests*. CAB International, Wallingford, UK, Map 677.

Anonymous (2014) *Dryocosmus kuriphilus*. In: *Invasive Species Compendium*. CAB International, Wallingford, UK. Available at: http://www.cabi.org/isc/datasheet/20005#toHostPlantsPlantsAffected (accessed 2 November 2014).

Avtzis, D.N. and Cognato, A.I. (2013) Genetic structure of *Curculio elephas* Gyll. (Coleoptera, Curculionidae) in Greece: an important pest of sweet chestnut. *Journal of Pest Science* 86, 491–497.

Ben Yehuda, S., Dunkelblum, E., Protasov, A., Zada, A. and Mendel, Z. (2007) Occurrence of *Synanthedon vespiformis* (Lepidoptera; Sesiidae) in stone fruit orchards and development of means for its monitoring. *Alon Hanotea* 61, 11–14. (in Hebrew with English abstract)

Blunck, H. (1953) *Tierishe Shadlinge an Nutapflazen Handbuch der Pflanzenkranheiten*. Band 4. Paul Parey, Berlin, Germany.

Bonnemaison, L. (1962) *Les Ennemis Animaux des Plantes Cultivées et des Forêts*, Volume 2. Editions Sep., Paris.

Bovey, P.A., Linder, A. and Müler, Q. (1975) Rechers sur les insectes des chataigners au Tessin (Suisse). *Journal Forestrier Suisse* 126, 781–820.

Burges, G. and Gal, T. (1981) Zur verbreitung und lebensweise des kastanienruesslers (*Curculio elephas* Gyll., Col.: Curculionidae) in ungarn 2. Verbreitung, Schaden, Schwarmen und Geschlechterverhaltnis. *Zeitschrift für Angewandte Entomologie* 92, 35–41.

Çetin, G., Orman, E. and Polat, Z. (2014) First record of the oriental chestnut gall wasp, *Dryocosmus kuriphilus* Yasumatsu (Hymenoptera: Cynipidae) in Turkey. *Bitki Koruma Bülteni* 54, 303–309.

Doganlar, M. (2014) First report on the natural enemy fauna of the chesnut gall wasp, *Dryocosmus kuriphilus* Yasumatsu (Hymenoptera: Cynipidae) in Yalova Turkey. *Turkish Journal of Biological Control* 5, 67–74. (in Turkish, with an English summary)

FAOSTAT (2014) Food and Agriculture Organization of the United Nations Statistics Division. Available at: http://faostat.fao.org (accessed 22 June 2017).

Hageneder, F. (2005) *The Meaning of Trees: Botany – History – Healing – Lore*. Chronicle Books, San Francisco, California.

Halperin, J. and Sauter, W. (1991–1992) An annotated list with new records of Lepidoptera associated with forest and ornamental trees and shrubs in Israel. *Israel Journal of Entomology* 25–26, 105–147.

Hoffmann, A. (1963) Tribu des Balanini: *Balaninus elephas*. In: Balachowsky, A.S. (ed.) *Entomologie Appliquee a l'Agriculture. Tome 2, Lepidoptera*. Mason et Cie Editeurs, Paris, pp. 1125–1128.

Jacobs, D.F. and Severeid, L.R. (2004) Dominance of interplanted American chestnut (*Castanea dentata*) in southwestern Wisconsin, USA. *Forest Ecology and Management* 191(1), 111–120.

Juma, C. (2014) *The Gene Hunters: Biotechnology and the Scramble for Seeds (No. 1)*. Princeton University Press, Princeton, New Jersey.

Kajiura, M. and Machida, Y. (1961) The breeding of resistant varieties to some insects and diseases in fruit trees. *Japanese Journal of Breeding* 11, 137–140. (in Japanese)

Karagöz, M. and Erincik, Ö. (2007) Distribution, damage and economic importance of the chesnut pests, *Cossus cossus* (Lep: Cossidae), *Synanthedon vespiformis* (Lep: Sessiidae) and *Capnodis* spp. (Col: Buprestidae), in Aydın Province. International Workshop on Chesnut Management in Mediterranean Countries: Problems and Prospects, 23–25 October, Bursa, Turkey, Abstract 11.

Karagöz, M., Gülcü, B., Hazır, S. and Kaya, H. (2009) Laboratory evaluation of Turkish entomopathogenic nematodes for suppression of the chestnut pests, *Curculio elephas* (Coleoptera: Curculionidae) and *Cydia splendana* (Lepidoptera: Tortricidae). *Biocontrol Science and Technology* 19, 755–768.

Kato, K. and Hijii, N. (1997) Effects of gall formation by *Dryocosmus kuriphilus* Yasumatsu (Hym., Cynipidae) on the growth of chestnut trees. *Journal of Applied Entomology* 121, 9–15.

Keith, N. and Alexandar, N. (2002) The invertebrates of living and decaying timber in Britain and Ireland – a provisional annotated checklist. *English Nature Research Reports* No. 467, p. 142. Available at: http://www.cerambycoidea.com/titles/alexander2002.pdf (accessed 9 December 2014).

Kepenekci, İ., Gokce, A. and Gaugler, R. (2004) Virulence of three species of entomopathogenic nematodes to the chestnut weevil, *Curculio elephas* (Coleoptera: Curculionidae). *Nematropica* 34, 199–204.

Levi-Zada, A., Ben-Yehuda, S., Dunkelblum, E., Gindin, G., Fefer, D., Protasov, A., Kuznetsowa, T., Manulis-Sasson, S. and Mendel, Z. (2011) Identification and field bioassays of the sex pheromone of the yellow-legged clearwing *Synanthedon vespiformis* (Lepidoptera: Sesiidae). *Chemoecology* 21, 227–233.

Manel, S. and Debouzie, D. (1997) Modeling insect development time of two or more larval stages in the field under variable temperatures. *Environmental Entomology* 26, 163–169.

Matosevic, D. and Melika, G. (2013) Recruitment of native parasitoids to a new invasive host: first results of *Dryocosmus kuriphilus* parasitoid assamblage in Croatia. *Bulletin of Insectology* 66, 231–238.

Menu, F., Roebuck, J.P. and Viala, M. (2000) Bet-hedging diapause strategies in stochastic environments. *The American Naturalist* 155, 724–734.

Moriya, S., Shiga, M. and Adachi, I. (2003) Classical biological control of the chestnut gall wasp in Japan. In: Van Driesche, R.G. (ed.) *Proceedings of the 1st International Symposium on Biological Control of Arthropods.* United States Department of Agriculture Forest Service, Honolulu, Hawaii, pp. 407–415.

Mudge, K.W. and Brennan, E.B. (1999) Clonal propagation of multipurpose and fruit trees used in agroforestry. In: Buck, L.E., Lassoie, J.P. and Fernandes, E.C.M. (eds) *Agroforestry in Sustainable Agricultural Systems.* CRC Press/Lewis Publishers, New York, pp.157–190.

Murakami, Y. (1981) The parasitoids of *Dryocosmus kuriphilus* Yasumatsu (Hymenoptera: Cynipidae) in Japan and the introduction of a promising natural enemy from China (Hymenoptera: Chalcidoidea). *Journal of the Faculty Agriculture, Kyushu University* 25, 81–87.

Nakamura, M. and Nakamura, K. (1977) Population dynamics of the chestnut gall wasp, *Dryocosmus kuriphilus* Yasumatsu (Hymenoptera: Cynipidae). V. Estimation of the effect of predation by spiders on the mortality of imaginal wasps based on the precipitin test. *Oecologia* 27, 97–116.

Nakanishi, T., Nakamuta, K., Mochizuki, F. and Fukumoto, T. (2013) Mating disruption of the carpenter moth, *Cossus insularis* (Staudinger) (Lepidoptera: Cossidae) with synthetic sex pheromone in Japanese pear orchards. *Journal of Asia-Pacific Entomology* 16, 251–255.

New Zealand Chestnut Council (2000) Chestnuts Worldwide and In New Zealand. Available online: http://www.nzcc.org.nz/factsheet.html (accessed 1 December 2015).

Önuçar, A. and Ulu, O. (1989) Faunistic survey studies in chestnut plantations and establishment of control measures against the infectious pest in nuts in İzmir. *Turkish Journal of Agriculture and Forestry* 13, 637–643. (In Turkish, with an English summary)

Paparatti, B. and Speranza, S. (2005) Management of chestnut weevil (*Curculio* spp.), insect key-pest in central Italy. *Acta Horticulturae* 293, 551–556.

Payne, J.A., Jayne, R.A. and Kays, S.J. (1983) Chinese chestnut production in the United States: practice, problems and possible solutions. *Economic Botany* 37, 187–200.

Ploye, H. and Menu, F. (2000) Clutch size manipulations in the chestnut weevil, *Curculio elephas*: fitness of oviposition strategies. *Oecologia* 122, 493–499.

Quacchia, A., Ferracini, C., Nicholls, J.A., Piazza, E., Saladini, M.A., Tota, F., Melika, G. and Alma, A. (2013) Chalcid parasitoid community associated with the invading pest *Dryocosmus kuriphilus* in north-western Italy. *Insect Conservation and Diversity* 6, 114–123. DOI: 10.1111/j.1752-4598.2012.00192.x.

Real, P. and Balachowsky, A.S. (1966) Famille des Aegeriidae (Sessiidae). In: Balachowsky, A.S. (ed.) *Entomologie Appliquee a l'Agriculture. Tome 2, Lepidoptera*. Mason et Cie Editeurs, Paris, pp. 303–304.

Robert, Y., Rabesse, J.M. and Rouze-Jouan, J. (1974) Sur l'utulisation des pieges jounes pour la capture de pucerons en culture de pomme de tere. Influence de la hauteur de piegeage. *Annales de Zoologie Ecologie Animale* 6, 349–372.

Robin, C., Morel, O., Vettraino, A.M., Perlerou, C., Diamandis, S. and Vannini, A. (2006) Genetic variation in susceptibility to *Phytophthora cambivora* in European chestnut (*Castanea sativa*). *Forest Ecology and Management* 226(1), 199–207.

Seçkin, E. (1981) Bursa ili kestanelerinde (*Castanea sativa* Mill.) zarar yapan Totricidae (Lepidoptera) familyası türleri, tanınmaları, zararları, kısa biyolojileri ve doğal düşmanları üzerinde araştırmalar. *T.C. Tarım Orman Bakanlığı Zirai Mücadele ve Zirai Karantina Genel Müdürlüğü Araştırma Eserleri* Seri No. 16, 122. (in Turkish)

Shimura, I. (1972) Studies on the breeding of chestnut, *Castanea* spp. II. Parasitic variation in the chestnut gall wasp, *Dryocosmus kuriphilus* Yasumatsu. *Bulletin of the Horticultural Research Station* A11, 1–13. (in Japanese, with English summary)

Shimura, I. (1973) Studies on the breeding behaviors of several characters in chestnut, *Castanea* spp. III. Damages variation in the chestnut cultivars by the chestnut gall wasp, *Dryocosmus kuriphilus* Yasumatsu. *Japanese Journal of Breeding* 23, 311–319. (in Japanese, with English summary)

Soula, B. and Menu, F. (2003) Variabilty in diapause duration chesnut weevil mixed ESS, genetic polymorphism or bet-hedging? *Oikos* 100, 574–580.

Stovin, G.H.T. (1958) *The Caterpillars of British Moths. The Eggs, Chrysalids and Food Plants*. Frederick Warne and Co., London, p. 327.

Szántóné-Veszelka, M., Poós, B. and Szőcs, G. (2010) Blackberry and raspberry, new hosts of the yellow legged clearwing moth, *Synanthedon vespiformis*: what can the recently developed sex attractant offer in monitoring and beyond? Paper presented at the 7th Workshop on Integrated Soft Fruit Production, of the International Organisation for Biological and Integrated Control (IOBC) Working Group, Integrated Plant Protection in Fruit Crops Subgroup 'Soft Fruits', 20–23 September 2010, Budapest, Hungary.

United States Department of Agriculture (USDA) (1983) European chesnut weevil. In: *Pests Not Known to Occur in the United States or of Limited Distribution*, No. 32. Animal and Plant Health Inspection Service (APHIS), Plant Protection and Quarantine (PPQ), Hyattsville, Maryland, pp. 1–10.

11 Pest Management in Organic Hazelnut Growing

Sebahat K. Ozman-Sullivan*
Ondokuz Mayis University, Samsun, Turkey

Introduction

Hazelnuts are a major crop that is commercially produced in many countries, including Turkey, Italy, Azerbaijan, Georgia, the USA and Spain. Turkey is the world's largest producer, with approximately 70% of the world's production and 80% of exports (Anonymous, 2014a). Insect and mite pests are a major problem of hazelnut production throughout growing areas worldwide (Viggiani, 1984; AliNiazee, 1997a, 1998; Tuncer *et al.*, 2001; Ozman-Sullivan and Sullivan, 2009). Nearly 20% of the potential crop is lost to insects and mites, despite millions of dollars being spent on control measures (AliNiazee, 1998). There are nearly 200 harmful species associated with hazelnut production in Europe, including Turkey, but only a few cause economically important damage and can be classified as major pests (Table 11.1). The remaining species are present at low levels and farmers do not attempt to control them (Ural *et al.*, 1973; Kurt, 1982; Viggiani, 1984; Işık *et al.*, 1987; Gantner, 2001; Tuncer *et al.*, 2001; Ozman-Sullivan and Sullivan, 2009). In the USA, 150 species were reported as causing damage, of which only 24 caused damage above the economic threshold (Messing and AliNiazee, 1985).

Despite the variability of the economic importance of the harmful species across hazelnut-growing countries, the big bud mite (*Phytoptus avellanae* Nal.), hazelnut weevil (*Curculio nucum* (L.)), pentatomid bug (*Palomena prasina* (L.)), ambrosia beetle (*Xyleborus dispar* (F.)), hazelnut aphid (*Myzocallis coryli* (Goeze)) and filbertworm (*Cydia latiferreana* (Walsingham)) are commonly reported as pests, causing damage that varies in severity from area to area and year to year. *C. nucum* and *P. avellanae* are the most important pests in Europe and *C. latiferreana* is the most important pest in the USA (Ural, 1957a; Ural *et al.*, 1973; Kurt, 1982; Viggiani, 1984; Tavamaishvili, 1990; AliNiazee, 1997a, b; Ozman and Toros, 1997a; Tuncer *et al.*, 2001; Ozman-Sullivan and Sullivan, 2009).

Most Turkish hazelnut growers use pesticides to control *C. nucum* because it causes direct damage and is well known as a key pest. However, very few Turkish farmers use pesticides against other economically important pests such as the big bud mite and twig borer. Cultural and mechanical control are also used by many growers to control hazelnut pests. Organic growers employ a range of

* Corresponding author: sozman@omu.edu.tr

Table 11.1. Economically important hazelnut pests.

Order	Family	Species
Trombidiformes	Phytoptidae	*Phytoptus avellanae* Nal.[a]
	Eriophyidae	*Cecidophyopsis vermiformis* (Nal.)
Coleoptera	Curculionidae	*Curculio nucum* (L.)[a]
		Anoplus roboris Sufr.
	Scolytidae	*Xyleborus dispar* (F.)[a]
	Cerambyciidae	*Obera linearis* (L.)
	Chrysomelidae	*Agelastica alni* (L.)
	Scarabaeidae	*Melolontha melolontha* (L.)
Hemiptera	Pentatomidae	*Palomena prasina* (L.)[a]
	Aphididae	*Myzocallis coryli* (Goeze)[b]
	Coccidae	*Parthenolecanium corni* (Bouche)
	Diaspididae	*Lepidosaphes ulmi* (L.)
Lepidoptera	Tortricidae	*Gypsonoma dealbana* (Frölich)
		Cydia latiferreana (Walsingham)[b]
		Archips rosana (L.)
	Erebidae	*Lymantria dispar* (L.)
		Hyphantria cunea (Drury)
Diptera	Cecidomyiidae	*Mikomya coryli* (Kieffer)

[a]Major pests in Europe.
[b]Major pests in the USA.

approved chemicals, and cultural, mechanical and physical treatments to manage pests, especially by collecting and destroying the pests, pruning dried and damaged shoots and branches, and cutting weeds. Methods such as pheromone disruption of mating, oviposition deterrents, habitat enhancement and trap crops are almost never used (Ozman-Sullivan and Sullivan, 2008, 2009). A low level of pesticide application, rich plant diversity and active management of orchards can provide very good habitat for natural enemies (Işık *et al.*, 1987; Ecevit *et al.*, 1996a; Ozman-Sullivan and Sullivan, 2009), but only one classical biological control programme for hazelnut pests has been implemented. In that case, *Trioxys pallidus* (Haliday) (Hymenoptera: Aphidiidae) was introduced from Europe to the USA in 1984 to control the filbert aphid, *M. coryli*. This wasp naturalized in Oregon and Washington plantations to such an extent that aphids have essentially become a non-pest (AliNiazee, 1995, 1997b, 1998, 2001).

Many beneficial species have been reported from hazelnut orchards in Turkey, including more than 150 insect species (Ural *et al.*, 1973; Işık *et al.*, 1987; Ecevit *et al.*, 1996a; Tuncer and Ecevit, 1996; Sullivan *et al.*, 2010a, 2011a, 2012a, b, c, 2015; Sullivan and Ozman-Sullivan, 2012), more than 70 beneficial mite species (Ozman and Cobanoglu, 2001; Cobanoglu and Ozman, 2002; Ozman-Sullivan *et al.*, 2005; Ozman-Sullivan, 2006a, b) and many entomopathogens (Ural, 1957a; İren, 1970; Işık *et al.*, 1983; Ozman, 1998; Ozman and Hatat, 1999; Yaman and Demirbağ, 2000; Sezen *et al.*, 2001, 2008; Nalcacioglu *et al.*, 2002; Ozman-Sullivan *et al.*, 2009; Sullivan *et al.*, 2010b, 2011b, 2015; Sullivan, 2011; Ozsahin *et al.*, 2014). This rich complex of organisms in the hazelnut ecosystem is of great importance in the suppression of hazelnut pests. However, much more research is needed in the evaluation of biocontrol agents and their inclusion in pest management programmes. More research is also needed in the evaluation of biopesticides and the development of resistant hazelnut varieties to control pests in organic hazelnut farming.

Economically Important Hazelnut Pests

Phytoptus avellanae Nal. (Trombidiformes: Phytoptidae) (hazelnut big bud mite, filbert bud mite)

Description

The adult female of *Phytoptus avellanae* is about 0.20–0.25 mm long, whitish and vermiform, with a prodorsal shield with a pair of frontal setae and a pair of scapular setae. The opisthosoma has subdorsal setae and about 70 annuli, the empodium is four to five rayed, and the genital coverflap is smooth, except for short basal lines. Males are similar to females but slightly smaller, without a genital coverflap. The nymph resembles the adult. The eggs are round and transparent (Keifer, 1940; Jeppson *et al.*, 1975; Ecevit *et al.*, 1992a; Ozman, 2000).

Previous studies referred to the two forms as 'gall' and 'vagrant' but the latest research on *P. avellanae* has confirmed that the vagrant form is a different species (Ozman, 2000; Ozman-Sullivan, 2014).

Life cycle and damage

P. avellanae has a simple life cycle. In spring, nymphs start migrating from old 'big buds' to healthy buds from mid-April. Intensive migration normally takes place between the last week of April and the last week of May in Turkey. Across cultivation zones, the migration period of *P. avellanae* varies, depending on differences in cultivars and its ecology, especially environmental conditions, which are very influential in the migration of mites (Viggiani and Bianco, 1973; Jeppson *et al.*, 1975; Krantz, 1979; Petanovic *et al.*, 1989; Beber, 1994; Ozman and Toros, 1997a, b). The migration time is reversed in the southern hemisphere, New Zealand and Australia. In New Zealand, *P. avellanae* emerges from the winter big buds from early spring until late spring, namely the end of August to the end of November (Weber, 2007). Thousands of second stage nymphs leave open, old big buds and disperse to the leaves and shoots but only a few reach and enter the new buds. Newly infested buds are observed in June in Turkey. Nymphs that enter the healthy buds settle on the meristematic apex. Generally, the first adults are seen in mid-June. The adults lay their eggs on the meristematic apex of the bud or the surrounding tissue. The infested buds can be distinguished from the healthy buds in July. By the end of July, the generations overlap and the population increases approximately ten times. *P. avellanae* individuals remain on the inner part of the big buds until October. Later, infestation spreads to the outer scales and the mite-induced big buds become very obvious. They continue their feeding and reproduction during winter. From the end of March, the adults of the last generation begin laying eggs and the nymphs that will migrate can be seen in the big buds. *P. avellanae* has about six generations/year. Since the population is stationary within the same bud, generations overlap. *P. avellanae* is seen alone or with *Cecidophyopsis vermiformis* (Nal.). Big buds with only *P. avellanae* are referred to as 'spring big bud', and big buds with both *P. avellanae* and *C. vermiformis* are termed 'summer big bud' or 'winter big bud', depending on the time of first occurrence (Keçeci Ozman, 1995; Ozman and Toros, 1997a, b).

P. avellanae feeds exclusively on generative buds, which means that each big bud equates to the loss of a cluster of nuts (Ozman and Toros, 1997c). Newly infested buds dry out if they are heavily infested during the migration period in spring or become a 'big bud' which are 0.5–3 cm in diameter, fleshy and contain 'warted' inner cells. All the generative buds on a shoot may become big buds. Big buds dry out and fall off during the next spring and the shoots may dry out (Keçeci Ozman, 1995; Ozman and Ecevit, 1996). It is a very important pest in hazelnut orchards that causes major yield loss. The damage caused by *P. avellanae* ranges from 18% to 90% big bud incidence in different countries (Viggiani, 1973; Krantz, 1979; Tavamaishvili, 1990; Keçeci Ozman, 1995; Ozman and Toros, 1997a, b, c; Stamenkovic *et al.*, 1997; AliNiazee, 1998; Gantner, 2001). Almost 260 big buds were counted on a stem of the sensitive variety 'Mincane' in Trabzon, Turkey (Keçeci Ozman, 1995).

There are secondary effects of infestation: *P. avellanae* is one of the chief transmitters of the fungal pathogen *Gloeosporium* sp. which causes twig desiccation (Pesante, 1973). The deformed big buds also provide an entry point for Eastern filbert blight which is a very important disease in the USA (Mehlenbacher and Miller, 1989).

Distribution

This species is widely distributed in Turkey and other countries in Europe. It is also present in North America, Asia, Australia and

New Zealand (Jeppson *et al.*, 1975; Viggiani, 1984; Petanovic *et al.*, 1989; Keçeci Ozman, 1995; Ozman and Toros, 1996; Milenkovic and Mitrovic, 2001; Snare, 2006; Weber, 2007).

Hosts

P. avellanae feeds only on hazelnut species, varieties and hybrids (Ourecky and Slate, 1969; Jeppson *et al.*, 1975; AliNiazee, 1980; Keçeci Ozman, 1995; Anonymous, 2008).

Management

CULTURAL CONTROL. Genetic resistance in hazelnuts is important for the control of this pest. The level of host plant susceptibility is related to the structure of its apical meristem and bud primordia (Burgess and Thompson, 1985). Loose bud cultivars are more sensitive to big bud infestation. The Turkish cultivars, 'Tombul', 'Mincane' and 'Uzunmusa' were most susceptible and 'Aci', 'Kus' and 'Palaz' most resistant (Keçeci Ozman, 1995; Ozman and Toros, 1997c). In Europe and the USA, the cultivars 'Butler', 'Daviana', 'Ennis', 'Negret' and 'Tonda Gentile delle Langhe' are considered to be highly susceptible and 'Barcelona', 'Tonda Romana', 'Riccia di Talanico' and 'Hall's Giant' are resistant (Thompson, 1977; Mehlenbacher and Miller, 1989; Snare, 2006).

MECHANICAL CONTROL. Removing big buds during winter and early March very effectively controls big bud mites. The big buds certainly should be placed on the plantation floor, not be burned or buried. While the big bud dries, the big bud mites die, but predatory insects and mites are mobile and can move to new habitat. This method protects predatory insects, mites and parasitoids that overwinter inside big buds (Keçeci Ozman, 1995; Ozman-Sullivan, 2006c).

BIOLOGICAL CONTROL. *Orius minitus* L. (Anthocoridae), a predatory midge *Arthrocnodax coryligallarum* (Targioni-Tozzetti) (Cecidomyiidae), and a chalcid parasitoid, *Tetrastichus eriophyes* Taylor (Eulophidae), were recorded as natural enemies of *P. avellanae* (Arzone, 1983; Keçeci Ozman, 1995). There are many beneficial mites living in hazelnut orchards (Ozman and Cobanoglu, 2001; Cobanoglu and Ozman, 2002; Ozman *et al.*, 2002, 2005). Predatory mites such as *Kampimodromus aberrans* (Oud.) and *Phytoseius plumifer* (Can. and Fanz.) are present in all seasons in big buds and can have high efficacy against *P. avellanae* in the Black Sea region of Turkey (Ozman-Sullivan, 2006a, b). In a laboratory study, a gravid *K. aberrans* female consumed an average of 20 *P. avellanae* adults/day. The same study reported an average of 21 *K. aberrans* per big bud (Ozman-Sullivan, 2006a). Predatory thrips and spiders in big buds also decrease the *P. avellanae* population (Keçeci Ozman, 1995; Ozman-Sullivan, 2006c).

The fungus *Verticillium lecanii* (Zimm.) Viégas, a potential biological control agent that was 99.5% effective against big bud mites, is distributed throughout the Black Sea region (Ozman, 1998; Ozman and Hatat, 1999).

CHEMICAL CONTROL. The economic threshold for *P. avellanae* damage is five big buds per stem. A single application of sulfur 80% WP (400 g/100 l) at the end of April or the beginning of May, just before the peak migration period of the mites from infested buds to new buds, was very effective in keeping big bud numbers low. During this period, the new shoots have between four and four-and-a-half leaves, newly forming buds are half a pin head in size and newly forming fruits with husks are approximately 3 mm in diameter (Ozman-Sullivan and Akca, 2005).

***Cecidophyopsis vermiformis* (Nal.) (Acari: Trombidiformes: Eriophyidae) (hazelnut big bud mite, filbert bud mite)**

Description

The *Cecidophyopsis vermiformis* adult is smaller than *P. avellanae*, being about 0.20 mm long. It is white and vermiform, lacking all prodorsal shield setae and subdorsal setae. It has an opisthosoma with about 80 annuli, the empodium is five rayed and the genital coverflap of the female is heavily ribbed. The nymph resembles the adult. The eggs are round and transparent (Keifer, 1944;

Jeppson *et al.*, 1975; Ecevit *et al.*, 1992a; Keçeci Ozman, 1995).

Life cycle and damage

C. vermiformis has a complex life cycle with two migration periods (spring and autumn) (Krantz, 1979; Ozman and Toros, 1997a). When the big buds open in spring, *C. vermiformis* last instar nymphs and adults migrate to the healthy buds with *P. avellanae* nymphs. The migration is intensive during May. If buds are heavily infested by both species, they dry out due to heavy feeding. The *C. vermiformis* that survive continue feeding and reproducing in the slightly infested buds and contribute to summer big bud formation. In summer big buds, the two populations are approximately equal in early summer whereas *C. vermiformis* becomes the dominant species from the end of July onwards and reaches its maximum population in autumn. Summer big buds become enlarged and begin to dry out around mid-August. *C. vermiformis* adults and nymphs then begin to migrate to spring big buds occupied by *P. avellanae* and form 'winter big bud'. The migration continues until mid-October, depending on the period of summer big buds' desiccation and falling. The *C. vermiformis* population in the winter big buds increases gradually and it becomes the predominant species in spring. The generations of both species overlap and the number of generations of both species cannot be accurately determined. *C. vermiformis* does not cause big bud formation alone but its activity with *P. avellanae* leads to the formation of 'summer' and 'winter' big buds (Ozman and Toros, 1997a, b, c). It lives inside the big buds created by *P. avellanae* (Keifer, 1944; Ozman and Toros, 1997a, c). Krantz (1979) stated that spring big buds reached 18–20% of the total buds but summer big buds rarely exceeded 3–5%. Similarly, Ozman and Toros (1996) reported that the number of big buds infested by only *P. avellanae* was greater than the number infested with both species in the Black Sea region of Turkey. During the spring and autumn migration periods, some *C. vermiformis* individuals migrate to the catkins (male flowers) and female flowers, and very rarely to the vegetative buds, where they continue feeding and reproduction. Infested female flowers do not produce nuts. Infested catkins are distorted, become brittle and produce little or no pollen. If the infestation on the flowers is too heavy, they dry out and fall (Ozman and Toros, 1997a, b, c).

Distribution

This species is widely distributed in Turkey and other countries in Europe. It is also present in North America and New Zealand (Keifer, 1944; Jeppson *et al.*, 1975; Ozman and Toros, 1996; Weber, 2007).

Hosts

The only known host is *Corylus* species, varieties and hybrids (Ourecky and Slate, 1969; Jeppson *et al.*, 1975; Krantz, 1979; AliNiazee, 1980; Keçeci Ozman, 1995).

Management

C. vermiformis is managed in the same way as *P. avellanae*.

Curculio nucum (L.) (Coleoptera: Curculionidae) (hazelnut weevil)

Description

The *Curculio nucum* adult is 6–8.5 mm in length, greyish brown in colour and covered with a strong pubescence. The rostrum is slender, strongly curved, longer than the body in the female, and shorter in the male. The eggs are oval, white and < 1 mm in length. The larva are 10–12 mm long, apodous, curved and creamy white with a brown head. The pupa is of the exarate type (Ural, 1957a; Anonymous, 1997a; Anonymous, 2008; Alford, 2014).

Life cycle and damage

C. nucum overwinters as full-grown diapausing larvae in the soil. In March, the overwintering larvae begin to pupate and adults appear. They feed on the same plant below 20°C (almost inactive below 16°C), and when the temperature increases, they start to fly. Adult emergence peaks in early May in Turkey and they can live for 3 months. Mating and oviposition begin in May and continue during July. Females produce an average of 42 eggs and they lay one egg under the shell of each nut. There is a brown line 4–5 mm in length on the surface of the shell above where the egg was laid. Embryonic development lasts 8–10 days. After egg hatching, the larva penetrates the kernel, feeds and completes development in 4–5 weeks. The larva comes out of a 1.5–2.0 mm diameter hole and enters the soil. It forms a hard, silken coccoon among the roots of the host plant at a depth of 10–15 cm and overwinters. Some overwintering larvae may stay in diapause for 2–3 years. There is only one generation/year. Damage of 20–70% of the crop has been reported (Ural, 1957a; Viggiani, 1984; Işık *et al.*, 1987; Tavamaishvili, 1990; Anonymous, 1997a; Gantner, 2001; Akca, 2003; Akca and Tuncer, 2005). The adults cause damage by feeding on opening buds, young foliage, leaves and flowers in early spring. When the new nuts start to develop, they feed on them and cause many to drop prematurely, or prevent full development, and their yellowish shells don't harden properly and shrink. Also, the feeding punctures of the adult on young hazelnuts cause deformation. If the larva feeds on the developed nut, the inside of the nut becomes darker and rots and then a black secretion exudes from the nut. One adult can damage 80 nuts through its feeding (Ural, 1957a; Anonymous, 2008). One male and female pair can cause damage to an average of 200 nuts, which equates to approximately 0.5 kg, by feeding and egg laying (Ural, 1957a). Milenkovic and Mitrovic (2001) stated that feeding punctures create conditions that favor infection by *Monilia* spp.

Distribution

C. nucum is a key pest in most hazelnut orchards in Turkey and in Europe (Ural, 1957a; Viggiani, 1984; Işık *et al.*, 1987; Tavamaishvili, 1990; Gantner, 2001). It is normally found at higher densities at higher elevations and in orchards close to forest areas than it is at low elevation because of the high water table in alluvial plains (Ural, 1957a; Kurt, 1982; Işık *et al.*, 1987; Tuncer and Ecevit, 1997).

Hosts

The adult feeds on hazelnut and on the fruits of various trees such as pear, peach and apple (Ural, 1957a; Anonymous, 1997a; Anonymous, 2008; Alford, 2014).

Management

CULTURAL CONTROL. Ploughing the soil without damaging roots is effective in decreasing *C. nucum* numbers. Collecting the adults with the 'beating sheet' method is also effective (Ural, 1957a; Anonymous, 2008).

Most of the thin-shelled nut varieties appear to be more susceptible. Preliminary studies indicated that the most common and marketable Turkish cultivars, 'Tombul', 'Palaz' and 'Çakıldak', are susceptible to nut weevil (Ecevit *et al.*, 1996b). The varieties 'Hall's Giant' and 'Lenka II' are more resistant to hazelnut weevil than the other tested hazelnut cultivars in Poland (Piskornik, 1994; Wojciechowicz-Żytko, 2005). Guidone *et al.* (2007) stated that early nut development is an important factor in resistance to the attacks of *C. nucum*.

BIOLOGICAL CONTROL. Because *C. nucum* spends most of its life in the nut, it is difficult for biological control agents to be effective against this pest. Ural (1957a) stated that *Beauveria densa* Link had some efficacy (5–18%) against *C. nucum* and that birds were predators. Sezen and Demirbağ (1999) isolated five bacteria, namely *Bacillus thuringiensis* Berliner, *Pseudomonas fluorescens* (Flügge), *Micrococcus luteus* (Schroeter), *Serratia marcescens* Bizio and *Escherichia coli* (Migula), from *C. nucum*, with *S. marcescens* having 100% biocidal activity against *C. nucum* larvae under laboratory conditions. Paparatti and Speranza (2005) reported 99.5% mortality for larvae treated with *Beauveria bassiana* (Bals.-Criv.) Vuill. while untreated larval mortality was 63.5%. Entomopathogenic nematodes also have the potential to be effective in the control of *C. nucum*. Peters *et al.* (2007) stated that *Heterorhabditis bacteriophora* (Poinar) is the most promising candidate for controlling the hazelnut weevil. Batalla-Carrera *et al.* (2013) reported that the *Steinernema feltiae* (Filipjev), *Steinernema* sp. and *H. bacteriophora* were used in spring and summer applications and reduced hazelnut weevil numbers in the range of 32–88%.

CHEMICAL CONTROL. Pesticides are applied from the end of April to the end of May at the beginning of fruit development (Milenkovic and Mitrovic, 2001; Anonymous, 2008). Spinosad 480 g/l, produced by the fermentation of a bacterium, *Saccharopolyspora spinosa* Mertz and Yao, is used in organic hazelnut growing at 150 ml/ha (Anonymous, 2015).

Anoplus roboris Suffr. (Coleoptera: Curculionidae) (hazelnut leaf holer)

Description

The adult body length of *Anoplus roboris* is about 2 mm, and it is black with short, white hair. The antennae have 11 segments and the elytra are marked with longitudinal lines. The larvae are cream in colour with a black head, and the last instar larva is about 3–4 mm long (Hoffman, 1954; Ecevit *et al.*, 1993).

Life cycle and damage

A. roboris overwinters as the adult in the soil or in other protected areas. They become active from March and feed on the buds and leaves, causing holes in the leaves. After 1–2 weeks feeding, at the beginning of April they start laying eggs, generally on the underside of the main vein and sometimes on a thick lateral vein. In total, they lay about 55 eggs. Larvae emerge in 9–15 days and open a gallery that runs towards the leaf margin between the upper and lower epidermis. Larval development takes 20–25 days. Mature larvae enter the soil and pupate. Pupal development takes 7–12 days and the new generation adults are seen in June. The adult lifespan is about 3 months and there is one generation/year. The leaf often tears as a result of feeding and egg deposition. It can cause economic damage that varies according to the year and location. If the population is high, they can cause 20% damage to the buds and leaves (Işık *et al.*, 1992; Anonymous, 2008).

Distribution

This species is found in Europe and Turkey (Hoffman, 1954; Işık *et al.*, 1992, Snezana, 2013).

Hosts

Hazelnut, alder and oak are the host plants for *A. roboris* (Işık *et al.*, 1992; Anonymous, 2008).

Management

BIOLOGICAL CONTROL. *Bacillus circulans* Jordan, *Bacillus polymyxa* (Prazmowski), *Bacillus sphaericus* Neide and *Enterobacter* sp. were isolated from *A. roboris*, with *B. sphaericus* having 67% efficacy under laboratory conditions (Demir *et al.*, 2002).

Xyleborus dispar (Fabricius) (Coleoptera: Scolytidae) (European shot-hole borer, ambrosia beetle)

Description

Xyleborus dispar has a short, broad body that is strongly convex and shiny, and mostly brownish black. Its head is almost black, while the antennae and legs are light brown. The thorax and elytra have a sparse coating of yellowish hairs; the thorax is rounded with numerous small projections anteriorly. The antennal club is articulated. There are morphological differences between the sexes: (i) the female body length is 3.0–3.5 mm, twice as long as it is wide, while the male is 2 mm long; (ii) the female is more elongate and cylindrical than the male; and (iii) the male thorax is relatively small and the abdomen is short, giving the body a hunch-backed appearance. The eggs are oval and 0.8–0.9 mm in length, pearly white and shiny. The larvae and pupae are 4.0–5.0 mm long and creamy white (Anonymous, 2008; David'yan, 2008a; Alford, 2014; CABI, 2014).

Life cycle and damage

X. dispar has one generation/year. It hibernates in the adult stage inside galleries tightly packed one behind the other in the stem or trunk. In the following spring, females leave the galleries through the parental entrance hole and begin flying from early March as the air temperature reaches about 18°C. Their emergence usually continues for 4–8 weeks. The males are unable to fly and mating usually takes place in the gallery. After emergence, females bore into the same tree or find another suitable host. The females bore a short, 1–3 cm deep radial entrance tunnel before excavating a transverse tunnel to both sides. From each of these, cylindrical breeding galleries are produced, directed perpendicularly both upwards and downwards. In smaller trunks or in branches, the galleries are often simpler. As they are boring, females passively inoculate the new gallery with the symbiotic ambrosia fungus, *Ambrosiella hartigii* Batra. When the ambrosia fungus has become established on the walls of the breeding passages, 10–15 days after opening the galleries, the female starts to lay eggs. Females lay about 50 eggs in total. Larvae emerge 8–10 days after oviposition and feed exclusively on the fungus growing in the galleries for 3–4 weeks and then pupate. The pupal period is 8–10 days and the new generation appears in summer but stays in the galleries until the next spring. Full development takes about 2 months. Due to the long flight and oviposition period of females, different developmental stages can be found together in a gallery. The average number of beetles in a gallery is about 25, with a maximum of about 40. The males are much rarer than the females, particularly in summer, comprising about 10% of populations (Işık, 1970; Mani *et al.*, 1992; Anonymous, 2008; Alford, 2014; CABI, 2014).

The beetles attack trunks and larger branches. Attacked trees have delayed growth and wilt and often die within a short time. In general, *X. dispar* attacks only stressed trees already damaged by frost, drought, disease or root feeders but when populations are high, it may become a primary species, attacking healthy trees, especially hazelnut, apple, pear and apricot (Kühnholz *et al.*, 2003; CABI, 2014). Ak *et al.* (2005a) stated that damage began to increase in April and the major damage occurred in July and August in hazelnut orchards in the Black Sea region of Turkey.

Distribution

X. dispar is a widespread species in Europe, Asia, North Africa and North America (CABI, 2014) that has become a serious pest in some coastal hazelnut-growing areas of the Black Sea region of Turkey in recent years (Ak *et al.*, 2005a, b; Saruhan and Akyol, 2013).

Hosts

X. dispar is a highly polyphagous species. It is a primary and secondary pest of a wide variety of fruit, nut and forest trees (Kühnholz *et al.*, 2003; Ak *et al.*, 2005a, b; Alford, 2014; CABI, 2014).

Management

CULTURAL CONTROL. Fertilization, pruning and burning of infested trees and branches, and using trap trees and trap branches for adults are effective in decreasing numbers of *X. dispar* (Borden *et al.*, 2003; Kühnholz *et al.*, 2003; CABI, 2014).

BIOLOGICAL CONTROL. There are natural enemies of *X. dispar* but they do not provide effective control. Adults of ambrosia beetles are predated by lizards, clerid beetles and ants as they attempt to bore into the host tree. *Colydium elongatum* (F.) (Coleoptera: Colydiidae) is another predator of *X. dispar* (David'yan, 2008a). A few species of Hymenoptera have been recorded from *X. dispar*: *Perniphora robusta* Ruschka, *Habritys brevicornis* (Ratzeburg) and *Vrestovia querci* Yang attack ambrosia beetles. *Eurytoma morio* Boheman is a polyphagous species that attacks both scolytids and their parasitoids (Noyes, 2003; CABI, 2014). Schvester (1950) found a nematode, *Parasitylenchus dispar xylebori* Schvester, in the body cavity of adult females. Canganella *et al.* (1994) collected *Pseudomonas chlororaphis* (Guignard and Sauvageau) and *Bacillus subtilis* (Ehrenberg) from *X. dispar* and in galleries. Sezen *et al.* (2008) isolated five bacteria from *X. dispar*: *Pseudomonas fluorescens* (Flügge), *Bacillus megaterium* de Bary, *B. thuringiensis* Berliner, *Pseudomonas rhizosphaerae* Peix *et al.* and *Pantoea cedenensis* Rios & Rojas.

BIOTECHNICAL CONTROL. Different kinds of traps are used to control *X. dispar*. The red wing trap with ethanol (Rebell rosso) is very useful for control and monitoring (Mani *et al.*, 1992). For monitoring, one or two traps/ha of orchard are placed in spring when maximum temperatures rise above 17°C (Ak *et al.*, 2005a; CABI, 2014). Speranza *et al.* (2009) stated that mass trapping should start at the beginning of March in Italy. Under favourable weather conditions, the lure, 250 ml 50% ethanol denatured with 1% toluene, has to be replaced every 2–3 days. When catches reach 20 beetles/trap/day, the risk of attack of economic importance is indicated. Eight traps/ha need to be placed in an 'at risk' orchard for control of *X. dispar*. The CSalomon® Palx trap is also used to capture *X. dispar*. The attractant-baited trap should be placed on the trunks of trees or hung from lower branches at a height of 1–1.5 m. Depending on the warmth of the weather, the 20% ethanol in the liquid bait remains attractive for at least 8–10 days before refilling is necessary (Anonymous, 2014b). The Lindgren funnel trap is used for monitoring adult flight (David'yan, 2008a). Ak *et al.* (2014) stated that the funnel-type and ribbed cage trap are more effective and practical than the red wing traps.

Obera linearis (L.) (Coleoptera: Cerambycidae) (longhorn beetle, twig borer)

Description

The adult *Obera linearis* is slender, 11–15 mm long, 3–4 mm wide, with a body that is entirely black and with yellowish legs. The antennae are long but not longer than the body. The female has shorter antennae than the male's antennae. The larva has the characteristic form of cerambycid larvae, apodous, yellowish or light brown, with a brown prothoracic shield. The final length is 20–25 mm. The eggs are 3 mm × 0.6 mm (Balachowsky, 1962; Kurt, 1982; Anonymous, 2008; Tzanakakis, 2008).

Life cycle and damage

O. linearis completes one generation every 2 years but in colder regions there is one generation every 3 years. The female lays eggs singly in slits she gnaws in the bark tissue of 1-year-old twigs, 10–15 cm from the tip. The first instar larva bores a gallery which kills the apical part of the shoot. The gallery, which reaches 40–60 cm in length in the first year, is directed towards the base of the twig, whereas in the second year it is directed towards the apex of the twig. It spends the first winter as a relatively young larva and in the second winter is a fully grown larva at the end of its gallery. It pupates in April and the adults appear in May–June. *O. linearis* infests 1–3-year-old shoots and causes damage by larval feeding in tunnels. Infested 1-year-old twigs wilt, dry up and split at the level of the egg-laying scar. Heavy infestations have an adverse effect on growth and may also reduce yield. It causes economic damage in some years and localities (Balochowsky, 1962; Kurt, 1982; AliNiazee, 1998; Milenkovic and Mitrovic, 2001; Saruhan and Tuncer, 2001; Anonymous, 2008; Tzanakakis, 2008; Marras *et al.*, 2009; Alford, 2014).

Distribution

O. linearis is found in Europe, Russia, the Caucasus and Turkey (Hoskovec and Rejzek, 1997). It is sometimes seen in high numbers in coastal hazelnut-growing areas in Turkey and also in Serbia (Kurt, 1982; Milenkovic and Mitrovic, 2001; Saruhan and Tuncer, 2001; Anonymous, 2008).

Hosts

Hazelnut is the preferred host. Rarely, it is found on walnut, elm, alder, hornbeam, poplar and ash (Balochowsky, 1962; Anonymous, 2008).

Management

CULTURAL CONTROL. Pruning and burning of the infested twigs, the best method of controlling this pest, has to be done from mid-July, when the damage starts, until the autumn, when infested twigs are obvious (Kurt, 1982; Milenkovic and Mitrovic, 2001; Anonymous, 2008; Marras *et al.*, 2009). Infested twigs should be be cut at least 5–10 cm below the dry part.

BIOLOGICAL CONTROL. Thirteen bacteria isolated from *O. linearis* were tested against its larvae, with *Serratia marcescens* Bizio causing 65% mortality. *Bacillus thuringiensis* Berliner isolated from *Melolontha melolontha* L. caused 90% mortality of *O. linearis* (Bahar and Demirbağ, 2007).

Agelastica alni (L.) (Coleoptera: Chrysomelidae) (alder leaf beetle)

Description

The adult *Agelastica alni* is 6–8 mm long, metallic dark blue to violet in colour, with black antennae, tibiae and tarsi. The body is rather bulbous, with elytra noticeably expanded towards the hind end. The antennae have 11 segments. The eggs are ovoid to cylindrical and yellow. The larvae are shiny black, up to 10–12 mm long (Ural, 1957b; Kurt, 1982; Alford, 2014).

Life cycle and damage

There is a single generation annually. Overwintering adults emerge in April. They mate after feeding for 2–3 weeks and lay eggs in large, scattered groups on the host's leaves. The oviposition period is 1–1.5 months and females lay 500–600 eggs in total. The eggs hatch in about 2 weeks. Larvae are present from June to July, moulting three times and becoming fully grown in about 1 month. They drop to the ground to pupate. Adults appear 1–2 weeks later, from mid-July onwards, mostly in August and September. They overwinter under fallen leaves and mosses and other protected places (Ural, 1957b; Kurt, 1982; Alford, 2014).

From April onwards, adults and larvae cause holes and gaps in the leaves while they feed together. At first, larvae graze the surface, but later they chew out holes between the major veins. Heavily infested older trees are weakened and their growth is retarded but younger trees can die (Ural, 1957b; Kurt, 1982; Alford, 2014).

Distribution

A. alni is found in Europe, Turkey, the Caucasus, Siberia and north-eastern Kazakhstan, and the USA where it was accidently introduced (Ural, 1957b; Kolk and Starzyk, 1996).

Hosts

It is an important and destructive pest of alder and other broadleaved trees, including fruit and ornamental trees (Ural, 1957b; Alford, 2014).

Management

BIOLOGICAL CONTROL. Laboratory studies have determined the efficacy of some biological control agents against *A. alni*. Sezen *et al.* (2004) reported that *Pseudomonas fluorescens* (Flügge) isolated from *A. alni* had the highest effect on both larvae and adults at 70% and 56%, respectively. Sezen and Demirbağ (2006) stated that *Bacillus thuringiensis* biovar. *tenebrionis*, *Bacillus sphaericus* (isolated from *Anoplus roboris*) and *B. thuringiensis* (isolated from *Melolantha melolontha*), caused 90% mortality of the larvae of *A. alni*. *B. thuringiensis* isolated from *Thaumetopoea pityocampa* (Den. and Schiff.) caused 100% mortality (Kati *et al.*, 2009). *Beauveria bassiana* (Bals.-Criv.) Vuill. isolates obtained from forest soils were pathogenic (Rumine *et al.*, 2007), and pre-pupal and pupal stages were very susceptible to *Heterorhabditis megidis* Poinar *et al.* in a laboratory and semi-field experiment (Tomalak, 2004). Bekircan *et al.* (2014) reported that the essential oil of *Thymus leucotrichus*

Hal. was an effective antifeedant against the larvae under laboratory conditions.

Melolontha melolontha (L.) (Coleoptera: Scarabaeidae) (cockchafer, May bug)

Description

The adult *Melolontha melolontha* is a large beetle, 25–30 mm long, chestnut-brown in colour, with a darker head and thorax, and partly coated in whitish hairs. The elytron has four ribs, and the abdomen terminates in a blunt, downwardly directed spine. There are five white triangular spots on the side of the abdomen. The head has a pair of fan-like antennae; in males, these are branched with seven leaf-like lamellae that are wider and two times larger than in females which have six lamellae. The eggs are oval, 2–3 mm long, and are whitish or yellowish. The larva is campodeid and C shaped, 40–45 mm long in the last instar, whitish yellow, with the head and legs being brown and shiny. The larva has three instars. The pupa is exarate, 25–35 mm long, whitish to brown (Ural, 1968; Anonymous, 2008; Malysh and Frolov, 2008; Alford, 2014).

Life cycle and damage

Adults usually emerge from the second half of April until the beginning of June and live for about 5–7 weeks. A few weeks after emergence, the females burrow into the soil to a depth of 15–25 cm and lay eggs in batches of 25–30, and about 60 in total. They usually hatch after about 4 weeks. Initially the larvae consume humus particles and then small roots of small seedlings and weeds. They become second instar larvae after 2 months and third instar larvae during the next summer. The third instar larval duration is about 1 year. The larvae attack plant roots, feeding for up to 3 years and overwintering at 50–60 cm depth in the soil. Pupation occurs in July of the third summer in an earthen cell 15–35 cm below the surface. Adults are formed by about 6 weeks later but they do not emerge from the soil until the following spring. The most notable damage is by second and third instar larval feeding on both small and large roots from May to September, seriously restricting growth; in severe cases, plants wilt and die. Adults are nocturnal, feeding at night on the leaves, buds, flowers and foliage of various trees and shrubs (Ural, 1968; Anonymous, 2008; Malysh and Frolov, 2008; Alford, 2014).

Distribution

M. melolontha is widely distributed in Europe. It is common in hazelnut-growing areas in Turkey but only causes economic damage locally in some areas (Ural, 1968; Anonymous, 2008; Malysh and Frolov, 2008; Alford, 2014).

Hosts

It is a common pest of the underground parts of various plants, including vegetables, herbaceous ornamentals, and trees and shrubs (Anonymous, 2008; Alford, 2014).

Management

CULTURAL CONTROL. Killing the larvae by cultivation of the soil is very important in reducing populations. Manure and hazelnut husks should not be kept around hazelnut orchards for a long period in order to prevent infestations and population increases (Anonymous, 2008).

MECHANICAL CONTROL. Adults should be killed when they are seen. If there is yellow azalea (*Rhododendron luteum* L.) around orchards, adults are attracted to its flowers (Anonymous, 2008). Adults may be caught and destroyed using light traps (Malysh and Frolov, 2008).

BIOLOGICAL CONTROL. Moles, badgers, bats, birds, ground beetles, large wasps and tachinid flies may notably decrease beetle numbers. The developmental stages of *M. melolontha* are naturally infected by the fungi *Beauveria brongniartii* (Saccardo) Petch and *B. bassiana*, the bacteria *Bacillus thuringiensis* Berliner and *Coccobacillus* sp., and the microsporidia *Pleistophora melolonthae* H. and *Thelohania* sp. *B. bassiana*-based mycoinsecticides produced good results (Malysh and Frolov, 2008). Keller *et al.* (2002) used *B. brongniartii* against *M. melolontha* adults. Sezen *et al.* (2007) isolated seven bacterial strains from *M. melolontha* and evaluated their pathogenicity against *M. melolontha* larvae. They reported that *B. thuringiensis* and *Bacillus weihenstephanensis* Lechner *et al.* may be useful as biological control agents. Sezen *et al.* (2008) stated that *B. thuringiensis* subsp. *tenebrionis* isolated from *X. dispar* had 100% efficacy against *M. melolontha* larvae under laboratory conditions. Yaman and Demirbağ (2000) isolated *B. thuringiensis* and *Bacillus sphaericus* Neide from *Hyphantria cunea* (Drury) and *Pierris brassicae* (L.), respectively. These isolates caused 44% and 32% mortality, respectively, of *M. melolontha.* Ugras and Demirbağ (2013) isolated *Serratia marcescens* from *M. melolontha*, and Paris and Segrétain (1975) reported that nine strains of *Beauveria tenella* (Sacc.) caused more than 32% mortality of *M. melolontha.* Fatu *et al.* (2015) used a bioproduct based on *B. brongniartii* against *M. melolontha* and it decreased the larval density below the economic threshold.

Palomena prasina (L.) (Hemiptera: Pentatomidae) (green stink bug, green shield bug)

Description

Adults of *Palomena prasina* are 11–14 mm long, with a dark wing membrane. The antennae are five segmented. The tarsus and fourth and fifth segments of the antennae are reddish. In the spring and summer, adults are bright green, becoming bronze-brown before winter hibernation. Later nymphs are often darker than earlier in the season. Newly emerged adults may have a pale wing membrane, leading to possible confusion with *Nezara viridula* (L). The eggs are light green and 1.3 mm long (Kurt, 1975; Anonymous, 2008; Bantock and Botting, 2013).

Life cycle and damage

P. prasina overwinters as an adult under fallen leaves and debris or in protected places. The adults emerge from hibernation at the end of March or in April when the temperature reaches 18–20°C. They feed for a month and then mate in June. The female lays her eggs in hexagonal batches of 14–28 on the underside of the hazelnut leaf. A single female can lay three to four egg batches and a total of around 100 eggs which hatch in 8–10 days. First stage nymphs remain together in sibling communities. This species passes through five nymphal stages. Each stage has a different coloration, and the final stage has short wings. The total duration of nymphal stages is about 2 months and they can be found from June to October. The first adults are seen in July and they hibernate in November. *P. prasina* produces one generation/year (Kurt, 1975; Saruhan, 2004; Anonymous, 2007; Saruhan and Tuncer, 2009).

The nymphs and adults feed on the foliage and nut clusters of many trees and shrubs. In Europe and Turkey, *P. prasina* is the most widespread of the bugs infesting hazelnut and sometimes causes substantial damage characterized by premature fruit drop and defects like blanks, and shrivelled or spotted kernels. This pest has caused damage to approximately 5% of kernels in Turkey which causes problems for exporters. In north-western Italy, the percentage of damaged kernels varied from 1.3% to 4.0% (Kiper and Yücetin, 1971; Kurt, 1975; Işık *et al.*, 1987; Tavella *et al.*, 2001; Saruhan, 2004; Tuncer *et al.*, 2005; Saruhan and Tuncer, 2009; Polajnar *et al.*, 2013).

Distribution

P. prasina has a Palearctic distribution and is common in Europe (Kurt, 1975; Alford, 2014).

Hosts

P. prasina is a polyphagous species that can be found in many habitats, including parks and gardens (Anonymous, 2008; Bantock and Botting, 2013; Alford, 2014).

Management

BIOLOGICAL CONTROL. *Gymnosoma rotundatum* L. (Tachinidae) and *Aridelus* sp. (Braconidae) are parasitoids of the nymph and adult of *P. prasina*, and *Trisolcus grandis* Thoms. (Scelionidae) is an egg parasitoid. These adult and nymph parasitoids cause 11–39% mortality of overwintering adults, depending on the year (Kurt, 1975). Furthermore, larvae of the tachinid flies *Cylindromyia brassicaria* (Fabricius) and *Phasia hemiptera* (Fabricius) have been reported as endoparasitoids of *P. prasina* (Anonymous, 2007). Ozsahin *et al.* (2014) stated that *Lysinibacillus sphaericus* (Meyer and Neide), *B. thuringiensis* and *Raoultella terrigena* (Izard *et al.*) isolated from *P. prasina* were 60%, 70% and 60% effective, respectively, against the adults of *P. prasina* under laboratory conditions.

Myzocallis coryli (Goeze) (Homoptera: Aphididae) (filbert aphid, European hazelnut aphid)

Description

The adults of *Myzocallis coryli* are small, soft-bodied, light to dark green and approximately 2.5 mm long; the nymphs are about half that size. The eggs are oval shaped and pale yellow, turning shiny black before hatching. Newly hatched nymphs are white to pale yellow (AliNiazee, 1980).

Life cycle and damage

The hatching of overwintered eggs begins in late February and early March and continues for 3–4 weeks. Nymphs moult four times and produce winged adults that give birth without sexual reproduction. In early spring, filbert aphids can be found on unfolding buds; later they occur on the undersides of leaves and occasionally on young nuts and husks. Due to the cool temperatures in early spring, the nymphs develop slowly and adults do not appear until late April or early May. Their population usually peaks in June and early July. Later in July and August, aphid numbers decline because of high temperatures. They complete six to eight parthenogenetic generations through the summer. The winged form of the filbert aphid, which develops in the autumn, produces oviparous adults. Males also appear in late autumn. The oviparous adults deposit eggs that overwinter in rows on twigs and branches (AliNiazee, 1980, 1983).

M. coryli prefers to feed on younger leaves. Damage is caused by large numbers sucking sap from the leaves which become yellow and drop. Feeding damage reduces both tree vigour and nut quality by causing reduction in nut size and empty nuts. Aphids also secrete large quantities of honeydew which may cause severe leaf burn, scorch or staining of the nuts. Leaves with honeydew become susceptible to the fungus known as fumagina (*Capnodium* sp.) which decreases photosynthesis (AliNiazee, 1980, 1983, 2001; AliNiazee and Messing, 1995). In Chile, *M. coryli* is the most important foliar feeder on hazelnuts (Aguilera *et al.*, 2014).

Distribution

M. coryli is found in hazelnut-growing areas in Europe, Turkey, south-west Asia, North and South America, Australia and New Zealand (AliNiazee, 1980; Viggiani, 1984; Işık *et al.*, 1987; Ecevit *et al.*, 1996b; Gantner, 2001; Snare, 2006; Yarahmadi and Rajabpour, 2012; Aguilera *et al.*, 2014).

Hosts

Corylus species are the main hosts of *M. coryli* (AliNiazee, 1983; Blackman and Eastop, 1985).

Management

CULTURAL CONTROL. Nitrogen, which can encourage excessive young growth which is attractive to aphids, should not be overused. In addition, weeds have to be controlled to avoid the build up of high aphid populations (Snare, 2006). Ecevit *et al.* (1996b) and Wojciechowicz-Żytko (2003) tested the susceptibility of hazelnut varieties to *M. coryli* and they recorded no differences.

BIOLOGICAL CONTROL. *Trioxys pallidus* Halliday (Hymenoptera: Braconidae) is an effective parasitoid of the filbert aphid. Mass rearing and release of *T. pallidus* has reduced the aphid population by nearly 75% across the hazelnut orchards in Oregon and Washington (AliNiazee and Messing, 1995; AliNiazee, 1997a, b, 2001).

In unsprayed orchards, aphids are controlled by their natural enemies and are rarely a problem (AliNiazee and Messing, 1995; AliNiazee, 1997b). Messing and AliNiazee (1985) reported 55 species of predators of aphids in hazelnut orchards in Oregon. The most important predators were *Adalia bipunctata* (L.) and *Cycloneda polita* Casey (Coccinellidae), *Deraeocoris brevis* (Uhler) (Miridae), *Hemerobius* spp. (Hemorobiidae) and *Chrysopa* spp. (Chrysopidae). A number of parasitoids were also found, including *Mesidiopsis* spp. (Hymenoptera: Aphelinidae). The pathogenic fungus, *Triplosporium fresenii* (Nowakowski) infects the filbert aphid but is only seen at very high aphid densities. More than 30 predators of *M. coryli*, including *A. bipunctata*, *Coccinella septempunctata* (L.), *Chrysoperla carnea* (Steph.) and *Orius* spp., and two parasitoids, *Apanteles pallipes* (Reinhard) and *Bracon* sp., have been reported from hazelnut orchards in Turkey (Işık *et al.*, 1987; Ecevit *et al.*, 1992b, 1996a; Tuncer *et al.*, 1997). *A. bipunctata* was also reported from Serbian orchards (Milenkovic and Mitrovic, 2001) and the larvae of *Adalia angulifera* Mulsant are a good predator of *M. coryli* in Chile (Aguilera *et al.*, 2014).

CHEMICAL CONTROL. Oil sprays can be used to smother eggs (AliNiazee, 1980; Snare, 2006). Azadirachtin was also effective on filbert aphid (Tuncer and AliNiazee, 1998). Mıcık and Akça (2011) reported that various concentrations of azadirachtin and spinosad had 100% efficacy against *M. coryli* nymphs under laboratory conditions.

***Parthenolecanium corni* (Bouche) (Hemiptera: Coccidae) (brown scale, European fruit scale, European fruit lecanium scale)**

Description

The non-mobile female of *Parthenolecanium corni* is chestnut brown, 4–6 mm long, 2–4 mm wide, 1.5–2 mm in height and oval, very convex with a roughened surface. The antennae are seven segmented. The adult male is brown with a single pair of wings, and the antennae are ten segmented. The first instar nymph (crawler) is 0.4 mm long, oval, flat, pale greenish to orange and light brown. The egg is about 0.3 mm long, oval, whitish and shiny (Kurt, 1982; Anonymous, 2008; David'yan, 2008b; Alford, 2014).

Life history and damage

P. corni generally has one generation/year but two to three generations may occur. The female lays 1000–3000 eggs under her body on twigs in April and May. Eggs hatch in approximately 30 days. First instar nymphs move to the young leaves where they settle near leaf veins and on twigs to begin feeding. The second instar nymphs appear in August. They continue feeding on the young growth but in the autumn, before the leaves fall, they migrate to the twigs and branches to overwinter. After overwintering, they move to young twigs, settle and feed when the sap flows in the spring. Female nymphs grow rapidly in March and become adult in April. Males begin flying at the beginning of May when they can be about 5% of the population on hazelnut and plum. Infestations disfigure and weaken host plants, and often cause premature leaf fall. Heavy infestations kill small branches and stunt tree growth. Young trees and bushes are more susceptible to injury. Infestation is high in shaded and unpruned plantations. Heavy infestations also produce large quantities of honeydew on which saprophytic fungi grow, preventing photosynthesis (Kurt, 1982; Ecevit *et al.*, 1987; Anonymous, 2008; David'yan, 2008b; Alford, 2014).

Distribution

The present distribution of *P. corni* is throughout Europe and also in Asia, North Africa, North America, Argentina, Australia and New Zealand (Kurt, 1982; David'yan, 2008b; Alford, 2014).

Hosts

P. corni is a polyphagous species causing damage to fruit and forest trees and shrubs, including ornamental plants (Anonymous, 2008; Alford, 2014).

Management

MECHANICAL CONTROL. Pruning of plantations and cutting and removing infested shoots and branches help control *P. corni* (Anonymous, 2008; David'yan, 2008b).

BIOLOGICAL CONTROL. *Coccinella septempunctata* (L.), *Vibidia duodecimguttata* (Poda), *Propylae quatuordecimpunctata* (L.) (Coccinellidae) and *Chrysoperla carnea* (Steph.) (Chrysopidae) are predators of scale insects in hazelnut orchards in Turkey (Ecevit *et al.*, 1996a). *Blastothrix confusa* Erd., *Microterus sylvius* (Dalm.), *Coccophagus lycimnia* (Walk.) and *Metaphycus insidiosus* Morc. are parasitoids of *P. corni* (David'yan, 2008b). Japoshvili and Karaca (2007) reported 21 encyrtid parasitoids of *P. corni* in Turkey, and Japoshvili *et al.* (2008) reported 32 species of chalcidoid parasitoids from *P. corni* in Georgia and south-eastern Europe. The most common parasitoid was *Blastothrix longipennis* Howard (Encyrtidae). *Verticillium lecanii* (Zimm.) Viegas and *Cordyceps clavulatus* (Schw.) El. et Ev. occur naturally in the Black Sea region of Turkey and can cause high mortality of scale insects under humid conditions. Pesticides should be selected to cause minimal harm to *V. lecanii* and *C. clavulatus* (İren, 1970; Işık *et al.*, 1983; Ecevit *et al.*, 1987; Anonymous, 2008; Ozman-Sullivan *et al.*, 2009). Işık *et al.* (1983) reported that a 10^8 suspension of *V. lecanii* was 90% effective against *P. corni* in a field trial in a hazelnut orchard in Turkey.

CHEMICAL CONTROL. Azadirachtin (10 g/l) is used at 300 ml/100 l against this pest in hazelnut orchards in Turkey (Tosun and Onan, 2014).

Lepidosaphes ulmi (L.) (Homoptera: Diaspididae) (mussel scale, oystershell scale)

Description

The adult female scale is about 2.5–3.5 mm long, elongate, often 'comma' shaped and grey to yellowish brown. The body under the scale is 1–2 mm long, milky white and not mobile. Eyes and legs are absent, and antennae are non-segmented. The scale of the male nymph is the same shape as the female but smaller. The adult male has eyes, antennae, three pairs of legs and one pair of wings. The body length is 0.5 mm. The first instar nymph (crawler) is 0.4 mm long, yellowish brown and mobile. Eggs are 0.3 mm long, elongated and oval shaped and white (Griswold, 1925; Anonymous, 2008; David'yan, 2008c; Alford, 2014).

Life cycle and damage

Lepidosaphes ulmi mostly has one generation/year. Eggs overwinter under the female's scale. Nymphs emerge at about 8°C at the beginning of May. The female and male have two and four nymphal stages, respectively. First instar nymphs are mobile but settle on twigs and branches 1–2 days after hatching. The first moult occurs after 10–15 days. The second instar nymph has no eyes or antennae, and development lasts about 1 month. Females appear in June and July, lay about 100 eggs in late August and September, then die. Not all colonies produce males and bisexual and parthenogenetic populations without morphological differences are known. *L. ulmi* lives on trunks, branches and to a lesser degree on fruits, causing defoliation and death of branches, and sometimes of the entire plant (Griswold, 1925; Kurt, 1982; Ecevit *et al.*, 1987; Anonymous, 2008; David'yan, 2008c).

Distribution

L. ulmi has almost worldwide distribution, including Southern America, Australia and New Zealand (Griswold, 1925; David'yan, 2008c; Alford, 2014).

Hosts

L. ulmi is a polyphagous species, having more than 150 hosts. It feeds on various fruit and forest trees, shrubs and ornamental plants, including hazelnut (Griswold, 1925; Anonymous, 2008; David'yan, 2008c; Alford, 2014).

Management

MECHANICAL CONTROL. Pruning of plantations, and cutting and removing infested branches before egg hatching are preventive controls against *L. ulmi* (Anonymous, 2008; David'yan, 2008c).

BIOLOGICAL CONTROL. *Chilocorus bipustulatus* (L.) (Coccinellidae), *Chrysoperla carnea* (Steph.) (Chrysopidae) and *Allotrombium* sp. (Trombiidae), which achieved 20% control in hazelnut orchards, are predators of *L. ulmi*. *Aphytis mytilaspidis* (Le Baron) (Aphelinidae), *Anabrolepis zetterstedtii* (Westwood), *Apterencyrtus microphagus* (Mayr) (Encyrtidae) and *Thysanus ater* Walker (Signiphoridae) are parasitoids of *L. ulmi* in hazelnut orchards in Turkey (Kurt, 1982; Ecevit *et al.*, 1996a; Anonymous, 2008). Birds and mites, especially *Hemisarcoptes malus* Shimer, are also important in controlling this pest (Griswold, 1925).

CHEMICAL CONTROL. An oil ('winter wash') can be applied in February and early March before buds open (Anonymous, 2008).

Gypsonoma dealbana (Frölich) (Lepidoptera: Tortricidae) (twig borer)

Description

The adult *Gypsonoma dealbana* is 4.0–4.5 mm long with a wingspan of 11–14 mm. It is distinguished from other *Gypsonoma* species by the creamy-white patch on the front of the head. The front wings are creamy brown and white while hind wings are greyish brown. The eggs are 0.5–0.7 mm long, white and not shiny. The final instar larvae are 7–9 mm long. The larval body is light yellow and covered by short hairs. The head is blackish brown, and the prothoracic shield pitch black and divided by a whitish medial line. The pupae are reddish brown and 7–8 mm long (Ural *et al.*, 1968; Anonymous, 2008; De Prins and Steeman, 2010; Kimber, 2015).

Life cycle and damage

Adults are seen in hazelnut orchards from around mid-May and fly from dusk onwards. Adult emergence continues for about 50 days. They lay eggs for 5–6 days after emergence and lay each egg separately, generally on the upperside of leaves. In total, they lay about 40 eggs that hatch after 11–14 days. The larvae feed on a range of deciduous trees, 'windowing' leaves in summer and autumn and eating the buds, catkins and young shoots and then spinning leaves together in the spring. Larvae feed for about 4–4.5 months on the undersurface of leaves where the lateral vein joins the mid-vein. Larvae also open a short gallery in the mid-vein on the leaf. The first damage, seen on leaves in July, characteristically is brown and triangular in shape. They move into the base of buds, catkins and big buds in autumn to overwinter. From the beginning of March, they move to new buds and shoots, commence feeding inside, and produce galleries. Young twigs and buds shrivel and dry up. One larva may cause damage to five shoots. Pupation occurs from early May in a cocoon on the tree or in the soil. There is one generation/year (Ural *et al.*, 1968; Anonymous, 2008; Kimber, 2015).

Distribution

This species is distributed across most of Europe, the Near East and the eastern part of the Palearctic ecozone (Aarvik, 2013). It is widespread in the hazelnut-growing area in Turkey but causes a serious problem only in some years and areas (Ural *et al.*, 1968; Işık *et al.*, 1987; Saruhan and Tuncer, 2001; Anonymous, 2008).

Hosts

G. dealbana occurs on hazelnut, willow, poplar, hawthorn, oak and many other trees and shrubs (Ural *et al.*, 1968; Anonymous, 2008; Kimber, 2015).

Management

MECHANICAL CONTROL. Cutting and removing infected, dry shoots in spring and summer is a very effective control method (Ural *et al.*, 1968).

BIOLOGICAL CONTROL. There are a number of braconid and ichneumonid parasitoids of *G. dealbana* in hazelnut orchards in Turkey. Most are larval parasitoids, with up to 15% parasitism (Ural *et al.*, 1973; Işık *et al.*, 1987; Ecevit *et al.*, 1996a; Anonymous, 2008). Ural *et al.* (1973) reported that *Pimpla didyma* Gravenhorst, *Evania fuscipes* Illiger, *Macrocentrus grandii* Goid., *Macrocentrus thoracicus* Nees and *Pristomerus* sp. are the most common parasitoids of *G. dealbana*. *Bacillus thuringiensis* Berliner isolated from *Hyphantria cunea* had 93% efficacy against *G. dealbana* larvae under laboratory conditions (Yaman and Demirbag, 2000).

Cydia latiferreana (Walsingham) (Lepidoptera: Tortricidae) (filbertworm)

Description

Cydia latiferreana has a wingspan of about 18 mm. The forewing colour ranges from pale tan to orange red and dark brown. There are two metallic transverse bands from the costa to the dorsum. The hind wings are almost black. Seven different forms (types A–G) are recognized, based on their morphology, with only forms A and C important on hazelnut. The last instar larvae are approximately 12–15 mm in length, with the abdomen being whitish grey and the head yellowish brown. The prothoracic shield is pale brown with faint, dark mottling. There is no anal comb (AliNiazee, 1980; Brown, 1983; Gilligan *et al.*, 2008).

Life cycle and damage

C. latiferreana overwinter as diapausing larvae in silk cocoons in leaves and debris on the ground, in sheltered places such as crevices, or in the soil. Most larvae pupate by the end of June. Adults emerge 2–5 weeks after pupation and fly from the middle of June until the end of October. Mating takes place soon after emergence and egg laying begins the next day. The eggs, laid singly on leaves and nuts, hatch in 8–10 days and the larvae search for nut clusters. Once a nut is selected, the larva tunnels through the husk and shell and feeds on the kernel while tunnelling to the centre. The larva develops in 3–4 weeks; occasionally, it moves to other nuts. Most damage is in July and August. The mature larva leaves the nut through the entrance hole or a new hole, then forms a cocoon where it spends the winter. A few early emerging larvae become adults in September (AliNiazee, 1980).

Damage is caused by internal feeding and symptoms include obvious exit holes, premature nut drop, visible frass and webbing. In addition to direct damage to nuts, considerable costs are incurred in separating damaged from undamaged nuts. In unsprayed orchards, filbertworm damage ranged from 12% to 37% (AliNiazee, 1980, 1983).

Distribution

C. latiferreana is widely distributed across North America and northern Mexico. It is the key pest in almost all hazelnut orchards in the Willamette Valley, Oregon, USA (AliNiazee, 1980, 2001; Powell and Opler, 2009).

Hosts

C. latiferreana is a pest of hazelnut, walnut, chestnut, almond, macadamia, pomegranate and cherry, as well as forest trees (AliNiazee, 1980; Powell and Opler, 2009).

Management

CULTURAL CONTROL. Sanitation plays an important role in the control *C. latiferreana*. Many prematurely dropped nuts contain the larvae. Destruction of these nuts eliminates infestation sources for the next year. Drying and packing sheds should be cleaned to avoid infestation (AliNiazee, 1980). Chambers *et al.* (2011) stated that a thicker basal scar in hazelnut cultivars contributes to resistance against filbertworm infestation; the cultivar 'Siciliana' had the thickest shell overall and lowest filbertworm infestation

(6%), whereas 'Daviana', 'Frango 2' and 'Casina' had the thinnest shells and 82%, 76%, and 55% nut infestation, respectively.

BIOTECHNICAL CONTROL. Black light traps are used for monitoring emergence and flight (AliNiazee, 1980). Commercial pheromone lures are available. Different trap designs and trap placement methods were tested in hazelnut orchards in Oregon, USA (AliNiazee, 1983). Hedstrom *et al.* (2014) stated that sex-pheromone-mediated mating disruption is promising and warrants further research.

BIOLOGICAL CONTROL. More than a dozen parasitoids have been reared from filbertworm, but none are effective in its control because the larvae live inside the nuts. The egg parasitoids, *Trichogramma minitum* Riley and *Trichogramma evanescens* Westwood, were effective in an experimental trial but they rarely provide economic control (AliNiazee, 1980, 2001).

Bruck and Walton (2007) determined that the entomopathogenic nematodes, *Heterorhabditis marelatus* (Liu and Berry) Pt. Reyes, *Steinernema carpocapsae* (Weiser) and *Steinernema kraussei* (Steiner) have the potential to control *C. latiferreana* in the field. Their infectivity ranged from 73% to 100% in laboratory bioassays and were similar in the field.

CHEMICAL CONTROL. *Bacillus thuringiensis* and neem are ineffective against high populations of the filbertworm. Traditional insecticides such as extracts of the pyrethrum plant are used against filbertworm in organic orchards but results are unpredictable (AliNiazee, 2001).

Archips rosana (L.) (Lepidoptera: Tortricidae) (European leafroller, hazelnut-filbert leafroller, rose leafroller)

Description

The wingspan of the female *Archips rosana* is 19–22 mm and that of the male is 17–19 mm. The head and thorax are brown and the abdomen brownish grey. The forewings are truncated at the top, light brown with three slanted, badly delimited spots while the hindwings are brownish grey with an orangy apex and fringes. The eggs are greenish when laid, then becoming brownish grey. The final instar larva is 18–22 mm in length, light green to olive green and with a shiny brown head and prothoracic plate. The pupa is 10–12 mm long, yellowish brown with a darker dorsum (Anonymous, 1997b; Ovsyannikova and Grichanov, 2008).

Life cycle and damage

A. rosana has one generation/year. Adults fly from the end of May until mid-August and live for 2–4 weeks. Egg laying starts 3–5 days after adult emergence. Most egg laying occurs in June and July. Females lay about 250 eggs, with 40–100 in one batch on the smooth surface of bark, main branches and trunk. Eggs are covered with a translucent, protective wax-like substance and remain in diapause until the next spring. Hatching takes place from the end of February to the beginning of May, depending on the temperature and the region. Larvae feed on buds, flowers, leaves and newly formed nuts. The young larvae penetrate buds and bind the young leaves together with silk. Then they roll up the leaves along the main vein, making a sort of cigar shape and gnawing the parenchyma, leaving only the veins. The tender terminal growth is often the most attacked part. The larval period lasts about 6–8 weeks. Pupation occurs in feeding places, especially in shrivelled leaves, and lasts 2–3 weeks. In heavily infested orchards in the USA, 50% of the total foliage can be rolled, causing heavy losses (AliNiazee, 1977, 1980, 1998; Al-Zabidi, 1994; Anonymous, 1997b; Ovsyannikova and Grichanov, 2008).

Distribution

A. rosana is native to the Palearctic but it was accidently introduced to North America (AliNiazee, 1980; Anonymous, 1997b; Ovsyannikova and Grichanov, 2008).

Hosts

A. rosana is highly polyphagous, feeding on more than 130 species, including fruit and forest trees as well as bushes, ornamental plants, grasses and some cereals and vegetables (Anonymous, 1997b; Ovsyannikova and Grichanov, 2008). In North America, it is an important pest of hazelnut orchards (AliNiazee, 1977, 1998).

Management

CULTURAL CONTROL. Removing old bark on the lower parts of trunks, and pruning old and damaged branches is a way of controlling *A. rosana* (Ovsyannikova and Grichanov, 2008).

BIOLOGICAL CONTROL. More than 100 parasitoids have been collected from *A. rosana* (AliNiazee, 1977; Yu *et al.*, 2005; Aydogdu, 2014). Natural enemies have been reported from hazelnut orchards in Oregon, including larval–pupal parasitoid wasps which reduce the population of *A. rosana* by as much as 70% (AliNiazee, 1977, 1980). The release of *Trichogramma* during the egg-laying period can control *A. rosana* (Ovsyannikova and Grichanov, 2008).

Bacillus thuringiensis is effective in controlling this pest when applied immediately after egg hatching begins (AliNiazee, 1974, 1980, 1998) and has been used in integrated pest management (IPM) programmes (DeFrancesco, 2006).

BIOTECHNICAL CONTROL. Pheromone traps are used to monitor and control this pest. Open traps placed at 0.6 m and 1.2 m above ground were significantly more effective than at other heights. It is active at night and responds positively to light traps (AliNiazee, 1976, 1998).

CHEMICAL CONTROL. Azadirachtin is used to control *A. rosana* in hazelnut orchards in the USA (AliNiazee, 1998; DeFrancesco, 2006). Dormant sprays of oil are effective in killing *A. rosana* eggs with application at the correct time (AliNiazee, 1997a, 1998).

Lymantria dispar **(L.) (Lepidoptera: Erebidae) (gypsy moth)**

There are several subspecies and races of gypsy moth: (i) the European gypsy moth *Lymantria dispar dispar* race Europe; (ii) the Asian gypsy moth *Lymantria dispar dispar* race Asian; and (iii) the Japanese gypsy moth, *Lymantria dispar japonica* (ISSG, 2005).

Description

The wingspan of the flightless adult female is about 50–60 mm. The wings and body are creamy white, with brownish grey, zigzag lines. The antennae are black and weakly bipectinate. The abdomen is thick and covered with dense black hairs. The eggs are greyish green and pellet-like. The adult male is much smaller than the female, with a wingspan of about 35–45 mm. Its abdomen is thin and the wings and body are greyish brown with black markings. The antennae are strongly bipectinate. The larvae are up to 55 mm long, with a pale bluish grey to pale creamy-grey body that is extensively marked with black, and with whitish sides and a pale dorsal line. There are five pairs of blue spots followed by six pairs of reddish spots on the thorax and abdomen, from head to rear. The larvae are covered with long hairs in clumps. The pupae are reddish brown and 18–28 mm long (Ural, 1957c; Kurt, 1982; Anonymous, 2008; Grichanov and Ovsyannikova, 2008; Alford, 2014).

Life cycle and damage

L. dispar has one generation/year. Adults are present from July to early September. Eggs are laid in large batches of several hundred on the bark of trees or on other surfaces. The female coats the egg batches with hairs from her anal tuft. Maximum fecundity is 1200 eggs and there are up to 700 eggs in one batch. The larvae do not emerge until the following April or May. They feed intensively on buds, leaves, young shoots and developing fruitlets and cause extensive defoliation. They become fully grown larvae in 2–3 months after passing through six larval stages. The larvae pupate in flimsy, silken cocoons spun among foliage, on the bark of trees and stones, or among debris on the ground. Adults emerge in 2–3 weeks (Ural, 1957c; Kurt, 1982; Anonymous, 2008; Grichanov and Ovsyannikova, 2008; Alford, 2014).

Distribution

It is a Eurasiatic species. It is also present in Northern Africa, North America and Japan (Grichanov and Ovsyannikova, 2008; Alford, 2014).

Hosts

L. dispar feeds on more than 600 forest, ornamental and fruit tree species (Anonymous, 2008; Grichanov and Ovsyannikova, 2008; Alford, 2014).

Management

MECHANICAL CONTROL. Collection and destruction of egg batches helps to reduce the population (Anonymous, 2008).

BIOLOGICAL CONTROL. *Bacillus thuringiensis* application as a microbial insecticide is effective against the larvae. There are also native predators, parasitoids and entomopathogens of *L. dispar*, including *Calosoma sycophanta* (L.), *Entomophaga maimaiga* Humber *et al.*, nuclear polyhedrosis virus (NPV), *Nosema* sp., birds and mice. Of these, *E. maimaiga*, which was introduced to the USA, and NPV, can cause massive mortality of larvae (Ural, 1957c; Tuncer *et al.*, 2001; Anonymous, 2008; Hajizadeh *et al.*, 2011; Demir *et al.*, 2012; Yaman *et al.*, 2012). In addition, the entomopathogenic nematodes, *Steinernema carpocapsae* (Weiser) and *S. feltiae* (Filipjev) have potential as control agents of this pest (Zamoum *et al.*, 2011; Papadopoulou *et al.*, 2012).

Plant extracts from *Aesculus hippocastanum* L., *Morus alba* L. and *Ocimum basilicum* L. are effective against *L. dispar* larvae (Gvozdenac *et al.*, 2012; Popovic *et al.*, 2013). Neem may also be suitable for IPM programmes (Zabel *et al.*, 2002).

***Hyphantria cunea* (Drury) (Lepidoptera: Erebidae) (fall webworm)**

Description

The adult *Hyphantria cunea* has a wingspan of 25–30 mm and the wings are mainly white, with the forewings being sometimes flecked with black. The male has noticeably bipectinate antennae. The male is 11 mm and the female 15 mm in length. The eggs are light green and 0.5–0.6 mm in diameter. The larva is up to 35 mm long, with the body varying from yellow or yellowish green to deep-reddish brown, with tufts of whitish hairs arising from black verrucae; the spiracles are white, ringed with black, and the head is shiny black. The pupa is 10–12 mm long and is shiny blackish brown (Anonymous, 2008; Alford, 2014).

Life history and damage

H. cunea has two generations/year but a third generation occurs in some years. It overwinters as a pupa in protected places such as under bark, logs and leaf litter and in cracked walls. Adults of the first and second generation occur between mid-May and mid-June and between late July and mid-August, respectively. Females lay eggs in clutches of several hundred on the underside of leaves, covering them with white abdominal hairs. Eggs hatch in about 1 week and the larvae pass through seven instars. The first three larval instars live in colonies under a web and then disperse. The larvae feed on leaves and young fruits and can consume almost all of the foliage. The first generation of larvae in June generally causes little damage as populations are low. Because of the high fecundity of *H. cunea*, most damage is caused by the second generation larvae in August and rarely a third generation in October (Tuncer, 1992; Anonymous, 2008; Sullivan, 2011; Alford, 2014).

Distribution

This species is found in North America, Europe, Asia and Japan (Sullivan, 2011; Alford, 2014).

Hosts

It is a highly polyphagous pest with more than 600 hosts that include forest, fruit and ornamental trees (Tuncer, 1992; Anonymous, 2008; Sullivan, 2011; Alford, 2014).

Management

MECHANICAL CONTROL. Collecting and destroying egg clusters and larval colonies of *H. cunea* are recommended. The procedure has to be repeated every 4–5 days to destroy larvae from later egg clusters. The fastening of straw belts around trees to provide artificial pupation sites can also be effective. The straw belts should not be burned after the end of pupation, but should instead be stored in containers and covered by a wire screen which allows the escape of emerged parasitoids but prevents escape of the moth (Szalay-Marzso, 1972; Anonymous, 2008).

BIOLOGICAL CONTROL. Predators, parasitoids and entomopathogens play an important role in the natural control of this pest. The eggs, larvae, pupae and adults are preyed on by at least 36 families of vertebrates and non-vertebrates including *Arma custos* (Fabr.), *Chrysoperla carnea* (Steph.) and birds in Europe and North America (Warren and Tadic, 1967; Szalay-Marzso, 1972; Anonymous, 2008; Sullivan, 2011). More than 130 species of the Ichneumonoidea, Chalcidoidea and Tachinidae, including *Pimpla instigator* (Fabr.), *Chouioia cunea* Yang and *Nemoraea pellucida* (Meigen) parasitize *H. cunea* (Warren and Tadic, 1967; Szalay-Marzso, 1972; Railyan, 1974; Johnson and Lyon, 1991; Tuncer and Ecevit, 1996; Yu *et al.*, 2005; Anonymous, 2008; Sullivan *et al.*, 2010a, 2011a, 2012a, 2015). The bacteria isolated from pupae include *B. thuringiensis* and *Pseudomonas* spp. (Yaman *et al.*, 2002; Sullivan *et al.*, 2010b). Other pathogenic organisms include fungi, a granulosis virus and an NPV (Jasinka, 1984; Yang *et al.*, 2005; Sullivan *et al.*, 2011b, 2015). Sullivan *et al.* (2011b) reported that isolates of *Paecilomyces fumosoroseus* (Wise) and *Beauveria bassiana* (Bals.-Criv.) Vuill. had very high efficacy against *H. cunea* larvae under laboratory conditions. Nordin and Maddox (1974) isolated the microsporidia *Nosema necatrix* Kramer and *Pleistophora schubergi hyphantriae* Weiser from *H. cunea*. The entomogenous nematode, *Steinernema feltiae* (Filipjev), had efficacy against *H. cunea* (Yamanaka *et al.*, 1986). In a laboratory study, azadirachtin, spinosad and *B. thuringiensis*, were used against third stage larvae, with the highest concentrations of all three causing 100% mortality (Saruhan *et al.*, 2014).

Biological control, including the use of bacterial, fungal and viral preparations, is effective on the larvae. The most used organisms are subspecies of *Bacillus thuringiensis* Berliner which are used when older larvae begin dispersing from their web to feed. *B. thuringiensis* subsp. *kurstaki* was reported to be effective in Hungary and Korea (Jasinka, 1984; Choi *et al.*, 1986). In Turkey, *B. thuringiensis* (16,000 IU/mg) at 50 g/100 l of water is very effective (Anonymous, 1996). Ecevit *et al.* (1994) reported 100% larval mortality 3 days after the application of the *B. thuringiensis* formulations, Thuricide Hp and Biobit. Railyan (1974) reported that Enterobactin, a preparation of *B. thuringiensis*, and a specific NPV, were used against the larvae in Moldavia.

Mikomya coryli **(Kieffer) (Diptera: Cecidomyiidae) (hazelnut gall midge)**

Description

Adults of *Mikomya coryli* are red and about 2 mm in length. Antennae have 14 segments in both sexes, but the male antennae are longer. The female has a long ovipositor. The eggs are 0.2–0.3 mm long, oval shaped and pale red. The larvae are white and 2.5–3 mm long when fully grown (Ural and Kurt, 1973).

Life cycle and damage

Adults begin emerging at the end of March, with 80–90% emerging in April. Just after mating, females lay 100–200 eggs between bud scales, on young leaves and in newly formed husks. The adult lifespan is 1–2 days. In the 3–4 weeks after the emergence of adults, galls appear along veins on leaves and husks. Each gall contains one larva. After rain, fully grown larvae begin to leave the galls and enter the soil. The larvae overwinter in a white cocoon about 1 cm deep in the soil. They pupate in March and the pupation period is 2 weeks. *M. coryli* has one generation/year.

Larvae feed on plant tissues and induce galls which are swollen and densely covered by long, red hairs. Damaged leaves dry after the larvae leave and drop earlier than normal leaves. Damaged fruits do not drop. Galls on the husk are of economic importance because they disfigure the nut and prevent its proper development (Ural and Kurt, 1973).

Distribution

M. coryli is widespread in Europe, including Turkey (Ural and Kurt, 1973; Pape and Thompson, 2013).

Hosts

Hazelnut is the host plant for *M. coryli* (Ural and Kurt, 1973; Anonymous, 2008).

Management

BIOLOGICAL CONTROL. Parasitoids, including *Amblyaspis angustula* Thoms., *Prosactogaster oebalus* (Walker) and *Inostemma* sp. (Platygastridae), *Torymus cultriventris* Ratz. (Torymidae) and three species of *Tetrastichus* (Eulophidae) attack the larvae and play an important role in natural control (Ural and Kurt, 1973).

Acknowledgement

The author thanks Gregory T. Sullivan of the University of Queensland in Brisbane, Australia for editing the chapter.

References

Aarvik, L.E. (2013) *Gypsonoma dealbana*. Fauna Europaea: Tortricidae. In: Karsholt, O. and Nieukerken, E.J. van, (eds) *Fauna Europaea: Lepidoptera, Moths. Fauna Europaea Version 2.6.2*. Available at: http://www.faunaeur.org (accessed 16 March 2015).

Aguilera, A., Neculman, R. and Rebolledo, R. (2014) Predatory capacity of *Adalia angulifera* (Coleoptera: Coccinellidae) larvae on *Myzocallis coryli* (Hemiptera: Aphididae) in Chile. *Ciencia e Investigacion Agraria* 41(1), 81–88.

Ak, K., Uysal, M. and Tuncer, C. (2005a) The injury level of bark beetles (Coleoptera: Scolytidae) in hazelnut orchards in Giresun, Ordu and Samsun provinces of Turkey. *Gaziosmanpaşa Üniversitesi Ziraat Fakültesi Dergisi* 22(1), 9–14. (abstract in English)

Ak, K., Uysal, M., Tuncer, C. and Akyol, H. (2005b) The bark beetle species (Col.: Scolytidae) harmful on hazelnut in middle and east Blacksea region of Turkey and their control strategies. *Selçuk Üniversitesi Ziraat Fakültesi Dergisi* 19(37), 37–40. (abstract in English)

Ak, K., Saruhan, I. and Akyol, H. (2014) Determination of performances of different trap types against *Xyleborus dispar* (Fabricius, 1792) and *Xyleborinus saxesenii* (Ratzeburg, 1837) (Coleoptera: Curculionoidea: Scolytidae). *Anadolu Journal of Agricultural Sciences* 29(1), 26–35. (abstract in English)

Akca, I. (2003) Orta karadeniz bölgesinde findik kurdu *Curculio nucum* L. 1758 (Coleoptera; Curculionidae) popülasyonlarinin biyolojisi ve zarari üzerine araştirmalar. Doktora tezi, Ondokuz Mayıs Üniversitesi, Fen Bilimleri Enstitüsü, Samsun.

Akca, I. and Tuncer, C. (2005) Biological control and morphological studies on nut weevil (*Curculio nucum* L. Col., Curculionidae). *Acta Horticulturae* 686, 413–419.

Alford, D.V. (2014) *Pests of Fruit Crops: A Colour Handbook*, 2nd edn. CRC Press, Boca Raton, Florida.

AliNiazee, M.T. (1974) Evaluation of *Bacillus thuringiensis* against *Archips rosanus* (Lepidoptera: Tortricidae). *The Canadian Entomologist* 106(4), 393–398.

AliNiazee, M.T. (1976) Field studies on sex pheromone trapping of the filbert leafroller, *Archips rosanus*, in Oregon. *Annals of the Entomological Society of America* 69(5), 820–824.

AliNiazee, M.T. (1977) Bionomics and life history of a filbert leafroller, *Archips rosanus* (Lepidoptera: Tortricidae). *Annals of the Entomological Society of America* 70(3), 391–401.

AliNiazee (1980) *Filbert Insect and Mite Pests*. Station Bulletin 643. Agricultural Experiment Station, Oregon State University, Corvallis, Oregon.

AliNiazee, M.T. (1983) Pest status of filbert insects in Oregon orchards: a ten year study. *The Canadian Entomologist* 12, 141–147.

AliNiazee, M.T. (1995) Recent developments in hazelnut pest control. *Proceedings of Nut Growers Society of Oregon, Washington and British Columbia* 80, 62–68.

AliNiazee, M.T. (1997a) Integrated pest management of hazelnut insects: a worldwide perspective. *Acta Horticulturae* 445, 469–476.

AliNiazee, M.T. (1997b) Biology, impact, and management of *Trioxys pallidus* in hazelnut orchards of Oregon, U.S.A. *Acta Horticulturae* 445, 477–482.

AliNiazee, M.T. (1998) Ecology and management of hazelnut pests. *Annual Review of Entomology* 43, 395–419.

AliNiazee, M.T. (2001) Hazelnut production without the use of broad spectrum disruptive insecticides: theory and practice. *Acta Horticulturae* 556, 489–494.

AliNiazee, M.T. and Messing, R.H. (1995) Filbert aphid. In: Nechols, J.R., Andres, L.A., Beardsley, J.W., Goeden, R.D. and Jackson, C.G. (eds) *Biological Control in the Western United States*. University of California, Division of Agriculture and Natural Resources, Publication 3361. University of California, Davis, California, pp. 123–126.

Al-Zabidi, A.H. (1994) Effects of temperature and photoperiod on growth and development of the filbert leafroller, *Archips rosanus* L. (Lepidoptera: Tortricidae) in the Willamatte Valley of Oregon. PhD thesis, Oregon State University, Corvallis, Oregon.

Anonymous (1996) *Zirai Mücadele Teknik Talimatları*, Cilt: 4. T.C. Tarım ve Köyişleri Bakanlığı Koruma ve Kontrol Genel Müdürlüğü, Ankara.

Anonymous (1997a) *Curculio nucum* (L.) Nut Weevil. Available at: http://www7.inra.fr/hyppz/RAVAGEUR/6curnuc.htm (accessed 4 December 2014).

Anonymous (1997b) *Archips rosana* (L.) Rose Leafroller. Available at: http://www7.inra.fr/hyppz/RAVAGEUR/6arcros.htm (accessed 17 April 2015).

Anonymous (2007) *Palomena Prasina*. *Encyclopedia of Life*. Available at: http://eol.org/pages/2869574/details (accessed 12 August 2014).

Anonymous (2008) *Zirai Mücadele Teknik Talimatları*, Cilt: 5. T.C. Tarım ve Köyişleri Bakanlığı Tarımsal Araştırmalar Genel Müdürlüğü, Başak Matbaacılık ve Tan. Hiz. Ltd. Şti., Ankara.

Anonymous (2014a) *2103 Yılı Fındık Sektörü Raporu*. Toprak Mahsulleri Ofisi Genel Müdürlüğü, Ankara.

Anonymous (2014b) Shot-hole borer – *Xyleborus dispar* F. Available at: http://csalomontraps.com/4listbylatinname/pdffajonkentik/xyleborusdisparang08.pdf (accessed 10 December 2014).

Anonymous (2015) The Dow Chemical Company (1995–2015). Dow AgroSciences. Ürünlerimiz. Available at: http://www.dowagro.com/tr-tr/turkiye/products/%C4%B0nsektisitler/laser (accessed 3 February 2015).

Arzone, A. (1983) Due fitomizi dannosi al nocciolo: l'acaro delle gemme e il Miride degli amenti. In: Tarantino, C., D'Agostino, G. and de Jorio, F. (eds) *Atti del Convegno Internazionale sul Nocciuolo*, 22–24 September 1983, Avellino, Italy. Regione Campania, Assessorato Agricoltura, Avellino, Italy, pp. 199–204.

Aydogdu, M. (2014) Parasitoid abundance of *Archips rosana* (Linnaeus, 1758) (Lepidoptera: Tortricidae) in organic cherry orchards. *North-Western Journal of Zoology* 10(1), 42–47.

Bahar, A.A. and Demirbağ, Z. (2007) Isolation of pathogenic bacteria from *Obera linearis* (Coleoptera: Cerambycidae). *Biologia, Bratislava* 62(1), 13–18.

Balochowsky, A.S. (1962) *Coleopteres. Entomologie Appliquée a l'Agriculture*. Tome I, Vol. 1. Masson et Cie, Paris.

Bantock, T. and Botting, J. (2013) British Bugs. Available at: http://www.britishbugs.org.uk/heteroptera/Pentatomidae/palomena_prasina.html (accessed 5 October 2014).

Batalla-Carrera, L., Morton, A. and Garcia-del-Pino, F. (2013) Field efficacy against the hazelnut weevil, *Curculio nucum* and short-term persistence of entomopathogenic nematodes. *Spanish Journal of Agricultural Research* 11(4), 1112–1119.

Beber, K. (1994) Studies on migration of big bud mite (*Phytoptus avellanae* Nal.) in northeastern Slovenia. *Acta Horticulturae* 351, 625–630.

Bekircan, Ç., Cüce, M. and Sökmen, A. (2014) Antifeedant activity of the essential oils from four different Lamiaceae species against *Agelastica alni* L. (Coleoptera: Chrysomelidae). *Advances in Zoology and Botany* 2(4), 57–62.

Blackman, R.L. and Eastop, V.F. (1985) *Aphids on the World's Crops: An Identification Guide*. John Wiley & Sons, New York.

Borden, J.H., Chong, L.J., Gries, R. and Pierce, H.D. Jr (2003) Potential for non-host volatiles as repellents in integrated pest management of ambrosia beetles. *Integrated Pest Management Reviews* 6, 221–236.

Brown, R.L. (1983) Taxonomic and morphological investigations of Olethreutinae: *Rhopobota, Griselda, Melissopus*, and *Cydia* (Lepidoptera: Tortricidae). *Entomography* 2, 97–120.

Bruck, D.J. and Walton, V.M. (2007) Susceptibility of the filbertworm (*Cydia latiferreana*, Lepidoptera: Tortricidae) and filbert weevil (*Curculio occidentalis*, Coleoptera: Curculionide) to entomopathogenic nematodes. *Journal of Invertebrate Pathology* 96, 93–96.

Burgess, J.E. and Thompson, M.M. (1985) Shoot development and bud infestation in hazel nut (*Corylus avellana*). *Annals of Applied Biology* 107, 397–408.

CABI (2014) *Xyleborus dispar* (pear blight beetle). In: Invasive Species Compendium. CAB International, Wallingford, UK. Available at: http://www.cabi.org/isc/datasheet/57157 (accessed 25 June 2014).

Canganella, F., Paparatti, B. and Natali V. (1994) Microbial species isolated from the bark beetle *Anisandrus dispar* F. *Microbiological Research* 149(2), 123–128.

Chambers, U., Walton, V.M. and Mehlenbacher, S.A. (2011) Susceptibility of hazelnut cultivars to filbertworm, *Cydia latiferreana*. *HortScience* 46(10), 1377–1380.

Choi, Y.C., Lee, S.W., Shin, Y.H. and Lee, K.H. (1986) Studies on the mass production of *Bacillus thuringiensis* as a microbial insecticide. *Research Reports of the Rural Development Administration, Plant Environment, Mycology & Farm Products Utilization, Korea Republic* 28(2), 56–59.

Cobanoglu, S. and Ozman, S.K. (2002) Beneficial mite species of hazelnut orchard ecosystems from the Black Sea region of Turkey. In: Žd'árková, E., Wakefield, M., Lukáš, J. and Hubert, J. (eds) *Biological Control of Pest Insects and Mites with Special Reference to Entomophthorales*. Proceedings of the 2nd Meeting of Working Group 4: Bio-control of Arthropod Pests in the Stored Products, 30–31 May 2002, Prague, Czech Republic. Research Institute of Crop Production, Prague, pp. 91–99.

David'yan, G.E. (2008a) *Xyleborus Dispar* (F.). In: Afonin, A.N., Greene, S.L., Dzyubenko, N.I. and Frolov, A.N. (eds) *Interactive Agricultural Ecological Atlas of Russia and Neighboring Countries. Economic Plants and Their Diseases, Pests and Weeds*. Available at: http://www.agroatlas.ru/en/content/pests/Xyleborus_dispar/ (accessed 21 June 2014).

David'yan, G.E. (2008b) *Parthenolecanium corni* Bouche. In: Afonin, A.N., Greene, S.L., Dzyubenko, N.I. and Frolov, A.N. (eds) *Interactive Agricultural Ecological Atlas of Russia and Neighboring Countries. Economic Plants and Their Diseases, Pests and Weeds*. Available at: http://www.agroatlas.ru/en/content/pests/Parthenolecanium_corni/ (accessed 25 July 2014).

David'yan, G.E. (2008c) *Lepidosaphes ulmi* L. In: Afonin, A.N., Greene, S.L., Dzyubenko, N.I. and Frolov, A.N. (eds) *Interactive Agricultural Ecological Atlas of Russia and Neighboring Countries. Economic Plants and Their Diseases, Pests and Weeds*. Available at: http://www.agroatlas.ru/en/content/pests/Lepidosaphes_ulmi/ (accessed 7 January 2015).

DeFrancesco, J. (2006) *Pest Management Strategic Plan for Hazelnuts in Oregon and Washington*. Available at: http://www.ipmcenters.org/pmsp/pdf/ORWA_Hazelnut.pdf (accessed 25 February 2015).

Demir, I., Sezen, K. and Demirbağ, Z. (2002) The first study on bacterial flora and biological control agent of *Anoplus roboris* Sufr. (Coleoptera). *The Journal of Microbiology* 40(2), 104–108.

Demir, İ., Eryüzlü, E. and Demirbağ, Z. (2012) A study on the characterization and pathogenicity of bacteria from *Lymantria dispar* L. (Lepidoptera: Lymantriidae). *Turkish Journal of Biology* 36, 459–468.

De Prins, W. and Steeman, C. (2010) *Gypsonoma dealbana* Frölich. Catalogue of the Lepidoptera of Belgium. Available at: http://uahost.uantwerpen.be/vve/Checklists/Lepidoptera/Tortricidae/Gdealbana.htm (accessed 25 August 2014).

Ecevit, O., Işık, M. and Yanılmaz, A.F. (1987) *Fındıklarda Zararlı Fındık Koşnili (*Parthenolecanium corni *Bouche. ve* Parthenolecaium rufulum *Ckll.) İle Virgül Kabuklu Biti (*Lepidosaphes ulmi *L.)'nin Biyoekolojik Özellikleri ve Fındık Koşnilinin Mücadele Metotları Üzerine Araştırmalar*. Ondokuz Mayıs Üniversitesi Yayınları, No. 19, Samsun, Türkiye.

Ecevit, O., Keçeci, S., Tuncer, C., Yanılmaz, A.F. and Işık, M. (1992a) Studies on Eriophyoidea (Acarina, Actinedida) of hazelnut groves in east Black Sea region. In: *Proceedings of the Second Turkish National Congress of Entomology*, 28–31 January 1992, Adana, Turkey. Entomoloji Derneği Yayınları No. 5, Izmir, Turkey, pp. 671–681. (abstract in English)

Ecevit, O., Işık, M., Keçeci, S. and Yanılmaz, A.F. (1992b) Studies on some bioecological characteristics of *Phylus coryli* L. (Hemiptera, Miridae) which is predator of *Myzocallis coryli* Goeze (Homoptera, Aphididae), a hazelnut pest in the east Black Sea region. In: *Proceedings of the Second Turkish National Congress of Entomology*, 28–31 January 1992, Adana, Turkey. Entomoloji Derneği Yayınları No. 5, Izmir, Turkey, pp. 217–226. (abstract in English)

Ecevit, O., Tuncer, C. and Keçeci, S. (1993) Studies on the description of *Anoplus roboris* Suffrian (Coleoptera: Curculionidae). *Türkiye Entomoloji Dergisi* 17(4), 235–238.

Ecevit, O., Tuncer, C., Hatat, G. and Keçeci, S. (1994) Studies on the efficiency of two *Bacillus thuringiensis* formulations (Thuricide HP and Biobit), azinphos methyl and triflumuron against fall webworm (*Hyphantria cunea* Drury, Lepidoptera: Arctiidae). In: *Proceedings of the Third Turkish National Congress of Biological Control*, 25–28 January 1994, İzmir, Turkey. Entomoloji Derneği Yayınları No. 7, Izmir, Turkey, pp. 519–528.

Ecevit, O., Tuncer, C., Ozman, S.K., Mennan, S. and Akca, I. (1996a) Natural enemies found in Black Sea hazelnut orchards and their potential in biological control. In: *Fındık ve Diğer Sert Kabuklu Meyveler Sempozyumu*, Ondokuz Mayıs Üniversitesi, Ziraat Fakültesi, 10–11 Ocak 1996, Samsun, Turkey, pp. 293–307. (abstract in English)

Ecevit, O., Ozman, S.K., Hatat, G., Okay, A.N., Kaya, A. and Mennan, S. (1996b) Determination of the susceptibility of some important hazelnut cultivars at the Black Sea region to pest and diseases. *Fındık ve Diğer Sert Kabuklu Meyveler Sempozyumu*, 10–11 Ocak 1996, Samsun, Turkey, pp. 77–93. (abstract in English)

Fatu, A.-C., Dinu, M.-M., Ciornei, C. and Andrei, A.-M. (2015) Biological control of *Melolontha melolontha* L. larvae with entomopathogenic bioinsecticide based on *Beauveria brongniartii*. *AgroLife Scientific Journal* 4(1), 64–69.

Gantner, M. (2001) Occurence of hazelnut pests in southeastern Poland. *Acta Horticulturae* 556, 469–477.

Gilligan, T.M., Wright, D.J. and Gibson, L.D. (2008) *Olethreutine Moths of the Midwestern United States, An Identification Guide*. Ohio Biological Survey, Columbus, Ohio.

Grichanov, I.Y. and Ovsyannikova, E.I. (2008) *Ocneria dispar* (L.). In: Afonin, A.N., Greene, S.L., Dzyubenko, N.I. and Frolov A.N. (eds) *Interactive Agricultural Ecological Atlas of Russia and Neighboring Countries. Economic Plants and Their Diseases, Pests and Weeds*. Available at: http://www.agroatlas.ru/en/content/pests/Ocneria_dispar/ (accessed 7 January 2015).

Griswold, G.H. (1925) A study of the oyster-shell scale, *Lepidosaphes ulmi* (L.) and one of its parasites, *Aphelinus mytilaspidis* Le B. *Memoir* 93. Cornell University Agricultural Experiment Station, New York.

Guidone, L., Valentini, N., Rolle, L., Me, G. and Tavella, L. (2007) Early nut development as a resistance factor to the attacks of *Curculio nucum* (Coleoptera: Curculionidae). *Annals of Applied Biology* 150(3), 323–329.

Gvozdenac, S., Indic, D., Vukovic, S. and Grahovac, M. (2012) Antifeeding activity of several plant extracts against *Lymantria dispar* L. (Lepidoptera: Lymantriidae) larvae. *Pesticides and Phytomedicine* 27(4), 305–311.

Hajizadeh, G., Kavosi, M.R. and Moshashaei, E. (2011) Natural enemies of the gypsy moth *Lymantria dispar* (L.) (Lepidoptera: Lymantriidae). *International Research Journal of Agricultural Sciences and Soil Sciences* 1(8), 301–306.

Hedstrom, C.S., Olsen, J., Walton, V.M. and Chambers, U. (2014) Pheromone mating disruption of filbert worm moth (*Cydia latiferreana*) in commercial hazelnut orchards. *Acta Horticulturae* 1052, 253–262.

Hoffman, A. (1954) *Faune de France, 59, Coleopteres, Curculionides*. Federation Francaise des Societes de Sciences Naturelles, Paris.

Hoskovec, M. and Rejzek, M. (1997) Cerambycidae. Longhorn Beetles (Cerambycidae) of the West Palaearctic Region. Available at: http://www.cerambyx.uochb.cz/obelin.htm (accessed 28 January 2015).

Invasive Species Specialist Group (ISSG) (2005) *Lymantria dispar*. Global Invasive Species Database. Available at: http://www.issg.org/database/species/ecology.asp?si=96&fr=1&sts=&lang=EN (accessed 8 February 2015).

İren, S. (1970) *Düzce ve Tirebolu'da Fındıklara Arız Olan* Parthenolecanium corni *(Bouche)'yi Parazitleyen* Cordyceps clavulatus *(Schw.) Ellis et Ev. ve* Verticillium lecanii *(Zimm.) Viégas Üzerinde Bir Araştırma*. Tarım Bakanlığı, Zirai Mücadele ve Zirai Karantina Genel Müdürlüğü. Yeni Desen Matbaası, Ankara.

Işık, M. (1970) *Karadeniz Bölgesi Fındık Bahçelerinde Zarar Yapan Dalkıran,* Xyleborus (Anisandrus) dispar *Fabr., (Col.: Scolytidae) Böceğinin Biyolojisi ve Mücadele Metotları Üzerine Araştırmalar*. T.C. Tarım, Orman ve Köyişleri Bakanlığı, Zirai Mücadele ve Zirai Karantina Genel Müdürlüğü Araştırma Eserleri, No. 30, Ankara.

Işık, M., Tuncdemir, M. and Yanılmaz, A.F. (1983) Study on the control possibilities of *Parthenolecanium rufulum* Ckll. by *Verticillium lecanii* (Zimm.) Viegas. *Turkiye Bitki Koruma Dergisi* 7(3), 167–175.

Işık, M., Ecevit, O., Kurt, M.A. and Yücetin, T. (1987) *Doğu Karadeniz Bölgesi Fındık Bahçelerinde Entegre Savaş Olanakları Üzerinde Araştırmalar*. Ondokuz Mayıs Üniversitesi Yayınları, No. 20, Samsun, Türkiye.

Işık, M., Ecevit, O., Keçeci, S. and Yanılmaz, A.F. (1992) Studies on some bioecological characteristics and control methods of *Anoplus roboris* Sufr., (Col., Curculionidae) a hazelnut pest which is harmful in east Black Sea region. In: *Proceedings of the Second Turkish National Congress of Entomology*, 28–31 January 1992, Adana, Turkey. Entomoloji Derneği Yayınları No. 5, Izmir, Turkey, pp. 307–318. (abstract in English)

Japoshvili, G. and Karaca, İ. (2007) Encyrtid (Hymenoptera: Chalcidoidea, Encyrtidae) parasitoits of Coccidae (Hemiptera: Coccoidea) in Turkey. *Turkish Journal of Entomology* 31(3), 175–188.

Japoshvili, G., Gabroshvili, N. and Japoshvili, B. (2008) The parasitoid complex of *Parthenolecanium corni* Bouche in the city of Tbilisi and its surroundings and comparison with some other European countries. *Bulletin of Entomological Research* 98, 53–56.
Jasinka, J. (1984) Farm studies on application techniques and pesticide efficiency against *Hyphantria cunea* Drury. *Növenyvedelem* 20(8), 368–372.
Jeppson, L.R., Keifer, H.H. and Baker, E.W. (1975) *Mites Injurious to Economic Plants*. University of California Press, Berkeley, California.
Johnson, W.T. and Lyon, H.H. (1991) *Insects that Feed on Trees and Shrubs*. Cornell University Press, New York.
Kati, H., Ince, I.A., Sezen, K., Isci, S. and Demirbag, Z. (2009) Characterization of two *Bacillus thuringiensis* ssp. *morrisoni* strains isolated from *Thaumetopoea pityocampa* (Lep., Thaumetopoeidae). *Biocontrol Science and Technology* 19(5), 475–484.
Keçeci Ozman, S. (1995) Ecological studies with an emphasis on the population fluctuations and host relations of big bud mites [(*Phytocoptella avellanae* (Nal.) and *Cecidophyopsis vermiformis* (Nal.) (Eriophyoidea: Acarina)] on filbert in the Black Sea region. PhD thesis, Ankara University, Graduate School of Natural and Applied Sciences, Ankara. (abstract in English)
Keifer, H.H. (1940) Eriophyid Studies, IX. *The Bulletin, Department of Agriculture, State of California* 29(2), 112–117.
Keifer, H.H. (1944) Eriophyid studies, XIV. *The Bulletin, Department of Agriculture, State of California* 33(1), 18–38.
Keller, S., Kessler, P., Jensen, D.B. and Schweizer, C. (2002) How many spores of *Beauveria brongniartii* are needed to control *Melolontha melolontha*? *IOBC (International Organisation for Biological and Integrated Control) WPRS Bulletin* 25(7), 59–63.
Kimber, I. (2015) *Gypsonoma dealbana*. ukmoths. Available at: http://ukmoths.org.uk/species/gypsonoma-dealbana/ (accessed 1 September 2015).
Kiper, G. and Yücetin, T. (1971) Doğu Karadeniz Bölgesi fındık bahçelerinde görülen fındık yeşil kokarcası (*Palomena prasina* L.) zararının depolanmış iç fındıklara intikal oranı üzerinde araştırmalar. *Bitki Koruma Bülteni* 11(4), 218–224.
Kolk, A. and Starzyk, J.R. (1996) *Atlas of Harmful Forest Insects*. MULTICO, Warsaw. (in Polish)
Krantz, G.W. (1979) The role of *Phytocoptella avellanae* (Nalepa) and *Cecidophyopsis vermiformis* (Nalepa) (Eriophyoidea) in big bud of filbert. In: Piffl, E. (ed.) *Proceedings of the 4th International Congress of Acarology*. Academiai Kiado, Budapest, Hungary, pp. 201–208.
Kühnholz, S., Borden, J.H. and Uzunovic, A. (2003) Secondary ambrosia beetles in apparently healthy trees: adaptations, potential causes and suggested research. *Integrated Pest Management Reviews* 6, 209–219.
Kurt, M.A. (1975) *Doğu Karadeniz Fındıklarında Zarar Yapan* Palomena prasina *L. (Hem.: Pentatomidae)'nin Biyo-Ökolojisi Üzerinde Araştırmalar*. T.C. Gıda-Tarım ve Hayvancılık Bakanlığı, Zirai Mücadele ve Zirai Karantina Genel Müdürlüğü, Samsun Bölge Zirai Mücadele Araştırma Enstitüsü Yayınları, No. 25, Samsun, Türkiye.
Kurt, M.A. (1982) *Doğu Karadeniz Bölgesinde Fındık Zararlıları, Tanınmaları, Yayılış ve Zararları, Yaşayışları ve Savaşım Yöntemleri*. T.C. Tarım ve Orman Bakanlığı, Zirai Mücadele ve Zirai Karantina Genel Müdürlüğü, Mesleki Kitaplar Serisi, No. 26, Ankara.
Malysh, J.M. and Frolov, A.N. (2008) *Melolontha melolontha* L. In: Afonin, A.N., Greene, S.L., Dzyubenko, N.I. and Frolov, A.N. (eds) *Interactive Agricultural Ecological Atlas of Russia and Neighboring Countries. Economic Plants and Their Diseases, Pests and Weeds*. Available at: http://www.agroatlas.ru/en/content/pests/Melolontha_melolontha/ (accessed 1 February 2015).
Mani, E., Remund, U. and Schwaller, F. (1992) Attack of the bark beetle, *Xyleborus dispar* F. (Coleoptera: Scolytidae) in orchards and vineyards. Importance, biology, flight observations, control, development and use of an efficient ethanol trap. *Acta Phytopathologica et Entomologica Hungarica* 27(1–4), 425–433.
Marras, P.M., Loru, L. and Pantaleoni, R.A. (2009) Observations on the biology and behaviour of *Obera linearis* (Coleoptera Cerambycidae) in Sardinia (Italy). *Acta Horticulturae* 845, 465–470.
Mehlenbacher, S.A. and Miller, A.N. (1989) 'Barcelona' hazelnut. *Fruit Varieties Journal* 43(3), 90–95.
Messing, R.H. and AliNiazee, M.T. (1985) Natural enemies of *Myzocallis coryli* in Oregon hazelnut orchards. *Journal of the Entomological Society of British Columbia* 82, 14–18.
Mıcık, M. and Akça, İ. (2011) Determination of the efficacy of some biopesticides against hazelnut filbert aphid [*Myzocallis coryli* (Goetze) Homoptera: Aphididae]. In: Book of abstracts, *Proceedings of the Fourth Plant Protection Congress of Turkey*, 28–30 June 2011, Kahramanmaraş, Turkey, p. 273.
Milenkovic, S. and Mitrovic, M. (2001) Hazelnut pests in Serbia. *Acta Horticulturae* 556, 403–406.
Nalcacioglu, R., Yaman, M., Dulger, S., Belduz, A.O. and Demirbag, Z. (2002) Isolation and characterization of *Bacillus thuringiensis* isolated from hazelnut fields in Turkey. *Fresenius Environmental Bulletin* 11(7), 337–341.

Nordin, G.L. and Maddox, J.V. (1974) Microsporidia of the fall webworm *Hyphantria cunea*: identification, distribution and comparison of *Nosema* sp. with similar species of *Nosema* spp. from other Lepidoptera. *Journal of Invertebrate Zoology* 24, 1–13.

Noyes, J.S. (2003) Universal Chalcidoidea Database. Available at: www.nhm.ac.uk/entomology/chalcidoids/about.html (accessed 14 December 2014).

Ourecky, D.K. and Slate, G.L. (1969) Susceptibility of filbert varieties and hybrids to the filbert bud mite, *Phytoptus avellanae* Nal. *Annual Report of the Northern Nut Growers Association* 60, 89–91.

Ovsyannikova, E.I. and Grichanov, I.Y. (2008) *Archips rosana* L. In: Afonin, A.N., Greene, S.L., Dzyubenko, N.I. and Frolov, A.N. (eds) *Interactive Agricultural Ecological Atlas of Russia and Neighboring Countries. Economic Plants and Their Diseases, Pests and Weeds*. Available at: http://www.agroatlas.ru/en/content/pests/Archips_rosana/ (accessed 23 January 2015).

Ozman, S.K. (1998) *Verticillum lecanii* (Zimm.) Viégas, a fungal pathogen of the big bud mites, *Phytoptus avellanae* Nal. and *Cecidophyopsis vermiformis* Nal. (Acarina: Eriophyoidea). In: Book of abstracts, *The VIth European Congress of Entomology*, 23–29 August 1998, Ceske Budejovice, Czech Republic, p. 610.

Ozman, S.K. (2000) Some biological and morphological differences between gall and vagrant forms of *Phytoptus avellanae* Nal. (Acari: Phytoptidae). *International Journal of Acarology* 26(3), 215–219.

Ozman, S.K. and Cobanoglu, S. (2001) Current status of hazelnut mites in Turkey. *Acta Horticulturae* 556, 479–487.

Ozman, S.K. and Ecevit, O. (1996) Relations between bud falls and big bud mites [*Phytoptus avellanae* Nal. and *Cecidophyopsis vermiformis* Nal. (Acarina: Eriophyoidea)] in hazelnut. In: *Proceedings of the Third Turkish National Congress of Entomology*, 24–28 September 1996, Ankara, Turkey. Ankara Üniversitesi Basımevi, Ankara, Turkey, pp. 337–345. (abstract in English)

Ozman, S.K. and Hatat, G. (1999) Biological control of *Phytoptus avellanae* Nal. and *Cecidophyopsis vermiformis* Nal. (Acarina: Eriophyoidea) by *Verticillium lecanii* (Zimm.) Viegas. In: *Proceedings of the Fourth Turkish National Congress of Biological Control*, 26–29 January 1999, Adana, Turkey. Entomoloji Derneği Yayınları No. 9, Izmir, Turkey, pp. 189–200. (abstract in English)

Ozman, S.K. and Toros, S. (1996) Distributions of big bud mites [*Phytoptus avellanae* Nal. and *Cecidophyopsis vermiformis* Nal. (Acarina: Eriophyoidea)] in the Black Sea region. In: *Proceedings of the Third Turkish National Congress of Entomology*, 24–28 September 1996, Ankara, Turkey. Ankara Üniversitesi Basımevi, Ankara, Turkey, pp. 328–336. (abstract in English)

Ozman, S.K. and Toros, S. (1997a) Life cycles of *Phytoptus avellanae* Nal. and *Cecidophyopsis vermiformis* Nal. (Acarina: Eriophyoidea). *Acta Horticulturae* 445, 493–501.

Ozman, S.K. and Toros, S. (1997b) Population fluctuations of *Phytoptus avellanae* Nal. and *Cecidophyopsis vermiformis* Nal. (Acarina: Eriophyoidea) in big buds of hazelnut. *Acta Horticulturae* 445, 511–520.

Ozman, S.K. and Toros, S. (1997c) Damage caused by *Phytoptus avellanae* Nal. and *Cecidophyopsis vermiformis* Nal. (Acarina: Eriophyoidea) in hazelnut. *Acta Horticulturae* 445, 537–543.

Ozman-Sullivan, S.K. (2006a) Life history of *Kampimodromus aberrans* (Oudemans) as a predator of *Phytoptus avellanae* Nalepa (Acari: Phytoseiidae, Phytoptidae). *Experimental and Applied Acarology* 38, 15–23.

Ozman-Sullivan, S.K. (2006b) *Phytoseius plumifer* as a predator of *Phytoptus avellanae* (Phytoseiidae, Phytoptidae). In: *Proceedings of the 12th International Congress of Acarology*, 21–26 August 2006, Amsterdam, The Netherlands, p. 152.

Ozman-Sullivan, S.K. (2006c) Harmful mites and their economic importance in hazelnut orchards. *Ondokuz Mayıs Üniversitesi Ziraat Fakültesi Dergisi* 21(2), 261–264. (abstract in English)

Ozman-Sullivan, S.K. (2014) Do the contradictory life cycles of the hazelnut big bud mite, *Phytoptus avellanae* (Nal.) (Acari: Phytoptidae) imply two species? In: Book of abstracts, *XIVth International Congress of Acarology*, 14–18 July 2014, Kyoto, Japan, p. 90.

Ozman-Sullivan, S.K. and Akca, I. (2005) Efficiency of pesticides against big bud mites [*Phytoptus avellanae* Nal. and *Cecidophyopsis vermiformis* Nal. (Acarina: Eriophyoidea)] on hazelnut. *Acta Horticulturae* 686, 393–399.

Ozman-Sullivan, S.K. and Sullivan, G.T. (2008) Pest management in organic hazelnut growing in Turkey. Paper presented at *XXIII International Congress of Entomology*, 6–12 July 2008, Durban, South Africa.

Ozman-Sullivan, S.K. and Sullivan, G.T. (2009) Strategies for improved pest management in Turkish hazelnut growing. *Acta Horticulturae* 845, 561–566.

Ozman-Sullivan, S.K., Kazmierski, A. and Cobanoglu, S. (2005) Alycina and Eupodina mites of hazelnut orchards in Turkey. *Acta Horticulturae* 686, 401–406.

Ozman-Sullivan, S.K., Öcal, H., Sullivan, G.T. and Mıcık, M. (2009) The effects on *Verticillium lecanii* (Zimmerman) Viégas of some pesticides used against *Parthenolecanium* spp. In: *Proceedings of the Third Plant Protection Congress of Turkey*, 15–18 July, 2009, Van, Turkey, p. 361.

Ozsahin, E., Sezen, K., Demir, I. and Demirbag, Z. (2014) Bacterial isolates from *Palomena prasina* (Hemiptera: Pentatomidae) include potential microbial control agents. *Biocontrol Science and Technology* 24(9), 1039–1051.

Papadopoulou, S., Chryssochoides, C. and Avtzis, D. (2012) Genetic diversity of *Lymantria dispar* Linnaeus (Lepidoptera: Lymantriidae) in northern Greece and evaluation of the effectiveness of novel insecticides. *Biotechnology and Biotechnological Equipment* 26(3), 2976–2980.

Paparatti, B. and Speranza, S. (2005) Biological control of hazelnut weevil (*Curculio nucum* L., Coleoptera, Curculionidae) using the entomopathogenic fungus *Beauveria bassiana* (Balsamo) Vuill. (Deuteromycotina, Hyphomycetes). *Acta Horticulturae* 686, 407–412.

Pape, T. and Thompson, F.C. (2013) *Mikomya coryli* (Kieffer). *Systema Dipterorum*. Available at: http://www.diptera.org/ (accessed 19 Feburary 2015).

Pesante, A. (1973) Disseccamento dei rametti di Nocciolo da *Gloeosporium* sp. [Twig dessication in hazel due to *Gleosporium* sp.]. *Phytopathologia Mediterranea* 12, 15–21.

Petanovic, R., Dobrivojevic, K. and Boskovic, R. (1989) Life cycle of hazelnut big bud mite *Phytoptus avellanae* (Nal.) (Acarida: Eriophyoidea) and the result of its control. *Zastita Bilja* 40(4), 433–441.

Peters, A., Sarraquigne, J.-P., Blum, B. and Kuske, S. (2007) Control of the hazelnut borer *Curculio nucum* with entomopathogenic nematodes. *Bulletin OILB/SROP* 30(1), 73–76.

Piskornik, Z. (1994) Relationship between the resistance of hazelnut cultivars to the hazelnut weevil and the content of sugars, aminoacids, and phenols in the endocarp tissue of growing nuts. *Acta Horticulturae* 351, 617–624.

Polajnar, J., Kavcic, A., Kosi, A.Z. and Cokl, A. (2013) *Palomena prasina* (Hemiptera: Pentatomidae) vibratory signals and their tuning with plant substrates. *Central European Journal of Biology* 8(7), 670–680.

Popovic, Z., Kostic, M., Stankovic, S., Milanovic, S., Sivcev, I., Kostic, I. and Kljajic, P. (2013) Ecologically acceptable usage of derivates of essential oil of sweet basil, *Ocimum basilicum*, as antifeedants against larvae of the gypsy moth, *Lymantria dispar*. *Journal of Insect Science* 13(161), 1–12.

Powell, J.A. and Opler, P.A. (2009) *Moths of Western North America*. University of California Press, Berkeley, California.

Railyan, N.N. (1974) The fall webworm in Moldavia. *Zashchita Rastenii* 10, 44–46. (Abstract in *The Review of Applied Entomology* (1976) 64(10), 5980.)

Rumine, P., Francardi, V. and de Silva, J. (2007) Laboratory trials of microbial control of *Agelastica alni* L. with *Beauveria bassiana* (Bals.) Vuill. *IOBC (International Organisation for Biological and Integrated Control) WPRS Bulletin* 30(1), 9.

Saruhan, I. (2004) Karadeniz bölgesi fındık üretim alanlarında görülen fındık kokarcası (*Palomena prasina* (Linneaus, 1761) Heteroptera: Pentatomidae)' nın biyolojisi, populasyon yoğunluğu ve zarar şekli üzerine araştirmalar. Doktora tezi, Ondokuz Mayıs Üniversitesi, Fen Bilimleri Enstitüsü, Samsun, Türkiye.

Saruhan, I. and Akyol, H. (2013) Monitoring population density and fluctuations of *Xyleborus dispar* and *Xyleborinus saxesenii* (Coleoptera: Scolytidae) with red winged sticky traps in hazelnut orchards. *African Journal of Agricultural Research* 8(19), 2189–2194.

Saruhan, İ. and Tuncer, C. (2001) Population densities and seasonal fluctuations of hazelnut pests in Samsun, Turkey. *Acta Horticulturae* 556, 495–502.

Saruhan, İ and Tuncer, C. (2009) Population density and fluctuations of the green shield bug (*Palomena prasina* L., Heteroptera: Pentatomidae) in hazelnut orchards of Turkey. *Acta Horticulturae* 845, 549–554.

Saruhan, I., Akca, I. and Kushiyev, R. (2014) Toxicity of some biopesticides to the fall webworm, *Hyphantria cunea* Durry (Lepidoptera: Arctiidae). *Egyptian Journal of Biological Pest Control* 24(1), 255–257.

Schvester, D. (1950) Sur un nematode du groupe du *Parasitylenchus dispar* Fuchs, parasite nouveau du xylebore disparate (*Xyleborus dispar* F.). *Annales des Epiphyties* 1, 48–53.

Sezen, K. and Demirbağ, Z. (1999) Isolation and insecticidal activity of some bacteria from the hazelnut beetle (*Balaninus nucum* L.). *Applied Entomology and Zoology* 34(1), 85–89.

Sezen, K. and Demirbağ, Z. (2006) Insecticidal effects of some biological agents on *Agelastica alni* (Coleoptera: Chrysomelidae). *Biologia, Bratislava* 61(6), 687–692.

Sezen, K., Yaman, M. and Demirbağ, Z. (2001) Insecticidal potential of *Serratia marcescens* Bn10. *Biologia, Bratislava* 56(3), 333–336.

Sezen, K., Demir, İ. and Demirbağ, Z. (2004) Study of the bacterial flora as a biological control agent of *Agelastica alni* L. (Coleoptera: Chrysomelidae). *Biologia, Bratislava* 59(3), 327–331.

Sezen, K., Demir, Y. and Demirbağ, Z. (2007) Identification and pathogenicity of entomopathogenic bacteria from common cockhafer, *Melolantha melontha* (Coleoptera: Scarabaeidae). *New Zealand Journal of Crop and Horticultural Science* 35(1), 79–85.

Sezen, K., Kati, H., Nalcacioğlu, R., Muratoğlu, H. and Demirbağ, Z. (2008) Identification and pathogenicity of bacteria from European shot-hole borer, *Xyleborus dispar* Fabricius (Coleoptera: Scolytidae). *Annals of Microbiology* 58(2), 173–179.

Snare, L. (2006) *Pest and Disease Analysis in Hazelnut.* New South Wales Department of Primary Industries, Orange, New South Wales, Australia.

Snezana, P. (2013) *Anoplus roboris.* Fauna Europaea: Curculionidae. In: Alonso-Zarazaga, M.A.(ed.) *Fauna Europaea: Coleoptera.* Fauna Europaea version 2.6.2, Available at: http://www.faunaeur.org (accessed 12 April 2015).

Speranza, S., Bucini, D. and Paparatti, B. (2009) New observation on biology of European shot-hole borer [*Xyleborus dispar* (F.)] on hazel in northern Latium (Central Italy). *Acta Horticulturae* 845, 539–542.

Stamenkovic, S., Milenkovic, S., Pesic, M. and Mitrovic, M. (1997) Population dynamics, harmfulness and control of *Phytoptus avellanae* (Nalepa) in western Serbia. *Acta Horticulturae* 445, 521–526.

Sullivan, G.T. (2011) Determination of biological control agents of overwintering *Hyphantria cunea* pupae and efficacy of entomopathogenic fungi. MSc thesis, Suleyman Demirel University, Graduate School of Applied and Natural Sciences, Isparta, Turkey.

Sullivan, G.T. and Ozman-Sullivan, S.K. (2012) Tachinid (Diptera) parasitoids of *Hyphantria cunea* (Lepidoptera: Arctiidae) in its native North America and in Europe and Asia – a literature review. *Entomologica Fennica* 23(4), 181–192.

Sullivan, G.T., Karaca, I., Ozman-Sullivan, S.K. and Kolarov, J. (2010a) Ichneumonid (Hymenoptera) parasitoids of overwintering *Hyphantria cunea* (Drury) (Lepidoptera: Arctiidae) in hazelnut plantations of the central Black Sea region of Turkey. *Zootaxa* 2608, 63–68.

Sullivan, G.T., Aksoy, H.M., Ozman-Sullivan, S.K. and Karaca, I. (2010b) Bacteria isolated from overwintering *Hyphantria cunea* (Lepidoptera: Arctiidae) pupae as potential entomopathogens. In: Book of abstracts, *43th Annual Meeting of Society for Invertebrate Pathology*, 11–15 July, 2010, Trabzon, Turkey, p. 154.

Sullivan, G.T., Karaca, I., Ozman-Sullivan, S.K. and Yang, Z.Q. (2011a) Chalcidoid parasitoids of overwintered pupae of *Hyphantria cunea* (Lepidoptera: Arctiidae) in hazelnut plantations of Turkey's central Black Sea region. *Canadian Entomologist* 143(4), 411–414.

Sullivan, G.T., Özman-Sullivan, S.K., Karaca, I., and Karaca, G. (2011b) Entomopathogenic efficacy of fungi isolated from overwintered *Hyphantria cunea* (Drury) (Lepidoptera: Arctiidae) pupae. In: Proceedings of the Fourth Plant Protection Congress of Turkey, 28–30 June, 2011, Kahramanmaraş, Turkey, p. 137.

Sullivan, G.T., Karaca, I., Ozman-Sullivan, S.K. and Kara, K. (2012a) Tachinid (Diptera: Tachinidae) parasitoids of overwintered *Hyphantria cunea* (Drury) (Lepidoptera: Arctiidae) pupae in hazelnut plantations in Samsun province, Turkey. *Journal of the Entomological Research Society* 14(1), 21–30.

Sullivan, G.T., Ozman-Sullivan, S.K. and Kesdek, M. (2012b) Carabids (Coleoptera: Carabidae) of hazelnut plantations in Turkey. In: Book of abstracts, *Proceedings of 8th International Congress on Hazelnut*, 19–22 March, 2012, Temuco, Chile, p. 53.

Sullivan, G.T., Ozman-Sullivan, S.K. and Kesdek, M. (2012c) Distributions and seasonal fluctuations of carabid species (Coleoptera: Carabidae) of hazelnut plantations in Turkey. Paper presented at the *XXIV International Congress of Entomology*, 19–25 August 2012, Daegu, Korea.

Sullivan, G.T., Karaca, I. and Ozman-Sullivan, S.K. (2015) Potential biological control agents collected from overwintering *Hyphantria cunea* (Drury) (Lepidoptera: Arctiidae) pupae in Turkey. In: *International Conference on Biopesticides* 7, 19–25 October, 2014, Side-Antalya, Turkey, p. 17.

Szalay-Marzso, L. (1972) Biology and control of the fall webworm (*Hyphantria cunea* Drury) in the Middle- and East European Countries. *Organisation Europeenne et Mediterranneene pour la Protection des Plantes* 3, 25–35. (Abstract in *The Review of Applied Entomology*, 1973, 61(3), 702.)

Tavamaishvili, L.E. (1990) The chief pests of hazelnuts in subtropical western Georgia and their control. *Subtropicheskie Kul'tury* 5, 119–125.

Tavella, L., Arzone, A., Miaja, M.L. and Sonnati, C. (2001) Influence of bug (Heteroptera, Coreidae and Pentatomidae) feeding activity on hazelnut in northwestern Italy. *Acta Horticulturae* 556, 461–467.

Thompson, M.M. (1977) Inheritance of big bud mite susceptibility in filbert. *Journal of the American Society for Horticultural Science* 102(1), 39–42.

Tomalak, M. (2004) Infectivity of entomopathogenic nematodes to soil-dwelling developmental stages of the tree leaf beetles *Altica quercetorum* and *Agelastica alni*. *Entomologia Experimentalis et Applic*ata 110(2), 125–133.

Tosun, N. and Onan, E. (2014) *Ruhsatlı Bitki Koruma Ürünleri 2014/2015.* Hasad Yayıncılık, İstanbul, Türkiye.

Tuncer, C. (1992) Amerikan beyaz kelebeği (*Hyphantria cunea* Drury, Lepidoptera: Arctiidae) 'nin Samsun ve çevresindeki biyolojisi ve özellikle konukçu bitkilerin değişik açılardan etkileri üzerinde araştırmalar. Doktora tezi, Ankara Üniversitesi, Fen Bilimleri Enstitüsü, Ankara.

Tuncer, C. and AliNiazee, M.T. (1998) Acute and chronic effects of neem on the filbert aphid, *Myzocallis coryli* Goetze (Hom.: Aphididae). *International Journal of Pest Management* 44(2), 53–58.
Tuncer, C. and Ecevit, O. (1996) Amerikan Beyaz Kelebeğinin Samsun ili fındık üretim alanlarındaki kısa biyolojisi ve doğal düşmanları. *Fındık ve Diğer Sert Kabuklu Meyveler Sempozyumu*, Ondokuz Mayıs Üniversitesi, Ziraat Fakültesi, 10–11 Ocak 1996, Samsun, Türkiye, pp. 134–145. (abstract in English)
Tuncer, C. and Ecevit, O. (1997) Current status of hazelnut orchards. *Acta Horticulturae* 445, 545–552.
Tuncer, C., Ecevit, O. and Akça, I. (1997) Observations on the biology of filbert aphid (*Myzocallis coryli*, Homoptera: Aphididae) in hazelnut orchards. *Acta Horticulturae* 445, 485–492.
Tuncer, C., Akça, I. and Saruhan, I. (2001) Integrated pest management in Turkish hazelnut orchards. *Acta Horticulturae* 556, 419–429.
Tuncer, C., Saruhan, İ. and Akça, İ. (2005) The insect pest problem affecting hazelnut kernel quality in Turkey. *Acta Horticulturae* 686, 367–375.
Tzanakakis, M.E. (2008) Hazelnut and walnut twig borer: *Obera linearis* L. (Coleoptera: Cerambycidae). In: Capinera, J.L. (ed.) *Encyclopedia of Entomology*, Volume 2. Springer Science+Business Media BV, Dordrecht, The Netherlands, pp. 1772–1774.
Ugras, S. and Demirbağ, Z. (2013) Screening antibacterial activity of entomopathogenic bacteria isolated from pests of hazelnut. *Biologia* 68(4), 592–598.
Ural, I. (1957a) *Doğu Karadeniz Fındıklarında Zarar Yapan* Balaninus (Curculio) nucum *L. Böceğinin Biyolojisi ve Mücadelesi Üzerine Araştırmalar*. Ankara Üniversitesi Ziraat Fakültesi Yayınları, 130, Çalışmalar 80, Ankara.
Ural, İ. (1957b) *Kızılağaç Böceği* (Agelastica alni). Samsun Bölge Zirai Mücadele Araştırma Enstitüsü Yayınları, No. 5, Çituri Biraderler Basımevi, İstanbul, Türkiye.
Ural, İ. (1957c) *Kırtırtılı (*Lymantria dispar*)*. Samsun Bölge Zirai Mücadele Araştırma Enstitüsü Yayınları, No. 6, Çituri Biraderler Basımevi, İstanbul, Türkiye.
Ural, I. (1968) Karadeniz fındıklarında zarar yapan Mayıs Böceği (*Melolontha melolontha*) üzerinde araştırmalar. *Bitki Koruma Bülteni* 8(1), 1–38.
Ural, İ. and Kurt, A. (1973) Research on the biology and chemical control of hazelnut gall midge (*Mikomyia coryli* Kieffer) which does damage to hazelnut in the east Black Sea region. *Bitki Koruma Bülteni* 13(1), 1–18. (abstract in English)
Ural, İ., Gökmen, N., Küçükarslan, N., Işık, M. and Kırtıloğlu, T. (1968) *Fındıklarda zarar yapan* Gypsonoma dealbana *Fröhl.'ün Biyolojisi ve Mücadelesi Üzerinde Araştırmalar*. T.C. Tarım Bakanlığı, Zirai Mücadele ve Zirai Karantina Genel Müdürlüğü, Araştırma Eserleri Serisi, Teknik Bülten No. 4, Yenilik Basımevi, İstanbul, Türkiye.
Ural, İ., Işık, M. and Kurt, A. (1973) Doğu Karadeniz Bölgesi fındık bahçelerinde tesbit edilen böcekler üzerinde bazı incelemeler. *Bitki Koruma Bülteni* 13(2), 55–66.
Viggiani, G. (1973) Nematodes, mites and insects injuries to hazel. *Annali della Facolta di Scienze Agrarie della Universita di Napoli* 6, 160–175. [Abstract in *The Review of Applied Entomology* (1975) 63(9), 3593.]
Viggiani, G. (1984) *Avversita, Malattie e Fitofagi del Nocciolo*. Istituto di Entomologia Agraria dell'Universita di Napoli-Portici, Italy.
Viggiani, G. and Bianco, M. (1973) Osservazioni biologiche sul *Phytoptus avellanae* Nal. in Campania e relative prove di lotta chimica. *Atti Giornate Fitopatologiche*, Aprile, 79–83.
Warren, L.O. and Tadic, M. (1967) The fall webworm, *Hyphantria cunea*, its distribution and natural enemies. A world list (Lepidoptera: Arctiidae). *Journal of the Kansas Entomological Society* 40(2), 194–202.
Weber, J.D. (2007) Phenology of hazelnut big bud mites in Canterbury and implications for management. MSc thesis, Lincoln University, Lincoln, New Zealand.
Wojciechowicz-Żytko, E. (2003) Development of *Myzocallis coryli* Goetze (Homoptera, Aphidodea) on the different hazel (*Corylus* L.) cultivars. *Journal of Plant Protection Research* 43(4), 369–374.
Wojciechowicz-Żytko, E. (2005) Infestation of hazel nuts by hazelnut weevil (*Curculio nucum* L., Coleoptera, Curculionidae) in Poland. *Journal of Plant Protection Research* 45(2), 59–61.
Yaman, M. and Demirbag, Z. (2000) Isolation, identification and determination of insecticidal activity of two insect-originated *Bacillus* spp. *Biologia, Bratislava* 55(3), 283–287.
Yaman, M., Nalcacioglu, R. and Demirbag, Z. (2002) Studies on bacterial flora in the population of the fall webworm, *Hyphantria cunea* Drury. (Lep., Arctiidae). *Journal of Applied Entomology* 126(9), 470–474.
Yaman, M., Bekircan, C., Radek, R. and Linde, A. (2012) The first record of nucleopolyhedrovirus isolated from the gypsy moth, *Lymantria dispar* (Lepidoptera: Lymantriidae) in Turkey. *Turkish Journal of Parasitology* 36(2), 92–95.

Yamanaka, S., Seta, K. and Yasuda, M. (1986) Evaluation of the use of entomogenous nematode, *Steinernema feltiae* (str. Mexican) for the biological control of the fall webworm, *Hyphantria cunea* (Lepidoptera: Arctiidae). *Japanese Journal of Nematology* 16, 26–31.

Yang, Z.Q., Wang, X.Y., Wang, C.Z., Qiao, X.R. and Pang, J.J. (2005) Studies on utilizing parasitoid *Chouioia cunea* Yang (Hymenoptera: Eulophidae) for sustainable control of fall webworm. *Scientia Silvae Sincae* 41(5), 72–80.

Yarahmadi, F. and Rajabpour, A. (2012) Seasonal population dynamics and spatial distribution of *Myzocallis coryli* Goetze on *Corylus avellana* in Iran. *Asian Journal of Biological Sciences* 5(1), 52–56.

Yu, D.S., Achterberg, C. van and Horstmann, K. (2005) *World Ichneumonoidea 2004. Taxonomy, Biology, Morphology and Distribution (Braconidae).* Taxapad 2005 (Scientific Names for Information Management) Interactive Catalogue on DVD/CDROM. Taxapad, Vancouver, Canada.

Zabel, A., Manojlovic, B., Rajkovic, S., Stankovic, S. and Kostic, M. (2002) Effect of neem extract on *Lymantria dispar* L. (Lepidoptera: Lymantriidae) and *Leptinotarsa decemlineata* Say. (Coleoptera: Chrysomelidae). *Journal of Pest Science* 75, 19–25.

Zamoum, M., Berchiche, S., Sai, K., Triggiani, O. and Tarasco, E. (2011) Preliminary survey of the occurence of entomopathogenic nematodes and fungi in the forest soils of Algeria. *Silva Lusitana* (Special issue), 141–145.

12 Pest Management in Organic Almond

Hüseyin Baspinar,[1]* David Doll[2] and Jhalendra Rijal[3]
[1]*Adnan Menderes University, Aydın, Turkey;* [2]*University of California Cooperative Extension, Merced County, California, USA;* [3]*University of California Cooperative Extension, Stanislaus County, California, USA*

Introduction

Almonds possess economic, medicinal and nutritional benefits and are consumed in nearly every country worldwide. Major production areas, however, are limited to Mediterranean-like climates, which are broadly categorized as hot, dry summers and mild, wet winters. Even though the almond tree is native to western Asia, the USA has the highest production of almonds in the world. In 2013, roughly 82% of the almond production was within the USA with an estimated 840.91 thousand t (Tables 12.1 and 12.2). Other major production areas include EU-27 (the 27 countries of the European Union) (predominantly Spain), Australia and Turkey. The export value of the almond crop for the USA is US$3387 billion in 2012 (Anonymous, 2013a).

Within agricultural systems, the almond tree is unique. Almond trees can grow on a variety of soil types, which include high pH and moderately saline soils, even though they perform best on well-drained, deep, fertile soils. Almond trees are able to survive on as little as 180 mm of water annually, and respond to increased water applications with increasing yield. Almonds bloom earlier than other *Prunus* spp., however, and therefore are susceptible to late spring frosts. They are also susceptible to a number of diseases and insects. These risks can be minimized by selecting later blooming or more resistant varieties.

Almonds are affected by a number of insect pests (Table 12.3). These pests attack the tree or kernel, reducing orchard vigour or yield. Pest pressure, however, tends to be lower than other crops due to the protection of the kernel/seed from the environment by a shell and hull, and the production areas being primarily in arid areas which tend to have low insect pressure. This provides an opportunity to grow almonds organically or without the requirement of a large amount of pesticides.

Organic production within almond is reliant on the use of variety selection or cultural practices to reduce insect infestation rates. Cultural methods include good sanitation practices in the orchard, which include removing or shredding old debris, dried nuts from the previous year's crop, and dead wood from the trees. Organically approved pesticides have been shown to be effective for some, but not all pests. Specific control measures for the pests listed in

* Corresponding author: hbaspinar@adu.edu.tr

Table 12.1. Production-wise ranking of different almond-growing countries of the world. (From Almond Board of California and International Nut and Dried Fruit Council (INC), 2013, cited in Anonymous, 2013a.)

Countries	Production (%)
USA	82
EU-27[a]	6
Australia	5
Turkey	2
Others	5

[a]EU-27, The 27 countries of the European Union.

Table 12.2. Forecasted world almond production in 2013–2014. (From Almond Board of California and International Nut and Dried Fruit Council (INC), 2013, cited in Anonymous, 2013a.)

Country	Production (thousands of kg)
USA	840.91
Australia	69.13
Spain	32.04
Turkey	15.04
Iran	15.04
Tunisia	13.04
Chile	10.00
Morocco	6.00
Greece	5.00
Italy	5.00
Others	30.04
World total	1041.24

Table 12.3 are outlined below. Please note that there are other pests which are not listed that are minor or limited to smaller areas of production.

Major Pests

Monosteira unicostata (Mulsant & Rey) (Hemiptera: Tingididae) (poplar lace bug)

Description

The body of *Monosteira unicostata* is light yellowish grey in colour and about 2.5 mm long. The hemelytra are divided in appearance (Lodos, 1982).

Life cycle

M. unicostata overwinters in the adult stage in hidden places, under debris or in the crevices on the trees in orchards. Females lay their eggs by inserting them in the plant tissue under the leaf. The surface of the inserted egg on the leaf tissue is covered with a dark-coloured liquid excreted by the female from its anus. It reaches high population numbers on poplar trees in August and September. Additionally, the other two species of Tingidae, *Monosteira lobulifera* Reuter and *Stephanitis pyri* (Fabricius) have also been found in almond orchards in Turkey (Bolu, 2007).

Damage

Adults and nymphs feed on the underside of leaves by sucking the sap which results in damage to the chlorophyll and in white patches on leaves. The leaves drop prematurely as a result of feeding by either adults or nymphs of the pest (Lodos, 1982). *M. unicostata* has three generations in 1 year (Russo *et al.*, 1994).

Distribution and host plants

The poplar lace bug is reported from Mediterranean countries, Turkistan, Hungary and the Caucasus. The host plants include poplar, *Salix*, apple, pear and almond (Lodos, 1982).

Management for organic farming

The efficacy of kaolin, azadirachtin and potassium salts of fatty acids combined with thyme essential oil against adults and fourth instar nymphs was evaluated in laboratory assays. It was concluded that the products tested have shown high and different efficacy on nymphs and adults of *M. unicostata*. This activity might be suitable for the practical application of these compounds to control its populations under real field conditions (Sanches-Ramos *et al.*, 2014).

The number of coccinellid and hemipteran predators feeding on *M. unicostata* has been determined in almond orchards (Bolu, 2007).

Table 12.3. Pests in almond orchards.

Species	Common name	Order: family
Amyelois transitella (Walker)	Navel orangeworm	Lepidoptera: Pyralidae
Quadraspidiotus perniciosus (Comstock)	San Jose scale	Hemiptera: Diaspididae
Monosteira unicostata (Mulsant & Rey)	Poplar lace bug	Hemiptera: Tingididae
Capnodis carbonaria (Klug)	Almond borer	Coleoptera: Buprestidae
Cerambyx dux (Faldermann)	Longhorn beetle	Coleoptera: Cerambycidae
Anarsia lineatella Zeller	Peach twig borer	Lepidoptera: Gelechiidae
Tropinota (= *Epicometis*) *hirta* (Poda)		Coleoptera: Scarabaeidae
Ectomyelois ceratoniae (Zeller)	Carob moth	Lepidoptera: Pyralidae
Cimbex quadrimaculatus (Müller)	Almond sawfly	Hymenoptera: Cimbicidae
Eurytoma amygdali Enderlein		Hymenoptera: Eurotomidae
Tetranychus pacificus McGregor, *Tetranychus urticae* Koch, *Tetranychus turkestani* Ugarov and Nikolski	Web-spinning spider mites (Pacific spider mite, two-spotted spider mite, strawberry spider mite)	Trombidiforma: Tetranychidae
Anthonomus amygdali Hustache[a]		Coleoptera: Curculionidae
Brachycaudus amygdalinus (Schouteden)[a]		Hemiptera: Aphididae
Panonychus ulmi (Koch)[a], *Bryobia rubrioculus* (Scheuten)[a]	European red mite, brown almond mite	Trombidiforma: Tetranychidae
Tetramorium caespitum (L.)[a], *Solenopsis xyloni* McCook[a], *Solenopsis molesta* (Say)[a]	Ants (pavement ant, southern fire ant, thief ant)	Hymenoptera: Formicidae

[a]Secondary pests.

Capnodis carbonaria (Klug) (Coleoptera: Buprestidae) (almond borer)

Description

Adults of *Capnodis carbonaria* are black or bronzed in colour and the pronotum is slightly shiny and ornamented in black and white. The body of adults becomes tapered from the anterior to the posterior. The length of the adult is 20–35 mm, and the forewings are very hard. The eggs are 1 mm in length and oval in shape. Larvae are flattened with 13 segments and yellow in colour. The young larva is very pubescent, but from the second instar on it changes and becomes hairless and smooth. The length of the developed larvae can reach up to 12 cm (Lodos and Tezcan, 1995), depending on geographic area and on which host they are feeding. The pupa is oval in appearance and resembles the adult in shape.

Life cycle

Almond borer beetles overwinter in the adult stage under debris or in the ground. They become active and copulate when the temperature increases over 25–26°C. Mated females deposit their eggs in the crevices of the bark very near to ground level, or on the ground near to the trunk of trees. Oviposition begins in May, but most of the eggs are laid in July and August (Lodos and Tezcan, 1995). A single female can deposit almost 2000 eggs during its lifespan. The eggs are wet when laid and are covered by debris and soil adhering to them, so that they are camouflaged. Hatching larvae from the eggs are quite active underground. They can move by using hairs on their body to reach for the roots. The hairs are lost when they tunnel into roots. Young larvae feed in cambium tissue of the root by tunnelling in 30–45 cm length. It takes 1–2 years to develop from larvae into pupae. They pupate near the root in the ground. Adults hatch from pupae after 4 weeks. Adults' emergence from pupae mostly occurs in July–August. This group of adults copulate and deposit their eggs in September–October. The second group of adults appears in October–November, and they become active in the next spring and oviposition takes place in July–August

(Anonymous, 2008b). The life cycle takes 12–15 months to be completed (Talhouk, 2009).

Damage

C. carbonaria adults feed on leaves and young shoots, but economic damage is rare. Young trees, between the ages of 1 and 4 years, are at the greatest risk due to tree collapse and death from larvae feeding in the roots. Older trees may die if many years of root feeding occur.

Distribution and host plants

The pest is known in Italy, the former Yugoslavia, Greece, Bulgaria, Cyprus, Lebanon, Israel, Syria, Iraq, Iran, south Caucasus and Afghanistan (Avidov and Harpaz, 1969; Lodos and Tezcan, 1995). Host plants include primarily almond and other fruit trees such as apricots, peaches, plums, cherries and sour cherries (Lodos and Tezcan, 1995) and additionally pistachios in Turkey (Nizamlıoglu, 1957).

Management for organic farming

CULTURAL PRACTICES. Weed control can help to destroy the adult habitat under the canopy. Tree trunks can be painted with whitewash on the bark to prevent egg laying by adult females. Early in the morning and late evening, adults can be hand collected when they are partly inactive on the trunk of trees. When damaged leaves drop from the trees, it is an indication of the damage by *Capnodis*, and *Capnodis* can be collected and removed by shaking the branches of young trees (Anonymous, 2008b).

BIOLOGICAL CONTROL. No information on biological control of *C. carbonaria* could be found in the literature, however, the nematode, *Steinernema carpocapsae* (Weiser) was reported to be very effective (96–100% efficacy) against neonate larvae of *Capnodis tenebrionis* (L.) in laboratory trials (Garcia Del Pino and Morton, 2005). *S. carpocapsae* in a chitosan formulation was found to be very effective against *C. tenebrionis* in field trials in apricot plantations in Spain (Martinez de Altub *et al.*, 2008).

There are very few natural enemies of *C. tenebrionis*. Only *Sclerodermus cereicollis* Kieffer (Hymenoptera: Bethylidae) and some entomopathogenic fungi were reported in south Italy. Additionally, two commercial formulations of *Bacillus thuringiensis* (Berliner) were found to be ineffective against this pest (Marannino and Lillo, 2007).

***Cerambyx dux* (Faldermann) (Coleoptera: Cerambycidae) (longhorn beetle)**

Description

The adult of *Cerambyx dux* is dark brown and 50–52 mm long. The antennae are longer than the body. The eggs are 4.5 mm long, oval and dirty white in colour. The newly hatched larvae are very small, around 4.5 mm long. The larvae are soft and creamy white in colour and cylindrical in shape. The larva may reach 9–10 cm just before it pupates (Talhouk, 1969). The pupae are initially dirty white, but become dark in colour over time (Avidov and Harpaz, 1969).

Life cycle

The longhorn beetle overwinters as the adult stage after hatching from the pupa in the late autumn within a tunnel inside the trunk of a tree. The following year, the beetle will emerge from the tunnel in the late spring or early summer. The adult female deposits eggs singly under the bark in crevices or cracks. A single female can deposit 30–40 eggs during its lifetime. The newly hatched larvae start boring into the trunk or main branches of the tree and feed on the wood (Avidov and Harpaz, 1969; Talhouk, 1969). Large amounts of sawdust or frass are evacuated from the gallery holes on the trunk. The larvae feeding in the wood produce tapping sounds. The larval development is completed in 15–17 months, from June until August or October in the Middle East (Talhouk, 1969).

Damage

The damage is caused by the larvae boring the tunnel in the wood tissue. This damage weakens the tree which may lead to the breaking of affected branches or scaffolds under the weight of heavy loads or the pressure of the wind (Talhouk, 1969).

Distribution and host plants

The longhorn beetle is reported in countries of the Middle East (Talhouk, 1969), and Mediterranean countries including Bulgaria, Crimea, Greece, Italy, Israel, Syria, the Lebanon, Macedonia, north-west Iran and Turkey (Avidov and Harpaz, 1969; Anonymous, 2011). The list of host plants includes stone fruits, such as peach, apricot, almond (Avidov and Harpaz, 1969) and plum (Talhouk, 1969).

Management for organic farming

CULTURAL PRACTICES. Wood-boring pests of trees are attracted to weak woody plants in order to deposit their eggs (Talhouk, 1969). Hence cultural practices such as regular irrigation, pruning and proper fertilization can help to maintain the trees in a healthy state so they are resistant to pest attack.

***Anarsia lineatella* Zeller (Lepidoptera: Gelechiidae) (peach twig borer moth)**

Description

Anarsia lineatella moths are grey in appearance with grey forewings. The wings may have darker and lighter spots and lines. The hindwings are lighter in colour than the forewings. The wingspan is 14–18 mm and the body length is 7–8 mm. The wings are fringed with long hairs (Avidov and Harpaz, 1969; Talhouk, 1969).

The eggs are oval and 0.5 × 0.3 mm in size, and when laid they are initially creamy white in colour, but later turn to orange and brown (Anonymous, 2008b). The newly hatched larva is light brown and later it turns to reddish brown in colour. Its head, pronotum and legs are black with whitish intersegmental areas giving the larvae a banded appearance. The body of the larva is covered with numerous hairs on the dorsal surface and when mature, the larval length is 10 mm. The pupa is elongate and 6 mm long with numerous hairs (Avidov and Harpaz, 1969; Talhouk, 1969).

Life cycle

The peach twig borer overwinters as a young larva in a cavity, which is termed a hibernaculum, approximately 2 mm beneath the bark of twigs and branches. In spring, the larva becomes active and leaves the cavity to feed on flower buds, leaves, nutlets, growing tips and the small buds. The larvae can change their feeding site and may attack several growing tips before they become mature. The larvae will burrow into a shoot tip travelling 2–5 cm down into the wood, killing the terminal bud. When fully grown, the larvae leave their tunnels to become pupae, spinning a cocoon on the branches (Avidov and Harpaz, 1969; Talhouk, 1969). Adults hatch from the pupae in early to mid-spring, depending on geographic area (Anonymous, 2008b). Females deposit their eggs, more than 140 in number (Avidov and Harpaz, 1969), on fruit or foliage after copulation. The peach twig borer produces four generations/year in Israel (Avidov and Harpaz, 1969), California (Strand, 2002) and in Syria (Talhouk, 1969), and three to five generations in Turkey (Anonymous, 2008b).

Damage

The first generation larvae hatching from the eggs prefer to feed on the fruits causing damage and fruit drop. Attacked fruits indicate a gummy point where the larva enters into the fruit, and larva can feed on the kernel or between the hull and the shell. The larvae of the peach twig borer may also attack both twigs and fruit during summer. In nurseries and young orchards, the larvae of the pest could cause severe damage on vigorous growing shoots and cause undesirable growth and lateral branching of the shoots (Anonymous, 2008b).

Distribution and host plants

The peach twig borer moth is reported to be found in North America, many European countries, the Mediterranean countries, Syria, Lebanon, Palestine, China, Japan, Australia (Avidov and Harpaz, 1969; Talhouk, 1969), Iraq (Ahmad and Khadhum, 1986) and Iran (Oloumi-Sadeghi and Esmaili, 1983). The list of host plants includes peach, nectarine, almond, apricot, plum, cherry and apple (Anonymous, 2008b).

Management for organic farming

CULTURAL PRACTICES. Infested shoots should be cut into 8–10 cm lengths weekly between March and September. They should be placed in cages covered with mesh so that once parasitoid adults hatch from the parasitized larvae the adults can escape and increase the parasitoid population in the orchard. During this practice, undesirable lateral shoots should also be cut off to prevent new infestations (Anonymous, 2008b). The infested fruits should be collected and destroyed, so that the population of the pest will be reduced in the next growing season.

BIOLOGICAL CONTROL. There are numerous parasitoids and predators controlling the peach twig borer populations. These include the following parasitoids that have been determined to date:

- *Apanteles anarsiae* Faure et Alab., *Apanteles glomeratus* L., *Ascogaster* sp., *Bracon gelechiae* Ashmead, *Macrocentrus ancylivorus* Rowher and *Spilochalcis* n.sp. *aff torvina* (Cresson) (Hymenoptera: Braconidae);
- *Paralitomastix pyralidis* (Ashmead) and *Paralitomastix varicornis* Nees. (Hymenoptera: Encyrtidae);
- *Ephialtes subglobiatus* L., *Aptesis* sp., *Mastrus* sp., *Phaeoganes rustigatus* Wesm., *Pimpla instigator* F. and *Pristomerus vulnelator* Panz. (Hymenoptera: Ichneumonidae);
- *Periclora gestroci* K. (Hymenoptera: Belulidae);
- *Brachymeria intermedia* Perk. and *Hyperteles lividus* (Ashmead) (Hymenoptera: Chalcididae);
- *Andreana* sp. (Hymenoptera: Apidae);
- *Dibrachys offinis* M. (Hymenoptera: Pteromalidae);
- *Haematopoda pluviallis* L. (Diptera: Tabanidae);
- *Erynnia tortricis* (Coquillett) (Diptera: Tachinidae);
- *Euderus cushmani* Crawford (Hymenoptera: Eulophidae); and
- *Pyemotes ventricosus* (Newport) (Acarina: Pyemotidae) (Daane *et al.*, 1993; Strand, 2002; Anonymous, 2008b).

The grey field ant, *Formica aerata* (Francoeur) (Hymenoptera: Formicidae) is reported to prey on the peach twig borer during spring and summer, but it was not able to keep the pest population below economically damaging levels (Strand, 2002).

It was also reported that dormant-season application of *Steinernema carpocapsae* (Weiser) (Rhabditida: Steinernematidae) and *Heterorhabditis* sp. (Rhabditida: Heterorhabditidae) reduced overwintering larval populations of the peach twig borer in hibernacula on almond trees in California orchards (Agudelo-Silva *et al.*, 1995).

MATING DISRUPTION. Mating disruption with sex pheromone can help to reduce the pest population. This was reported to reduce the peach twig borer moth populations in plum orchards; however, it was not reliable when used alone. It is effective in orchards when the moth population is low (Strand, 2002).

CHEMICAL TREATMENTS. Sprays of microbial product such as *Bacillus thuringiensis* and the Entrust formulation of spinosad at bloom can control the peach twig borer. Mid-spring sprays of Entrust should be timed to the hatching larvae of the first generation (Strand, 2002).

***Tropinota* (= *Epicometis*) *hirta* (Poda) (Coleoptera: Scarabaeidae) (flower chafers)**

Description

The adult of *Tropinota hirta* is dark brown and 8–12 mm long. The body is covered with dense and long yellowish-white hairs. There are white patches on the elytra. Eggs are 2.0–2.5 mm in diameter and spherical in shape and white in colour. Larvae are a brown-coloured grub (Anonymous, 2008b).

Life cycle

T. hirta overwinters as the adult stage in the soil. Adults become active in the spring during the blossoming period of the fruit trees. The adult population reaches its peak by the end of the spring. Adults feed on the blossoms, young leaves and buds and even fruits, and deposit their eggs into the soil. The grubs feed on the roots of the weeds after hatching from the eggs, and they develop on decomposing plant matter, and do not cause any damage to the almond trees (Avidov and Harpaz, 1969; Anonymous, 2008b). The grubs complete their development in 6–9 weeks in the soil and pupate. Adults hatching from the pupae overwinter in the soil (Anonymous, 2008b).

Damage

The damage is caused to the flowers, young leaves, buds and even fruits which are attacked by the adults, which then lay their eggs into the soil. The grubs do not cause any damage to the almond trees, as they feed on the roots of weeds after hatching from the eggs, and they develop on decomposing plant matter (Avidov and Harpaz, 1969; Anonymous, 2008b).

Distribution and host plants

Flower chafers are known to occur in Europe, the Near East and North Africa. Its list of host plants includes apple, apricot, cherry, sour cherry, peach, pear, plum and many other plants (Anonymous, 2008b).

Management for organic farming

CULTURAL PRACTICES. The adults, grubs and adults could be destroyed by tillage of the soil. The trees can be shaken during the morning hours when the adults are motionless on the plants, and the adults that are dropped can be picked off by hand (Anonymous, 2008b).

BIOTECHNICAL CONTROL. Traps combined with visual (blue colour) and chemical (1:1 cinnamyl alcohol/*trans*-anethole mixture, known as flower scent volatiles) play an important role in mass trapping of the pest (Knudsen *et al.*, 1993; Toth *et al.*, 2004).

***Ectomyelois ceratoniae* (Zell.) (Lepidoptera: Pyralidae) (carob moth)**

Description

Forewings of the adult *Ectomyelois ceratoniae* are narrow, dull and dark grey in colour; two 'w'-shaped light stripes stand out on the forewings when the adult is at rest. The hindwings are white in colour with distinctive veins. The body length and the wingspan are 8–11 mm and 16–28 mm, respectively (Avidov and Harpaz, 1969; Anonymous, 2008a), depending on the geographical area where they live.

The eggs are oval and 0.7 × 0.5 mm in size, and when laid they are initially white in colour, but later they turn to red-brown. The larvae are 15–18 mm long, and the larval body is pinkish with a brown head and pronotum. The pupa is 3 × 10 mm in size, and pupation takes place in a light grey cocoon (Avidov and Harpaz, 1969; Anonymous, 2008a).

Life cycle

The carob moth overwinters as larvae within the almond 'mummies' (mummified fruit), under the bark or in the crevices of the trees. First adults appear between April and June depending on the geographical area. One female deposits 100–350 eggs on the fruits during its lifespan. Larvae start to feed on fruits right after they emerge from the eggs. The carob moth produces four to five generations in a year (Avidov and Harpaz, 1969; Anonymous, 2008a).

Damage

The first generation develops from mid-April until late June (Avidov and Harpaz, 1969) and is harmful to the almond. It can be considered that from late June onwards the almond fruits become rigorous so that the second generation larvae are not capable of penetrating into the fruits as long as the fruit skin is not injured or split. The second and following generations develop mainly in carob, citrus (Avidov and Harpaz, 1969) and pomegranate.

Distribution and host plants

The carob moth is distributed in Africa, central and southern Europe, Central and South America and the Near East, and it is likely to be introduced into many temperate countries inside food consignments (Avidov and Harpaz, 1969). The list of host plants includes carob, orange, grapefruit, pistachio, pomegranate, apple, pear, hazelnut, almond, walnut, chestnut, date, fig, grape, olive, persimmon and quince (Avidov and Harpaz, 1969; Anonymous, 2008a).

Management for organic farming

CULTURAL PRACTICES. All infested fruits both on trees and on the ground should be collected regularly from the almond orchards and destroyed. The almond orchards preferably should be established in areas that are free from the other hosts of the carob moth.

BIOLOGICAL CONTROL. The predator *Orius minutus* L. (Hem.: Anthocoridae) and the parasitoids *Phanerotoma flavitestacea* Fish., *Habrabracon hebetor* Say., *Habrabracon brevicornis* (Wesmael), *Bracon lactus* Wesmael, *Apanteles* sp. (Hym.: Braconidae), *Pristomerus vulnerator* Panz. (Hym.: Ichneumonidae) and *Trichogramma* spp. (Hym.: Trichogrammatidae) are common natural enemies in Turkey (Anonymous, 2008a). It was reported that *P. flavitestacea*, *Clausicella suturata* Rond. (Dip.: Tachinidae) and the ectoparasitic mite *Pyemotes* (= *Pediculoides*) *ventricosus* (Newp.) (Acarina: Pyemotidae) were common natural enemies in Israel (Avidov and Harpaz, 1969). *Apanteles myeloenta* Wilkinson (Hymenoptera: Braconidae) was found to be very common in Iran (Kishani-Farahani *et al.*, 2012). *Apanteles* spp. group ultor (Hym.: Braconidae) was reported as a very common parasitoid species of the carob moth in Iraq (Al-Maliki and Al-Izzi, 1986). Enhancement of the natural enemies could help to reduce the carob moth population in almond orchards.

In addition to natural enemies, *Bacillus thuringiensis* can control carob moth populations when sprayed regularly at intervals of every 10–15 days from the first larval emergence (Anonymous, 2008a).

MATING DISRUPTION. There are some commercial mating disruption products to control the carob moth in dates, which could be tested in almond orchards.

Amyelois transitella (Walker) (Lepidoptera: Pyralidae) (navel orangeworm)

Description

Adult moths have silvery grey and black patterns on the forewings and legs. The hindwings are light, darkening at the apex and along the veins. A pair of palps in front of the head forms a snout-like projection. The body length and the wingspan are 8–12 mm and 19–23.5 mm, respectively (Wade, 1961). Females have a larger wingspan, with a range from 18 to 27 mm.

Eggs are oval and 0.5–1 mm in diameter, and when laid are initially white in colour. As the eggs mature, they turn pink and then to red-brown. Newly hatched larvae are reddish brown in colour, but change to pink or white depending upon diet. The head and pronotum are dark in colour in all instars, and a pair of crescent-shaped marks on the second segment helps distinguish the moth from other larvae. Larvae grow to 15–18 mm long, and the larval body is pinkish, and head and pronotum are brown. The pupa ranges in length from 7.25 mm to 12 mm, is light to dark brown, and is often found within shells or between the shell and the hull.

Life cycle

The navel orangeworm moth (NOW) overwinters as pupae and larvae within dried, shrivelled mummified fruit that remain on the tree after the previous year's harvest. Larvae do not enter diapause, so adult emergence may occur during warm periods within the winter. Pupation occurs within the mummies in early spring, and emergence marks the beginning of the first flight. First generation eggs are laid on mummy nuts, and these serve as the only food source for the developing larvae. First generation female moths emerge in late spring or early summer for the second flight and lay eggs on mummy nuts or fruit damaged by other moth pests. Developing larvae will feed on almond hulls and kernels, but develop faster on almond kernels. Successive flights will increase egg-laying female populations. As the almond-ripening process begins, and hull-split is initiated, female moths will lay eggs on the exposed shell and kernel. On average, a female deposits 84.6 eggs, with as many as 250 being observed (Wade, 1961). NOW typically produces four generations a year in California with the second, third and fourth flight potentially causing damage to the almond crop (Strand, 2002).

Damage

Kernel feeding by NOW larvae causes economic losses, especially in areas of higher population densities (Strand, 2002).

Distribution and host plants

The NOW is commonly found in Mexico and throughout the south-western USA (Wade, 1961). The list of host plants includes citrus, apples, apricots, figs, nectarine, peach, pear, plum, quince, almonds, pecans and walnuts. It is commonly found within trees or shrubs that produce seed pods. These include carob pods, bottle-tree seeds, dates, jujube, loquat, pomegranate, *Acacia farnesiana*, *Genipa americana*, Texas ebony and yucca pods (Wade, 1961).

Management for organic farming

CULTURAL PRACTICES. Cultural practices that can help with control of this pest include:

- Sanitation – all infested fruits both on trees and on the ground should be collected regularly from the almond orchards and destroyed. Mummy nuts can be removed from the tree by mechanically shaking or by hand pulling.
- Early harvest – almonds should be harvested as soon as feasible. An earlier timed harvest can reduce the exposure to NOW flights, leading to a reduction in damage.
- Varietal selection – hard-shelled or other varieties that have a tight shell seal are more resistant to infestation by NOW.

BIOLOGICAL CONTROL. There are two parisitoid wasps introduced into California to manage this pest. The encyrtid wasp, *Copidosoma plethorica* (Caltagirone) lays its eggs inside the NOW larva, and each egg develops into a large number of larva that consume the host and pupate inside the exoskeleton. The bethylid wasp *Goniozus legneri* Gordh lays eggs on the surface of the NOW larva and the egg hatches into a larva that consumes the NOW larva from the outside. Both wasps can occur within the same orchard, and control is greater when both species are present. Even with high densities, natural populations do not provide reliable control of NOW.

Flocks of birds that move into the orchard during the dormant period will often feed upon mummy nuts. This feeding assists the sanitation process and effectiveness is determined by the type of bird and proximity of the orchard to bird flight patterns.

In addition to natural enemies, *Bacillus thuringiensis* can control NOW, but since this bacterium must be ingested, coverage is critical. Sprays must be made regularly at intervals of 10–15 days once hull-split has begun.

MATING DISRUPTION. There are some commercial mating disruption products to control the NOW being tested within almonds in California.

Quadraspidiotus perniciosus **(Comstock) (Hemiptera: Diaspididae) (San Jose scale)**

Description

The adult male of *Quadraspidiotus perniciosus*, the San Jose scale is a yellow-brown two winged insect that is very small in size, between 1 mm and 2 mm. The female lives under a scale covering and when scraped away, may reveal a bright yellow body. There is no visible egg stage and nymphs emerge as 'crawlers' during the first instar and migrate to other feeding sites.

Life cycle

There are three stages during the first instar, which include the 'crawler', 'white-cap' and 'black-cap'. The bright yellow crawler is about 0.2 mm in length and will relocate through animal, wind or human intervention. Within 8–24 h of emergence, it will insert mouthparts into the tree, and begin to feed on the tree's sap. As it feeds, a white waxy covering begins to form ('white-cap' stage), and after a week of feeding, the cap will begin to turn black ('black cap' stage). The development of the male requires three moults, and upon emergence it is short lived. Females live under a scale, and emit a sex pheromone to attract males. The male flight and female receptivity tend to peak in the early spring, and within 5–6 weeks the first crawlers emerge. Crawlers that emerge in mid-spring will give rise to the first generation male flight in the summer, and two more generations will follow. Crawlers produced in the late autumn will overwinter as black caps and produce the overwintering flight.

Damage

Damage is from sucking of plant juices and injection of a toxin, which leads to death of twigs, limbs and overall decline in productivity. Red halos often appear around the feeding on green 1-year-old wood, and damage is often visible as necrotic spots when the scale-infested bark is scraped away.

Distribution and host plants

The San Jose scale has a worldwide distribution. The list of host plants includes apple, pear, sweet cherry, peach, prune, other tree fruits and nuts, berry bushes and many kinds of shade trees and ornamental shrubs.

Management for organic farming

BIOLOGICAL CONTROL. San Jose scale has many natural enemies that can keep the pest under control. Within Californian orchards, two predaceous beetles have been identified (*Chilocorus orbus* Casey and *Cybocephalus californicus* Horn) as well as several wasps. The most important wasps are the Encyrtidae species including *Encarsia perniciosi* (Tower) and *Aphytis* spp.

CHEMICAL CONTROL. If large populations are detected, applications of narrow-range oil when the trees are dormant in winter are effective in reducing all the stages of San Jose scale. Spring sprays timed to crawler emergence are also effective, but later sprays are not.

Cimbex quadrimaculata (Müller) (Hymenoptera: Cimbicidae) (almond sawfly)

Description

The adult of *Cimbex quadrimaculata* is 22–24 mm long; its head is dark brown and thorax black in colour. The abdomen is yellow with black crossing, narrow lines. The egg is greenish in colour and about 2.75 mm long. The general colour of the larvae is grey and there are many black dots on the body. The body of the larvae is about 38 mm long when full grown. The pupae are light brown and 25 mm long (Talhouk, 1969).

Life cycle

C. quadrimaculata overwinters as the mature larval stage in the soil. It pupates in March–April depending on the geographical area. Adults emerge in late March and April. They deposit their eggs on the foliage of the trees. The larvae hatching from the eggs feed greedily on the leaves, and can cause severe damage on young trees. The larvae leave the trees usually moving into the soil in May when they attain full size, and they remain in diapause in their cocoons until the following spring (Talhouk, 1969).

Damage

It is not considered as a serious pest, but in some years the larvae can defoliate lonely trees (Talhouk, 1969).

Distribution and host plants

The almond sawfly is known to be found in Cyprus, Lebanon, Palestine, Syria, Turkey and parts of Western Europe (Talhouk, 1969). Avidov and Harpaz (1969) have recorded it in Israel. The pest attacks almond and pear (Talhouk, 1969).

Management for organic farming

C. quadrimaculata is not a serious pest. The parasitization rate in the larval and pupal stage of the pest is quite high. *Listrognathus mactator* (Thunberg) (Hymenoptera: Ichneumonidae: Cryptinae) (Özgen *et al.*, 2010) and *Opheltes glaucopterus* (Linnaeus) and *Phobetes nigriceps* (Gravenhorst) (Hymenoptera: Ichneumonidae: Ctenopelmatinae) (Özbek, 2014) were determined as larva–pupa parasitoids of *C. quadrimaculata*.

Eurytoma amygdali Enderlein (Hymenoptera: Eurotomidae) (almond fruit wasp)

Description

Eurytoma amygdali wasp is black in colour, and forewings are transparent, metallic and shiny and triangular in shape. The tibiae and the connecting other leg segments are yellow in colour. Females bear a distinctive ovipositor (Anonymous, 2008a). The body length is between 4 mm and 8 mm (Avidov and Harpaz, 1969).

The eggs are minute and milky white in colour and bear two prolonged appendages, one being longer than the other. Larvae are without legs, 7–8 mm in length and white in colour. The body of the larvae is covered with scattered hairs. The pupa is white in colour in the beginning, but later turns dark during its development (Anonymous, 2008a).

Life cycle

E. amygdali overwinters as a developed larva inside almond fruits, and before almond blossom time the larvae pupate, and the adults emerge in late February to early March in the Middle East (Talhouk, 1969). However, the majority of the overwintering larvae develop into the pupal stage in the next spring. The duration of the pupal stage takes about 18–51 days depending on the temperature. The first adults appear between March and April in Israel, and also in mid-April and June, depending on climatic and geographic conditions (Avidov and Harpaz, 1969; Anonymous, 2008a). After the adults emerge from the fruit, an emergence hole is visible which is about 2 mm in diameter. After copulation, females can deposit 47–88 eggs into the endosperm of the fresh fruit. The incubation period of eggs is variable; it takes place between 24 days and 27 days. It was reported that emergence usually starts in March in Israel (Plaut, 1971) and in April and May in Greece (Katsoyannos *et al.*, 1992). Males emerge earlier than females, and females deposit up to five eggs per fruit under normal conditions (Plaut, 1971). Laboratory studies have shown that the females of this species use a host-marking pheromone, immediately after oviposition. Therefore, the pheromone enables the females to distinguish the infested and uninfested fruit and to select uninfested fruits for depositing eggs during the oviposition period (Kouloussis and Katsoyannos, 1991). Drilling and deposition of the egg into the nuclear tissue of the young fruits takes about 9–34 min, up to five eggs per fruit being laid under natural conditions (Talhouk, 1977a). It has one generation/year. Mentjelos and Atjemis (1970) stated that when larval development was completed by the end of June or the beginning of July, then the larva enters diapause and remains in this stage for one to three winters in Greece.

Damage

After the eggs hatch, the young larva starts to tunnel into the middle of the fruit. *E. amygdali* is one of the only pests reported to feed on almond fruits in the east Mediterranean countries (Talhouk, 1977a). It can damage up to 50% of the almond orchards in Bulgaria (Ivanov, 1960) and 71% in Macedonia (Cakar, 1980).

Distribution and host plants

The geographical distribution of *E. amygdali* includes the Middle East and east

Mediterranean countries (Plaut, 1972; Talhouk, 1977a). The pest develops on almond, apricot and plum (Anonymous, 2008a).

Management for organic farming

CULTURAL PRACTICES. All infested fruits both on trees and on the ground should be collected from the almond orchards after harvesting, and destroyed.

BIOLOGICAL CONTROL. Many natural enemies of *E. amygdali* have been reported from almond-growing countries. The parasitoids, *Aprostocetus bucculentus* (Kostjukov) (Hymenoptera: Eulophidae), *Gugolzia bademia* Doganlar (Hymenoptera: Pteromalidae) and *Adontomerus amygdali* (Boucek) (Hymenoptera: Torymidae) are common in Turkey (Bolu and Özgen, 2007; Anonymous, 2008a). *Adontomerus amygdali* (Hymenoptera: Chalcidoidea: Torymidae) and *Aprostocetus bucculentus* (Hymenoptera: Chalcidoidea: Eulophidae) are gregarious ectoparasitoids on the larvae of *E. amygdali. Pyemotes amygdali* Cobanoglu and Doganlar (2006) (Acarina: Pyemotidae) is a gregarious ectoparasitoid on prepupae, pupae and newly hatched adults of all of the hymenopterous insects (Doganlar *et al.*, 2006). *Thanasimus* sp. (Coleoptera: Cleridae) is a predator of hymenopterous insects in almond fruits. The natural parasitism on *E. amygdali* by *A. amygdali* reached 0.38–35.2% in places where the parasitoid was present. But, in the case of *A. bucculentus* it was less than 5%. Parasitism/predation rates by *P. amygdali* and *Thanasimus* sp. which have been found in Hatay province (Turkey) ranged from 7.56% to 44.53% and 0.38% to 11.2%, respectively (Doganlar *et al.*, 2006). It is likely that the natural enemies help to keep the pest population at a lower level, so the habitat in almond orchards should be amended and protected to ensure the survival of populations of parasitoids.

CHEMICAL CONTROL. Determination of the first emergence of adults is very important in the spring. Cages covered with mesh cloth can help to determine the first adult emergence by observing the infested fruits from the previous year placed in the cages. Chemical spraying that is acceptable in organic production can be started after emergence of the adult in the spring. The emergence period of adults can vary between 24 days and 45 days. If the emergence period were extended, one more spray may be needed to control the pest effectively.

Web-spinning spider mites

Web-spinning spider mites that are pests on almond include the following species:

- *Tetranychus pacificus* McGregor (Acari: Tetranychidae) (Pacific spider mite);
- *Tetranychus urticae* Koch (Acari: Tetranychidae) (two-spotted spider mite); and
- *Tetranychus turkestani* Ugarov and Nikolski (Acari: Tetranychidae) (strawberry spider mite).

Description

Mites are tiny arthropods (< 1 mm) belonging to the class Arachnida (other examples: spiders, ticks). The web-spinning mite species described in this section are more or less similar in morphology (adult stages), life cycle, feeding habit and nature of damage to the plants. Therefore, the same management strategy applies to all three species.

Adult mites are pale green to black in colour, which changes into red or orange during the winter. Male mites are smaller than females and they do not overwinter.

Life cycle

Adult females overwinter under bark, leaf litter and winter weeds on the orchard floor. Upon reaching the conducive environmental conditions in spring, mites migrate from their overwintering sites to the trees for egg laying. Mites deposit eggs on the underside of the leaf surface, and upon hatching first instar larvae start feeding on leaves. At least three moults occur. Early in the season, mites are abundant in the bottom half of the trees, but will become widespread throughout the tree later in the season depending on temperature and the degree of infestation.

Temperature plays a significant role in increased mite reproduction, thus maximum population increase occurs between June and September. Spider mites can complete their life cycle within 7 days under high temperature conditions, and can have between eight and ten generations/year in California (Strand, 2002).

Damage

Spider mite infestations often begin on the underside of the leaves. All stages of mites feed on almond leaves by sucking the cell contents. Spider mite infestation is characterized by the presence of webbing covering tree leaves and twigs. The webbing has several biological and ecological functions including dispersal and reproduction of mites, and protection from natural enemies (Gerson, 1985; Kennedy and Smitley, 1985). In the beginning, damage by spider mite feeding results in stippled leaves, which advances to yellowing and dropping of leaves as the infestation progresses. The degree of mite infestation is negatively correlated with chlorophyll content and photosynthetic activity of the leaves (Andrews and La Pré, 1979), and this eventually affects tree health and productivity. Mite damage in the current year translates into the reduction in growth and productivity of the trees in the following years (Barnes and Andrews, 1978).

Distribution and hosts plants

The three spider mite species possess a wide geographical distribution, and they are one of the most widely distributed pests of many wild, ornamental and cultivated plants. In the USA, two-spotted spider mites feed on over 300 host plants, and one-third of them are cultivated crops.

Pest monitoring

Spider mites favour a dry and low-moisture type of environment, thus water-stressed orchards are often at risk of high infestation. Properly irrigated orchards may not require treatment for mites in most cases as almond trees can tolerate low to moderate mite pressure without affecting tree productivity.

Another important aspect of effective mite control is judicial use of available control measures. The mite population in several crops including almond is often well controlled by natural enemies, and the use of broad-spectrum insecticides can disrupt the natural control system resulting in elevated levels of spider mite population. A high natural enemy:mite ratio does not require treatment intervention in almond.

Monitoring of orchards for predators and spider mites is critical. Sampling at least once every 2 weeks during the early part of the growing season and weekly thereafter until harvest is recommended. If the orchard has a history of heavy mite infestation or water-stressed trees, monitoring every few days may be necessary. During the early phase of the growing season, sampling should focus on areas with a greater likelihood of early infestation such as areas near to dirt roads and areas with water-stressed trees. Once infestation has reached the economic threshold, sampling is necessary for the rest of the orchard. Dividing orchards into sampling areas is helpful to determine whether the spot treatment in a high mite infestation area is sufficient. For each sampling area, 15 random leaves should be selected from each of five selected trees, and these should be examined with a hand lens on both sides of each leaf for the presence of spider mites and eggs, predatory mites or eggs, and other predators. The treatment decision can be made based on presence/absence sampling for mite and predator. Details of the sampling protocol are described in Strand (2002).

Management for organic farming

USE OF OIL. Several types of organic oil are available commercially to use both in conventional and in organic productions, although all oil types may not be acceptable for organic use. Since oil works by contact action (including smothering and barrier effects), good spray coverage is crucial for its effectiveness. Due to the potential risk of phytotoxicity, it is important to apply oil to well-watered trees. Oil also kills beneficial arthropods that come into contact during the spray application, but there are minimal risks on remaining beneficials due to low residual activities. More than one application may be necessary to control a large pest population.

CULTURAL CONTROL. Irrigating the orchard properly to reduce water-stressed trees is critical to reduce overall mite populations in the orchard. In addition, reducing dusty conditions by oiling or watering dirt roads and maintaining a good ground cover in the orchard are preventative measures to minimize mite infestations.

BIOLOGICAL CONTROL. There are several species of biocontrol agents that are effective in reducing the spider mite population in almond. The abundance and effectiveness of species can vary with the geographic region and other environmental factors. The western predatory mite, *Galendromus occidentalis* (Nesbitt), six-spotted thrips, *Scolothrips sexmaculatus* (Pergande) and a black-coloured ladybird beetle species, also called the spider mite destroyer (*Stethorus* sp.) are reported in almond orchards in the USA (Strand, 2002). The western predatory mite is the most widespread and effective predator. Similar in size to a spider mite, the western predatory mite lacks black spots on its body, and is highly mobile. Six-spotted thrips can quickly migrate among leaves and prey on spider mites efficiently. Spider mite destroyer beetles are good fliers and can concentrate their feeding on spider-mite-aggregated areas of the orchard. These natural enemies are available commercially to use as an augmentative release in almond orchards to boost natural populations.

Secondary Pests

Ants

Ants that are pests on almond include the following species:

- *Tetramorium caespitum* (L.) (pavement ant);
- *Solenopsis xyloni* McCook (southern fire ant); and
- *Solenopsis molesta* (Say) (thief ant).

Description

The workers of the pavement ants are dark brown to black in colour with body size ~ 3.5 mm long, consisting of parallel furrows or ridges on the head and thorax (Bruder and Gupta, 1972). Reproductive ants (swarmers) have wings, and are twice as big as workers with similar other morphological structures (Jacobs, 2013). Pavement ants prefer sandy or loamy soil for nesting. Not much information is available about the colony biology for this ant.

Southern fire ants are stinking ants native to the southern parts of the USA. The southern fire ant workers vary from 1.8 mm to 6.4 mm in size. This ant has an amber-coloured head and thorax with a black abdomen. The eyes are noticeably big and the body is covered with golden hairs. Similar to pavement ants, fire ants also have a two-segmented pedicel, a structure that connects the abdomen with the thorax. The distribution of this ant ranges from California to South Carolina (southern part) and Florida (north-west corner) (Smith, 1965; Taber, 2000).

Thief ants are slightly smaller than the fire ants. These ants are present in relatively small numbers and nest in proximity to other ant nests, from which they often steal food.

Ant nests are in small mounds or patches of loose soil, commonly found close to wetted areas in orchards. These nests are closer to berms in orchards with flood irrigation and with clay soil, but are also found in other areas of the orchard that have conditions of loose soil. Fire ants swarm upon disturbance. Southern fire ant nests are often associated with clumps of weeds, such as nuts edge (*Cyperus esculentus* L.) or spotted spurge (*Euphorbia maculata* (L.)). Ants are active as pests in orchards with peak activities in the morning and just before sunset (Strand, 2002).

Damage

The southern fire ants are a more widespread problem in almond, although pavement ants are more problematic in the northern part of the Central Valley of California. Damage on almond by ants is due to direct feeding on the nut kernels; feeding on kernels results in chewing marks and white dust in the nuts. Ants can completely hollow out the meat from the kernel leaving only parts of the pellicle (i.e. the outer skin of the kernel). Ants can damage nuts still attached to the young trees; however, the major damage occurs to harvested nuts that are on the orchard floor as a part of the harvesting process (Zalom and Bentley, 1985). Orchards with a sprinkler or a drip irrigation system and with cover crops are more at risk of infestation. Almond varieties with a tight shell seal or with minimal splits (< 0.75 mm) experience less damage, and the shell seal can vary according to other factors such as the year, the crop size, the nut size and horticultural practices.

Distribution and host plants

T. caespitum is native to Europe but is also actually reported in North America. *S. xyloni* is found in the USA and Mexico, and *S. molesta* is reported in North America, Mexico and recently in Malaysia. They are omnivorous, and *S. xyloni* feeds on various plant parts such as fruits, seeds, honeydew, plant sap, stems, buds and tubers of several plants including althea, dahlia, citrus, okra, pecan, walnut, almond, tomato, melons, potato, strawberry, yukka, maize and aubergine (Smith, 1937, 1965; Zalom and Bentley, 1985; Taber, 2000). *S. molesta* are present in relatively small numbers and nest in proximity to other ant nests (Thomson, 1989), from which they often steal food.

Pest monitoring

In spring, surveying the orchard floor for ant colonies 2–3 days after irrigation is very important. For sampling, the orchard block should be divided into five survey areas (each survey area ~ 93 m^2 (= 1000 ft^2)) including the area from mid-alley to mid-alley beneath the trees. Active ant colonies should be surveyed and counted from individual survey areas. Based on total colony counts from five survey areas (i.e. 5 × 93 m^2 = 465 m^2) in spring, and the number of days in which nuts are on the ground after the harvest provides estimates of the percentage nut damage caused by ants (Table 12.4). In addition, inspecting a sample of 500 harvested nuts for ant damage provides information on the effectiveness of current pest management practices, and therefore provides guidelines for future pest management planning. Full details of the sampling protocol are explained in Strand (2002).

Management for organic farming

Insect growth-regulator-based baits are effective methods for managing ant population in almonds. Baits are more effective than insecticide sprays to control ants because receiver ants (worker ants) carry baits inside the colony and the whole colony can be destroyed. Since baits are relatively slow-acting products, application should be made several weeks before the harvest. Higher moisture reduces the effectiveness of the baits, so it is recommended to avoid use of baits 1–2 days before and after irrigation. Some of these baits are registered for use in organic production.

CULTURAL CONTROL. Flood irrigating can reduce ant populations. Damage is significantly higher in orchards that harvest the nuts off the ground. Nuts should be removed

Table 12.4. Percentage damage caused by ants to almonds on the ground in an almond orchard. (From Strand, 2002.)

No. of colony entrances/ 465 m^2 (5000 ft^2)[a] in spring	Days nuts are on the ground				
	4	7	10	14	21
15	0.9%	1.6%	2.1%	3.1%	4.9%
45	1.4%	2.3%	3.2%	4.7%	7.0%
185	2.0%	3.6%	5.0%	7.0%	11.1%

[a]Value of 5000 ft^2 is the value according to Strand (2002). This has been converted into square metres (5000 ft^2 = ~ 465 m^2).

from the orchard floor as rapidly as possible following tree shaking to minimize ant damage on harvested nuts. Table 12.4 shows the risks of potential damage by ants depending on the time between tree shaking and the removal of nuts from the ground.

Brachycaudus amygdalinus (Schouteden) (Hemiptera: Aphididae) (short-tailed almond aphid)

Description

The body of the apterous female *Brachycaudus amygdalinus* on almond is dark green, but pale green on the posterior part of the abdomen, and 1.6–2.1 mm long. It has short legs and antennae. The antenna is six segmented, the siphunculi are short and pale green, the apices dark and the cauda is very short. The male is winged, the head and thorax are black, and the abdomen is dark brown. The body is 1.1–1.8 mm long, the cauda and siphunculi are black, and the genitalia are dark brown (Avidov and Harpaz, 1969; Lodos, 1982; Blackman and Eastop, 2000).

Life cycle

B. amygdalinus overwinters in the egg stage, eggs having been laid by gamic females within bark crevices and in bud axils of the almond tree. The eggs hatch in the next spring when the almond trees are in leaf, and reproduce virginoparously. The population of *B. amygdalinus* can reach high levels on the underside of the leaves in the spring as a result of rapid colonization of foundresses (Avidov and Harpaz, 1969). The colonies become overcrowded, and the winged form exists by the summer. They migrate back to the alternative hosts during summer, and the almond trees are free from the aphid during summer until autumn (Swirskii, 1954).

Damage

Infested leaves roll up and drop prematurely, and new growth is stunted as a result of feeding by the aphid pest (Avidov and Harpaz, 1969).

Distribution and host plants

The geographical distribution of the short-tailed almond aphid includes central and western Asia, Crimea, Israel, Europe, South Africa, the Middle East, Ukraine, Pakistan and Turkey (Bodenheimer and Swirski, 1957; Avidov and Harpaz, 1969; Lodos, 1982; Blackman and Eastop, 2000, 2006). The pest develops on almond and peach (Blackman and Eastop, 2000).

Management for organic farming

CULTURAL PRACTICES. Growers should avoid application of excess nitrogenous fertilizer and irrigation to control shoot flushing and leaf formation where there are high population numbers of colonizing aphids.

BIOLOGICAL CONTROL. Enhancing the numbers of natural enemies in the spring may help to control *B. amygdalinus* populations. There are numerous natural enemies that either feed or breed on the aphids. Many species of Aphidiidae, Braconidae, Eulophidae, Encyrtidae and Pteromalidae are parasitoids of aphids; and many species of Chrysopidae, Coccinellidae, Lygaeidae, Miridae, Nabidae, Anthocoridae, Cecidomyiidae, Syrphidae and Trombidiidae are predators of aphids (Anonymous, 2008a).

In addition to *B. amygdalinus* on almond the following aphid species are also reported: (i) *Hyalopterus amygdale* (Blanchard) (Russo *et al.*, 1994); (ii) *Brachycaudus amygdalinus* (Schouteden) (Avidov and Harpaz, 1969; Talhouk, 1977b; Sekkat, 1984; Russo *et al.*, 1994); (iii) *Brachycaudus helichrysi* (Kaltenbach); (iv) *Pterochloroides persicae* (Cholodkovsky) (Talhouk, 1977b); (v) *Hyaleptorus pruni* (Geoffroy); and (vi) *Myzus persicae* (Sulzer) (Sekkat, 1984).

Anthonomus amygdali Hustache (Coleoptera: Curculionidae) (almond weevil)

Description

The adult of *Anthonomus amygdali* is brown in colour and 3.0–4.2 mm long. The egg is milky white, oval in shape and 0.8 × 0.5 mm in size. The larva has a cylindrical body and is 4.8–5.5 mm long, and the head of the larva is shiny reddish brown.

Life cycle

A. amygdali overwinters as a larva and feeds on the buds throughout winter. It pupates in the spring months. Adults of the pest are situated in the shelter during the summer and become active in the autumn. They feed on the buds of the almond trees and deposit their eggs on the buds which generate in the next spring. They produce a single generation annually.

Damage

The infestation rate of the pest on the blossom of almond trees was estimated to be 1–5% in west Turkey (Önuçar and Zümreoğlu, 1985).

Distribution and host plants

Distribution includes many countries in Europe and the Middle East (Anonymous, 2013b). The pest develops on almond, apple, cherry, peach, plum, quince, walnut and *Pyracantha coccinea* Roem. (Anonymous, 2008b).

Management for organic farming

CULTURAL PRACTICES. It may help to reduce damage by pruning off the damaged shoots and branches of the trees. Additionally, adults can be picked off when they drop on a sheet placed on the ground by shaking the trees. Also damage can be reduced by picking off the damaged blossoms on the ground (Anonymous, 2008b).

BIOLOGICAL CONTROL. *Scambus pomorum* (Ratzeburg) (Hymenoptera: Ichneumonidae), *Bracon disdiscoidens* Weems and *Syrrhizus delusorius* Foraty (Hymenoptera: Braconidae) are known as common and effective parasitoids (Anonymous, 2008b).

Other pests

Fifty-four species from the superfamily Curculionoidea (Rhynchitidae – two species, Brentidae – 20 species, Curculionidae – 30 species and Scolytidae – two species) were collected from almond trees (Bolu and Legalov, 2008). There were many other pests in almond orchards, namely *Tatianaerhynchites aequatus* (Linnaeus) and *Epirhynchites smyrnensis* (Desbrochers des Loges) (Coleoptera: Rhynchitidae), *Diloba caeruleocephala* (L.) (Lepidoptera: Noctuidae), *Nordmannia acacia* (F.) (Lepidoptera: Lycaenidae), *Polydrosus roseiceps* Pesarini (Coleoptera: Curculionidae), *Hedya nubiferana* (Haworth) (Lepidoptera: Tortricidae), *Aporia crataegi* (L.) (Lepidoptera: Pieridae), *Agrilus roscidus* Kiesenweter (Coleoptera: Curculionidae) and *Capnodis tenebricosa* (Oliver) (Coleoptera: Buprestidae) (Bolu *et al.*, 2011). However, the number of studies on the biology, damage and control methods of these other pests are not sufficient, and they need to be developed.

Other mites in the almond orchard

Other mites in the almond orchard include:

- *Panonychus ulmi* (Koch) (Tetranychidae) (European red mite); and
- *Bryobia rubrioculus* (Scheuten) (Tetranychidae) (brown mite).

Description

Both European red mites and brown mites are not considered a major problem in almond orchards. These mites overwinter as eggs on tree parts such as fruit spurs, buds and twigs. They have red eggs which look similar except the European red mite egg which has a typical spine-like projection (i.e. a stipe) arising from the centre of the egg. Newly hatched larvae are green, which changes into red after feeding. Stipes are lacking in brown mite eggs.

Life cycle

Egg hatching of brown mites coincides with the leaf and flower bud opening time in almonds. Freshly hatched larvae which are red in colour with six legs, eventually change to a brown colour with eight legs resembling the adults. Brown mites are not active during the hot summer time, and have two to three generations/year, while European red mites are active for a longer part of the growing season and have five to ten generations in California (Strand, 2002).

Damage

Feeding by European red mites causes leaf stippling. Under prolonged feeding, leaf margins initially look yellowish brown, which eventually turns into a burned type of symptom. Healthy trees can tolerate high infestations (up to 50 mites per leaf). Brown mite feeding can cause leaf chlorosis, but leaf dropping is rare. Feeding activities on leaves occur only during the cool parts of the day. Infestation by brown mites is often confined to a few trees in the orchard.

Sampling and management

Generally these mite species are under natural control. In fact, they serve as food sources for important beneficials during the early part of the season. One of the beneficials, the western predatory mite is effective in reducing European red mite and brown mite populations. Although not prevailing in all orchards, the brown lacewing, *Hemerobius* sp., is an effective predator against mite pests. Spur sampling to look at mite egg presence between late autumn and early January is recommended to guide treatment decisions. Late dormant application of oil targeting mite eggs is suggested if infestation exceeds 20% of the sampled spurs. Occasional infestations of brown mite can be seen in a cool spring if the dormant treatment is inadequate. Biological control and certain oil sprays are available for use in organic production.

References

Agudelo-Silva, F., Zalom, F.G., Hom, A. and Hendricks, L. (1995) Dormant season application of *Steinernema carpocapsae* (Rhabditida: Steinernematidae) and *Heterorhabditis* sp. (Rhabditida: Heterorhabditidae) on almond for control of overwintering *Amyelois transitella* and *Anarsia lineatella* (Lepidoptera: Gelechiidae). *Florida Entomologist* 78, 516–523.

Ahmad, T.R. and Khadhum, A.A. (1986) Influence of pheromone trap design and placement on capture of peach twig borer, *Anarsia lineatella* (Zeller) (Lepidoptera: Gelechiidae). *International Journal of Tropical Insect Science* 7, 637–640.

Al Maliki, S.K. and Al Izzi, M.A.J. (1986) Parasites of *Ectomyelois ceratoniae* with biological studies on *Apanteles* sp. *group ultor* in Iraq. *Entomophaga* 31, 313–319.

Andrews, K.L. and La Pré, L.F. (1989) Effects of Pacific spider mite on physiological processes of almond foliage. *Journal of Economic Entomology* 72, 651–654.

Anonymous (2008a) *Technical Instruction for Plant Protection*, Volume 5. Turkish Ministry of Food Agriculture and Livestock, Başak Press, Ankara.

Anonymous (2008b) *Technical Instruction for Plant Protection*, Volume 4. Turkish Ministry of Food Agriculture and Livestock, Başak Press, Ankara.

Anonymous (2011) *Cerambyx dux*. Available at: http://www.cerambyx.uochb.cz/cerambyxdux.htm. (accessed 20 November 2014).

Anonymous (2013a) Bureau of the Census US Department of Commerce Foreign Trade Statistics. Available at: http://www.almonds.com/sites/default/files/content/attachments/2013_almanac.pdf (accessed 20 November 2014).

Anonymous (2013b) *Technical Instruction for Plant Protection*. Turkish Ministry of Food Agriculture and Livestock, Volume 4. Başak Press, Ankara, Turkey.

Avidov, Z. and Harpaz, I. (1969) *Plant Pests of Israel*. Israel Universities Press, Jerusalem, p. 549.

Barnes, M.M. and Andrews, K.L. (1978) The effects of spider mites on almond tree growth and productivity. *Journal of Economic Entomology* 71, 555–558.

Blackman, R.L. and Eastop, V.F. (2000) *Aphid's on the World's Crops: An Identification and Information Guide*, 2nd edn. Wiley, New York.

Blackman, R.L. and Eastop, V.F. (2006) *Aphids of the World's Herbaceous Plants and Shrubs: An Identification and Information Guide*. Wiley, New York.

Bodenheimer, F.S. and Swirski, E. (1957) *The Aphidoidea of the Middle East.* Weizmann Science Press of Israel, Jerusalem.

Bolu, H. (2007) Population dynamics of lacebugs (Heteroptera: Tingidae) and its natural enemies in almond orchards of Turkey. *Journal of the Entomological Research Society* 9, 33–37.

Bolu, H. and Legalov, A.A. (2008) On the Curculionoidea (Coleoptera) fauna of almond (*Amygdalus communis* L.) orchards in south-eastern and eastern Anatolia in Turkey. *Baltic Journal of Coleopterology* 8, 75–85.

Bolu, H. and Özgen, İ. (2007) Distribution areas, infestation rates and parasitoids of the almond seed wasp *Eurytoma amygdali* Enderlein (Hymenoptera: Eurytomidae). *Journal of Agricultural Faculty of Harran University* 11, 59–65.

Bolu, H., Özgen, İ. and Ayaz, T. (2011) Insect pests in almond orchards in the South East Anatolia. In: *Proceedings of IVth Turkish Plant Protection Congress*, 28–30 June, 2011, Kahramanmaraş, Turkey. Sutcu İmam University, Faculty of Agriculture, Kahramanmaraş, Turkey, p. 295.

Bruder, K.W. and Gupta, A.P. (1972) Biology of the pavement ant, *Tetramorium caespitum* (Hymenoptera: Formicidae). *Annals of Entomological Society of America* 62, 258–367.

Cakar, L. (1980) *Eurytoma amygdali* End. – (Hymenoptera, Chalcidoidea, Eurytomidae) – a pest of almond in Macedonia. *Zastita Bilja* 31, 263–272.

Daane, K.M., Yokota, G.Y. and Dlott, J.W. (1993) Dormant-season sprays affect the mortality of peach twig borer (Lepidoptera: Gelechiidae) and its parasitoids. *Journal of Economic Entomology* 86, 1679–1685.

Doganlar, O., Yıldırım, A.E. and Doganlar, M. (2006) Natural enemy complex of *Eurytoma Amygdali* Enderlein, 1907 (Hymenoptera, Eurytomidae) in Eastern Mediterranean region of Turkey: notes on their interaction and effectiveness. *Research Journal of Agriculture and Biological Sciences* 2, 282–286.

Garcia Del Pino, F. and Morton, A. (2005) Efficacy of entomopathogenic nematodes against neonate larvae of *Capnodis tenebrionis* (L.) (Coleoptera: Buprestidae) in laboratory trials. *BioControl* 50, 307–316.

Gerson, U. (1985) Webbing. In: Helle, W. and Sabelis, M.W. (eds) *Spider Mites: Their Biology, Natural Enemies and Control*, Vol. 1A. Elsevier, New York, pp. 223–232.

Ivanov, S. (1960) *Eurytoma amygdali* End. in Bulgaria and its control. *Rastit Zasht, Sofia* 8, 41–61.

Jacobs, S.B. (2013) Pavement ant, *Tetramorium caespitum* (L). Entomological Notes, Penn State Cooperative Extension. Available at: http://ento.psu.edu/extension/factsheets/pdf/pavementAnt.pdf (accessed 17 November 2015).

Katsoyanos, B.L., Kouloussis, N.A. and Bassillio, A. (1992) Monitoring population of the almond seed wasp, *Eurytoma amygdali* with sex pheromone traps and other means, and optimal timing of chemical control. *Entomologia Experimentalis et Applicata* 62, 9–16.

Kennedy, G.G. and Smitley, D.R. (1985) Dispersal. In: Helle, W. and Sabelis, M.W. (eds) *Spider Mites: Their Biology, Natural Enemies and Control*, Vol. 1A. Elsevier, New York, pp. 233–242.

Kishani-Farahani, H., Goldansaz, S.H. and Sabahi, Q. (2012) A survey on the overwintering larval parasitoids of *Ectomyelois ceratoniae*. *Crop Protection* 36, 52–57.

Knudsen, J.T., Tollsten, L. and Bergström, L.G. (1993) Floral scents – a checklist of volatile compounds isolated by head-space techniques. *Phytochemistry* 33, 253–280.

Kouloussis, N.A. and Katsoyannos, B.I. (1991) Host discrimination and evidence for a host marking pheromone in the almond seed wasp, *Eurytoma amygdali*. *Entomologia Experimentalis et Applicata* 58, 165–174.

Lodos, N. (1982) *Türkiye Entomolojisi* II. *Genel, Uygulamalı, Faunistik*. Ege University Agricultural Faculty Books, No. 429. Ege University Press, Bornova-İzmir, Turkey. (in Turkish)

Lodos, N. and Tezcan, S. (1995) *Turkish Entomology V, Buprestidae*. Ege University Press, Bornova-İzmir, Turkey. (in Turkish)

Marannino, P. and Lillo, E. (2007) The peach flatheaded rootborer, *Capnodis tenebrionis* (L.), and its enemies. *Insect Pathogens and Insect Parasitic Nematodes, IOBC (International Organisation for Biological and Integrated Control) WPRS Bulletin* 30, 197–200.

Martinez de Altube, M.M., Strauch, O., Castro, G.F. and Pena, A.M. (2008) Control of the flat-headed root borer *Capnodis tenebrionis* (Linne) (Coleoptera: Buprestidae) with the entomopathogenic nematode *Steinernema carpocapsae* (Weiser) (Nematoda: Steinernematidae) in a chitosan formulation in apricot orchards. *BioControl* 53, 531–539.

Mentjelos, J. and Atjemis, A. (1970) Studies on the biology and control of *Eurytoma amygdale* in Greece. *Journal of Economic Entomology* 63, 1934–1936.

Nizamlıoglu, K. (1957) *Fruit Pests and Their Control in Turkey*. Koruma Pesticide Company Press, No. 5. Koruma Pesticide Company, Istanbul.

Oloumi-Sadeghi, H. and Esmaili, M. (1983) The moth population study of peach twig borer (*Anarsia lineatella* Zeller) in Ghazvin and Karadj from 1975–80. *Journal Entomologie et Phytopathologie Appliquees* 50, 1–16.

Önuçar, A. and Zümreoğlu, A. (1985) Preliminary studies on blossom weevils (*Anthonomus* spp. Col.: Curculionidae) harmful on fruit trees in Aegean region. *Plant Protection Bulletin* 25, 139–149.

Özbek, H. (2014) Ichneumonid parasitoids of the sawfly *Cimbex quadrimaculata* (Müller) feeding on almonds in Antalya, along with a new parasitoid and new record. *Turkish Journal of Zoology* 38, 657–659.

Özgen, İ., Yurtcan, M., Bolu, H. and Kolarov, J. (2010) *Listrognathus mactator* (Thunberg, 1824) (Hymenoptera: Ichneumonidae): a new recorded parasitoid of *Cimbex quadrimaculatus* (Müller, 1776) (Hymenoptera: Cimbicidae) in Turkey. *Entomological News* 121, 391–392.

Plaut, H.N. (1971) On the biology of the adult of the almond wasp, *Eurytoma amygdali* End. (Hym., Eurytomidae), in Israel. *Bulletin of Entomological Research* 61, 275–281.

Plaut, H.N. (1972) On the biology of the immature stages of the almond wasp, *Eurytoma amygdale* End. (Hym. Eurytomidae) in Israel. *Bulletin of Entomological Research* 61, 681–687.

Russo, A., Siscaro, G. and Spampinato, R.G. (1994) Almond pests in Sicily. *Acta Horticulture (ISHS)* 373, 309–316.

Sanches-Ramos, I., Pascual, S., Marcotegui, A., Fernandez, C.E. and Gonzalez-Nunez, M. (2014) Laboratory evaluation of alternative control methods against the false tiger, *Monosteira unicostata* (Hemiptera: Tingidae). *Journal of Pest Management Science* 70, 454–461.

Sekkat, A. (1984) Biological control of almond aphids in the Saïss region. *Actes de l'Institut Agronomique et Veterinaire Hassan II* 4, 105–111.

Smith, M.R. (1937) Consideration of the fire ant *Solenapsis xyloni* as an important southern pest. *Journal of Economic Entomology* 29, 120–122.

Smith, M.R. (1965) House-infesting ants of the eastern United States. *Technical Bulletin, US Department of Agriculture* 1326, 38–40.

Strand, L. (2002) *Integrated Pest Management for Almonds*, 2nd edn. University of California Agricultural and Natural Resources Publication 3308. University of California Agricultural and Natural Resources, Oakland, California, 199 pp.

Swirski, E. (1954) Fruit tree aphids of Israel. *Bulletin of Entomological Research* 45, 623–638.

Taber, S.W. (2000) Fire ants native to the United States. In: Taber, S.W. (ed.) *Fire Ants*. Texas A&M University Press, College Station, Texas, pp. 87–93.

Talhouk, A.S. (1969) *Insects and Mites Injurious to Crops in Middle Eastern Countries*. Monographien Zur Angew. Entomologie, Beihefte zur Zeitschrift für angewante Entomologie. Verlag Paul Parey, Hamburg, Germany.

Talhouk, A.S. (1977a) Contributions to the knowledge of almond pests in East Mediterranean countries. V. The fruit-feeding insects, *Eurytoma amygdali* End., and *Anarsia lineatella*. *Zeitschrift für Angewandte Entomologie* 83, 145–154.

Talhouk, A.S. (1977b) Contribution to the knowledge of almond pests in East Mediterranean countries. *Zeitschrift für Angewandte Entomologie* 83, 248–257.

Talhouk, A.S. (2009) Contribution to the knowledge of almond pests in East Mediterranean countries. *Journal of Applied Entomology* 80, 162–169.

Thomson, C.R. (1989) The thief ants, *Solenopsis molesta* (Hymenoptera: Formicidae). *Florida Entomologist* 72, 268–283.

Toth, M., Schmera, D. and Imrei, Z. (2004) Optimization of a chemical attractant for *Epicometis* (*Tropinota*) *hirta* Poda. *Zeitschrift für Naturforschung* 59c, 288–292.

Wade, W.H. (1961) Biology of the navel orangeworm, *Paramyelois transitella* (Walker), on almonds and walnuts in northern California. *Hilgardia* 31, 71.

Zalom, F.G. and Bentley, W.J. (1985) Southern fire ant (Hymenoptera: Formicidae) damage to harvested almonds in California. *Journal of Economic Entomology* 78, 339–341.

13 Pest Management in Organic Rice: Latin America and the Caribbean

Alberto Pantoja,[1*] Edgar A. Torres,[2] Anamaria Garcia,[1] Eduardo J. Gaterol,[3] Gustavo A. Prado[2] and Maribel Cruz[3]

[1]Food and Agriculture Organization of the United Nations (FAO), Regional Office for Latin America and the Caribbean, Santiago, Chile; [2]Centro Internacional de Agricultura Tropical (CIAT), Cali, Colombia; [3]Fondo Latinoamericano para Arroz de Riego (FLAR), Cali, Colombia

Introduction

Rice (*Oryza sativa* L.) is the staple food for half of the world population and is considered the single largest food source for the poor in coastal areas of Latin America and the Caribbean (LAC) countries (Zorrilla *et al.*, 2013). Rice is also the fastest growing food source in sub-Saharan Africa (GRiSP, 2014).

Although, LAC only produces 4.5% of the world's rice, the crop is a staple for many coastal communities in the region (Pantoja *et al.*, 1997; Zorrilla *et al.*, 2013). In 2013, about 5.3 million ha of rice was planted in LAC (FAO, 2015). In descending order, Brazil, Colombia, Ecuador, Peru and Venezuela are the main rice producers in LAC (Fig. 13.1). South America represents 87% of the total production area in LAC, while the Caribbean and Central America accounts for 8% and 5% of the rice area in LAC, respectively. Information on organic rice production in LAC is not readily available. A survey conducted by the authors with the national rice programme for countries in the region indicated that Brazil (5000 ha), Argentina (874 ha) and Costa Rica (25 ha) are the leading organic rice producers in the region.

The demand for rice is increasing and organic rice is acquiring importance worldwide. In spite of the increasing demand for organic crops, little information is available on organic rice farming systems in LAC. Pantoja *et al.* (1997) reviewed the state of the art in pest management for rice in LAC, providing non-chemical alternatives for insect, weeds and diseases management. However, this is the only comprehensive review for rice in LAC, and the review does not provide specific recommendations for organic rice farming systems or organic plant–pest interactions.

This chapter focuses on organic rice pest-management options for LAC. Information on biological control agents, cultural practices, resistant varieties and other non-chemical means of pest control are discussed. The chapter presents general principles and practical methods of integrated pest management (IPM) for organic rice production systems in LAC, focusing on arthropod management, but where appropriate reference is made to diseases and weed management tactics.

* Corresponding author: Alberto.pantoja@ars.usda.gov

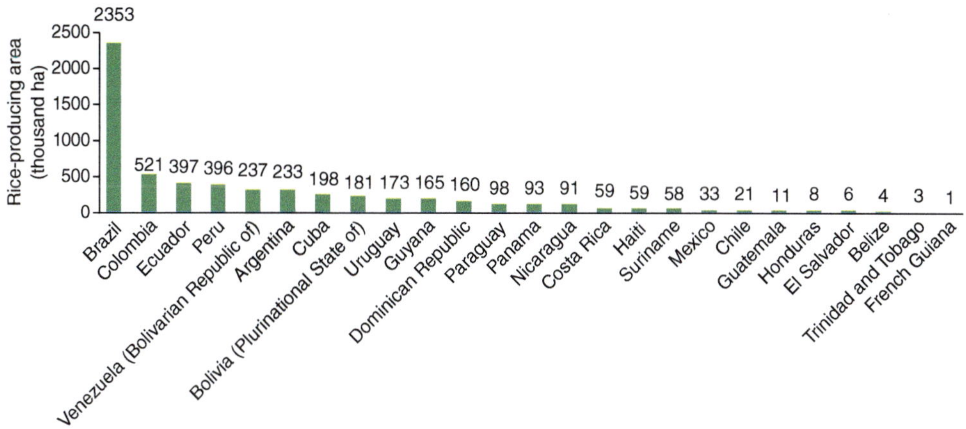

Fig. 13.1. Rice-producing countries in Latin America and the Caribbean (LAC). (Source FAO, 2015)

Pest Management

The term organic rice production represents a holistic production management system that avoids the use of synthetic fertilizers, pesticides, plant growth regulators and genetically modified organisms. Organic rice production is considered an interesting option for sustainable agriculture because it offers a unique combination of low external input technology, environmental conservation and input/output efficiency. Additionally it can provide access to premium price markets through labelling. The organic labels can originate from both third-party certification and participatory guarantee systems. Organic rice growers rely on crop rotations, crop residues management, composting, mechanical cultivation and IPM as means of providing fertilization and pest-management options. Growers are increasingly adopting organic agriculture techniques as a method of improving productivity, sustainability, income and food security. However, in spite of the potential access to premium markets and agroecological options, organic rice production has many challenges and represents a small portion of the total rice production in LAC.

Pest management is one of the biggest challenges to rice production in tropical production systems such as LAC (Pantoja *et al.*, 1997), where there is a paucity of information on organic-rice-based systems. In agricultural systems, a pest is defined as any organism that disturbs or affects the development and growth of a crop. In this chapter, we define rice pests as arthropods, pathogens and weeds, but the chapter focuses on arthropod management.

Globally, pest management in conventional rice production systems represents approximately 35% of production costs (12% insect pests, 10% weeds, 12% pathogens and 1% vertebrates) (Pantoja *et al.*, 1997). In conventional rice production systems, chemical control is the primary pest management strategy. However, there is paucity of information on pest control costs for organic pest management, especially in LAC, where few non-chemical alternatives are available for organic pest management.

The development of arthropod pest management programmes for organic systems has been reviewed by Zehnder *et al.* (2007) and Wyss *et al.* (2005), but there is no specific review on arthropod management in organic rice. Wyss *et al.* (2005) proposed four phases for progressive pest management in organic systems. Phase one includes cultural practices compatible with natural process (crop rotation, soil management, farm location, etc.). Phase two includes vegetation management to improve natural enemies' impact on the pest, while in phase three inundative and inoculative releases of biological control agents are considered. The use of organic-approved insecticides and mating disruption techniques are the last resort and implemented in phase four.

Farm location has been used as a means to avoid (isolation) or affect pest incidence and development (unfavourable climate) in organic systems (Wyss *et al.*, 2005; Zehnder *et al.*, 2007). Although, rice is a highly adaptable crop that can be produced under flooded or upland conditions, as well as in areas prone to flooding, the location of an organic rice farm is most likely determined by water availability and soil type, rather than pest management considerations.

Since planting dates are important in managing insect pest in LAC, especially disease vectors, the length of the growing season and water management are important factors in the implementation of IPM schemes. In general, varieties with a short growing season benefit from lower pest infestations as compared with varieties with a long growing season. Depending on the variety, temperature, soil, and water availability, the average growing season in LAC can be from 3 to 7 months. In addition to varieties that are resistant to major pests, growers might elect to plant early maturing varieties to avoid pest damage (Pantoja *et al.*, 1997).

The concept of 'healthy soil' has been cited as a main factor for pest management in organic farming (Oelhaf, 1978; Merrill, 1983; Altieri *et al.*, 2005). Most rice planted in the world is under irrigated or flooded conditions that alters soil chemistry (nutrient availability) and microbial properties. Flooding and drainage limits the use of habitat management and soil fertility, which are considered to be the pillars of improving soil health (Altieri *et al.*, 2005). Research is needed to better understand the relationship between pest and disease management and the changes in chemical and physical properties induced by flooding/drainage systems in organic rice production.

The concept of IPM has several definitions. We follow the Food and Agriculture Organization of the United Nations (FAO) that defines IPM as an ecosystem approach to crop production and protection that combines different management strategies and practices to grow healthy crops and minimize the use of pesticides (FAO, 2015). The FAO promotes IPM as the preferred approach to crop protection and regards it as a pillar of both sustainable intensification of crop production and pesticide risk reduction. As such, IPM is being mainstreamed in organic crop production programmes.

Host plant resistance has been a pillar for IPM in rice since the 1970s. The International Rice Research Institute (IRRI) in the Philippines, the International Center for Tropical Agriculture (CIAT) in Colombia, the Latin American Fund for Irrigated Rice (FLAR) in Colombia, and Africa-Rice in Côte d'Ivoire provide growers with rice germplasm with resistance to major diseases and abiotic factors affecting rice production in their respective regions. Zehnder *et al.* (2007) discuss the need to increase plant diversity to provide shelter and food sources to natural enemies when host plant resistance is the main pest-management option. Research is needed to better integrate and understand the use of natural enemies conservation strips, intercropping, companion/trap crops and other options for natural enemies' enhancement under flooded soil conditions.

The use of natural enemies is probably the most viable option for pest management in organic rice systems. Information on rice pests' natural enemies is widely available (Reissing *et al.*, 1986; Pantoja *et al.*, 1997). However, in most cases the literature reports on the presence and abundance of the natural enemies (Pantoja *et al.*, 1997) and a limited number of programmes have focused on natural enemies inundation or inoculation methods for organic production (Zehnder *et al.*, 2007). The cost and availability of biocontrol agents as compared with conventional control, is another limiting factor in the use of biocontrol programmes in organic rice farming systems.

To develop an IPM programme the organic rice grower needs to know: (i) the main pests affecting the crop; (ii) their feeding habits; (iii) their life history; (iv) the damaging stage; and (v) the part of the plant affected. Information on the agronomic and climatic factors influencing the development of the pest, the relationship between plant density and the plant growth stage and the time of insect appearance are factors that can be used to develop IPM programmes in organic settings. This chapter presents rice pests classified by their economic importance (Table 13.1). Key pests are defined as those present most of the time or those causing significant yield losses. Secondary pests are those that under specific circumstances can cause economic damage.

Table 13.1. Common rice pests in Latin America and the Caribbean (LAC). (Adapted from Pantoja *et al.*, 1997.)

Taxonomic group	Economic importance[a]	Damage	Management[b]	References
COLEOPTERA				
Curculionidae				
Lissorhoptrus bosqui Kuschel	M	Root	WM	Pantoja *et al.* (1997)
Lissorhoptrus brevirostris Suffrian	M	Root	WM	Pantoja *et al.* (1997)
Lissorhoptrus isthmicus Kuschel	M	Root	WM	Pantoja and Medina (1998)
Lissorhoptrus kuscheli O'Brien	M	Root	WM	Pantoja *et al.* (1999b)
Lissorhoptrus oryzophilus Kuschel	M	Root	WM	Pantoja *et al.* (1997)
Lissorhoptrus venezolanus Kuschel	M	Root	WM	Pantoja *et al.* (1997)
Scarabeidae				
Eutheola bidentata (Burmeister)	S	Root	SP, WM	Pantoja *et al.* (1997)
Phyllophaga spp.	S	Root	SP, WM	Pantoja *et al.* (1997)
HEMIPTERA				
Cercopidae				
Aeneolamia spp.	S	Stem	PD, RV	Pantoja *et al.* (1997)
Cicadellidae				
Draeculacephala soluta Gibson	S	Stem	WM, BC, PD	Arciniegas *et al.* (1999)
Hortensia similis Walker	S	Stem	WM, BC, PD	Arciniegas *et al.* (1999)
Delphacidae				
Tagosodes orizicolus (Müir) (= *Sogatodes oryzicola*)	M	Stem	RV, PD, BC	Zeigler *et al.* (1994)
		Vector		Arias *et al.* (1993), Pantoja and Hernández (1993)
Tagosodes cubanus (Crawford)	M	Stem	RV, PD, BC	Zeigler *et al.* (1994)
		Vector		Pantoja *et al.* (1997)
Pentatomidae				
Oebalus insularis (Stal)	M	Panicle	WC, BC	Gutiérrez *et al.* (1982, 1985, 1987), Franqui *et al.* (1998), Pantoja *et al.* (2000)
Oebalus ypsilon-griseus (De Geer)	M	Panicle	WC, BC	Pantoja *et al.* (1995)
Oebalus ornatus (Sailer)	M	Panicle	WC, BC	Pantoja *et al.* (1992a, 1995, 2000)
Oebalus pugnax-torridus (Sailer)	M	Panicle	WC, BC	Pantoja *et al.* (2000)
Tibraca limbativentris (Stal)	S	Stems	WC, WM, BC	Pantoja *et al.* (2000, 2005)
Tibraca obscurata (Bergoth)	S	Stems	WC, WM, BC	Pantoja *et al.* (2000, 2007)
Mormidea maculata Dallas	S	Panicle	WC, BC	Pantoja *et al.* (2000)
Mormidea pictiventris Stål	S	Panicle	WC, BC	Pantoja *et al.* (2000)
ORTHOPTERA				
Gryllotalpidae				
Neocurtilla hexadactyla (Perty)	S	Roots	WM, BC, PD	Pantoja *et al.* (1997)
LEPIDOPTERA				
Pyralidae				
Diatraea saccharalis (Fabricius)	S	Stems	WM, BC	Pantoja (1994)
Elasmopalpus lignosellus (Zeller)	S	Stems	WC, BC	Pantoja *et al.* (1997)
Rupela albinella (Cramer)	S	Stems, foliage	WC, WM, BC, PD	Pantoja and Matta (1993)
Noctuidae				
Spodoptera frugiperda (J.E. Smith)	S	Foliage	WC, WM, BC, PD	Pantoja *et al.* (1986a, 1994, 1997)

Continued

Table 13.1. Continued.

Taxonomic group	Economic importance[a]	Damage	Management[b]	References
Mocis spp.	S	Foliage	WC, WM, BC, PD	Pantoja *et al.* (1997)
Nymphula depunctalis Guenée	S	Foliage, stems	WC, WM, BC, PD	Pantoja *et al.* (1997)
DIPTERA				
Ephydridae				
Hydrellia wirthi (Korytkowski)	S	Foliage	WC, WM, BC, PD	Pantoja and Salazar (1992a, b), Pantoja *et al.* (1993a)
HYMENOPTERA				
Formicidae				
Atta laevigata F. Smith	S	Foliage, stems	SP, PD, CR	Pantoja *et al.* (1992a, b, 1997)
Acromyrmex landoltii Forel	S	Foliage, stems	SP, PD, CR	Pantoja *et al.* (1992a, b, 1997)
ACARI				
Acaridae				
Schizotetranychus oryzae (Rossi de Simons)	S	Foliage, vector	WC, WM, PD	Mejia *et al.* (1998a, b)

[a]M, Major pest; S, secondary pest.
[b]BC, Biological control; CR, crop rotation; PD, plant density; RV, resistant varieties; SP, soil preparation, land levelling; WC, weed control; WM, water management.

Major Rice Pests in LAC

Tagosodes orizicolus (Müir) and *Tagosodes cubanus* (Crawford) (Homoptera: Delphacidae) (sogata) and rice hoja blanca virus

Tagosodes orizicolus and *Tagosodes cubanus*, commonly known as sogata (Pantoja *et al.*, 1997) are considered the most important pests of rice in tropical LAC. Both species are present during all stages of plant development, although insects are more frequently collected at the beginning of the rice culture. Both adults and nymphs feed on plant stems and foliage causing a condition known as hopper burn. However, the economic importance of sogata is mostly associated with their ability to transmit rice hoja blanca virus (RHBV) (Zeigler *et al.*, 1994).

The symptoms of RHBV are chlorotic areas (pale yellow stripes) parallel to the midrib, which usually range from the apex of the blade to the base of the leaf sheath. RHBV causes plant stunting, reduced yield and plant death. In addition to affecting the plant, the virus has a deleterious effect on the insect. Carrier sogata females lay fewer eggs and for a shorter period of time than virus-free females (Zeigler *et al.*, 1994). However, the effect of the deleterious effect in controlling insect populations under field conditions is unknown.

Varietal resistance is the main control component of IPM programmes against both mechanical damage and RHBV on rice. CIAT, in Cali, Colombia, has been releasing rice varieties that are resistant to RHBV and to mechanical damage since the 1970s (Zeigler *et al.*, 1994). In recent years FLAR and CIAT provided national programmes in the region of advanced lines and varieties with resistance to RHBV, resistance to mechanical damage, or resistance to both. CIAT also assists national programmes with varietal evaluation and IPM programmes at the national level. The mechanism of resistance and inheritance to sogata mechanical damage and RHBV has been studied by Zeigler *et al.* (1994), Pardey *et al.* (2000) and summarized by Zorrilla *et al.* (2013).

Sogata is parasitized by *Haplogonatopus hernandezae* Olmi and *Elenchus* sp. In Norte de Santander, Colombia more than 40% of the adults and nymphs and up to 15% of the eggs of *T. orizicolus* are parasitized (Arias *et al.*, 1993). Parasitism is higher in females than in male sogata. The levels of parasitism observed in Colombia provided adequate control, even if varieties susceptible to mechanical damage by sogata were used; however, the relationship between parasitism, sex and their implications in controlling field sogata populations is not well understood. In addition to the level of parasitism, growers should consider the presence of weeds in fields. A weed species common in rice fields, *Echinochloa* sp. is a known host for sogata and is also considered a reservoir for RHBV (Pantoja *et al.*, 1997). Growers should keep fields free of weeds to reduce sogata incidence and reduce the probability of RHBV incidence. The combination of resistant varieties (to mechanical damage and RHBV) and appropriate weed management tactics coupled with parasitism should provide adequate control under organic production systems.

Spiders are important regulators of sogata adults and nymphs. Bastidas *et al.* (1993, 1994a, b) suggest sampling for spiders in rice fields and adjusting sogata action thresholds based on the number of spiders per sample. A reduction of three sogatas from the threshold for each spider collected in the sample is recommended. Action thresholds for sogata and the interaction with spiders are discussed by Pantoja *et al.* (1997) and Bastidas *et al.* (1993, 1994a, b). However, there is no IPM programme specifically developed for organic rice production in LAC.

Humidity and rainfall affect sogata incidence (Pantoja *et al.*, 1997; Vivas *et al.*, 2009). In areas where sogata is endemic, planting should be avoided during times of low rainfall (dry season). The grower should use multi-year weather data and insect incidence data to synchronize planting during periods of the year when planthopper populations are low. For example in the Cauca Valley of Colombia, planthopper population density is high during January and July. Rice fields established during those months will be exposed to severe sogata attack. However, in Cuba the best planting time coincides with low precipitation because it coincides with the arrival of cold temperatures. The low temperatures contribute to reduced insect development and damage (Gutierrez *et al.*, 1987).

Lissorhoptrus spp. (Coleoptera: Curculionidae) (rice water weevil)

Worldwide, species of insects belonging to the *Lissorhoptrus* species complex are considered the most limiting factor affecting flooded rice (Reissing *et al.*, 1986). Insects belonging to this species complex are commonly known as rice water weevils (RWW). In LAC, several species or RWW are reported. In Colombia, *Lissorhoptrus bosqui* Kuschel, *Oryzophagus* sp. and *Lissorhoptrus oryzophilus* Kuschel have been reported (Pantoja *et al.*, 1999b). In Venezuela *Lissorhoptrus venezolanus* Kuschel is the most common species; *L. venezolanus* was first reported in Venezuela in 1975 as a secondary pest, since then their population has increased and it is now considered an economically important pest (Pantoja *et al.*, 1997, 1999b). In Cuba, *Lissorhoptrus brevirostris* Suffrian is the main RWW species affecting flooded rice and is considered a key pest (EcuRed, 2015). In Cuba cultural and biological control of the RWW is commonly used. In Puerto Rico, *Lissorhoptrus isthmicus* Kuschel is the main RWW present in rice fields (Pantoja and Medina, 1998). In Asia and the USA, *L. oryzophilus* is considered a key pest in rice production (Reissing *et al.*, 1986).

Two other weevils, *Onychylis secundus* Burke and *Ochetina uniformis* Pascoe, similarly shaped and coloured to *Lissorhoptrus* spp. can be collected from rice fields and can be misidentified as RWW (Pantoja *et al.*, 1997). However, the areas where these two species (*O. secundus* and *O. uniformis*) occur is mainly associated with the presence of weeds in rice fields and their economic importance has not been determined. Keeping rice fields free of weeds reduces the incidence of RWW as well as related species such as *O. secundus* and *O. uniformis* (Pantoja *et al.*, 1997).

The RWW is a hydrophilic pest and they are attracted to the water layer established in flooded fields. Draining rice fields reduces the appearance of adult RWW which is reflected in reduced oviposition and reduced foliar damage. Damage to foliage is used as an indicator of the presence of RWW adults, but foliar damage is not considered of economic importance. Draining also increases larval mortality. However, draining fields causes nitrogen losses and affects (reduces) weed control. Additionally the RWW larva can survive long periods of time under moist conditions. The grower should consider the benefits and constraints of drainage and water management to control this pest.

Two biological control agents are known against RWW. The fungi *Beauveria bassiana* (Balsamo-Crivelli) Vuillemin and *Metarhizium anisopliae* (Metchnikoff) Sorokin are commonly used in Cuba against *L. brevirostris* (EcuRed, 2015). However, the persistence of the fungus in the field is short lived, and there is little information on the use of these biological agents in organic rice farming. RWW larvae can also be attacked by aquatic predators and nematodes, but there is little information on the efficiency of these natural enemies in controlling pests under field conditions. The literature does not report on augmentative management programmes for these insects in organic systems.

Some weed species host both larvae and adults of the RWW. Adult RWW are often found in Johnson grass (*Sorghum halepense* L.). Removing weeds from rice fields, levies and neighbouring areas reduces the incidence of RWW adults. Removing weeds also reduces competition for water and soil nutrients. Weed and water management are probably the best options for RWW management in organic settings (Pantoja *et al.*, 1997).

Stink bugs (Hemiptera: Pentatomidae)

Stink bugs are considered serious pests of rice throughout the world. Nymphs and adults feed on various parts of the rice plant, but tend to prefer the developing grain. Depending on the species, damage can be to the panicle, kernels or stems.

Several species of stink bugs are considered key pests of rice in LAC (Pantoja *et al.*, 1992a, 1995, 2000, 2005, 2007; Daza and Pantoja, 1993; Vivas and Astudillo, 2010). The presence of a species complex (Pantoja *et al.*, 1992a, 1995, 2000, 2005, 2007) rather than a single species, and the difference in damage potential among species in LAC make the implementation of management practices difficult. Several stink bug species can be collected simultaneously, but an action threshold only exists for *Oebalus poecilus* (Sailer) and *Oebalus ornatus* (Sailer) (Pantoja *et al.*, 1992a). The species composition or proportion of insects from each species needs to be determined before corrective measures are taken. For example in Colombia growers are familiar with *O. poecilus*, but not with *O. ornatus* that can represent up to 93% of the stink bugs collected with sweep nets (Pantoja *et al.*, 1995).

The symptoms of damage and its severity depend on when feeding occurs and what part of the plant is affected. Grains attacked at the milk stage do not fill properly, and the panicles often remain erect as the crop matures. Stink bug damage can result in discoloration and structural weaknesses that frequently cause the grain to shatter during milling (Pantoja *et al.*, 1992a). Damaged grains that survive the milling process often have small black or brown spots called pecky rice that represents a serious quality issue affecting the price. Insect feeding during the flowering and milk stages of grain development cause more damage than during the soft dough stage. In Cuba, *Oebalus insularis* is also a serious pest during the milk stage of grain development but causes less damage as the grain hardens (Meneses *et al.*, 1990).

In south-east Colombia, *O. ornatus* adults peak when plants are about 60 days old. In Latin America, most varieties initiate panicle emergence at this age. A similar behaviour has been reported for stink bugs in Florida and Cuba (Gutiérrez *et al.*, 1982). Stink bugs collected during the vegetative stage of growth (0–30-day-old plants) are mainly associated with the presence of weeds. Daza and Pantoja (1992, 1993) reported higher stink bug populations in weedy rice fields than in weed-free fields. Although *O. ornatus* can cause damage to rice yield in some regions of Colombia, the low population levels encountered in the region suggest that economic damage will rarely occur (Pantoja *et al.*, 1992a, 1995, 1998, 1999a, 2000).

The natural enemies of stink bugs in LAC have been studied by Pantoja and Daza (1993) and Franqui *et al.* (1988). Although several species of parasitoids have been reported, their effect on natural stink bug populations is still unknown and requires research before they can be used to manage natural pest infestations in organic rice farming systems.

Two other pentatomid species, *Tibraca limbativentris* (Stal) and *Tibraca obscurata* (Bergoth) are common pests in upland rice and occasionally attack irrigated fields (Pantoja *et al.*, 2005, 2007). In contrast to the *Oebalus* spp. complex, these two species feed on the stem affecting panicle emergence. The geographical distribution and abundance in rice fields of these two species is poorly documented. Pantoja *et al.* (2005, 2007) documented their incidence, damage and biology in Colombia, Venezuela and Ecuador. Both species feed on rice stems causing a condition known as white heads. Damage is caused by both adults and nymphs that can be found at the base of the stem, usually with the head down.

These two species can be distinguished by size differences: *T. limbativentris* is larger than *T. obscurata*. Nymphs and adults introduce their stylet into the stem damaging the developing panicle. The damage puncture causes panicle constriction that distinguishes it from dead hearts caused by stem borers (dead hearts are also characterized by white or discoloured panicles). The damage (presence of white panicles) is most evident during the flowering and boot stage of plant development (i.e. when the leaf stem bulges concealing the developing panicle), when the white panicles appear. However, damage occurs early in plant establishment, and management tactics should not be implemented using white panicles as the indicator of insect presence or damage.

In spite of the reported agricultural significance of *Tibraca* spp. in the region, few data are available on the economic importance and damage functions of these two species in rice. In Colombia and Venezuela farmers apply insecticide for *Tibraca* spp. control, but action thresholds are based on the presence of white panicles (Aponte *et al.*, 1992), not on insect density. However, the presence of white heads in the field is a symptom of damage that occurred earlier, during the filling stage of grain development, and is not a reliable indicator of insect presence or population density (Pantoja *et al.*, 1997). Sampling should be based on egg counts; unfortunately, there are no efficient sampling methods to inspect and count *T. limbativentris* and *T. obscurata* eggs on rice plants. Sampling and management tactics in areas of high *T. limbativentris* and *T. obscurata* incidence should be implemented early during plant development (25–40-day-old plants), not at detection of white panicles (Pantoja *et al.*, 1997, 2005, 2007).

There is a paucity of information on biological control agents and management tactics for these two pests in rice fields. Insect incidence is affected by plant density, but there is no published information on the correlation between plant density and damage reduction in rice fields. The fungus *Paecilomyces* sp. affects both adults and nymphs, but the controlling effect is unknown under field conditions. Grasses growing at the edge and on ridges of rice fields and debris from previous crops (ratoon) are alternate hosts of these insects. Keeping rice fields weed free should reduce the appearance and incidence of the pests. There are no specific organic control measurements for these two species on rice.

Spodoptera frugiperda (J.E. Smith) (Lepidoptera: Noctuidae) (fall armyworm)

The fall armyworm, *Spodoptera frugiperda*, is among the most serious pests that limit rice production in LAC. This pest is considered an economically important pest in Puerto Rico, Panama, Colombia and Brazil (Navas, 1966; Gallego, 1967; Machado, 1978; Pantoja *et al.*, 1986a, c). In spite of the reported economic importance, damage to rice is sporadic and has been associated with the presence of weeds in fields and climatic conditions (Pantoja *et al.*, 1997).

Although varietal resistance has been identified (Pantoja *et al.*, 1986b) management is usually accomplished by means of insecticides. Up to 70% of larvae can be affected or destroyed by parasites, but the 30% surviving larva can be of economic importance (Pantoja *et al.*, 1985; Pantoja and Fuxa, 1992).

Natural enemies include the fungus *Nomuraea rileyi* (Farl.) Samson and parasitoids *Chelonus insularis* (Cresson), *Euplectrus furnius* Walker, *Cotesia marginiventris* (Cresson), *Lespesia* sp., *Euplectrus platyhypenae* Howard, *Archytas marmoratus* (Townsend) and *Rogas laphygmae* (Viereck). However, the effects of these natural enemies in controlling field populations are still unclear (Pantoja and Fuxa, 1992; Pantoja *et al.*, 1997).

Cultural practices to manage *Spodoptera* include weeding, flooding and pheromone trapping (Pantoja *et al.*, 1997). Sampling rice fields for oviposition is an effective practice to establish possible infestation and apply control methods (Pantoja *et al.*, 1986c; Pantoja *et al.*, 1997). However, there are no action thresholds for this pest in organic rice production systems.

Secondary Pests

Depending on the geographical area and climate, some insect pests that usually exist as stable populations in balance with the environment can cause economic damage or might require intervention. The main secondary pests for the LAC area are presented here.

Hydrellia wirthi Korytkowski (Diptera: Ephydridae) (rice leaf miner)

The rice leaf miner, *Hydrellia wirthi*, is a hydrophilic pest common in LAC, although the geographic distribution is not clearly defined (Pantoja and Salazar, 1993a, b; Salazar *et al.*, 1993; Pantoja and Salazar, 1994). Female flies are attracted to water and oviposit on foliage. The larva bores into foliage to pupate on the leaf sheaths. Growers use leaf damage to estimate *H. wirthi* damage, but sampling should be directed to the egg stage.

Since adults are attracted by water, cultural practices are effective against this pest (Pantoja and Salazar, 1993a, b; Salazar *et al.*, 1993; Pantoja and Salazar, 1994). For example, damage by *H. wirthi* can be reduced by draining rice fields. Drainage affects the numbers of adults that appear in the fields resulting in reduced oviposition and reduced larval survival, but drainage may increase the incidence of weeds (Pantoja and Salazar, 1993a, b; Pantoja *et al.*, 1993; Pantoja and Salazar, 1994). Additionally, drainage increases the loss of nitrogen from fields. The grower has to balance the benefits of insect control against possible negative side effects from the implementation of cultural techniques such as drainage.

Transplanting is another cultural practice effective at reducing *H. wirthi* incidence in rice fields. Apparently the mechanical damage caused during the hardening and transplanting makes plants less attractive for oviposition. There is no evidence that this practice is used in organic farming systems for rice leaf miner management in LAC. Varieties with a short life cycle or early maturity can affect insects with a short development cycle such as *H. wirthi*. Crop rotation is undoubtedly an effective tactic for controlling arthropods, weeds and diseases in rice. Unfortunately, most of the land used in LAC for rice production is not rotated with other crops, especially irrigated rice lands. Wherever possible rotation of rice with legumes and other broadleaved crops is recommended for weed control to disrupt insect and weed cycles.

Inadequate seed density and poor seed distribution contribute to the incidence of hydrophilic pests such as *Hydrellia* sp. Adults of these species prefer to oviposit in areas where the water surface is visible. When planting density increases the visibility of the water surface is reduced, consequently decreasing the incidence of *H. wirthi* (Pantoja and Salazar, 1993a, b; Salazar *et al.*, 1993; Pantoja and Salazar, 1994; Pantoja *et al.*, 1997). High seeding densities, in turn, reduce weed incidence, but favour the incidence of rice blast (*Pyricularia oryzae* Cavara) if the crop is grown in an area of high relative humidity.

Rice blast, caused by *P. oryzae* (teleomorph: *Magnaporthe oryzae*), is considered the most limiting disease for rice production in LAC. Losses by *P. oryzae* can reach 100% incidence on highly susceptible cultivars under appropriate environmental conditions. The fungus produces spots or lesions on leaves, kernels and panicles. On leaves, early lesions appear as small brown spots. As the lesions develop, the lesion takes a diamond shape with a grey-coloured centre and a chlorotic halo edge that runs along the lesion. In susceptible cultivars lesions coalesce causing plant death especially in seedlings. At the internodes straw-coloured rings are formed, reducing growth. The most visible damage is to the panicle, where brown spots can be visible. Constriction at the internodes prevents nutrient flow, resulting in empty or partially filled kernels. The environmental factors that favour the development of the fungus are: (i) high relative humidity (greater than 80%); (ii) temperatures between 25°C and 28°C; (iii) dew periods between 12 h and 14 h; (iv) low average temperature during cold nights followed by warm days; (v) prolonged humidity; (vi) low luminosity; and (vii) light wind. Factors that favour the pathogen incidence include nitrogen excess or deficiencies, susceptible varieties and soil type (Pantoja *et al.*, 1997).

Strategies for rice blast management include a combination of resistant varieties, cultural management and chemical management. The use of resistant varieties is the most effective management method. Resistance to blast is unstable; varieties released by CIAT in LAC last up to 3 years, but in some cases resistance has been broken in 2 years. The exception in LAC is a variety 'Fedearroz 50' that is still demonstrating field resistance 15 years after its release. Alternating resistant varieties and soil preparation are the most common cultural practices used in the region (Pardey *et al.*, 2000).

Leafhoppers (Hemiptera: Cicadellidae)

Several species of leafhoppers can be collected in rice fields. Leafhoppers can be collected in high densities, but usually do not cause economic damage to rice. The population dynamics and damage functions by the two most common species in rice fields in Colombia (*Draeculacephala* spp. and *Hortensia* spp.) were studied by Arciniegas and Pantoja (1993) and Arciniegas *et al.* (1999). Keeping fields weed free is probably the most effective way to reduce the incidence of these two pests in rice fields.

Other secondary pests are presented in Table 13.1. For more information on management options, see Pantoja *et al.* (1997).

The Way Forward

Despite the growth of organic production, the information and options for pest management in organic rice systems are limited. In LAC, CIAT and FLAR have provided national programmes with resistant varieties to sogata, RHBV and *Pyricularia* leaf spot disease; however, there is no breeding programme in the region targeting other important pests such as the rice weevil. Pantoja *et al.* (1997) provided comprehensive pest-management options for rice producers in LAC. The review focused on non-chemical means of pest control on rice, but did not include specific recommendations for organic rice production systems.

Although, the demand for organic rice is increasing worldwide, rice producers in LAC are not synchronized with the expanding market. Based on interviews with national experts in the region, insect pest problems are not the most limiting factor for organic rice expansion in the region. The problems include: (i) the lack of organic-certified seed; (ii) the lack of machinery and mills dedicated to organic rice production; (iii) the different ecological practices specified by international and national organic production standards; (iv) the cost of certification; and (v) marketing issues. Research into these issues is required for the rapid expansion of certified organic rice production in LAC.

Despite the growth of organic agriculture, there has been a lack of research to understand the complex mechanisms operating in organic rice systems, including plant–pest interactions under flooded and upland rice systems. Research priority should be directed to preventative and curative measures approved by the various international and national organic production standards.

References

Altieri, M.A., Ponti, L. and Nicholls, C. (2005) Enhanced pest management through soil health: toward a belowground habitat management strategy. *Biodynamics* (Summer), 33–40.

Aponte, O., Escalona, L.E., Vivas, L.E., Ramírez, L.M. and Freitez, F.P. (1992) *Manejo Integrado de Artrópodos Plaga en el Cultivo de Arroz en Venezuela.* Centro Internacional de Agricultura Tropical (CIAT), Cali, Colombia.

Arciniegas, I.C. and Pantoja, A. (1993) Hospederos alternos de *Draeculacephala soluta* Gibson y *Hortensia similis* Walker (Homoptera: Cicadellidae) en el arroz del Valle del Cauca, Colombia. *Manejo Integrado de Plagas (Costa Rica)* 26, 34–35.

Arciniegas, I.C., Pantoja, A., Garcia, C. and Duque, M.C. (1999) Umbral de acción de *Hortencia similis* Walker y *Draeculacephala soluta* Gibbons (Homoptera: Cicadellidae) en el cultivo de arroz en el valle del Cauca, Colombia. *Journal of Agriculture of the University of Puerto Rico* 83, 65–74.

Arias, M., Vivas, M., Cuevas, A. and Pantoja, A. (1993) Parasitism of *Tagosodes orizicolus* and *Tagosodes cubanus* in northeastern Colombian rice fields. *International Rice Research Newsletter* 18(2), 32.

Bastidas, H., Pantoja, A., Zuluaga, I. and Murillo, A. (1993) Colombian rice fields' spiders. *International Rice Research Newsletter* 18(2), 32–33.

Bastidas, H., Pantoja, A. and Hernández, M.P. (1994a) Consumo de presas por *Argiope argentata* (Araneae: Araneidae) y *Pleisometa argira* (Araneae: Tetragnathidae) en arroz irrigado en Colombia. *Manejo Integrado de Plagas (Costa Rica)* 32, 30–32.

Bastidas, H., Pantoja, A., Murillo, A., Zuluaga, J.I. and Duque, M.C. (1994b) Identificación, dinámica poblacional y consumo de presas por arañas en arroz en el Valle del Cauca. *SOCOLEM* 20, 149–160.

Daza, E. and Pantoja, A. (1992) Hospederos alternos de Pentatómidos: implicaciones en el manejo de plagas. *Turrialba* 42, 408–410.

Daza, E. and Pantoja, A. (1993) Hospederos alternos de *Oebalus ornatus*. *Mormidea pictiventris* y *Mormidea maculata*: implicaciones en el manejo integrado de plagas. *Turrialba* 42, 408–410.

EcuRed (2015) *El gorgojo acuático del arroz.* Available at: http://www.ecured.cu/index.php/Gorgojo_acu%C3%A1tico_del_Arroz (accessed 6 January 2015).

Food and Agriculture Organization of the United Nations (FAO) (2015) *Integrated Pest Management.* Available at: http://www.fao.org/agriculture/crops/thematic-sitemap/theme/pests/ipm/en/' (accessed 8 February 2015).

Franqui, R.A., Pantoja, A. and Medina, S. (1998) Natural enemies of Pentatomids affecting rice fields in Puerto Rico. *Journal of Agriculture of the University of Puerto Rico* 72, 371–373.

Gallego, M.F.L. (1967) Lista preliminar de insectos de importancia economica y secundarios, que afectan los principales cultivos, ani males domesticos y al hombre, en Colombia. *Revista de la Facultad de Agronomia de Colombia* 65, 32–66.

Global Rice Science Partnership (GRiSP) (2014) Global Rice Science Partnership. Available at: http://library.cgiar.org/bitstream/handle/10947/3085/GRiSP%202013%20Annual%20Report.pdf?sequence=1 (accessed 8 February 2015).

Gutiérrez, A., Meneses, R. and Corona, R. (1982) Pérdidas ocasionadas por la alimentación de *Oebalus insularis* en la fase lechosa del grano de arroz. *Ciencia y Tecnologia Agricola Arroz* 5(1), 71–79.

Gutiérrez, A., Arias, A., García, A. and Corona, R. (1985) Evaluación del nivel de daño causado por diferentes índices de población de *Oebalus insularis* en el cultivo del arroz. *Ciencia y Tecnologia Agricola Arroz* 8(1), 63–74.

Gutiérrez, A., Meneses, R., Arias, E., García, A. and Hernández, A. (1987) Estimación de las poblaciones de *Oebalus insularis* en el cultivo del arroz. *Ciencia y Tecnologia Agricola Arroz* 10(1), 43–53.

Machado, S.S. (1978) Alguns fatores limitantes da produtividade das lavoras de arroz. *Lavoura Arrozeira* 307, 52–54.

Mejia, O.I., Pantoja, A., Messa, N.C. and Duque, M.C. (1998a) Evaluation of rice germplasm for resistance to *Schizotetranichus oryzae* (Rossi de Simons). *Journal of Agriculture of the University of Puerto Rico* 82, 217–220.

Mejia, O.I., Ramírez, L.M., Escalona, L.E., Messa, N.C., Duque, M.C. and Pantoja, A. (1998b) Development of *Schizotetranichus oryzae* (Rossi de Simons) on rice. *Journal of Agriculture of the University of Puerto Rico* 82, 187–192.

Meneses, R., Gutiérrez, A., Arias, E., Hernández, A., García, A. and Amador, M. (1990) Resultados de los estudios realizados en Cuba para el manejo de *Sogatodes oryzicola* Muir, *Oebalus insularis* Stal, *Lissorhopterus brevirostis* (Sulf) e Hydrellia sp., en el cultivo de arroz. Instituto de Investigaciones del Arroz, Ministerio de Agricultura, La Habana, Cuba.

Merrill, M.C. (1983) Bio-agriculture: a review of its history and philosophy. *Biological Agriculture and Horticulture* 1, 181–210.

Navas, D. (1966) Evaluacion del dano que ocasionan algunas plagas del arroz en Panama. In: *Programa Cooperativo Centroamericano para el Mejoramiento de Cultivos Alimenticios, XII Reunion Anual,* 28 March–2 April, Managua, Nicaragua, pp. 77–78.

Oelhaf, R.C. (1978) *Organic Farming: Economic and Ecological Comparisons with Conventional Methods.* Allanheld Osmun, Montclair, New Jersey.

Pantoja, A. (1992) *Hydrellia wirthi* Korytkowski damage to rice in Colombia. *International Rice Research Newsletter* 17(6), 30.

Pantoja, A. (1994) *Diatraea saccharalis* on rice plants in Puerto Rico. *Journal of Agriculture of the University of Puerto Rico* 78, 195.

Pantoja, A. and Daza, E. (1993) Natural enemies of *Oebalus ornatus* (Sailer) (Hemiptera: Pentatomidae) recovered from rice fields in Southwestern Colombia. *Trends in Agricultural Science (Entomology)* 1, 41–43.

Pantoja, A. and Fuxa, J.R. (1992) Prevalence of biotic agents in fall armyworm infesting rice in Puerto Rico. *Folia Entomologica Mexicana* 84, 79–84.

Pantoja, A. and Hernández, M.P. (1993) *Sogatodes* o *Tagosodes*: sinonimia y evaluación de daño mecánico. *Arroz (Colombia)* 42, 30–31.

Pantoja, A. and Matta, J. (1993) *Rupela albinella* (Lepidoptera: Pyralidae) oviposition behavior and egg distribution on rice plants. *Journal of Agricultural Science* 28(3), 249–253.

Pantoja, A. and Medina, S. (1998) *Lissorhoptrus isthmicus* Kuschel (Coleoptera: Curculionidae) a new rice pest for Puerto Rico. *Journal of Agriculture of the University of Puerto Rico* 72, 333.

Pantoja, A. and Salazar, A. (1993a) *Hydrellia wirthi* Korytkowsky: a new rice pest in Colombia. *International Rice Research Newsletter* 18(2), 32.

Pantoja, A. and Salazar, A. (1993b) Ovipositional preference of the rice leaf miner, *Hydrellia wirthi* Korytkowsky (Diptera: Ephydridae), on selected rice varieties. *Tropical Agriculture* 70, 378–379.

Pantoja, A. and Salazar, A. (1994) Sampling *Hydrellia wirthi* with sticky traps. *Journal of Agriculture of the University of Puerto Rico* 78, 191.

Pantoja, A., Smith, C.M. and Robinson, J.F. (1985) Natural control agents affecting *Spodoptera frugiperda* (Lepidoptera: Noctuidae) infesting rice in Puerto Rico. *Florida Entomologist* 68, 488–490.

Pantoja, A., Smith, C.M. and Robinson, J.F. (1986a) Effects of the fall armyworm (Lepidoptera: Noctuidae) on rice yields. *Journal of Economic Entomology* 79, 1324–1329.

Pantoja, A., Smith, C.M. and Robinson, J.F. (1986b) Evaluation of rice germplasm for resistance to the fall armyworm (Lepidoptera: Noctuidae). *Journal of Economic Entomology* 79, 1319–1323.

Pantoja, A., Smith, C.M. and Robinson, J.F. (1986c) Fall armyworm oviposition and egg distribution on rice. *Journal of Agricultural Entomology* 3, 114–119.

Pantoja, A., Daza, E. and Duque, M.C. (1992a) Efecto de *Oebalus ornatus* (Sailer) y *Oebalus insularís* (Stal) (Hemíptera: Pentatomidae) sobre el arroz: una comparación entre especies. *Manejo Integrado de Plagas (Costa Rica)* 26, 31–33.

Pantoja, A., García, C. and Mejía, O. (1992b) Prácticas agronómicas como alternativa al control químico de *Acromyrmex landolti* Forel en el establecimiento de arroz. En: Cuevas, F. (ed.) *Arroz en América Latina: Mejoramiento, manejo y comercialización.* Centro Internacional de Agricultura Tropical (CIAT), Cali, Colombia, pp. 273–274.

Pantoja, A., Salazar, A., Mejía, O.I., Velásquez, J.G. and Duque, M.C. (1993) Cultural practices to manage the rice leaf miner. *Hydrellia wirthi* Korytkowski (Diptera: Ephydridae) in Colombia. *Journal of Economic Entomology* 86(6), 1820–1823.

Pantoja, A., García, C. and Mejía, O.I. (1994) Flight activity of *Spodoptera frugiperda* (J.L. Smith) in northeastern Colombia acid sabana rice fields. *International Rice Research Newsletter* 19(4), 19–20.

Pantoja, A., Daza, E., García, C., Mejia, O.I. and Raider, D.A. (1995) Relative abundance of stink bugs (Hemiptera: Pentatomidae) in southwestern Colombia rice fields. *Journal of Entomological Science* 30, 463–467.

Pantoja, A., Fischer, A., Correa, F., Sanint, L. and Ramirez, A. (1997) *Manejo Integrado del Cultivo del Arroz en Latino América y el Caribe*. Centro Internacional de Agricultura Tropical (CIAT) Publication 292. CIAT, Cali, Colombia.

Pantoja, A., García, C.A., Mejia, O.I., Ramírez, L.M., Escalona, L.E. and Duque, M.C. (1998) Disminución del rendimiento y calidad del arroz de secano por *Oebalus ypsilon-griseus*. *Manejo Integrado de Plagas (Costa Rica)* 47, 37–40.

Pantoja, A., Daza, E., Mejia, O.I., Duque, M.C. and Escalona, L.E. (1999a) Development of *Oebalus ornatus* (Sailer) and *Oebalus insularis* (Stal) on rice. *Journal of Entomological Science* 34, 335–338.

Pantoja, A., Ramirez, L.M., Escalona, L.E., Bastidas, H. and Duque, M.C. (1999b) Number of instars of *Lissorhoptrus kuscheli* O'Brien (Coleoptera: Curculionidae). *Annals of the Entomological Society of America* 92, 236–238.

Pantoja, A., García, C. and Duque, M.C. (2000) Population dynamics and effects of *Oebalus ornatus* (Hemiptera: Pentatomidae) on rice yield and quality in southwestern Colombia. *Journal of Economic Entomology* 93, 276–279.

Pantoja, A., Triana, M., Bastidas, H., García, C. and Duque, M.C. (2005) Development of *Tibraca obscurata* and *Tibraca limbativentris* (Hemiptera: Pentatomidae) on rice in southwestern Colombia. *Journal of Agriculture of the University of Puerto Rico* 89, 221–228.

Pantoja, A., Triana, M., Bastidas, H., García, C., Mejia, O.I. and Duque, M.C. (2007) Damage by *Tibraca limbativentris* (Hemiptera: Pentatomidae) to rice in southwestern Colombia. *Journal of Agriculture of the University of Puerto Rico* 91, 11–18.

Pardey, C., Cuevas, F., Martínez, C.P. and Pantoja, A. (2000) Herencia de la resistencia al daño mecánico por *Tagosodes orizicolus* (Muir) (Homoptera: Delphacidae) en arroz. *Journal of Agriculture of the University of Puerto Rico* 84, 47–52.

Reissing, W.H., Heinrichs, E.A., Litsinger, J.A., Moody, K., Fiedler, L., Mew, T.W. and Barrion, A.T. (1986) *Illustrated Guide to Integrated Pest Management in Rice in Tropical Asia.* International Rice Research Institute (IRRI), Los Baños, Filipinas.

Salazar, A., Pantoja, A. and Duque, M.C. (1993) Aspectos biológicos del minador de la hoja del arroz en el Valle del Cauca. *Arroz (Colombia)* 42, 38–40.

Vivas, L. and Astudillo, D. (2010) Plantas hospederas de chinche vaneadora en el cultivo del arroz en Calabozo, Estado Guarico, Venezuela. *Agronomía Tropical* 60(4), 370–374.

Vivas, L.E., Astudillo, D. and Poleo, J. (2009) Monitoreo de *Tagosodes orizicolus* e incidencia del virus de la hoja blanca 'VHB' en el cultivo de arroz en calabozo, estado Guárico, Venezuela. *Agronomía Tropical* 59(4), 57–67.

Wyss, E., Luka, H., Pfiffner, L., Schlatter, C., Uehlinger, G. and Daniel, C. (2005) Approaches to pest management in organic agriculture: a case study in European apple orchards. *Organic Research* (May), 33–36.

Zehnder, G., Gurr, G.M., Kühne, S., Wade, M.R., Wratten, S.D. and Wyss, E. (2007) Arthropod pest management in organic crops. *Annual Review of Entomology* 52, 57–80.

Zeigler, R.S., Pantoja, A., Duque, M.C. and Weber, G. (1994) Characteristics of resistance in rice to rice hoja blanca virus (RHBV) and its vector, *Tagosodes orizicolus* (Muir). *Annals of Applied Biology* 124, 429–440.

Zorrilla, G., Martinez, C., Berrio, L., Corredor, E., Carmona, L. and Pulver, E. (2013) Improving rice production systems in Latin America and the Caribbean. In: Hershey, C.H. and Neate, P. (eds) *Eco-efficiency: From Vision to Reality.* Centro Internacional de Agricultura Tropical (CIAT) Publication No. 381. CIAT, Cali, Colombia, pp. 161–170.

14 Wheat Production in Organic Farming

Paolo Bàrberi* and Ambrogio Costanzo
Institute of Life Sciences, Scuola Superiore Sant'Anna, Pisa, Italy

Introduction

Wheat is the cereal crop with the largest acreage under organic management. The most recent figures available show that in 2012 the total area grown worldwide with organic cereals totalled 2,652,864 ha, of which wheat represented *c.*42% (Willer and Lernoud, 2014).

A wheat field is an agroecosystem that can harbour a variety of insects, but in major wheat production areas only a few of them are considered to be serious pests, thereby downgrading the importance of pest management as relative to, for example, weed or nutrient management (Birzer and Badgery, 2006; Weisz *et al.*, 2014).

Nevertheless, there may be situations in which pest management is necessary or advisable. For example, in fragile agroecosystems of West Asia and North Africa, where drought is recurrent, insect pests and diseases often contribute to yield instability (Ceccarelli *et al.*, 1992). Increased instability of weather patterns consequent to climate change may also exacerbate pest problems by altering the complex biological interactions affecting pests (Chakraborty and Newton, 2011). Furthermore, climate change is expected to increase problems due to invasive alien pests (Masters and Norgrove, 2010), which are also likely to occur as a consequence of increased global trade (Hulme, 2009). Wheat, as a major commodity, may be particularly exposed to these risks. Climate change may also change the pest status of arthropod species not formerly known as wheat pests. For example, this has been suggested for *Australiodillo bifrons* (Budde-Lund), a slater (Isopoda) species endemic to New South Wales, Australia. *A. bifrons* used to feed on decaying organic matter but has recently been shown to cause significant damage to wheat seedlings, especially in fields surrounded by windbreaks that can harbour large populations (Paoletti *et al.*, 2008).

Organic farming, through the reduction in pesticide use, can substantially increase agricultural sustainability by preserving key organisms and the agroecological services associated with them. In wheat, it has been estimated that halving the treatment frequency index (TFI) would increase the abundance of earthworms by a factor of 1.5 to 4, contributing to soil fertility conservation (Pelosi *et al.*, 2013). Interestingly, a reduction in insecticide use would be more beneficial to earthworms than a corresponding reduction in herbicide or fungicide use, due to the

* Corresponding author: paolo.barberi@santannapisa.it

more environmentally harmful profile of the former. Besides soil fertility, organic farming would be expected to also improve the biological regulation of insect pests in wheat, not only through the reduction in insecticide use but also through the creation of a more favourable habitat (Puech *et al.*, 2014).

In this chapter the strategies and methods used for insect pest management in organic wheat are summarized. The first thing that is tackled is the general issue of where and when pests need to be controlled in organic wheat. Then preventive approaches to pest management are considered, such as strategies based on habitat diversity, on conservation biological control, and on cropping system or management diversity. Afterwards, some direct methods of pest control applied to organic wheat are presented. Lastly, the perspectives of pest management in organic wheat are considered, taking into account emerging trends such as the application of functional biodiversity in wheat-based systems.

Do Insect Pests Need to be Controlled in Organic Wheat?

In an integrated pest management (IPM) strategy, regular pest monitoring is of utmost importance to determine whether or not farmers should intervene to control biotic aggressors during the crop growing cycle. This is particularly relevant for organic farming, where pest management should mainly rely on preventive rather than curative (direct) measures (IFOAM, 2005). Therefore, any methods aimed to provide accurate estimates of early pest occurrence in organic wheat should be encouraged. In the case of organic wheat, the scientific literature clearly shows that aphids are to be considered the most serious insect pest to be monitored and possibly controlled.

Optimization of sampling efforts in early wheat growth stages may be crucial for timely application of IPM measures, as it has been demonstrated for the English grain aphid, *Sitobion avenae* (F.) in Iran, where good estimates of population densities could be attained with relatively little sampling effort when values were > 3 aphids per tiller but not when they were < 0.5 aphid per tiller (Shahrokhi and Amirmaafi, 2012).

Yang *et al.* (2005) showed that use of remote sensing techniques (a hand-held radiometer) allowed detection of early wheat stress caused by greenbug (*Schizaphis graminum* (Rondani)) infestation in the US Great Plains through analysis of reflectance data and derived vegetation indices from 16 wavelength bands. The band centred at 694 nm and the vegetation indices derived from the bands centred at 694 and 800 nm were identified as those most sensitive to damage detection. Subsequent work by the same team (Yang *et al.*, 2009) showed that with the same method it is also possible to discriminate between the stress induced in wheat by two different aphid species (greenbug and Russian wheat aphid, *Diuraphis noxia* (Kurdjumov)). Ratio-based vegetation indices (based on 800/450 nm and 950/450 nm wavelength bands) were the most promising, but they need further testing at canopy level and under field conditions.

Models and decision support systems can be instrumental in fine tuning predictions of population densities of wheat pests and associated natural enemies. Freier *et al.* (2002) estimated the regulation effect of natural enemies on three wheat aphid species (*S. avenae*, *Rhopalosiphum padi* (L.) and *Metopolophium dirhodum* (Walker)) with the simulation model GETLAUS01 using 9-year insect counting values at two sites. The model was first used to make the simulated infestation development fit actual field density data. Then, simulations were repeated but without natural enemies (their density was set to zero). The difference between aphid densities without and with predator influence allowed the authors to estimate the effect of predators. Simulation results indicated that aphid infestation without predators' influence was at least double that of actual field data, where predation occurs.

Generic crop simulation models can be useful for predicting the effect of major pests on wheat growth and yield based on a set of environmental and management parameters. Aggarwal *et al.* (2006) validated the InfoCrop model for wheat under contrasting tropical agroenvironmental conditions based on data from 11 field experiments differing in site, season, nitrogen (N) and irrigation management, variety, organic matter and pest incidence. The model simulated wheat total biomass and grain yield losses to different pests and their populations satisfactorily and with good sensitivity. Interestingly, this model was shown to adequately quantify multiple pests damage through iso-loss curves.

Models incorporating a spatial distribution component are important to provide dynamic information for pest control. A model based on the aggregation–diffusion behaviour of aphids was developed by Li *et al.* (2010) to simulate the spatio-temporal distribution dynamics of four aphid populations setting their patchiness index as a dependent variable. Simulation results showed, for example, that the dynamics of wheat aphid populations were different from that of maize aphid populations. This model can be used for comparative studies of the spatial dynamics of aphids and of their natural enemies, to: (i) identify the dominant species of natural enemies; (ii) highlight the occurrence of interguild competition; and (iii) screen wheat cultivars for resistance to aphids.

Recently, a novel modelling approach based on the butterfly catastrophe theory was used to build a wheat aphid population dynamics model as a function of four controlling factors (natural enemies, weather, pesticides and carrying capacity) with satisfactory results (Wu *et al.*, 2014). Sensitivity analysis of the four factors allowed prediction of the situations in which a catastrophic increase in wheat aphid populations can be prevented.

To improve monitoring and modelling performance, increased knowledge of the population ecology of wheat pests is necessary. In organic farming, where either weeds or wild plants living in semi-natural habitats adjacent to the field are expected to be more abundant than in conventional farming, one important issue is to understand if wild plant species and habitats may favour pest establishment on wheat. Vialatte *et al.* (2005) showed that the genetic structure of *S. avenae* populations found on winter cereal crops and on associated wild plants in western France was highly diverse, indicating limited gene flow between the field and its margins. However, plants belonging to the same tribe (*Triticeae*, *Poeae* and *Aveneae*) showed closer genetic relatedness, indicating that aphids tend not to specialize on a single host but rather on a group of related plant species. This suggests that reducing populations of grass weeds in and around a wheat field, which is more likely to occur in diversified organic crop rotations than in simplified conventional ones, would in turn reduce aphid outbreaks in the crop.

Field studies on the spatial distribution of wheat pests are expected to improve the estimates of spatially explicit models and decision support systems. Winder *et al.* (2013) studied the within-canopy spatial distribution of the aphids *S. avenae* and *M. dirhodum* in southern England, with a special focus on estimating their availability to ground-active predators. Aphid spatial pattern was associated with crop nitrogen uptake or crop yield, but the two species showed opposite trends: *S. avenae* was negatively associated with yield or plant nitrogen whereas *M. dirhodum* showed positive association. For both aphid species, the spatial pattern of individuals falling to the ground may modi ate the effectiveness of ground-active predators as pest control agents.

Approaches to Pest Management in Organic Wheat

Prevention is recognized as the core element of crop protection against weeds, diseases and pests in organic farming. Compared with other biotic aggressors, approaches to pest control are characterized by a stronger focus on the landscape scale, due to the intrinsic higher mobility of arthropods. Prevention of pest damage to wheat can be pursued either by: (i) enhancing the crop ability to suppress, resist and/or tolerate pests attacks; or (ii) enhancing the activity of natural enemies through optimized cropping system design and landscape configuration. Integration between these two approaches can increase the effectiveness of prevention strategies in organic wheat, thus facilitating successful crop protection.

Hereafter in this chapter the main domains of pest management in organic wheat will be reviewed. First, habitat and landscape diversity related to conservation biological control are discussed. Second, the effect of cropping system design on pests in wheat is addressed. Within cropping systems, particular attention is dedicated to management diversity (i.e. the diversity of management tactics and their effects on wheat pests) and to the inclusion of host resistance in breeding and host genetic diversity in cropping systems. Finally, an insight on direct pest control measures applicable in organic wheat is provided.

Habitat diversity and conservation biological control

A wheat field is expected to host several arthropods that can act as natural enemies of wheat pests. These natural enemies can act as predators or parasitoids and can show a ground-dwelling or flying habit. Spiders (Araneae) and carabid and staphylinid beetles (Coleoptera: Carabidae and Staphylinidae) are the most important taxa of ground-dwelling generalist predators. Flying predators are represented by, for example, coccinellid beetles (Coleoptera: Coccinellidae) and syrphids (Diptera: Syrphidae). Flying parasitoids generally belong to the Diptera and Hymenoptera orders, and special importance in cereal aphids control is attributed to parasitic wasps, especially Hymenoptera belonging to the Apidiinae subfamily of Braconidae. Among all these groups, Schmidt *et al.* (2003) showed that parasitic wasps have the strongest effect on the three aphid species *S. avenae*, *M. dirhodum* and *R. padi* in insecticide-free winter wheat fields. They also proved that co-occurring parasitoids and predators' activity do not overlap but seem to act in synergy. The role of polyphagous predators, in turn, can significantly reduce aphid infestations in wheat, as demonstrated on the spider *Pardosa lugubris* (Walckenaer) (Araneae: Lycosidae) against *R. padi*, especially early in the growing season and in the presence of alternative preys (Oelbermann and Scheu, 2009).

Conservation of natural enemies of wheat pests is a crucial aspect of effective pest control in organic wheat. It is widely accepted that diversified agricultural landscapes are more likely to support biological pest control than simplified landscapes (Bianchi *et al.*, 2006). However, this statement is far too generic to be translated into concrete management options that can ensure a higher and more stable pest control service with greater buffer capacity. As a matter of fact, at least two aspects have to be considered in detail: (i) which aspect of landscape diversification is concretely correlated with improved pest control; and (ii) the effect that landscape diversification can have not only on natural enemies but also on pests and on, for example, hyperparasitoids (that may counteract the positive effect of parasitoids of wheat aphids).

Plećaš *et al.* (2014) addressed these two issues by comparing the effect of three contrasting landscape configurations on cereal aphid–parasitoid–hyperparasitoid interactions: (i) landscape complexity addressed as the proportion of crop to non-crop area in circular-shaped portions of land; (ii) configurational heterogeneity addressed as small versus large field landscapes; and (iii) edge effects of different field-margin types. They found a positive effect of landscape complexity, a variable effect of field-margin vegetation and a neutral effect of landscape configuration heterogeneity on the biological control of aphids.

Roschewitz *et al.* (2005) suggested that high landscape complexity with the presence of undisturbed perennial vegetation can support parasitoids but also cereal aphid populations, as the alternative hosts of the main wheat aphids were mainly present in non-cropped areas. According to these authors, a key issue to disentangle these complex interactions is the spatial scale at which landscape complexity is relevant. The lower dispersal ability of parasitoids compared with aphids made them favoured by landscape complexity within a 0.5–2 km radius from the focal field. Instead, wheat aphids showed a significant relationship with landscape complexity at higher scales (1–6 km), suggesting that effective landscape manipulation can be driven by the identification of a 'functional spatial scale' (Thies *et al.*, 2005). Zhao *et al.* (2014) extended this concept to the hyperparasitoid trophic level, showing that parasitoid and hyperparasitoid diversity and activity increased with landscape complexity, but hyperparasitoids were more scale-sensitive than primary parasitoids.

Habitat diversity can be manipulated both at a landscape and at a field scale to improve conservation biological control through well-targeted agrienvironmental schemes. For instance, Holland *et al.* (2012) showed that increasing the proportion of linear grass margins – one of the most widespread agrienvironmental options available to UK farmers – enhanced the activity of both ground-dwelling and flying predators against cereal aphids. Similarly, Levie *et al.* (2005) demonstrated the effectiveness of red clover (*Trifolium pratense* L.) + perennial ryegrass (*Lolium perenne* L.) grass strips sown along wheat fields as reservoirs of parasitoids and predators and their consequent effect in reducing aphid populations. Maintaining overwintering habitats for polyphagous predators – the so-called 'beetle banks' – was proved to reduce *S. avenae* populations in a wheat field experiment hosting a grass strip sown with a mixture of the grass species *Dactylis glomerata* L. and *Holcus lanatus* L. (Collins *et al.*, 2002).

Cropping systems

Compared with conventional systems, organic arable systems are commonly characterized by the lack of use of chemical insecticides and by more diversified crop rotations in space and time. When it comes to pests, there are two common expectations in organic cropping systems. On the one hand, the absence of chemical pest control is perceived as a risk of pest proliferation. While on the other hand, both the absence of chemical control and the intrinsic higher crop and plant diversity harboured by organic fields is expected to enhance the activity of natural enemies and hence to reduce pest abundance. The first expectation has been contradicted by Gosme *et al.* (2012), who showed that the presence of organic fields in an agricultural landscape actually decreased aphid abundance in both organic and conventional neighbouring fields. Lohaus *et al.* (2013) showed that an organic and a conventional winter wheat field differed in the food web structure linking aphids with their primary and secondary parasitoids, despite small differences in total richness of each trophic group. Surprisingly, aphids and primary parasitoids showed greater intensity and evenness of interactions in conventional wheat fields, with cascade interactions at the next trophic level. Evidence from this study does not support the general hypothesis that organic farming practices should encourage species richness and food web complexity.

It should be noted that not all the aphids living in cereal fields are crop pests, since some live on weeds that may serve as alternative hosts or prey for aphid parasitoids or predators. This makes aphid–natural enemies interactions even more complex. A study carried out in Catalonia (Caballero-López *et al.*, 2012) showed that conventional wheat fields, where weeds were scarce, were dominated by grass aphids (mainly pests), while organic wheat fields, where weeds and legumes were more abundant, showed considerable presence of forb aphids (not to be considered pests in this context). In conventional fields, aphid parasitoids were encouraged by the presence of aphid pests, while in organic fields aphid parasitoids and predators were mainly encouraged by the indirect effect of legumes. Abundance and typology of plant species – including weeds – occurring in agroecosystems are therefore important drivers of pest–natural enemies interactions.

Further insights into the distribution of basal resources affecting pest–natural enemies interactions in wheat-based conventional and organic systems were provided by a stable isotope study carried out in Switzerland by Birkhofer *et al.* (2011) within the context of the DOK long-term trial (a trial looking at the long-term effects of biodynamic, bio-organic and conventional farming systems on soil conditions, yield and product quality).

These authors found that management system effects were species specific. In the organic system, generalist predators consumed greater proportions of herbivore prey. Starvation and intraguild predation rates increased in some predator species with time. The species specificity of effects made these authors suggest that the best 'insurance' strategy to conserve the biological pest control service in the long term would be to maintain a diverse natural enemy community including species with different phenology and sensitivity to management practices. In this respect, organic management was more promising than conventional management because it enhanced the presence of microbial and faunal decomposers, whose effect propagated into the aboveground system via the abundance of generalist predators increasing conservation biological control (Birkhofer *et al.*, 2008). It should be noted that in the DOK trial wheat grain and straw yields were 23% higher in conventional systems receiving mineral fertilizers and herbicides, clearly showing the trade-off between productivity and agroecosystem health.

Increasing the diversity of crops in wheat fields through the inclusion of adequate companion species can enable mechanisms that prevent pests from attacking wheat. An effective example is the intercropping of wheat with garlic (*Allium sativum* L.), that was able to reduce *S. avenae* infestations thanks to both garlic repellent-emitting properties and an increased density of natural enemies, suggesting the existence of a 'push–pull'-like effect (Zhou *et al.*, 2013). A trap-crop effect of intercropped lucerne (*Medicago sativa* L.) in wheat was demonstrated by Saeed *et al.* (2013), as lucerne was able to feed aphid populations earlier than wheat, thus attracting aphid natural enemies. As early as 1928, Seamans developed a detailed management protocol for establishing grass trap-crop boundaries alongside wheat fields using, for example, smooth brome (*Bromus inermis* Leyss.) and rye (*Secale cereale* L.). These plants extrude ears earlier than wheat and can then easily attract adults of the wheat stem sawfly, *Cephus cinctus* (Norton), as they 'become suitable for oviposition earlier than the main crop of wheat and therefore receive the bulk of the eggs' (Seamans, 1928), and can subsequently be removed before larvae can mature in the stubble.

Management diversity

Besides being based on more diversified crop rotations in space and time, organic systems usually also have a greater level of management diversity than conventional systems. Management diversity is due partly to the different agronomic requirements needed by the higher number of crops present in organic rotations, and partly to the more frequent management adaptations required in a less standardized system. It can be hypothesized that a greater level of management diversity should turn into reduced pest pressure in wheat. However, a comprehensive review by Glen (2000) actually showed a very complex picture in which different management practices may have contradictory effects on different wheat pests. As an example, variation in wheat sowing dates may have opposite effects on aphid infestation and on the wheat bulb fly, *Delia coarctata* (Fallen). Therefore, the link between management diversity and reduced pest pressure needs to be better circumstantiated by clear experimental evidence.

Crop fertilization is one of the management practices most likely to influence pest pressure due to its effect on the concentration of plant elements, especially N. In continental Europe, Lohaus and Vidal (2013) tested the hypothesis of lower aphid abundance under organic fertilization in a 4-year study. Two N intensities (legume-based fertilization and legume-based plus organic liquid manure), typical of organic systems, were compared with a high (mineral) N intensity, typical of conventional systems. The authors found that aphid species were selected by the different N fertilization treatment. Legume-based organic wheat showed clear dominance of *S. avenae* (89–96% of the total aphid community), whereas manure-fertilized organic wheat and conventional wheat were mainly infested by *M. dirhodum* and *R. padi*. Total aphid abundance was positively correlated with wheat grain N content in the latter two systems but not in legume-based organic wheat. Despite lower total aphid abundance, this system may result in greater crop yield reduction because *S. avenae* is usually the most troublesome aphid species in wheat.

Evidence from Ke and Scheu (2008) confirmed that insect pest pressure in wheat can be indirectly affected by the community of decomposers resulting from the effect of management practices on soil organic matter distribution in soil. In a greenhouse study, these authors compared the effect of three types of 15N-labelled litter distribution – simulating mulching, ploughing and disking – on a wheat–aphid model system. Earthworms, Collembola and litter distribution interacted to affect wheat growth, N uptake and aphid (*R. padi*) development. Interestingly, by altering the distribution of litter, earthworms reduced infestation of crops by aphids via reduced wheat capture of litter N, especially when litter was concentrated deeper in the soil (simulated ploughing). The main take-home message from this study is that management practices promoting a continuous moderate release of nutrients from soil organic matter should result in maximum plant growth with minimum plant pest infestation.

Enhanced biological pest control can also occur via appropriate compost applications. Results from a 4-year field-scale manipulative study (Bell *et al.*, 2008) indicated that compost may induce both direct and indirect effects on wheat aphids and their predators. Although not in all years, compost application significantly increased the abundance of predators, with consequent lower aphid presence. Instead, alternative preys always responded strongly to compost application. However, using compost as a management practice that consistently fosters biological pest control would only be possible once the finer mechanisms linking organic matter type, compost application time and rate, and environmental conditions are fully elucidated.

Tillage is also likely to influence the occurrence of pests and natural enemies in wheat. In Slovakia, Gallo and Pekár (2001) found that while most wheat pests were favoured by deep ploughing, the majority of natural enemies occurred in higher densities under shallow ploughing. The type of preceding crop (silage maize or pea), instead, did not affect the abundance of either pests or natural enemies. However, organically managed wheat reduced the overall abundance of insect pests as compared with integrated wheat.

Sometimes, simple practices can prove effective in enhancing biological pest control in wheat. A wheat crop grown on bare soil may be more exposed to pest attack due to reduced abundance of ground-dwelling predators. This problem may easily be solved by mulching, as demonstrated by Schmidt *et al.* (2004), since a surface straw layer can provide a more suitable habitat for, and thus enhance the density of, spiders.

Pests other than aphids may at times become problematic in organic wheat. Management measures to keep them under control have been reviewed by Weisz *et al.* (2014) for the typical growing conditions of North Carolina. For example, reasonable control of the cereal leaf beetle, *Oulema melanopus* (L.), can be expected by ensuring good tillering of the crop through timely sowing, use of high quality seed, adequate pre-plant fertility, and split N applications in late winter.

The Hessian fly, *Mayetiola destructor* (Say), has become more frequent in North Carolina due to the increased use of no-till and of wheat as either a cover crop or a habitat for hunted birds (Weisz *et al.*, 2014). Reducing the presence of wheat stubble across the crop rotation is highly likely to keep this pest at bay. Since the Hessian fly is a weak flier, increasing the distance between nearby wheat fields to > 400 m would further contribute to reducing pest infestations. Furthermore, stubble cultivation by disking would kill this pest. Stubble burning, instead, is not thought to be so effective because many pupae located underneath the soil surface would survive. Use of cereal cover crops other than wheat (e.g. oats, rye and/or triticale) would reduce pest populations because these crops are not hosts to this pest. In some areas of the USA it is suggested that farmers sow wheat only after the first seasonal frost has occurred, because this would kill the Hessian fly adults. In case of expected high pest pressure, sowing a wheat variety that is resistant to the Hessian fly biotype-L would be an important preventive measure.

Although in most cases it would be difficult to find clear-cut cause–effect relationships between individual management practices and pest control, adequate cropping system planning based on management diversity can be crucial to ensure in-field expression of the biological pest control agroecosystem service. Further research is undoubtedly needed to unravel the effects of individual management practices on pest occurrence and pest–natural enemies interactions in organic wheat.

Host plant resistance and genetic diversity

Prevention of insect pest outbreaks and subsequent damage through host resistance is considered one of the most cost-effective approaches in both organic and conventional agriculture. Genetic variation for host resistance against pests was identified in vast collections of wheat germplasm, and molecular bases for antixenosis- and antibiosis-based host resistance have been identified in common wheat against, for example, the wheat Russian aphid, *Diuraphis noxia* (Mordvilko)) and the greenbug (*S. graminum*) (Castro *et al.*, 2005; El Bouhssini *et al.*, 2011), thus enabling breeding for pest-resistant wheat cultivars.

However, breeding for pest-resistant wheat is subject to unpredictability driven by genetic diversity of pest populations and, consequently, by the variability in pest traits related to pest–host interactions. Therefore, breeding for insect-resistant wheat cultivars for organic agriculture should also explore the genetic basis of phenotypic plasticity linked to resistance and its interaction with the biotic environment (Tétard-Jones and Leifert, 2011). Variable pest population response to host resistance is a further disturbing element. As shown by Hawley *et al.* (2003), antibiosis and antixenosis can effectively reduce wheat Russian aphid damage on wheat in the short term, but care must be taken in the long term as resistant biotypes may be selected within aphid populations.

The effect of preventive approaches based on homogeneous resistant host crops appears limited since it can be overcome by the complex interactions occurring between a homogeneous host population and genetically heterogeneous pest populations. Research should also better explore the effect of host genetic diversity on pest reduction, based for example on the remarkable advantages of using wheat cultivar mixtures in low-input agriculture (Kaut *et al.*, 2009) and for disease reduction (Finckh, 2008). This effect was clearly highlighted by Tooker and Frank (2012), who focus on the relationships between genetically heterogeneous hosts, pests and their natural enemies.

Many studies have tested the hypothesis that the use of pest-resistant wheat cultivars results in reduced activity of natural enemies, finding interesting results. For example, it was observed that aphid-resistant wheat cultivars can enhance the parasitization of *S. avenae* by *Aphidius* spp. (Cai *et al.*, 2009). Fuentes-Contreras and Niemayer (1998) studied host acceptance and suitability of wheat cultivars containing high or low levels of hydraxomic acids (Hx). Aphids (*S. avenae*) feeding on high-Hx plants were smaller, showed a slower growth rate and were more exposed to parasitoid (*Aphidius rhopalosiphi* (De Stefani-Perez)) oviposition. In a more comprehensive experiment, Fuentes-Contreras and Niemayer (2000) obtained significant reduction of *S. avenae* population growth rates combining wheat resistance, parasitoids (*A. rhopalosiphi*) and enthomopathogenic fungi (*Pandora neoaphidis* (Remaudiere *et* Hennebert)). Fuentes-Granados *et al.* (2001) found no negative effects on the parasitoid *Lysiphlebus testaceipes* (Cresson) (Hymenoptera: Aphidiidae) oviposition parameters and a little delay in time for adult emergence (unlikely to alter pest–parasitoid interaction) when the host *Schizaphis graminum* was feeding on a resistant cultivar.

Evidence from these studies show that use of host resistance and host genetic diversity is a promising strategy to improve pest management in organic wheat. Nevertheless, the potential of this strategy is fully expressed when it is integrated in a more comprehensive approach encompassing all components of the agroecosystem.

Direct methods

Direct pest control methods have less importance in organic than in conventional agriculture, as chemical insecticides are excluded. However, direct measures to reduce pest incidence and damage can be useful when embedded in an appropriate system. Available measures rely on the use of insecticides permitted in organic farming and/or on employing compounds (semiochemicals) able to alter the pest's behaviour towards host plants.

Weisz *et al.* (2014) provided examples of how approved insecticides can be useful in particular situations, such as significant infestation of the North American armyworm, *Mythimna unipuncta* (Haworth), that can infest small-grain crops, especially wheat, from late April to mid-May. Organic growers have the choice of accepting armyworms feeding on their crop or using an insecticide such as a spinosad in emergency situations. However, it should be stressed that in the case of the cereal leaf beetle (*O. melanopus*) spinosad can only provide adequate control of light infestations, but is not that effective when pest populations are high.

The use of semiochemicals for direct pest control is an emerging method that relies on mimicking the production of plant defences against pests. Piesik *et al.* (2010) observed that: (i) mechanically damaged wheat plants or plants damaged by *O. melanopus* released increased amounts of volatile organic compounds (VOCs); and (ii) uninjured wheat plants increased the emission of VOCs when positioned near other plants which were mechanically or insect damaged, thus demonstrating the validity of this approach. For example, it has been proved that the plant semiochemical *cis*-jasmone induces plant resistance that deters herbivores and attracts natural enemies (Delaney *et al.*, 2013). This compound was studied in detail by Bruce *et al.* (2003), who demonstrated its repellent properties against winged forms of *S. avenae*. Wheat seedlings sprayed with formulated *cis*-jasmone were less susceptible to attack by *S. avenae* than control plants. Furthermore, *cis*-jasmone field applications resulted in reduced cereal aphid populations infesting wheat.

Results of some experiments also suggest that the use of semiochemicals can be more effective when included in a systems approach. For example, Prinsloo *et al.* (2007) found an interaction between wheat cultivar and semiochemicals (volatile substances from plant essential oils, methyl salicylate, 1,8-cineole and menthol) application in reducing *D. noxia* density, since the application was successful on an aphid-resistant cultivar but not on a susceptible one. Dong *et al.* (2012) found that the combination of wheat intercropping with oilseed rape (*Brassica napus* L. var. *oleifera* DC) and the release of methyl salicylate was more effective in reducing the density of *S. avenae* and increasing the number of ladybird beetles than each of the two practices applied alone.

To date, direct pest control in organic wheat offers a limited range of potentially effective solutions, which are mainly useful in case of emergency or peculiar situations. These measures can be fully effective when adequate landscape and cropping system design have been able to reduce pest pressure on wheat, thus it can play a role in an 'ultimate' IPM perspective, where application of a systems approach is to be considered a 'must' (Ratnadass and Barzman, 2014).

Conclusions and Perspectives

Although pest management is unlikely to be the most important technical problem in many organic wheat-growing areas, appropriate management of wheat-based cropping systems would contribute to minimize the risk of serious pest outbreaks. In this respect, (re)design of crop management systems upon the agroecological approach by making full use of functional biodiversity across multiple spatial and temporal scales seems a promising approach.

In entomology, functional biodiversity is often considered a synonym of conservation biological control (e.g. in an IOBC (International Organization for Biological and Integrated Control) perspective), but a more comprehensive definition (see e.g. Moonen and Bàrberi, 2008) would actually encompass many more solutions than just habitat management to encourage the presence and activity of autochthonous natural enemies. These solutions span across the three domains of: (i) *genetic diversity* (e.g. resistant cultivars); (ii) *species diversity* (e.g. appropriate crop rotations, cover crops and intercrops); and (iii) *habitat diversity* (e.g. semi-natural areas and their spatial arrangement at field, farm and/or landscape scale or the presence of deterring crops like rape planted close to wheat) (Li *et al.*, 2011).

The actual mechanisms leading to effective reduction or management of pest populations in organic wheat are likely to differ among contexts. These mechanisms can be grouped in three categories that may facilitate agroecosystem design (Costanzo and Bàrberi, 2014):

- *functional identity*, i.e. the presence of a set of homogeneous phenotypic traits that are related to the expression of the biological pest control service (e.g. a resistant cultivar, a repellent cover crop or a nettle, *Urtica dioica* L., strip) (Alhmedi *et al.*, 2006);
- *functional composition*, i.e. the complementary effect of different traits, expressed by co-occurring elements, on the provision of the biological pest control service (e.g. in variety mixtures or intercrops like wheat–garlic) (Zhou *et al.*, 2013);
- *functional diversity* (*sensu stricto*), i.e. the direct effect of heterogeneity within the crop stand on the expression of the biological pest control service (e.g. in wheat composite cross populations).

This is an interesting and novel field of research which is likely to provide a better understanding of crop–pest–natural enemies interactions and consequently to highlight the most appropriate solutions for pest management in organic wheat in different agroclimatic zones.

References

Aggarwal, P.K., Banerjee, B., Daryaei, M.G., Bhatia, A., Bala, A., Rani, S., Chander, S., Pathak, H. and Kalra, N. (2006) InfoCrop: a dynamic simulation model for the assessment of crop yields, losses due to pests, and environmental impact of agro-ecosystems in tropical environments. II. Performance of the model. *Agricultural Systems* 89, 47–67.

Alhmedi, A., Haubruge, E., Bodson, B. and Francis, F. (2006) Inter- and intra-guild interactions related to aphids in nettle (*Urtica dioica* L.) strips closed to field crops. *Communications in Agricultural and Applied Biological Sciences* 71, 413–423.

Bell, J.R., Traugott, M., Sunderland, K.D., Skirvin, D.J., Mead, A., Kravar-Garde, L., Reynolds, K., Fenlon, J.S. and Symondson, W.O.C. (2008) Beneficial links for the control of aphids: the effects of compost applications on predators and prey. *Journal of Applied Ecology* 45, 1266–1273.

Bianchi, F.J., Booij, C.J. and Tscharntke, T. (2006) Sustainable pest regulation in agricultural landscapes: a review of landscape composition, biodiversity and natural pest control. *Proceedings of the Royal Society B: Biological Sciences* 273, 1715–1727.

Birkhofer, K., Bezemer, T.M., Bloem, J., Bonkowski, M., Christensen, S., Dubois, D., Ekelund, F., Fließbach, A., Gunst, L., Hedlund, K., Mäder, P., Mikola, J., Robin, C., Setälä, H., Tatin-Froux, F., Van der Putten, W.H. and Scheu, S. (2008) Long-term organic farming fosters below and aboveground biota: implications for soil quality, biological control and productivity. *Soil Biology and Biochemistry* 40, 2297–2308.

Birkhofer, K., Fließbach, A., Wise, D.H. and Scheu, S. (2011) Arthropod food webs in organic and conventional wheat farming systems of an agricultural long-term experiment: a stable isotope approach. *Agricultural and Forest Entomology* 13, 197–204.

Birzer, T. and Badgery, W. (2006) Organic wheat farming in Australia: issues farmers face. *Australian Farm Business Management Journal* 3, 1–13.

Bruce, T.J.A., Martin, J.L., Pickett, J.A., Pye, B.J., Smart, L.E. and Wadhams, L.J. (2003) *Cis*-jasmone treatment induces resistance in wheat plants against the grain aphid, *Sitobion avenae* (Fabricius) (Homoptera: Aphididae). *Pest Management Science* 59, 1031–1036.

Caballero-López, B., Blanco-Moreno, J.M., Pérez-Hidalgo, N., Michelena-Saval, J.M., Pujade-Villar, J., Guerrieri, E., Sánchez-Espigares, J.A. and Sans, F.X. (2012) Weeds, aphids, and specialist parasitoids and predators benefit differently from organic and conventional cropping of winter cereals. *Journal of Pest Science* 85, 81–88.

Cai, Q.N., Ma, X.M., Zhao, X., Cao, Y.Z. and Yang, X.Q. (2009) Effects of host plant resistance on insect pests and its parasitoid: a case study of wheat-aphid-parasitoid system. *Biological Control* 49, 134–138.

Castro, A.M., Vasicek, A., Manifiesto, M., Giménez, D.O., Tacaliti, M.S., Dobrovolskaya, O., Röder, M.S., Snape, J.W. and Börner, A. (2005) Mapping antixenosis genes on chromosome 6A of wheat to greenbug and to a new biotype of Russian wheat aphid. *Plant Breeding* 124, 229–233.

Ceccarelli, S., Valkoun, J., Erskine, W., Weigand, S., Miller, R. and Van Leur, J.A.G. (1992) Plant genetic resources and plant improvement as tools to develop sustainable agriculture. *Experimental Agriculture* 28, 89.

Chakraborty, S. and Newton, A.C. (2011) Climate change, plant diseases and food security: an overview. *Plant Pathology* 60, 2–14.

Collins, K.L., Boatman, N.D., Wilcox, A. and Holland, J.M. (2002) Influence of beetle banks on cereal aphid predation in winter wheat. *Agriculture, Ecosystems and Environment* 93, 337–350.

Costanzo, A. and Bàrberi, P. (2014) Functional agrobiodiversity and agroecosystem services in sustainable wheat production. A review. *Agronomy for Sustainable Development* 34, 327–348.

Delaney, K.J., Wawrzyniak, M., Lemańczyk, G., Wrzesińska, D. and Piesik, D. (2013) Synthetic *cis*-jasmone exposure induces wheat and barley volatiles that repel the pest cereal leaf beetle, *Oulema melanopus* L. *Journal of Chemical Ecology* 39, 620–629.

Dong, J., Liu, Y.J., Li, P.L., Lin, F.J., Chen, J.L. and Liu, Y. (2012) Ecological effects of wheat–oilseed rape intercropping combined with methyl salicylate release on *Sitobion avenae* and its main natural enemies. *Chinese Journal of Applied Ecology* 23, 2843–2848.

El Bouhssini, M., Ogbonnaya, F.C., Ketata, H., Mosaad, M.M., Street, K., Amri, A., Keser, M., Rajaram, S., Morgounov, A., Ihawi, F., Dabus, A. and Smith, C.M. (2011) Progress in host plant resistance in wheat to Russian wheat aphid (Hemiptera: Aphididae) in North Africa and West Asia. *Australian Journal of Crop Science* 5, 1108–1113.

Finckh, M.R. (2008) Integration of breeding and technology into diversification strategies for disease control in modern agriculture. *European Journal of Plant Pathology* 121, 299–409.

Freier, B., Roßberg, D., Gosselke, U. and Triltsch, H. (2002) Evaluation of beneficial effects in wheat fields by using computer simulation. *Gesunde Pflanzen* 54, 188–193.

Fuentes-Contreras, E. and Niemeyer, H.M. (1998) Dimboa glucoside, a wheat chemical defense, affects host acceptance and suitability of *Sitobion avenae* to the cereal aphid parasitoid *Aphidius rhopalosiphi*. *Journal of Chemical Ecology* 24, 371–381.

Fuentes-Contreras, E. and Niemeyer, H.M. (2000) Effect of wheat resistance, the parasitoid *Aphidius rhopalosiphi*, and the entomopathogenic fungus *Pandora neoaphidis*, on population dynamics of the cereal aphid *Sitobion avenae*. *Entomologia Experimentalis et Applicata* 97, 109–114.

Fuentes-Granados, R.G., Giles, K.L., Elliott, N.C. and Porter, D.R. (2001) Assessment of greenbug-resistant wheat germplasm on *Lysiphlebus testaceipes* Cresson (Hymenoptera: Aphidiidae) oviposition and development in greenbug over two generations. *Southwestern Entomologist* 26, 187–194.

Gallo, J. and Pekár, S. (2001) Effect of ploughing and previous crop on winter wheat pests and their natural enemies under integrated farming system in Slovakia. *Anzeiger fur Schadlingskunde* 74, 60–65.

Glen, D.M. (2000) The effects of cultural measures on cereal pests and their role in integrated pest management. *Integrated Pest Management Reviews* 5, 25–40.

Gosme, M., de Villemandy, M., Bazot, M. and Jeuffroy, M.H. (2012) Local and neighbourhood effects of organic and conventional wheat management on aphids, weeds and foliar diseases. *Agriculture Ecosystems & Environment* 161, 121–129.

Hawley, C.J., Peairs, F.B. and Randolph, T.L. (2003) Categories of resistance at different growth stages in Halt, a winter wheat resistant to the Russian wheat aphid (Homoptera: Aphididae). *Journal of Economic Entomology* 96, 214–219.

Holland, J.M., Oaten, H., Moreby, S., Birkett, T., Simper, J., Southway, S. and Smith, B.M. (2012) Agri-environment scheme enhancing ecosystem services: a demonstration of improved biological control in cereal crops. *Agriculture Ecosystems & Environment* 155,147–152.

Hulme, P.E. (2009) Trade, transport and trouble: managing invasive species pathways in an era of globalization. *Journal of Applied Ecology* 46, 10–18.

International Federation of Organic Agriculture Movements (IFOAM) (2005) *Principles of Organic Agriculture*. IFOAM, Bonn, Germany.

Kaut, A.H.E.E., Mason, H.E., Navabi, A., O'Donovan, J.T. and Spaner, D. (2009) Performance and adaptability of performance of spring wheat variety mixtures in organic and conventional management systems in western Canada. *Journal of Agricultural Science* 147, 141–153.

Ke, X. and Scheu, S. (2008) Earthworms, Collembola and residue management change wheat (*Triticum aestivum*) and herbivore pest performance (Aphidina: *Rhophalosiphum padi*). *Oecologia* 157, 603–617.

Levie, A., Legrand, M.-A., Dogot, P., Pels, C., Baret, P.V. and Hance, T. (2005) Mass release of *Aphidius rhopalosiphi* (Hymenoptera: Aphidiinae), and strip management to control of wheat aphids. *Agriculture Ecosystems & Environment* 105, 17–21.

Li, C., Wu, W.Q., Zhu, L., Zhang, Q.W. and Liu, X.X. (2011) Effects of wheat planted adjacent to rape on the major pests and their natural enemies in wheat field. *Chinese Journal of Applied Ecology* 22, 3371–3376.

Li, D., Zhao, H. and Hu, X. (2010) A model to describe the spatio-temporal distribution dynamics of aphid populations. *Shengtai Xuebao/Acta Ecologica Sinica* 30, 4986–4992.

Lohaus, K. and Vidal, S. (2013) Nitrogen supply in conventional versus organic farming systems: effects on the performance of cereal aphids. *Organic Agriculture* 3, 129–139.

Lohaus, K., Vidal, S. and Thies, C. (2013) Farming practices change food web structures in cereal aphid–parasitoid–hyperparasitoid communities. *Oecologia* 171, 249–259.

Masters, G. and Norgrove, L. (2010) Climate change and invasive alien species. *CABI Working Paper* 1, 30 pp.

Moonen, A.C. and Bàrberi, P. (2008) Functional biodiversity: an agroecosystem approach. *Agriculture, Ecosystems & Environment* 127, 7–21.

Oelbermann, K. and Scheu, S. (2009) Control of aphids on wheat by generalist predators: effects of predator density and the presence of alternative prey. *Entomologia Experimentalis et Applicata* 132, 225–231.

Paoletti, M.G., Tsitsilas, A., Thomson, L.J., Taiti, S. and Umina, P.A. (2008) The flood bug, *Australiodillo bifrons* (Isopoda: Armadillidae): a potential pest of cereals in Australia? *Applied Soil Ecology* 39, 76–83.

Pelosi, C., Toutous, L., Chiron, F., Dubs, F., Hedde, M., Muratet, A., Ponge, J.F., Salmon, S. and Makowski, D. (2013) Reduction of pesticide use can increase earthworm populations in wheat crops in a European temperate region. *Agriculture Ecosystems & Environment* 181, 223–230.

Piesik, D., Łyszczarz, A., Tabaka, P., Lamparski, R., Bocianowski, J. and Delaney, K.J. (2010) Volatile induction of three cereals: influence of mechanical injury and insect herbivory on injured plants and neighbouring uninjured plants. *Annals of Applied Biology* 157, 425–434.

Plećaš, M., Gagić, V., Janković, M., Petrović-Obradović, O., Kavallieratos, N.G., Tomanović, T., Thies, C., Tscharntke, T. and Ćetković, A. (2014) Landscape composition and configuration influence cereal aphid–parasitoid–hyperparasitoid interactions and biological control differentially across years. *Agriculture Ecosystems & Environment* 183, 1–10.

Prinsloo, G., Ninkovic, V., Van Der Linde, T.C., Van Der Westhuizen, A.J., Pettersson, J. and Glinwood, R. (2007) Test of semiochemicals and a resistant wheat variety for Russian wheat aphid management in South Africa. *Journal of Applied Entomology* 131, 637–644.

Puech, C., Baudry, J., Joannon, A., Poggio, S. and Aviron, S. (2014) Organic vs conventional farming dichotomy: does it make sense for natural enemies? *Agriculture Ecosystems & Environment* 194, 48–57.

Ratnadass, A. and Barzman, M. (2014) Ecological intensification for crop protection. In: Ozier-Lafontaine, H. and Lesueur-Jannoyer, M. (eds) *Sustainable Agriculture Reviews 14: Agroecology and Global Change*. Springer, New York, pp. 53–81.

Roschewitz, I., Hücker, M., Tscharntke, T. and Thies, C. (2005).The influence of landscape context and farming practices on parasitism of cereal aphids. *Agriculture Ecosystems & Environment* 108, 218–227.

Saeed, Q., Zaka, M., Saeed, S. and Bakhtawar, M. (2013) Lucerne as trap crop in wheat for development of predators population against wheat aphids (Aphididae: Homoptera). *Pakistan Journal of Zoology* 45, 193–196.

Schmidt, M.H., Lauer, A., Purtauf, T., Schaefer, M. and Tscharntke, T. (2003) Relative importance of predators and parasitoids for cereal aphid control. *Proceedings of the Royal Society B: Biological Sciences* 270, 1905–1909.

Schmidt, M.H., Thewes, U., Thies, C. and Tscharntke, T. (2004) Aphid suppression by natural enemies in mulched cereals. *Entomologia Experimentalis et Applicata* 113, 87–93.

Seamans, H.L. (1928) The value of trap crops in the control of the wheat stem sawfly in Alberta. *59th Annual Report of the Entomological Society of Ontario* pp. 59–64.

Shahrokhi, S. and Amirmaafi, M. (2012) Optimum sampling plan for the population of English grain aphid, *Sitobion avenae* (F.) (Hem.: Aphididae) on wheat in Iran. *Journal of Food, Agriculture and Environment* 10, 401–403.

Tétard-Jones, C. and Leifert, C. (2011) Plasticity of yield components of winter wheat in response to cereal aphids. *NJAS – Wageningen Journal of Life Sciences* 58, 139–143.

Thies, C., Roschewitz, I. and Tscharntke, T. (2005) The landscape context of cereal aphid–parasitoid interactions. *Proceedings of the Royal Society B: Biological Sciences* 272, 203–210.

Tooker, J.F. and Frank, S.D. (2012) Genotypically diverse cultivar mixtures for insect pest management and increased crop yields. *Journal of Applied Ecology* 49, 974–985.

Vialatte, A., Dedryver, C.A., Simon, J.C., Galman, M. and Plantegenest, M. (2005) Limited genetic exchanges between populations of an insect pest living on uncultivated and related cultivated host plants. *Proceedings of the Royal Society B: Biological Sciences* 272, 1075–1082.

Weisz, R., Couger, C. and Reisig, D. (2014) Crop production management – organic wheat and small grains. In: *North Carolina Organic Grain Production Guide*, Chapter 4. Available at: http://content.ces.ncsu.edu/chapter-4-crop-production-management-organic-wheat-and-small-grains/ (accessed 31 March 2015).

Willer, H. and Lernoud, J. (eds) (2014) *The World of Organic Agriculture. Statistics and Emerging Trends 2014*, revised version of 24 February 2014. FiBL-IFOAM Report. Research Institute of Organic Agriculture (FiBL) and International Federation of Organic Agriculture Movements (IFOAM), Bonn, Germany.

Winder, L., Alexander, C.J., Woolley, C., Perry, J.N. and Holland, J.M. (2013) The spatial distribution of canopy-resident and ground-resident cereal aphids (*Sitobion avenae* and *Metopolophium dirhodum*) in winter wheat. *Arthropod–Plant Interactions* 7, 21–32.

Wu, W., Piyaratne, M.K.D.K., Zhao, H., Li, C., Hu, Z. and Hu, X. (2014) Butterfly catastrophe model for wheat aphid population dynamics: construction, analysis and application. *Ecological Modelling* 288, 55–61.

Yang, Z., Rao, M.N., Elliott, N.C., Kindler, S.D. and Popham, T.W. (2005) Using ground-based multispectral radiometry to detect stress in wheat caused by greenbug (Homoptera: Aphididae) infestation. *Computers and Electronics in Agriculture* 47, 121–135.

Yang, Z., Rao, M.N., Elliott, N.C., Kindler, S.D. and Popham, T.W. (2009) Differentiating stress induced by greenbugs and Russian wheat aphids in wheat using remote sensing. *Computers and Electronics in Agriculture* 67, 64–70.

Zhao, Z.H., Hui, C., Hardev, S., Ouyang, F., Dong, Z. and Ge, F. (2014) Responses of cereal aphids and their parasitic wasps to landscape complexity. *Journal of Economic Entomology* 107, 630–637.

Zhou, H.B., Chen, J.L., Liu, Y., Francis, F., Haubruge, E., Bragard, C., Sun, J.R. and Cheng, D.F. (2013) Influence of garlic intercropping or active emitted volatiles in releasers on aphid and related beneficials in wheat fields in China. *Journal of Integrative Agriculture* 12, 467–473.

15 Pest Management in Organic Vegetable Greenhouses

Carmelo Peter Bonsignore* and Vincenzo Vacante
Mediterranean University, Salita Melissari, Italy

Introduction

The ecological factors acting inside the greenhouse exert a variable influence on the bio-ecology of crops and on the animals associated with them. Usually, the development and the productivity of cultivated plants are positively stimulated, while the animal populations (pests and natural enemies) are influenced in various ways depending on the taxa and their evolutionary adaptations. So, different pests find the inside of the greenhouse a suitable environment in which to live and their populations can develop at high levels creating serious problems such that they need to be controlled, while that of their natural enemies, whether indigenous and/or released, may adopt different models, to the point that sometimes the development of populations of important beneficials is inconsistent. The phenomenon depends on various ecological factors (e.g. temperature, relative humidity (RH), short cultural cycle, homogeneity of crops, high density of plants, agronomic practices, choice of varieties, high use of fertilizers and high availability of food for pests) and defines a simple agroecosystem that is ecologically unstable, which from a practical point of view encourages the application of radical methods of control, such as the generalized use of pesticides.

Unfortunately, the risk of toxic residues on foodstuffs due to the shorter growing cycles and the scale of production of greenhouse crops, the resistance of pests to various active ingredients, and the high cost of chemical control does not encourage the use of pesticides (which is alarming on certain crops and world areas where the crops are food staples on which people depend) and highlights serious hygienic, ecological and economic contraindications. This scenario suggests the application of alternative methods of control, such as integrated pest management, biological control and organic farming. Among these strategies, organic farming offers greater assurance from an ecotoxicological and economic viewpoint.

The technical rules governing the application of pest control in organic farming and the practical means available have been extensively discussed in Chapters 1 and 2 of this volume, respectively. In this chapter the technical choices of pest control in organic vegetable greenhouses are discussed. The topic is introduced by a brief discussion on the general features of greenhouses

* Corresponding author: cbonsignore@unirc.it

(metal-glass, wood-plastic, etc.), and on the relationship of these features with the microclimate. This is to help the reader to better evaluate the factors that influence the development of pests and their natural enemies in this particular environment, with the prior aim to hinder the development of pests and to promote – when and where possible – that of beneficials.

Greenhouses: General Features and Microclimate

General features

The term greenhouse refers in general to a cultivation environment protected with glass or other material of a synthetic nature in which the average temperature level is higher than outdoors. The greenhouse is therefore a protective structure that allows farmers to manage the production cycles in order to mitigate the effects of the external climate conditions and to protect plants from various kinds of adversity, including in the broadest sense animals, weeds and fungi. Hence, in an agronomic sense, the greenhouse is a more suitable environment than open ground. However, a protective structure is not always capable of preventing crops from being attacked by pests and pathogens; although, often, the internal physical conditions and technical choices can facilitate the introduction and development of various natural enemies (predators and parasitoids) which can limit the action of such pests.

The greenhouse structure is generally made of metal or wood. The cover material that creates the greenhouse can be glass or different kinds of plastic. In the first case, the glass cover defines the type of greenhouse called a 'glasshouse', while the plastic cover defines 'plastic houses' or 'protected crops'. A further type of protected environment are 'tunnels', which are similar to plastic houses and are from 1 m to 2 m in height, but their thermo-hygrometric conditions are not always comparable to those of a greenhouse. Glasshouses are common in temperate areas and their thermal and humidity conditioning can maintain complete physical isolation. These structures can be covered by netting, and openings (e.g. windows and vents) can ensure ventilation and cooling in warmer environments. Meshes of selected hole size and the shape of the netting should be viewed in relation to the need to ventilate but also to limit insect access through the covering (Fatnassi *et al.*, 2006; Rigakis *et al.*, 2015). The 'plastic houses' are more prevalent in tropical and subtropical regions (e.g. the Mediterranean area). This type of greenhouse does not assure optimal control of temperature and RH. The choice of the type of greenhouse depends on: (i) the country; (ii) the external climatic conditions; (iii) the type of crop; (iv) the material available; and (v) the economic and sociocultural position of the human populations involved.

As regards the growing substrate, it is possible to use the open ground with or without a plastic mulch, the outside soil with the aid of technical facilities and various substrates or a hydroponic system for irrigation. This last technical choice circumvents the risk of nematodes attacking the plant roots.

The relationships between pests and their natural enemies living in greenhouses are affected in the same way as in outdoor crops, but also by the morphological and genetic characters of the cultivated plant (habit, hairiness, strength, tolerance, etc.). The different influence on fertility and/or development of a pest exerted by different species and varieties of plants varies and is often conditioned by these factors. So, for example, different tomentose production in tomato leaves has different effects on the action of parasitoids on whitefly (McAuslane *et al.*, 1995). Over the past decades, during the selection process carried out by geneticists to create new varieties, plant resistance to insects has often been overlooked and this has resulted in greater susceptibility to one or more biotic factors. Only recently has the problem been better evaluated in terms of genetic selection, with the introduction of resistance characters or also through the use of grafts on to species that are more tolerant of the soil-dwelling pest.

According to the aspects reported above, in greenhouses pest populations have a greater opportunity to establish and this is linked to the thermo-hygrometric parameters and to the rich concentration of food provided, even if only for a short time, supported by the massive use of technical means (fertilizers, mulching, number and distribution of plants, etc.).

It can be concluded that from an ecological point of view, the greenhouse is a habitat sometimes characterized by large spatial and temporal variations in the population density of pests. The process of aggregation of pests is not unique to the whole area of the greenhouse and often it relates to the distribution of a particular pest on just a part of the plant (vegetative apex, buds, leaves, etc.). Populations of pests are formed by several individuals which are equipped with the typical features of a species (i.e. structure between different stages, sex ratio, mortality and migration). The size of a population determines when control actions should be introduced or the introduction of natural enemies. Some pest species need to be contained even when the infestation is at a low intensity, given the possibility of different species transmitting viral diseases to the plants that are grown. From a practical point of view, in addition to macroscopic actions of random monitoring for pests, the use of trap plants and/or botanical varieties that attract the pests allows verification of the existence and density of different pest populations. Also partial shelters can be provided for the predators and/or natural enemies of pests consisting of weed flora or of parts of the greenhouse (e.g. external or internal rows of the crop) where the pest may occur but be kept at a low density by predation or parasitism activity of natural enemies.

Microclimate

As discussed above, the greenhouse is a specialized, artificial agricultural environment, characterized in general by unique ecological conditions. Although these conditions are generally fairly uniform throughout a particular greenhouse there are sometimes exceptions, as in the plastic houses of the Mediterranean area, where different factors (different side openings, variable soil level, variable height of the same greenhouse, etc.) may affect the ventilation of the crop differently and produce small microclimatic areas that affect the development of pest populations in the crops differently. Regardless of these aspects, the main resources involved in the ecology of a greenhouse are light, soil and water. These resources play a lead role in the bio-ecology of pests and of their natural enemies. The temperature depends on the climate with the mediation of the greenhouse structure and its ventilation, etc.; it is the main climatic factor that influences the activity of the animal communities in the greenhouse.

Depending on the geographic location, solar radiation and temperature may affect the microclimate of the greenhouse differently. In temperate regions, greenhouses have less heat input, increasing the average temperature at lower latitudes. The utilization of solar radiation by plants varies with the orientation of the greenhouse and this influences crop growth, which is also influenced by the arrangement of the plants within the greenhouse and the nature of the greenhouse cover (e.g. glass or plastic).

The physical nature of the soil relates to the temperature and influences the ecology of a greenhouse. Loose and sandy soils absorb greater amounts of heat during the day than other soils and the release of this heat during the night affects the levels of RH. The organic component of the soil supports the terrestrial fauna, including the predators (e.g. some stigmeid and phytoseiid mites) which under appropriate conditions can move on to the foliage and attack important pests such as spider mites, true bugs and whiteflies. The general increase in predator species richness in the greenhouse environment is fundamental for more precise pest control, although the effects of species richness are density dependent (Griffiths *et al.*, 2008). The structure and composition of the soil affects the development of specific pests, such as nematodes.

Trends and changes in the climate of the external environment affect the internal greenhouse microclimate. These effects are less evident in temperature-controlled greenhouses. The microclimate of greenhouses and the daily variations of temperature are closely related to the population dynamics of pests and their natural enemies, influencing the relationship in a complex way and directly affecting the success of biological control in greenhouses. The spread of harmful species and the daily activities of these populations are influenced by temperature. The relationship between temperature and the efficiency of predation may play a significant role in pest control in protected crops and has been extensively investigated: for example, variations and deviations from the optimum temperature affect the times when predation events occur (Logan and Wolesensky, 2007). This is the case for the greenhouse whitefly, *Trialeurodes vaporariorum* (Westwood), in which the daily thermal performance and plant health influence the number of flights, and this same relationship between environmental factors and flight activity has been observed in the predator of *T. vaporariorum*, the muscid *Coenosia attenuata* Stoin (Boncignoro, 2015, 2016). Similarly temperature affects the parasitoid–host relationship between the wasp *Encarsia formosa* (Hymenoptera: Aphelinide) and *T. vaporariorum* in that temperatures below 18°C limit the activity of *E. formosa* (Hulspas-Jordaan *et al.*, 1987). Many other examples are documented in the literature.

Less obvious relationships with populations of injurious insects are linked to RH in the greenhouse. RH is closely related to the type of irrigation, the mulching, the type of greenhouse cover, the soil and the species of cultivated plant. The interconnections of RH with daily temperatures often determine how much condensation occurs on the walls of the greenhouse cover and subsequent effects on plants, for example the greatly feared development of cryptogamic fungi. Controlling the temperature and humidity inside the greenhouse requires daily management of the side openings or vents, exposing the crops at different times of the year to the action of pests (and pathogens) moving actively towards the greenhouse or variously conveyed from one environment to another. In addition to being dictated by the internal environment, the presence in the greenhouse of one or more important pests such as whiteflies and cutworms also depends on the external environment, which varies according to the seasons and the cultural cycles. The influence of RH on the development of the phytoseiid mite *Phytoseiulus persimilis* Athias Henriot (Pralavorio and Almaguel-Rojas, 1980) is well known. There are many other examples documented in the literature. From a practical point of view, knowledge of the effects of changes in temperature and RH help to curb the negative effects; extremes of temperature and RH in greenhouses can be restrained by applying different techniques with the primary purpose of prejudicing the development of some pests and/or facilitating that of their natural enemies, or to hamper the development of pathogenic fungi and/or to delay the phase of the harvest of certain plant species (e.g. rose, carnation). So, the whitening of the roof with lime in the summertime lowers the temperature by a few degrees. The abolition of mulch elevates RH, as well as surface irrigation. This method facilitates, for example, the use of the predatory mite *P. persimilis*. The rational opening of the lateral sides favours ventilation and lowers the temperature and the RH (Vacante and Benuzzi, 2007).

The List of Pests

The list of pests of greenhouse crops includes mainly generalist species (polyphagous) and secondarily specialist species (monophagous and oligophagous). The most injurious species are nematodes, mites and insects, and exhibit a generalist polyphagous adaptation, such as aphids, aleirodids, leafminers, spider mites, etc. A restricted number of species play a specialist role, usually infesting few host plants, such as the tomato russet mite, *Aculops lycopersici* (Tryon) (Eriophyidae) (Vacante, 2016). According to their bio-ecological adaptations two groups of pests are recognized, referred to here as 'major pests' and 'minor pests', respectively. Moreover, some arthropods, called 'occasional pests' (e.g. Isopoda or Diplopoda; Vacante and Benuzzi, 2007), occasionally damage vegetables cultivated in the greenhouse, but reasons of space do not allow extensive discussion of these species.

Major Pests

Nematodes

The nematodes have a fusiform body that is usually 0.2–0.4 mm long, light coloured and often transparent; the females may differ from males by having a subspheric or saccular body. The group includes several species adapted to various environments, with a typical terrestrial phytophagy, causing direct and indirect damage, sometimes associated with severe crop losses. The most important species belong to the order Tylenchida and to the families Anguinidae, Heteroderidae and Aphelenchoididae.

Anguinidae

This family includes relatively few injurious species, among which are: (i) the potato tuber nematode, *Ditylenchus destructor* (Thorne), which is characterized by a wide geographical distribution and is strongly polyphagous (Sikora and Greco, 1990; Douda, 2005; Mwaura *et al.*, 2015); and (ii) *Ditylenchus dipsaci* (Kuhn), which is also reported in greenhouses.

DITYLENCHUS DIPSACI (STEM AND BULB NEMATODE)

Description. The adult is 1.5 mm long; the female is robust and curved and the male is thinner and smaller. The stylet is very short and delicate, and the glandular area of the oesophagus is adjacent to the intestine.

Life cycle. This nematode does not seem to establish itself in tropical regions except at higher altitudes with a temperate climate. The moisture content of the soil and temperatures over 12°C promote the development of this nematode and a better temperature for its development is 20°C (Tenente and Evans, 1998). The survival of *D. dipsaci* in adverse conditions is entrusted to cryptobiosis, which ensures its survival even in the absence of plants. The species overwinters in soil as pre-adults and adults but not as eggs (Lewis and Mai, 1960). The spread is also linked to the presence of the grey field slug, *Deroceras reticulatum* (Müller) (Agriolimacidae) (Cook *et al.*, 1989). Like other terrestrial nematodes, *D. dipsaci* undergoes complex relationships with other soil microorganisms such as bacteria and fungi antagonists, reported as its natural enemies.

Damage. Attacked plants have deformed leaves and stems which are clear green and pale yellowish in colour and according to plant species may cause cracking and stunting of the plants. The extent of damage is related to the pest population density and depends also on the soil structure and the general condition of the plants.

Distribution and host plants. The nematode occurs locally in most temperate areas of the world (Europe and the Mediterranean region, North and South America, northern and southern Africa, Asia and Oceania) (EPPO, 2013a, b). According to its polyphagous adaptation it infests a number of different plant species in greenhouses, especially bulbous flowering plants.

Management for organic farming. The control of this nematode is relegated to prevention through the reduction of factors that favour its development, such as humidity and temperature. The soil solarization and the use of resistant varieties limit the density of populations of the nematode. In the UK, narcissus bulbs are routinely treated with hot water for the control of *D. dipsaci* prior to planting (Gratwick and Southey, 1986).

Heteroderidae

Members of the family Heteroderidae that are injurious to greenhouse crops are usually called root knot nematodes. The most important species belong to the genus *Meloidogyne*, which includes more than 90 species,

23 of which are reported for the European continent (Karssen, 2002; Wesemael *et al.*, 2011). The most common species associated with greenhouse crops are *Meloidogyne incognita* (Kofoid *et* White), *Meloidogyne javanica* (Treub), *Meloidogyne arenaria* (Neal) and *Meloidogyne hapla* Chitwood. The similarity of their bio-ecology, damage and control suggests a common treatment of these species.

MELOIDOGYNE SPP. These include:

- *M. incognita* (southern root knot nematode or cotton root knot nematode);
- *M. javanica* (sugarcane eelworm); and
- *M. arenaria* (peanut root knot nematode).

Description. The root knot nematodes possess a marked sexual dimorphism, with the female being usually globose and sedentary at maturity. The male is vermiform. They are long, from 400 μm to 1000 μm, and possess a stylet for injecting secretions as well as ingesting nutrients from the host plant cells of roots. Species differentiation includes: (i) the morphology of perineal patterns; (ii) the head of females, males and second-stage juveniles; and (iii) the stylet of both sexes (Eisenback *et al.*, 1981).

Life cycle. The *Meloidogyne* spp. are sedentary endoparasites. Females lay eggs into a gelatinous matrix referred to as 'egg masses' or 'egg sacs', each containing over 1000 eggs, and appearing larger than the female body. The first larval stage shows a visible stylet and is not very active. The second larval stage leaves the egg mass, and moves through the soil in search of a root on which to feed. Juveniles penetrate the root tips, and occasionally attack the roots in the area of root elongation. They initiate the development of giant cells in the meristematic, cortical and xylem tissues of the root and consequently galling of roots occurs. The third- and fourth-stage juveniles and the young females occur after about 6–8 days and 15 days, respectively. The adult females appear after 20 days and egg laying starts after 25 days (Ibrahim and El-Saedy, 1987). The biological cycle is faster with soil temperatures above 20°C, and with suitable temperatures the populations are active throughout the year. In the Mediterranean region they develop four to five generations/year, while in warmer areas there are up to seven generations. Although the spread of root knot nematodes is an active process, their dispersal occurs mostly passively, with no spread via seeds but possibly by their presence in vegetative planting material such as corms, bulbs, tubers or roots (Karssen and Moens, 2006).

Damage. Root knot nematodes live intimately associated with plant roots and their feeding activity on younger tissues induces the formation of hypertrophic areas visible macroscopically as galls. The number of galls on the roots depends on the level of infestion and their shape varies according to the species of nematode. Attacked plants typically show a reduced growth, short internodes, light-green-coloured leaves, and low productivity. Crop losses may be variable and may affect over 60% of yield with a serious impact on qualitative and quantitative standards. Early infestions are very much feared and can lead to the death of young seedlings. Plants attacked by nematodes have lower resistance and are likely to be attacked by plant pathogenic fungi (e.g. *Fusarium* spp. and *Verticillium* spp.). Nematode attacks can sometimes break down the resistance of plants to diseases caused by other organisms (Castillo *et al.*, 2000; Back *et al.*, 2002). *Meloidogyne* spp. can occur in a wide range of soil types but their association with crop damage is markedly evident on sandy soils or sandy patches within fields (Van Gundy, 1985).

Distribution and host plants. Root knot nematodes are more common in tropical and subtropical areas, in the warmer conditions of southern Europe, and also in the glasshouses of temperate areas of central and northern Europe. Host plants number hundreds of species, including fruits, grasses, vegetables and numerous weeds. The greenhouse crops attacked by root knot nematodes primarily include members of the families *Solanaceae* and *Cucurbitaceae*

(Barbary *et al.*, 2015) and also ornamental plants (e.g. rose) (Amsing, 2004).

Management for organic farming. The control of root knot nematodes must first be assigned to preventive actions (e.g. purchase of certified plant material and soil substrates free from nematodes). Inspection of propagation material is focused on quarantine organisms but also it could be important to avoid problems with other root knot nematodes. For example, in warmer climates *Meloidogyne enterolobii* Yang *et* Eisenback (Rodriguez *et al.*, 2003) and *M. javanica* (Vovlas *et al.*, 2005) can compromise potato culture, whereas in cooler climates *Meloidogyne minor* Karssen *et al.* is a threat (Lammers *et al.*, 2006).

Differences between temperate and tropical species of *Meloidogyne* and their occurrence in Europe imply the need for different management strategies in southern and northern Europe. Possible crop rotations for the control of root knot nematodes are limited by the wide host range of several important species. The banning of methyl bromide and other restrictions on fumigant pesticides in the European Union (EU) has significantly increased the application of biofumigation in southern Europe (Wesemael *et al.*, 2011). In any case, any action of control must reduce the pre-transplantation population density of these pests. Prior knowledge of the nematode charge per gram of soil is important in predicting the damage. Threshold values in the Mediterranean are 2–2.5 and 4–5 eggs and larvae/g of soil for pepper and tomato, respectively.

Crop rotations, such as alternating between sesame and aubergine crops, impedes the development of nematodes. Even deep tillage and proper soil drainage hinder the development of *Meloidogyne* populations. It is good practice not to leave parts of plants and/or whole plants that have been attacked by nematodes in the ground. The use of soil containing *Ulva fasciata* (Delila), *Eichhornia crassipes* (Martius), *Glycyrrhiza lepidota* (Nutt), *Artemisia cana* Pursh, *Brassica oleracea* L., etc. suppresses root knot nematode populations. Moreover, correct application of fertilizers favours the development of soil fungi and hinders that of nematode populations. The use of resistant varieties of plants or the use of rootstocks can reduce the negative effects of nematode attack but considerations of the cost of grafted plants are important and it must be assessed in each individual case.

Physical means include the use of soil solarization in warmer areas. Moreover, using fungi and bacteria in the soil against soil-borne nematodes is a possibility (Trudgill *et al.*, 2000; Davies, 2009). However, the application on a commercial scale of species of bacteria such as *Pasteuria penetrans* (Thorne) Sayre *et* Starr is still not widespread, while that of fungi *Pochonia chlamydosporia* (Tzortzakakis and Petsas, 2003), *Myrothecium verrucaria* (Albertini *et* Schweinitz) and *Paecilomyces lilacinus* (Thom) Samson RA and other species of fungi have begun to find application in various greenhouse districts (Kiewnick and Sikora, 2006). Arbuscular mycorrhizal fungi (AMF) and endophytic fungi growing within plant tissues without causing disease can play a protective role against parasitic nematodes (Talavera *et al.*, 2001).

Moreover, a containment action against plant pathogenic nematodes is exercised by the extracts of *Azadirachta indica* (L.) A. Juss., applied by fertigation during the transplantation of seedlings.

Aphelenchoididae

The most important pest of this family reported from greenhouse crops is *Aphelenchoides fragariae* (Ritzema Bos), sometimes associated with *Aphelenchoides ritzemabosi* Schwartz.

APHELENCHOIDES FRAGARIAE (STRAWBERRY CRIMP NEMATODE)

Description. The body is slender, with the cuticle patterned with fine transverse striae; the lateral field shows two incisures appearing as a plain narrow band. The cephalic region is smooth and anteriorly flattened. The tail is elongate-conoid, and the distal peg is simple and spike-like (Allen, 1952).

Life cycle. A. fragariae is a polyphagous species. The nematodes infect young leaves;

nematodes can spread directly from infected plants to healthy plants via contact between healthy and infected leaves (Jagdale and Grewal, 2006). This nematode spreads with the planting material. The species overwinters as juveniles and adults in soil, dry leaves and dormant buds, but not in roots (Jagdale and Grewal, 2006).

On the strawberry it can be found associated with the blackcurrant nematode or chrysanthemum foliar nematode, *A. ritzemabosi*, which is also polyvoltine and polyphagous, and overwinters as juveniles which may live for a long time in the ground in a quiescent manner.

Damage. A. fragariae attacks the above-ground parts of plants and may be endo- or ectoparasitic. The symptoms on strawberry consist of: (i) malformations of the shoot such as twisting and puckering of leaves; (ii) discoloured areas with a hard and rough surface; (iii) undersized leaves with crinkled edges; (iv) reddening of petioles; (v) short internodes of runners; (vi) reduced flower trusses with only one or two flowers; and (vii) death of the crown bud (Dicker, 1948; Franklin, 1950; CABI, 2017). Ectoparasitic feeding on the folded crown and runner buds causes small dry, brown feeding areas which can be observed on expanded leaves usually near the midrib; endoparasitic feeding within the leaf tissue produces typical leaf-blotch symptoms.

In begonias, the nematode feeds on and destroys the mesophyll cells of the leaves and may cause reddening along the veins causing the entire leaf blade to appear red; severe necrosis may result in the presence of *Xanthomonas begoniae* (Takimoto) Dowson (Riedel and Larsen, 1974). On flowering plant leaves, the feeding areas appear as irregular, water-soaked patches later turning brown, violet or purple. Red plant symptoms are reported on strawberry (Goodey, 1933). The presence of the nematode and bacterial pathogens produce a typical symptom, and the interaction of *A. fragariae* and *Corynebacterium* (*Rhodococcus*) *fascians* (Tilford) Dowson is necessary to produce 'cauliflower' disease of strawberry while neither pathogen inoculated separately reproduces the disease (Siddiqui *et al.*, 2012).

Distribution and host plants. A. fragariae has a worldwide distribution (Jagdale and Grewal, 2002, 2004), and is reported on over 250 plants in 47 families (Sturhan, 1962).

Management for organic farming. The general lines of control of the foliar nematodes are similar to those discussed for other nematode species. Alternative methods that can be safely used by nursery managers and homeowners are briefly described here. They include: (i) host plant resistance, e.g. on strawberry (Naumova, 1972); (ii) biological control with nematophagous fungi (Jansson and Nordbring-Hertz, 1980; Cayrol *et al.*, 1986); and (iii) cultural control (use of healthy well-adapted cultivars, manuring based on soil analysis with special attention to boron, draining or planting on ridges to avoid waterlogging, and irrigating at planting, during the summer of planting and again in the following spring) (Clerjeau *et al.*, 1983). The cultivation of non-host crops such as wheat in soil reduced the populations of nematode (Yamada and Takakura, 1987). Hot water treatment has been used to disinfest plant materials of insect pests (Hara *et al.*, 1993, 1994) and plant-parasitic nematodes (Birchfield, 1954; Birchfield and van Pelt, 1958; Tsang *et al.*, 2001).

Acari

The Acari are small arthropods, usually called mites, that vary in length from 300 μm to 500 μm, with exception of the eriophyoids, whose body is shorter and vermiform. The body of mites is usually ovoid, with or without abdominal segmentation, and divided into different regions. The life cycle of mites develops through the biological stages of egg, larva and nymph, the latter equipped with four pairs of legs. The species that injure protected crops belong to the families Tetranychidae, Eriophyidae and Tarsonemidae. Some Tenuipalpidae and Acaridae occur sometimes in protected crops, but their role is usually secondary. In the greenhouses of the Mediterranean the acarid *Tyrophagus neiswanderi* Johonsonn *et* Bruce is sporadically injurious to cucumber plants in greenhouses causing abnormalities of leaf edges and the heads of plants species *Phalenopsis*, *Cymbidium* and *Freesia* (Vacante, 1988, 2016).

Tetranychidae

The most important tetranychid injurious to protected vegetable crops is *Tetranychus urticae* Koch. In some countries the tomato red spider mite, *Tetranychus evansi* Baker, may cause severe damage to solanaceous plants, especially the tomato (Migeon, 2007), but its attacks mainly occur outdoors, and only a few reports concern its presence in glasshouses (EPPO, 2007), probably due to different varieties and control strategies applied to protected crops. Occasionally, in some countries the bean spider mite, *Tetranychus ludeni* Zacher is reported (Zhang, 2002; Vacante, 2016). Zhang (2002) lists some *Eotetranychus*, *Panonychus* and *Oligonichus* species as occasional pests of greenhouse crops. On protected crops of Mediterranean vegetables Vacante (2016) has collected *Panonichus citri* (McGregor), *Bryobia praetiosa* Koch, *Bryobia vasiljevi* Reck and *Petrobia tunisiae* Manson, but none of these species produced severe damage. Only *T. urticae* and *T. evansi* are discussed here.

TETRANYCHUS URTICAE (TWO-SPOTTED SPIDER MITE)

Description. The body is ovoid and variable in colour; there is a green form (GF) and a red form (RF). Summer females of the RF are similar in colour to the strawberry spider mite but with one conspicuous black spot on each side, while overwintering forms are yellowish orange. The peritreme is hooked. The opisthosomal dorsal striae are lobate, and the lobes vary from triangular to semicircular. The male aedeagus has a small knob, with the axis of the knob parallel to the axis of the shaft and the anterior and posterior angulations of the knob small and similar (Vacante, 2016).

Life cycle. The two-spotted spider mite is polyphagous and polyvoltine. Males develop from haploid eggs and females from diploid eggs. In the Mediterranean region and in tropical areas the species is active throughout the year while in the greenhouses of continental areas the species overwinters as fertilized females (Veerman, 1977; Kroon *et al.*, 2004). In Mediterranean greenhouses (e.g. in southern Italy) the irrational use of mite-stimulant pesticides, chemically unbalanced fertilization and the positive influence of microclimatic factors may stimulate the development of severe infestations of *T. urticae* even in open fields (James and Price, 2002). The main limiting factors are: (i) phytoseiid mites (*P. persimilis*, *Amblyseius californicus* (McGregor) and other species); (ii) cecidomids (*Feltiella acarisuga* (Vallot)); (iii) Coccinellidae (*Stethorus punctillum* Weise); (iv) green lacewings (*Chrysoperla* sp.); and (v) anthocorids (*Orius* sp.) (Vacante, 2016).

Damage. The mite populations feed on tender tissues. Motile forms develop initially on the undersides of leaves and only at high density spread on to the top of the plant, on the flowers and fruits (in various stages of development). The mite secretes abundant silky strands that at high density form a dense silky canvas and may cover the entire canopy. The feeding activity of motile forms of *T. urticae* produces mesophyll that loses its chlorophyll becoming empty and pale in colour, which at a high density of attack appears on leaves and fruits as irregular areas of variable colour from light green to rusty brown. The plant tissues lose water and nutrients, with very serious effects on the vegetative development and plant productivity and leaves can dry up and fall off. Young fruits undergo arrested development and show russeting associated with morphological abnormalities. High levels of attack seriously affect the yield and heavily infested plants may die (Vacante, 2016).

Distribution and host plants. The two-spotted spider mite has a worldwide geographical distribution and has been collected on 1059 wild and cultivated plants belonging to 124 families, including all plant species cultivated in greenhouses (Vacante, 2016).

Management for organic farming. Control requires rigid prevention based on the reduction of mite populations in the environment. In warmer climates and in the summer months the closure of the greenhouse openings at the end of the cycle of cultivation allows temperatures to reach above 50°C and to disinfest the environment of any motile forms of the two-spotted spider mite that remain in the environment. Moreover, in the next crop cycle before the transplant of young plants it is necessary to eliminate the weeds inside the greenhouse and in the adjacent areas (which are probable hosts of mites). Biological control involves the release of the phytoseid mites *P. persimilis* and *A. californicus*. The latter species has more ecological limits, and therefore the first predatory mite is the more suitable for release. According to the density and frequency of attack, *P. persimilis* is generally released at densities from 1.5 to 30 mobile forms/m^2 of crop. However, in some areas (e.g. the Mediterranean region) the release of the predator is hindered by the high temperatures and the low RH. These obstacles can be overcome by netting and openings (vents) that can ensure good ventilation and cooling in warmer environments. In addition, it is necessary to use genetically healthy predator strains. This need has been suggested after the mass release of strains produced by insectaries located far from the area of release of the predator. In the absence of non-selective chemical treatments various natural enemies, including the two predators mentioned above, develop useful populations in greenhouses (Vacante, 2016).

Eriophyidae

The eriophyid mites have a worm-like appearance, are provided with two pairs of legs at all stages and are inconspicuous in colour. The most important pest of this family is *Aculops lycopersici* (Tryon), a vagrant leaf species living on both wild and cultivated annual solanaceous plants; the cultivated plants may be protected crops and/or those grown outdoors.

ACULOPS LYCOPERSICI (TOMATO RUSSET MITE)

Description. The body is robust, fusiform, from 150 μm to 180 μm long and whitish to yellowish in colour. The prodorsal shield has a broad and short frontal lobe that is precipitous anteriorly, and topped by a transverse line that extends back along the sides to the rear margins. The opisthosoma has about 27 dorsal annuli and 60 ventral annuli. The microtubercles are pointed and located on the ventral annuli margins. The genitalia have faint basal granules and between eight and ten longitudinal ribs (Vacante, 2016).

Life cycle. The tomato russet mite lives on members of the *Solanaceae*, among which it prefers tomato, cultivated outdoors and/or grown as a protected crop, and to a lesser extent aubergine, potato and pepper. In the Mediterranean region, the species is active throughout the year although more are found in spring–summer (Vacante, 1982, 2016). Males develop from haploid eggs and females from diploid eggs. It is a warm-weather species and has no dormant stages. Winter cold may kill all mite stages in the field but not in protected crops. In southern areas of North America and in greenhouses, the mite commonly survives during the winter on black nightshade growing near tomato crops. Morning glory and bindweed can also harbour the mites, and leaves of the latter develop a silvery sheen and may fall without russeting (Rice and Strong, 1962). In California, continuous reproduction has been observed in the field from early May to November or until the first frost occurs. Reproduction is continuous as long as conditions permit, and only slows down at minimum survival temperatures for the host plants.

Damage. The attack usually starts from the collar to the neck of the plants and continues up to infest the entire canopy. The changes are evident with browning on the stalk and cortical fissures. The leaves lose their green colour and at an advanced stage have a bronze colour, associated with crumpled edges and a cartilaginous consistency. The attack stops fruit development as shown on the epicarp as irregular areas that are suberized and brownish–russet in colour and with a mesh of irregular cracks. High levels of attack lead to death of the plants (Vacante, 1982, 2016).

Distribution and host plants. The tomato russet mite has a worldwide geographical distribution. It has undergone a host expansion, having lived originally on wild solanaceous plants native to the Americas and later being associated with tomato (Oldfield, 1996). It is now reported on different plant families, but it is commonly associated with tomato grown outdoors and as a protected crop, and less frequently on potato, aubergine and pepper (Vacante, 2016).

Management for organic farming. The control requires preventive and curative actions. Preventative action consists mainly of weeding the crop and the surrounding areas, paying particular attention to wild and cultivated solanaceous plants. The tomato russet mite has a patchy distribution in the field and the level of group aggregation is correlated with different factors, such as its habits, environmental variables, etc. This facilitates detection and monitoring of its populations and assessment of the population density. Lines resistant to tomato russet mite have been investigated in various countries, but no line appears to be currently available commercially.

Irrigation management early in the first phase of tomato growth could help to prevent mite development later in the growing phase (Gispert *et al.*, 1989).

Curative action includes use of natural enemies and a number of natural enemies, such as fungi and predators (insects and mites) have been reported in various countries. In the last three decades, most studies have concerned the phytoseiid mites, such as *Amblyseius concordis* (Chant) and other species, and some predatory insects, but biological control of the tomato russet mite awaits adequate solutions. The control is entrusted fundamentally to prevention and use of sulfur, which despite its negative impact on beneficials must be favoured because of its reduced toxicological impact on mammals (Vacante, 1982, 2016).

Tarsonemidae

The Tarsonemidae are small mites, from 100 μm to 300 μm long, that are translucent, pale or whitish, with the colour influenced by the ingested food (e.g. greenish, depending on the green leaves attacked). The males usually differ from females in size and morphological characteristics. The body of the female is usually ovoid and sometimes elongate while that of the male is smaller and has a characteristic subterminal 'genital capsule' and a strong leg IV. The most important species infesting vegetables cultivated in the greenhouse are the polyphagous *Polyphagotarsonemus latus* (Banks) and *Phytonemus pallidus* (Banks) that are injurious to strawberries and other economic plants (Vacante, 2016).

POLYPHAGOTARSONEMUS LATUS (BROAD MITE)

Description. The female has an oval body which is about 200 μm long and is white amber to yellow or pale greenish in colour, sometimes with an intense white median spot and with faint contours. The body setae are short, the pseudostigmatic organs are spherical and the ventral metapodosoma has four pairs of setae. The tibiotarsus of leg I has one claw. The male has three pairs of setae on the prodorsum and the ventral metapodosoma four pairs of setae; the tibia and tarsus IV are fused into a tibiotarsus ending with a knob-like claw (Vacante, 2016).

Life cycle. The broad mite feeds on tender tissues of new leaves, blossoms and young fruits. Females oviposit in hollows of the lower leaf surface and on younger fruit. Reproduction is by arrhenotokous parthenogenesis and the development typically requires tropical climatic conditions. Experimental studies using lemons show that optimum development occurs at 25°C and an RH between 90% and 100% (Jones and Brown, 1983).

In greenhouse crops the broad mite is generally active throughout the year with a reproductive rate slowing during the colder months (Jeppson *et al.*, 1975; Schwartz, 1977). The broad mite exhibits similar behaviour in cold Mediterranean greenhouses (Vacante and Benuzzi, 2007).

The natural enemies include different predatory mites such as *Amblyseius ovalis* (Evans), *A. californicus* and *Neoseiulus barkeri* Hughes.

Damage. The broad mite feeds on tender tissues of young leaves, flowers and immature fruits of a large variety of economic plants. Infested leaves become bronzed with down-curling margins, buds and flowers are aborted and distorted, shoots grow twisted and fruit may be misshapen and russeted (Gerson, 1992). On pepper seedlings, infestation by the broad mite prevents flower and fruit development, later causing flower dropping and silvering of the fruits (Kulkarni, 1923; Vacante and Benuzzi, 2007); similar damage occurs on aubergine plants, both outdoors and in greenhouses (Vacante and Benuzzi, 2007). The stems of terminal shoots and the lower surfaces of the young leaves of tomato become shiny bronze or brownish. The first injury is browning of the epidermal cells, but later the cells of developing tissues collapse and die. The young expanding leaves become narrow, stiff, twisted or crumpled, fail to develop, and dry. The stems of young plants appear swollen, roughened or russeted and greyish green in colour (Vacante, 2016).

Distribution and host plants. The mite has a worldwide distribution and is common in tropical and subtropical areas and in greenhouses of temperate and subtropical countries (Vacante, 2016).

Management for organic farming. Control measures for this mite also consist of preventive and curative actions. Prevention consists mainly of the control of weeds inside and outside the greenhouse and on transplanting seedlings that are free from attack. Because of their small size, broad mites are not noticeable until they cause serious damage to apical leaves in pepper seedlings (Jovicich *et al.*, 2004), suggesting that control measures should be started at the first symptoms of an attack. The use of resistant varieties and of biological control does not seem to offer concrete guarantees of success. As for *A. lycopersici* the use of sulfur can successfully resolve an attack.

PHYTONEMUS PALLIDUS (CYCLAMEN MITE)

Description. The body of the female is about 250 μm long and yellowish brown in colour. The prodorsal shield does not extend over the gnathosoma and has two pairs of setae, the first pair of which is shorter than the second pair; the latter is much longer than the dorsal hysterosomal setae. The trichobotria are capitate. The male is smaller than the female. The prodorsum has four pairs of setae, the fourth pair of which is shorter than the third pair and set laterally of the line formed by the first three pairs of setae. The tibia and the tarsus of leg IV are fused, and the femurogenu has a conspicuous and rounded flange (Vacante, 2016).

Life cycle. The cyclamen mite lives on a wide range of wild and cultivated plants, belonging to almost 26 botanical families, among which it severely infests the strawberry. The mite is negatively phototropic and their development requires high levels of humidity. This adaptation means that the mite prefers to live in sites such as unopened leaflets in the crown of the host plant, between the young leaves in the bud or in the cavities of flower buds. In these sites, the cyclamen mite lays its eggs and reproduces. *P. pallidus* usually feeds on the upper side of young and not completely developed leaves. In temperate climates, the cyclamen mite overwinters as females, but during mild winters or in mild climates, it can continue to reproduce. The mite develops 3.5 generations/year, and at the end of August and in September the young females start to search for wintering places (Łabanowska, 2006). In cool temperate regions, *P. pallidus* develops both in greenhouses and outdoors (Fjelddalen and Stenseth, 1958; Stenseth and Nordby, 1976).

Damage. The cyclamen mite infests various economic plants, among which are strawberries, watercress, and many ornamental flowers and shrubs (Van Eyndhoven and Groenewold, 1959). Infested strawberry leaves have a roughened and wrinkled upper leaf surface, with irregular folding and fluting of the leaf margins. The veins bulge upward like blisters, and the petioles fail to elongate, and bear leaflets that fail to unfold completely. These small leaflets become pale yellowish green with a hard brittle texture and turn brown or silvered on exposure to the sun. The flowers and fruits become brown near the inner bases of the sepals, and may turn black and dry (Dustan and Matthewman, 1932).

Distribution and host plants. The geographical distribution is worldwide and the list of hosts plants includes 26 botanical families and several plant species, including the strawberry (Vacante, 2016).

Management for organic farming. Control of the cyclamen mite includes preventive and curative actions. Cyclamen mites can easily be transferred from one crop to another by animals, pickers and equipment, including strawberry freezer trays. Early plant inspection allows detection of mite presence on young plants before and after transplantation, and optimizing the application of control strategies.

The immersion of strawberry plants in water at 43.5°C for 30 min results in full mortality of the cyclamen mite, but treatment of the plants with water-saturated air at 43.5°C for 1 h is more efficient. The plants must be loose on a screen or slatted boxes, or stacked to allow the penetration of the water vapour, but after treatment the plants should be dried before packing, and planted as soon as possible (Munger, 1933; Smith, 1939). The University of California Pest Management Guidelines for the cyclamen mite on strawberry (UC IPM, 2014) recommend dipping trays of long-term cold-stored (–2.22°C) transplants into a hot water bath for 7 min before planting. The plants should be washed to remove all dirt, and then placed in a circulating water bath that is held at a constant temperature of 48.88°C. Afterwards, the plants should be submerged in very cold water and then planted as soon as possible. This treatment is not recommended for freshly dug transplants that have only been stored at 0.55°C.

No cultivar has been found to be completely resistant to mite attacks.

It is good practice not to transplant – where and when possible – into areas that have previously been affected by the presence of the mite. Because infested nursery plants are the major source of *P. pallidus* in annual plantings, it is necessary to use uninfested nursery stock. The removal of infested plants as soon as symptoms are detected also slows the spread of infestations in the crop (UC IPM, 2014). Releases of the phytoseiid mites *Neoseiulus cucumeris* and *N. barkeri* on to strawberry produce encouraging results (Tuovinen and Lindqvist, 2010), but tests with a different set of predatory mites (*Anthoseius rhenanus* (Oudemans), *Euseius finlandicus* (Oudemans), *N. cucumeris*, *N. barkeri*) indirectly demonstrated that the biological control of cyclamen mite on this crop awaits a definitive solution.

Insecta

Insects reported from vegetables cultivated in greenhouses include several orders and numerous species. However, a restricted number of species cause constant damage and may be included among the key pests, while other species are occasional and play a secondary role. There is not enough space in this chapter to allow extensive treatment of all these insects and so only essential characteristics for major species are reported here. The interested reader can find further updates in specialized texts.

Thripidae

This family belongs to the order Tysanoptera and includes a number of different pest species infesting vegetable crops in greenhouses. The most important species is *Frankliniella occidentalis* (Pergande), followed by the onion thrips, *Thrips tabaci* (Lindeman), and the melon thrips, *Thrips palmi* Karny.

The onion thrips has a cosmopolitan distribution and possesses a strong polyphagia. Among the vegetables cultivated in greenhouses the pest usually infests members of the *Cucurbitaceae* and the *Liliaceae*.

Immediately after transplantation or during the growing season adults move from weeds associated with the vegetable crops or growing near to the greenhouses on to the crop. The species is considered to be responsible for the transmission of pathogens including tomato spotted wilt virus (TSWV) and iris yellow spot virus (IYSV) (Srinivasan *et al.*, 2012). For biological control of this pest see the discussion on the western flower thrips.

The melon thrips is a polyphagous species, spread throughout the southern hemisphere, where it is considered an important pest on different plant species that originate from this part of the world. In the greenhouse it prefers sweet pepper, cucumber and aubergine. Also for this species see the discussion on the biological control of the western flower thrips.

It is possible to find sporadically in greenhouses the thrips *Heliothrips haemorroidales* (Bouché) or *Thrips nigropilosus* Uzel on basil. However, in this chapter only the western flower thrips are discussed in more detail.

FRANKLINIELLA OCCIDENTALIS (WESTERN FLOWER THRIPS)

Description. The body is tiny, less than 2 mm long, slender and with narrow fringed wings. The female has a spindle-shaped abdomen, and varies in colour from yellow to brown to nearly black. The male is smaller than the female, it has a narrow abdomen with a rounded end and is pale yellow (almost white) (Moulton, 1948).

Life cycle. The life cycle of the western flower thrips develops through the egg, larva, nymph (two instars) and adult instars. The female lays the eggs in the tender tissues of young leaves, flowers and fruits, and larvae, nymph and adults feed on these organs and also on pollen. The thrips populations aggregate inside the flowers of host plants and are constantly active during the year on different vegetables in protected crops. On aubergine and pepper, the western flower thrips is more common on flowers and leaves and on cucumber it is more frequent on the leaves. Various natural enemies hinder the development of populations of this pest, including different predators (mites and insects) and entomophatogenic fungi.

Damage. The western flower thrips causes direct damage by feeding on the leaves, flowers and fruit, and indirect damage by acting as a vector for plant viruses (Lewis, 1997). The direct damage is due to the action of the nymphs and adults that pierce the young epigeal tissues injecting toxic saliva and then sucking out the juices. The leaves show silvery areas devoid of pigment, which tend to become necrotic, associated with deformations and necrosis of the leaf blades. The attached petals show discoloration and necrosis; if the attack affects the ovary the flower falls. Areas of fruits that are attacked may become corky and the epicarp may be deformed. The pest also causes damage during the deposition of eggs in the tender tissues.

Distribution and host plants. F. occidentalis is indigenous to North America (Canada, Mexico and continental USA) and actually has a worldwide distribution. It infests more than 500 host plants belonging to about 50 families (Waterhouse and Norris, 1989), including a number of economically important crops (Cho *et al.*, 1989) such as pepper, tomato, aubergine, cucumber, strawberry and ornamental plants in greenhouses and also fruit crops (grapes, nectarines) outdoors.

Management for organic farming. Prevention of attack plays a key role, including prior control of wild populations of western flower thrips inside the greenhouse and outdoors and ensuring that only young plants that are free from the pest are transplanted. The cryptic behaviour of *F. occidentalis* in flower buds makes it less easy to treat with insecticides and hinders the control of its population. In this context, biological control is highly desirable and provides economic and eco-toxicological benefits. Parasitic wasps, predatory mites, entomopathogenic nematodes (Premachandra *et al.*, 2003; Loomans, 2003; Lim and Van Driesche, 2004; Arthurs and Heinz, 2006; Arthurs *et al.*, 2009), generalist predators including minute pirate bugs (*Orius* spp.) (Stoltz and Stern, 1978; Waterhouse and Norris, 1989; Bahsi and Tunc, 2008), Nabidae (*Nabis* spp.) (Benedict and Cothran, 1980) and lacewing larvae of *Chrysoperla carnea* Stephens have been used for biological control of *Frankliniella* spp. (Waterhouse and Norris, 1989; Obrist *et al.*, 2005; Atakan, 2006). Among the generalist predators reported from the Mediterranean area, the anthocorid *Orius laevigatus* (Fieber) plays an important role in the control of western flower thrips (Vacante and Tropea Garzia, 1993; Bonsignore and Vacante, 2012) as well as of aphids, mites and whitefly populations (Alvarado *et al.*, 1997; Montserrat *et al.*, 2000) in unheated greenhouses. *Orius* sp. feeds more easily on flowers where there are western flower thrips, so they are considered the main biological control agents in the greenhouse. The use of different species of *Orius* often involves preventive release although in the Mediterranean area the climate and alternate host plants mean that these predators are often permanently present in the greenhouse (Isenhour and Yeargan, 1981; Nakashima Hirose, 1997; Bahşi and Tunç, 2008; Shakaya *et al.*, 2009). The phytoseiids *Amblyseius cucumeris* Oudemans, *Amblyseius degenerans* (Berlese) and *Amblyseius swirskii* Athias Henriot offer an alternative method of control which may be more suitable in certain geographical areas (Calvo *et al.*, 2015). Some entomopathogenic fungi, such as *Neozygites parvispora* (McLeod *et* Carl) (Vacante *et al.*, 1994), *Verticillum lecani* (Zimmerman *et* Végas) and *Beauveria bassiana* (Balsamo) Vuill. exert an antagonistic action against thrips.

Among physical means to control this pest, the use of plastic net to obstruct the entry of whiteflies in the greenhouse also reduces the presence of thrips by 50%, probably due to the optical interference exerted by the structure.

Aleyrodidae

The Aleyrodidae are small insects, varying in length from 1 mm to 2 mm, usually with a wingspan of less than 3 mm. The adults have well-developed antennae that have seven segments. Two ocelli are set at the anterior margins of the compound eyes. The two most important species of this family, *Trialeurodes vaporariorum* (Westwood) and *Bemisia tabaci* (Gennadius) are discussed here.

TRIALEURODES VAPORARIORUM (GREENHOUSE WHITEFLY)

Description. The adult is 1–2 mm long and yellowish in colour. The upper and lower parts of the compound eyes are completely separate. The greenhouse whitefly has four wax-coated wings that are pale yellow and held near parallel to the leaf surface.

Bio-ecology. The greenhouse whitefly is an important pest in many regions of the world (Lourenção *et al.*, 2008) and despite the increasing distribution of the whitefly *B. tabaci* in different areas of the world, it remains the major insect pest of greenhouse crops in Europe. The life cycle of the greenhouse whitefly develops through the egg, larva (the only mobile form), three larval instars, pupa and adult. The pupa is whitish in colour. The pest populations live on the undersides of young leaves, where they feed, excrete honeydew and reproduce. The females lay many pale-yellow eggs mainly on young leaves; the eggs turn grey prior to hatching. The pupa shows the compound eyes and other body tissues become visible as the larvae thicken and rise from the leaf surface. In tropical and subtropical areas the populations of the greenhouse whitefly are steadily active, as well as in protected agricultural settings such as greenhouses of temperate areas, where it can occur even during winter. Populations increase usually in spring–summer, and sometimes also during the winter months (e.g. in the protected crops of the Mediterranean area). The pest attacks ornamental and vegetable crops (Mound and Halsey, 1978; Oliveira *et al.*, 2003; Erdogan *et al.*, 2012) mainly cultivated in greenhouses.

Greenhouse whitefly spread as a result of: (i) transplant of infested young plants; (ii) by wild populations developing inside the greenhouse; or (iii) by introduction of adults from outdoors. A number of natural enemies (predators, parasitoids, pathogens) develop on populations of the greenhouse whitefly.

Damage. The feeding activity of *T. vaporariorum* can reduce the plant productivity and longevity, as well as act as a vector of plant viruses, a few of which are responsible for significant economic damage (Duffus *et al.*, 1996; Wisler *et al.*, 1997; Wintermantel, 2004). In the Mediterranean area, *T. vaporariorum* is widely distributed in greenhouses that are used for off-season production for zucchini squash (*Cucurbita pepo* L.) and tomato (*Solanum lycopersicum* L.). Adults and larvae (not pupae) feed on phloem sap, producing large amounts of honeydew, which results in the formation of sooty mould. In general, traits of the host plant often affect the activity and performance of *T. vaporariorum*. Leaf morphostructural variables, age and the chemical characteristics of the plant host are crucial for whitefly fitness (Inbar and Gerling, 2008).

Distribution and host plants. The total world record of greenhouse whitefly host plants is approximately 859 species, belonging to 469 genera in 121 families (CABI, 2016).

Management for organic farming. Regular field monitoring is important for the detection and management of *T. vaporariorum*. Visual sampling of adult and larvae by leaf examination is the most common and accurate method of monitoring whiteflies. Yellow sticky traps help to monitor the flights of whiteflies into the greenhouse or outdoors immediately after the transplantation of the crops. Agronomic and environmental aspects of actively growing plants act dynamically on whitefly life-history traits: so, a reduction in fitness of *T. vaporariorum* larvae coincided with an increase in the number of daylight hours (Johansen, 2009), and with variation in the metabolite composition of the host (Park *et al.*, 2009). Such aspects might facilitate the spread and persistence of *T. vaporariorum* in an environment, and vary the degree of infestation even for the same growing area. The density of whitefly in the greenhouse is unpredictable in relation to the entry of whiteflies from the outside (Gabarra *et al.*, 2004); the spread of the species within a single plant through arrival of additional flights of the pest is also influenced by the host plant conditions (Bonsignore, 2015).

The natural enemies of *T. vaporariorum* include predators, parasitoids and pathogens. The most important predators are the mirids *Nesidiocoris tenuis* (Reuter), *Macrolophus caliginosus* Wagner and *Dycyphus errans* (Wolff), the whitefly predator *Delphastus pusillus* (LeConte) (Coccinellidae) and the muscid flies *Coenosia attenuata* Stein (Muscidae). In the Mediterranean region (especially in the western Mediterranean) *N. tenuis* and *C. attenuata* develop spontaneously in greenhouses not treated with pesticides (or treated with selective pesticides) and they usually significantly hinder the development of populations of *T. vaporariorum*. The biological pollination of some crops (e.g. tomato) has facilitated the development of their populations. Also various parasitoids develop spontaneously in greenhouses not treated with pesticides (e.g. the aphelinids *Encarsia formosa*, *Encarsia pergandiella* Howard and *Encarsia lutea* (Masi)), but their action is usually insufficient and in temperate greenhouses the interaction between *T. vaporariorum* and its parasitoid *E. formosa* must be created by artificial releases; the parasitoid is introduced on the crop at the stage of the mature pupa (black in colour). The first appearance of *T. vaporariorum* requires between four and six releases of four to six pupae of *E. formosa*/m^2 at 15-day intervals in spring and about 7-day intervals during the summer, in order to obtain a 60–70% level of parasitism. In temperate regions the parasitoid is now well acclimatized.

Even the phytoseiid *Amblyseius swirskii* Athias Henriot and some pathogens, such as *Verticillium lecanii* (Zimmermann) Viegas and *Beauveria bassiana* (Balsamo) Vuillemin perform a control action. Further information on biopesticides can be found in Chapter 2, this volume.

Placing mesh over the side apertures of greenhouses hinders the introduction of

T. vaporariorum populations from outside. Recent studies have demonstrated the effectiveness of ultraviolet absorbent plastic films at reducing *T. vaporariorum* infestations on protected crops (Mutwiwa *et al.*, 2005).

From a practical point of view, there is no single pest control strategy, the choice of control measures being influenced by geographic location, microclimate, type of greenhouse and sociocultural level of the human populations involved. Thus, in the greenhouses of temperate regions the use of *E. formosa* is prevalent and provides positive results, while in subtropical areas other means of control are used, for example the use of biopesticides and natural enemies that spontaneously develop within the greenhouse. Usually a concrete strategy of control must integrate a variety of different methods.

BEMISIA TABACI (SILVERLEAF WHITEFLY)

Description. *Bemisia tabaci* is a species complex including about 24 species that are morphologically indistinguishable from each other (De Barro *et al.*, 2011). The pest has been reported from the USA since the end of 1800s, and in the mid-1980s a virulent strain was found on ornamental plants (poinsettia) in Florida. This strain was called strain B (biotype B) to distinguish it from the earlier known strain A. Successively, strain B has been also collected on tomatoes and other economic plants.

The adult of *B. tabaci* has a body up to 0.8 mm long which is snow-white in colour, according to the secretion of wax across its wings and body. The upper and lower parts of the compound eyes are connected by a single ommatidium. During feeding or resting stages the adult covers its body with its wings.

Life cycle. Populations of *B. tabaci* live on the under surface of young and tender leaves, where they feed and reproduce. Females are diploid and originate from fertilized eggs whereas the males are haploid and originate from unfertilized eggs. The fecundity ranges from 50 to 400 eggs. The eggs are laid in groups on the under leaf surface, starting as whitish in colour and after progressing in development becoming brown in colour; hatching occurs after 5–7 days. The life cycle develops through the egg, larva (only mobile stage), three larval instars, pupa and adult. The pupa is yellow in colour. In Mediterranean greenhouses high levels of populations occur in spring–summer; and sometimes severe attacks develop during the winter months. Spread occurs by transplant of infested young plants, by wild populations developing inside the greenhouse or by introduction of adults from outdoors. Although this species is particularly active on tomatoes and cucurbits and its populations may be associated with that of *T. vaporariorum*, they do not occupy the same ecological niche, *B. tabaci* being largely present on the lower third of the canopy and *T. vaporariorum* on the apical part of the plant (Tsueda and Tsuchida, 1998; Vacante and Benuzzi, 2006). A number of natural enemies develop on populations of this pest, the most important of which are *Eretmocerus mundus* Mercet, *Eretmocerus hayati* Zolnerowich *et* Rose and *Eretmocerus emiratus* Zolnerowich *et* Rose (Aphelinidae). Populations of *E. mundus* that developed naturally on protected crops in Sicily produced over 60% of insect pupae control on pepper (Vacante *et al.*, 1995). Different pathogens develop on *B. tabaci*, such as *Paecilomyces fumosoroseus* (Wize) A.H.S. Br. and G. Sm., *Aschersonia aleyrodis* Webber, *V. lecanii* and *B. bassiana* (Hoddle, 1999; Vacante *et al.*, 2001a).

Damage. The feeding activity of adult and larval instars of the silverleaf whitefly causes direct and indirect damage. With its mouth the silverleaf whitefly pierces the phloem or the lower leaf surfaces of the host plant and removes the nutrients. Infested areas show chlorotic spots, are whiter in colour, and at high levels of attack loss of leaves may occur. Adult and larval instars excrete honeydew over the host foliage and fruits (Brown *et al.*, 1995). On the honeydew sooty moulds develop, which hinder photosynthesis and induce less growth, lower yield and poor quality of the plants. Moreover, the silverleaf whitefly is a vector for plant disease, transmitting lettuce infectious yellows virus, tomato yellow leaf curl virus and the African cassava mosaic virus (Brown *et al.*, 1995); recently, it has been found that the pest is a vector for cassava brown streak virus disease (Ntawuruhunga and Legg (2007). The ability of whitefly to adapt to various plants facilitates the spread of plant viruses, notoriously transmitted by these insect pests.

Distribution and host plants. The whitefly has spread to all continents, especially in tropical and subtropical areas. It feeds on over 500 hosts, among which are various economic plants, such as tomatoes, squash, broccoli, cauliflower, cabbage, melons, cotton, carrots, sweet potato, cucumber and pumpkin, and different ornamental plants (poinsettia, etc.). In the greenhouse it causes considerable damage to tomatoes and cucurbits.

Management for organic farming. Preventive actions are similar to those discussed for *T. vaporariorum*, especially using mesh on the side apertures of greenhouses and the control of weeds. Also in this case control requires integration of different methods, such as natural enemies, physical means, biopesticides and agronomic choices.

Biological control can be carried out with releases of *E. mundus*. Adult females of this parasitoid live in optimal conditions at about 25°C, and may lay up to 50 or more eggs below the nymphs of the pest. The young larva of *E. mundus* enters later into the host body until it kills it when it becomes a pupa. The release programmes provide a sequence of introductions of the parasitoid, repeated just to reach a total quantity of 12–20 individuals/m^2 or more depending on the situation, in order to allow a progressive settlement of the parasitoid and the containment of the pest. In many cases the use of active ingredients with low environmental impact favours the development of natural populations.

Aphididae

Aphids have soft bodies and antennae with as many as six segments provided with an 'end process'. The abdomen has a pair of 'cornicles' or 'siphunculi', through which they exude droplets of a quick-hardening defensive fluid. Aphids have a tail-like protrustion called a cauda above the rectal apertures. The family includes species injurious to herbaceous and arboreal plants. A symmetrical stinging apparatus adapted to sucking causes direct and indirect damage and possesses a high capacity to transmit viruses. Aphid populations are able to grow exponentially if they are able to optimize their nutritional and reproductive processes by directing most of the nutrients to the reproduction of the species. The possibility of forming winged forms allows this group of insects to quickly move towards sources of food. In fact, the presence in greenhouse crops is basically due to the ability of the winged forms to penetrate inside the greenhouse through the lateral openings. The main aphid pests of the *Solanaceae*, *Cucurbitaceae* and other salad crops are *Aphis gossypii* Glover, *Aphis fabae* Scopoli, *Myzus persicae* Sulzer, *Macrosiphum euphorbiae* Linnaeus and *Hyperomyzus lactucae* (Linnaeus). Other species, belonging to different families (e.g. the Lacnidae) sporadically infest vegetables in greenhouses.

APHIS GOSSYPII GLOVER (COTTON APHID)

Description. Adults of *A. gossypii* range from under 1 mm to 1.5 mm in body length. The cornicles or siphunculi are uniformly sclerotized from the tip to the base, darkly pigmented, longer than the cauda and gradually tapering towards the apex with a small dilation there. The dorsal abdominal segments are uniformly sclerotized and unpigmented. The cauda usually have four to seven hairs and are paler than the cornicles. The metafemoral hairs are all shorter than the basal width of the metafemur. *A. gossypii* lacks a stridulatory apparatus, and the antennal tubercles are weakly developed. The terminal process is more than twice the length of the last antennal segment, but less than 3.5 times as long (Blackman and Eastop, 1984).

Life cycle. Adults and nymphs of the cotton aphid live and feed on the underside of leaves or on the growing tips of shoots, sucking juices from the plant. Reproduction is mostly asexual with either alate or apterous females. In warmer environments, *A. gossypii* exhibits an anholocyclic life cycle, while in cooler areas it exhibits either a heteroecious or autoecious holocyclic life cycle (Slosser *et al.*, 1989; Zhang and Zhong, 1990). The heteroecious cycle involves migration from a winter host to a summer host in the spring and a return to a winter host in the autumn for laying eggs.

So, in some countries (e.g. the southern USA) the females of the cotton aphid continue to produce offspring without mating so long as the weather is favourable, while in other areas the aphid displays a holocyclic adaptation characterized by two host plant species, with *Catalpa*, *Rhamnus* or *Hibiscus* as primary host plants. In Europe *A. gossypii* reproduces exclusively by asexual reproduction and can develop nearly 50 generations/year. In some countries (e.g. in Russia) overwintering eggs are laid on various wild plants and the winged forms migrate to secondary host plants of various families, such as *Rosaceae*, *Chenopodiaceae*, *Malvaceae*, *Cruciferae*, *Cucurbitaceae*, *Solanaceae* and *Compositae*, where high levels of populations develop parthenogenetically. Parthenogenic females have a lifespan of about 20 days during which they can generate up to 85 nymphs. As autumn approaches, the winged forms migrate back to the primary hosts. Here, both males and sexual females are produced, mating takes place and the females lay eggs which overwinter, ready to repeat the life cycle the following year. The natural enemies include various species, among which the most important are included in the Aphidiinae and Aphelinidae families.

Damage. The initial symptom of *A. gossypii* attack is a yellowing of the leaves. As the aphids increase in number, the leaves become puckered and curled. As the populations develop, the aphids move to younger leaves, stems and flowers (sepals mostly). Honeydew excreted by the aphids facilitates sooty mould development, causing a reduction in photosynthesis and inducing a decrease in the quantity and quality of the crop yield. Honeydew can also act as an attractant to other crop pests, and insects such as bees, wasps and ants, that may provide protection for the aphids from their natural enemies (Slosser *et al.*, 1989). Plants also become stunted and (particularly in the cucurbits) the stems become twisted. At high population densities *A. gossypii* may kill the host plant.

Populations of *A. gossypii* infest most parts of the plant if their density is high. The cotton aphid transmits various virus diseases, and it is the most important vector of cucumber mosaic virus (CMV) in cucurbits. CMV has one of the widest host ranges of any plant virus. It can be acquired in 5–10 s and be transmitted in less than 1 min. The ability of CMV to be transmitted declines after about 2 min and is usually lost within 2 h (Francki *et al.*, 1985).

Distribution and host plants. The cotton aphid has a worldwide distribution and has about 700 host plants. Vegetables cultivated in greenhouses that may be attacked include watermelons, cucumbers, cantaloupe melons, squash and pumpkin. It also infests pepper and aubergine.

Management for organic farming. Cultural control has been applied on some crops, for example potatoes (Potts and Gunadi, 1991), but it has not been tested on vegetables in greenhouses. Resistance to the cotton aphid has been well documented on some cultured crops, for example in aubergine (Sambandam and Chelliah, 1970) or in melons, whose resistance was conferred by the *Vat* gene (Chen *et al.*, 1997), but its application in the field has not been documented.

There are few records available of biological control being practised in the field, and it is not known as a general strategy for the concrete biological control of this aphid. One successful study using biological control was in Egypt. Two field releases of the green lacewing *C. carnea* at a ratio of 1:5 (predator:aphids) eliminated the aphid in 12 days, whereas it took a single release of the eleven-spot ladybird beetle *Coccinella undecimpunctata* (Linnaeus) at a ratio of 1:50 to get 99.7% control on okra (Zaki *et al.*, 1999). According to suppliers of beneficials the biological control of this aphid may be carried out with releases of the braconid *Aphidius colemani* Viereck, introduced into greenhouses at the first appearance of the aphid on the crops. The parasiotid can be hampered in summer by the presence of hyperparasites, and is best suited to temperatures between 18°C and 30°C; sustained temperatures over 30°C may reduce its effectiveness. From a commercial point of view it is dispatched as 'mummies' packed in a Petri dish or vial. The release rate is 0.2 mummies/m^2 as a preventative measure applied preferably weekly or bi-weekly before aphids are found on the crop. Conversely, one to five mummies/m^2 can be applied as a light to moderate curative measure, over several weeks, as aphids are found on the crop. It is also possible to release the braconid *Lysiphlebus testaceipes* (Cresson), with repeated introductions following the appearance of the first aphids, and preventive release in the most severe situations, even with the use of banker plants. The use of pathogenic fungi also produces positive results.

MYZUS PERSICAE SULZER (PEACH POTATO APHID)

Description. The adult females are wingless and parthenogenetic, with oval bodies from 1.2 mm to 2.1 mm long and variable in colour, from whitish green to pale yellow green, grey green, mid-green, dark green, pink or red. Apart from genetically determined colour variation, any one genotype will be more deeply pigmented green or magenta in cold conditions. Winged morphs have a black central dorsal patch on the abdomen. Immatures of the winged females are often pink or red, especially in autumn, and immature males are yellowish (Blackman and Eastop, 1984).

Life cycle. M. persicae alternates between hosts, with sexual reproduction taking place during part of the life cycle (heteroecious holocyclic) on *Prunus* (usually peach) and a number of summer host plants; however, it is anholocyclic during the summer on secondary hosts in many parts of the world where peach is lacking, and a mild climate allows overwintering of the active stages. In the tropics and subtropics the aphid is usually anholocyclic. Blackman (1974) discussed the life cycle variability of *M. persicae* on a worldwide basis. An extensive list of known natural enemies of *M. persicae* includes over 30 species of primary parasitoid, and various predators, such as coccinellids (*Adonia* spp., *Coccinella* spp., *Hippodamia* spp. and *Scymnus* spp.) and syrphids (Van Emden *et al.*, 1969). Infections of *V. lecanii*, *B. bassiana* and *Conidiobolus* sp. are well documented (Kish *et al.*, 1994).

Damage. The direct feeding damage of high-density populations may cause stunting and reduced root weight in host plants; however, on most crops populations of this aphid do not reach levels that cause obvious symptoms such as chlorosis or leaf curling, and the excretion of abundant honeydew associated with the development of sooty mould. On potatoes direct damage can cause high yield losses (Sexson *et al.*, 2005) and on peppers and flower crops cultivated in greenhouses visible distortion of leaves can occur. Moreover, *M. persicae* is the most important aphid virus vector, transmitting well over 100 plant viruses that cause diseases in about 30 different families, including many major crops.

Distribution and host plants. The peach potato aphid has a worldwide distribution. In winter its primary host is almost invariably the peach, *Prunus persica* (L.) Batsch, sometimes other *Prunus* spp. and peach-almond hybrids. It is not clear whether the sexual part of its life cycle is completed on other species. *M. persicae* is highly polyphagous on summer hosts which are members of over 40 different families, including the *Brassicaceae, Solanaceae, Poaceae, Leguminosae, Cyperaceae, Convolvulaceae, Chenopodiaceae, Compositae, Cucurbitaceae* and *Umbelliferae.* Summer hosts include many economically important plants.

Management for organic farming. In temperate areas the release of natural enemies is routine in the control of *M. persicae* on greenhouse crops (Powell and Pell, 2007).

Here lacewing larvae have given good control on aubergines, and the midge *Aphidoletes aphidimyza* (Rondani) has proved an effective predator of the aphid on peppers. However, most releases are of parasitoids of *M. persicae*, the main species being *Aphidius colemani*, *Aphidius matricariae* (Haliday) and *Aphelinus abdominalis* (Dalman). In subtropical areas the natural development of *A. matricariae* hinders the development of this aphid (Vacante, 2000).

Agromyzidae

This family includes over 2000 species, 400 of which are reported from the Palearctic region. Most species possess endophytic larvae, which feed and develop in the leaves where they hollow out typical mines. Different species are adapted to a limited number of host plants and only a restricted number of them are very polyphagous. Members of the family Agromyzidae that infest vegetables cultivated in greenhouses are commonly referred to as 'leafminers' and the most common species are *Liriomyza trifolii* (Burgess), *Liriomyza bryoniae* (Kaltenback), *Liriomyza strigata* (Meigen), *Liriomyza huidobrensis* (Blanchard) and *Chromatomya horticola* (Goureau). Only *L. trifolii* is discussed here, taking into account that all species infesting vegetables in greenhouses exhibit similar behaviour and may be controlled in the same way.

LIRIOMYZA TRIFOLII (AMERICAN SERPENTINE LEAFMINER)

Description. Adults of *L. trifolii*, *L. bryoniae*, *L. huidobrensis* and *Liriomyza sativae* can be confused with each other, and in addition to morphological characters they may be distinguished by using variations in allozyme patterns as revealed by gel electrophoresis (Menken and Ulenberg, 1986). The adult of *L. trifolii* has a small body, from 1 mm to 1.7 mm long. The mesonotum is grey-black with a yellow blotch at each hind corner. The scutellum is bright yellow; the face, frons and third antennal segment are bright yellow in colour. Both sexes are usually similar. The larvae and puparia have a pair of posterior spiracles terminating in three cone-like appendages (Spencer, 1972).

Life cycle. Females of *L. trifolii* puncture the leaves of the host plants causing wounds for feeding or laying eggs. Feeding punctures destroy a large number of cells and are visible to the naked eye. About 15% of oviposition punctures host viable eggs (Parrella *et al.*, 1981). Males of *L. trifolii* are unable to puncture the leaves but feed on the punctures produced by females. According to the temperature, the eggs hatch in 2–5 days, and at 24°C hatching requires from 4 to 7 days (Harris and Tate, 1933). The endophytic larvae feed at the expense of the foliar mesophyll tissue, producing a specific mine characterized by its own size and shape. The larval development depends on temperature and probably on host plant, and several generations may occur per year (Spencer, 1972). The puparia may be observed outside the leaf and/or in the soil beneath the plant. Also for this biological instar the development depends on season and temperature. Between 20°C and 30°C the adult emergence occurs 7–14 days after pupariation (Leibee, 1984). According to temperature and host plants the American serpentine leafminer develops different cycles throughout the year, and the highest levels of populations occur during spring–summer. At 35°C, 20°C and 15°C on celery *L. trifolii* completes its life cycle in 12 days, 26 days and 54 days, respectively (Leibee, 1984), while on *Phaseolus lunatus* L. at 20°C it takes 20 days (Poe, 1981). Various natural enemies develop on the larvae of the American serpentine leafminer.

Damage. The feeding punctures of the American serpentine leafminer appear on the leaf tissue as white speckles, and the oviposition punctures are usually smaller and more uniformly round. The shape of the leaf mines can vary according to the host plant but they are commonly long, linear, narrow and do not widen greatly towards the end. Their colour is usually greenish white. In very small leaves the limited area for feeding results in the formation of a secondary blotch at the end of the mine, before pupariation (Spencer, 1972).

Distribution and host plants. The American serpentine leafminer has a worldwide distribution, and its host range includes over 400 species of plants in 28 families including both ornamental crops (Bogran, 2005) and vegetables, among which are various members of the *Cucurbitaceae* and *Solanaceae* (EFSA, 2012).

Management for organic farming. The action of the eulophid *Diglyphus isaea* Walker, active ectoparasitoid of the larval stages of leafminers and common in the Palearctic region, allows greater control of different members of the Agromyzidae than that of pesticides. Like the coccinellid *Rodolia cardinalis* (Mulsant), it probably represents the most successful case of biological control of insect pests. The success of this parasitoid is explained by its remarkable adaptation and its reliable functional response. The species is active at a range of temperatures between 15°C and 30°C. In greenhouses untreated with pesticides populations of *D. isaea* develop spontaneously and assure radical control of leafminers (Vacante *et al.*, 1988; Vacante, 2000). Sometimes, it is necessary to introduce the eulophid into greenhouses, at amounts of 200–400 adults/1000 m^2. For *L. trifolii* the use of sterile males (sterile insect technique or SIT) could improve the effectiveness of the parasitoid *D. isaea* (Kaspi and Parrella, 2006). The use of *D. isaea* in greenhouse crops of temperate areas, where it is naturally present (Van de Veire and Vacante, 1984), has been held back by commercial interests that have wrongly privileged the release of *Dacnusa sibirica* Telenga (Braconidae) for years.

Gelechiidae

This family includes various injurious species. Of these, *Tuta absoluta* (Meyrick) is a serious pest of tomato. Other occasional species are the tomato pinworm, *Keiferia lycopersicella* (Walsingham) and the potato tuber moth, *Phthorimaea operculella* (Zeller). For their biological control see the Crambidae section in this chapter.

TUTA ABSOLUTA (MEYRICK) (TOMATO LEAFMINER)

Description. The adult is about 10 mm long, with silverish-grey scales, filiform antenae, alternating light or dark segments and recurved labial palps, which are well developed. The first instar larvae are whitish soon after eclosion (hatching), becoming greenish or light pink in the second to fourth instars according to food (leaflet or ripe fruit, respectively). There are usually four instars (Imenes *et al.*, 1990).

Life cycle. The female of *T. absoluta* lays about 260 eggs during its lifetime. The peak of oviposition occurs on the first and second day after adult mating, when around 92% of the total number of eggs are laid. The female releases a sex pheromone 1 or 2 days after emergence that lures males to exhibit mating behaviour and copulation. The average number of matings for the female in the laboratory was about 10.4 (Imenes *et al.*, 1990; Uchoa-Fernandes *et al.*, 1995a). At 26–30°C and 60–75% RH the eclosion of the eggs requires about 5–7 days. In these conditions, the four larval instars are completed in around 20 days. After this period the larva eliminates all material of the gut and builds a cocoon for pupation. *T. absoluta* has different behaviour of pupating according to whether it occurs in processing tomato or fresh market tomato plants. In the first case, it pupates on the soil (1–2 cm deep) and in the latter case the larva builds a cocoon and pupates on the leaf surface or inside the mines (Uchoa-Fernandes *et al.*, 1995b). In favourable weather conditions eight to ten generations can occur in a single year.

Damage. The tomato leafminer is well known as a serious pest of tomato crops. The larva feeds upon tomato plants, producing wide mines in the leaves, burrowing in stalks and consuming apical buds and green and ripe fruits. It can cause a yield loss of 100% (Apablaza, 1992).

Distribution and host plants. This moth was first reported as a tomato pest in different South American countries, and today it is known in Central America, the Caribbean, South America, Europe, Asia and Africa. The main host plant is the tomato, but the pest also attacks other crop plants of the *Solanaceae* family such as potato, aubergine, pepino, pepper and tobacco (Galarza, 1984; Desneux *et al.*, 2010), and other *Solanaceae* (Clarke, 1962; Vargas, 1970; Coelho and Franta, 1987) including various wild species (Garcia and Espul, 1982).

Management for organic farming. The control needs the early detection of the pest on tomato plants, especially on the apical buds, flowers or new fruits, where the black frass is visible. In severe attacks the moth infests the leaves on other parts of the plant; mines are evident on infested leaves (Imenes *et al.*, 1990). Significant advances have been made with sex pheromones, important male attractants (Desneux *et al.*, 2010); the use of pheromone-based strategies has become an effective control technique (Cocco *et al.*, 2013) to the point that the pheromone traps are retained as the first means of defence against this pest in greenhouses (and in open fields), where they may be employed both for monitoring and male annihilation purposes.

Cultural practices may hinder population development of this moth. Such practices include ploughing, manuring, irrigation, crop rotation, solarization, elimination of infested leaves and destruction of infested tomato plants. The removal of alternative reservoir hosts (e.g. the nightshades) is strongly recommended before and during the cropping cycle. In greenhouses, one of the management tactics used to reduce the initial level of populations is to keep infested greenhouses closed after harvest to prevent the migration of adults to open-field crops. Alternating host crops, mainly tomato and potato, with non-host cultures can ensure a long-term reduction in pest pressure.

Host-plant resistance has been investigated by developing tomato accessions with high zingiberene and/or acylsugar contents resulting in low ovipostion rates and larval feeding of *T. absoluta* (de Azevedo *et al.*, 2003; Maluf *et al.*, 2010).

Different natural enemies are used to control the tomato leafminer in open-field and greenhouse tomato cultivation. The most common predators are the mirid bugs *N. tenuis* and *Macrolophus pygmaeus* (Rambur) that are commercially available and widely used in North Africa and Europe. Different trials have demonstrated the efficacy of *Bacillus thuringiensis* in controlling *T. absoluta*. The first instar larvae are the most susceptible target and, on this basis, various commercially available formulations have been recommended for use without side effects on beneficial arthropods (Mollá *et al.*, 2011).

Laboratory and field trials revealed high larval mortality (78.6–100%) and low pupal mortality (10%) caused by the entomopathogenic nematode *Steinernema feltiae* (Filipjev) (Garcia del Pino *et al.*, 2013). However, the entomopathogenic fungi *Metarhizium anisopliae* (Metchnikoff) Sorokin and *B. bassiana* develop on the eggs, larvae and adults of the pest. Recent research has revealed up to 54% mortality of *T. absoluta* adults by *M. anisopliae* (Pires *et al.*, 2009, 2010).

Noctuidae

The family Noctuidae currently includes about 20,000 species, of which more than 1100 are known only for Europe. Several species infest vegetables in greenhouses, among which are: (i) the black cutworm, *Agrotis ipsilon* (Hufnagel); (ii) the turnip moth, *Agrotis segetum* (Denis *et* Sciffermüller); (iii) *Chrysodeixis chalcites* (Esper); (iv) *Spodoptera littoralis* (Boisduval); (v) the cabbage looper, *Trichoplusia ni* (Hübner); (vi) the setaceous Hebrew character, *Xestia c-nigrum* (Linneus); and (viii) *Helicoverpa armigera* (Hübner) and other species. Polyphagia appears as a widespread adaptation, and some species possess a worldwide distribution. The polyvoltine adaptation is generalized and the number of generations per year varies according to the species and the environmental conditions. Although some noctuid also develop on flower plants, such as chrysanthemum, roses and carnation, usually species of this family are more common on vegetables (peppers, tomatoes, aubergine, strawberries, zucchini, etc.). The most important species infesting vegetable crops in greenhouses are *C. chalcites*, *S. littoralis* and *H. armigera*.

CHRYSODEIXIS CHALCITES (GOLDEN TWIN SPOT MOTH)

Description. The adult wingspan is about 40 mm. The forewing is 15–17 mm, usually gold, although some individuals have forewings that are more of a bronze colour. Two oval silver spots are set on the forewing, and in some individuals these are united. The hindwing is paler. Two prominent crests are set on the thorax (Pinhey, 1975; Bretherton *et al.*, 1983). Mature larvae are 34–38 mm long, pale yellow–green in colour and with a glossy green to grey head edged with a black streak (Haggett, 1980).

Life cycle. C. chalcites is a polyvoltine species, developing from eight to nine generations/year in Egypt (Rashid *et al.*, 1971; Harakly and Farag, 1975). After emergence the females mate and begin to lay the eggs within 2 or 3 days (Gasim and Younis, 1989). The female briefly touches the leaf to deposit one, two or a few eggs at a time on the upper and lower leaf surfaces at night (Harakly and Farag, 1975). At 20°C, the egg incubation lasts between 5 days and 26 days (Gaumont and Moreau, 1961). The first instar larvae graze on the underside of leaves feeding on parenchyma. They can be quite difficult to detect. A larva will drop from the leaf and hang on a silken thread if disturbed (Goodey, 1991). The mature larva stops feeding and enters a prepupal stage. It spins a cocoon within which it pupates. The cocoon is usually attached to the underside of a leaf but can be in the soil (Harakly and Farag, 1975). As regards natural enemies, the hymenopteran eulophid *Eulophus pennicornis* (Nees) was found in glasshouse sweet peppers in Belgium parasitizing *C. chalcites* (Van de Veire, 1993). Two other parasitoids have been recorded from Spain, *Cotesia kazak* (Telenga) and *Meteorus pulchricornis* (Wesmael) (Braconidae), and these parasitized 31.4% of larvae (Cabello, 1989).

Damage. C. chalcites is a polyphagous moth feeding on the foliage and fruit of vegetable, fruit and ornamental crops. In protected crops, it can occur at any time of the year (Lempke, 1982; Vos and Rutten, 1995) and can reach high levels of infestation on vegetables and ornamental plants. The moth is reported as a serious pest in Bulgaria (Loginova, 1992) and Turkey (Uygun and Ozgur, 1980) where it affects tomato, cucumber and peppers. In Sicily it represents one of the four main noctuid pests of protected crops (Vacante, 2000; Vacante and Benuzzi, 2007) and it is a continual pest in glasshouses in The Netherlands (de Vos and Rutten, 1995) and Belgium (Van de Veire, 1993).

Distribution and host plants. The distribution of *C. chalcites* primarily occurs between 45°N and 35°S, from southern Europe and the Mediterranean and the Middle East to southern Africa. Immigrant adults from North Africa or southern Europe, borne on strong southerly winds, are sometimes reported in central and northern Europe (Austria, Denmark, Germany, Sweden, Switzerland and the UK) in the late summer or autumn (Jor, 1973; Bretherton *et al.*, 1983; Hachler *et al.*, 1998; Palmqvist, 1998, 2002). About 50 records of *C. chalcites* as a migrant to the UK are documented between 1943 and 1990 (Bretherton *et al.*, 1983). Outdoor breeding populations occur in Europe as far north as northern Spain and northern Italy. Lempke (1982) and de Vos and Rutten (1995) found that *C. chalcites* is present all year round in glasshouses in The Netherlands. Van de Veire (1993) reported populations established in glasshouses in Belgium. However, it is not known whether *C. chalcites* can overwinter outdoors in The Netherlands (Lempke, 1982) or elsewhere in northern Europe.

C. chalcites is highly polyphagous, feeding on many fruit, vegetable and ornamental crops and weeds belonging to different plant families, including the *Acanthaceae*, *Asteraceae*, *Bignoniaceae*, *Boraginaceae*, *Brassicaceae*, *Convolvulaceae*, *Crassulaceae*, *Lamiaceae*, *Fabaceae*, *Malvaceae*, *Orchidaceae*, *Rosaceae*, *Scrophulariaceae*, *Solanaceae*, *Verbenaceae* and *Violaceae*. It can be a pest of crops grown outdoors and under protection, including both shade and glasshouses.

Management for organic farming. Visual inspections are challenging and not always decisive to understand the degree of infestation. Monitoring adult populations can be carried out with pheromone traps or with light traps; both means are able to provide guidance on the presence of adults in the greenhouse and therefore to detect the first oviposition and presence of the early larval stages. However, it is necessary to take into account that migratory populations from North Africa and/or the Middle East do not show a constant relationship between the number of males captured in the traps and the number of larvae infesting plants. The detection of volatile substances (herbivore-induced plant volatiles or HIPVs) emitted by plants subjected to biotic stress due to the attack of this moth and other species may enhance the presence of such species, as occurs in *T. ni* (Miresmailli *et al.*, 2010).

From a practical point of view, the early detection of eggs and larvae of the first larval stage allows the successful use of microbiological control, for example with *B. thuringiensis*, distributed one or more times by uniformly wetting the plant and operating preferably in the afternoon, or the use of alternative technical means such as use of azadirachtin.

In Italian glasshouses, the predatory pentatomid heteropterans *Podisus maculiventris* and *Podisus nigrispinus*, both from North America, have been tested as control agents, but without success (Vacante *et al.*, 1996). Different strains of *B. thuringiensis* gave full control (100% efficacy) of *C. chalcites* when sprayed on tomatoes grown under net protection or in non-heated greenhouses in Sicily, Italy (Vacante *et al.*, 2001b).

SPODOPTERA LITTORALIS (COTTON LEAFWORM)

Description. The adult is 15–20 mm long and grey-brown in colour. The wingspan is 30–38 mm, and the forewings are grey to reddish brown with paler lines along the veins (in males, bluish areas occur on the wing base and tip); the ocellus is marked by two or three oblique whitish stripes. The hindwings are greyish white and irridescent with grey margins and usually lack darker veins (CABI/EPPO, 1997). The mature larvae are from 40 mm to 45 mm long, hairless, cylindrical, tapering towards the posterior and variable in colour (blackish grey to dark green, becoming reddish brown or whitish yellow) (Bishari, 1934; Brown and Dewhurst, 1975).

Life cycle. Adults of *S. littoralis* fly at night, mostly between 10 p.m. and midnight (Nasr *et al.*, 1981). The flight activity depends on atmospheric conditions: increases in RH and decreases in air temperature induce flight (Hassan *et al.*, 1960). The flight range during a 4-h period can be up to 1.5 km (Salama and Shoukry, 1972). The females lay most of their egg masses (20–1000 eggs) on the lower surface of younger leaves or upper parts of the plant (Khalifa *et al.*, 1982). The eggs are spherical, somewhat flattened, and laid in clusters arranged in more-or-less regular rows in one to three layers, with hair scales derived from the tip of the abdomen of the female (Pinhey, 1975). The first three larval instars feed mainly in the dark (this behaviour pattern may be less noticeable in early instars) on the lower surface of the leaves, whereas later instars feed on both surfaces. Approximately 50% of mated females lay their eggs on the same night of mating, before sunrise (Hassan *et al.*, 1960). The minimum constant temperature for normal development in all stages is 13–14°C. Resistance to cold generally increases through the larval stages and is greatest in the pupal stage (Miller, 1977). In Egypt seven overlapping generations of *S. littoralis* were observed when feeding on cotton, and three peak infestation periods occur (El-Shafei *et al.*, 1981; Khalifa *et al.*, 1982). The natural enemies number dozens of parasitoids and predators. Generalist predators include ladybird beetles, which feed on egg masses and young larvae.

Damage. The damage arises from extensive feeding by larvae, leading to complete stripping of the plants. Symptoms on fruits consist of visible frass, internal feeding, an obvious exit hole and premature drop. Leaves show external feeding and shredding.

Distribution and host plants. The distribution of *S. littoralis* includes various countries of Asia, Africa and Europe. The northern distribution limit of this species in Europe corresponds to the climatic zone in which winter frosts are infrequent. It occurs throughout Africa and extends eastwards into Turkey and north into eastern Spain, southern

France and northern Italy.The host range covers over 40 families, containing at least 87 plant species of economic importance (Salama *et al.*, 1970).

Management for organic farming. The management for organic farming involves different technical choices that may be used together to work in harmony. The hand collection of egg masses, the use of microbial pesticides and that of slow-release pheromone formulations for mating disruption hinder the development of populations of the moth. Pheromones have also been used for mass trapping using a lure-and-kill strategy (McVeigh and Bettany, 1987) and for monitoring populations.

Parasitoids (braconids, encyrtids, tachinids and ichneumonids) and predators have been extensively investigated for biological control of *S. littoralis.* A nuclear polyhedrosis virus has been evaluated against *Spodoptera* (Elnagar and El-Sheikh, 1990; Jones *et al.*, 1994), whereas fungi and microsporidia have also been recorded as pathogens. Parasitic nematodes such as *Neoaplectana carpocapsae* Weiser have also been evaluated. *B. thuringiensis* is effective against the moth (Navon *et al.*, 1983), but it is necessary to take into account that *S. littoralis* is resistant to different strains (Salama *et al.*, 1989).

HELICOVERPA ARMIGERA (COTTON BOLLWORM)

Description. The wingspan is 3.5–4 cm and wings are broad across the thorax and then tapering, and they are 14–18 mm long. The male is usually greenish grey and the female orange-brown in colour. The forewings have a line of seven to eight blackish spots on the margin and a broad, irregular, transverse brown band. The hindwings are pale-straw colour with a broad dark-brown border containing a paler patch; they have yellowish margins and strongly marked veins and a dark comma-shaped marking in the middle. The antennae are covered with fine hairs (Cayrol, 1972; Delattre, 1973). The mature larvae are about 30–40 mm long, with the head brown and mottled, and the prothoracic and supra-anal plates and legs pale brown, the claws and the spiracles are black, and the skin surface consists of close-set, minute tubercles.

Life cycle. According to geographical location, *H. armigera* develops from two complete generations/year to a variable number of generations (from a partial third to six generations)/year. The species overwinters in the pupal stage in the soil. The adults are present on crops (or on weeds) from May to October, and the females may lay up to 3180 eggs (up to 457 in 24 h), singly and mainly at night. A high number of pathogens, predators and parasitoids of *H. armigera* are reported in the world. The worldwide distribution, abundance and potential for biocontrol by the natural enemies of economically important *Heliothis* and *Helicoverpa* spp. have been reviewed by King *et al.* (1982), King and Coleman (1989) and King and Jackson (1989).

Damage. Young fruits of tomato are invaded and fall; larger larvae may bore into older fruits. Secondary infections by other organisms lead to rotting.

Distribution and host plants. This moth is reported from many countries of Asia, Africa, Europe, Oceania and also from Central America, the Caribbean and South America (EPPO, 2014). It is a polyphagous species infesting a wide range of cultivated plants; for the whole list of host plants see Matthews (1991) and Manjanuth *et al.* (1989). Among vegetables, the pest is common on tomato (and in some countries the pepper) cultivated in the greenhouse.

Management for organic farming. Although the trapping of adult moths has sometimes been applied to assess the need for control, usually it is applied as a qualitative measure indicating the start of an infestation or a migratory 'wave front', highlighting the time to begin scouting for eggs and larvae in the crop. The relationship between catch and later larval populations is often poor (Rothschild *et al.*, 1982), and the adult captures differ markedly between traps separated by only a few tens of metres, although it was closest when moth densities were low and at the beginning of the seasonal cycle.

The success of biological control based on inoculations and/or mass release of beneficials appears to be uncertain. In fact, the number of encouraging results of research studies carried out in various countries of the world are associated with various unsuccessful attempts. Better results are produced by microbiological control, for example with *B. thuringiensis* and/or *H. armigera* nuclear polyhedrosis virus (HaNPV). These means are present and active under natural conditions, and particularly HaNPV exerts some impact on the pest populations, although seldom reaching the epizootic proportions necessary to achieve effective control (King and Jackson, 1989).

Minor Pests

Grillidae

Gryllus bimaculatus De Geere is sometimes harmful to the fruits of melon and pepper in Mediterranean greenhouses (Vacante and Benuzzi, 2007). The species sometimes causes rounded or curved erosion on the apical portion of the fruit and in that adjacent to the pedicle. However, the origin of this damage is uncertain being also attributed to the action of mice or birds. The natural enemies include nematodes, Diptera and Hymenoptera. The control of weeds inside the greenhouse and outdoors is a good technique of prevention.

Gryllotalpidae

This family includes a few dozen species. Of these, the mole cricket *Gryllotalpa gryllotalpa* (Linneus) is common. The species has a large body and is adapted to life underground in sufficiently deep soils rich in organic matter. Its feeding regime is omnivorous with a prevailing entomophagy against other insects (eggs of grasshoppers and other mole crickets) and other invertebrates. However, its action of excavation of tunnels may completely cut the roots of wild and cultivated plants, such as vegetables. In some cases, the mole cricket produces direct damage to tubers and erodes the plant at the collar. In greenhouses the presence of this pest is uncommon, and it relates to the failure to disinfect the substrate and/or the use of previously uncultivated fields for the production of vegetables in a greenhouse. Solanaceous plants and cucurbits cultivated in tunnels are frequently damaged.

Control of the mole cricket includes deeply ploughing the soil before installation of the greenhouse structure, and ascertaining the presence of the pest in the soil in order to mechanically collect its nests. Appreciable results have been obtained with the use of the nematode *Steirnema scapterisci* (Nguyen *et* Smart). The natural enemies also include the mite *Mycterothrombium neglectum* (Bruyant) (Microtrombidiidae) (Smart *et al.*, 1990).

Pentatomidae

This family includes insects of large size that are often brightly coloured with stinging/sucking mouthparts. The most common species infesting vegetables in greenhouses are the green vegetable bug, *Nezara viridula* (Linneus), the bug *Aelia rostrata* Boheman and *Acrosternum acutum* (Dallas). Their presence is occasional, and usually affects a few plants in the outer rows. The pest may be controlled – when and where necessary – with the use of pyrethrins.

Sciaridae

The Sciaridae have a worldwide geographical distribution. Different species, such as *Pryxia scabiei* (Hopkins), *Bradysia paupera* Tuomikoski, *Lycoriella* sp. and *Platisciara* sp. damage fungi and cultivated plants. Damage to seedlings of vegetables and flowers (e.g. poinsettia) are reported both in the greenhouse and in the nursery. Biological control involves the use of predatory mites such as *Hypoaspis miles* Berlese and *Hypoaspis aculeifer* (Canestrini) (Laelapidae) and entomopathogenic nematodes.

Crambidae

The Crambidae include different species of agricultural importance, among which the most common on vegetables cultivated in the greenhouse are: (i) the European maize borer, *Ostrinia nubilalis* (Hübner); (ii) the rusty-dot pearl, *Udea ferrugalis* (Hübner); (iii) the melonworm moth, *Diaphania ialinata* (Linneus); (iv) the pickleworm *Diaphania nitidalis* (Stoll); and (v) the European pepper moth, *Duponchelia fovealis* (Zeller).

The European maize borer has a wide geographical distribution (Europe, North Africa, Asia Minor and North America); it is commonly harmful to maize and may damage vegetables (beans, peppers, asparagus and other species) in greenhouses, especially when adjacent to maize fields. On pepper the larvae penetrate and undermine the stems of the berries, and on bean they feed on the developing pods. The use of insect screens limits the introduction of the pest into the greenhouse. The biological control includes the release of the tricogrammatid *Trichogramma maidis* (Pintureau *et* Vogele) integrated with treatments with *B. thuringiensis* var. *kurstaki*.

For controlling the pickleworm *D. nitidalis* a push–pull cropping approach (Cook *et al.*, 2007) could prove to be effective with a possible reduction in the number of eggs laid when squash (*Cucurbita pepo* L.) is used as a trap crop and watermelon (*Citrullus lanatus* [Thunb.] Matsum. & Nakai) is used as a deterrent intercrop to manage pickleworm on cantaloupe (*Cucumis melo* L.) (Leiner and Spafford, 2016).

The European pepper moth, *D. fovealis*, is polyphagous and polyvoltine, and has a wide geographical distribution. The moth is native to southern Europe and North Africa and is common in most Italian regions, including the large islands (Bassi *et al.*, 1995; Bonsignore and Vacante, 2010). Since the 1980s its distribution area has gradually expanded, probably due to the importation of infested ornamental plants, so that over a few decades it now affects several central European countries and some in the north (Belgium, Denmark, Finland, Germany, England, The Netherlands, the Czech Republic, Sweden and Hungary). The same spread was recorded in the Mediterranean, where the species has been reported in Portugal, Spain (including the Canary Islands), Gibraltar, France, Italy (including Sicily) and Sardinia, Greece, Macedonia, Malta and Cyprus; moreover, the moth was found in Canada (CABI, 2008), the USA (Epstein, 2005), Turkey (Efil *et al.*, 2014), Iran (Mozafary, 1993), South Africa (Solis, 2006) and recently in Brazil (Zawadneak *et al.*, 2016). The natural enemies include the laelapid mites *H. miles* and *H. aculeifer* and the rove beetle *Atheta coriaria* Kraatz (Staphylinidae), predators of its eggs and young larvae. Soil-borne organisms must be properly inoculated in the substrate, paying attention to the influence that it can have on their bio-ecology and consequently on their action of control (Messelink and van Wensveen, 2003; Blok and Messelink, 2009). A study of *D. fovealis* infestations in greenhouses in The Netherlands showed that the most effective biocontrol agent is the soil-dwelling predatory mite *H. miles*, commercially available in the USA (Brambila and Stocks, 2010).

Tortricidae

This family includes various pests of greenhouse crops, some of which possess a wide and expanding geographical distribution (Wilkerson *et al.*, 2005). The most important species are *Pandemis dumetana* (Treitschke) and the strawberry leafroller, *Ancylis comptana* (Fröhlich), but their attacks are occasional.

Coccinellidae

This family mainly includes predatory species of small arthropods (insects and mites) and only a narrow range of species are adapted to phytophagy. The latter includes the African melon ladybird beetle, *Henosepilachna elateri* (Rossi), which is sometimes injurious to *Cucurbitaceae* cultivated in the greenhouse. The geographical distribution of this species includes the Mediterranean region, North Africa and Turkestan, where it attacks mainly wild curcurbitaceous plants (e.g. *Ecballium elaterium* Linneus). In milder areas the pest may develop from two to five generations/year, and in the greenhouse its attacks generally occur on the external rows. The coccinellid lives on the undersides of leaves where it erodes the surface, saving the veins and the upper epidermis. Usually greenhouse attacks are not severe but the control of this pest needs to be assessed from time to time. The control of cucurbits that appear spontaneously around greenhouses is a good prevention practice.

Crysomelidae

This family includes many species that are largely phytophagous in the adult stage and partly in the larval stage. In the greenhouses of the northern Mediterranean area the Colorado potato beetle, *Leptinotarsa decemlineata* (Say) is reported. The species is native to North America and lives on wild and cultivated *Solanaceae* (potato, aubergine, pepper, tomato). Although the pest prefers potato in the greenhouse, it infests aubergine and secondarily tomato. Greenhouse-grown plants, however, appeared to be less suitable for *L. decemlineata* growth than field-grown plants, and host plant suitability declined with host plant age (Hare and Andreadis, 1983). The damage caused consists of foliar erosion produced mainly by larvae, which can also devour stems and branches of young plants. The natural enemies include beetles, ground beetles (*Carabus* sp., *Calosoma* sp.), spider beetles, Hemiptera pentatomids (*Zicrona caerulea* Linneus and *Picromes bidens* Linneus) and the chrysopid *C. carnea*.

Methods of biological control that are commercially available include the use of strains of *B. thuringiensis* var. *tenebrionis* and formulations that are azadirachtin based (Zehnder and Warthen, 1988). The use of a synthetic plant attractant and an antifeedant as components of a stimulo-deterrent diversion strategy for management of the Colorado potato beetle has been proposed (Martel *et al.*, 2005).

Curculionidae

The Curculionidae vary in size, and bear typical distinctive long snouts and geniculate antennae with small clubs. The family is very large and the species that afflict greenhouse crops are relatively few. The most common pests of vegetables in the greenhouse are *Otiorrhynchus rugosustriatus* Goeze, *Otiorrhynchus cribicollis* Gyllen and sometimes other *Otiorrhynchus* spp. They are polyphagous and usually develop one generation/year, overwintering as larvae of different stages. Among vegetables, these curculionids primarily infest strawberry and in some environments also various ornamental plants in nurseries. The damage is due to erosion produced by the feeding activity of the larvae on the roots and the area of the collar where they produce typical notches that may induce the death of the plant, while the adults erode the leaf edges producing typical circular or lunate cuts. On strawberry a high density population may induce extensive damage with the loss of 100% of the crop; the annual (rather than biennial) crop of strawberry significantly reduces the extent of the attacks.

Biological means of control include the use of entomophagous nematodes, such as *S. feltiae*, *Steinernema kraussei* (Steiner) and *Heterorhabditis* sp.

Conclusions

What has been discussed here in this chapter has shown that pest management in organically farmed vegetables grown in the greenhouse is a challenge but it can have a positive conclusion. The strategy that needs to be implemented is of a protocol based primarily on prevention, associated, when and where possible, with curative measures. In this sense, the application of the threshold concept is largely unproductive and misleading, resulting in the control allowed by an holistic concept of defence, where all technical choices, first of all the agronomic choices, must contribute together.

Each analysis takes place in the context of highly specialized agriculture that is notoriously subject to extensive changes in trade between different areas of the world, and to rapid technical changes. The list of pests reported here is not exhaustive, and neither is the list of natural enemies. From this point of view, the contribution represents a first approach to a complex topic, where some solutions to problems are yet to come. This deficiency suggests the need for more studies on the biocenosis of the agroecosystem, especially the ecological interrelationships between the different trophic levels.

Finally, from a practical point of view, it cannot be underestimated that greenhouse crops in addition to the problem of pests are subjected to attacks from various pathogens, which are technically possible to control (see Chapter 2, this volume).

References

Allen, M.W. (1952) Taxonomic status of the bud and leaf nematodes related to *Alphelenchoides fragariae* (Ritzema Bos, 1891). *Proceedings of the Helminthological Society* 19(2), 108–120.

Alvarado, P., Balta, O. and Alomar, O. (1997) Efficiency of four heteroptera as predators of *Aphis gossypii* and *Macrosiphum euphorbiae* (Hom.: Aphididae). *Entomophaga* 42(1–2), 215–226.

Amsing, J. (2004) Wortelknobbelaaltjesproblematiek in de glastuinbouw. *Gewasbescherming* 35, 260–263.

Apablaza, J. (1992) La polilla del tomate y su manejo. *Tattersal* 79, 12–13.

Arthurs, S. and Heinz, K.M. (2006) Evaluation of the nematodes *Steinernema feltiae* and *Thripinema nicklewoodi* as biological control agents of western flower thrips *Frankliniella occidentalis* infesting chrysanthemum. *Biocontrol Science and Technology* 16(2), 141–155.

Arthurs, S., McKenzie, C.L., Chen, J., Dogramaci, M., Brennan, M., Houben, K. and Osborne, L. (2009) Evaluation of *Neoseiulus cucumeris* and *Amblyseius swirskii* (Acari: Phytoseiidae) as biological control agents of chilli thrips, *Scirtothrips dorsalis* (Thysanoptera: Thripidae) on pepper. *Biological Control* 49, 91–96.

Atakan, E. (2006) Associations between *Frankliniella* spp. and *Orius niger* populations in cotton. *Phytoparasitica* 34(3), 221–234.

Azevedo, S.M. de, Faria, M.V., Maluf, W.R., Oliveira, A.C.B. de and Freitas, J.A. de (2003) Zingiberene-mediated resistance to the South American tomato pinworm derived from *Lycopersicon hirsutum* var. *hirsutum*. *Euphytica* 134(3), 347–351.

Back, M.A., Haydock, P.P.J. and Jenkinson, P. (2002) Disease complexes involving plant parasitic nematodes and soil borne pathogens. *Plant Pathology* 51, 683–697.

Bahşi, Ş.Ü. and Tunç, İ. (2008) Development, survival and reproduction of *Orius niger* (Hemiptera: Anthocoridae) under different photoperiod and temperature regimes. *Biocontrol Science and Technology* 18, 767–778.

Barbary, A., Djian-Caporalino, C., Palloix, A. and Castagnone-Sereno, P. (2015) Host genetic resistance to root-knot nematodes, *Meloidogyne* spp., in Solanaceae: from genes to the field. *Pest Management Science* 71(12), 1591–1598. doi: 10.1002/ps.4091.

Bassi, G., Passerin d'Entreves, P., Speidel, W. and Zangheri, S. (1995) Lepidoptera Pyraloidea. In: Minelli, A., Ruffo, S. and La Posta, S. (eds) *Checklist delle Specie della Fauna Italiana*. Calderini, Bologna, Italy, p. 87.

Benedict, J.H. and Cothran, W.R. (1980) Damsel bugs useful as predators but need help. *California Agriculture* 34(8–9), 11–12.

Birchfield, W. (1954) The hot water treatment of nematode-infested nursery stock. *Proceedings of the Florida State Horticultural Society* 67, 94–96.

Birchfield, W. and van Pelt, H.M. (1958) Thermotherapy for nematodes of ornamental plants. *Plant Disease Report* 42, 451–455.

Bishari, I. (1934) The cotton worm *Prodenia litura* F. in Egypt. *Bulletin of the Entomological Society of Egypt* 18, 223–404.

Blackman, R.L. (1974) Life-cycle variation of *Myzus persicae* (Sulz.) (Hom., Aphididae) in different parts of the world, in relation to genotype and environment. *Bulletin of Entomological Research* 63(4), 595–607.

Blackman, R.L. and Eastop, V.F. (1984) *Aphids on the World's Crops. An Identification and Information Guide*. Wiley, Chichester, UK.

Blok, C. and Messelink, G.J. (2009) Improving control of *Duponchelia fovealis* (Lepidoptera: Pyralidae) by rooting media related strategies. *Acta Horticulture* 819, 203–208.

Bogran, C.E. (2005) *Biology and Management of* Liriomyza *Leafminers in Greenhouse Ornamental Crops*. Agrilife Extension Texas A&M System. Texas A&M University, Texas. Available at: http://extentopubs.tamu.edu/eee_00030.html (accessed 9 December 2016).

Bonsignore, C.P. (2015) Effect of environmental factors on the flight activity of *Trialeurodes vaporariorum* (Westwood) under greenhouse conditions. *Entomological Science* 18, 207–216.

Bonsignore, C.P. (2016) Environmental factors affecting the behavior of *Coenosia attenuata* Stein, a predator of *Trialeurodes vaporariorum* (Westwood) in tomato greenhouses. *Entomologia Experimentalis et Applicata* 158, 87–96.

Bonsignore, C.P. and Vacante, V. (2010) *Duponchelia fovealis* (Zeller). Una nuova emergenza per la fragola? *Protezione delle Colture* 3, 40–43.

Bonsignore, C.P. and Vacante, V. (2012) Influences of botanical pesticides and biological agents on *Orius laevigatus – Frankliniella occidentalis* dynamics under greenhouse conditions. *Journal of Plant Protection Research* 52, 15–23.

Brambila, J. and Stocks, I. (2010) The European pepper moth, *Duponchelia fovealis* Zeller (Lepidoptera: Crambidae), a Mediterranean pest moth discovered in central Florida. FDACS – Division of Plant Industry. Available at: http://www.freshfromflorida.com/content/download/23893/486212/duponchelia-fovealis.pdf (accessed 10 October 2016).

Bretherton, R.F., Goater, B. and Lorimer, R.I. (1983) Noctuidae (Part II) and Agaristidae. In: Heath, J. and. Emmet, M.A. (eds) *The Butterflies and Moths of Great Britain and Ireland*, Vol. 10. Harley Books, Colchester, UK, pp. 36–402.

Brown, E.S. and Dewhurst, C.F. (1975) The genus *Spodoptera* (Lepidoptera, Noctuidae) in Africa and the Near East. *Bulletin of Entomological Research* 65(2), 221–262.

Brown, J.K., Frohlich, D.R. and Rosell, R.C. (1995) The sweetpotato or silverleaf whiteflies: biotypes of *Bemisia tabaci* or a species complex? *Annual Review of Entomology* 40(1), 511–534.

Cabello, I. (1989) Natural enemies of noctuid pests (Lep., Noctuidae) on alfalfa, corn, cotton and soybean crops in southern Spain. *Journal of Applied Entomology* 108(1), 80–88.

CABI (2008) Map 704. In: *Distribution Maps of Plant Pests*. June 2008. CAB International, Wallingford, UK.

CABI (2016) *Trialeurodes vaporariorum* (Whitefly, Greenhouse). Available at: http://www.cabi.org/isc/datasheet/54660 (accessed 9 December 2016).

CABI (2017) *Aphelenchoides fragariae* (strawberry crimp nematode). In: CABI Invasive Species Compendium. Available at: http://www.cabi.org/isc/datasheet/6381 (accessed 30 June 2017).

CABI/EPPO (European and Mediterranean Plant Protection Organization) (1997) *Quarantine Pests for Europe*, 2nd edn. CAB International, Wallingford, UK.

Calvo, F.J., Knapp, M., van Houten, Y.M., Hoogerbrugge, H. and Belda, J.E. (2015) *Amblyseius swirskii*: what made this predatory mite such a successful biocontrol agent? *Experimental and Applied Acarology* 65, 419–433.

Castillo, P., Di Vito, M., Vovlas, N. and Jimenez-Diaz, R.M. (2000) Host–parasite relationship in root-knot disease of white mulberry. *Plant Disease* 85, 277–281.

Cayrol, J.C., Castet, R. and Samson, R.A. (1986) Comparative activity of different *Hirsutella* species towards three plant parasitic nematodes. *Revue de Nematologie* 9(4), 412–414.

Cayrol, R.A. (1972) Famille des Noctuidae. Sous-famille des Melicleptriinae. *Helicoverpa armigera* Hb. In: Balachowsky, A.S. (ed.) *Entomologie Appliquée à l'Agriculture*, Vol. 2. Masson et Cie, Paris, pp. 1431–1444.

Chen, J.Q., Rahbé, Y., Delobel, B., Sauvion, N., Guillaud, J. and Febvay, G. (1997) Melon resistance to the aphid *Aphis gossypii*: behavioural analysis and chemical correlations with nitrogenous compounds. *Entomologia Experimentalis et Applicata* 85(1), 33–44.

Cho, J.J., Mau, R.F.L., German, T.L., Hartmann, R.W. and Yudin, L.S. (1989) A multidisciplinary approach to management of tomato spotted wilt virus in Hawaii. *Plant Disease* 73(5), 375–383.

Clarke, J.F. (1962) New species of microlepidoptera from Japan. *Entomology News* 73, 102.

Clerjeau, M., Rancillac, M. and Veschambre, D. (1983) The position regarding strawberry decline in France. *Pepinieristes Horticuteurs Maraichers – Revue Horticole* 237, 39–42.

Cocco, A., Deliperi, S. and Delrio, G. (2013) Control of *Tuta absoluta* (Meyrick) (Lepidoptera: Gelechiidae) in greenhouse tomato crops using the mating disruption technique. *Journal of Applied Entomology* 137(1–2), 16–28.

Coelho, M.C.F. and Franta, F.H. (1987) Biologia, quetotaxia da larva e descritao da pupa e adulto da trata-do-tomateiro. *Pesquisa Agropecuária Brasileira* 22(2), 129–135.

Cook, R., Thomas, B.J. and Mizen, K.A. (1989) Dissemination of white clover mosaic virus and stem nematode, *Ditylenchus dipsaci*, by the slug, *Deroceras reticulatum*. *Monograph British Crop Protection Council* 41, 107–112.

Cook, S.M., Khan, Z.R. and Pickett, J. (2007) The use of push–pull strategies in integrated pest management. *Annual Review of Entomology* 52, 375–400.

Davies, K.G. (2009) Understanding the interaction between an obligate hyperparasitic bacterium, *Pasteuria penetrans* and its obligate plant-parasitic host, *Meloidogyne* spp. *Advances in Parasitology* 68, 211–245.

De Barro, P.J., Liu, S.-S., Boykin, L.M. and Dinsdale, A.B. (2011) *Bemisia tabaci*: a statement of species status. *Annual Review of Entomology* 56(1), 1–19.

Delattre, R. (1973) *Parasites et Maladies en Culture Cotonniere. Manuel Phytosanitaire*. [*Pests and Diseases in Cotton Growing. Phytosanitary Handbook*.] Institut de Recherches du Coton et des Textiles Exotiques, Paris, p. 146.

Desneux, N., Wajnberg, E., Wyckhuys, K.A.G., Burgio, G., Arpaia, S., Narvaez-Vasquez, C.A., Gonzalez-Cabrera, J., Ruescas, D.C., Tabone, E., Frandon, J., Pizzol, J., Poncet, C., Cabello, T. and Urbaneja, A. (2010) Biological invasion of European crops by *Tuta abosluta*: ecology, geographic expansion and prospects for biological control. *Journal of Pest Science* 83, 197–215.

Dicker, G.H.L. (1948) A preliminary report on the strawberry eelworm (*Aphelenchoides fragariae* Ritzema Bos). In: *Report of the East Malling Research Station, England 1947*. East Malling Research Station, East Malling, Kent, UK, pp. 144–147.

Douda, O. (2005) Host range and growth of stem and bulb nematode (*Ditylenchus dipsaci*) populations isolated from garlic and chicory. *Plant Protection Science* 41, 104–108.

Duffus, J.E., Liu, H.Y. and Wisler, G.C. (1996) Tomato infectious chlorosis virus – a new clostero-like virus transmitted by *Trialeurodes vaporariorum*. *European Journal of Plant Pathology* 102, 219–226.

Dustan, A.G. and Matthewman, W.G. (1932) Some notes on the cyclamen mite, *Tarsonemus pallidus* (Banks), a pest of strawberry plants. *Annual Report of the Entomological Society of Ontario (1931)* 62, pp. 34–37.

Efil, L., Özgür, O. and Efil, F. (2014) A new pest, *Duponchelia fovealis* Zeller, on strawberries in Turkey – damage, distribution and parazitoid. *Journal of Entomology and Zoology Studies* 2(4), 328–334.

Eisenback, J.D., Hirschmann, H., Sasser, J.N. and Triantaphyllou, A.C. (1981) *A Guide to the Four Most Common Species of Root-Knot Nematodes (Meloidogyne spp.)*, With a Pictorial Key. North Carolina State University and the United States Agency for International Development, Raleigh, North Carolina, p. 48.

Elnagar, S. and El-Sheikh, M.A. (1990) Baculovirus, as a promising biocontrol element in the pest management of the cotton leafworm in Egypt. In: *Integrated Pest Management in Tropical and Subtropical Cropping Systems '89*, 8–15 February 1989, Bad Durkheim, Frankfurt, Germany. Deutsche Landwirtschafts-Gesellschaft, Frankfurt, Germany, pp. 669–683.

El-Shafei, S.A., Iss-hak, R.R. and Nasr, E.S.A. (1981) Seasonal abundance of the cotton leafworm moths, *Spodoptera littoralis* (Boisd.), in relation to the accumulated heat. *Research Bulletin, Faculty of Agriculture, Ain Shams University* 1613, 5 [+7] pp.

EPPO (European and Mediterranean Plant Protection Organization) (2007) Report of a Pest Risk Analysis for *Tetranychus evansi*. Document 08-14562. EPPO, Paris. Available at: http://www.eppo.int/QUARANTINE/Pest_Risk_Analysis/PRAdocs_insects/08-14562%20PRA%20report%20T%20evansi%20final%2008-10%20for%20website.doc (accessed 6 May 2015).

EPPO (European and Mediterranean Plant Protection Organization) (2013a) PQR: EPPO Plant Quarantine Data Retrieval System. Available at: http//www.eppo.orgDATABASES/pqr/pqr.htm (accessed 10 December 2016).

EPPO (European and Mediterranean Plant Protection Organization) (2013b) Diagnostic protocols for regulated pests: pictorial glossary of morphological terms in nematology. *EPPO Technical Document* No. 1056 (Rev. 4). Available at: http://www.eppo.int/QUARANTINE/diag_activities/EPPO_TD_1056_Glossary.pdf (accessed 10 December 2016).

EPPO (European and Mediterranean Plant Protection Organization) (2014) *PQR Database*. EPPO, Paris.

Epstein, M.E. (2005) Lepidoptera, *Duponchelia fovealis* Zeller (Pyraloidea: Crambidae). *California Plant Pest and Disease Report* 22(1), 20–21.

Erdogan, C., Velioglu, A.S., Gurkan, M.O., Denholm, I. and Moores, G.D. (2012) Chlorpyrifos ethyl-oxon sensitive and insensitive acetylcholinesterase variants of greenhouse whitefly *Trialeurodes vaporariorum* (Westw.) (Hemiptera: Aleyrodidae) from Turkey. *Pesticide Biochemistry and Physiology* 104, 273–276.

European Food Safety Authority (EFSA) (2012) Scientific opinion on the risks to plant health posed by *Liriomyza huidobrensis* (Blanchard) and *Liriomyza trifolii* (Burgess) to the EU territory with the identification and evaluation of risk reduction options. *EFSA Journal* 10(12), 190 pp. Available at: www.efsa.europa.eu/efsajournal (accessed 30 June 2017).

Fatnassi, H., Boulard, T., Poncet, C. and Chave, M. (2006) Optimisation of greenhouse insect screening with computational fluid dynamics. *Biosystems Engineering* 93, 301–312.

Fjelddalen, J. and Stenseth, C. (1958) Nye kjemiske midler for bekjempelse av jordbaermidd (*Tarsonemus pallidus* Banks). *Frukt og Baer* 1958, 37–42. (in Norwegian)

Francki, R.I.B., Milne, R.G. and Hatta, T. (1985) *Atlas of Plant Viruses*, Volume II. CRC Press, Boca Raton, Florida, p. 284.

Franklin, M.T. (1950) Two species of Aphelenchoides associated with strawberry bud disease in Britain. *Annual of Applied Biology* 37, 1–10.

Gabarra, R., Alomar, Ò., Castañé, C., Goula, M. and Albajes, R. (2004) Movement of greenhouse whitefly and its predators between in and outside of Mediterranean greenhouses. *Agricultural Ecosystem Environment* 102, 341–348.

Galarza, J. (1984) Laboratory assessment of some solanaceous plants. Possible food-plants of the tomato moth *Scrobipalpula absoluta* (Meyr.) (Lepidoptera: Gelechiidae). *IDIA* 421(424), 30–32.

Garcia, M.F. and Espul, J.C. (1982) Bioecology of the tomato moth (*Scrobipalpula absoluta*) in Mendoza, Argentine Republic. *Revista de Investigaciones Agropecuarias* 17(2), 135–146.

Garcia del Pino, F., Alabern, X. and Morton, A. (2013) Efficacy of soil treatments of entomopathogenic nematodes against the larvae, pupae and adults of *Tuta absoluta* and their interaction with the insecticides used against this insect. *BioControl* 58(6), 723–731.

Gasim, G.Y. and Younis, H.T. (1989) Biological studies on tomato leafworm *Plusia chalcytes* L. (Noctuidae: Lepidoptera) under effect of constant temperatures. *Mesopotamia Journal of Agriculture* 21(2), 325–334.

Gaumont, R. and Moreau, R. (1961) Observations on the bionomics of *Plusia chalcites* Esp. (Lepidoptera, Noctuidae). *Agricultural Zoology Reviews* 60(1–3), 31–36. (in French)

Gerson, U. (1992) Biology and control of the broad mite, *Polyphagotarsonemus latus* (Banks) (Acari: Tarsonemidae). *Experimental and Applied Acarology* 13, 16–178.

Gispert Galvan, M.C., Perring, T.M., Zarate de Lara, G. and Llanderal Cazares, C. (1989) Efecto del riego en la fluctuacion poblacional del acaro del tomate (*Aculops lycopersici* Massee). *Agrociencia* 76, 153–165.

Goodey, B. (1991) *Chrysodeixis chalcites* – observations on the life cycle in captivity. *Entomologist's Record and Journal of Variation* 103, 111–118.

Goodey, T. (1933) *Plant Parasitic Nematodes and the Diseases They Cause*. Methuen, London, p. 306.

Gratwick, M. and Southey, J.F. (eds) (1986) *Hot-water Treatment of Plant Material*, 3rd edn. Her Majesty's Stationery Office (HMSO) Reference book 201. HMSO, London.

Griffiths, G.J.K., Wilby, A., Crawley, M.J. and Thomas, M.B. (2008) Density-dependent effects of predator species-richness in diversity–function studies. *Ecology* 89, 2986–2993.

Hachler, M., Jermini, M. and Brunetti, R. (1998) Two new harmful noctuids on tomatoes in glasshouse in south and western Switzerland. *Revue Suisse Viticulture de Arboriculture Horticulture* 30(5), 281–285. (in French)

Haggett, G.M. (1980) Larvae of the British Lepidoptera not figured by Buckler, Part XI. *Proceedings and Transactions of the British Entomological and Natural History Society* 13(3/4), 95–105.

Hara, A.H., Hata, T.Y., Hu, B.K.S. and Tenbrink, V.L. (1993) Hot water immersion as a potential quarantine treatment against *Pseudaulapsis cockerelli* (Homoptera: Diaspididae). *Journal of Economic Entomology* 86, 1167–1170.

Hara, A.H., Hata, T.Y., Hu, B.K.S., Kaneko, R.T. and Tenbrink, V.L. (1994) Hot-water immersion of Cape jasmine cuttings for disinfestations of green scale (Homoptera: Coccidae). *Journal of Economic Entomology* 87, 1569–1573.

Harakly, F.A. and Farag, S.S. (1975) Biological studies on the tomato looper *Chrysodeixis chalcytes* (Esper) in Egypt. *Bulletin de la Société Entomologique d'Égypte* 59, 295–299.

Hare, J.D. and Andreadis, T.G. (1983) Variation in the susceptibility of *Leptinotarsa decemlineata* (Coleoptera: Chrysomelidae) when reared on different host plants to the fungal pathogen, *Beauveria bassiana* in the field and laboratory. *Environmental Entomology* 12, 1891–1896.

Harris, H.M. and Tate, H.D. (1933) A leafminer attacking the cultivated onion. *Journal of Economic Entomology* 26, 515–516.

Hassan, A.S., Moussa, M.A. and, Nasr, E.A. (1960) Behaviour of larvae and adults of the cotton leaf worm, *Prodenia litura*. *Bulletin de la Société Entomologique d'Égypte* 44, 337–343.

Hoddle, M.S. (1999) The Biology and Management of the Silverleaf Whitefly, *Bemisia argentifolii* Bellows and Perring (Homoptera: Aleyrodidae) on Greenhouse Grown Ornamentals. Available at: http://biocontrol.ucr.edu/bemisia.html (accessed 1 September 2016).

Hulspas-Jordaan, P.M., Christochowitz, E.E., Woets, J. and van Lenteren, J.C. (1987) The parasite–host relationship between *Encarsia formosa* (Hymenoptera: Aphelinide) and *Trialeurodes vaporariorum* (Homoptera: Aleyrodidae). XXIV. Effectiveness of *Encarsia formosa* in the greenhouse at low temperatures. *Journal of Applied Entomology* 103, 368–378.

Ibrahim, I.K.A. and El-Saedy, M.A. (1987) Development of *Meloidogyne incognita* and *M. javanica* in soybean roots. *Nematologica Mediterranea* 15, 47–50.

Imenes, S.D.L., Uchôa-Fernandes, M.A., Campos, T.B. and Takematsu, A.P. (1990) Aspectos biológicos e comportamentais da trata do tomateiro *Scrobipalpula absoluta* (Meyrick, 1917), (Lepidoptera-Gelechiidae). *Arquivos do Instituto Biológico (Sao Paulo)* 57(1–2), 63–68.

Inbar, M. and Gerling, D. (2008) Plant-mediated interactions between whiteflies, herbivores, and natural enemies. *Annual Review of Entomology* 53, 431–448.

Isenhour, D.J. and Yeargan, K.V. (1981) Effect of temperature on the development of *Orius insidiosus,* with notes on laboratory rearing. *Annals of the Entomological Society of America* 74(1), 114–116.

Jagdale, G.B. and Grewal, P.S. (2002) Identification of alternatives for the management of foliar nematodes in floriculture. *Pest Management Science* 58, 451–458.

Jagdale, G.B. and Grewal, P.S. (2004) Effectiveness of a hot water drench for the control of foliar nematodes, *Aphelenchoides fragariae,* in floriculture. *Journal of Nematology* 36, 49–53.

Jagdale, G.B. and Grewal, P.S. (2006) Infection behavior and overwintering survival of foliar nematodes, *Aphelenchoides fragariae,* on hosta. *Journal of Nematology* 38, 130–136.

James, D.G. and Price, T.S. (2002) Fecundity in twospotted spider mite (Acari: Tetranychidae) is increased by direct and systemic exposure to imidacloprid. *Journal of Economic Entomology* 95, 729–732.

Jansson, H.B. and Nordbring-Hertz, B. (1980) Interactions between nematophagous fungi and plant-parasitic nematodes: attraction, induction of trap formation and capture. *Nematologica* 26(4), 383–389.

Jeppson, L.R., Keifer, H.H. and Baker, E.W. (1975) *Mites Injurious to Economic Plants.* University of California Press, Berkeley, California.

Johansen, N.S. (2009) Effect of continuous light on the biology of the greenhouse whitefly, *Trialeurodes vaporariorum* on roses. *Entomologia Experimentalis et Applicata* 133, 244–250.

Jones, K.A., Irving, N.S., Grzywacz, D., Moawad, G.M., Hussein, A.H. and Fargahly, A. (1994) Application rate trials with a nuclear polyhedrosis virus to control *Spodoptera littoralis* (Boisd.) on cotton in Egypt. *Crop Protection* 13(5), 337–340.

Jones, V.P. and Brown, R.D. (1983) Reproductive responses of the broad mite, *Polyphagotarsonemus latus* (Banks) (Acari: Tarsonemidae) to constant temperature-humidity regimes. *Annals of the Entomological Society of America* 76, 466–469.

Jor, J. (1973) *Plusia chalcites* Esp., found on Funen (Lep., Noctuidae). *Entomologiske Meddelelser* 41(3), 187–188.

Jovicich, E., Cantliffe, D.J., Osborne, L.S. and Stoffella, P.J. (2004) Mite population and damage caused by broad mites (*Polyphagotarsonemus latus* [Banks]) infesting bell pepper (*Capsicum annuum* L.) at different seedling developmental stages. *Acta Horticulture* 659(1), 339–344.

Karssen, G. (2002) *The Plant-Parasitic Nematode Genus* Meloidogyne *Göldi, 1892 (Tylenchida) in Europe.* Brill, Leiden, The Netherlands, p. 161.

Karssen, G. and Moens, M. (2006) Root-knot nematodes. In: Perry, R.N. and Moens, M. (eds) *Plant Nematology.* CAB International, Wallingford, UK, pp. 59–90.

Kaspi, R. and Parrella, M.P. (2006) Improving the biological control of leafminers (Diptera: Agromyzidae) using the sterile insect technique. *Journal of Economic Entomology* 99, 1168–1175.

Khalifa, A., Iss-hak, R.R. and Foda, M.E. (1982) Vertical and horizontal distribution of the Egyptian cotton leafworm eggmasses in cotton fields in Egypt. *Research Bulletin, Faculty of Agriculture, Ain Shams University* 1749, 6 [+8] pp.

Kiewnick, S. and Sikora, R.A. (2006) Biological control of the root-knot nematode *Meloidogyne incognita* by *Paecilomyces lilacinus* strain 251. *Biological Control* 38, 179–187.

King, E.G. and Coleman, R.J. (1989) Potential for biological control of *Heliothis* species. *Annual Review of Entomology* 34, 53–75.

King, E.G. and Jackson, R.D. (eds) (1989) *Proceedings of the Workshop on Biological Control of* Heliothis*: Increasing the Effectiveness of Natural Enemies,* November 1985, New Delhi. Far Eastern Regional Research Office, United States Department of Agriculture, New Delhi, India.

King, E.G., Powell, J.E. and Smith, J.W. (1982) Prospects for utilization of parasites and predators for management of *Heliothis* spp. In: Reed, W. and Kumble, V. (eds) *Proceedings of the International Workshop on* Heliothis *Management*, 15–20 November 1981, International Crops Research Institute for the Semi-Arid Tropics (ICRISAT) Center, Patancheru, India. ICRISAT, Patancheru, Andhra Pradesh, India, pp. 103–122.

Kish, L.P., Majchrowicz, I. and Biever, K.D. (1994) Prevalence of natural fungal mortality of green peach aphid (Homoptera: Aphididae) on potatoes and non-solanaceous hosts in Washington and Idaho. *Environmental Entomology* 23(5), 1326–1330.

Kroon, A., Veenendaal, R.L., Bruin, J. and Sabelis, M.W. (2004) Predation risk affects diapause induction in the spider mite *Tetranychus* urticae. *Experimental and Applied Acarology* 34, 307–314.

Kulkarni, G.S. (1923) The 'murda' disease of chilli (*Capsicum*). *Indian Journal of Agriculture* 17, 51–54.

Łabanowska, B.H. (2006) Potential agents for controlling the strawberry mite (*Phytonemus pallidus* ssp. *fragariae* Zimm.) after the withdrawal of endosulfan. *Journal of Fruit and Ornamental Plant Research* 14(3), 67–72.

Lammers, W., Karssen, G., Jelleman, P., Baker, R., Hockland, S., Fleming, C. and Turner, S. (2006) Pest Risk Assessment *Meloidogyne minor*. Available at: https://www.eppo.int/QUARANTINE/Pest_Risk_Analysis/PRAdocs_nematodes/08-14648%20PRA%20%20MELGMI.pdf (accessed 24 January 2017).

Leibee, G.L. (1984) Influence of temperature on development and fecundity of *Liriomyza trifolii* (Burgess) (Diptera: Agromyzidae) on celery. *Environmental Entomology* 13, 497–501.

Leiner, R.C. and Spafford, H. (2016) Oviposition preferences of pickleworm (Lepidoptera: Crambidae) in relation to a potential push–pull cropping management approach. *Environmental Entomology* 45(3), 677–684.

Lempke, B.J. (1982) [Interesting observations and catches of Lepidoptera in 1980]. *Entomologische Berichten (Amsterdam)* 42(8), 116–119. (in Dutch)

Lewis, G.D. and Mai, W.F. (1960) Overwintering and migration of *Ditylenchus dipsaci* in organic soils of New York. *Phytopathology* 50, 341–343.

Lewis, T. (ed.) (1997) *Thrips as Crop Pests*. CAB International, Wallingford, UK, 740 pp.

Lim, U.T. and Van Driesche, R.G. (2004) Assessment of augmentative releases of parasitic nematode *Thripinema nicklewoodi* for control of *Frankliniella occidentalis* in *Impatiens* bedding plants. *Environmental Entomology* 33(5), 1344–1350.

Logan, J.D. and Wolesensky, W. (2007) Accounting or temperature in predator functional responses. *Natural Resource Model* 20, 549–574.

Loginova, E. (1992) Some new pests of glasshouse crops in Bulgaria and their control by an IPM programme. *Bulletin OEPP/EPPO [European and Mediterranean Plant Protection Organization*] 22, 357–361.

Loomans, A. (2003) Parasitoids as biological control agents of thrips pest. PhD thesis, Wageningen University, Wageningen, The Netherlands, 198 pp.

Lourenção, A.L., Alves, A.C., Fugi, C.G.Q. and Matos, E.S. (2008) Outbreaks of *Trialeurodes vaporariorum* (West.) (Hemiptera: Aleyrodidae) under field conditions in the State of São Paulo, Brazil. *Neotropical Entomology* 37, 89–91.

Maluf, W.R., Silva, V. de F., Cardoso Mdas, G., Gomes, L.A.A., Gonçalves Neto, Â.C., Maciel, G.M. and Nízio, D.A.C. (2010) Resistance to the South American tomato pinworm *Tuta absoluta* in high acylsugar and/or high zingiberene tomato genotypes. *Euphytica* 176(1), 113–123.

Manjanuth, T.M., Bhatnagar, V.S., Pawa, R.C.S. and Sitanatham, S. (1989) Economic importance of *Heliothis* spp. in India and an assessment of their natural enemies and host plants. In: King, E.G. and Jackson, R.D. (eds) *Proceedings of the Workshop on Biological Control of* Heliothis*: Increasing the Effectiveness of Natural Enemies*, November 1985, New Delhi. Far Eastern Regional Research Office, United States Department of Agriculture, New Delhi, India, pp 196–228.

Martel, J.W., Alford, A.R. and Dickens, J.C. (2005) Laboratory and greenhouse evaluation of a synthetic host volatile attractant for Colorado potato beetle *Leptinotarsa decemlineata* (Say). *Agriculture for Entomology* 7, 71–78.

Matthews, M. (1991) Classification of the Heliothinae. *NRI Bulletin* No. 44. Natural Resources Institute (NRI), Chatham, UK.

McAuslane, H.J., Johnson, F.A., Colvin, D.L. and Sojack, B. (1995) Influence of foliar pubescence on abundance and parasitism of *Bemisia argentifolii* (Homoptera: Aleyrodidae) on soybean and peanut. *Environmental Entomology* 24, 1135–1143.

McVeigh, L.J. and Bettany, B.W. (1987) The development of lure and kill technique for control of the Egyptian cotton leafworm, *Spodoptera littoralis*. *Bulletin OILB-SROP* [*Cette Organisation Internationale de Lutte Biologique (OILB)-Section régionale ouest-paléarctique (SROP)*] 10(3), 59–60.

Menken, S.B.J. and Ulenberg, S.A. (1986) Allozymatic diagnosis of four economically important *Liriomyza* species (Diptera, Agromyzidae). *Annals of Applied Biology* 109(1), 41–47.

Messelink, G. and Wensveen, W. van (2003) Biocontrol of *Duponchelia fovealis* (Lepidoptera: Pyralidae) with soil-dwelling predators in potted plants. *Communications in Agricultural and Applied Biological Sciences* 68(4a), 159–165.

Migeon, A. (2007) Acarien rouge de la tomate: nouvelles observations et perspectives. *PHM Revue Horticole* 488, 20–24.

Miller, G.W. (1977) Mortality of *Spodoptera littoralis* (Boisduval) (Lepidoptera: Noctuidae) at non-freezing temperatures. *Bulletin of Entomology Research* 67(1), 143–152.

Miresmailli, S., Gries, R., Gries, G., Zamar, R.H. and Isman, M.B. (2010) Herbivore-induced plant volatiles allow detection of *Trichoplusia ni* (Lepidoptera: Noctuidae) infestation on greenhouse tomato plants. *Pest Management of Science* 66(8), 916–924.

Mollá, O., Gonzalez-Cabrera, J. and Urbaneja, A. (2011) The combined use of *Bacillus thuringiensis* and *Nesidiocoris tenuis* against the tomato borer *Tuta absoluta*. *BioControl* 56, 883–891.

Montserrat, M., Albajes, R. and Castañe, C. (2000) Functional response of four heteropteran predators preying on greenhouse whitefly (Homoptera: Aleyrodidae) and western flower thrips (Thysanoptera: Thripidae). *Environmental Entomology* 29(5), 1075–1082.

Moulton, D. (1948) The genus *Frankliniella* Karny, with keys for the determination of species. *Review of Entomology* 10, 55–114.

Mound, L.A. and Halsey, S.H. (1978) *Whitefly of the World: A Systematic Catalogue of the Aleyrodidae (Homoptera) with Host Plant and Natural Enemy Data*. British Museum (Natural History). Wiley, Chichester, UK.

Mozafary, M. (1993) The introduction of *Duponchelia fovealis* (Zeller) on corn plants in Khuzetan province (Lep.: Pyralidae). In: *Proceedings of the 11th Plant Protection Congress of Iran*, 28 August–2 September 1993, Rasht, Iran, p. 90. (in Persian)

Munger, F. (1933) *Investigations in the Control of the Cyclamen Mite (*Tarsonemus pallidus *Banks)*. Technical Bulletin No. 93. Minnesota Agricultural Experiment Station, St Paul, Minnesota.

Mutwiwa, U.N., Borgemeister, C., Elsner, B.-von and Tantau, H.J. (2005) Effects of UV-absorbing plastic films on greenhouse whitefly (Homoptera: Aleyrodidae). *Journal of Economic Entomology* 98(4), 1221–1228.

Mwaura, P., Niere, B. and Vidal, S. (2015) Resistance and tolerance of potato varieties to potato rot nematode (*Ditylenchus destructor*) and stem nematode (*Ditylenchus dipsaci*). *Annals of Applied Biology* 166(2), 257.

Nakashima, Y. and Hirose, Y. (1997) Temperature effects on development of *Orius tantillus* (*Het.: Anthocoridae*), a predator of *Thrips palmi* (*Thys.: Thripidae*). *Entomophaga* 42, 337–342.

Nasr, E.S.A., El-Shafei, S.A. and Iss-hak, R.R. (1981) The hourly activity of *Spodoptera littoralis* moths during the night as indicated by light trap catches. *Research Bulletin* No. 1610. Faculty of Agriculture, Ain Shams University, Cairo, Egypt, p. 5.

Naumova, G.A. (1972) *Introduction and Results of Studying Introduced and New Soviet Strawberry Varieties*. Kul'tura zemlyaniki v SSSR, USSR, Moscow, pp. 294–298.

Navon, A., Wysoki, M. and Keren, S. (1983) Potency and effect of *Bacillus thuringiensis* preparations against larvae of *Spodoptera littoralis* and *Boarmia* (*Ascotis*) *selenaria*. *Phytoparasitica* 11(1), 3–11.

Ntawuruhunga, P. and Legg, J. (May 2007) New Spread of Cassava Brown Streak Virus Disease and its Implications for the Movement of Cassava Germplasm in the East and Central African Region (Crop Crisis Control Project), p. 6. Available at: http://c3project.iita.org/Doc/A25-CBSDbriefMay6.pdf (accessed 24 January 2017).

Obrist, L.B., Klein, H., Dutton, A. and Bigler, F. (2005) Effects of *Bt* maize on *Frankliniella tenuicornis* and exposure of thrips predators to prey-mediated *Bt* toxin. *Entomologia Experimentalis et Applicata* 115(3), 409–416.

Oldfield, G.N. (1996) Diversity and host plant specificity. In: Lindquist, E.E., Sabelis, M.W. and Bruin, J. (eds) *Eriophyoid Mites: Their Biology, Natural Enemies and Control*. *World Crop Pests*, Volume 6. Elsevier Science, Amsterdam, pp. 199–216.

Oliveira, M.R.V., Amâncio, E., Laumann, R.A. and Gomes, L.O. (2003) Natural enemies of *Bemisia tabaci* (Gennadius) B biotype and *Trialeurodes vaporariorum* (Westwood) (Hemiptera: Aleyrodidae) in Brasília, Brazil. *Neotropical Entomology* 32, 151–154.

Palmqvist, G. (1998) Remarkable records of Macrolepidoptera in Sweden 1997. *Entomology Tidskr* 119(1), 13–27.

Palmqvist, G. (2002) Remarkable records of Macrolepidoptera in Sweden 2001. *Entomology Tidskr* 123(1/2), 53–63.

Park, M.K., Kim, J.G., Song, Y.H., Lee, J.H., Shin, K. and Cho, K. (2009) Effect of nitrogen levels of two cherry tomato cultivars on development, preference and honeydew production of *Trialeurodes vaporariorum* (Hemiptera: Aleyrodidae). *Journal of Asia-Pacific Entomology* 12, 27–32.

Parrella, M.P., Allen, W.W. and Morishita, P. (1981) Leafminer species causes California mum growers new problems. *California Agriculture* 35(9/10), 28–30.

Pinhey, E.C.G. (1975) *Moths of Southern Africa*, descriptions and colour illustrations of 1183 species. A. Balkema, Rotterdam, The Netherlands, p. 273.

Pires, L.M., Marques, E.J., Wanderley-Teixeira, V., Teixeira, Â.A.C., Alves, L.C. and Alves, E.S.B. (2009) Ultrastructure of *Tuta absoluta* parasitized eggs and the reproductive potential of females after parasitism by *Metarhizium anisopliae*. *Micron* 40(2), 255–261.

Pires, L.M., Marques, E.J., Oliveira, J.V. de and Alves, S.B. (2010) Selection of isolates of entomopathogenic fungi for controlling *Tuta absoluta* (Meyrick) (Lepidoptera: Gelechiidae) and their compatibility with insecticides used in tomato crop. *Neotropical Entomology* 39(6), 977–984.

Poe, S.L. (1981) Miner notes. *Society of American Florists* 2, 1–10.

Potts, M.J. and Gunadi, N. (1991) The influence of intercropping with *Allium* on some insect populations in potato (*Solanum tuberosum*). *Annals of Applied Biology* 119(1), 207–213.

Powell, W. and Pell, J.K. (2007) Biological control. In: Emden, H.F. and van and Harrington, R. (eds) *Aphids as Crop Pests*. CAB International, Wallingford, UK, pp. 469–513.

Pralavorio, M. and Almaguel-Rojas, L. (1980) Influence de la temperature et de 1'humidite relative sur le developpement et la reproduction de *Phytoseiulus persimilis*. *IOBC (International Organisation for Biological and Integrated Control) WPRS Bulletin* 3(3), 157–162.

Premachandra, W.T.S.D., Borgemeister, C., Berndt, O., Ehlers, R.U. and Poehling, H.M. (2003) Combined releases of entomopathogenic nematodes and the predatory mite *Hypoaspis aculeifer* to control soil-dwelling stages of western flower thrips *Frankliniella occidentalis*. *BioControl* 48(5), 529–541.

Rashid, F.F., Hammad, S.M. and Hassan, S.M. (1971) The biology of *Autographa chalcites* L. in Alexandria region (Lepidoptera: Noctuidae). *Bulletin de la Société Entomologique d'Égypte* 55, 419–426.

Rice, R.E. and Strong, F.E. (1962) Bionomics of the tomato russet mite, *Vasates lycopersici* (Massee). *Journal of Economic Entomology* 55, 431–435.

Riedel, R.M. and Larsen, P.O. (1974) Interrelationship of *Aphelenchoides fragariae* and *Xanthomonas begoniae* on Rieger begonia. *Journal of Nematology* 6(4), 215–216.

Rigakis, N., Katsoulas, N., Teitel, N., Bartzanas, T. and Kittas, C. (2015) A simple model for ventilation rate determination in screenhouses. *Energy and Buildings* 87, 293–301.

Rodriguez, M.G., Sanchez, L. and Rowe, J. (2003) Host status of agriculturally important plant families to the rootknot nematode *Meloidogyne mayaguensis* in Cuba. *Nematropica* 33, 125–130.

Rothschild, G.H.L., Wilson, A.G.L. and Malafant, K.W. (1982) Preliminary studies on the female sex pheromones of *Heliothis* species and their possible use in control programs in Australia. In: Reed, W. and Kumble, V. (eds) *Proceedings of the International Workshop on* Heliothis *Management*, 15–20 November 1981, International Crops Research Institute for the Semi-Arid Tropics (ICRISAT) Center, Patancheru, India. ICRISAT, Patancheru, Andhra Pradesh, India, pp. 319–327.

Salama, H.S. and Shoukry, A. (1972) Flight range of the moth of the cotton leaf worm *Spodoptera littoralis* (Bois.). *Zeitschrift für Angewandte Entomologie* 71(2), 181–184.

Salama, H.S., Dimetry, N.Z. and Salem, S.A. (1970) On the host preference and biology of the cotton leaf worm *Spodoptera littoralis*. *Zeitschrift für Angewandte Entomologie* 67, 261–266.

Salama, H.S., Foda, M.S. and Sharaby, A. (1989) A proposed new biological standard for bioassay of bacterial insecticides vs. *Spodoptera* spp. *Tropical Pest Management* 35(3), 326–330.

Sambandam, C.N. and Chelliah, S. (1970) Evaluation of certain *Solanum* spp. for resistance to *Aphis gossypii* Glover. *Indian Journal of Entomology* 32(3), 270–271.

Schwartz, A. (1977) *Citrus Silver Mite,* Hemitarsonemus latus *(Banks)*. Leaflet, Citrus Series H 14, Department of Agricultural Technical Services of the Republic of South Africa. Government Printer, Pretoria.

Sexson, D.L., Wyman, J., Radcliffe, E.B., Hoy, C.J., Ragsdale, D.W. and Dively, G.P. (2005) Potato. In: Foster, R. and Flood, B. (eds) Vegetable Insect Management. Meister Publishing, Willoughby, Ohio, pp. 92–107.

Shakaya, S., Weintraub, P.G. and Coll, M. (2009) Effect of pollen supplement on intraguild predatory interactions between two omnivores: the importance of spatial dynamics. *Biological Control* 50, 281–287.

Siddiqui, Z.A., Nesha, R., Singh, N. and Alam, S. (2012) Interactions of plant-parasitic nematodes and plant-pathogenic bacteria. In: Maheshwari, D.K. (ed.) *Bacteria in Agrobiology: Plant Probiotics*. Springer, Berlin, pp. 251–267.

Sikora, R.A. and Greco, N. (1990) Nematode parasites of food legumes. In: Luc, M., Sikora, R.A. and Bridge, J. (eds) *Plant Parasitic Nematodes in Subtropical and Tropical Agriculture*. CAB International, Wallingford, UK, pp. 181–235.

Slosser, J.E., Pinchak, W.E. and Rummel, D.R. (1989) A review of known and potential factors affecting the population dynamics of the cotton aphid. *Southwest Entomology* 14(3), 302–313.

Smart, G.C. Jr, Nguyen, K.B., Parkman, J.P. and Frank, J.H. (1990) Biological control of mole crickets in the genus *Scapteriscus* with the nematode *Steinernema scapterisci* Nguyen and Smart. In: INRA (ed.) *Rencontres Caraibes en lutte Biologique*, 5–7 November 1990, Guadeloupe. Institut National de la Recherche Agronomique (INRA), Paris.

Smith, F.F. (1939) Control of cyclamen and broad mite on gerbera. *United States Department of Agriculture Circular* 516, 1–14.

Solis, M.A. (2006) Key to selected Pyraloidea (Lepidoptera) larvae intercepted at US ports of entry: revision of Pyraloidea. In: Weisman, D.M. (ed.) *Keys to Some Frequently Intercepted Lepidoptorous Larvae*. Miscellaneous Entomology Collections, United States Department of Agriculture (USDA) Systematic Entomology Laboratory, Washington, DC, pp. 1–58.

Spencer, K.A. (1972) *Agromyzidae (Diptera) of Economic Importance*. Dr W. Junk BV, The Hague, The Netherlands, 418 pp.

Srinivasan, R., Sundaraj, S., Pappu, H.R., Diffie, S., Riley, D.G. and Gitaitis, R.D. (2012) Transmission of iris yellow spot virus by *Frankliniella fusca* and *Thrips tabaci* (Thysanoptera: Thripidae). *Journal of Economic Entomology* 105(1), 40–47.

Stenseth, C. and Nordby, A. (1976) Damage, and control of the strawberry mite *Steneotarsonemus pallidus* (Acarina: Tarsonemidae), on strawberries. *Journal of Horticulture Science* 51, 49–54.

Stoltz, R.L. and Stern, V.M. (1978) The longevity and fecundity of *Orius tristicolor* when introduced to increasing numbers of the prey *Frankliniella occidentalis*. *Environmental Entomoloy* 7(2), 197–198.

Sturhan, D. (1962) Fber neue Wirtspflanzen der BlattSlchen *Aphelenchoides fragariae* und *A. ritzemabosi*, mit Bemerkungen zu den Wirtspflanzenkreisen beider Nematodenarten. *Anz für Schädlingskunde* 35(5), 65–67.

Talavera, M., Itou, K. and Mizukubo, T. (2001) Reduction of nematode damage by root colonisation with arbuscular mycorrhiza (*Glomus* spp.) in tomato-*Meloidogyne incognita* (Tylenchida: *Meloidogynidae*) and carrot-*Pratylenchus penetrans* (Tylenchida: *Pratylenchidae*) pathosystems. *Applied Entomology and Zoology* 36, 387–392.

Tenente, R.V. and Evans, A.A.F. (1998) Life cycle of *Ditylenchus dipsaci* teasel race on onion seedlings under different temperature conditions. *Nematologia Brasileira* 22, 87–96.

Trudgill, D.L., Bala, G., Blok, V.C., Daudi, A., Davies, K.G., Gowen, S.R., Fargette, M., Madulu, J.D., Mateille, T., Mwageni, W. *et al.* (2000) The importance of tropical root-knot nematodes (*Meloidogyne* spp.) and factors affecting the utility of *Pasteuria penetrans* as a biocontrol agent. *Nematology* 2, 823–845.

Tsang, M.M.C., Hara, A.H. and Sipes, B.S. (2001) A hot water drenching system for disinfecting roots and media of potted plants of the burrowing nematodes. *Applied Engineering in Agriculture* 17, 533–538.

Tsueda, H. and Tsuchida, K. (1998) Differences in spatial distribution and life history parameters of two sympatric whiteflies, the greenhouse whitefly *Trialeurodes vaporariorum* Westwood and the silverleaf whitefly *Bemisia argentifolii* Bellows and Perring, under greenhouse and laboratory conditions. *Applied Entomology and Zoology* 33, 379–383.

Tuovinen, T. and Lindqvist, I. (2010) Maintenance of predatory phytoseiid mites for preventive control of strawberry tarsonemid mite *Phytonemus pallidus* in strawberry plant propagation. *Biological Control* 54, 119–125.

Tzortzakakis, E.A. and Petsas, S.E. (2003) Investigation of alternatives to methyl bromide for management of *Meloidogyne javanica* on greenhouse grown tomato. *Pest Management Science* 59, 1311–1320.

Uchoa-Fernandes, M.A., Della Lucia, T.M.C. and Vilela, E.F. (1995a) Mating, oviposition and pupation of *Scrobipalpuloides absoluta* (Meyr.) (Lepidoptera: Gelechiidae). *Anais da Sociedade Entomológica do Brasil* 24(1), 159–164.

Uchoa-Fernandes, M.A., Vilela, E.F. and Della Lucia, T.M.C. (1995b) Ritmo diário de atratao sexual em *Scrobipalpuloides absoluta* (Meyrick, 1917) (Lepidoptera: Gelechiidae). *Revista Brasileira de Biologia* 55(1), 67–73.

University of California Integrated Pest Management Program (UC IPM) (2014) UC Pest Management Guidelines. Strawberry, Cyclamen Mite. Available at: http://www.ipm.ucdavis.edu/PMG/r734400211.html (accessed 26 August 2014).

Uygun, N. and Ozgur, F. (1980) Identification of pests of greenhouse vegetables in the Icel and Adana regions, and the effects of endosulfan smoke tablets and pirimicarb on *Myzus persicae* (Sulz.). *Turkiye Bitki Koruma Dergisi* 4(3), 185–192.

Vacante, V. (1982) La difesa del pomodoro in serra dall'eriofide rugginoso (*Aculops lycopersici* (Massee)) (Acarina, Eriophyidae). *Colture Protette* 11(6), 29–34.

Vacante, V. (1988) Note su *Tyrophagus neiswanderi* Johnston & Bruce (Acarina, Acaridae). I. Descrizione morfologica degli adulti. In: *Atti XV Congresso Nazionale Italiano di Entomologia*, 13–17 Giugno 1988, L'Aquila, Italy. Società Entomologica Italiana, L'Aquila, Italy, pp. 213–222.

Vacante, V. (2000) Animali dannosi alle ortive da serra. In: Baccetti, B., Barbagallo, S., Süss, L. and Trembaly, E. (eds) *Manuale di Zoologia Agraria*. Antonio Delfino, Rome, pp. 429–448.

Vacante, V. (2016) *The Handbook of Mites of Economic Plants*. CAB International, Wallingford, UK, 872 pp.

Vacante, V. and Benuzzi, M. (2007) *Difesa delle Colture in Serra*. Edagricole, Bologna, Italy, p. 437. (in Italian)

Vacante, V. and Tropea Garzia, G. (1993) Prime osservazioni sulla risposta funzionale di *Orius laevigatus* (Fieber) nel controllo di *Frankliniella occidentalis* (Pergande) su peperone in serra fredda. *Colture Protette* 22(1), 33–36.

Vacante, V., Nucifora, A. and Calabretta, C. (1988) Considerations sur les possibilitiées de lutte biologique et integrée contre la mouche mineuse *Liriomyza trifolii* (Burg.) dans la culture protegée de gerbera en Italie. In: *Proceedings of the XVIII International Congress of Entomology*, 3–9 July 1988, Vancouver, British Columbia, Canada. Entomological Society of Canada, Vancouver, Canada, p. 330.

Vacante, V., Cacciola, S.O. and Pennisi, A.M. (1994) Epizootiological study of *Neozygites parvi spora* (Zygomycota: Entomophthoraceae) in a population of *Frankliniella occidentalis* (Thysanoptera: Thripidae) on pepper in Sicily. *Entomophaga* 39, 123–130.

Vacante, V., Tropea Garzia, G., Onillon, J.C. and Pucci, C. (1995) Observations sur la dynamique des populations de l'aleurode *Bemisia tabaci* (Gennadius) en serre de poivron. *Mededelingen van de Faculteit Landbouwwetenschappen Rijksuniversiteit Gent* 60(3a), 635–643.

Vacante, V., Cocuzza, G., Pucci, C. and Clercq, P. de (1996) Note on *Podisus maculiventris* and *P. nigrispinus* (Heteroptera Pentatomidae). *Informatore Fitopatologico* 46(6), 11–16.

Vacante, V., Benuzzi, M., Palmeri, V., Gilioli, G. and Brafa, G. (2001a) Impiego sperimentale di *Beauveria bassiana* contro *Bemisia tabaci* su poinsezia (*Euphorbia pulcherrima*) nel ragusano. *Notiziario Sulla Protezione delle Piante* 13, 169–173.

Vacante, V., Benuzzi, M., Palmeri, V. and Brafa, G. (2001b) Experimental trials of microbiological control of the Turkey moth (*Chrysodeixis chalcites* (Esper)) in Sicilian greenhouse crops. *Informatore Fitopatologico* 51(7/8), 73–76.

Van de Veire, M. (1993) First observations in glasshouse sweet peppers in Belgium and laboratory rearing of the parasitic wasp *Eulophus pennicornis* (Hym.: Eulophidae). *Entomophaga* 38(1), 61–62.

Van de Veire, M. and Vacante, V. (1984) Greenhouse whitefly and leaf miners control by the combined use of the colour attraction system with the parasite wasp, *Encarsia formosa* (Gahan). *Mededelingen van de Faculteit Landbouwwetenschappen Rijksuniversiteit Gent* 49/1, 107–114.

Van Emden, H.F., Eastop, V.F., Hughes, H.D. and Way, M.J. (1969) The ecology of *Myzus persicae*. *Annual Review of Entomology* 14, 197–270.

Van Eyndhoven, G.L. and Groenewold, H. (1959) On the morphology of *Steneotarsonemus pallidus* and *S. fragariae* (Acarina, Tarsonemidae). *Entomologische Berichten (Amsterdam)* 19, 123–124.

Van Gundy, S.D. (1985) Ecology of *Meloidogyne* spp. – emphasis on environmental factors affecting survival and pathogenicity. In: Sasser, J.N. and Carter, C.C. (eds) *An Advanced Treatise on Meloidogyne. Volume I: Biology and Control*. North Carolina State University Graphics, Raleigh, North Carolina, pp. 177–182.

Vargas, H.C. (1970) Observaciones sobre la biologfa y enemigos naturales de la polilla del tomate, *Gnorimoschema absoluta* (Meyrick) (Lepidoptera: Gelechiidae). *Idesia (Chile)* 1, 75–110.

Veerman, A. (1977) Aspects of the induction of diapause in a laboratory strain of the mite *Tetranychus urticae*. *Journal of Insect Physiology* 23, 703–711.

Vos, R. de and Rutten, A.L.M. (1995) Migrating Lepidoptera in 1992 (fifty-third report). *Entomologische Berichten (Amsterdam)* 55(3), 37–46.

Vovlas, N., Lucarelli, G., Sasanelli, N., Troccoli, A., Papajova, I., Palomares-Rius, J.E. and Castillo, P. (2008a) Pathogenicity and host–parasite relationships of the root-knot nematode *Meloidogyne incognita* on celery. *Plant Pathology* 57, 981–987.

Vovlas, N., Troccoli, A., Minuto, A., Bruzzone, C., Sasanelli, N. and Castillo, P. (2008b) Pathogenicity and host–parasite relationships of *Meloidogyne arenaria* in sweet basil. *Plant Disease* 92, 1329–1335.

Waterhouse, D.F. and Norris, K.R. (1989) *Frankliniella occidentalis* (Pergande). In: *Biological Control Pacific Prospects* – Supplement 1. Australian Centre for International Agriculture Research, Canberra, pp. 24–35.

Wesemael, W.M.L., Viaene, N. and Moens, M. (2011) Root-knot nematodes (*Meloidogyne* spp.) in Europe. *Nematology* 13(1), 3–16.

Wilkerson, J.L., Webb, S.E. and Capinera, J.L. (2005) *Vegetable Pests III: Lepidoptera*. University of Florida Institute of Food and Agricultural Sciences (UF/IFAS) CD-ROM. SW 182. UF/IFAS, Gainsville, Florida.

Wintermantel, W.M. (2004) Emergence of greenhouse whitefly (*Trialeurodes vaporariorum*) transmitted criniviruses as threats to vegetable and fruit production in North America. The American Phytopathological

Society (APS). Available at: http://www.apsnet.org/publications/apsnetfeatures/Pages/GreenhouseWhitefly.aspx (accessed 1 June 2016).

Wisler, G.C., Duffus, J.E., Liu, H.Y., Li, R. and Falk, B.W. (1997) New whitefly transmitted closterovirus identified in tomatoes. *California Agriculture* 51, 24–26.

Yamada, E. and Takakura, S. (1987) Ecological investigations on the strawberry nematode, *Aphelenchoides fragariae* on lilies. *Japanese Journal of Nematology* 17(12), 1–7.

Zaki, F.N., El-Shaarawy, M.F. and Farag, N.A. (1999) Release of two predators and two parasitoids to control aphids and whiteflies. *Anz Schädlingskd* 72(1), 19–20.

Zawadneak, M.A.C., Gonçalves, R.B., Pimentel, I.C., Schuber, J.M., Santos, B., Poltronieri, A.S. and Solis, M.A. (2016) First record of *Duponchelia fovealis* (Lepidoptera: Crambidae) in South America. *IDESIA* 34, 91–95.

Zehnder, G. and Warthen, J.D. (1988) Feeding inhibition and mortality effects of neem-seed extract on the Colorado potato beetle (Coleoptera: Chrysomelidae). *Journal of Economic Entomology* 81, 1040–1044.

Zhang, G.X. and Zhong, T.S. (1990) Experimental studies on some aphid life-cycle patterns and the hybridization of two sibling species. In: Campbel, R.K. and Eikenbary, R.D. (eds) *Aphid–Plant Genotype Interactions*. Elsevier Press, New York, pp. 37–50.

Zhang, Z.Q. (2002) Taxonomy of *Tetranychus ludeni* (Acari: Tetranychidae) in New Zealand and its ecology on *Sechium edule*. *New Zealand Entomology* 25, 27–34.

16 Pests and Natural Enemies in Organic Field Vegetables in Tropical and Subtropical Areas

Dominique Bordat*
Ex CIRAD, Montferrier, France

Introduction

In tropical conditions, up to 70% of vegetable damage is caused by arthropods and the remaining damage is caused by diseases (fungi, bacteria) and weeds. The damage caused by pests is often not identified by local growers and the natural enemy fauna is unknown to them. The growers are often uneducated and their knowledge of vegetable crops comes from traditional methods passed down from their parents.

The majority of growers use chemical control to manage pest populations. Most chemicals used are ineffective against pest populations, their use resulting in the appearance of resistant populations. Many pesticides used for cotton pest control which are not recommended for use on vegetables enter the horticultural system when farmers obtain them on the black market. As there are no regulations for their use on vegetables, overuse and non-compliance with application timing before harvesting occur. Subsequent overuse of pesticides leaves farmers worse off and consumers at risk from pesticide residues on the crops. In Senegal the most frequently used pesticides have been detected in 97% of vegetable samples, with concentrations higher than the allowed maximum limits (AML) (Ngom *et al.*, 2013).

Two types of farmers producing organic field vegetables in the tropics and subtropics exist: (i) the small farmer, supplying the traditional market; and (ii) the farmer associated with private firms which export their produce. The small farmer has little or no understanding of crop damage caused by pests and of the beneficial interrelations between live species present in his/her field. These farmers use pesticides and, when financially able, use products found on the local market or proposed by retailers. By contrast, the farmer associated with private firms is often trained in horticulture and helped by foreign experts who advise him/her. He/she uses planned chemical control following specific programmes and formulations adapted to pests, which he/she is able to recognize. He/she understands the agroecological concept but does not take risks in using it to manage the pest populations, because his/her production and career in the firm depend on an agreement based on profitability.

The agroecological concept should be applied to produce healthy vegetables

* dominique.bordat@cirad.fr

sustainably. However, important commercial and social pressures restrain its application in the tropics. The facts mentioned above impede the application of pest control in organic farming in these regions of the world. However, the implementation of this strategy is a desirable solution and it should be scientifically and practically encouraged. To contribute to this aim, this chapter presents the preliminary knowledge on the main pests (and their natural enemies) on field vegetables in tropical and subtropical areas. This knowledge is fundamental to the application of organic farming but the list of pests provided in this chapter is not exhaustive.

Major Pests

About ten of the 150 pest species observed on vegetable crops cause very important recurrent damage (Bordat and Arvanitakis, 2004). Under some favourable conditions (climate, significant population of pest) these species can destroy the total crop. About 20 other species are sporadic, but they cause important damage too. Among these different species, a few can be managed by biological control (biopesticides, local natural enemies).

Wet and hot tropical climates and especially the absence of winter favour large persistent pest populations resulting in extensive damage. The damage level accepted on vegetables is subjective, and depends on human dietary custom. For example the melon (*Cucumis melo* L.) is cultivated commercially for its fruits, and damage caused by 'fruits flies' (Diptera; Tephritidae) is very important; however, it is also widely cultivated in the tropics for its young shoots which are consumed boiled. In this case, the pests that eat leaves in the families Coccinellidae and Chrysomelidae will be the most important pests. Similarly, *Amaranthus* spp. which are grown and consumed in Central Africa, are judged as weeds in West Africa. In the first case, the two Lepidoptera Pyralidae that eat leaves are *Hymenia recurvalis* (F.) and *Psara basalis* (Walker), and in Central Africa these will be judged as pests, while in West Africa these two species can be judged as beneficial.

Vegetable crops are grown all year round and everywhere when water is easy to reach. Vegetable production is high yield and constant with no breaks in the cultural cycle and damaging pests are present all the year round. Traditional crops such as okra (*Abelmoschus esculentus* (L.) Moench), roselle (*Hibiscus sabdariffa* L.), the African aubergine and the morelles (*Solanum* spp.), the leafy vegetables, *Corchorus* spp. and *Amaranthus* spp., which are adapted to the tropical climate conditions are often consumed by pests but always produce. The majority of vegetables grown in the tropics, such as cabbages, melons, tomatoes, cucumbers and aubergines, which are not adapted to a tropical climate come from temperate areas of the world, and are becoming more sensitive to the pests. The hot, wet, tropical climate restricts these temperate plant species and favours pests, which are not affected by winter conditions, found in mid-latitude countries (23–63°), except for high altitude areas in the tropics such as the Cameron Highlands in Malaysia, the South West of Cameroon, Kenya, and the high altitude areas of Chile and Argentina. Generally, the climate is rather selective for the natural onomy fauna (dry season species differ from rainy season species) but it does not affect the pest species, which are present as soon as the host plant is cultivated. In productive fields, crops are exposed to pest populations selected by natural selection, which owing to their phenotypic plasticity are at their maximum fitness (Pichon *et al.*, 2006). Individuals from the same species often have behavioural differences dependent on different localities in the same country (Roux *et al.*, 2007).

It is very difficult to classify the pests in relation to the damage they cause. Some species cause a great deal of damage in some areas while they cause less damage in others. The species found and collected from vegetables are not all noted. Only the pests that cause most damage or are the most common are described in this chapter.

The pest species found on open field vegetables in Africa (West Africa and Central Africa), Reunion and the Mayotte islands are shown in Table 16.1, those from South America

Table 16.1. Common pests species found on vegetable fields in West and Central Africa, Reunion and the Mayotte islands.

Order	Family	Species	Host plants	Scale of damage	Presence	Site of damage
Acari	Eriophyidae	*Aculops lycopersici* Massée	*Solanaceae*	++	Permanent	Le, Fr, St
	Tarsonemidae	*Polyphagotarsonemus latus* (Banks)	*Solanaceae*, *Cucurbitaceae*	+++	Permanent	Le, Fr
	Tetranychidae	*Tetranychus evansi* Baker & Pritchard	Tomato	+++(+)	Sporadic	Le, Fl, Fr, St
	Tetranychidae	*Tetranychus urticae* (Koch)	*Solanaceae*, bean, onion, *Cucurbitaceae*	+++(+)	Permanent	Le, Fl, Fr, St
Coleoptera	Chrysomelidae	*Aspidomorpha cincta* (F.)	Sweet potato	++	Permanent	Le
	Chrysomelidae	*Aulacophora foveicollis* Lucas	*Cucurbitaceae*	++(+)	Sporadic	Le, Fl
	Chrysomelidae	*Cassida liquefacta* Spaeth	Amaranth, *Celosia* spp.	++	Sporadic	Le
	Chrysomelidae	*Epithrix* sp.	Aubergine	++	Sporadic	Le
	Chrysomelidae	*Laccoptera corrugata* (Sahlberg)	Sweet potato	+(+)	Sporadic	Le
	Chrysomelidae	*Lamprocopa occidentalis* (Weise)	*Cucurbitaceae*	+(+)	Sporadic	Le, Fl
	Chrysomelidae	*Podagrixina decolorata* (Duvivier)	Aubergine, okra	++	Sporadic	Le, Fl, Fr
	Chrysomelidae	*Phyllotreta cheiranthi* Weise	Aubergine, cabbage	++	Sporadic	Le
	Chrysomelidae	*Rhytidocassis scutellaris* (Klug)	Sweet potato	+(+)	Sporadic	Le
	Coccinellidae	*Epilachna nigeriana* Mader	Vernonia	++	Sporadic	Le
	Coccinellidae	*Epilachna pavonia* (Olivier)	Aubergine, *Solanum macrocarpon*	+(+)	Sporadic	Le
	Coccinellidae	*Henosepilachna elaterii* (Rossi)	*Cucurbitaceae*	++	Sporadic	Le, St, Fr
	Coccinellidae	*Henosepilachna reticulata* (Olivier)	*Cucurbitaceae*	++	Sporadic	Le, St, Fr
	Curculionidae	*Cylas puncticollis* Boheman	Sweet potato	++(++)	Sporadic	St, tuber
	Curculionidae	*Gasteroclisus rhomboïdalis* (Boheman)	Amaranth, *Celosia* spp., aubergine	++	Permanent	Le, St
	Curculionidae	*Hypolixus nubilosus* (Boheman)	Amaranth, *Celosia* spp., aubergine, okra	+(+)	Sporadic	Le, St
	Meloidae	*Hycleus argentata* F.	Okra	++	Sporadic	Le, Fl
	Meloidae	*Hycleus hermaniae* (F.)	Okra, bean	++	Sporadic	Le, Fl

Continued

Table 16.1. Continued.

Order	Family	Species	Host plants	Scale of damage	Presence	Site of damage
	Meloidae	*Hycleus senegalensis* (Voigts)	Okra, bean	++	Sporadic	Le, Fl
	Meloidae	*Hycleus trifasciatus* (Thunberg)	Okra	++	Sporadic	Le, Fl
	Meloidae	*Hycleus vestita* Reiche	Okra, bean	++	Sporadic	Le, Fl
	Meloidae	*Lydomorphus westermanii* (Mäklin)	Okra	+(+)	Sporadic	Fr
	Scarabeidae	*Diplognatha gagates* Forster	Okra, bean, vernonia	++	Sporadic	Le, Fr
	Scarabeidae	*Pachnoda interrupta* (Olivier)	Okra	+(+)	Sporadic	Fr, Fl
	Tenebrionidae	*Lagria vilosa* F.	Sweet potato	+(+)	Sporadic	Le
Diptera	Agromyzidae	*Liriomyza huidobrensis* (Blanchard)	Polyphagous	++(++)	Permanent	Le
	Agromyzidae	*Liriomyza sativae* Blanchard	Polyphagous	++(++)	Permanent	Le
	Agromyzidae	*Liriomyza trifolii* (Burgess)	Polyphagous	++(++)	Permanent	Le
	Agromyzidae	*Melanagromyza cleomae* Spencer	Cabbage	++	Permanent	Le
	Agromyzidae	*Ophiomya phaseoli* Tryon	Bean	++	Permanent	Le, St
	Cecidomyiidae	*Asphondyla* sp.	African aubergine	++	Sporadic	Fr
	Tephritidae	*Bactrocera cucurbitae* (Coquillet)	*Cucurbitaceae*	++++	Permanent	Fr
	Tephritidae	*Ceratitis capitata* (Wiedemann)	Hot pepper	++++	Permanent	Fr
	Tephritidae	*Dacus bivittatus* (Bigot)	*Cucurbitaceae*	++++	Permanent	Fr
	Tephritidae	*Dacus ciliatus* Loew	*Cucurbitaceae*	++++	Permanent	Fr
	Tephritidae	*Dacus vertebratus* Bezzi	*Cucurbitaceae*	++++	Permanent	Fr
	Tephritidae	*Neoceratitis cyanescens* (Bezzi)	Hot pepper, tomato	++++	Permanent	Fr
Hemiptera	Aleyrodidae	*Bemisia tabaci* (Gennadius)	Polyphagous	++++	Permanent	Le
	Aleyrodidae	*Trialeurodes vaporariorum* (Westwood)	Polyphagous	++	Permanent	Le
	Aphididae	*Aphis gossypii* Glover	Polyphagous	++(+)	Permanent	Le, Fl, Fr
	Aphididae	*Lipaphis pseudobrassicae* (Davis)	Cabbage	+++(+)	Permanent	Le

Continued

Table 16.1. Continued.

Order	Family	Species	Host plants	Scale of damage	Presence	Site of damage
	Aphididae	*Myzus persicae* (Sulzer)	Cabbage	+++(+)	Permanent	Le
	Cicadellidae	*Jacobiasca hybrida* Bergevin & Zanon	Aubergine	+++	Sporadic	Le
	Coreidae	*Leptoglossus australis* (F.)	*Cucurbitaceae*, tomato	+(+)	Permanent	Fl
	Lygaeidae	*Spilostethus longulus* Dallas	*Cucurbitaceae*, tomato	+(+)	Sporadic	Fr
	Lygaeidae	*Spilostethus rivularis* (Germar)	*Cucurbitaceae*, tomato	++	Sporadic	Fr
	Pentatomidae	*Acrosternum acutum* (Dallas)	Okra, bean, *Solanaceae*	++	Sporadic	Le, Fl, Fr
	Pentatomidae	*Nezara viridula* (L.)	Polyphagous	+++	Permanent	Fl, Fr, St
Lepidoptera	Crambidae	*Psara basalis* (Walker)	Amaranth, *Celosia* spp.	+(+)	Sporadic	Le
	Crambidae	*Crocidolomia binotalis* Zeller	Cabbage	+++	Sporadic	Le
	Crambidae	*Diaphana indica* (Saunders)	*Cucurbitaceae*	++(+)	Sporadic	Le, Fl, Fr
	Crambidae	*Hellula undalis* (F.)	Cabbage	+++(+)	Permanent	Le
	Crambidae	*Leucinodes orbonalis* Guénée	Aubergine, *S. macrocarpon*	+++	Sporadic	Fr, St
	Crambidae	*Maruca testulalis* (Geyer)	Bean	+(+)	Sporadic	Fr
	Crambidae	*Sceliodes laisalis* (Walker)	Aubergine, *S. macrocarpon,*	+++	Sporadic	Fr
	Gelechiidae	*Scrobipalpa ergasima* (Meyrick)	Aubergine	+++(+)	Sporadic	Fl
	Noctuidae	*Agrotis ipsilon* (Hufnagel)	Polyphagous	+++	Sporadic	St
	Noctuidae	*Eublemma admota* Felder	Aubergine	+(+)	Sporadic	Le
	Noctuidae	*Helicoverpa armigera* (Hübner)	Polyphagous	+++(+)	Permanent	Le, Fl, Fr
	Noctuidae	*Spodoptera exigua* (Hübner)	Onion	+++	Sporadic	Le
	Noctuidae	*Spodoptera littoralis* Boisduval	Polyphagous	++	Sporadic	Le
	Nolidae	*Earias biplaga* Walker	Okra	++	Sporadic	Fr
	Nolidae	*Earias insulana* Boisduval	Okra	++	Sporadic	Fr
	Nolidae	*Selepa docilis* Butler	Aubergine	+(+)	Sporadic	Le
	Plutellidae	*Plutella xylostella* (L.)	*Brassicaceae*	+++(+)	Permanent	Le
	Pyralidae	*Hymenia recurvalis* (F.)	Amaranth	+++	Permanent	Le
	Pyralidae	*Phycita melongena* Aina	Aubergine, okra	+(+)	Sporadic	Le
	Pyralidae	*Sillepte derogata* (F.)	Okra	+(+)	Sporadic	Le
	Tortricidae	*Cryptophlebia leucotreta* (Meyrick)	Hot pepper	+++	Sporadic	Fr
Orthoptera	Pyrgomorphidae	*Zonocerus variegatus* (L.)	Polyphagous	+++	Sporadic	Le, Fl, Fr

Continued

Table 16.1. Continued.

Order	Family	Species	Host plants	Scale of damage	Presence	Site of damage
Thysanoptera	Thripidae	*Frankliniella occidentalis* (Pergande)	Polyphagous	++(+)	Sporadic	Fe, Fl, Fr
	Thripidae	*Thrips palmi* Karny	Polyphagous	+++	Sporadic	Le
	Thripidae	*Thrips tabaci* Lindeman	Onion	+++	Permanent	Fe
Virus		Papaya ring spot virus (PRSV)	*Cucurbitaceae*	+++(+)	Sporadic	Le
		Tomato spotted wilt virus (TSWV)	Lettuce	+++(+)	Sporadic	Le
		Tomato yellow leaf curl virus (TYLCV)	Tomato	+++(+)	Sporadic	Le
		Zucchini yellow mosaic virus (ZYMV)	*Cucurbitaceae*	+++(+)	Sporadic	Le, Fr

[a]Scale of damage:
+, unimportant;
++, important;
+++, very important;
++++, can destroy the crop.
[b]Site of damage:
Le, leaf;
Fl, flower;
Fr, fruit;
St, stem.

(Brazil and French Antilles) in Table 16.2, those from South-east Asia in Table 16.3 and those from New Caledonia in Table 16.4.

Acari

Eriophyidae

ACULOPS LYCOPERSICI (TRYON) (TOMATO RUSSET MITE). This species causes a lot of damage on tomato especially in the dry and hot seasons, where mites breed quickly. Vermiform, they are invisible to the naked eye. The leaves curl up and become bronze coloured. The fruits mummify and are unmarketable.
Host plants: *Solanaceae*.
Distribution: Africa, Asia, New Caledonia, South America.

Tarsonemidae

POLYPHAGOTARSONEMUS LATUS (BANKS) (BROAD MITE). Adults and larvae are rarely visible to the naked eye. They attack mainly young leaves and flower buds, causing dwarf leaves with sinuous and prominent veins. The adults and larvae sting causing fruit deformations. The fruits are unmarketable.
Host plants: *Solanaceae*, *Cucurbitaceae*.
Distribution: Africa, Asia, New Caledonia, South America.

Tetranychidae

TETRANYCHUS EVANSI BAKER & PRITCHARD (TOMATO RED SPIDER MITE). The *Tetranychus* are the only Acari species which weave webs. This particular species completely covers the attacked plant with webbing and consequently adults and larvae are protected against natural enemies and pesticides. Populations increase rapidly with high temperatures.
Host plants: tomato.
Distribution: Africa, South America.

TETRANYCHUS URTICAE KOCH (TWO-SPOTTED SPIDER MITE). This species is often present during the hot

Table 16.2. Common pests found on vegetable fields in South America (including the French Antilles).

Order	Family	Species	Host plants	Scale of damage	Presence	Site of damage
Acari	Eriophyidae	*Aculops lycopersici* Massée	*Solanaceae*	++	Sporadic	Le, Fr, St
	Tarsonemidae	*Polyphagotarsonemus latus* (Banks)	*Solanaceae*, *Cucurbitaceae*	+++	Sporadic	Le, Fr
	Tetranychidae	*Tetranychus evansi* Baker & Pritchard	*Solanaceae*	+++(+)	Sporadic	Le, Fr
	Tetranychidae	*Tetranychus urticae* (Koch)	*Solanaceae*, bean, *Cucurbitaceae*	+++(+)	Permanent	Le, Fr
Coleoptera	Chrysomelidae	*Acalymma separata* Baley	*Cucurbitaceae*	+(+)	Sporadic	Le, St, Fr
	Chrysomelidae	*Cerotoma arcuata* Olivier	*Cucurbitaceae*, bean	+(+)	Sporadic	Le, St, Fr
	Chrysomelidae	*Cerotoma ruficornis* (Olivier)	Aubergine	+(+)	Sporadic	Le, Fl, Fr
	Chrysomelidae	*Diabrotica fucata* F.	Aubergine, *Cucurbitaceae*	+(+)	Sporadic	Le, St, Fr
	Chrysomelidae	*Diabrotica speciosa* (Germar)	Polyphagous	+(+)	Sporadic	Le, St, Fr
	Chrysomelidae	*Diabrotica sinuata* Olivier	Cabbage, *Cucurbitaceae*, bean	+++	Sporadic	Le
	Chrysomelidae	*Systena testaceovi* Heta	Turnip	+(+)	Sporadic	Le
	Tenebrionidae	*Lagria villosa* L.	Cabbage	+(+)	Permanent	Le
Diptera	Agromyzidae	*Amauromyza maculosa* (Malloch)	Lettuce	++(++)	Permanent	Le
	Agromyzidae	*Liriomyza huidobrensis* (Blanchard)	Polyphagous	++	Permanent	Le
	Agromyzidae	*Liriomyza sativae* Blanchard	Polyphagous	++(++)	Permanent	Le
	Agromyzidae	*Liriomyza trifolii* (Burgess)	Polyphagous	++(++)	Permanent	Le
	Tephritidae	*Ceratitis capitata* (Wiedemann)	Hot pepper	++++	Permanent	Fr
Hemiptera	Aleyrodidae	*Bemisia tabaci* (Gennadius)	Polyphagous	++++	Permanent	Le
	Aphididae	*Brevicoryne brassicae* (L.)	Cabbage	++	Permanent	Le
	Cicadellidae	*Empoasca fabae* Harris	Sweet potato	+(+)	Sporadic	Le
	Coreidae	*Spartocera fusca* (Thunberg)	Aubergine	+(+)	Permanent	Fl, Fr
	Lygaeidae	*Lygaeus bicrucis* Say	Bean	++	Permanent	Fe, Fl
	Pentatomidae	*Edessa* sp.	Sweet potato, aubergine	+(+)	Sporadic	Le
	Pentatomidae	*Thyanta perditor* (F.)	Cabbage, bean	+(+)	Permanent	Le
	Pentatomidae	*Nezara viridula* (L.)	Polyphagous	++(+)	Permanent	Le,Fr
	Pyrrhocoridae	*Dysdercus andreae* (L.)	Okra	+(+)	Sporadic	Fl, Fr
	Pyrrhocoridae	*Dysdercus flavolimbatus* Stal	Okra	+(+)	Permanent	Fl, Fr
	Tingidae	*Stephanitis* sp.	Aubergine	++	Sporadic	Le

Continued

Table 16.2. Continued.

Order	Family	Species	Host plants	Scale of damage	Presence	Site of damage
Lepidoptera	Crambidae	*Diaphana hyalinata* L.	Melon	++	Permanent	Le, Fr
	Crambidae	*Hellula undalis* (F.)	Cabbage	+++(+)	Permanent	Le
	Crambidae	*Maruca testulalis* (Geyer)	Bean	+(+)	Sporadic	Fr
	Gelechiidae	*Tuta absoluta* (Meyrick)	Tomato	+++(+)	Permanent	Le, Fr, St
	Noctuidae	*Helicoverpa armigera* (Hübner)	Polyphagous	+++(+)	Permanent	Fr
	Noctuidae	*Trichoplusia ni* (Hübner)	Tomato	++	Sporadic	Le
	Nymphalidae	*Chlosyne lacinia saundersii* Doubleday	Sweet potato	+(+)	Sporadic	Le
	Pieridae	*Ascia monuste* (L.)	Cabbage	++	Permanent	Le
	Pieridae	*Ascia monuste orseis* (Latreille)	Cabbage	++	Sporadic	Le
	Plutellidae	*Plutella xylostella* (L.)	*Brassicaceae*	+++(+)	Permanent	Le
Thysanoptera	Thripidae	*Frankliniella occidentalis* (Pergande)	Polyphagous	++(+)	Sporadic	Fe, Fl, Fr
	Thripidae	*Thrips palmi* Karny	Polyphagous	++(+)	Sporadic	Fe, Fl, Fr
	Thripidae	*Thrips tabaci* Lindeman	Onion	+++	Permanent	Fe

[a]Scale of damage:
+, unimportant;
++, important;
+++, very important;
++++, can destroy the crop.
[b]Site of damage:
Le, leaf;
Fl, flower;
Fr, fruit;
St, stem.

dry season. The damage begins with yellowish or bleached streaks on the foliage which dry out. The plants may become brown and shrivelled and the mummified fruits are unfit for consumption.
Host plants: bean, onion, *Cucurbitaceae*, *Solanaceae*.
Distribution: Africa, Asia, New Caledonia, South America.

Insecta, Thysanoptera

Thripidae

FRANKLINIELLA OCCIDENTALIS (PERGANDE) (WESTERN FLOWER THRIPS). The orange adults (1–2 mm in length) and yellow larvae are visible with difficulty to the naked eye. They clear the plant cells causing a silver appearance to the foliage. China cabbages are very sensitive. This species can transmit viruses such as tomato spotted wilt virus (TSWV) and destroy the crop.
Host plants: cabbage, bean, lettuce, onion, *Solanaceae*.
Distribution: Africa, Asia.

THRIPS PALMI KARNY (MELON THRIPS). Adults and larvae are similar to *F. occidentalis*. The first damage always appears on the young leaves causing: (i) shrivelled leaves on members of the *Cucurbitaceae*; (ii) deformed leaves and veins on sweet pepper; and (iii) silver streaks along the veins in aubergine, cucumber and watermelon. The attacked fruits have some silver areas, which reduce their commercial value. When populations are very large, the farmed crop is totally destroyed.
Host plants: bean, onion, *Cucurbitaceae*, *Solanaceae*.

Table 16.3. Common pests found on vegetable fields in South-east Asia.

Order	Family	Species	Host plants	Scale of damage	Presence	Site of damage
Acari	Eriophyidae	*Aculops lycopersici* Massée	*Solanaceae*	++	Sporadic	Le, Fr, St
	Tarsonemidae	*Polyphagotarsonemus latus* (Banks)	*Solanaceae*, *Cucurbitaceae*	+++	Sporadic	Le, Fr
	Tetranychidae	*Tetranychus urticae* (Koch)	*Solanaceae*, bean, *Cucurbitaceae*	+++(+)	Permanent	Le, Fr
Coleoptera	Chrysomelidae	*Aulacophora femoralis* (Melsheimer)	Polyphagous	+(+)	Sporadic	Le, Fl
	Chrysomelidae	*Aulacophora lewisii* Baley	Cucumber	+(+)	Sporadic	Le, Fl
	Chrysomelidae	*Monolepta signata* Olivier	Aubergine, bean	++	Sporadic	Le
	Chrysomelidae	*Phyllotreta vittata* F.	Cabbage, tomato	++	Permanent	Le
	Chrysomelidae	*Sagra femorata* ssp. *tonkinensis* (Drury)	Bean	+(+)	Sporadic	Le, Fl
	Coccinellidae	*Henosepilachna sparsa vigintisexpunctata* (Boisduval)	Aubergine, tomato, cucumber	++	Permanent	Le
	Cucurlionidae	*Hypomeces squamosus* (F.)	Zucchini	+(+)	Sporadic	Le
	Meloidae	*Epicauta maklini* Haag	*Cucurbitaceae*	+(+)	Sporadic	Fr, Fl
	Scarabeidae	*Heteroprotaetia fusca* (Herbst)	Aubergine	+(+)	Sporadic	Fr, Fl
	Scarabeidae	*Urbania acuminata* (F.)	Aubergine	+(+)	Sporadic	Fr, Fl
	Scarabeidae	*Hemicoryphocera nigrotestaceus* (Wallace)	Aubergine	+(+)	Sporadic	Fr, Fl
Orthoptera	Acrididae	*Cyrtacanthacris tatarica* (L.)	*Cucurbitaceae*	++(++)	Sporadic	Le, St, Fl
Diptera	Agromyzidae	*Liriomyza huidobrensis* (Blanchard)	Polyphagous	++(++)	Sporadic	Le
	Agromyzidae	*Liriomyza sativae* Blanchard	Polyphagous	++(++)	Sporadic	Le
	Agromyzidae	*Liriomyza trifolii* (Burgess)	Polyphagous	++(++)	Sporadic	Le
	Tephritidae	*Bactrocera cucurbitae* (Coquillet)	*Cucurbitaceae*	++++	Permanent	Fr
	Tephritidae	*Ceratitis capitata* (Wiedemann)	Hot pepper	++++	Permanent	Fr
	Tephritidae	*Dacus dorsalis* (Hendel)	*Cucurbitaceae*	++++	Permanent	Fr
Hemiptera	Aleyrodidae	*Bemisia tabaci* (Gennadius)	Polyphagous	++++	Permanent	Le
	Coreidae	*Anoplocnemis phasiana* (F.)	Aubergine	+(+)	Sporadic	Fl, Fr
	Coreidae	*Cletus* sp.	Tomato, cabbage	+(+)	Sporadic	Fl, Fr
	Coreidae	*Leptoglossus australis* (F.)	Cucumber	++(+)	Permanent	Le, Fe, Fr
	Dinidoridae	*Coridius fuscus* (Westwood)	Cucumber	++	Sporadic	Le, Fr
	Pentatomidae	*Cyclopelta obscura (*Lepeletier)	Bean	+(+)	Sporadic	Fl, Fr
	Pentatomidae	*Eurydema pulchrum* (Westwood)	Chinese cabbage	+(+)	Sporadic	Le
	Pentatomidae	*Nezara viridula* (L.)	Polyphagous	+++	Permanent	Fe, Fl, Fr
Lepidoptera	Crambidae	*Crocidolomia binotalis* Zeller	Cabbage	+++	Sporadic	Le
	Crambidae	*Diaphana indica* (Saunders)	*Cucurbitaceae*	++(+)	Sporadic	Le, Fl, Fr
	Crambidae	*Hellula undalis* (F.)	Cabbage	+++(+)	Permanent	Le

	Crambidae	*Leucinodes orbonalis* Guénée	Aubergine	+++	Sporadic	Fr
	Noctuidae	*Helicoverpa armigera* (Hübner)	Polyphagous	+++(+)	Permanent	Fr
	Noctuidae	*Spodoptera litura* (F.)	Cabbage	++	Permanent	Le
	Plutellidae	*Plutella xylostella* (L.)	*Brassicaceae*	+++(+)	Permanent	Le
Thysanoptera	Thripidae	*Frankliniella occidentalis* (Pergande)	Polyphagous	++(+)	Sporadic	Fe, Fl, Fr
	Thripidae	*Thrips palmi* Karny	Polyphagous	+++	Permanent	Fe, Fl, Fr
	Thripidae	*Thrips tabaci* Lindeman	Onion	+++	Permanent	Fe

[a]Scale of damage:
+, unimportant;
++, important;
+++, very important;
++++, can destroy the crop.
[b]Site of damage:
Le, leaf;
Fl, flower;
Fr, fruit;
St, stem.

Table 16.4. Common pests found on vegetable fields in Oceania (New Caledonia).

Order	Family	Species	Host plants	Scale of damage	Presence	Site of damage
Acari	Eriophyidae	*Aculops lycopersici* Massée	*Solanaceae*	++	Sporadic	Le, Fr, St
	Tarsonemidae	*Polyphagotarsonemus latus* (Banks)	*Solanaceae*, bean	++(+)	Sporadic	Le, Fr
	Tetranychidae	*Tetranychus urticae* (Koch)	*Solanaceae*, bean, carrot	+++	Permanent	Le, Fr
Coleoptera	Chrysomelidae	*Aulacophora indica* (Gmelin)	*Cucurbitaceae*	++	Permanent	Le, Fl, Fr
	Chrysomelidae	*Monolepta palustris* (Perroud)	Polyphagous	++	Permanent	Le, Fl, Fr, St
	Chrysomelidae	*Monolepta semiviolacea* Fauvel	Polyphagous	++	Sporadic	Le, Fl, Fr, St
	Chrysomelidae	*Psylliodes brettinghami* Baley	Aubergine, potato, turnip	++	Permanent	Le
	Coccinellidae	*Henosepilachna sparsa vigintisexpunctata* (Boisduval)	*Solanaceae*	++	Sporadic	Le
Diptera	Agromyzidae	*Liriomyza huidobrensis* (Blanchard)	Polyphagous	++(+)	Permanent	Le
	Agromyzidae	*Liriomyza sativae* Blanchard	Polyphagous	++(+)	Permanent	Le
	Agromyzidae	*Liriomyza trifolii* (Burgess)	Polyphagous	++(+)	Permanent	Le
	Tephritidae	*Bactrocera cucurbitae* (Coquillet)	*Cucurbitaceae*	+++(+)	Permanent	Fr
	Tephritidae	*Bactrocera curvipennis* (Froggatt)	Sweet pepper	+++(+)	Permanent	Fr
Hemiptera	Aleyrodidae	*Bemisia tabaci* (Gennadius)	Polyphagous	+++(+)	Permanent	Le
	Aleyrodidae	*Trialeurodes vaporariorum* (Westwood)	Polyphagous	+++	Permanent	Le
	Aphididae	*Aphis crassivora* Koch	Polyphagous	+++(+)	Permanent	Le, Fl, Fr, St
	Aphididae	*Aphis gossypii* Glover	Polyphagous	+++(+)	Permanent	Le, Fl, Fr, St
	Aphididae	*Myzus persicae* (Sulzer)	Polyphagous	+++(+)	Permanent	Le, Fl, Fr, St
	Coreidae	*Leptoglossus australis* (F.)	*Cucurbitaceae*, tomato	++	Permanent	Fl, Fr
	Lygaeidae	*Graptostetus* sp.	Sweet pepper	+(+)	Sporadic	Fl, Fr
	Lygaeidae	*Oxycarenus luctuosus* (Montrouzier)	Aubergine	+(+)	Sporadic	Fl, Fr
	Pentatomidae	*Nezara viridula* (L.)	Polyphagous	++(+)	Permanent	Fl, Fr, St
Lepidoptera	Crambidae	*Crocidolomia binotalis* Zeller	Cabbage	+++(+)	Permanent	Le
	Crambidae	*Diaphana indica* (Saunders)	*Cucurbitaceae*	+(+)	Sporadic	Le, Fr

Continued

Table 16.4. Continued.

Order	Family	Species	Host plants	Scale of damage	Presence	Site of damage
	Crambidae	*Hellula undalis* (F.)	Cabbage	++(+)	Permanent	Le
	Crambidae	*Sceliodes cordalis* (Doubleday)	Aubergine, tomato	++	Sporadic	Fr
	Gracillaridae	*Acrocercops* sp.	Bean	+(+)	Sporadic	Le
	Noctuidae	*Agrotis ipsilon* (Hufnagel)	Polyphagous	++(+)	Sporadic	St
	Noctuidae	*Chrysodeixis* sp.	Cabbage, lettuce, tomato	++	Sporadic	Le
	Noctuidae	*Helicoverpa armigera* (Hübner)	Polyphagous	++(+)	Permanent	Le Fl, Fr, St
	Noctuidae	*Trichoplusia orichalcea* (F.)	Carrot, cabbage	+(+)	Sporadic	Le
	Noctuidae	*Spodoptera litura* (F.)	Polyphagous	++(+)	Sporadic	Le
	Plutellidae	*Plutella xylostella* (L.)	*Brassicaceae*	+++(+)	Permanent	Le
	Pyralidae	*Hymenia recurvalis* F.	Beet, spinach	+(+)	Sporadic	Le
Thysanoptera	Thripidae	*Frankliniella occidentalis* (Pergande)	Polyphagous	++(+)	Sporadic	Fe, Fl, Fr
	Thripidae	*Thrips palmi* Karny	Polyphagous	++(+)	Sporadic	Le, Fl, Fr, St
	Thripidae	*Thrips tabaci* Lindeman	Onion, garlic	++(+)	Permanent	Le, Fl
Virus		Papaya ring spot virus (PRSV)	*Cucurbitaceae*	+++(+)	Sporadic	Le
		Zucchini yellow mosaic virus (ZYMV)	*Cucurbitaceae*	+++(+)	Sporadic	Le, Fr

[a]Scale of damage:
+, unimportant;
++, important;
+++, very important;
++++, can destroy the crop.
[b]Site of damage:
Le, leaf;
Fl, flower;
Fr, fruit;
St, stem.

Distribution: Africa, Asia, New Caledonia, South America.

THRIPS TABACI (LINDEMAN) (ONION THRIPS). This species is specific to the family *Aliaceae*. The attacked leaves shrivel and become dry. During the hottest part of the day, adults and larvae gather around the central leaf nodes to suck up the cell contents.
Host plant: onion.
Distribution: Africa, Asia, New Caledonia, South America.

Insecta, Hemiptera

Aleyrodidae

BEMISIA TABACI (GENNADIUS) (SWEET POTATO WHITEFLY) Adults are white and measure 2 mm in length. They live fixed under the foliage where they suck sap from the plant. Biotype B, causes silvering of the attacked foliage mainly on *Cucurbitaceae*. Many biotypes can transmit a dangerous virus on tomato, the tomato yellow leaf curl virus (TYLCV).

Host plants: cabbage, *Solanaceae*, *Cucurbitaceae*, *Fabaceae*, *Malvaceae*, *Asteraceae*.
Distribution: Africa, Asia, New Caledonia, South America.

TRIALEURODES VAPORARIORUM (WESTWOOD) (GLASSHOUSE OR GREENHOUSE WHITEFLY). Adults are much larger than *B. tabaci*. Adults are often localized on the top of the plant, where the leaves are softest. The honeydew from feeding favours mildew formation. This species does not seem to transmit TYLCV on tomato.
Host plants: bean, *Cucurbitaceae*, *Solanaceae*.
Distribution: Africa, Asia, New Caledonia, South America.

Aphididae

Numerous species of aphid are found in vegetables including: (i) *Aphis crassivora* Koch; (ii) *Aphis gossypi* Glover; (iii) *Brevicoryne brassicae* (L.); (iv) *Lipaphis pseudobrassicae* (Davis); and (v) *Myzus persicae* (Sulzer). The adults live in colonies with their larvae. Their colour is variable: yellow, orange, greenish grey or black. They cause leaf rolling and young fruit deformation. Their damage is very important as the females transmit viruses such as:

- papaya ring spot virus (PRSV) which causes bleaching and deformation of cucurbit foliage; and
- zucchini yellow mosaic virus (ZYMV) which is sometimes present with PRSV, and causes bleaching of leaves. Young zucchini fruits may be severely malformed with raised protuberances that can make the fruit unmarketable.

In the field, it is very difficult to diagnose these viruses because symptoms are not typical and often the two viruses coexist on the same plant.

On cabbage, no viruses are transmitted but heavy attack by adults and larvae of *L. pseudobrassicae* and *M. persicae* obstruct plant development. The cabbage foliage becomes greyish and the plant becomes dwarfed.
Host plants: *Cucurbitaceae*, *Asteraceae*, *Solanaceae*, *Brassicaceae*.
Distribution: Africa, South America, New Caledonia.

Coreidae

LEPTOGLOSSUS AUSTRALIS (F.) (LEAF-FOOTED BUG). Adults are black, measure about 20 mm in length, and have black legs with flattened posterior tibia. The eggs typically look like lines of small kegs. Larvae are reddish to black. Adults and larvae suck flowers causing flower and immature fruit drop.
Host plants: tomato, *Cucurbitaceae*.
Distribution: Africa, Asia, New Caledonia.

Dinidoridae

CORIDIUS FUSCUS (WESTWOOD). Adults are dark brownish with a dark lower forewing. They measure 15–20 mm in length. Larvae and adults bite all parts of the host plant. Heavy infestation results in fruits becoming distorted and the plants stop growing.
Host plant: cucumber.
Distribution: Asia.

Jassidae

JACOBIASCA HYBRIDA BERGEVIN AND ZANON. The spindle-shaped adults measure about 5 mm in length and are light green. They suck the cell contents of the foliage, causing discoloration, drying out and eventually death of the plant.
Host plant: aubergine.
Distribution: Africa.

Lygaeidae

SPILOSTETHUS SPP. (SEED BUGS). Two species are often present on vegetables. *Spilostethus longulus* Dallas adults measure about 10 mm in length and are orange red with several black triangular markings on the body. Feeding by adults and larvae cause distortion and arrested development of young fruits. *Spilostethus rivularis* (Germar) adults have similar coloration but are larger (about 15 mm) and are often present on hot and sweet pepper as well as on cucurbits where their feeding leads to flower and fruit drop.
Host plants: *Cucurbitaceae*, *Solanaceae*.
Distribution: Africa.

Pentatomidae

ACROSTERNUM ACUTUM (DALLAS). The females are green and have two yellow spines on the sides of the thorax. They measure between 10 mm and 12 mm. The males are shorter in length. Eggs are laid in a group on the leaves in geometric form. After hatching, young larvae stay with the egg group for several hours before dispersing on to the plant. The mobile pupae are very colourful. All stages bite all parts of the host plant causing bud drop and fruit deformation.
Host plants: okra, bean, *Solanaceae*.
Distribution: Africa.

NEZARA VIRIDULA (L.) (GREEN VEGETABLE BUG). On tomato, a few individuals can cause damage. Adults are about 15 mm in length and are green without a yellow spine. Similar to *A. acutum*, females lay eggs in a geometrical pattern. Adults, larvae and pupae bite all parts of the host plant. The young bean pod becomes distorted and on tomato fruits some discoloured marks appear which reduce the commercial sale value.
Host plants: amaranth, cabbage, okra, bean, *Cucurbitaceae*, *Solanaceae*.
Distribution: Africa, New Caledonia, South America, Asia.

Insecta, Lepidoptera

Crambidae

PSARA BASALIS (WALKER) (LEAF CATERPILLAR). The adults are yellow-ochre with black spots. The wingspan measures between 16 mm and 20 mm. Larvae are greenish white and eat the foliage from under a previously made fold that they form. Often found with *H. recurvalis*, this species causes less damage and the larvae are less agile.
Host plants: amaranth, *Celosia* spp.
Distribution: Africa.

MARUCA TESTULALIS (GEYER) (MARUCA POD BORER, BEAN POD BORER, MUNG MOTH). The forewings are brownish with white areas. The adult wingspan measures 20 mm. The females oviposit cream, oval eggs near floral buds. The larvae are yellowish white with black spots. They bore into bean pods and eat into young seeds. They attack both flowers and leaves and join them by silk. They pupate in a cocoon in the pod.
Host plants: pigeon pea, cowpea, mung bean and soybean.
Distribution: Africa, New Caledonia, South America.

CROCIDOLOMIA BINOTALIS ZELLER (CABBAGE CLUSTER CATERPILLAR). The males are yellow-ochre with black markings on the forewings. Females are uniformly ochre with a wingspan of about 23 mm. They lay eggs in clusters on the stem or on the leaves. Larvae are about 20–25 mm in length and have a variable greenish to white coloration with two dark-brown longitudinal lines. They always occur in colonies and they consume the cabbage heart. Pupation is in the soil or in plant debris.
Host plants: *Brassicaceae*, including cultivated cabbage.
Distribution: Africa, New Caledonia, Asia.

HELLULA UNDALIS (F.) (CABBAGE WEBWORM). This is a small moth with a wingspan of 23 mm and off-white brownish freckled forewings. Females lay single eggs under the leaves. The whitish larvae have two light-brown longitudinal lines on the body. The head is dark brownish to black. They damage the major veins and the plant axial bud. As a result of this the secondary buds on cabbage start to develop, producing several little heads which are not marketable.
Host plants: *Brassicaceae*.
Distribution: Africa, South America, New Caledonia, Asia.

DIAPHANIA INDICA (SAUNDERS) (CUCUMBER MOTH OR COTTON CATERPILLAR). Adults have translucent whitish wings with broad, dark brown borders. The wingspan is about 25–30 mm. There is a tuft of light brown 'hairs' on the abdomen tip, more developed in the females than in the males. Females lay eggs under the leaves near the veins. The larvae are translucent at the beginning and then are green with two longitudinal white lines on the body. Feeding damage leads to drying

out of leaves and young fruit. Pupation is on the foliage in a leaf fold made by the larva.
Host plants: *Cucurbitaceae* and other plants.
Distribution: Africa, Asia, Oceania.

DIAPHANIA HYALINATA (L.) (MELON WORM MOTH). The wingspan is 27–30 mm. The wings are pearly white centrally and slightly iridescent, but are edged with a broad band of dark brown. There is a tuft of light brown 'hairs' on the abdomen tip too. The larvae are similar to those of *D. indica* and the biology of this species is similar too.
Host plants: melon, cucumber.
Distribution: South America.

LEUCINODES ORBONALIS GUÉNÉE (EGGPLANT FRUIT AND SHOOT BORER OR BRINJAL FRUIT AND SHOOT BORER). The adults are white with two light brown and two black freckles on the forewings, with a wingspan of 20 mm. Its abdomen is typically raised in the rest position. The larvae are pinkish and bore into the fruit. This species is also a stem borer, destroying the terminal shoots.
Host plants: aubergine.
Distribution: Africa, Asia.

SCELIODES LAISALIS (WALKER) (EGGPLANT FRUIT BORER). The adults are white with greenish and dark brown freckles on the forewings. The wingspan, larval coloration, biology and rest position are similar to *L. orbonalis*, except its geographical distribution differs. This species prefers drier areas than *L. orbonalis*.
Host plants: aubergine, African aubergine, *Solanum macrocarpon*.
Distribution: Africa.

SCELIODES CORDALIS (DOUBLEDAY) (POROPORO FRUIT BORER OR EGGFRUIT CATERPILLAR). This species is similar to *S. laisalis* described above.
Host plants: aubergine, tomato.
Distribution: New Caledonia.

Gelechiidae

SCROBIPALPA ERGASIMA (MEYRICK). The adults have a 13 mm wingspan and are light brown. The translucent larvae have a black head and are solitary; they bore into the floral buds. No symptom of damage is visible from the exterior, except the exit hole made by the larva leaving to pupate. Pupation is in the soil where the larvae build shells for protection.
Host plant: aubergine.
Distribution: Africa.

Gracillariidae

ACROCERCOPS SP. The larvae from this little moth bores into the foliage causing whitish cells as a result of air breaking into the damaged cells. Pupation occurs on the plant. This species can cause serious damage when beans are young.
Host plant: bean.
Distribution: New Caledonia.

Noctuidae

AGROTIS IPSILON (HUFNAGEL) (BLACK CUTWORM). The nocturnal adults are greyish to black in colour. Adults have a wingspan of 25 mm. The nocturnal larvae cut young plant stems at the neck.
Host plants: cabbage, okra, *Solanaceae*.
Distribution: Africa, Asia, New Caledonia, South America.

SPODOPTERA EXIGUA (HÜBNER) (BEET ARMYWORM OR SMALL MOTTLED WILLOW MOTH). The adults are dark brownish with whitish designs on the forewings. They have a wingspan of 20–25 mm. The larvae have two longitudinally white bands on the sides. They feed inside the onion leaves.
Host plant: onion.
Distribution: Africa.

SPODOPTERA LITTORALIS BOISDUVAL (AFRICAN COTTON LEAFWORM OR EGYPTIAN COTTON LEAFWORM). Adults with a wingspan of 30 mm are blackish brown with designs and lines on the forewings. The males have bluish streaks on each forewing. Several hundred eggs are oviposited on to leaves in groups which are covered with scales. The larval colour varies from yellow-green to brown but is recognizable by two black triangles located at the front and at the extremity of each side of the abdomen. After hatching, larvae stay in a group and consume the leaf parenchyma under the foliage. Seedlings are

often destroyed by larvae which pupate in the soil.
Host plants: okra, bean, *Brassicaceae*, *Solanceae*, *Fabaceae*.
Distribution: Africa.

SPODOPTERA LITURA (F.) (ORIENTAL LEAFWORM MOTH, TARO CATERPILLAR OR TOBACCO CATERPILLAR). Adults measure between 15 mm and 20 mm in length and have a wingspan of 30–38 mm. Forewings are grey to reddish brown, with a complex pattern of creamy streaks and paler lines. The males have a blue-grey band at the apex of each wing. Larvae have bright yellow stripes along the back and the sides. Larval colours vary from pale green to dark green. Females lay eggs in masses of 200–300 eggs. Egg masses are covered with scales and are laid on the underside of the host plant leaf. Larvae eat leaves, and even flowers and fruits. Pupation is carried out in the soil.
Host plants: lettuce, cabbage, beet, amaranth, tomatoes, aubergine, bean.
Distribution: Asia, New Caledonia.

TRICHOPLUSIA NI (HÜBNER) (CABBAGE LOOPER). The forewings are mottled grey-brown in colour. They bear silvery white spots centrally. The adults have a wingspan of 33–38 mm. Eggs are deposited singly on either the upper or lower surface of the leaf. The larvae are green, but marked with a distinct white stripe on each side. They consume the foliage. At pupation a thin white cocoon is formed on the underside of the foliage, in plant debris or among clods of soil. The pupa is initially green, but soon turns dark brown or black.
Host plants: *Brassicaceae*, bean, *Solanaceae*, *Cucurbitaceae*, sweet potato, celery, lettuce.
Distribution: South America.

HELICOVERPA ARMIGERA (HÜBNER) (COTTON BOLLWORM, CORN EARWORM OR OLD WORLD (AFRICAN) BOLLWORM). Recently observed in Brazil, the adults from this species have a wingspan of 35 mm and are active at night. The forewings are light brown for the male but more greenish for females. Females oviposit single eggs towards the top of the plant. Larvae are brown to green in colour with two dark lines on the whitish ventral side of the abdomen. Pupation is in the soil, and sometimes into the fruit. The most important damage is on tomato. After a defoliation period, the young larvae eat the flowers and then attack the fruits, which later rot. One larva can bore several fruits during this larval cycle.
Host plants: amaranth, cabbage, lettuce, onion, *Solanaceae*, *Cucurbitaceae*.
Distribution: Africa, Asia, New Caledonia, South America.

Nolidae

SELEPA DOCILIS BUTLER. The adults are light brown with a wingspan varying from 18 mm to 20 mm. Eggs are laid singly on the leaf edges. The larvae, always in groups, have long hairs. They consume the foliage and leave the major veins. They can attack fruits. Pupation is made on the plant in a conical cocoon.
Host plant: aubergine.
Distribution: Africa.

EARIAS BIPLAGA (WALKER) (SPINY BOLLWORM) AND *EARIAS INSULANA* (BOISDUVAL) (EGYPTIAN STEM-BORER, EGYPTIAN BOLLWORM, SPINY BOLLWORM OR COTTON SPOTTED BOLLWORM). Adults are greenish yellow with dark-brown bordered forewings. They measure about 20 mm in length. The *E. insulana* adult has two forms: (i) in the dry season, they are totally ochre in colour; and (ii) in the rainy season they become green. The females lay single bluish eggs on young fruits. Of the two species, the larvae are more spiny in *E. biplaga* which feeds inside the fruit. Pupation occurs outside the fruit. The cocoon is globular in *E. biplaga* and more flattened in *E. insulana*.
Host plant: okra.
Distribution: Africa.

Pieridae

ASCIA MONUSTE (L.) (GREAT SOUTHERN WHITE). The adult is white with a black freckle on the top of each forewing. The wingspan is of 63–86 mm. The light yellow eggs are laid in groups on the foliage. The larvae are

blue with yellow lines and occur in groups consuming the foliage. The pupae are light yellow and black. Pupation is on the plant.
Host plants: *Brassicaceae* (including cultivated cabbage, radish and *Lepidium* species) and plants of the *Capparidaceae* (e.g. nasturtium).
Distribution: South America.

Plutellidae

PLUTELLA XYLOSTELLA (L.) (DIAMONDBACK MOTH). Adults are very small moths (wingspan of 10 mm). Their colour varies from light brown (females) to dark brown (males). Females lay single eggs or in groups of around ten under leaves or on stems, near the leaf veins. After hatching, larvae feed on the foliage making 'comma'-like holes. After the second stage, they fall from the host plant and stay hanging on a thread. They mine the foliage making typical transparent 'windows'. Heavy larval attack can destroy the cabbage crop. Recently, two researchers, after a DNA barcoding study, discovered in Australia two cryptic species, *P. xylostella* (L.) and *Plutella australiana* Landry & Herbert, (Lepidoptera, Plutellidae), which were present on cabbage crops in the same localities and on the same plots (Landry and Hebert, 2013). This discovery causes complications in the evaluation of the pest level of these two species for their respective population control.
Host plants: *Brassicaceae*.
Distribution: Australia, New Zealand, Europe, Africa, North and South America, Asia, New Caledonia and the Hawaiian Islands.

Pyralidae

HYMENIA RECURVALIS (F.) (AMARANTH CATERPILLARS). The adult measures 15 mm, the forewings are dark brown with a white line across them. The larvae are very agile, and are green with two white longitudinal lines on the body. They eat the foliage and leave the veins.
Host plants: spinach, beet, cotton, maize, soybean.
Distribution: Africa.

Insecta, Coleoptera

Chrysomelidae

The Chrysomelidae discussed in this chapter include three groups of species, referred to as the 'flea beetles', the 'leaf beetles' and the 'tortoise beetles'.

THE FLEA BEETLES. The adults jump when disturbed. The females lay eggs in the soil. The larvae eat into the roots, but damage is never very important. The principal damage is caused by adults which create shot holes in the foliage of the plant host, reducing plant photosynthesis.

Epithrix sp. The adults (2 mm in length) are dark blue to metallic black. They are always numerous and can make a lot of damage, principally on young plants. Usually species from this genus are specific to plants in the family *Solanaceae*.
Host plant: aubergine.
Distribution: Africa.

Phyllotreta cheiranthi Weise. The small adults (1.5–2 mm in length) are metallic blue-green and numerous. They can create considerable damage principally on young plants.
Host plants: aubergine, cabbage, *S. macrocarpon*.
Distribution: Africa.

Phyllotreta vittata (F.). The small adults have a black body and the elytra are yellow with longitudinal black lines. They can create considerable damage principally on young plants.
Host plants: cabbage, tomato.
Distribution: Asia.

Podagrixina decolorata (Duvivier). The adults (3–4 mm in length) are uniformly orange red. They create shot holes in the leaves and erode flowers and fruits. They can cause important damage on leafy vegetables (e.g. *Corchorus* sp.) consumed by people.

Host plants: aubergine, *Corchorus* sp., *Malvaceae*.
Distribution: Africa.

Psylliodes brettinghami Baley. The adults (2–3 mm in length) are a metallic blue-green colour.
Host plants: aubergine, potato.
Distribution: New Caledonia.

THE LEAF BEETLES. Only the adults cause important damage. They consume leaves and flowers and cause considerable damage in young plants. The eggs are laid in the soil and whitish larvae enter the roots and bore the stem of the host plant causing little damage. Pupation is in the soil in a shell.

Aulacophora foveicollis Lucas and *Lamprocopa occidentalis* (Weise). These species often occur in large numbers on the same plant. Adults of *A. foveicollis* are red and those of *L. occidentalis* are yellowish, they are about 8 mm in length and they consume leaves and flowers.
Host plants: *Cucurbitaceae*.
Distribution: Africa.

Aulacophora indica (Gmelin). This species is identical to *A. foveicollis*, except that the body colour is orange yellow.
Host plants: *Cucurbitaceae*.
Distribution: New Caledonia.

Diabrotica sinuata Olivier. The adults are 5–8 mm in length. The head is black, the thorax yellow. The elytra are yellow with two or three black bands. A lot of varieties exist with different coloration.
Host plants: cabbage, bean, *Cucurbitaceae*.
Distribution: South America.

Monolepta palustris (Perroud) and *Monolepta semiviolacea* Fauvel. These two species cohabit, *M. semiviolacea* being less important. Adults (5–8 mm in length) are light orange-yellow. They can attack stem and fruits reducing their value for market.
Host plants: aubergine, beet, cabbage, carrot, bean, tomato, *Cucurbitaceae*.
Distribution: New Caledonia.

Monolepta signata Olivier. The elytra are dark blue with two whitish spots on each. The thorax is orange and legs yellow. Adults measure about 6 mm in length.
Host plants: aubergine, bean.
Distribution: Asia.

THE TORTOISE BEETLES. Several species are present on sweet potato foliage. A few species cause important damage.

Aspidomorpha cincta (F.). Adults measure 7–10 mm in length and have a circular form. Their body has golden sparkles and a black 'X' form on the back. The legs are masked by the body. Young larvae are always in groups and consume the foliage, spreading out as they age. While often spectacular, the damage caused does not threaten the sweet potato crop, indeed sweet potatoes tolerate this pest.
Host plant: sweet potato.
Distribution: Africa.

Cassida liquefacta Spaeth. Adults are 4 mm in length and green in colour. Larvae and adults consume the leaves. When there are large infestations, damage can be significant because amaranth and *Celosia* spp. are grown for their edible foliage.
Host plants: amaranth, *Celosia* spp.
Distribution: Africa.

Coccinellidae

The species discussed here are members of the Hepilachninae subfamily, which are all phytophagous species. It is essential to distinguish these species from similar looking predator species (Coccinellinae). They are usually referred to as ladybird beetles.

Adult phytophagous species measure between 6 mm and 10 mm in length and are more or less orange red in colour with some black spots. Their body is always mat (i.e. dull and without a shine), while predator species' bodies are always bright and shiny with variable coloration (red, yellow, orange or black, with black spots, lines or freckles). The yellow eggs are laid in a group under leaves or on the fruits. The phytophagous larvae have numerous spines, a spherical body without recognizable legs, move little, and are white to yellow in colour. By contrast, predatory larvae have no spines, are slender with visible legs, are very agile and their coloration is always black with sometimes two black spots on the thorax. Pupation is on plant leaves. Adults and larvae scrape the foliage.

EPILACHNA NIGERIANA MADER. The adults measure 4–7 mm in length. The body and spines of the larvae are white.
Host plants: *Vernonia* spp.
Distribution: Africa.

EPILACHNA PAVONIA (OLIVIER). The adults are about 8 mm in length and have a transverse dark grey band on the thorax and on the elytra edge. The elytra are punctuated with dark grey spots.
Host plants: aubergine, *S. macrocarpon*.
Distribution: Africa.

HENOSEPILACHNA ELATERII (ROSSI). Adults (6–9 mm in length) have an orange red body punctuated with black spots. The larval body and spines are light yellow. Larvae attack flowers and fruits causing great damage on young plants.
Host plants: *Cucurbitaceae*.
Distribution: Africa.

HENOSEPILACHNA RETICULATA (OLIVIER). The adults are larger than *H. elaterii* (7–10 mm in length). The larvae are cream with black spines, and can attack flowers and fruits too.
Host plants: *Cucurbitaceae*.
Distribution: Africa.

HENOSEPILACHNA SPARSA VIGINTISEXPUNCTATA (BOISDUVAL). Adults are 7–10 mm in length with a body that is orange red with 28 black spots in transverse lines.
Host plants: aubergine, potato.
Distribution: New Caledonia, Asia.

Cucurlionidae

CYLAS PUNCTICOLLIS BOHEMANN (SWEET POTATO WEEVIL). Females are mat black and measure 5–8 mm in length and bore into stems and tubers to oviposit. The larvae make galleries in stems, which dry out as a result. The more important damage comes when the tuber harvest is delayed. Females lay eggs in the uncovered tuber and the larvae begin making their galleries and continue during the postharvest period. The harvest can be destroyed during tuber storage.
Host plant: sweet potato.
Distribution: Africa.

GASTEROCLISUS RHOMBOIDALIS (BOHEMAN.). Adults measure 11–15 mm in length. Their body is covered with yellow or white pubescence. A black network and two designs in 'V' form are more and less visible on the back. Adults consume leaves and females oviposit into stems, which are bored by larvae causing the plant to dry out.
Host plants: amaranth, aubergine, *Celosia* spp.
Distribution: Africa.

Meloidae

The species in this family reported here are usually referred to as 'chafers'. Only the adults cause damage. They consume leaves and flowers and sometimes chew the fruit. The larvae parasitize bees.

HYCLEUS ARGENTATA F. The adults are 8–12 mm in length. The head, thorax and body are black with more-or-less large black spots on the silver grey elytra.
Host plant: okra.
Distribution: Africa.

HYCLEUS HERMANIAE (F.). The adults are 7–14 mm in length. The head, thorax, body and elytra are black with six yellow spots.
Host plants: okra, bean.
Distribution: Africa.

HYCLEUS SENEGALENSIS (VOIGTS). The adults are 20–27 mm in length. The head, thorax, body and elytra are black with six red spots.
Host plants: okra, bean.
Distribution: Africa.

HYCLEUS TRIFASCIATUS (THUNBERG). The adults are 25–30 mm in length. The head, thorax, body and elytra are black with six big yellow spots.

Host plant: okra.
Distribution: Africa.

HYCLEUS VESTITA REICHE. Adults are 12 mm in length. The head, thorax, body and elytra are black with eight red spots.
Host plants: okra, bean.
Distribution: Africa.

Scarabeidae

The species reported in this chapter are usually referred to as 'long-legged chafers' or 'chafers'.

DIPLOGNATHA GAGATES (FORSTER). The adults are 18–23 mm in length with metallic dark brown coloration. They chew okra pistil flowers, *Vernonia* leaves, stems and pods. On bean, one adult attacks several pods. The larvae are saprophytic.
Host plants: okra, bean, *Vernonia*.
Distribution: Africa.

PACHNODA INTERRUPTA (OLIVIER) (SORGHUM CHAFER). Numerous *Pachnoda* species are present on okra, but make little damage. *P. interrupta* adults (12–16 mm in length) are shiny black with red spots distributed on the thorax and on the elytra. They chew flowers and fruits. The larvae are saprophytic.
Host plant: okra.
Distribution: Africa.

Diptera

Agromyzidae

The species discussed here are usually reported as 'leafminers'. The females lay cream-coloured eggs on the leaf parenchyma. After hatching, larval boring causes large serpentine (i.e. winding and twisting) mines, which form whitish plaques formed when air enters the damaged cells. At the end of the larval cycle, larvae fall on the soil and pupate. For some species the pupation is on the plant. The pupae are orange, dark red or brown depending on the species. The adult females feed on the foliage making small spots. Females split the parenchyma and suck out the cell contents.

LIRIOMYZA TRIFOLII (BURGESS) (AMERICAN SERPENTINE LEAFMINER) AND *LIRIOMYZA SATIVAE* BLANCHARD (VEGETABLE LEAFMINER). These two species cause similar damage and often cohabit in the same fields and on the same crop. These species (which are often confused) have different responses in relation to the host plants (Bordat *et al.*, 1987), to pesticides and to natural enemies (Bordat *et al.*, 1988).

The adults are very small (1.5–2 mm in length) and larvae are yellow and black. Pupal coloration varies with age from light orange to brown. These species are referred to as 'serpentine leafminers'. Heavy attack causes drying out of foliage and whole plants.
Host plants: *Solanaceae*, *Cucurbitaceae*, *Asteraceae*, *Malvaceae*, *Fabaceae*, *Brassicaceae*.
Distribution: Africa, Asia, South America, New Caledonia.

LIRIOMYZA HUIDOBRENSIS (BLANCHARD) (PEA LEAFMINER). This species is also referred to as 'serpentine leafminers' and produces mines that are more localized near the leaf veins. The adults (2 mm in length) and larvae are white and black. The pupae vary from whitish to dark brown and often stay hanging under the host plant leaves (particularly on bean).
Host plants: *Solanaceae*, *Cucurbitaceae*, *Asteraceae*, *Malvaceae*, *Fabaceae*, *Brassicaceae*.
Distribution: Africa, Asia, South America, New Caledonia.

MELANAGROMYZA CLEOME SPENCER. The adults (2 mm in length) are black. Larvae bore into the petiole and the mine is visible outside. Pupation is in the stem. Plants become dwarfed and do not grow.
Host plant: cabbage.
Distribution: Africa, Asia.

AMAUROMYZA MACULOSA (MALLOCH) (BLOTCH LEAFMINER). The adults (2–3 mm in length) are black. Females lay eggs in the leaves. The larvae consume the parenchyma causing

whitish lenses caused by air entering into the damaged cells. This species is a 'plaque miner'. Pupation is in the interior of the leaves.
Host plant: lettuce.
Distribution: South America.

OPHIOMYIA PHASEOLI TRION (BEAN FLY). The adults (2–3 mm in length) are black. Eggs are laid in the parenchyma and larvae bore into the petiole and the stem. Pupation is in the neck (the point where the petiole meets the stem), which splits. The plants affected are sickly.
Host plant: bean.
Distribution: Africa, Asia.

Tephritidae

The species discussed here are usually reported as 'fruit flies'. Often different species causing similar damage cohabit in the same fields and on the same crop.

The females with their boring ovipositor lay several white eggs in the fruit. After some days, the oviposition area rots. The numerous larvae bore into the fruits which rot. Pupation is in the soil. The adult size varies between 6 mm and 8 mm in length. The males are smaller than females.

BACTROCERA CUCURBITAE (COQUILLET) (MELON FLY). Head, thorax and abdomen are dirty orange, the escutcheon is light yellow and the wings are translucent with black trails.
Host plants: melon, cucumber, tomato.
Distribution: Africa, Asia, New Caledonia.

BACTROCERA CURVIPENNIS (FROGGATT). The head is dark red, the thorax blackish, the abdomen orange yellow and the escutcheon yellow. The wings are translucent with a thick black vein at the edge.
Host plants: This species is polyphagous, reported from at least 41 host plant species in 30 genera and 21 families.
Distribution: New Caledonia.

DACUS BIVITTATUS (BIGOT) (PUMPKIN FLY). The head, thorax and abdomen are black brown and the escutcheon is yellow. The wings are translucent with an opaque area.
Host plants: *Cucurbitaceae.*
Distribution: Africa.

DACUS CILIATUS LOEW. The head, thorax and abdomen are light orange brown, the escutcheon is yellow and the wings are translucent.
Host plants: *Cucurbitaceae.*
Distribution: Africa, Asia.

DACUS VERTEBRATUS BEZZI. The head, thorax and abdomen are garnet red, the escutcheon is yellow and the wings are translucent with a black spot at the end of each wing.
Host plants: *Cucurbitaceae.*
Distribution: Africa.

CERATITIS CAPITATA (WIEDEMANN) (MEDITERRANEAN FRUIT FLY OR MEDFLY). The head is white, the thorax grey and white and the escutcheon is black. The wings have orange bands and black designs. The adults are smaller in relation to the other species, measuring 4–7 mm in length.
Host plant: hot pepper.
Distribution: Africa, Asia, South America.

NEOCERATITIS CYANESCENS (BEZZI) (TOMATO FRUIT FLY). The head is red, the thorax grey with black lines, the abdomen red with two transverse white lines and the escutcheon yellow. The wings are translucent with a black trail-like comb on each wing.
Host plants: tomato, hot and sweet pepper.
Distribution: Africa.

Means of Controlling Pests in Organic Production

For good pest control in vegetables, in a sustainable production project, the local natural enemy fauna must be known qualitatively (diversity of species) and quantitatively (scale of individuals).

The main natural enemies found in Africa (West, Central, Reunion and Mayotte islands) are shown in Table 16.5, those from South America (Brazil and French Antilles) in Table 16.6, and those from South-east Asia and New Caledonia in Table 16.7.

Generally, habitats are poor and lacking in diversity for the primary parasitoid

Table 16.5. Natural enemy species found in vegetable fields in West and Central Africa, Reunion and the Mayotte islands.

Type	Order	Family	Species	Host	Family	Host stage
Predators	Diptera	Syrphidae	*Eristalinus tabanoides* (Jaennieke)	Aphids	Aphidae	La
		Syrphidae	*Ischiodon aegyptius* (Weidemann)	*Lipaphis pseudobrassicae*	Aphidae	Ad, La
	Hemiptera	Pentatomidae	*Afrius purpureus* Westwood	Various	-	Ad, La
		Pentatomidae	*Dorycoris pavoninus* (Westwood)	Various	-	Ad, La
		Pentatomidae	*D. pavoninus* var. *miniatus* (Westwood)	Various	-	Ad, La
		Reduvidae	*Coranus aegyptius* (F.)	Various	-	Eg, Ad, La
		Reduvidae	*Coranus pallidus* Reuter	Various	-	Eg, Ad, La
		Reduvidae	*Coronopsis vittata* Horvath	Various	-	Eg, Ad, La
		Reduvidae	*Cosmolestes pictus* (Klug)	Various	-	Eg, Ad, La
		Reduvidae	*Haematochares obscuripennis* Stal	Various	-	Eg, Ad, La
		Reduvidae	*Hediocoris fasciatus* (Reuter)	Various	-	Eg, Ad, La
		Reduvidae	*Hediocoris fasciatus* var. *reuteri* Villiers	Various	-	Eg, Ad, La
		Reduvidae	*Rhinocoris albopilosus* Signoret	Various	-	Eg, Ad, La
		Reduvidae	*Rhinocoris bicolor* F.	Various	-	Eg, Ad, La
		Reduvidae	*Rhinocoris rapax* Stal	Various	-	Eg, Ad, La
		Reduvidae	*Rhinocoris segmentarius* Germar	Various	-	Eg, Ad, La
		Reduvidae	*Sphedanolestes pictumellus* Schoulgedey	Various	-	Eg, Ad, La
	Coleoptera	Carabidae	*Stenocallida fasciata* Dejean	Various	-	Eg, La
		Carabidae	*Stenocallida* sp.	Various	-	Eg, La
		Coccinellidae	*Cheilomenes propinqua triangulifera* (Mulsant)	Aphids	-	Ad, La
		Coccinellidae	*Cheilomenes sulphurea* Olivier	Aphids	-	Ad, La
		Coccinellidae	*Cheilomenes vicina* Mulsant	Aphids	-	Ad, La
		Coccinellidae	*Exochomus laeviuseulus* Weise	Aphids	-	Ad, La
		Coccinellidae	*Exochomus troberti* Mulsant	Aphids	-	Ad, La
		Coccinellidae	*Hippodamia variegata* (Goeze)	Aphids, thrips	-	Ad, La
		Coccinellidae	*Xanthadalia effusa* (Erichson)	Aphids	-	Ad, La
Parasitoids	Hymenoptera	Braconidae	*Apanteles bredoi* De Saeger	*Psara basalis*	Pyralidae	La
		Braconidae	*Apanteles litae* Dixon	*Plutella xylostella*	Plutellidae	La
		Braconidae	*Apanteles sagax* Wilkinson	*Sylepta derogata* (F.)	Pyralidae	La
		Braconidae	*Apanteles syleptae* Ferrière	*S. derogata*, *P. basalis*	Pyralidae	La
		Braconidae	*Apanteles* sp.	*Hymenia recurvalis*	Pyralidae	La
		Braconidae	*Apanteles* sp.	*Scrobipalpa ergasima*	Gelechiidae	La

Continued

Table 16.5. Continued.

Type	Order	Family	Species	Host	Family	Host stage
		Braconidae	*Austroopius insignipennis* (Granger)	*Neoceratitis cyanescens*, *Dacus ciliatus*	Tephritidae	La
		Braconidae	*Bracon* sp. 1	*Asphondyla* sp.	Cecidomyiidae	La
		Braconidae	*Bracon* sp. 2	*Asphondyla* sp.	Cecidomyiidae	La
		Braconidae	*Bracon* sp. 1	*S. ergasima*	Gelechiidae	La
		Braconidae	*Bracon* sp. 2	*S. ergasima*	Gelechiidae	La
		Braconidae	*Cardiochiles* sp.	*P. basalis*	Pyralidae	La
		Braconidae	*Cotesia* cf. *hyperion* De Saeger	*P. xylostella*	Plutellidae	La
		Braconidae	*Cotesia vestalis* (Haliday)	*P. xylostella*	Plutellidae	La
		Braconidae	*Cotesia* sp.	*Spodoptera* spp.	Noctuidae	La
		Braconidae	*Dacnusa* sp.	*Liriomyza huidobrensis*	Agromyzidae	La
		Braconidae	*Diolcogaster* aff. *semirufa* (De Saeger)	*Selepa docilis*	Noctuidae	La
		Braconidae	*Diolcogaster* sp.	*S. docilis*	Noctuidae	La
		Braconidae	*Dolichogenidea* sp.	*P. melongena*	Pyralidae	La
		Braconidae	*Macrocentrus* sp.	*P. melongena*	Pyralidae	La
		Braconidae	*Meteoridea testacea* Granger	*H. recurvalis*	Pyralidae	La
		Braconidae	*Opius dissitus* Muesebeck	*Liriomyza* spp.	Agromyzidae	La
		Braconidae	*Orgilus* sp.	*H. recurvalis*	Pyralidae	La
		Braconidae	*Protomicroplitis* spp.	*P. melongena*	Pyralidae	La
		Chalcididae	*Brachymeria citrea Steffan*	*P. xylostella*	Plutellidae	Pu
		Eulophidae	*Callitula* sp.	*Melanagromyza cleome*	Agromyzidae	La
		Eulophidae	*Chrysocharis caribea* Boucek	*Liriomyza* spp.	Agromyzidae	La
		Eulophidae	*Euplectrus hargreavesi* Ferrière	*Helicoverpa armigera*	Noctuidae	La
		Eulophidae	*Euplectrus laphygmae* (Ferrière)	*Spodoptera* spp.	Noctuidae	La
		Eulophidae	*Hemiptarsenus varicornis* Guirault	*Liriomyza* spp.	Agromyzidae	La
		Eulophidae	*Oomyzus sokolowskii* (Kurdjumov)	*P. xylostella*	Plutellidae	La
		Eulophidae	*Sigmophora* sp. 1	*Asphondyla* sp.	Cecidomyiidae	La
		Eulophidae	*Sigmophora* sp. 2	*Asphondyla* sp.	Cecidomyiidae	La
		Eulophidae	*Tetrastichus* sp.	*S. docilis*	Noctuidae	Pu
		Eupelmidae	*Neanastatus turneri* Ferrière	*Asphondyla* sp.	Cecidomyiidae	La
		Eurytomidae	*Eurytoma* sp. 1	*Asphondyla* sp.	Cecidomyiidae	La
		Eurytomidae	*Eurytoma* sp. 2	*Asphondyla* sp.	Cecidomyiidae	La
		Ichneumonidae	*Charops* sp. 1	*P. basalis*	Pyralidae	La

	Ichneumonidae	*Charops* sp. 2	*S. littoralis*	Noctuidae	La
	Ichneumonidae	*Charops* sp. 3	*H. armigera*	Noctuidae	La
	Ichneumonidae	*Ceratocryptina* sp.	*Autoba admota* (Felder & Rogenhofer, 1874)	Noctuidae	La.
	Ichneumonidae	*Diadegma mollipla* Holmgren.	*P. xylostella*	Plutellidae	La
	Ichneumonidae	*Diadegma insulare* (Cresson)	*P. xylostella*	Plutellidae	La
	Pteromalidae	*Catolaccus* sp.	*Asphondyla* sp.	Cecidomyiidae	La
	Pteromalidae	*Halticoptera circulus* Walter	*Ophiomyia phaseoli*	Agromyzidae	La
	Pteromalidae	*Halticoptera* sp.	*L. huidobrensis*	Agromyzidae	La
Diptera	Tachinidae	*Blepharella* sp.	Various		La
	Tachinidae	*Pesibaea* sp.	*Spodoptera littoralis*	Noctuidae	La

[a]Host stage:
Ad, adult;
Eg, egg;
La, larva;
Pu, pupa.

Table 16.6. Natural enemy species found in vegetable fields in South America (including the French Antilles).

Type	Order	Family	Species	Host	Family	Host stage
Predators	Coleoptera	Coccinellidae	*Cladis nitidula* F.	Aphids	Aphidae	Ad, La
		Coccinellidae	*Cycloneda sanguinea limbifer* (L.)	Aphids	Aphidae	Ad, La
		Coccinellidae	*Eriopsis connexa* (Germar)	Aphids	Aphidae	Ad, La
		Coccinellidae	*Exocomus* sp.	Aphids	Aphidae	Ad, La
		Coccinellidae	*Hippodamia convergens* Guérin	Aphids	Aphidae	Ad, La
		Coccinellidae	*Megilla maculata* (De Geer)	Aphids	Aphidae	Ad, La
	Hemiptera	Pyrrochoridae	*Dysdercus andreae* L.	Various	-	Eg, Ad, La
Parasitoids	Hymenoptera	Braconidae	*Opius* sp.	*Liriomyza* spp.	Agromyzidae	La
		Eulophidae	*Chrysocharis caribea* Boucek	*Liriomyza* spp.	Agromyzidae	La
		Eulophidae	*Chrysonotomia* sp.	*Liriomyza trifolii*	Agromyzidae	La
		Braconidae	*Apanteles piceotrichosus* Blanchard	*Plutella xylostella*	Plutellidae	La
		Braconidae	*Cotesia vestalis* (Haliday)	*P. xylostella*	Plutellidae	La
		Braconidae	*Meteorus* sp.	*Trichoplusia ni*	Noctuidae	La
		Chalcididae	*Conura fulvovariegata* (Cameron)	*P. xylostella*	Plutellidae	La
		Eulophidae	*Oomyzus sokolowskii* (Kurdjumov)	*P. xylostella*	Plutellidae	La
		Ichneumonidae	*Diadegma leontiniae* (Brethes)	*P. xylostella*	Plutellidae	La

[a]Host stage:
Ad, adult;
Eg, egg;
La, larva.

species (Sow *et al.*, 2013b), but conversely often show a high diversity for hyperparasitoids (Table 16.8). For example, seven species have been found on *Cotesia vestalis* (Haliday) cocoons, (Hymenoptera; Braconidae) an endoparasitoid from *P. xylostella* larvae present on cabbage crops grown in the same locality in Benin (Arvanitakis *et al.*, 2014). The specialist parasitoids that the author found on one host species are not abundant compared with generalist predators, which are less efficient because they attack a large number of different prey species. Often, these specialist parasitoids are often less efficient under natural conditions (Labou *et al.*, 2016a), even though they have a big potential when they are studied in laboratory conditions (Sow *et al.*, 2013c, d).

Introduced natural enemies can vary in fitness depending on their origin. Fitness can be reduced in some populations because of the presence of endosymbiotic bacteria (e.g. *Wolbachia* spp.). On the contrary, fitness can be better, as in the case of *C. vestalis* where females native to Benin can parasitize about 80 larvae in 24 h, when those native to Martinique island can parasitize a maximum of 40 larvae (Rincon *et al.*, 2006). In the open field in Benin, despite an important percentage parasitism, often varying between 50% and 80% depending on the month, *C. vestalis* populations cannot control

Table 16.7. Natural enemy species found in vegetable fields in South-east Asia and in Oceania (New Caledonia).

Type	Order	Family	Species	Host	Family	Host stage	Continent
Predators	Coleoptera	Coccinellidae	*Coccinella transversalis* F.	Aphids	Aphididae	Ad, La	Oceania
		Coccinellidae	*Harmonia octomaculata* F.	Aphids	Aphididae	Ad, La	Oceania
		Coccinellidae	*Lemnia mulsanti* Montrouzier	Aphids	Aphididae	Ad, La	Oceania
		Coccinellidae	*Micraspis discolor* (F.)	Aphids	Aphididae	Ad, La	Asia
		Coccinellidae	*Micraspis lineata* (Thunberg)	Aphids	Aphididae	Ad, La	Oceania
		Coccinellidae	*Olla nigrum* Mulsant	Aphids	Aphididae	Ad, La	Oceania
Parasitoids	Hymenoptera	Braconidae	*Snellenius* sp.	*Spodoptera litura*	Noctuidae	La	Asia
		Eulophidae	*Hemiptarsenus varicornis* Guirault	*Liriomyza* spp.	Agromyzidae	La	Asia, Oceania
		Eulophidae	*Chrysocharis pentheus* (Walker)	*Liriomyza* spp	Agromyzidae	La	Oceania

[a]Host stage:
Ad, adult;
La, larva.

diamondback moth (DBM) populations on cabbage year round (Arvanitakis *et al.*, 2014). Sometimes a new parasitoid species appears in the vegetable field. This is the case for *Diadegma insulare* (Cresson), (Hymenoptera; Ichneumonidae) which has been collected several times since 2014 in Senegal. This is the first record of this parasitoid from Senegal and West Africa (Labou *et al.*, 2016b) and the list of the beneficial fauna is thus increased.

The large programmes of natural enemy releases in South-east Asia in the 1980s to control DBM have not obtained efficient results in the long term, despite the assistance and support of 'farm field schools', while those in Kenya have. However, in cabbage fields in the Eastern Cape (South Africa), DBM populations are relatively controlled by the local natural enemy fauna. It is true, that seven hymenopteran endoparasitoid species, five endo-larval species (*C. vestalis*, *Apanteles eriophyes* Nixon (Braconidae); *Diadegma mollipla* (Holmgren), *Mesochorus* sp. (Ichneumonidae), and *Oomyzus sokolowskii* (Kurdjumov) (Eulophidae)) and two endo-pupal species (*Diadromus collaris* Gravenhorst and *Itoplectis* sp., Ichneumonidae) are present on cabbage cultured all the year round (Smith and Villet, 2002).

In temperate areas, the hedges and floral borders around the culture plots provide a precious help because of their capacity to pull in and protect natural enemies. In Sahelian areas, there are few flowers and hedges are most often the refuge of numerous pests (fruit flies, bugs, locusts, rodents, etc.) which shelter from the wind and sun during the hottest time of day.

New Trends for the Future

One solution may be 'the hut garden' where production is on a small scale (between 100 m^2 and 500 m^2), very diversified at the vegetal species level (trees, fruit trees, vegetables, aromatic plants, etc.), little attacked by pests,

Table 16.8. Hymenopteran hyperparasitoid species found on natural enemies in vegetable crops in Africa, Asia, South America and Oceania.

Family	Species	Host	Order	Family	Continent
Ceraphronidae	*Aphanogmus fijiensis* (Ferrière)	*Cotesia vestalis*	Hymenoptera	Braconidae	Africa
	A. fijiensis (Ferrière)	*Diadegma* sp.	Hymenoptera	Ichneumonidae	Asia
	Aphanogmus reticulatus (Fouts)	*C. vestalis*	Hymenoptera	Braconidae	Africa
Chacididae	*Conura fulvovariegata* (Cameron)	*Diadegma leontiniae*	Hymenoptera	Ichneumonidae	South America
	Conura sp. gr. *immaculata*	*D. leontiniae*	Hymenoptera	Ichneumonidae	South America
	Hockeria sp.	*C. vestalis*	Hymenoptera	Braconidae	Africa
Elasmidae	*Elasmus* sp.	*C. vestalis*	Hymenoptera	Braconidae	Africa
Eulophidae	*Nesolynx phaosoma* (Waterston)	*Apanteles sagax*	Hymenoptera	Braconidae	Africa
	Notanisomorphella borborica (Giard)	*C. vestalis*	Hymenoptera	Braconidae	Africa
	Pediobius spp.	*C. vestalis*	Hymenoptera	Braconidae	Africa
Figitidae	*Anacharoïdes* sp.	*Ischiodon aegyptus*	Diptera	Sirphidae	Africa
	Proaspicea sp.	*I. aegyptus*	Diptera	Sirphidae	Africa
Ichneumonidae	*Charops* sp.	*I. aegyptus*	Diptera	Sirphidae	Africa
	Diplazon laetatorius F.	*I. aegyptus*	Diptera	Sirphidae	Africa
	Diplazon sp.	-	Diptera	Sirphidae	South America
	Mesochorus sp.	*Diolcogaster* sp.	Hymenoptera	Braconidae	Africa
	Stictopisthus sp.	*C. vestalis*	Hymenoptera	Braconidae	Africa
Pteromalidae	*Pachyneuron nelsoni* Girault	*I. aegyptus*	Diptera	Sirphidae	Africa
Pteromalidae	*Pteromalus* sp.	*C. vestalis*	Hymenoptera	Braconidae	Africa
Pteromalidae	*Trichomalopsis lasiocampa* (Graham)	*C. vestalis*	Hymenoptera	Braconidae	Africa
Pteromalidae	*Trichomalopsis orizae* (Risbec)	*C. vestalis*	Hymenoptera	Braconidae	Africa
Pteromalidae	*Trichomalopsis* sp.	*D. leontiniae*	Hymenoptera	Ichneumonidae	South America
Pteromalidae	*Trichomalopsis* sp.	*C. vestalis*	Hymenoptera	Braconidae	Africa

but with a low yield at harvest. This efficient crop system is not enough to feed a city population.

The solutions for a sustainable and acceptable harvest exist, but for that, the current chemical control system used now, must be excluded. Growers are now more likely to change their cultural habit as the use of non-chemical control methods has become accepted. Nets (soaked with pesticides) put on to crops may have a good future, but currently few local farmers have an interest in them (due to cost, more work to put them on to crops, theft in the fields, etc.). However, ecologically they are not efficient, because: (i) they destroy natural enemies which touch down on the nets (the nets are often soaked with pyrethroids); and (ii) the nets must have a very fine mesh to prohibit pest outbreaks of micro-insects (whiteflies, thrips, aphids, etc.) and such fine nets reduce ventilation which increases the temperature about 2–3°C minimum at the crop level, preventing predators coming in and restricting rain absorption. In addition, in a lot of cases, nets serve as a refuge for snakes and rodents, which protect themselves from predators.

It is very important to drive biological control by preservation of natural enemies. In a lot of localities, even if the natural fauna is not efficient enough to control pest damage (Labou *et al.*, 2016a), it is always present and limiting pesticide use will favour an increase in their populations. Several cases in Africa can be reported:

- The minimal damage caused by *P. basalis* amaranth caterpillar which is parasitized by three Braconidae and one Ichneumonidae species, when *H. recurvalis*, often associated with it on the plants, is not parasitized and causes heavy damage.
- *S. docilis* aubergine defoliators seem to be controlled by two Braconidae and two Eulophidae species. Damage is sporadic and of little importance.
- Similarly for *Asphondyla* sp. (Cecidomyiidae), the aubergine gall midge, which is an occasional pest; six parasitoids species (two Braconidae, two Eulophidae and two Eurytomidae) have been found on the pest.
- Unfortunately, in some cases when the pest populations increase, for example *S. ergasima* (aubergine flower buds borer), damage can be very important despite two Braconidae species being present causing about 85% parasitism.

The use of 'friendly plants' and the association of more cultivated species (tomato, cabbage, aubergine, melon, onion, etc.) with species that are little attacked (e.g. carrot, lettuce, aromatic plants) can be tested. Several plants and bio-product associations (Sow *et al.*, 2013a) that disturb pest populations are studied under field conditions at the University of Cheick Anta Diop (UCAD) Dakar, Senegal with satisfactory results and these findings are actually confirmed in farmers' fields. These plants associations are: (i) lettuce/cabbage and cabbage/mint against aphids; (ii) cabbage/lettuce, cabbage/Chinese cabbage, cabbage/carrot against DBM; and (iii) tomato/lettuce against *Tuta absoluta* (Meyrick) (Lepidoptera; Gelechiidae), a tomato pest recently introduced (in 2013).

In West Africa, many farmers know the benefit provided by Mexican marigold (*Tagetes erecta* L.) to limit nematodes in the soil.

This type of control (plant associations) is very well received by farmers, because they understand its place in the horticultural techniques for growing vegetables.

Conclusions

The list of pests in this chapter is not exhaustive, a pest is absent in a country until the time it is found! Under tropical conditions, vegetable crops grow in organic matter, with phosphorus, potassium and nitrogen. They have a shorter vegetative cycle than in temperate climates. The tropical climate, with hot temperatures and high humidity, discriminates against the 'European vegetables' such as cabbage, cucumber, tomato, melon and lettuce. These species are more sensitive to the pest populations, which are favoured by the climate.

A large diversity of vegetable species are the prey of many regular and sporadic pests destroying all parts of the plants (leaves, flowers, fruits, stems, etc.). The life histories of many species are unfamiliar or unknown to the farmers, causing indiscriminate use of chemical control to manage pests.

It is likely that a better knowledge within the tropical vegetable growers' environment of pest damage, the natural enemy fauna, the beneficial effect of vegetable species associations (allelopathy), and the great public health risks of pesticide use will lead to the sustainable use of the agroecological concept. During field surveys it was clear that many farmers are looking for solutions to their pest problems using non-chemical means.

Tropical pest–host plant interactions in low latitude areas (0–23°) differ from those in medium latitude areas (23–63°), where interactions depend on seasonal factors (Dobzhansky, 1950; Bates, 1953).

Currently, pest control problems in vegetable crops grown in tropical areas are not dealt with specifically in relation to the tropical conditions in which they are grown. Often and nearly always in vain, attempts are made to implant in the tropics the same process and dynamics that are applied in, for example, temperate areas and the failure comes as a reminder that the specificity of the environment should not be ignored (Francis Hallé, 2014). Now, particularly in Africa, a lot of researchers who are specialized in pest control and have studied in temperate areas are often not in touch with field conditions in their country of origin. Future research is required to improve understanding of biocontrol processes and the agroecological concept as applied specifically to organic vegetable production in tropical and subtropical areas.

References

Arvanitakis, L., David, J.-F. and Bordat, D. (2014) Incomplete control of the diamondback moth, *Plutella xylostella,* by the parasitoid *Cotesia vestalis* in a cabbage field, under tropical conditions. *BioControl* 59, 671–679.

Bates, M. (1953) *Where Winter Never Comes: A Study of Man and Nature in Tropics*. Payot, Paris.

Bordat, D. and Arvanitakis, L. (2004) *Arthropodes des Cultures Légumières d'Afrique de l'Ouest, Centrale, Mayotte et Réunion*. CIRAD (Centre de coopération internationale en recherche agronomique pour le développement), Montpellier, France.

Bordat, D., Laterrot, H., Renand, M. and Moretti, A. (1987) Phénomènes de résistance de *Lycopersicon cheesmanii* Riley à *Liriomyza trifolii* (Burgess) et *L. sativae* Blanchard. *L'Agronomie Tropicale* 42(1), 43–46.

Bordat, D., Robert, P. and Renand, M. (1988) Sensibilité de *Liriomyza trifolii* (Burgess) et de *L. sativae* Blanchard (Dipt. Agromyzidae) à onze souches de champignons entomopathogènes. *L'Agronomie Tropicale* 43(1), 68–73.

Dobzhansky, T. (1950) Evolution in the tropics. *American Scientist* 38, 209–221.

Hallé, F. (2014) *Un Monde sans Hiver*. Le Seuil, Paris.

Labou, B., Brévault, T., Bordat, D. and Diarra, K. (2016a) Determinants of parasitoid assemblages of the diamondback moth, *Plutella xylostella,* in a cabbage farmer's field in Senegal. *Crop Protection* 89, 6–11.

Labou, B., Bordat, D., Niang, A.A. and Diarra, K. (2016b) First record of the larval parasitoid *Diadegma insulare* (Cresson) (Hymenoptera: Ichneumonidae; Campopleginae) from Senegal. *African Entomology* 24(2), 533–535.

Landry, J.-F. and Hebert, P.D.N. (2013) *Plutella australiana* (Lepidoptera, Plutellidae), an overlooked diamondback moth revealed by DNA barcodes. *Zookeys* 327, 43–63.

Ngom, S., Manga, A., Diop, M., Thiam, M.B., Rousseau, J., Cisse, I. and Traore, S. (2013) Étude de l'évolution des résidus de pesticides dans les produits horticoles de grande consommation au Sénégal. *Revue Ivoirienne des Sciences et Technologie* 21&22, 31–44.

Pichon, A., Arvanitakis, L., Roux, O., Kirk, A.A., Alauzet, C.D., Bordat, D. and Legal, L. (2006) Genetic differentiation among world populations of the diamondback moth *Plutella xylostella* (Lepidoptera: Plutellidae). *Bulletin of Entomological Research* 96, 137–144.

Rincon, C., Bordat, D., Löhr, B. and Dupas, S. (2006) Reproductive isolation and differentiation between five populations of *Cotesia plutellae* (Hymenoptera: Braconidae), parasitoid of *Plutella xylostella* (Lepidoptera: Plutellidae). *Biological Control* 36(2), 171–182.

Roux, O., Gevrey, M., Arvanitakis, L., Bordat, D. and Legal, L. (2007) ISSR-PCR: Tool for discrimination and genetic structure analysis of *Plutella xylostella* populations native to different geographical areas. *Molecular Phylogenetics and Evolution* 43, 240–250.

Smith, T.J. and Villet, M.H. (2002) Biological control of the diamondback moth, *Plutella xylostella* (L.) in cabbage fields in the Eastern Cape, South Africa. In: Kirk, A.A. and Bordat, D. (eds) *Improving Biocontrol of* Plutella xylostella, *Proceedings of the International Symposium*. CIRAD, Montpellier, France, pp. 242–245.

Sow, G., Niassy, S., Arvanitakis, L., Bordat, D. and Diarra, K. (2013a) Effect of timely application of alternated treatments of *Bacillus thuringiensis* and neem on agronomical particulars of cabbage. *African Journal of Agricultural Research* 8(48), 6164–6170.

Sow, G., Diarra, K., Arvanitakis, L. and Bordat, D. (2013b) Relationships between the diamondback moth, climatic factors, cabbage crops and natural enemies in a tropical area. *Folia Horticulturae* 25(1), 3–12.

Sow, G., Arvanitakis, L., Niassy, S., Diarra, K. and Bordat, D. (2013c) Performance of the parasitoid *Oomyzus sokolowskii* (Kurdjumov), Hymenoptera; Eulophidae, on its host *Plutella xylostella* (L.), Lepidoptera; Plutellidae under laboratory conditions. *International Journal of Tropical Insect Sciences* 33(1), 38–45.

Sow, G., Arvanitakis, L., Niassy, S., Diarra, K. and Bordat, D. (2013d) Life history traits of *Oomyzus sokolowskii* (Kurdjumov), Hymenoptera; Eulophidae, a parasitoid of the diamondback moth. *African Entomology* 21(2), 1–8.

17 Pest Management in Organic Field Vegetables in Temperate Areas

Claudia Daniel,[1]* Rosemary Collier,[2] Urs Niggli[1] and Martin Koller[1]
[1]*Research Institute of Organic Agriculture, Forschungsinstitut für biologischen Landbau (FiBL), Frick, Switzerland;* [2]*Warwick Crop Centre, The University of Warwick, Warwick, UK*

Introduction

Preventative measures and the creation of resilient growing systems are key parts of pest management practice in organic field vegetable production (see also Chapter 1, this volume). Soil improvement, crop rotation, site selection, habitat management, variety choice, timing of planting and plant spacing are all considered before planting in order to avoid the most detrimental pest problems. Due to the huge variety of vegetable crops and their associated pest insects, specific tailored approaches are necessary. The effects of cultural practices to reduce pest outbreaks, such as weed control, appropriate irrigation, suitable fertilization, the use of mulches, and adjusted harvest times have been investigated during the last few years. For some crops (e.g. cabbage and carrot; Finch and Collier, 2000), a holistic pest control strategy, which combines preventative and direct control measures, is available, whereas there are still huge knowledge gaps for other crops. For example, in lettuce production, pest management under organic conditions still relies heavily on direct pest control measures. However, direct control methods using bioinsecticides can have adverse side effects on beneficial arthropods and thus destabilize the system and lead to outbreaks of secondary pests. In this chapter strategies and methods for pest control in organic field vegetable production that are currently available are described. The chapter is mainly based on information published in scientific journals and in technical leaflets, as well as on the experiences of advisors. After some general remarks about approaches to pest management, the description of pest management strategies is organized by crop. At the end of the chapter, in Table 17.1, there is a summary of strategies and control methods for the most important pests.

Preventative Measures

The creation of **healthy, biologically active soils** through the addition of organic matter using cover crops, compost or manure is the basis of organic farming. In vegetable production, however, freshly applied organic matter can increase the incidence of certain pests (e.g. bean seed fly). As a consequence, manuring schemes need to be adapted to the following crop and to local pest pressure.

* claudia.daniel@fibl.org

Most field vegetables need well-drained soils for optimal plant development. The level of **nitrogen fertilization** can affect pest incidence: at lower nitrogen levels, reduced thrips infestations have been observed in onion crops and similar observations were made for aphids in cabbage crops (Mutiga *et al.*, 2010). The size of the aphid infestation in cabbage crops was not only influenced by the level of fertilization (Buckland *et al.*, 2013), but also by the type of fertilizer: the use of vermicompost reduced the incidence of aphids (Little *et al.*, 2011).

The prompt **removal of harvest residues** is another important preventative measure. Many pest insects can complete their development or overwinter in crop residues. Thus, incorporating harvest residues into the soil, or collecting and composting of crop debris, are options to break the life cycle of many pest insects. These measures can also reduce the inoculum of many diseases. On the other hand, intensive tillage can have a negative impact on beneficial soil organisms, such as earthworms and carabid beetles.

Site selection has a huge impact on the abundance of pest insects and their antagonists. Several species of pest insect are very weak fliers (e.g. *Contarinia nasturtii*). Damage is reduced in exposed locations, whereas hedges and forest boundaries can favour these insects by providing sheltered immigration pathways. The distribution of insects by wind may lead to increased pest densities in locations that are downwind of sources of infestation. It is advisable to avoid growing crops in close proximity to established crops which are susceptible to the same pest species. For polyphagous pests, the close proximity of new crops to unrelated hosts can lead to increased damage. For example, growing brassica vegetables next to allium or cereal crops increases the risk of damage by onion thrips (*Thrips tabaci*), because onion thrips disperse from these crops after harvest. Close proximity between brassica vegetables and oilseed rape fields can also increase pest problems, because winter oilseed rape can act as a green bridge for many brassica pests during winter.

Crop rotation is another essential part of organic farming. Pest insects which overwinter in the soil, together with pest insects with a low ability for dispersal (e.g. *C. nasturtii*, *Psila rosae*) can be managed by crop rotation. However, some polyphagous pest insects, as well as pest insects with a good ability for dispersal, cannot be managed by crop rotation. In vegetable production, rotation between plant families is mainly focused on the prevention of diseases and root nematodes. The main plant families are: (i) brassica vegetables (*Brassicaceae*); (ii) legumes (*Fabaceae*); (iii) alliums (*Liliaceae*); (iv) lettuce/endive (*Asteraceae*); and (v) carrot (*Apiacaeae*). A break of 4–5 years between crops of the same type is necessary to reduce disease pressure. Careful planning of a cropping sequence also considers the inclusion of cover crops for soil improvement and possibly the inclusion of grain crops to extend the rotation period. If brassica vegetables are grown in a crop rotation, brassica cover crops should be avoided in order to keep the break between brassica species. Legume cover crops with their nitrogen fixing ability or 'neutral' crops such as phacelia or different *Poaceae* might provide an alternative. Cover crops are helpful for increasing soil nitrogen or to break pest or weed cycles, but might increase the abundance of pest insects such as *Tipula* sp. or *Agrotis* sp.

Habitat manipulations which increase plant diversity in the agricultural landscape favour populations of beneficial insects by supplying nectar, pollen, alternative hosts and sheltered habitats. In order to tailor habitat manipulations to specific pest problems, a detailed knowledge of the biology of the key pests, beneficials and plants, as well as their interactions, is necessary. At the field level, flower strips and companion plants can be used to enhance the abundance of naturally occurring antagonists of pests. The selection of these companion plants needs to be tailored to the needs of the specific growing system: an example for cabbage crops is described in Chapter 1, this volume. For some cropping systems, selective companion plants (i.e. companion plants that attract and feed beneficial insects, but not the pest insects) may be identified, but the adults of some pest insects such as *Delia radicum* (Nilsson *et al.*, 2011) and *Plutella xylostella* (Winkler *et al.*, 2009) are attracted to many different species of flower to feed and might therefore also be favoured by cropping systems using such companion plants. The design of an appropriate cropping system needs therefore to be adapted to local pest incidence.

Intercropping describes the simultaneous production of more than one type of crop in a field, whereas **companion plants** are non-crop plants grown within a field. Both strategies have similar purposes: to reduce colonization by pest insects and the attraction and support of natural enemies by providing pollen, nectar and shelter (Parolin *et al.*, 2012). Pest insects use chemical and visual stimuli to detect host plants. Intercropping has been shown to interfere with the pest's ability to find the host plant in the case of the cabbage pests *Brevicoryne brassicae* (Vidal, 1997), *P. xylostella* (George *et al.*, 2013) and *D. radicum* (Finch, 1993; Finch and Kienegger, 1997; Parsons *et al.*, 2007). Intercropping of peas with wheat significantly reduced colonization by pea aphid (Ndzana *et al.*, 2014). For cucumber beetles (*Acalymma vittatum*, *Diabrotica undecimpunctata howardi*) it has been shown that the presence of non-host plants, as well as plants attractive for beneficial insects, can reduce the numbers of beetles (Cline *et al.*, 2008). However, intercropping can complicate harvest procedures and result in additional labour costs. **Trap cropping** (i.e. growing a crop in certain parts of the field which is more attractive to particular pests than the main crop) was shown to reduce colonization by the leek moth *Acrolepiopsis assectella*, which preferred larger plants of a more advanced growth stage for oviposition, and the diamondback moth *P. xylostella*, which preferred Indian mustard to white cabbage (Asman, 2002). Perimeter trap cropping (i.e. surrounding the main crop with an attractive trap crop) was shown to reduce colonization by cucumber beetles in squash (Adler and Hazzard, 2009; Cavanagh *et al.*, 2009). The use of **mulches** has also been shown to reduce pest incidence (e.g. *Delia* sp. and aphids) (Singh, 1992; Heimbach *et al.*, 2002; Hommes *et al.*, 2003), but additional labour is needed for the application and removal of the mulch.

In areas where certain pest insects are known to be prevalent, farmers adapt their **selection of crops**. For example, cauliflower is planted instead of broccoli during the summer months in areas with high populations of swede midge (*C. nasturtii*) and cucumber is planted instead of squash to avoid damage by squash bug (*Anasa tristis*). The **choice of tolerant/resistant varieties** is sometimes another option to reduce damage. Many disease resistant varieties are available, whereas resistance to pest insects in vegetables is still underexploited. Aphid-resistant lettuce varieties are frequently used to manage *Nasonovia ribisnigri* and *Pemphigus bursarius* (Liu and McCreight, 2006) although some European populations of *N. ribisnigri* are now able to colonize 'resistant' varieties (Cid *et al.*, 2012). Cabbage varieties tolerant to thrips and carrot varieties tolerant to *P. rosae* (Ellis *et al.*, 1987b) are available.

Carefully selecting the timing of sowing, planting and harvesting is used to reduce the phenological coincidence between crop and pest. This strategy can be used for carrot production: by sowing carrots towards the end of May, the peak flight of the first generation of the carrot fly *P. rosae* is avoided (Finch, 1993). With an early harvest, oviposition by the third generation of this pest can be reduced (Finch, 1993). In cabbage production, carefully selected sowing/planting times can reduce damage to brassicas by cabbage root fly (*D. radicum*) and damage by leek moth (*A. assectella*) can be reduced by early harvest of onions (Rahn, 1982).

Targeted **irrigation** can be another approach as part of a holistic pest control strategy. Increased irrigation can mitigate damage by lettuce root aphid (*P. bursarius*), black bean aphid (*Aphis fabae*) or flea beetles (*Phyllotreta* sp.), whereas reduced irrigation in cabbage can prevent egg hatch of cabbage root fly (*D. radicum*) because its eggs are sensitive to drought (Lepage *et al.*, 2012). After egg hatch, however, increased irrigation can mitigate *D. radicum* damage. The abundance of *T. tabaci* and two-spotted spider mite (*Tetranychus urticae*) can be reduced by overhead irrigation (Kannan and Mohamed, 2001). Irrigation leads to increased thrips mortality due to epizootics of entomopathogenic fungi (Maniania *et al.*, 2003).

Direct Pest Control

Floating row covers and **fine-mesh netting covers** can be used to provide physical protection for crops, for example against *D. radicum* or *P. rosae* (Finch, 1993), flea beetles and many other pests. However, farmers are reluctant to use crop covers unless no other means of control are available, because: (i) use of covers is labour intensive; (ii) mechanical weed control is impaired; (iii) disease pressure is increased due to increased humidity under the covers; and (iv) secondary pests might be favoured because of the exclusion of antagonists. In order to avoid some of the disadvantages of covers, vertical barriers have been developed. These are usually 1–2 m high fences made of insect-proof netting with an outer overhang of 25–30 cm. They have been shown to exclude low-flying insects (e.g. *D. radicum*, *P. rosae* or *C. nasturtii*) from crops (Vernon and Mackenzie, 1998; Bomford *et al.*, 2000; Wyss and Daniel, 2004; Blackshaw *et al.*, 2012). However, their vulnerability in high winds and complicated installation and handling procedures have impeded introduction of this strategy very widely.

Application of **entomopathogenic organisms**, for example *Bacillus thuringiensis* (*Bt*) against lepidopteran larvae, or **insecticides approved for organic crops**, such as spinosad or horticultural soaps, is used widely to reduce insect damage, especially to manage insects that feed directly on the marketable parts of plants, where only a little damage can be tolerated. Most insecticides approved for organic crops are not systemic so direct contact of the product with the pest insect is needed for a good efficacy. Targeted application technology can improve the efficacy. For example, droplet sprayers for spraying the lower surfaces of the foliage of vegetable and field crops have been developed by the Federal Research Station Agroscope in Switzerland (Rüegg *et al.*, 2006) and are marketed through collaboration between Amazon, Lechler and Syngenta. Droplet sprayers consist of plastic drag hoses with spray nozzles fixed on their tip. With this technology, pests such as aphids, thrips or cabbage whitefly larvae, which are 'hidden' among the foliage, are controlled more effectively than with standard sprayers. In addition, the efficacy of ultraviolet (UV)-sensitive biocontrol products (e.g. *Bt*) is improved by application to the lower surfaces of leaves. However, even the insecticides used in organic farming (e.g. spinosad, pyrethrum and rotenone) can have detrimental side effects on non-target organisms (Jansen *et al.*, 2010). After the application of spinosad against *C. nasturtii* or Lepidoptera, adverse side effects on aphid parasitoids often lead to an increase in aphid abundance (Hommes and Herbst, 2014). Parasitoids of Lepidoptera are also affected negatively. To avoid the negative impact of direct control measures on ecosystem functionality, selective methods of control, combined with specific prevention strategies are preferable (see Table 17.1 at the end of this chapter) and use of non-selective pesticides should be limited to a minimum. Among the non-selective insecticides available for use on organic crops, spinosad probably has the most pronounced and prolonged effect on parasitic Hymenoptera, predatory bugs, syrphids and many other beneficial insects (Biondi *et al.*, 2012). Adverse side effects have also be seen following application of pyrethrin insecticides (Jansen *et al.*, 2010), whereas kaolin primarily has a repellent effect; it can disrupt the behaviour of parasitoids, but usually does not kill them (Markó *et al.*, 2008). Horticultural soaps and oils only have adverse side effects on insects living directly on the leaf surface and only for as long as the application layer remains wet. The use of fine-mesh netting covers can also have side effects by excluding natural enemies, but this effect ceases as soon as the nets are removed. Quassia, neem and *Bt* products are harmless to most beneficial insects (El-Wakeil *et al.*, 2006) and are therefore considered as selective control methods.

In the following sections pest insects and their management in specific crops are described.

Allium Crops

Leek, onion, garlic, chives and shallots are the main allium crops produced in temperate zones.

Allium leaf miner

Allium leaf miner (*Napomyza gymnostoma*) can cause considerable damage to leeks, onions and chives due to the feeding of larvae inside the young leaves. Pupae of this pest found within harvested onions and leeks lead to rejections by retailers and consumers. A further species of leaf miner of allium crops (*Liriomyza nietzkei*) occurs in some regions, but usually causes less damage. Covering most of the crop with fine-mesh netting covers while leaving some outer rows uncovered as a trap crop provides reliable control. Once leaf miners have completed oviposition, the border rows should be removed in order to prevent the emergence of the subsequent generation. Harvest residues and crop debris should also be removed. High populations of these pests can develop in chives and 'spill over' into adjacent onion and leek fields. Care should therefore be taken to separate susceptible crops spatially. Differences in susceptibility to leaf miner infestation between different leek varieties have been observed in some experiments. Applications of neem or spinosad give good control.

Onion maggot

The onion maggot (*Delia antiqua*) oviposits its eggs at the base of onion plants, the larvae feed on the roots and at the base of the plants. One larva can kill several young plants, older plants can tolerate a certain amount of damage, but larval feeding activity leads to damaged, misshapen bulbs. Red onions, leeks and garlic can also be attacked, but are usually less attractive to these flies. The flies are attracted to the odour of rotting onions (Judd and Borden, 1992; Gouinguene *et al.*, 2005). Therefore, proximity of new crops to more established onion crops, as well as to piles of compost, should be avoided; volunteer onion plants should be removed. Crop debris and harvest residues should be ploughed deeply to destroy pupae and a crop rotation strategy to avoid growing susceptible crops in the same location in successive years should be implemented. Mulches of different material (e.g. straw) can reduce egg laying, because female flies walk from the leaves downwards to oviposit at the base of plants. If possible, the planting of onions should be delayed until after the flight period of the first generation in spring (Nault *et al.*, 2011). Larvae of *D. antiqua* are attacked by several predatory insects, such as staphylinid or carabid beetles, and by parasitoids. Flower strips, untilled refuge strips for predatory ground beetles and the use of selective pesticides can favour these antagonists and thus lower the pest population. Optimal control is achieved using fine-mesh netting covers immediately after planting, seed treatments with spinosad (Nault *et al.*, 2006; Wilson *et al.*, 2015) or by making applications of spinosad during the early evening. A biocontrol strategy based on the use of entomopathogenic nematodes has been developed (Morris, 1985), but does not always provide reliable control.

Leek moth

Leek moth (*Acrolepiopsis assectella*) larvae mine inside the hollow leaves of onions or tunnel into the centre of leek plants. They overwinter as pupae or adults in plant debris and in sheltered areas on the field. Therefore, a 3-year crop rotation and the removal of harvest residues can reduce the population size. By delaying planting, oviposition by the first generation can be avoided and, through early harvesting, oviposition by the last generation can be avoided (Rahn, 1982). After harvest, any infested outer leaves can be removed in order to obtain a marketable crop. The leek moth is attacked by parasitoid wasps and deployment of flower strips and the use of selective insecticides can help to maintain these antagonists in the agricultural landscape. The biocontrol agent *Bt aizawai* has good efficacy against young larvae if applied before the larvae enter the leaf tissue. Crops covered with fine-mesh netting are also well protected. In addition, spinosad applications can be used to control this pest.

Onion thrips and western flower thrips

Onion thrips (*T. tabaci*) and western flower thrips (*Frankliniella occidentalis*) suck on tender plant tissue leading to silvery spots and black faecal residues. They can also transmit viruses. Large infestations can lead to reduced plant development or even plant death. However, in most cases, no yield reduction is observed (Kannan and Mohamed, 2001; Mautino *et al.*, 2012) and damage is primarily a cosmetic problem leading to rejections by multiple retailers. In leek, only the green parts of the leaves show the damage and marketing of leeks without most of the foliage is an option if plants are infested. Tolerant varieties, which show less damage, are available. Reduced nitrogen fertilization (Buckland *et al.*, 2013), the use of straw mulch (Larentzaki *et al.*, 2008a; Schwartz *et al.*, 2009), under-sowing crops with clover (Theunissen and Schelling, 1996) or intercropping onions with celery (Baumann *et al.*, 2000) can all reduce thrips abundance. The most severe damage occurs under hot and dry weather conditions, because more generations of thrips are produced at higher temperatures. Damage may be reduced by overhead irrigation. For example, Kannan and Mohamed (2001) indicated that 10 min of irrigation around noon on hot and dry days is sufficient to decrease pest populations. However, experimental work in the UK on the use of irrigation was inconclusive (Burnstone, 2009). It is likely that irrigation and rainy weather conditions lead to increased thrips mortality because they provide favourable conditions for infection by entomopathogenic fungi (Maniania *et al.*, 2003). Predatory thrips (Thysanoptera: Aeolothripidae) are other important antagonists of pest thrips and as long as the predatory thrips are not killed by the use of broad-spectrum insecticides, they occur in sufficient densities to keep onion thrips infestations at a tolerable level in most onion-producing areas (Bosco and Tavella, 2010). Other antagonists, such as predatory bugs (*Orius* sp. and *Anthocoris* sp.), lacewings and predatory mites, can be enhanced by flower strips. Thrips can feed on crops from several other plant families, including brassica vegetables and cereals. Close proximity of allium crops to these potential sources of infestation should be avoided. Thrips are unable to fly far independently and the main method of dispersal is when they are carried by the wind. Deep ploughing of postharvest debris can reduce the abundance of the overwintering stages in plant debris and soil. The direct control of thrips is possible using spinosad or pyrethrum, although it is possible that certain populations of *T. tabaci* are resistant to pyrethrum through the development of resistance to synthetic pyrethroid insecticides (Foster *et al.*, 2010) and frequent application of spinosad also leads to resistant thrips populations (Lebedev *et al.*, 2013). Thorough wetting of all plant parts is essential and can be improved by using adapted targeted application techniques (e.g. droplet sprayers) and high water volumes. Kaolin has been shown to have a good efficacy if applied frequently (Larentzaki *et al.*, 2008b).

Brassica Vegetables

The most important brassica vegetables are cauliflower, broccoli and several types of cabbage, but there are a number of other brassica crops (kale, Brussels sprout, swede, turnip, radish, Chinese cabbage, Chinese mustard). Brassica vegetables are attacked by a range of specialized pest insects.

Cabbage maggot

The adult cabbage maggot (*Delia radicum*) overwinters as a pupa, emerges in April and oviposits in the soil at the base of its host plant. Larvae tunnel in the roots and stems causing plants to wilt and, in some cases, to die. Young plants are particularly susceptible. The fly completes between one and four generations/year depending on temperature and, in locations where the weather is particularly warm, it will aestivate in the pupal stage (Finch and Collier, 1985). In addition, the eggs may be subject to desiccation when conditions are very hot and dry. Simulation models have been developed in the UK and Germany to predict periods of oviposition (Phelps *et al.*, 1993; Hommes and Gebelein, 1996). High populations of *D. radicum* are observed in areas where there is intensive production of brassica vegetables and oilseed rape. Although the female flies are able to fly more than 3 km (Finch and Skinner, 1975; Helenius, 1997), increasing the distance between new crops and sources of cabbage maggot may reduce the risk of infestation. Postharvest residues and plant debris in infested fields should be ploughed or disked immediately after harvest in order to reduce emergence of the new generation (Dosdall *et al.*, 1996). Since eggs are sensitive to desiccation, reducing irrigation during oviposition periods may prevent egg hatch (Lepage *et al.*, 2012). In addition, delaying planting until the soils are warmer might reduce the risk of damage because it fosters faster plant growth which reduces the time span of highly susceptible young plants. More damage is observed in soils with high organic matter content (Koštál *et al.*, 2000); decomposing organic matter (harvest residues of previous crops and manure) is especially attractive to flies. Therefore, the previous crop needs to be thoroughly ploughed and a waiting period of 2–3 weeks before cabbage planting is advisable. Damage to the roots of leafy brassica plants can be mitigated by irrigation and stimulating the growth of additional roots by deep planting and earthing up. Host finding and egg laying of *D. radicum* can be disrupted by undersowing or intercropping with non-host plants (Kostal and Finch, 1994; Theunissen *et al.*, 1995; Finch *et al.*, 2003; Morley *et al.*, 2005; Finch and Collier, 2012) as well as by high weed populations (Dosdall *et al.*, 2003). Straw mulch can prevent egg laying (Humphreys and Mowat, 1994) and mulching with grass clippings can increase egg predation (Hellqvist, 1996). Planting flower strips close to brassica fields, as well as growing particular companion plants, can support naturally occurring predators of eggs and larvae (Meyling *et al.*, 2013). The use of turnip rape as a trap crop, to attract *D. radicum* and its parasitoids, has been suggested (Rousse *et al.*, 2003). The application of products based on entomopathogenic fungi may be an option for control (Razinger *et al.*, 2014; Rannback *et al.*, 2015) but currently no products are available commercially for control of cabbage maggot. The application of entomopathogenic nematodes is another option (Vänninen *et al.*, 1999). There have been various attempts to repel or kill cabbage maggot adults or larvae with extracts of garlic (Prowse *et al.*, 2006). Garlic extracts have been shown to have insecticidal activity against the larvae, providing the concentration is sufficiently high (A. Jukes, 2003, unpublished data). Efficacy under field conditions may be harder to achieve. Evidence that garlic treatments repel adult flies is limited. Adults can be excluded from crops using fine-mesh netting (Blackshaw *et al.*, 2012). A mesh size of 1.3 mm is commonly used in the UK. However, recent experimental studies have shown that adult flies will lay eggs on and through the mesh, provided they have contact with the foliage (R. Collier, 2011, unpublished data). It is not clear how frequently they do this in commercial crops. Research has been undertaken to determine the efficacy of vertical barriers to exclude cabbage maggot adults from susceptible crops (Vernon and Mackenzie, 1998). These have also been used commercially on a small scale by growers in Norway, although in this instance the netting is impregnated with insecticide (R. Meadow, Norwegian Institute for Agricultural and Environmental Research (Bioforsk), Høgskoleveien, Norway, 2009, personal communication). Seed treated with spinosad (Ester *et al.*, 2003) or pre-planting module drench applications of spinosad or neem are also possible for transplanted brassica crops.

Flea beetles

Flea beetles (*Phyllotreta* sp.) overwinter in the soil surrounding a previous host crop or in leaf litter in hedgerows and can actively fly several kilometres. Nevertheless isolating new crops from potential sources of infestation (cabbage, rapeseed or other brassica crops used as a green manure) can reduce the risk of attack. Feeding damage caused by adult beetles to cotyledons and young leaves can cause severe damage to young plants. Healthy, well-developed young plants in good growing conditions can sustain higher levels of damage. Eggs are deposited in the soil, close to the plants and the larvae feed on the roots. Damage is observed especially during periods of dry weather and can be mitigated by irrigation. In infested crops postharvest residues and plant debris should be ploughed or disked in immediately after harvest in order to prevent emergence of the new adult generation. Exclusion of flea beetles is possible using nets with a maximum mesh size of 0.8 mm. Dust or spray applications of silicate rock dusts or kaolin can protect young plants during the early susceptible stages. In addition, applications of pyrethrin and spinosad can be used to control the beetles. Spinosad has the greatest efficacy but also causes most side effects on non-target beneficial insects (Andersen *et al.*, 2006).

Lepidopteran species

Several lepidopteran species (diamond back moth *Plutella xylostella*, small white butterfly *Pieris rapae*, large white butterfly *Pieris brassicae*, cabbage looper *Trichoplusia ni*, cabbage moth *Mamestra brassicae* and garden pebble moth *Evergestis forficalis*) are specialist pests of brassica crops. Damage is caused by larval feeding on leaves and contamination by frass. Most lepidopteran pests have two or more generations/year. Removal of crop debris and cruciferous weeds is important to reduce the size of the local population. Overwintering pupae within cabbage fields can be destroyed by cultivation before planting. However, *P. xylostella* in particular is a migrant species and may cover large distances at high altitude (Chapman *et al.*, 2002). Wildflower strips and companion plants can attract and sustain naturally occurring parasitoids and predators of Lepidoptera. Trap crops located on field margins have been shown to have an effect on *P. xyllostella* (Asman, 2002), whereas sheltering crops using tall barriers of non-host crops (sorghum, maize) can prevent immigration of *P. xyllostella* and *T. ni*. Direct control of lepidopteran larvae is possible using different *Bt* products. *Bt kurstaki* has good efficacy against young larvae of *P. xylostella* and *Pieris* sp. If noctuid species (*M. brassicae*; *T. ni*) are present, *Bt aizawai* provides better control. *Bt* is rapidly inactivated by UV radiation, therefore application should be conducted when the weather is overcast or during the evening. Thorough wetting of all plant parts is essential and can be improved by using certain application techniques. For example, with droplet sprayers, products are also sprayed on to the lower leaf surfaces, which provides better UV protection and therefore a prolonged period of efficacy. In addition, most lepidopteran larvae feed on the lower leaf surface. The addition of 1% sugar or vinasse (byproduct of the sugar industry) increases larval feeding and ensures rapid uptake of lethal concentrations. For good efficacy of *Bt* products, a minimum temperature of 12°C is necessary in order to have sufficient larval feeding activity. At lower temperatures, the use of spinosad is more effective, but frequent spinosad applications can result in the development of resistance (Zhao *et al.*, 2002). In addition, the entomopathogenic fungus *Beauveria bassiana* can be used to manage *P. xylostella*, *Pieris* sp. and *T. ni*. Neem may also be effective.

Aphids

Aphids that are pests of brassica crops include the cabbage aphid *Brevicoryne brassicae* and the green peach aphid *Myzus persicae*. Aphid infestations can lead to stunted plants with yellow-coloured curled leaves.

Young plants are often more susceptible. Aphids, particularly *M. persicae*, also transmit plant viruses. Covers made of fine-mesh netting can be used in seedling production in order to obtain uninfested seedlings. The creation of good growing conditions after planting (e.g. additional irrigation during dry weather) is important to foster plant growth. Older plants can tolerate small infestations. In most cases, aphids are sufficiently controlled by the plethora of aphid antagonists (lacewings, ladybird beetles, syrphids, *Aphidoletes aphidimyza*, parasitoids, earwigs). Flower strips and flowering companion plants (e.g. *Alyssum*; Brennan, 2013) within a field will attract and sustain naturally occurring parasitoids and predators of aphids (Theunissen *et al.*, 1995). The use of reflective or straw mulch can reduce host recognition and immigration of aphids (Döring *et al.*, 2004). Aphid infestations can be exacerbated when non-selective measures (such as spinosad, pyrethrins, fine-mesh netting) are used to control other pests: the reduction in the numbers of antagonists reduces natural mortality. Postharvest residues and plant debris from infested crops should be ploughed or disked in immediately after harvest. Direct aphid control is possible using quassia, neem, horticultural oils and soaps, as well as pyrethrin. Good wetting of all plant parts is essential and can be improved by: (i) using certain application techniques (e.g. droplet technology); (ii) adding wetting agents and adhesives (e.g. polyterpenes such as Nu-film or Heliosol); and (iii) applying treatments before the leaves are heavily curled.

Thrips tabaci

T. tabaci can cause cosmetic damage to headed cabbage, especially under hot and dry weather conditions. Application of overhead irrigation can reduce infestations (Kannan and Mohamed, 2001). The thrips may immigrate from cereal or onion fields. Isolation from such fields can therefore reduce the risk of attack. The choice of resistant/tolerant varieties is a key option for control (Shelton *et al.*, 1998). Direct control is also possible using entomopathogenic fungi (*B. bassiana*, *Isaria fumosorosea*), spinosad or pyrethrin. The addition of a wetting agent improves the efficacy of these products against thrips.

Swede midge

Swede midge (*Contarinia nasturtii*) can infest most brassica vegetables and causes distorted growing tips. Considerable damage is caused to broccoli, although there are differences in susceptibility among different varieties. In areas with high pest pressure, farmers tend to grow cauliflower instead of broccoli during the summer months (Hallett, 2007). The tiny, short-lived adult midges are very weak flyers and are easily translocated by wind. Infestation is usually higher in moist fields and close to sheltered hedges. As such, site selection and crop rotation are the most important control strategies (Chen *et al.*, 2009). Isolation from established brassica vegetable and oilseed rape fields (> 100 m) and cultivation in windy, dry locations can reduce damage. The reduction of brassica weeds within the field, as well as in its surroundings, is also important (Hallett, 2007; Chen *et al.*, 2009). In addition, postharvest residues and plant debris from infested crops should be ploughed or disked in immediately after harvest in order to prevent emergence of the new adult generation (Chen and Shelton, 2007). Direct control is possible using insect-proof netting with a mesh size of 0.8 mm (especially in seedling production). Insect fences, 1.0–1.5 m high, with an outside overhang can prevent immigration of midges into the crops. Spray applications of spinosad have good efficacy. Thorough wetting of all plant parts is essential and can be improved by using targeted application techniques (e.g. droplet technology) and the addition of a wetting agent. However, the timing of applications is a challenge for farmers: pheromone traps for swede midges are available, but identification of adult midges is difficult and farmers need training to do this.

Cabbage whitefly

Normally the cabbage whitefly (*Aleyrodes proletella*) does not reduce the yield of brassica vegetables. Crop quality is reduced by the production of honeydew which leads to sooty moulds, especially on Brussels sprout. Harvested crops of kale contaminated with larvae and adults are rejected by retailers. Large infestations of whitefly occur particularly in areas with intensive production of oilseed rape: when the oilseed rape crop matures and desiccates in summer, adult whiteflies migrate to nearby vegetable crops. Control of whiteflies in organic crops is challenging due to their high rate of reproduction. Partially resistant varieties are available (Nebreda *et al.*, 2005; Pelgrom *et al.*, 2015). Plants should be inspected thoroughly to detect developing infestations. The combination of fine-mesh netting (mesh size 0.8 mm) and applications of neem using a droplet sprayer provides good results. In instances where whiteflies move into the crop during temporary removal of the net covers for mechanical weed control, releases of parasitoids (*Encarsia tricolor*) can keep pest populations at a low level under the netting (Saucke *et al.*, 2011). Unfortunately, *E. tricolor* is currently not available commercially.

Carrots and other *Apiaceae*

Carrot, parsnip, celery, fennel, celeriac and parsley belong to this group of crops.

Carrot rust fly

Carrot rust fly (*Psila rosae*) can damage carrot, parsnip, celeriac and celery crops. Eggs are laid in the soil, close to the host plant and newly hatched larvae feed on the small lateral roots at first. Older larvae tunnel into the tissue of the main roots which leads to cosmetic damage and can result in root rot due to pathogens. Depending on the climate, carrot rust flies can overwinter both as pupae in the soil and as larvae which continue to feed in infested roots; therefore the removal of crop debris after harvest can help to reduce population levels in an area. A break of 3 years between carrot crops is recommended to reduce damage (Ellis *et al.*, 1987b). Sheltered locations close to hedges, maize fields, forest borders and other wind-sheltering structures are at higher risk of infestation (Ellis *et al.*, 1987a) and wind-exposed fields show less damage. Adult carrot flies do not travel over great distances and crops which are separated from sources of infestation by a distance of at least 1 km have a low risk of being colonized (Finch and Collier, 2004). However, this may be difficult to achieve in many vegetable production areas. As flies colonize the carrot crops nearest to their emergence sites, strips of carrots sown between sources of infestation and new fields may act as trap crops and arrest migrating adults (Herrmann *et al.*, 2009). In Central Europe, three generations of *P. rosae* occur per year (Davies and Collier, 2000). Adult numbers can be monitored using orange-yellow sticky traps (Collier and Finch, 1990; Collier, 2009). In addition, simulation models are available to predict flight periods (Phelps *et al.*, 1993; Hommes and Gebelein, 1996). Larvae of the first generation can kill young carrot plants if the flight period of the first generation coincides with the germination of young plants (Collier and Finch, 2000). Early sown carrots can be protected from immigration of first generation flies by floating row covers (Davies and Collier, 2000) which may already be in use on early grown carrots to accelerate growth. The most severe damage, however, is caused by larvae of the second/third generation tunnelling into the tap root of carrots destined for winter storage and winter harvest. Therefore, carrots for winter storage should be grown at a greater distance from sources of carrot fly to minimize the risk of infestation. In addition, carrots grown for winter storage should be sown earlier (Berry *et al.*, 1997) in order to harvest them within 4 weeks of the start of the third flight period, before larvae start tunnelling into the main root (Ellis *et al.*, 1987b). Nets with a mesh size of 1.4 mm can prevent immigration of flies. Insect fences of 1–2 m height with an external overhang may prevent immigration of flies into crops (Collier and Finch, 2009). Frequent hoeing and earthing up can disturb the feeding of the young larvae. Intercropping of carrots with clover can reduce damage (Theunissen and Schelling, 2000). Intercropping carrots with onions also gave good results (Uvah and Coaker, 1984), but the different growing seasons of these two crops make this combination challenging.

Carrot psyllid

Carrot psyllid (*Trioza apicalis*) overwinters as an adult on conifers, especially spruce (Kristoffersen and Anderbrant, 2007). Migration of adults to carrot crops starts in May and ceases in approximately mid-June (Laska, 2011). Females lay eggs on the leaves. Severe damage (curled leaves and misshapen roots) is caused by feeding by adults on young plants (younger than the five-leaf stage) (Nissinen *et al.*, 2007) especially under warm and dry weather conditions. This pest can also act as a vector of bacterial diseases (Munyaneza, 2010). In areas where this pest is known to occur, sowing should be delayed until after the first flight period. Covers made of fine-mesh netting give good control. Mulches using spruce sawdust (Rämert, 1993; Nehlin *et al.*, 1996) and living mulches of hairy vetch (Meadow, 2010) can reduce egg laying and support antagonists.

Other sucking pests

Thrips tabaci (see allium crops) and **two-spotted spider mite** (*Tetranychus urticae*) can in some cases reach damaging levels in celery. Both pests cause damage mainly during hot and dry weather conditions and can be markedly reduced by overhead irrigation. In addition, **aphids** (*Cavariella aegopodii*) can occur. Damage is sometimes caused in parsley, while in other crops these aphids are normally controlled sufficiently by natural enemies. They are vectors of virus diseases which can reduce yield significantly. Direct control is possible using pyrethrum, and horticultural oils and soaps.

Cucurbits

Melon, zucchini, pumpkin, squash, gherkin and slicing cucumbers are field-grown cucurbit vegetables in temperate areas.

Aphids

Several aphid species (e.g. *Aphis gossypii*) can infest cucurbit plants. The most severe damage is caused through virus transmission. In order to prevent virus transmission, volunteer plants in the vicinity of crops, which are likely to be hosts of virus, should be removed. Direct damage by aphids includes severely distorted foliage and contamination with honeydew leading to sooty moulds. Aphids are often well controlled by natural enemies and the use of selective control measures against other pests can preserve these enemies. Isolation of new crops from older crops can reduce the immigration of aphids and reflective mulches can be used to reduce colonization during immigration. Direct control is possible using neem, horticultural oils and soaps or pyrethrin (Pinto *et al.*, 2013).

Cucumber beetles

Cucumber beetles (*Acalymma vittatum*, *Diabrotica undecimpunctata howardi*) can be vectors of bacterial wilt (Caudle *et al.*, 2013) and can kill small seedlings directly through their feeding activity. By establishing the crop from transplants instead of by direct sowing, the most sensitive growth stage is exposed to these pests for a shorter period of time. Cultivars with a lower concentration of cucurbitacin are less susceptible to damage. Immigration of pests into new crops can be delayed by using reflective row covers (Cline *et al.*, 2008) or perimeter trap cropping (Gardner *et al.*, 2015). However, the efficacy of the latter strategy depends on the right choice of cultivars; trap crop plants need to be larger than the main crop (Adler and Hazzard, 2009). Crop rotation, removal of crop debris, premature destruction of heavily infested crops followed by cultivation, to prevent development of a second generation, and mowing of headlands are preventative measures to reduce pest pressure in an area. Companion plants to attract natural enemies (buckwheat) and repellent non-host companion plants (radish) can reduce populations due to their properties as physical barriers which deter beetle movement (Cline *et al.*, 2008; Hinds and Hooks, 2013). Direct control of cucumber beetles is possible using fine-mesh netting, pyrethrin, neem or kaolin (Caudle *et al.*, 2013). The efficacy of kaolin applications can be increased by combining its use with attractive trap crops.

Squash bug

Squash bug (*Anasa tristis*) and other stink bugs can transmit viruses. Direct feeding damage by adults and nymphs leads to wilting and delayed flowering, increased fruit abortion, resulting in smaller and fewer fruit (Biernacki and Lovett-Doust, 2002). Bugs overwinter under crop debris so removal of harvest residues, field sanitation and crop rotation can reduce the risk of infestation. Young plants are more susceptible than older plants so establishment of the crop from transplants instead of by direct sowing exposes the susceptible growth stages for a shorter period. Squash bugs aggregate in sheltered places (i.e. under boards, large leaves) and, in large fields, shelters should be removed. In smaller fields, bugs can be removed from their shelters during the day and destroyed. Squash bugs prefer pumpkins, squash and watermelons to cucumbers. Therefore, squash plants can be used in trap crops around cucumber fields to prevent immigration of bugs into the main crop. Crop covers made from fine-mesh netting provide good control, but bugs invade the fields as soon as the covers are removed to manage weeds or to allow pollination. A combination of row covers with the artificial introduction of bees under the covers can solve this problem (Minter and Bessin, 2014). Applications of neem and pyrethrin can reduce infestations. Dust applications of diatomaceous earth to the base of plants also showed promising results (Cranshaw *et al.*, 2001).

Seedcorn maggot

Seedcorn maggot (*Delia platura*), described for legumes (see next section), can also damage cucurbit crops.

Legumes

Peas, broad beans and dwarf beans are legume vegetables.

Seedcorn maggot

The seedcorn maggot (*D. platura*) oviposits on germinating seeds of bean plants (Gouinguene and Staedler, 2006) and other crops. Damage (plant death) is caused by larval feeding on seeds, roots or the growing point of young plants, especially under cool and wet weather conditions where plant development is slow. Three or four generations/year are possible, but the first generation (flight period in April/May) usually causes the most severe damage. This pest is closely related to other *Delia* sp. (i.e. *Delia florilega*) which cause similar types of damage. Adult flies are attracted to the odour of decaying organic material, so no fresh manure should be applied before planting and plant residues from previous crops should be ploughed in. Sowing legumes where potato or spinach were grown previously should be avoided. Repeated soil cultivations before sowing can help to reduce pest pressure. Shallow sowing (3 cm) in warm soil can reduce damage, because it encourages rapid germination and development of young plants. Early sown crops can be protected with covers of fine-mesh netting, floating row covers or by seed treatments with spinosad. Destruction (ploughing) of heavily infested crops can prevent development of a further generation. Following intense soil cultivation to destroy any larvae in the soil, the crop can be re-sown.

Pea and bean weevil

The larvae of the pea and bean weevil (*Sitona lineatus*) overwinter in the soil and feed on the roots of many cultivated and wild leguminous plant species, delaying growth. Adult beetles feed on the leaves. Usually damage remains below the economic level and can be reduced further by an appropriate crop rotation, early sowing, additional hoeing, application of silicate rock dusts or the use of fine-mesh netting covers.

Pea moth

The larvae of the pea moth (*Cydia nigricana*) feed on the seeds of pea plants. Tolerance levels demanded by the processing industry are very low: if more than 0.5% of peas are infested, the whole batch is rejected. The pea moth overwinters as a pupa in the soil where pea crops were grown previously and adults emerge in May/June. By using a combination of early sowing, fast-maturing cultivars (Schultz and Saucke, 2005) and isolating new crops from sources of infestation (> 500 m separation) (Huusela-Veistola and Jauhiainen, 2006; Thoeming *et al.*, 2011), it is possible to avoid colonization by this pest. This is currently the only strategy available to minimize damage by pea moth in organic production systems. For late sown pea crops, even fields that are more than 10 km away from sources of infestation can be damaged. Several insecticides approved for organic production (neem, baculovirus, pyrethrin) have been tested. At low population densities, pyrethrin applications have some efficacy. Mating disruption techniques have been developed (Witzgall *et al.*, 1996) but the migratory behaviour of gravid females has been shown to limit the success of this strategy (Saucke *et al.*, 2014).

Pea aphid

Pea aphid (*Acyrthosiphon pisum*) damages pea crops especially during flowering and pod filling and by transmission of viruses. It also infests other legumes (i.e. soybean) and lucerne, vetch and clover are overwintering hosts. The close proximity of new crops to soybean fields can lead to increased pest pressure, especially late in the season. In order to prevent migration from overwintering hosts to new pea crops, nearby lucerne, vetch and clover crops should be mown or harvested in the previous autumn. Intercropping peas with wheat can reduce colonization of pea plants by aphids (Ndzana *et al.*, 2014). Predators and parasitoids of pea aphids can be attracted using flowering strips. Direct control is possible using horticultural soaps, pyrethrin or neem.

Black bean aphid

The black bean aphid (*Aphis fabae*) can cause damage when the weather is warm and dry. Damage occurs mainly along the edges of fields. The level of damage is usually below economic threshold levels. The application of irrigation can sometimes mitigate damage. The selection of tolerant varieties and intercropping can reduce damage. Control is only necessary when infestation levels are high and quassia, horticultural soaps and pyrethrin can be applied. Intercropping legumes with cereals can reduce damage (Hansen *et al.*, 2008). Black bean aphid infestations may not be regulated effectively by natural enemies, because *A. fabae* is of lower quality as a food source for natural enemies, such as coccinellids, than other species of aphid (Volkl and Stechmann, 1998; Hinkelman and Tenhumberg, 2013).

Pea weevil

Pea weevil (*Bruchus pisorum*) larvae feed inside the seeds of many legume crops. Damage is caused mainly in crops grown for seed production. Early harvest, before the new generation of adults leave the peas, is the most effective way to prevent populations increasing.

Lettuce

This section considers the pests of the *Compositae*: leaf lettuce (*Lactuca* sp.), chicory, radicchio and endive (*Cichorium* sp.).

Aphids

The main pests are several species of aphid (*Nasonovia ribisnigri*, *M. persicae*, *Macrosiphum euphorbiae*, *Uroleucon sonchi*, *P. bursarius* and others). By using clean, uninfested seedlings, the transfer of aphids into newly planted crops can be avoided. The application of fine-mesh netting covers after transplanting can help to prevent aphid immigration. Most species of aphid infest only the outer leaves of the lettuce head and do not cause economic damage. Infestations by these aphids are often reduced by a range of parasitoids and predators. Wildflower strips close to lettuce crops or companion plants, such as *Alyssum*, within the field can help to establish high populations of aphid antagonists (Pascual-Villalobos *et al.*, 2006; Brennan, 2013; Barriere *et al.*, 2014). In addition, species that only infest the outer leaves can be controlled by contact insecticides, such as pyrethrin, horticultural oils and soaps, quassia or neem. Only *N. ribisnigri* infests the inner leaves of lettuce heads (Liu, 2004). Due to the low tolerance of retailers and consumers to aphid infestations in lettuce heads, this species can cause considerable economic damage. Because of its inaccessibility, this aphid species is rarely attacked by parasitoids. However, some predators can reduce the numbers of aphids within lettuce heads (Hopper *et al.*, 2011). Control of *N. ribisnigri* is difficult with contact insecticides. However, varieties of lettuce resistant to *N. ribisnigri* are available (Liu and McCreight, 2006), although there are now biotypes of *N. ribisnigri* which are able to overcome the resistance (Sauer-Kesper *et al.*, 2011; Cid *et al.*, 2012). Ongoing breeding programmes aim at introducing new resistance genes from wild *Lactuca* species (ten Broeke *et al.*, 2013).

Lettuce root aphid (*P. bursarius*) causes damage to the roots of lettuce plants, which leads to yellowing, wilting and impaired head formation. Most damage is caused in summer following immigration of winged aphids from their overwintering hosts and during dry weather conditions. The application of irrigation can mitigate damage and help plants to compensate for reduced water uptake by the roots. The primary (winter) hosts of this species are poplar trees (*Populus nigra*) (Miller *et al.*, 2005). Crops grown at some distance from poplar trees may avoid immigration by winged root aphids in early summer. A degree day forecast is available to predict the timing of the migration from poplar trees to lettuce crops (Collier *et al.*, 1994). Varieties resistant to *P. bursarius* are available (Ellis *et al.*, 2002).

Lepidopteran pests

In addition, polyphagous lepidopteran pests (e.g. cutworms and noctuids, listed in the next section) can cause feeding damage on lettuce.

Polyphagous Pest Insects of Different Crops

Most pest insects of field vegetable crops are specialists which colonize a small number of closely related crops (e.g. *Brassicaceae*). However, a small number of polyphagous pest species can colonize a range of crops from different plant families. These include soil-dwelling pests such as crane flies, cutworms and slugs.

Crane flies

Crane flies (*Tipula* sp.) oviposit in autumn in densely growing plant stands (i.e. green manure, grass–clover mixture). Damage is caused in the following spring when the larvae feed on plant roots. Crops that are sown directly are more susceptible, because germinating plants can be killed. Therefore, following the use of green manures, sowing of new crops should be delayed until the pupation of larvae in mid-May. Intense superficial cultivation of the soil can reduce the numbers of larvae.

Wireworms

Wireworms (*Agriotes* sp.) also live in the soil and feed on the roots of many plants. Damage is likely only when vegetable crops are grown 1 or 2 years after a perennial grass–clover mixture. Damage can be prevented by using an appropriate crop rotation strategy.

Cutworm

Cutworm (*Agrotis* sp.) larvae are also hidden in the soil during the day and damage plants by feeding on the hypocotyl and leaves during the night. For some species, such as *Agrotis segetum*, pheromone traps are available to monitor the activity of male moths. Crops sowed directly are most susceptible, especially in early spring (Central Europe) and June/July (UK). Repeated soil cultivations can expose the larvae to birds and other predators. In addition, repeated cultivation reduces the availability of food plants for cutworm larvae. The young larvae of some species, such as *A. segetum*, are susceptible to high soil moisture (Esbjerg, 1988) and numbers can be reduced by the application of irrigation. A model for forecasting the abundance of *A. segetum* larvae has been developed in the UK (Bowden *et al.*, 1983) and is used by farmers and growers. Biological control of these pests may be possible using *Bt* or entomopathogenic nematodes (*Steinernema carpocapse*). For best efficacy, entomopathogenic nematodes should be applied to moist soil; during periods of dry weather, the application of nematodes should be combined with the application of irrigation. *Bt* is most effective if mixed with mollases and lucerne meal as bait and spread along the plant rows in the evening. Spinosad applied in bait formulations may also have good efficacy.

Noctuids

Several species of noctuids (*Autographa gamma*, *Mamestra oleracea*) and other Lepidoptera can colonize many different vegetable crops. Most damage is caused to leafy vegetables (lettuce, spinach) and consumers do not like to find larvae in harvested produce. The presence of wildflower strips in the vicinity of crops or flowering companion plants within the field can help to establish high populations of parasitoids and predators. *Bt aizawai* has good efficacy against young larvae if applied at temperatures > 12°C in the evening or when weather conditions are overcast. Pyrethrin has some efficacy against young larvae. Applications of spinosad are also effective against older larvae.

Polyphagous aphids

Polyphagous aphids (*M. persicae*, *Aphis fabae*, *Macrosiphum euphorbiae*, *Aulacorthum solani*) can occur on most crops. With the exception of lettuce, they normally do not cause economic damage in open field production, because aphids are attacked by a plethora of natural enemies. The numbers of aphid antagonists can be enhanced and maintained by restrained use of broad-spectrum insecticides (Smith *et al.*, 2008; Fagan *et al.*, 2010) and by the presence of wildflower strips close to vegetable crops. Applications of neem or horticultural soaps can reduce aphid infestations in most crops.

Conclusions

For most of the pests which infest organic vegetable crops, a number of control strategies are available. However, some of the strategies (e.g. the use of fine-mesh netting) are considerably more expensive or more labour intensive than the strategies used in conventional farming. Mulching, intercropping and undersowing also increase the amount of work required when cultivating and harvesting crops. As a result most farmers implement these techniques very reluctantly despite the encouraging results provided by experimental studies. Information about the impact of certain cultural practices (e.g. timing of sowing and harvesting, effect of irrigation) is still very scarce and is based mainly on farmers' experience. More research is needed to determine the impact of different production systems on pest abundance and subsequent damage. In addition, better understanding of the biology and dispersal and host finding behaviours of pest insects would be useful for improved targeting of preventative measures. For some pests (e.g. *D. radicum*) detailed information is available, while there are huge gaps in our knowledge of other pests (e.g. *A. proletella*). Approaches such as intercropping and companion planting have shown promise for the management of certain pest species, but more research is required to extend these approaches to other crops. Approaches which enhance the numbers of predators, parasitoids and naturally occurring pathogens can be very effective for pest species where relatively large infestations can be tolerated (e.g. aphids) without there being a negative effect on crop quality and yield. For this approach, too, further research is needed to extend it to other crops.

References

Adler, L.S. and Hazzard, R.V. (2009) Comparison of perimeter trap crop varieties: effects on herbivory, pollination, and yield in butternut squash. *Environmental Entomology* 38, 207–215.

Andersen, C.L., Hazzard, R., Van Driesche, R. and Mangan, F.X. (2006) Alternative management tactics for control of *Phyllotreta cruciferae* and *Phyllotreta striolata* (Coleoptera: Chrysomelidae) on *Brassica rapa* in Massachusetts. *Journal of Economic Entomology* 99, 803–810.

Asman, K. (2002) Trap cropping effect on oviposition behaviour of the leek moth *Acrolepiopsis assectella* and the diamondback moth *Plutella xylostella*. *Entomologia Experimentalis et Applicata* 105, 153–164.

Barriere, V., Lecompte, F., Nicot, P.C., Maisonneuve, B., Tchamitchian, M. and Lescourret, F. (2014) Lettuce cropping with less pesticides. A review. *Agronomy for Sustainable Development* 34, 175–198.

Baumann, D.T., Kropff, M.J. and Bastiaans, L. (2000) Intercropping leeks to suppress weeds. *Weed Research* 40, 359–374.

Berry, N.A., Wratten, S.D. and Frampton, C. (1997) Effects of sowing and harvest dates on carrot rust fly (*Psila rosae*) damage to carrots in Canterbury, New Zealand. *New Zealand Journal of Crop and Horticultural Science* 25, 109–115.

Biernacki, M. and Lovett-Doust, J. (2002) Developmental shifts in watermelon growth and reproduction caused by the squash bug, *Anasa tristis*. *New Phytologist* 155, 265–273.

Biondi, A., Mommaerts, V., Smagghe, G., Viñuela, E., Zappalà, L. and Desneux, N. (2012) The non-target impact of spinosyns on beneficial arthropods. *Pest Management Science* 68, 1523–1536.

Blackshaw, R.P., Vernon, R.S. and Prasad, R. (2012) Reduction of *Delia radicum* attack in field brassicas using a vertical barrier. *Entomologia Experimentalis et Applicata* 144, 145–156.

Bomford, M.K., Vernon, R.S. and Pats, P. (2000) Importance of collection overhangs on the efficacy of exclusion fences for managing cabbage flies (Diptera: Anthomyiidae). *Environmental Entomology* 29, 795–799.

Bosco, L. and Tavella, L. (2010) Population dynamics and integrated pest management of *Thrips tabaci* on leek under field conditions in northwest Italy. *Entomologia Experimentalis et Applicata* 135, 276–287.

Bowden, J., Cochrane, J., Emmett, B.J., Minall, T.E. and Sherlock, P.L. (1983) A survey of cutworm attacks in England and Wales, and a descriptive population model for *Agrotis segetum* (Lepidoptera: Noctuidae). *Annals of Applied Biology* 102, 29–47.

Brennan, E.B. (2013) Agronomic aspects of strip intercropping lettuce with *Alyssum* for biological control of aphids. *Biological Control* 65, 302–311.

Buckland, K., Reeve, J.R., Alston, D., Nischwitz, C. and Drost, D. (2013) Effects of nitrogen fertility and crop rotation on onion growth and yield, thrips densities, iris yellow spot virus and soil properties. *Agriculture Ecosystems & Environment* 177, 63–74.

Burnstone, J. (2009) Investigations into the biology and behaviour of *Thrips tabaci* L. PhD thesis, University of Warwick, Warwick, UK. Available at: http://wrap.warwick.ac.uk/3124/1/WRAP_THESIS_Burnstone_2009.pdf (accessed 10 July 2017).

Caudle, J.R., Coolong, T., Williams, M.A., Vincelli, P. and Bessin, R. (2013) Development of an organic muskmelon production system against bacterial wilt disease. *Acta Horticulturae* 1001, 249–254.

Cavanagh, A., Hazzard, R., Adler, L.S. and Boucher, J. (2009) Using trap cops for cotrol of *Acalymma vittatum* (Coleoptera: Chrysomelidae) reduces insecticide use in butternut squash. *Journal of Economic Entomology* 102, 1101–1107.

Chapman, J.W., Reynolds, E.R., Smith, A.D., Riley, J.R., Pedgley, D.E. and Woiwod, I.P. (2002) High-altitude migration of the diamondback moth *Plutella xylostella* to the UK: a study using radar, aerial netting, and ground trapping. *Ecological Entomology* 27, 641–650.

Chen, M. and Shelton, A.M. (2007) Impact of soil type, moisture, and depth on swede midge (Diptera: Cecidomyiidae) pupation and emergence. *Environmental Entomology* 36, 1349–1355.

Chen, M., Li, W. and Shelton, A.M. (2009) Simulated crop rotation systems control swede midge, *Contarinia nasturtii*. *Entomologia Experimentalis et Applicata* 133, 84–91.

Cid, M., Avila, A., Garcia, A., Abad, J. and Fereres, A. (2012) New sources of resistance to lettuce aphids in *Lactuca* spp. *Arthropod–Plant Interactions* 6, 655–669.

Cline, G.R., Sedlacek, J.D., Hillman, S.L., Parker, S.K. and Silvernail, A.F. (2008) Organic management of cucumber beetles in watermelon and muskmelon production. *HortTechnology* 18, 436–444.

Collier, R. (2009) Review of carrot fly control in Northern Europe – 2009. *OEPP/EPPO* [European and Mediterranean Plant Protection Organization] *Bulletin* 39, 116–120.

Collier, R.H. and Finch, S. (1990) Some factors affecting the efficiency of sticky traps for capturing the carrot fly *Psila rosae*. *Bulletin of Entomological Research* 80, 153–158.

Collier, R.H. and Finch, S. (2000) Strategies for reducing carrot fly (*Psila rosae* F.) damage in organic crops. In: *52nd International Symposium on Crop Protection*, 9 May, Gent, Belgium. Mededelingen-Faculteit Landbouwkundige en Toegepaste Biologische Wetenschappen, Universiteit Gent, Belgium, 65(2a), pp. 227–233.

Collier, R. and Finch, S. (2009) A review of research to address carrot fly (*Psila rosae*) control in the UK. *OEPP/EPPO* [European and Mediterranean Plant Protection Organization] *Bulletin* 39, 121–127.

Collier, R.H., Davies, J., Roberts, M., Leatherland, M., Runham, S. and Blood-Smyth, J. (1994) Monitoring and forecasting the times of attack of the lettuce root aphid, *Pemphigus bursarius* L. *IOBC (International Organisation for Biological and Integrated Control) WPRS Bulletin* 17, 31–40.

Cranshaw, W., Bartolo, M. and Schweissing, F. (2001) Control of squash bug (Hemiptera: Coreidae) injury: management manipulations at the base of pumpkin. *Southwestern Entomologist* 26, 147–150.

Davies, J. and Collier, R. (2000) Strategies for controlling carrot fly while minimizing pesticide inputs. *Acta Horticulturae* 533, 575–582.

Döring, T.F., Kirchner, S.M., Kühne, S. and Saucke, H. (2004) Response of alate aphids to green targets on coloured backgrounds. *Entomologia Experimentalis et Applicata* 113, 53–61.

Dosdall, L.M., Herbut, M.J., Cowle, N.T. and Micklich, T.M. (1996) The effect of tillage regime on emergence of root maggots (*Delia* spp.) (Diptera: Anthomyiidae) from canola. *Canadian Entomologist* 128, 1157–1165.

Dosdall, L.M., Clayton, G.W., Harker, K.N., O'Donovan, J.T. and Stevenson, F.C. (2003) Weed control and root maggots: making canola pest management strategies compatible. *Weed Science* 51, 576–585.

El-Wakeil, N.E., Gaafar, N.M. and Vidal, S. (2006) Side effect of some neem products on natural enemies of *Helicoverpa* (*Trichogramma* spp.) and *Chrysoperla carnea*. *Archives of Phytopathology and Plant Protection* 39, 445–455.

Ellis, P.R., Freeman, G.H., Dowker, B.D., Hardman, J.A. and Kingswell, G. (1987a) The influence of plant density and position in field trials designed to evaluate the resistance of carrots to carrot fly (*Psila rosae*) attack. *Annals of Applied Biology* 111, 21–31.

Ellis, P.R., Hardman, J.A., Cole, R.A. and Phelps, K. (1987b) The complementary effects of plant-resistance and the choice of sowing and harvest times in reducing carrot fly (*Psila rosae*) damage to carrots. *Annals of Applied Biology* 111, 415–424.

Ellis, P.R., McClement, S.J., Saw, P.L., Phelps, K., Vice, W.E., Kift, N.B., Astley, D. and Pink, D.A.C. (2002) Identification of sources of resistance in lettuce to the lettuce root aphid, *Pemphigus bursarius*. *Euphytica* 125, 305–315.

Esbjerg, P. (1988) Behaviour of 1st- and 2nd-instar cutworms (*Agrotis segetum* Schiff.) (Lep., Noctuidae): the influence of soil moisture. *Journal of Applied Entomology* 105, 295–302.

Ester, A., de Putter, H. and van Bilsen, J. (2003) Filmcoating the seed of cabbage (*Brassica oleracea* L. convar. *capitata* L.) and cauliflower (*Brassica oleracea* L. var. *botrytis* L.) with imidacloprid and spinosad to control insect pests. *Crop Protection* 22, 761–768.

Fagan, L.L., McLachlan, A., Till, C.M. and Walker, M.K. (2010) Synergy between chemical and biological control in the IPM of currant-lettuce aphid (*Nasonovia ribisnigri*) in Canterbury, New Zealand. *Bulletin of Entomological Research* 100, 217–223.

Finch, S. (1993) Integrated pest-management of the cabbage root fly and the carrot fly. *Crop Protection* 12, 423–430.

Finch, S. and Collier, R. (1985) Laboratory studies on aestivation in the cabbage root fly (*Delia radicum*). *Entomologia Experimentalis et Applicata* 38, 137–143.

Finch, S. and Collier, R.H. (2000) Integrated pest management in field vegetable crops in northern Europe – with focus on two key pests. *Crop Protection* 19, 817–824.

Finch, S. and Collier, R.H. (2004) A simple method – based on the carrot fly – for studying the movement of pest insects. *Entomologia Experimentalis et Applicata* 110, 201–205.

Finch, S. and Collier, R.H. (2012) The influence of host and non-host companion plants on the behaviour of pest insects in field crops. *Entomologia Experimentalis et Applicata* 142, 87–96.

Finch, S. and Kienegger, M. (1997) A behavioural study to help clarify how undersowing with clover affects host-plant selection by pest insects of brassica crops. *Entomologia Experimentalis et Applicata* 84, 165–172.

Finch, S. and Skinner, G. (1975) Dispersal of the cabbage root fly. *Annals of Applied Biology* 81, 1–19.

Finch, S., Billiald, H. and Collier, R.H. (2003) Companion planting – do aromatic plants disrupt host-plant finding by the cabbage root fly and the onion fly more effectively than non-aromatic plants? *Entomologia Experimentalis et Applicata* 109, 183–195.

Foster, S.P., Gorman, K. and Denholm, I. (2010) English field samples of *Thrips tabaci* show strong and ubiquitous resistance to deltamethrin. *Pest Management Science* 66, 861–864.

Gardner, J., Hoffmann, M.P. and Mazourek, M. (2015) Striped cucumber beetle (Coleoptera: Chrysomelidae) aggregation in response to cultivar and flowering. *Environmental Entomology* 44, 309–316.

George, D.R., Collier, R.H. and Whitehouse, D.M. (2013) Can imitation companion planting interfere with host selection by *Brassica* pest insects? *Agricultural and Forest Entomology* 15, 106–109.

Gouinguene, S.P. and Staedler, E. (2006) Oviposition in *Delia platura* (Diptera, Anthomyiidae): the role of volatile and contact cues of bean. *Journal of Chemical Ecology* 32, 1399–1413.

Gouinguene, S., Buser, H.R. and Stadler, E. (2005) Host-plant leaf surface compounds influencing oviposition in *Delia antiqua*. *Chemoecology* 15, 243–249.

Hallett, R.H. (2007) Host plant susceptibility to the swede midge (Diptera: Cecidomyiidae). *Journal of Economic Entomology* 100, 1335–1343.

Hansen, L.M., Lorentsen, L. and Boelt, B. (2008) How to reduce the incidence of black bean aphids (*Aphis fabae* Scop.) attacking organic growing field beans (*Vicia faba* L.) by growing partially resistant bean varieties and by intercropping field beans with cereals. *Acta Agriculturae Scandinavica Section B-Soil and Plant Science* 58, 359–364.

Heimbach, U., Eggers, C. and Thieme, T. (2002) Weniger Blattläuse durch Mulchen? *Gesunde Pflanzen* 54, 119–125.

Helenius, J. (1997) Spatial scales in ecological pest management (EPM): importance of regional crop rotations. *Biological Agriculture & Horticulture* 15, 163–170.

Hellqvist, S. (1996) Mulching with grass-clippings in cauliflower: effects on yield and brassica root flies (*Delia* spp). *International Journal of Pest Management* 42, 39–46.

Herrmann, F., Buck, H., Liebig, N., Hommes, M. and Saucke, H. (2009) Vermeidung und reduktion von möhrenfliegenschäden im Ökolandbau. In: Mayer, J., Alföldi, T., Leiber, F., Dubois, D., Fried, P., Heckendorn, F., Hillmann, E., Klocke, P., Lüscher, A., Riedel, S., Stolze, M., Strasser, F., van der Heijden, M. and Willer, H. (eds) *Proceedings of the 10. Wissenschaftstagung Ökologischer Landbau ETH Zürich, 11–13 Februar 2009*, Vol. 1. Verlag Dr. Köster, Berlin, Germany.

Hinds, J. and Hooks, C.R.R. (2013) Population dynamics of arthropods in a sunn hemp–zucchini interplanting system. *Crop Protection* 53, 6–12.

Hinkelman, T.M. and Tenhumberg, B. (2013) Larval performance and kill rate of convergent ladybird beetles, *Hippodamia convergens*, on black bean aphids, *Aphis fabae*, and pea aphids, *Acyrthosiphon pisum*. *Journal of Insect Science* 13, 46. doi: 10.1673/031.013.4601.

Hommes, M. and Gebelein, D. (1996) Simulation models for the cabbage root fly and the carrot fly. *IOBC (International Organisation for Biological and Integrated Control) WPRS Bulletin* 19, 60–65.

Hommes, M. and Herbst, M. (2014) Supervised control of aphids and caterpillars on white cabbage and impact of flower strips. *IOBC (International Organisation for Biological and Integrated Control) WPRS Bulletin* 107, 43–49.

Hommes, M., Siekmann, G., Piepenbrock, O., Baur, U., Fricke, A. and Thieme, T. (2003) *Reducing Aphid Infestations in Selected Field Vegetables with Mulches of Different Materials and Colours*. Final report. Geschäftsstelle Bundesprogramm Ökologischer Landbau in der Bundesanstalt für Landwirtschaft und Ernährung (BLE), Bonn, Germany. p. 83.

Hopper, J.V., Nelson, E.H., Daane, K.M. and Mills, N.J. (2011) Growth, development and consumption by four syrphid species associated with the lettuce aphid, *Nasonovia ribisnigri*, in California. *Biological Control* 58, 271–276.

Humphreys, I.C. and Mowat, D.J. (1994) Effects of some organic treatments on predators (Coleoptera, Carabidae) of cabbage root fly, *Delia radicum* (L) (Diptera, Anthomyiidae), and on alternative prey species. *Pedobiologia* 38, 513–518.

Huusela-Veistola, E. and Jauhiainen, L. (2006) Expansion of pea cropping increases the risk of pea moth (*Cydia nigricana*; Lep., Tortricidae) infestation. *Journal of Applied Entomology* 130, 142–149.

Jansen, J.P., Defrance, T. and Warnier, A.M. (2010) Effects of organic-farming-compatible insecticides on four aphid natural enemy species. *Pest Management Science* 66, 650–656.

Judd, G.J.R. and Borden, J.H. (1992) Influence of different habitats and mating on olfactory behavior of onion flies seeking ovipositional hosts. *Journal of Chemical Ecology* 18, 605–620.

Kannan, H.O. and Mohamed, M.B. (2001) The impact of irrigation frequency on population density of thrips, *Thrips tabaci* Rom (Thripidae, Thysanoptera) and yield of onion in El Rahad, Sudan. *Annals of Applied Biology* 138, 129–132.

Kostal, V. and Finch, S. (1994) Influence of background on host-plant selection and subsequent oviposition by the cabbage root fly (*Delia radicum*). *Entomologia Experimentalis et Applicata* 70, 153–163.

Koštál, V., Baur, R. and Städler, E. (2000) Exploration and assessment of the oviposition substrate by the cabbage root fly, *Delia radicum* (Diptera: Anthomyiidae). *European Journal of Entomology* 97, 33–40.

Kristoffersen, L. and Anderbrant, O. (2007) Carrot psyllid (*Trioza apicalis*) winter habitats – insights in shelter plant preference and migratory capacity. *Journal of Applied Entomology* 131, 174–178.

Larentzaki, E., Plate, J., Nault, B.A. and Shelton, A.M. (2008a) Impact of straw mulch on populations of onion thrips (Thysanoptera: Thripidae) in onion. *Journal of Economic Entomology* 101, 1317–1324.

Larentzaki, E., Shelton, A.M. and Plate, J. (2008b) Effect of kaolin particle film on *Thrips tabaci* oviposition, feeding and development on onions: a lab and field case study. *Crop Protection* 27, 727–734.

Laska, L. (2011) Biology of *Trioza apicalis* – a review. *Plant Protection Science* 47, 68–77.

Lebedev, G., Abo-Moch, F., Gafni, G., Ben-Yakir, D. and Ghanim, M. (2013) High-level of resistance to spinosad, emamectin benzoate and carbosulfan in populations of *Thrips tabaci* collected in Israel. *Pest Management Science* 69, 274–277.

Lepage, M.P., Bourgeois, G., Brodeur, J. and Boivin, G. (2012) Effect of soil temperature and moisture on survival of eggs and first-instar larvae of *Delia radicum*. *Environmental Entomology* 41, 159–165.

Little, A.G., Arellano, C., Kennedy, G.G. and Cardoza, Y.J. (2011) Bottom-up effects mediated by an organic soil amendment on the cabbage aphid pests *Myzus persicae* and *Brevicoryne brassicae*. *Entomologia Experimentalis et Applicata* 139, 111–119.

Liu, Y.B. (2004) Distribution and population development of *Nasonovia ribisnigri* (Homoptera: Aphididae) in iceberg lettuce. *Journal of Economic Entomology* 97, 883–890.

Liu, Y.B. and McCreight, J.D. (2006) Responses of *Nasonovia ribisnigri* (Homoptera: Aphididae) to susceptible and resistant lettuce. *Journal of Economic Entomology* 99, 972–978.

Maniania, N.K., Sithanantham, S., Ekesi, S., Ampong-Nyarko, K., Baumgartner, J., Lohr, B. and Matoka, C.M. (2003) A field trial of the entomogenous fungus *Metarhizium anisopliae* for control of onion thrips, *Thrips tabaci*. *Crop Protection* 22, 553–559.

Markó, V., Blommers, L.H.M., Bogya, S. and Helsen, H. (2008) Kaolin particle films suppress many apple pests, disrupt natural enemies and promote woolly apple aphid. *Journal of Applied Entomology* 132, 26–35.

Mautino, G.C., Bosco, L. and Tavella, L. (2012) Integrated management of *Thrips tabaci* (Thysanoptera: Thripidae) on onion in north-western Italy: basic approaches for supervised control. *Pest Management Science* 68, 185–193.

Meadow, R. (2010) The carrot psyllid, *Trioza apicalis* – biology and control. *Bioforsk Report* 5, 1–14.

Meyling, N.V., Navntoft, S., Philipsen, H., Thorup-Kristensen, K. and Eilenberg, J. (2013) Natural regulation of *Delia radicum* in organic cabbage production. *Agriculture Ecosystems & Environment* 164, 183–189.

Miller, N.J., Kift, N.B. and Tatchell, G.M. (2005) Host-associated populations in the lettuce root aphid, *Pemphigus bursarius* (L.). *Heredity* 94, 556–564.

Minter, L.M. and Bessin, R.T. (2014) Evaluation of native bees as pollinators of cucurbit crops under floating row covers. *Environmental Entomology* 43, 1354–1363.

Morley, K., Finch, S. and Collier, R.H. (2005) Companion planting-behaviour of the cabbage root fly on host plants and non-host plants. *Entomologia Experimentalis et Applicata* 117, 15–25.

Morris, O.N. (1985) Susceptibility of 31 species of agricultural insect pests to the entomogenous nematodes *Steinernema feltiae* and *Heterorhabditis bacteriophora*. *The Canadian Entomologist* 117, 401–407.

Munyaneza, J.E. (2010) Psyllids as vectors of emerging bacterial diseases of annual crops. *Southwestern Entomologist* 35, 471–477.

Mutiga, S.K., Gohole, L.S. and Auma, E.O. (2010) Effects of integrating companion cropping and nitrogen application on the performance and infestation of collards by *Brevicoryne brassicae*. *Entomologia Experimentalis et Applicata* 134, 234–244.

Nault, B.A., Straub, W. and Taylor, A.G. (2006) Performance of novel insecticide seed treatments for managing onion maggot (Diptera: Anthomyiidae) in onion fields. *Crop Protection* 25, 58–65.

Nault, B.A., Werling, B.P., Straub, R.W. and Nyrop, J.P. (2011) Delaying onion planting to control onion maggot (Diptera: Anthomyiidae): efficacy and underlying mechanisms. *Journal of Economic Entomology* 104, 1622–1632.

Ndzana, R.A., Magro, A., Bedoussac, L., Justes, E., Journet, E.P. and Hemptinne, J.L. (2014) Is there an associational resistance of winter pea–durum wheat intercrops towards *Acyrthosiphon pisum* Harris? *Journal of Applied Entomology* 138, 577–585.

Nebreda, M., Nombela, G. and Muniz, M. (2005) Comparative host suitability of some *Brassica* cultivars for the whitefly, *Aleyrodes proletella* (Homoptera: Aleyrodidae). *Environmental Entomology* 34, 205–209.

Nehlin, G., Valterová, I. and Borg-Karlson, A.-K. (1996) Monoterpenes released from Apiaceae and the egg-laying preferences of the carrot psyllid, *Trioza apicalis*. *Entomologia Experimentalis et Applicata* 80, 83–86.

Nilsson, U., Rannback, L.M., Anderson, P., Eriksson, A. and Ramert, B. (2011) Comparison of nectar use and preference in the parasitoid *Trybliographa rapae* (Hymenoptera: Figitidae) and its host, the cabbage root fly, *Delia radicum* (Diptera: Anthomyiidae). *Biocontrol Science and Technology* 21, 1117–1132.

Nissinen, A., Vanhala, P., Holopainen, J.K. and Tiilikkala, K. (2007) Short feeding period of carrot psyllid (*Trioza apicalis*) females at early growth stages of carrot reduces yield and causes leaf discolouration. *Entomologia Experimentalis et Applicata* 125, 277–283.

Parolin, P., Bresch, C., Desneux, N., Brun, R., Bout, A., Boll, R. and Poncet, C. (2012) Secondary plants used in biological control: a review. *International Journal of Pest Management* 58, 91–100.

Parsons, C.K., Dixon, P.L. and Colbo, M. (2007) Relay cropping cauliflower with lettuce as a means to manage first-generation cabbage maggot (Diptera: Anthomyiidae) and minimize cauliflower yield loss. *Journal of Economic Entomology* 100, 838–846.

Pascual-Villalobos, M.J., Lacasa, A., Gonzalez, A., Varo, P. and Garcia, M.J. (2006) Effect of flowering plant strips on aphid and syrphid populations in lettuce. *European Journal of Agronomy* 24, 182–185.

Pelgrom, K.T.B., Broekgaarden, C., Voorrips, R.E., Bas, N., Visser, R.G.F. and Vosman, B. (2015) Host plant resistance towards the cabbage whitefly in *Brassica oleracea* and its wild relatives. *Euphytica* 202, 297–306.

Phelps, K., Collier, R.H., Reader, R.J. and Finch, S. (1993) Monte Carlo simulation method for forecasting the timing of pest insect attacks. *Crop Protection* 12, 335–342.

Pinto, E.S., Barros, E.M., Torres, J.B. and dos Santos Neves, R.C. (2013) The control and protection of cotton plants using natural insecticides against the colonization by *Aphis gossypii* Glover (Hemiptera: Aphididae). *Acta Scientiarum-Agronomy* 35, 169–174.

Prowse, G.M., Galloway, T.S. and Foggo, A. (2006) Insecticidal activity of garlic juice in two dipteran pests. *Agricultural and Forest Entomology* 8, 1–6.

Rahn, R. (1982) Effects of cultural practices on insect pests – the case of onion (*Allium cepa*) and of the leek moth, *Acrolepiopsis assectella* (Lepidoptera: Plutellidae). *Agronomie* 2, 695–699.

Rämert, B. (1993) Sawdust can be used for control of the carrot psyllid (*Trioza apicalis*). *Vaxtskyddsnotiser* 57, 34–38.

Rannback, L.-M., Cotes, B., Anderson, P., Ramert, B. and Meyling, N.V. (2015) Mortality risk from entomopathogenic fungi affects oviposition behavior in the parasitoid wasp *Trybliographa rapae*. *Journal of Invertebrate Pathology* 124, 78–86.

Razinger, J., Lutz, M., Schroers, H.-J., Urek, G. and Grunder, J. (2014) Evaluation of insect associated and plant growth promoting fungi in the control of cabbage root flies. *Journal of Economic Entomology* 107, 1348–1354.

Rousse, P., Fournet, S., Porteneuve, C. and Brunel, E. (2003) Trap cropping to control *Delia radicum* populations in cruciferous crops: first results and future applications. *Entomologia Experimentalis et Applicata* 109, 133–138.

Rüegg, J., Eder, R. and Anderau, V. (2006) Improved application techniques: ways to higher efficacy of fungicides and insecticides. *Outlooks on Pest Management* 17, 80–84. DOI: 10.1564/1516apr1507.

Saucke, H., Schultz, B., Wedemeyer, R., Liebig, N., Zimmermann, O. and Katz, P. (2011) Biotechnical control of cabbage whitefly in brassica vegetables – current status and perspectives. *Gesunde Pflanzen* 63, 183–189.

Saucke, H., Balasus, A., Finckh, M.R., Formowitz, B., Schmidt, R. and Kratt, A. (2014) Mating disruption of pea moth (*Cydia nigricana*) in organic peas (*Pisum sativum*). *Entomologia Experimentalis et Applicata* 150, 199–207.

Sauer-Kesper, C., Lucia, N., Buser, H. and Vogler, U. (2011) The new biotype Nr. 1 of the currant lettuce aphid: its distribution and impact on Swiss lettuce production. *Agrarforschung Schweiz* 2, 462–469.

Schultz, B. and Saucke, H. (2005) Einfluss verschiedener Saattermine auf den Erbsenwicklerbefall (*Cydia nigricana* Fabr.) in ökologischen Gemüseerbsen. In: Heß, J. and Rahmann, G. (eds) *Proceedings of the 8. Wissenschaftstagung Ökologischer Landbau.* Kassel University Press GmbH, Kassel, Germany, pp. 105–108.

Schwartz, H.F., Gent, D.H., Fichtner, S.M., Hammon, R., Cranshaw, W.S., Mahaffey, L., Camper, M., Otto, K. and McMillan, M. (2009) Straw mulch and reduced-risk pesticide impacts on thrips and iris yellow spot virus on western-grown onions. *Southwestern Entomologist* 34, 13–29.

Shelton, A.M., Wilsey, W.T. and Schmaedick, M.A. (1998) Management of onion thrips (Thysanoptera: Thripidae) on cabbage by using plant resistance and insecticides. *Journal of Economic Entomology* 91, 329–333.

Singh, S.P. (1992) Studies on mulching of vegetable crops – a review. *Advances in Horticulture and Forestry* 2, 115–143.

Smith, H.A., Chaney, W.E. and Bensen, T.A. (2008) Role of syrphid larvae and other predators in suppressing aphid infestations in organic lettuce on California's Central Coast. *Journal of Economic Entomology* 101, 1526–1532.

ten Broeke, C.J.M., Dicke, M. and van Loon, J.J.A. (2013) Resistance to a new biotype of the lettuce aphid *Nasonovia ribisnigri* in *Lactuca virosa* accession IVT280. *Euphytica* 193, 265–275.

Theunissen, J. and Schelling, G. (1996) Pest and disease management by intercropping: suppression of thrips and rust in leek. *International Journal of Pest Management* 42, 227–234.

Theunissen, J. and Schelling, G. (2000) Undersowing carrots with clover: suppression of carrot rust fly (*Psila rosae*) and cavity spot (*Pythium* spp.) infestation. *Biological Agriculture & Horticulture* 18, 67–76.

Theunissen, J., Booij, C.J.H. and Lotz, L.A.P. (1995) Effects of intercropping white cabbage with clovers on pest infestation and yield. *Entomologia Experimentalis et Applicata* 74, 7–16.

Thoeming, G., Poelitz, B., Kuehne, A. and Saucke, H. (2011) Risk assessment of pea moth *Cydia nigricana* infestation in organic green peas based on spatio-temporal distribution and phenology of the host plant. *Agricultural and Forest Entomology* 13, 121–130.

Uvah, I.I.I. and Coaker, T.H. (1984) Effect of mixed cropping on some insect pests of carrots and onions. *Entomologia Experimentalis et Applicata* 36, 159–167.

Vänninen, I., Hokkanen, H. and Tyni-Juslin, J. (1999) Screening of field performance of entomopathogenic fungi and nematodes against cabbage root flies (*Delia radicum* L. and *D. floralis* (Fall.); Diptera, Anthomyiidae). *Acta Agriculturae Scandinavica B-Plant Soil Sciences* 49, 167–183.

Vernon, R.S. and Mackenzie, J.R. (1998) The effect of exclusion fences on the colonization of rutabagas by cabbage flies (Diptera: Anthomyiidae). *Canadian Entomologist* 130, 153–162.

Vidal, S. (1997) Factors influencing the population dynamics of *Brevicoryne brassicae* in undersown Brussels sprouts. *Biological Agriculture & Horticulture* 15, 285–295.

Volkl, W. and Stechmann, D.H. (1998) Parasitism of the black bean aphid (*Aphis fabae*) by *Lysiphlebus fabarum* (Hym., Aphidiidae): the influence of host plant and habitat. *Journal of Applied Entomology* 122, 201–206.

Wilson, R.G., Orloff, S.B. and Taylor, A.G. (2015) Evaluation of insecticides and application methods to protect onions from onion maggot, *Delia antiqua*, and seedcorn maggot, *Delia platura*, damage. *Crop Protection* 67, 102–108.

Winkler, K., Wäckers, F.L., Kaufman, L.V., Larraz, V. and van Lenteren, J.C. (2009) Nectar exploitation by herbivores and their parasitoids is a function of flower species and relative humidity. *Biological Control* 50, 299–306.

Witzgall, P., Bengtsson, M., Karg, G., Backman, A.C., Streinz, L., Kirsch, P.A., Blum, Z. and Lofqvist, J. (1996) Behavioral observations and measurements of aerial pheromone in a mating disruption trial against pea moth *Cydia nigricana* F (Lepidoptera, Tortricidae). *Journal of Chemical Ecology* 22, 191–206.

Wyss, E. and Daniel, C. (2004) Die Wirksamkeit von Einflugbarrieren gegen die Besiedelung von Broccoli und Kohlrabi durch die Kohldrehherzgallmücke *Contarinia nasturtii* (Diptera: Cecidomyiidae). *Mitteilungen der Deutschen Gesellschaft für Allgemeine und Angewandte Entomologie* 14, 387–390.

Zhao, J., Collins, H.L., Gusukuma-Minuto, L., Mau, R.F.L., Thompson, G.D. and Shelton, A.M. (2002) Monitoring and characterization of diamondback moth resistance to spinosad. *Journal of Economic Entomology* 95, 430–436.

Appendix

Table 17.1. Summary of strategies and control methods for the most important pests of organic field vegetables in temperate areas.

Crop	Pest	Prevention	Antagonists	Selective control	Non-selective control
Allium crops	Leaf miners *Napomyza gymnostoma*	Removal of postharvest residues and crop debris; isolation of onion and leek fields from chive crops; variety choice; trap cropping			Fine-mesh netting (0.8 mm), spinosad
	Onion maggot *Delia antiqua*	Flower strips; untilled refuge strips; choice of selective pesticides; removal of crop debris and volunteer onions; crop rotation; isolation from older crops and compost piles; delayed planting; mulches to prevent egg laying	Parasitoids, staphylinid and carabid beetles, pathogens, nematodes	Entomopathogenic nematodes	Fine-mesh netting, spinosad seed treatments, spinosad
	Leek moth *Acrolepiopsis assectella*	Crop rotation; removal of postharvest residues; flower strips; delayed planting; early harvest; removal of infested leaves from the harvested crop	Parasitoids	*Bacillus thuringiensis* (*Bt*) *aizawai*	Fine-mesh netting (1.4 mm), spinosad
	Thrips tabaci	Overhead irrigation; tolerant varieties; flower strips to attract natural enemies; avoidance of non-selective insecticides; mixed cropping of leek and celery; removal of postharvest onion debris	Predators (thrips, bugs, mites, lacewings), entomopathogenic fungi, entomopathogenic nematodes		Spinosad, pyrethrin, kaolin
Brassica vegetables	Cabbage root maggot *Delia brassicae*	Adapted irrigation; removal of decaying organic material; low density of cabbage and oilseed rape production in an area; stimulating growth of lateral roots; flower strips; destroy harvest residues	Ground-dwelling predators (carabids, staphylinids, spiders)	Drench application of neem	Fine-mesh netting (2 mm), drench application of spinosad, seed treatments with spinosad, pyrethrin
	Flea beetles *Phyllotreta* sp.	Increased irrigation; low density of cabbage and oilseed rape production in an area; destruction of postharvest residues; undersowing with green manure	Parasitoids, entomopathogenic nematodes	Neem	Fine-mesh netting (0.8 mm), silicate rock dusts, kaolin, spinosad, pyrethrin

	Lepidoptera *Plutella xylostella*, *Pieris* sp., *Trichoplusia ni*, *Mamestra brassicae*	Soil cultivation in early spring to destroy pupae; flower strips and companion plants; destruction of harvest residues	Parasitoids, predators (carabids, staphylinids, coccinelids)	*Bt*, *Beauveria bassiana*, neem	Fine-mesh netting (2 mm), pyrethrin, spinosad
	Aphids *Brevicoryne brassicae*, *Myzus persicae*	Avoidance of non-selective insecticides; flower strips; pest-free seedlings; irrigation after planting; destruction of postharvest residues	Predators (lacewings, ladybird beetles, syrphids, aphid midges, earwigs), parasitoids	Quassia, neem (against *M. persicae*)	Horticultural oils and soaps, pyrethrin, kaolin
	Thrips *Thrips tabaci*	Resistant/tolerant varieties; isolation from cereal and onion fields; overhead irrigation; destruction of postharvest residues	Predators (thrips, bugs, mites, lacewings), entomopathogenic fungi	*B. bassiana*, *Isaria fumosorosea*, neem	Spinosad, pyrethrin, horticultural oils and soaps
	Swede midge *Contarinia nasturtii*	Crop rotation; site selection; control of brassica weeds; tolerant varieties; destruction of postharvest residues	Parasitoids	Vertical barriers (fences)	Fine-mesh netting (0.8 mm), spinosad
	White fly *Aleyrodes proletella*	Isolation from other brassica and rapeseed crops	Parasitoids (*Encarsia tricolor*), *Harmonia axyridis*	*Encarsia tricolor*, neem	Fine-mesh netting (0.8 mm), pyrethrin
Carrots and other *Apiaceae*	Carrot rust fly *Psila rosae*	Isolation from sources of infestation; crop rotation; destruction of postharvest residues; early sowing; adapted harvest times; trap crops		Vertical barriers (fences)	Fine-mesh netting (1.4 mm)
Carrots	Carrot psyllid *Trioza apicalis*	Late sowing	Carabids, staphylinids, spiders		Fine-mesh netting (2.4 mm)
Celery	Two-spotted spider mite *Tetranychus urticae*	Overhead irrigation	Predatory mites		Horticultural oils and soaps
Cucurbits	Aphids *Aphis gossypii* and others	Remove volunteer virus host plants; avoid broad-spectrum insecticides; isolation from sources of infestation; reflective mulches	Predators (lacewings, ladybird beetles, syrphids, aphid midges, earwigs), parasitoids	Neem	Pyrethrin, horticultural oils and soaps

Continued

Table 17.1. Continued.

Crop	Pest	Prevention	Antagonists	Selective control	Non-selective control
	Cucumber beetles *Acalymma vittatum, Diabrotica undecimpunctata howardi*	Use transplants instead of direct seeding; perimeter trap cropping; tolerant cultivars, crop rotation; mowing of headlands	Tachinid flies, parasitoids, nematodes, carabid beetles, ground-dwelling spiders, ladybird beetles		Kaolin, fine-mesh netting, pyrethrin
	Squash bug (*Anasa tristis*) and other stink bugs	Remove crop debris; crop rotation; using transplants instead of direct seeding; create habitats for natural predators; perimeter trap cropping	Tachinid flies, carabid beetles, parasitoids, pathogens	Neem	Fine-mesh netting, pyrethrin
Legumes (and other crops)	Seedcorn maggot (*Delia platura, Delia florilega* and others)	Avoid decomposing organic matter (no fresh compost, no manure before planting, no recently incorporated harvest residues or cover crops); avoid potatoes or spinach as previous crops; repeated soil cultivation before sowing; shallow sowing in warm soil (delayed planting); re-seeding after intense soil cultivation	Ground-dwelling predators (carabids, staphylinids, spiders), parasitoids		Fine-mesh netting (2 mm), spinosad seed treatments
Legumes	Pea and bean weevil *Sitona lineatus*	Crop rotation; early sowing; additional hoeing	Ground-dwelling predators (carabids, staphylinids, spiders)		Fine-mesh netting, silicate rock dusts
Pea	Pea moth *Cydia nigricana*	Early sowing of rapidly maturing cultivars; isolation from sources of infestation	Parasitoids		Pyrethrin, spinosad
Legumes, Pea	Pea aphid *Acyrthosiphon pisum*	Isolation from overwintering fields of lucerne, vetch or clover; isolation from soybean fields; flower strips	Predators (lacewings, ladybird beetles, syrphids, aphid midges, earwigs), parasitoids	Neem	Horticultural soaps
Bean	Black bean aphid *Aphis fabae*	Increased irrigation	Polyphagous predators do *not* reduce numbers	Quassia	Horticultural soaps, pyrethrin

Lettuce	Aphids *Nasonovia ribisnigri*, *M. persicae*, *Macrosiphum euphorbiae*, *Uroleucon sonchi* and others	Resistant varieties; pest-free seedlings; flower strips	Parasitoids, predators	Quassia, neem	Pyrethrin, horticultural oils and soaps, fine-mesh netting
	Lettuce root aphid *Pemphigius bursarius*	Resistant varieties; remove poplar trees (winter host) in proximity; increased irrigation			
Nearly all vegetable crops	Crane fly *Tipula* sp.	Delayed seeding after green manuring; soil cultivation	Soil-dwelling predators, birds		
	Cutworms *Agrotis* sp.	Repeated soil cultivation; irrigation	Ground-dwelling predators, parasitoids, enthomopathogenic nematodes	*Steinernema carpocapse*, *Bt*	Spinosad (bait formulations)
Many vegetables (lettuce, spinach, beans)	Noctuids *Autographa gamma*, *Mamestra oleracea* and others	Wild flower strips; companion plants	Parasitoids, predators	*Bt aizawai*	Pyrethrin, spinosad

18 Pest Management in an Organic Tea Plantation

Qiang Xiao*
Tea Research Institute, Chinese Academy of Agricultural Sciences, Hangzhou, China

Introduction

Tea (*Camellia sinensis* (L.) O. Kuntze) is an important global beverage, which originated in China, the largest tea-producing country in the world. In 2013, more than 1.945 million ha of tea was planted in China producing about 1.89 million t (FiBL and IFOAM, 2014). Organic tea originated in 1983 and was first certificated in 1990 in China (Xu *et al.*, 2000). So far, the acreage of certificated organic tea in China is more than 53,000 ha (FiBL and IFOAM, 2014). Tea is an intensively managed perennial monoculture crop cultivated in large- and small-scale plantations located between latitudes 41°N and 16°S (Ye *et al.*, 2014). The warm and humid climate in this area is very conducive to the occurrence of pests of tea. According to incomplete statistics, more than 1000 arthropod species are associated with tea around the world (Hazarika *et al.*, 2009). These pests have influenced tea yields and quality. Therefore, the management of tea pests is an important measure to ensure the quality of organic tea production.

Principles of Pest Control in Organic Tea

Tea is a perennial plantation crop of a shrub that remains evergreen throughout the seasons. Utilizing current cultivation techniques and management models, a relative steady microclimate, which makes its biological community structure more complex and the kinds and number of species more varied than any other ecosystem, can be created in a tea plantation. To some extent, these traits of a tea plantation, such as the complex biological community structure and abundant species, benefit the conservation of its biodiversity as well as its ecological balance. During a long cultivation, various pest management techniques including cultural, mechanical and biological control have consciously or unconsciously been successfully applied to pest control in tea gardens (Ye *et al.*, 2014). Meanwhile, these techniques have been improved according to different types of tea production, various pest characteristics and outbreak patterns, which can in turn

* xqtea@vip.163.com

help to maintain the sustainable development of tea production.

Although there are hundreds of pests in a tea garden, there are only a few dozen of them which are considered to be key pests that can cause damage up to economic threshold levels (ETLs). These pests always share some similar traits. First, the period of pest outbreak is usually consistent with the developmental stage of tea buds. Secondly, pest damage often exceeds the compensating abilities and tolerance of tea. Thirdly, the population scale of pest species often fluctuates around ETLs. In the pest management system of a tea garden, the economic loss caused by plant insect pests can be reduced as long as these key pests are controlled effectively.

Therefore, the principle of plant insect pest management in the organic tea garden is based on a good understanding of the particular ecological environment of the tea plantation while respecting nature. Pest control in a tea plantation utilizes the basic methods of ecology, making good use of the naturally regulated system that exists in which tea trees are the subjects and the tea plantation environment is the background. Cultural control methods are used as the main measure while physical and biological control methods are adopted as auxiliary ones. Additionally, none of the chemical pesticides that are allowed according to organic agricultural production standards are used for pest control in a tea plantation, in order to ensure the healthy growth of the tea tree.

Pests and their Damage in Tea

One thousand and thirty-four arthropods are associated with the tea plant and 430 of them are common species in China (Chen and Chen, 1989). According to the feeding mode and part of the plant that is injured, insects can be divided into four major groups: (i) sap-sucking pests; (ii) leaf-eating pests; (iii) wood-boring pests; and (iv) underground pests. The sap-sucking pests and leaf-eating pests have the biggest influence on tea production.

Sap-sucking pests

Sap-sucking pests suck juices up from the tea plant using their specially adapted mouthparts, leading to atrophy of new shoots, leaf etiolation and growth of the plant being arrested. The main species include the tea leafhopper, mites and scale insects.

The tea leafhopper (*Empoasca* spp.) is a serious insect pest that is widely distributed in tea-growing areas. Adults and nymphs lead to the dehydration and wilting of tea shoots, slow growth and edge-burnt leaves. The tea leafhopper has more than ten generations a year in most tea gardens. All stages of the tea leafhopper are present simultaneously in the emergence period, and these generations obviously overlap. Adults and nymphs mostly inhabit the underside of tender leaves, and then move to the immature stems. The pest overwinters as adults on the tea plants, weeds or other crops. The tea leafhopper severely reduces tea production in summer and autumn, resulting in crop losses from 10% to 15%, and even exceeding 50%.

Mites, as a group, are one of the most serious and persistent pests of tea. The main species of mites include *Acaphylla theae* Watt, *Polyphagotarsonemus latus* (Banks) Ewing, *Calacarus carinatus* Green and *Oligonychus coffeae* Nietner. *A. theae* and *P. latus* both destroy tender leaves. The former produces a rusty spot on the back of leaves, the latter produces brown stripes and makes the underside of leaves rough. *C. carinatus* destroys mature leaves, making leaves turn from green to purple bronze. *O. coffeae* destroys mature and old leaves, changing leaves to dark red and white crumbs. Mites are more prone to occurring in hot and humid seasons. Picking and pruning can inhibit the occurrence of mites.

Scale insects have between one and three generations in a year, easily occurring in tea gardens in the dense bushes. More than 30 kinds of scale insects have been recorded and the main species include *Lopholeucaspis japonica* Cockerell, *Ceroplastes pseudoceriferus* Green and *Ceroplastes rubens* Maskell. Nymphs and female adults of scale insects stay on the stems and leaves, secreting waxes and sucking the tea plant juice. The damage they cause includes reducing tree vigour and affecting tree growth.

Leaf-eating pests

Leaf-eating pests, a more threatening and fulminant pest, damage by feeding on the leaves of tea, and mainly include Lepidoptera and Coleoptera.

Lepidoptera

More than 300 species of Lepidoptera have been recorded, of which the vast majority are leaf-eating pests. Lepidoptera pests pass through four different developmental stages (the egg, larva, pupa and adult) and feed on tea leaves and tender buds, severely harming all leaves. Among the common kinds are the tea geometrid (*Ectropis obliqua* Wehrli), *Euproctis pseudoconspersa* Strand, *Iragoides fasciata* Moore, *Adoxophyes orana* Fischer von Roslerstamm and *Caloptilia theivora* Walsingham. The tea geometrid feeds on tender leaves of tea bushes, and then feeds on mature leaves as the larval age increases. *E. pseudoconspersa* and *I. fasciata* feed on mature tea leaves, and their bodies have toxic hairs that cause skin redness and itchiness that affects picking and other field operations. The larvae of *A. orana* and *C. theivora* roll up the leaves and feed on the leaves inside.

Coleoptera

Myllocerinus aurolineatus is a major kind of coleopteran pest. The larvae of *M. aurolineatus* feed on the roots of tea trees and weeds in the soil while the adults feed on tender leaves seriously affecting tea production and quality. One generation of *M. aurolineatus* occurs in a year, the adults coming out of the ground one after another, and they fall to feign death if the tea trees are shaken; the adults feed on the surface of tea bushes in the early morning and after dusk. Tea garden cultivations, climatic conditions and natural enemy species have an effect on the occurrence of *M. aurolineatus*.

Wood-boring pests

Wood-boring pests damage tea plants by drilling into stems and fruits. The main stem-boring pests are *Casmara patrona* Meyrick, *Linoclostis gonatias* Meyrick, *Zeuzera coffeae* Nietner, *Parametriotes theae* Kusnetzov and *Chreonoma atritarsis* Picard, while the main fruit-boring pest is *Curculio chinensis* Chevrolat. Wood-boring pests have one or two generations/year and overwinter as aged larvae in tea stems or in the soil. As tea is an evergreen shrub grown in order to harvest mainly fresh leaves, the damage caused to tea production by wood-boring insects is not obvious.

Underground pests

These insect pests live underground and feed on the roots of tea plants. Familiar ones are *Anomala corpulenta* Motschulsky, *Agrotis ypsilon* Rott and *Odontotermes formosanus* Schiraki. Such pests are mostly omnivorous insect pests and as well as damaging tea in the tea garden they can also damage a variety of other trees and crops.

Control Methods for Tea Pests

Cultural control

Cultural control refers to the method of prevention and control of tea pests through various cultivation management measures, which is the foundation of the prevention and control of pests in organic tea production. It is a main measure during the process of tea production based on the cultivation management of the tea tree, as well as the key method of prevention and control of tea pests. It decreases pest species by changing their living environment and forming adverse conditions for pests to exist and reproduce offspring. The specific measures for cultural control include: (i) improving the agroecological environment; (ii) using tea varieties that are resistant to pest attack; (iii) pruning and picking; and (iv) overall management of the tea plantation.

Agroecological environment

The tea plantation and the ecological environments around it determine the biodiversity of the tea plantation and the degree of occurrence of tea pests. A good ecological environment is one that is suitable to maintain biodiversity and increase the regulating ability to resist pests. However, large areas of tea are grown as a monoculture which definitely simplifies the biological structure and decreases the community diversity, leading to a higher incidence of pest occurrence and spread. For the organic tea plantation, it is very necessary to increase the vegetation in the neighbourhood of the tea plantation by adopting afforestation – planting street trees, shade trees and windbreak trees. For some tea plantations, it is recommended to move tea plantations back to the forests and to redistribute crops so that a more complex ecological system is created in the neighbourhood of the tea plantation, improving the ecological environment of the tea plantation and making the regulating ability of nature more stable.

Host plant resistance

Host plant resistance could reduce the pest damage to tea trees. Different tea varieties differ in leaf morphology and structure, tree shade and amino acid content that would affect the occurrence and damage caused by tea pests such as *Empoasca vitis* Gothe and mites (Chen *et al.*, 1996; Huang and Zhang, 1998; Liu *et al.*, 1999). A series of resistant tea varieties have been bred. Resistance to pests in tea breeding is the result of natural adaptation to pathogenic microorganisms and pest communities over a long period of evolution. The best varieties of tea trees are cultivated using methods of selection, hybridization and selective breeding (Wang and Yang, 2003). The variety with the strongest resistance against the main local pests should be selected when planting the new tea garden or changing tea varieties. Additionally, it would be selected and be collocated with other different cloned tea varieties when cultivating a large new tea plantation, which could prevent the breakout and prevalence of tea pests due to the change of the resistance of tea varieties or the adaptation of pests (Gong *et al.*, 2008).

Pruning and picking

As the material of picked tea, the bud of tea trees is also the nutrient source of tea pests. Due to this fact, timely pruning and picking can ensure the quality of tea, and meanwhile can also reduce the nutritional condition of tea pests and destroy their eggs. As tea leaves are picked, pest eggs in the buds can be removed, which obviously reduces the damage caused by tea pests such as *Toxoptera aurantii* Boyer, *Empoasca pirisuga* Matumura, *C. theivora*, *P. latus* and *A. theae*. In practical operation, buds with pests should be prioritized to be picked and pruned. When spring comes early, pruning and picking should be started early accordingly, while few bugs and leaves should be removed in summer due to the frequent occurrence of pests, and removing the top of tea trees and postponing shutting up the tea plantation are adopted if large amounts of pests appear in autumn.

Proper pruning and picking can promote the growth of tea trees, enlarge the area of plucking, and control the pests effectively as well. Pruning is an effective method to prevent and control the pests on the leaves such as aphid, *E. pirisuga* and *E. obliqua* (Du *et al.*, 2003; Zeng *et al.*, 2010b). Measures of heavy pruning should be taken when the tea plantation is seriously damaged by pests or growth is not good. When the tree plants become too close to one another the leaves and stems are cut right back to ensure good ventilation and inhibit the occurrence of pests.

Physical control

Physical control based on pest behavioural habits of aggregation, orientation and feeding preference has been developed to monitor and control pests in the tea plantation. The following are measures of physical control that would be applied to control tea pests: (i) lamp traps; (ii) artificial capture;

(iii) coloured sticky plates; and (iv) sex pheromone traps.

Lamp traps

Some insects are strongly positively phototactic, which can be used in pest management by using different kinds of lamp traps. There were three main types of lamp traps: (i) black-light traps; (ii) frequency-vibrated lamps; and (iii) solar lamps (Xu *et al.*, 2013). Luo *et al.* (2006) showed that the application of frequency-vibrated lamps in a tea plantation could successfully trap 68 kinds of pests distributed across 26 families of seven orders, especially lepidopteran and hemipteran pests such as the tea tussock moth, *E. pseudoconspersa*, the tea geometrid, *E. obliqua*, the tea lesser leafhopper, *E. pirisuga* and the citrus spiny whitefly, *Aleurocanthus spiniferus*. Using different types of lamps could attract different kinds of pests (Zeng *et al.*, 2010a). Lamp trapping could also be used to monitor and forecast the pest population in an organic tea plantation (Song *et al.*, 2005). However, lamp trapping cannot differentiate between pests and their natural enemies, so it seems extremely important to avoid the peak density of natural enemies while using lamp trapping (Zeng *et al.*, 2010a).

Artificial capture

Artificial capture is also an effective method of reducing the population of pests in an organic tea plantation. It is suitable to control pests such as large and sluggish pests, pests that aggregate or pests that have death-feigning behavioural habits. For example, the larvae of *E. pseudoconspersa* or *Andraca bipunctata* which have aggregation behavioural habits could be controlled by pruning the branch where they have aggregated and putting them into 1% soap solution. Searching for and removing the bursae of bagworm moths and rolled leaves containing moths are also essential for reducing their populations. Placing plastic film under the tea crop and patting the tea bushes can collect and kill death-feigning weevils.

Coloured sticky plates

Coloured sticky plates were devised according to the preference of insects for different colours. Yellow sticky plates are the most commonly used in an organic tea plantation. Using jasmine yellow sticky plates can efficiently attract and stick the tea leafhopper (*Empoasca* spp.), the citrus spiny whitefly (*Aleurocanthus spinifeus*) and *Ricania speculum* (Lin *et al.*, 2009). Zhao *et al.* (2001) showed that the tea leafhopper had a preference for yellow green and pale green.

Sex pheromone traps

Sex pheromones, that are generally biosynthesized and released by females and perceived by the conspecific males, play an important role in mediating insect mating behaviour. Thus taking full advantage of species-specific sex pheromone to disrupt the ratio of females and males in the field could make a large contribution to pest management. So far, the main components of sex pheromones of some serious moths found in tea gardens such as *E. obliqua*, *E. pseudoconspersa* and *Adoxophyes honmai* have been identified (Table 18.1). These specific, ecologically friendly sex pheromone products not only play a significant role in trapping male adults (Ge *et al.*, 2002), but they also function as isotropic agents involved in disturbing mating communication of conspecific males and females (Wang *et al.*, 2006). Monitoring and forecasting of devastatingly fulminic tea pests such as the tea geometrid, *E. obliqua* is a crucial aspect of the pest management system. Making full use of species-specific sex pheromones could forecast the exact occurrence period and amount of specific pests and effectively control the damage caused by them.

Biological control

Biological control is the method that controls pests with entomophagous insects, parasitic insects, pathogenic microorganisms or biological metabolites. Compared with other agricultural ecosystems, the tea

Table 18.1. Major component of sex pheromones of the main lepidopteran pests in a tea garden.

Pest	Main components of sex pheromone	Reference
Adoxophyes honmai Yasuada	Z9-TDA, Z11-TDA, *E*11-TDA, 10-Me-DDA	Tamaki and Noguchi (1984)
Homona magnanima Diakonoff	Z11-TDA, Z9-DDA, 11-DDA	Noguchi *et al.* (1979)
Ectropis obliqua Wehrli	Z3Z6Z9-18Hy, Z3Z9-6, 7epo-18Hy, Z3Z6Z9-22Hy, Z3Z6Z9-19Hy, Z3Z6Z9-24Hy	Yin *et al.* (1993), Liu *et al.* (1994), Yang (2009)
Euproctis pseudoconspersa Strand	(R)-10Me14Me-15:iBu, 14Me-15:iBu	Wakamura *et al.* (1994)
Caloptilia theivora Walsingham	*E*11-HDDA, Z11-HDDA	Noguchi *et al.* (1979)

plantation with its relatively stable ecological environment has more significant advantages, whether using its own natural enemy resources or releasing natural enemies in the field. Therefore, biological control is the preferred method used in organic tea pest management and includes: (i) protection and utilization of natural enemy resources; (ii) release of pathogenic microorganisms; and (iii) application of botanical and mineral pesticides.

Protection and utilization of natural enemy resources

As it is a relatively closed ecological environment with a wide variety of insect groups, the tea plantation is a beneficial environment for natural enemies in which to settle and reproduce. According to surveys, there are approximately 1100 species of natural enemies in the tea plantation, including predatory and parasitic insects, predatory spiders, parasitic microorganisms and beneficial birds (Cao *et al.*, 2013). Predatory natural enemies such as spiders, ladybird beetles and lacewings are the largest and most important species of natural enemies in the tea plantation. Of the total number of predatory natural enemies, 80–90% are made up of these types of natural enemy, of which there are about 290 kinds of spider species (Chen *et al.*, 2000, 2004). Research shows that the occurrence of spider species follows a spatial relationship with the primary pests of the tea plant (i.e. spiders that predate a particular pest are found in greater numbers where there are high densities of the pest) (Zhou *et al.*, 2013). Similarly, the occurrence of parasitic natural enemies and entomogenous fungi in the tea plantation ecosystem could parasitize the main pests of tea such as *E. obliqua* and *Empoasca flavescens*, effectively controlling their occurrence and development in the long term (Liang, 1981; Chen and Huang, 2001). Therefore, protecting natural enemy resources, which could play a full role in maintaining the ecological balance, should be the main method of biological control in the tea plantation.

There are various measures that could be taken to protect the natural enemies and these are described as follows. First, some measures, for example planting shelter trees and roadside shade trees (e.g. firs, palms and neem trees) around the tea plantation, intercropping the tea plant with fruit trees or interplanting with plants that can be used as a green manure in the young tea garden, are indispensable in creating favourable habitats for natural enemies of tea pests. In addition, laying grass between tea tree lines in summer and winter could also create suitable habitats and breeding sites for natural enemies. Secondly, damage caused to natural enemies should be reduced while undertaking agronomic operations. Clipped branches from pruning the tea tree should be first piled up near the tea plantation before removal, as this helps the natural enemies return to the tea plantation. Thirdly, providing some necessary nutrient resources for natural enemies can prolong their lifespan and improve their reproductive success. For example, cultivating nectariferous plants around the tea plantation that bloom at a different period could supply nutrition for some parasitic and predatory insects which feed by sucking the nectar, effectively improving their chance of survival (Lu *et al.*, 2005).

Release of pathogenic microorganisms

The stable ecological environment and suitable temperature and humidity of the tea garden are favourable for the survival and breeding of pathogenic microorganisms, so the application of pathogenic microorganisms to control tea pests has made great progress. For example, mortality of the tea leafhopper has reached about 49.5–68.4% by releasing *Beauveria bassiana* or *Isaria javanicus* in the field to control this leafhopper pest (Pu and Feng, 2004; Zhan *et al.*, 2012). As for pathogenic bacteria, *Bacillus thuringiensis* (*Bt*) is a good biological agent controlling larvae of lepidopteran pests and has been widely used in tea plantations.

Insect viruses are a very promising group of pathogenic microorganisms used in the biological control of pests in the tea plantation. Currently, there have been 81 kinds of virus isolated from tea pests, of which 45 species are the nuclear polyhedrosis virus (Hong, 1998). Among these kinds of virus, *Adoxophyes orana* Fischer von Roslerstamm granulosis virus (AoGV), *Ectropis oblique hypulina* Wehrli nuclear polyhedrosis virus (EoNPV) and *Euproctis pseudoconspersa* nuclear polyhedrosis virus (EpNPV) are widely applied, with the control mortality between 93% and 100% in the field (Du *et al.*, 1984; Yin *et al.*, 2003). A field survey indicated that the virus still plays a role in the natural balance in the field several years after its initial release (Du *et al.*, 1984). Now EoNPV and EpNPV mixed with *Bt*, respectively, have been registered as biological control agents in China and are being produced commercially in factories (Xu *et al.*, 2005; Leng *et al.*, 2007).

Application of botanical and mineral pesticides

Botanical or mineral pesticides, which have active ingredients that were isolated from plants or minerals, have been registered as products for pest control. These products can be applied to control pests in an organic tea plantation after the products have been audited and certified by the organic certification body. There are several kinds of botanical or mineral pesticides used in organic tea plantations, such as azadirachtin, rotenone, matrine, mineral oil and lime sulfur. Botanical pesticides generally can be used to control the lepidopteran pests such as the black tussock moth and the tea caterpillar, with the control mortality around 80% (Wang *et al.*, 2013). Mineral oil is mainly used to control mites in the tea tree, with its control mortality against *A. theae* reaching 90% (Ning *et al.*, 2004). Normally lime sulfur would be used in winter to reduce the occurrence of pests in the coming year.

Conclusion

With the worldwide popularity of the beverage of tea, the production of tea will develop constantly. The consumption of organic tea, which has developed as a symbol of 'safe' tea, will increase year by year. The development of pest control in an organic tea plantation is closely related to organic tea production. To accommodate the development of the organic tea plantation, the control of tea pests in the future will be preceded by the following measures.

The first measure is to take full advantage of the ecological function of natural factors in the tea garden ecological system, protecting and utilizing natural resources such as natural enemies, pathogenic microorganisms and so forth, enhancing the control effect of natural disaster, and establishing a sustainable and effective ecological control mechanism.

The second measure is to use modern information technologies such as remote sensing, geographic information system and GPS (Global Positioning System) to dynamically research the damage caused by tea pests, construct information recognition patterns, achieving the aim of real-time pest monitoring and forecasting, and improving the overall level of pest monitoring.

The third measure is to research and develop biological pesticides such as microbial pesticides, botanical pesticides, mineral pesticides and other new products and technologies with the purpose of meeting the demands of the organic tea garden pest management.

References

Cao, D., Zhu, L. and Gai, X. (2013) Research advances on resources and function of natural enemies in tea garden. *Shandong Agricultural Sciences* 45(12), 117–122.

Chen, H., Xu, M. and Chen, X. (1996) On the resistance mechanisms of tea clones to pink rust mite. *Acta Phytophylacica Sinica* 23, 137–142.

Chen, Y. and Huang, M. (2001) Biological control progress of *Ectropis obliqua* Prout. *Natural Enemies of Insects* 23, 181–184.

Chen, Y., Song, C. and Liu, L. (2000) Studies on species of spiders in tea garden in China. *Journal of Tea Science* 20, 59–66.

Chen, Y., Chen, Z. and Song, C. (2004) Review on the investigation and protection measurement of spiders in Chinese tea gardens. *Acta Archaeologica Sinica* 13(2), 125–128.

Chen, Z. and Chen, X. (1989) An analysis on the world tea pest fauna. *Journal of Tea Science* 9, 13–22.

Du, D., Wu, Y. and Gan, Z. (1984) Study on the utilization of *Adxophyes orana* granulosis virus in tea plantation. *Journal of Tea Science* 4(1), 39–44.

Du, X., Dong, M. and Zhang, Y. (2003) The effect of pruning tea trees on controlling *Empoasca vitis* in the production of organic oolong tea. *Fujian Tea* 2, 22–23.

FiBL and IFOAM (2014) *The World of Organic Agriculture. Statistics and Emerging Trends.* Research Institute of Organic Agriculture (FiBL), Frick, Switzerland and International Federation of Organic Agriculture Movements (IFOAM), Bonn, Germany.

Ge, F., Chen, X. and Wang, C. (2002) The control effectiveness of tea tussock moth (*Euproctis pseudoconspersa*) with synthetic sex pheromone by mass trapping. *Journal of Tea Science* 22(2), 115–118.

Gong, X., Yu, Y. and Xiao, B. (2008) Effects of different cultivating modes of tea gardens on environment and tea quality. *Acta Botanica Boreali-Occidentalia Sinica* 28, 2485–2491.

Hazarika, L.K., Bhuyan, M. and Hazarika, B.N. (2009) Insect pests of tea and their management. *Annual Review of Entomology* 54, 267–284.

Hong, B. (1998) Pest virus resources in Chinese tea gardens. *Journal of Tea* 24, 82–84.

Huang, Y. and Zhang, J. (1998) Anatomical characteristics of leaf structure of tea plant resistant to leafhopper (*Empoasca vitis* Gothe). *Journal of Tea Science* 18, 35–38.

Leng, Y., Xiao, Q. and Yin, K. (2007) The effect specialty of *Ep*NPV-Bt preparation. *Acta Phytophylacica Sinica* 34, 177–181.

Liang, Z. (1981) Entomogenous fungi from the pests of tea plants. *Acta Phytopathologica Sinica* 11, 9–16.

Lin, J., Han, B. and Zhuo, X. (2009) Comparison of trapping efficacy of various colours for insects in tea gardens. *Acta Ecologica Sinica* 29, 4303–4316.

Liu, T., Li, Z., Luo, Z., Jiang, Y. and Yao, E. (1994) Synthesis of some bioactive components on *Ectropic obliqua* sex pheromone. *Acta Scientiarum Naturalium Uniwersitatis Nankaiensis* 4, 82–62.

Liu, Y., Xu, Z. and Zhou, Z. (1999) Morphological and biochemical parameters of tea varieties resistant to *Polyphagotarsonemus latus* Banks. *Journal of Sichuan Agricultural University* 17, 187–191.

Lu, Y., Shi, X., Zhong, C., Wang, H., Chen, J., Yu, Y. and Yang, Y. (2005) Impacts of honeydew on the growth, fecundity and foraging behavior of natural enemies. *Chinese Bulletin of Entomology* 4, 379–385.

Luo, X.M., Li, S.H., Ren, Q., Luo, W. and Tang, Y. (2006) Analysis on the effect of frequency – vibrating insecticidal lamps on trapping and controlling pests in tea plantation of hilly tea area. *Journal of Hunan Agricultural University (Natural Sciences)* 2006(5), 86–88.

Ning, W., Pan, X. and Hu, M. (2004) Effects of petroleum oil for control *Acaphylla theae*. *Pesticide Science and Administration* 25(6), 21–23.

Noguchi, H., Tamaki, Y. and Yushima, T. (1979) Sex pheromone of the tea tortrix moth: isolation and identification. *Applied Entomology and Zoology* 2, 225–228.

Pu, X. and Feng, M. (2004) Efficacy of emulsifiable formulations of two entomopathogenic fungi against small green leafhoppers on tea plant. *Chinese Journal of Applied Ecology* 15(4), 619–622.

Song, C., Lan, J. and Xu, H. (2005) Using the jiaduo insect killer lamp to monitor and control the tea pests. *Chinese Bulletin of Entomology* 42, 324–325.

Tamaki, Y. and Noguchi, H. (1984) Biological activities of analogues of 10-methyldodecyl acetate, a sex pheromonal component of the smaller tea tortrix moth (*Adoxophyes* sp., Lepidoptera: Tortricidae). *Applied Entomology and Zoology* 19, 245–251.

Wakamura, S., Yasuda, T., Ichikawa, A. and Wasuda, B. (1994) Sex attractant pheromone of the tea tussock moth, *Euproctis pseudoconspersa* Strand (Lepidoptera: Lymantriidae): identification and field attraction. *Applied Entomology and Zoology* 29, 403–411.

Wang, Q., Wang, D. and Wu, G. (2013) Research advances on *Empoasca vitis* (Gothe) in tea trees in China. *Fujian Journal of Agricultural Sciences* 28, 615–623.

Wang, X. and Yang, Y. (2003) Research progress on resistance breeding of tea plant. *Journal of Tea Science* 23(2), 94–98.

Wang, Y., Ge, F. and Liu, X. (2006) Field experiments for controlling the tea tussock moth, *Euproctis pseudoconspersa,* by mating disruption with sex pheromone. *Chinese Bulletin of Entomology* 43(1), 60–63.

Xu, J., Xiao, Q., Liu, Q., Tang, M., Yin, K., Zhang, A. and Zhu, S. (2005) The research and application of biological pesticide *Eo*NPV-Bt. *Jiangsu Agricultural Science* 6, 56–58.

Xu, Y., Han, W. and Shi, Y. (2000) Key technologies and development prospect of organic tea. *Journal of Tea* 26, 11–13.

Xu, Y., Wang, H. and Wang, H. (2013) The technology of tea plants pruning and its reasonable application. *Journal of Anhui Agricultural Sciences* 41, 7439–7441.

Yang, Y. (2009) *Extraction and Identification of Pheromone of Ectiopis obliqua Prout and Correlative Studies on its Biology*. Anhui Agricultural University, Hefei, Anhui, China.

Ye, G., Xiao, Q., Chen, M., Chen, X., Yuan, Z., Stanley, D.W. and Hu, C. (2014) Tea: biological control of insect and mite pests in China. *Biological Control* 68, 73–91.

Yin, K., Hong, B., Shang, Y., Yao, E. and Li, Z. (1993) The biology research of sex pheromone of *Ecotripis obliqua*. *Natural Science Progress* 3, 332–338.

Yin, K., Chen, H. and Tang, M. (2003) Studies on application techniques of *Eo*NPV preparations. *Virologica Siniga* 18, 492–495.

Zeng, M., Liu, F. and He, Y. (2010a) Population monitoring of the major tea pests under pest forecast lamp and evaluation of killing effects under frequency vibrational lamp and light emitting diode lamp. *Chinese Agricultural Science Bulletin* 26(19), 246–252.

Zeng, M., Liu, F. and Wang, D. (2010b) Effects of light pruning on arthropod population and community diversity in tea plantations. *Fujian Journal of Agricultural Sciences* 25, 623–626.

Zhan, M., He, L. and Chen, M. (2012) Screening for a highly virulent entomogenous fungal strain to *Empoasca vitis* (Homoptera: Cicadellidae) and its field efficacy. *Chinese Journal of Biological Control* 28(1), 41–46.

Zhao, D, Chen, Z. and Chen, J. (2001) Study on preference of green leafhopper *Empoasca vitis* (Göthe) for different colors. *Journal of Tea Science* 21(01), 78–80.

Zhou, X., Bi, S., Huang, B., Ke, S., Zou, Y., Shi, X., Ke, L., Yang, L., Guo, H. and Lin, Y. (2013) Spatial relationships among main natural enemies and four insect pests in tea plantations. *Journal of South China Agricultural University* 4, 489–498.

19 Insect Pests of Coffee and their Management in Nature-friendly Production Systems

Juan F. Barrera*
El Colegio de la Frontera Sur (ECOSUR), Tapachula, Mexico

Introduction

It is speculated that at least 700 years ago, coffee (*Coffea arabica* L.) came out of the Abyssinian Mountains (present day Ethiopia) in merchant and slave-trader caravans heading for the Arabian Peninsula. There, in Yemen, the first coffee plantations flourished. However, it was not until the end of the 17th and beginning of the 18th century that the Dutch and French began coffee cultivation in their colonies abroad, putting an end to the Arabians' monopoly, and *C. arabica* commenced the adventure of colonizing the world. Much later, in the 19th century, another coffee species, robusta (*Coffea canephora* Pierre ex A. Froehner), was discovered on the plains of western Africa (Coste, 1964; Haarer, 1964; Smith, 1985; Wrigley, 1988).

Of more than 100 species of the genus *Coffea*, all of which originated in Africa and Madagascar, *C. arabica* and *C. canephora* are the most important and are cultivated commercially. From roasted ground *C. arabica* beans, an aromatic, mild, digestive drink is obtained. Because the drink made from *C. canephora* is stronger, more bitter, less aromatic and has twice as much caffeine, it is used in mixtures or to make soluble instant coffee. Today, coffee is one of the world's most important agricultural products. For many developing countries, coffee has been the second most important product, only after oil, because of the value in foreign currency it generates. Coffee is the main source of income for more than 125 million people worldwide, many of whom belong to very low-income farming families of indigenous ethnic groups (Osorio, 2002). These families of small farmers produce 75% of the world's coffee. *C. arabica* is cultivated at middle and high altitudes in most of the producer countries, particularly in Latin America, while *C. canephora* is mostly cultivated at low altitudes in Africa and Asia where the climate and coffee rust (*Hemileia vastatrix* Berkeley & Broome) make it difficult to cultivate *C. arabica*. From an environmental standpoint, coffee cultivated under shade is considered very important for maintaining biodiversity (de Graaff, 1986; Perfecto *et al.*, 1996; Maestri *et al.*, 2001; Gaitán *et al.*, 2015).

According to the statistics of the International Coffee Organization (ICO), the world production of coffee in 2015 was

* jbarrera@ecosur.mx

143.3 million 60-kg sacks, of which 58% were *C. arabica* and 42% was *C. canephora* (ICO, 2016a). Of this production, 43% was produced in South America, 33% in Asia and Oceania, 12% in Mexico and Central America, and 12% in Africa. The top ten producer countries are (thousands of 60-kg sacks) Brazil (43,235), Vietnam (27,500), Colombia (13,500), Indonesia (12,317), Ethiopia (6700), India (5833), Honduras (5750), Uganda (4755), Guatemala (3400) and Peru (3300).

Unpredictable price fluctuation is one of the major problems of the coffee industry. Low prices drastically affect the incomes of the growers with repercussions on the welfare of their households and management of the plantations. Since the breakdown of the International Coffee Agreement in 1989, serious international price crises have forced producers to abandon their plantations and out-migrate (Osorio, 2004; Goodman, 2008). Price drops can be even more devastating in a context of adverse weather and high incidences of pests and diseases (Eakin *et al.*, 2013, 2014). Growers have adopted and adapted strategies to palliate the adverse effects of price crises. Among these strategies are selling certified coffee in special markets that pay premiums and better prices, such as Fair Trade, organic coffee and other forms of nature-friendly and socially responsible production (Giovannucci and Koekoek, 2003; Martínez-Torres, 2006; Bacon *et al.*, 2008; Blackman and Naranjo, 2012). In the 2007 International Coffee Agreement, the ICO encourages these production initiatives because they assure the crop's sustainability (ICO, 2016b).

Coffee (*Coffea* spp.) is a perennial bush of the family *Rubiaceae* cultivated in the tropical and subtropical regions of the world, in a belt between approximately 25° N and 25° S of the equator (Coste, 1964). Optimum conditions for the crop depend on the coffee species. *C. arabica* requires temperatures of 18–22°C and 1400–2000 mm of annual precipitation. For *C. canephora*, optimal conditions are 22–28°C and 2000–2200 mm precipitation (Gaitán *et al.*, 2015). Under these conditions, the coffee plants are evergreen, with one annual cycle of growth and fruiting in regions farther from the equator, such as India, Ethiopia, Hawaii, Central America, south and central Brazil and Zimbabwe. Regions closer to the equator, such as Kenya and Colombia, have up to two cycles per year, typically having a faster rate of initial flowering and a slower shoot growth rate in the dry and/or cold months of 'winter', while flowering and fast shoot growth is 'triggered' by 'spring' rains (Cannell, 1985).

Coffee cultivation under diversified shade is the most traditional cropping system and is the basis of nature-friendly production systems. In these agroforestry production systems, typically managed by small growers, tall *C. arabica* varieties such as 'Typica' and 'Bourbon' predominate and are highly regarded in the market. Besides coffee, growers obtain many benefits from the shade trees: lumber, firewood and fruit (Herzog, 1994; Beer *et al.*, 1998). The plantation itself benefits from leaf litter from the shade trees. First, the litter protects the soil from erosion and improves crop resilience during drought; it also limits proliferation of weeds (Guharay *et al.*, 2001). Later, the litter is incorporated into the soil as organic matter through processes of decomposition mediated by a myriad of macro and microorganisms, among which an abundant, diverse community of predators prospers and participates in regulating populations. Litter's nutrient contribution to the crop reduces dependence on chemical fertilizers, and natural control exerted by predator fauna that inhabits the soil can diminish the need for chemical control of pests and diseases. Aboveground, the shade also benefits the plantation by creating a microclimate that attenuates the impact of extreme temperatures and precipitations and by providing a favourable habitat for the flora and fauna of surrounding ecosystems. Shade also has a positive effect on the quality of brewed coffee, especially in suboptimal low-altitude regions (Muschler, 2001). Yield in these agroforestry systems is lower, but the low yield is compensated by increased longevity of the coffee plants (Beer *et al.*, 1998).

As Wrigley (1988) states, it is not surprising that the evergreen nature of the coffee plant, its broad geographic distribution and the favourable environmental conditions for development and reproduction of arthropods, has propitiated an abundant community of phytophagous species – up to 3000 worldwide, according to Waller *et al.* (2007) – despite the caffeine in leaves, flowers and seeds. Caffeine is an alkaloid that inhibits insect feeding and acts as an insecticide (Nathanson, 1984; Guerreiro-Filho and Mazzafera, 2003; Green *et al.*, 2015). What is surprising, in a superficial analysis, is that only a few dozen of these organisms are economically important, and of these, most are geographically very localized problems and few are cosmopolitan pests (Oduor and Simons, 2003; Barrera, 2008). A more detailed analysis can explain that the same conditions that have favoured phytophagous pests have also been beneficial for their natural enemies. This is especially true in coffee plantations under shade trees where organisms of different trophic levels coexist in a complex network of interactions and contribute to regulating phytophagous populations (Oduor and Simons, 2003; Vandermeer *et al.*, 2010; Perfecto *et al.*, 2014). Nevertheless, these interactions are highly vulnerable to the effect of chemical pesticides, which are very harmful for natural enemies. On the other hand, it is necessary to take into account that while increasing shade can be unfavourable for some pests and diseases, it can also favour others (Beer *et al.*, 1998; Guharay *et al.*, 2001; Oduor and Simons, 2003).

The objective of this chapter is to present the most important insect pests of coffee (*Coffea* spp.) worldwide, emphasizing their management in nature-friendly productive systems. The first part refers to these insects based on the classification of Waller *et al.* (2007), who propose four groups defined by the part of the plant they attack: (i) stem borers and branch borers; (ii) berry-feeding insects; (iii) insects that feed on buds, leaves, shoots and flowers; and (iv) root- and collar-feeding insects. The second part broaches some considerations on management of these insect pests.

Coffee Pests

Stem borers and branch borers

Coffee stem and branch borers form a group made up mostly of insects of the order Coleoptera, among which are outstanding members of the Bostrichidae (Bostrichinae), Cerambycidae (Cerambycinae, Lamiidae and Prioninae) and Curculionidae (Scolytinae) families. Other orders that include insects that attack stems and branches are Orthoptera (Gryllidae) and Lepidoptera (Cossidae). However, the families Cerambycidae and Curculionidae include the most economically important species.

The coffee tree attracts many species of Cerambycidae, especially in Africa; however, most of them are not specific to coffee (Vayssière, 1964). Among the species of Cerambycidae, *Xylotrechus quadripes* (Chevrolat) (Cerambycinae) is outstanding; it is distributed from South east Asia to India where it is the most important pest of *C. arabica*, although it also attacks *C. canephora* and *Coffea liberica* (Venkatesha and Dinesh, 2012; Thapa and Lantinga, 2016). *Bixadus sierricola* White and *Monochamus leuconotus* (Pascoe) (Lamiidae) are widely distributed in Africa (Oduor and Simons, 2003; Crowe, 2004; Waller *et al.*, 2007; Kutywayo *et al.*, 2013); and *Plagiohammus maculosus* (Bates) (Lamiidae) is distributed in Central America and Mexico (Barrera, 2008; Constantino *et al.*, 2014). Some Cerambycidae, such as *B. sierricola*, cause more damage by ringing branches and young plants, which causes their death, while others, such as the larva of *P. maculosus* penetrates the trunk at ground level and makes tunnels that can extend several tens of centimetres above ground, while others such as *M. leuconotus* bore underground through the root, killing young plants. The plants attacked by *P. maculosus* can be easily identified by the mound of sawdust at the base of the trunk. In Africa, larvae of *Neonitocris princeps* (Jordan), a less important insect pest of *C. arabica*, *C. canephora* and *Coffea excelsa*, make several holes along the primary branches of the coffee tree that look like the holes of a flute before penetrating the trunk. In addition, in Africa, *Dirphya nigricornis* Olivier causes similar damage but attacks only *C. arabica*. Besides coffee, most of the Cerambycidae attack other woody plants, but except for very few, such as *N. princeps*, other host plants are unknown (Waller *et al.*, 2007).

Insects of the Curculionidae family that attack coffee stems and branches belong to the subfamily Scolytinae. Distributed worldwide, *Xylosandrus compactus* (Eichhoff) is outstanding for its economic importance, while *Xylosandrus morigerus* (Blandford) is of lesser importance and can be found in South-east Asia, where it originated; in America, it has been reported from Mexico to Brazil. Both species of *Xylosandrus* perforate young coffee stems and branches. They are polyphagous and xylomycetophagous beetles whose attack is associated with branch breaking and rot. Their host is *C. canephora*, but attacks on *C. arabica* have also been reported (Vélez-Angel, 1972; Wood, 1982; Barrera *et al.*, 2002a; Waller *et al.*, 2007).

A relatively recent problem in Honduran coffee plantations is 'flute disease' caused by at least two species of crickets, *Paroecanthus* spp. (Orthoptera: Gryllidae) (Muñoz, 1990; Padilla and Rodríguez, 2000; Barrera, 2008). These crickets are nocturnal and during the day they hide in litter and weeds. The damage is caused by the female when she oviposits on stems and branches, making several orifices a few centimetres apart, generally in a straight line, so that the branch or stem looks like a flute, the reason for the common name of this problem. A female lays up to eight eggs in each orifice making an 'X', with two eggs at each end of the four incisions it makes under the bark. In Honduras, these insect populations can become very dense during the dry season in unshaded coffee plantations located between 900 m and 1600 m above sea level. During this time, many orifices can be observed on a single branch or stem causing physiological disorders resulting in yellowing foliage, which later falls together with the fruits. Very serious attacks can kill the entire plant. Among the cricket's natural enemies are egg parasitoids of the genus *Acmopolynema* (Hymenoptera: Mymaridae) in Honduras and Mexico.

Berry-feeding insects

The most outstanding insect of this group, the coffee berry borer *Hypothenemus hampei* (Ferrari) (Curculionidae: Scolytinae), is also the major coffee pest worldwide (Le Pelley, 1973; Barrera, 1994; Vega *et al.*, 2015). After more than 115 years of reports as a coffee pest in Gabon, *H. hampei* has invaded the world's most important coffee-producing countries. The coffee berry borer is a small black beetle that attacks the fruits of *C. arabica*, *C. canephora*, *C. liberica* and *C. excelsa*, but it prefers the first two of these. The females, whose wings – unlike the male's wings – are functional, are responsible for colonizing coffee berries. Generally, it attacks the tip of the berry where it makes a perforation to reach the coffee bean. Consequently, the young perforated berries can fall and the developing seed rots. When the coffee bean reaches approximately 20% of its dry weight (as of 110–140 days, depending on the altitude), the female enters, digs irregular galleries and periodically oviposits small groups of eggs (Barrera, 1994). When the eggs hatch, the larvae expand the galleries, voraciously eating the seed. After reaching the end of its development, the larva undergoes a brief pre-pupa stage before becoming a pupa. Its entire development occurs inside the fruit colonized by the mother, lasting 28–34 days, at 25.9–19.1°C. Adult females of the progeny are more numerous and almost twice the size of the males. Before abandoning the fruit where they were born, the females mate with their brothers. Mated females leave the fruit and search for new fruits to begin a new generation. In Mexico and Central America, towards the end of the coffee harvest, it stops raining and the dry season begins. In response to these environmental conditions, adult females modify their behaviour. They avoid leaving the fruits where they were born and a major accumulation of borers can be seen in unharvested black fruits.

The first rains of the year trigger their mass emergence (Baker *et al.*, 1992; Baker and Barrera, 1993).

In Africa, and to a lesser degree in South-east Asia, a group of bug species of the genera *Antestiopsis* (Hemiptera: Pentatomidae) is found that cause major damage to the flower buds, green branches and berries – especially green berries – of *C. arabica*, although they also attack *C. canephora* and *C. liberica* (Vayssière, 1964; Le Pelley, 1973, Waller *et al.*, 2007). Large quantities of insecticides are used against these pests (Oduor and Simons, 2003). On the African continent, their distribution is differential. For example, *Antestiopsis intricata* (Ghesquiere and Carayon) and *Antestiopsis orbitalis ghesquierei* Carayon are more important in West Africa, while *A. orbitalis bechuana* (Kirkcaldy) is more important in East Africa (Crowe, 2004). Other species present in Africa are *A. orbitalis orbitalis* (Westwood), *Antestiopsis facetoides* Greathead, *Antestiopsis crypta* Greathead, *Antestiopsis falsa* (Schouteden) and *Antestiopsis clymeneis galtei* (Frappa). *Antestiopsis cruciata* (F.), *Antestiopsis semiviridis* (Walker) and *Antestia partita* (Walker) are found in South-east Asia. When they feed, adult bugs can transmit fungi of the genus *Nematospora*, which are responsible for coffee bean rot (Le Pelley, 1973). Proliferation of shoots on the branch tips is another kind of damage associated with attack by these insects. In Uganda, besides coffee, other plants have been reported as hosts (Waller *et al.*, 2007).

Several species of fruit flies (Diptera: Tephritidae) attack coffee berries. The Mediterranean fly *Ceratitis capitata* (Widemann), a polyphagous pest, is distributed worldwide and is of major importance to fruit production, particularly attacking *C. arabica*. In Africa *Trirhithrum inscriptum* (Graham) is present in *C. arabica* and *Trirhithrum coffeae* Bezzi in *C. canephora* (Crowe, 2004). In America, three species of *Anastrepha* have been reported in coffee: (i) *Anastrepha fraterculus* (Widemann) (Waller *et al.*, 2007) in Central and South America; (ii) *Anastrepha obliqua* (Macquart), the mango fly, in the Caribbean and south/south-eastern Brazil (Le Pelley, 1973); and (iii) *Anastrepha sosorcula* Zucchi, reported only in Brazil (Aguiar-Menezes *et al.*, 2007). The larvae of these flies develop between the pulp and the seed in ripe berries. It does no damage to the coffee bean, but it is assumed that they can give coffee an unpleasant taste ('potato taste') by introducing microorganisms, such as bacteria, into the fruit (Crowe, 2004). In Brazil, flies of the genera *Neosilba* and *Dasiops* (Diptera: Lonchaeidae), such as *Neosilba bifida* Strikis and Prado, *Neosilba certa* (Walker), *Neosilba glaberrima* (Widemann), *Neosilba pendula* (Bezi), *Neosilba pseudopendula* (Korytkowsky and Ojeda) and *Dasiops rugifrons* Henning have been reported in coffee berries. However, of all these species, only *N. bifida* is a primary invader (Aguiar-Menezes *et al.*, 2007).

The berry moth *Prophantis smaragdina* (Butler) (Lepidoptera: Pyralidae) is widely distributed across sub-Saharan Africa, Madagascar, the Indian Ocean islands and some parts of Asia (Waller *et al.*, 2007). The larvae emerge from eggs that the adult oviposits individually, usually on green berries. Recently emerged larvae feed on small flowers and fruits, and as they grow they feed on larger berries. Finally, the larvae penetrate the berries where they complete their development. Attacked berries turn brown and the complete cluster of fruits is covered by a net of silk threads the larva excretes (Gaitán *et al.*, 2015).

Insects that feed on buds, leaves, green shoots and flowers

This is the most numerous group of insects associated with coffee, and of these, the species belonging to the orders Lepidoptera, Hemiptera and Coleoptera are the most numerous (Waller *et al.*, 2007). However, most of these species are not relevant as pests. Some members of this group are pests that appear sporadically and are limited to certain regions. As in the case of other insect groups that affect coffee, populations of these insects boom when environmental conditions are favourable or when management of the plantation harms the natural enemies that keep them under control. Among the pests of this group that deserve attention are leaf miners, particularly the species of the genus *Leucoptera* (Lepidoptera: Lyonetiidae), such as *Leucoptera meyricki* Ghesquiere, *Leucoptera caffeina* Washbourn and *Leucoptera coma* Ghesquiere in Africa and Asia (Yemen), and *Leucoptera coffeella* (Guerin-Meneville) in tropical America. Apparently, the only species that is strictly monophagous is *L. coffeella* (Waller *et al.*, 2007). The adult is a micro-lepidopteran, silver coloured with a strip of brown on the outer edge of the forewings that have dark spots on their distal tips. They lay eggs on coffee leaves in groups (*L. caffeina*) or dispersed (*L. meyricki* and *L. coffeella*). The larvae emerge from the egg and penetrate the leaf where they dig an untidy gallery or mine that can be confused with damage from certain diseases. Once it reaches its maximum development, the larva exits the gallery and spins a cocoon to pupate on the edge of a leaf or in the litter. These insects prefer coffee plantations located in low areas exposed to the sun. In many regions of Brazil, *L. coffeella* is often the most important pest of the crop, especially in the driest season of the year. The presence of few or no shade trees and intensive use of fungicides against *H. vastatrix* and insecticides against the miner, characterize these regions. In addition to the favourable abiotic conditions, the outbreaks of miners in this country can be explained by the pest's development of resistance to insecticides and by the chemicals that have decimated populations of natural enemies, which form the basis for their natural control (Reis and Souza, 1986; Matiello, 1991; Guedes and Oliveira, 2002). Numerous parasitoid species and predators have been identified as natural enemies of the coffee leaf miner, many acting as hyperparasitoids (Aranda-Delgado, 1986; Campos *et al.*, 1989; Waller *et al.*, 2007; Lomelí-Flores *et al.*, 2009). In East Africa at least 18 species of primary parasitoids of *L. meyricki* have been reported (Waller *et al.*, 2007), while in Mexico 22 species or morpho-species of *L. coffeella* parasitoids of the families Eulophidae and Braconidae have been collected. Of these, more than 83% of the specimens belong to five genera of Eulophidae: *Neochrysocharis*, *Stiropius*, *Closterocerus*, *Pnigalio* and *Zagrammosoma* (Lomelí-Flores *et al.*, 2009). In Colombia seven species of parasitoids of the Eulophidae family have been identified, among which *Closterocerus coffeellae* (Ihering) was the most abundant (David-Rueda *et al.*, 2016).

Several species of scales, mealybugs and aphids (Hemiptera) that inhabit the foliage can reach populations large enough to cause damage to coffee plants. These insects form dense colonies; their mouth apparatus is adapted for piercing and sucking sap, weakening the plant. They can cause the death of branches and even of the entire plant during drought or nutrient deficits (Waller *et al.*, 2007). Most are polyphagous, highly sedentary insects, often associated with ants, such as *Crematogaster*, *Camponotus* and *Azteca*, which benefit from the honeydew these hemipterans excrete abundantly. In return, the ants provide protection against their natural enemies (Waller *et al.*, 2007; Vandermeer *et al.*, 2010). Among the most cosmopolitan species that feed on coffee plants are *Coccus viridis* (Green) (Coccidae) (Crowe, 2004; Lan and Wintgens, 2004; Waller *et al.*, 2007; Barrera, 2008), *Planococcus citri* (Risso) (Pseudococcidae) (Waller *et al.*, 2007; Barrera, 2008), *Pseudaonidia trilobitiformis* (Green) (Diaspididae) (Raga *et al.*, 2003; Culik *et al.*, 2008) and *Toxoptera aurantii* (Boyer de Fonscolombe) (Aphididae) (Waller *et al.*, 2007; Barrera, 2008). Some species of this group of insects live in the foliage but colonize the roots, for example, *P. citri*, *Planococcus lilacinus* (Cockerell) (Pseudococcidae) and *Saissetia coffeae* (Walker) (Coccidae). Among the most harmful species are *Asterolecanium coffeae* Newstead (Asterolecaniidae), a scale insect found in Central and East Africa that can kill coffee plants (Crowe, 2004; Waller *et al.*, 2007). Natural enemies usually maintain populations of these hemipterans at low numbers (Le Pelley, 1973; Murphy, 1991; Vandermeer *et al.*, 2010; Perfecto *et al.*, 2014).

Leafhoppers (Hemiptera: Cicadellidae) associated with coffee are a group that has recently acquired importance because some species have been found to be vectors of

Xylella fastidiosa (Wells, Raju, Hung, Weisburg, Mandelco-Paul, and Brenner), a Gram-negative bacteria of the xylem belonging to the family Xanthomonadaceae (Wells *et al.*, 1987; Janse and Obradovic, 2010). This bacterium was reported in *C. arabica* for the first time in Brazil in 1995 and has been confirmed as the causal agent of coffee leaf scorch (CLS). The bacterium has also been found in robusta (*C. canephora*) coffee plants and specific strains have been reported causing diseases in orange, grape, olive and almond crops, among others (Jacques *et al.*, 2016). Although most of the infected plants are asymptomatic, diseased coffee plants exhibit very short internodes, loss of older leaves, pale-green terminal shoots, small deformed yellow leaves, scorching at the leaf apex and edges, regressive death of lateral shoots, reduced size and number of fruits, and general delay in growth. This disease caused a 30% reduction in yield in some coffee plantations of northern São Paulo, Brazil. It has been shown that citrus variegated chlorosis (CVC) caused by *X. fastidiosa* is pathogenic for coffee plants. This may explain why CLS is more frequent in coffee plantations located near citrus plantations (Li *et al.*, 2001). In Costa Rica, where the disease was reported in 2001, it is known as 'crespera' because the plants exhibit irregular growth with an atypical curly aspect. The symptoms of 'crespera' in the *C. arabica* varieties *Caturra* and *Catuaí* differ to a certain degree from those reported in Brazil: (i) reduced leaf size; (ii) malformation of leaves; (iii) curly leaf edges; (iv) shortened internodes; and (v) chlorotic leaf mosaic that can become necrotized (Rodríguez *et al.*, 2001). In Brazil, the following leafhoppers have been confirmed as vectors of *X. fastidiosa* in coffee and citruses: *Bucephalogonia xanthophis* (Berg), *Dilobopterus costalimai* Young, *Homalodisca ignorata* Melichar and *Oncometopia facialis* (Signoret) (Marucci *et al.*, 2008). In Costa Rica, *Erythrogonia sonora* Melichar, *Graphocephala permagna* Nielson and Godoy, *Graphocephala crusa* Godoy and *Kapateira coffea* Godoy (Rodríguez *et al.*, 2001; Godoy and Villalobos, 2006; Godoy *et al.*, 2006) have been named as potential vectors of *X. fastidiosa* in coffee.

Another group of insects that feed on coffee foliage is ants of the genera *Atta* and *Acromyrmex* (Hymenoptera: Formicidae). The species reported inhabit tropical America. These ants can cause considerable defoliation to several plant species since their colonies can be enormous. They use the leaves to cultivate a fungus, which they use as food, deep in their underground nests. Most of the species of *Acromyrmex*, such as *Acromyrmex asperus* (F. Smith), *Acromyrmex niger* (F. Smith), *Acromyrmex octospinosus* (Reich) and *Acromyrmex subterraneus* (Forel), have been reported in South America, particularly in Brazil. The species of *Atta*, such as *Atta cephalotes* (L.), *Atta insularis* Guerin-Meneville and *Atta mexicana* (F. Smith), have been reported mostly in Mexico, Central America and the Caribbean. Because of their wide distribution, *A. cephalotes* and *Atta sexdens* (L.) are the most common species in coffee plantations (Waller *et al.*, 2007; Barrera, 2008).

Many species of caterpillars (Lepidoptera) of the families Drepanidae, Geometridae, Limacodidae, Lymantriidae, Psychidae, Pyralidae, Sphingidae and Tortricidae have been reported worldwide to be associated with coffee and can cause very severe defoliation. An extreme case occurred on the Malayan peninsula where an attack by *Cephalotes hylas* (L.) (Sphingidae), a caterpillar that can reach a length of 7 cm, caused failure of the coffee crop (Waller *et al.*, 2007). However, most of these caterpillars are of little economic consequence because they are kept under control by their natural enemies. As in other instances, population outbreaks of these insects have been related to the use of insecticides. Although most of the species of caterpillars reported on coffee plantations are polyphagous, some, such as *Epicampoptera* spp. (Drepanidae), have only been observed attacking *Coffea* (Crowe, 2004; Waller *et al.*, 2007) and have caused heavy defoliation in Africa. The caterpillars *Parasa lepida* (Cramer) and *Latoia vivida* (Walker) (Limacodidae) are found in Asia and Africa, respectively. The damage caused by pricks from their stinging hairs can be more important than that caused by their feeding on the coffee plants (Waller *et al.*, 2007).

A group of beetles (Coleoptera) of the families Chrysomelidae, Curculionidae and Scarabaeidae has been reported to feed on leaves of *Coffea* spp., preferring tender shoots. Some, such as *Macrostylus* (Curculionidae), can damage the growing tips of stems causing their bifurcation. The larvae of these insects inhabit the soil and feed on roots; they can cause severe damage to the root system of coffee plants. Around 40 species of Chrysomelidae, most Eumolpinae and Galerucinae, have been reported to attack coffee (Waller *et al.*, 2007). In Colombia, there are reports of species of *Homophoeta*, *Cerotoma*, *Diabrotica* and *Colaspis* affecting new plantations (Cárdenas and Posada, 2001). In East Africa, three species of *Dactylispa* (Waller *et al.*, 2007) have been reported. In southern Chiapas, Mexico, *Rhabdopterus jansoni* (Jacoby) was observed attacking mature plantations of *C. arabica* (Barrera *et al.*, 2008b; Kuesel *et al.*, 2014). According to Waller *et al.* (2007), about 70 species of weevils (Curculionidae) associated with coffee have been reported worldwide. Some, such as *Lachnopus coffeae* Marshall and *Lachnopus buchanani* Marshall (Curculionidae), are major pests in Puerto Rico and Cuba, respectively. In Papua, New Guinea, five species of *Apirocalus* have been reported (Waller *et al.*, 2007). In Colombia, weevils of the genera *Compsus*, *Epicaerus* and *Macrostylus* have been observed attacking foliage of newly planted coffee trees in the field (Cárdenas and Posada, 2001). Scarabaeidae, which defoliate coffee trees, are a group of some 45 species; most are Melolonthinae and Rutelinae, which have been reported all over the world (Waller *et al.*, 2007). The larvae of these insects are more important as soil pests than as adults feeding on leaves. In Africa, Asia and America species of *Adoretus*, *Ancistrosoma*, *Anomala*, *Platycoelia*, *Phyllophaga* and *Popillia* have been reported (Cárdenas and Posada, 2001; Waller *et al.*, 2007; Barrera, 2008).

Species of Orthoptera, mainly Tettigoniidae and Pyrgomorphidae, are reported as defoliators of coffee plants (Waller *et al.*, 2007). Of the Tettigoniidae, *Idiarthron subquadratum* De Saussure and Pictet is outstanding; the nymphs and adults of this species are nocturnal and have been reported in Mexico and Central America causing severe damage to foliage, shoots, fruits, stems and young branches, mainly of *C. arabica* (Barrera *et al.*, 2002b; Zúñiga *et al.*, 2002; Zavala-Olalde *et al.*, 2005). Of the Pyrgomorphidae, two species of *Zonocerus* are mentioned in tropical Africa, one distributed more towards the north (*Zonocerus variegatus* (L.)) than the other ((*Zonocerus elegans* (Thunberg)) (Crowe, 2004; Waller *et al.*, 2007).

Some bugs (Hemiptera) of the Tingidae family, such as *Dulinius unicolor* (Signoret) in Madagascar and *Habrochila* spp. in continental Africa, and of the family Miridae, such as *Ruspoliella coffeae* (China) and *Volumnus obscurus* Poppius in Central Africa, feed on coffee leaves and flower buds. The Tingidae feed on the underside of leaves, on buds and fruits while even low populations of Miridae can be very harmful because they can cause flowers to abort (Crowe, 2004; Waller *et al.*, 2007). *Monalonion velezangeli* Carvalho and Costa (Miridae) is a pest that has recently appeared in coffee plantations in Colombia. The damage they cause to young leaves, new shoots, non-lignified stems and flowers is seen as scorch-like spots, which negatively affect the plant's development and, consequently, production (Giraldo-Jaramillo *et al.*, 2010).

Waller *et al.* (2007) state that some 30 species of thrips (Thysanoptera) have been reported to affect coffee worldwide. These species belong to the families Aeolothripidae, Panchaetothripidae, Phlaeothripidae and Thripidae. Thrips are small insects that have scraping mouthparts; they live on the underside of the leaves puncturing the epidermis and sucking out the cell contents, causing stippling, discoloured flecking (silvering of the leaf surface) and leaf distortion, and in flowers they feed on pollen and other floral parts (Bethke *et al.*, 2014). They are not considered important because of the short duration of the flowering stage. However, they can be particularly harmful during the dry season. In Africa, *Diarthrothrips coffeae* Williams (Aeolothripidae) can cause defoliation and flower abortion (Crowe, 2004; Waller *et al.*, 2007). Some thrips, such as *Heliothrips haemorhoidalis* Bouche and *Selenothrips rubrocinctus* (Giard) (Thripidae), attack coffee but prefer cacao (*Theobroma cacao* L.). Others, such as *Scirtothrips bispinosus* (Bagnall) (Thripidae), are also pests of tea (Waller *et al.*, 2007), while still others, such as *Taeniothrips xanthocerus* (Hood) and *Taeniothrips antennatus* Bagnall (Thripidae), feed on spores of *H. vastatrix*, the reason they are considered to have certain importance in dispersing the disease (Crowe, 2004).

Root- and collar-feeding insects

Control of insects that inhabit or pass part of their life cycle in the soil and feed on roots is often more complicated than controlling those that live aboveground feeding on plant foliage. This is largely due to the lack of information concerning their identity and habits, but also because of the difficulty in assessing the damage they cause (Morón and Rodríguez-del-Bosque, 2010). The insects associated with the root system of coffee trees are sucking insects, crickets, termites and larvae of Coleoptera, Lepidoptera and Diptera. In general, these insects cause damage by sucking sap from the roots, gnawing on the roots or the collar of the plants. They can be found attacking plants in nurseries or in the field.

The group of insects that feed on the collar and root includes major coffee pests, such as mealybugs (Hemiptera: Pseudococcidae). Waller *et al.* (2007) names at least 40 species of mealybugs associated with coffee worldwide belonging to the following genera (number of species in parentheses): *Benedictycoccina* (1), *Capitisetella* (1), *Cataenococcus* (1), *Coccidella* (1), *Dysmicoccus* (7), *Formicoccus* (2), *Geocuccus* (1), *Neochavesia* (2), *Paraputo* (2), *Planococcus* (5), *Pseudocuccus* (3), *Pseudorhizoecus* (1), *Puto* (3), *Rhizoecus* (8) and *Ripersiella* (2). Many of the reported species are found in tropical America. It is worth mentioning that some mistaken identifications have occurred in the past. Female and immature mealybugs are sucking, sedentary, apterous insects, with a soft oval body covered with a waxy powder, which, depending on the species, can appear on lateral filaments. Males are smaller, winged, and do not feed. Plants attacked by these pests have a corky layer that covers the roots, under which their colonies take shelter. This layer is the compressed mycelia of fungi that live in association with the mealybugs; one of these fungi is *Diacanthodes novoguineensis* (Hennings) O. Fidalgo, which is found in association with *Planococcus fungicola* Watson and Cox in East Africa (Watson and Cox, 1990). Attacked plants weaken and can die during the dry season from the combined action of the feeding mealybugs and the fungal layer that impedes water and nutrient absorption by the roots. Mealybugs are also associated with ants, mainly of the genera *Acropyga*, *Pheidole* and *Solenopsis*, which consume the honeydew the mealybugs excrete abundantly. In exchange, the ants protect the mealybugs from natural enemies and impede proliferation of microorganisms by cleaning the mealybug colony (Waller *et al.*, 2007).

Another group of hemipterans found in coffee roots is scales of the families Coccidae and Ortheziidae, although they are less important than the members of the Pseudococcidae. Among the Coccidae, *Coccus brasiliensis* Da Fonseca and *S. coffeae* stand out. The latter is also very commonly found in the foliage (Waller *et al.*, 2007). Part of this family is *Toumeyella coffeae* Kondo, apparently the only species of this genus that inhabits coffee roots (*C. arabica*); the specimens described were collected in Colombia and Venezuela (Kondo, 2013). Of the scale family Ortheziidae associated with coffee roots, there are at least two species of *Mixorthezia* reported in the Caribbean and South America (Waller *et al.*, 2007).

Several species of cicadas (Hemiptera: Cicadidae) have been observed attacking roots, trunks and branches of *C. arabica* in South America, especially in Brazil (Reis and de Souza, 1986; Matiello, 1991; Martinelli and Zucchi, 1997), and more recently in Guatemala (ANACAFÉ, 2014). Adult female cicadas cause damage to the branches when they oviposit under the bark causing the branches to dry up. The cicada nymphs go underground for 3–4 years feeding on the root sap. Besides coffee, damage in cacao and avocado have been reported. At least 20 species of cicadas of the genera *Baeturia*, *Carineta*, *Dorisiana*, *Dundubia*, *Fidicina*, *Quesada*, *Ueana*, *Yanga* and *Zammara* have been reported attacking coffee (Waller *et al.*, 2007). Of these *Quesada gigas* (Olivier) is outstanding for its broad distribution and the damage it causes (Matiello, 1991; Martinelli and Zucchi, 1997). It is believed that cicada attacks on coffee are a consequence of deforestation, which has eliminated their natural host plants (Matiello, 1991).

Several species of the order Coleoptera that damage coffee roots and the collar have been reported. Outstanding for their importance are white grubs (Scarabaeidae), particularly the larvae of species belonging to the genus *Phyllophaga*, which have been reported in coffee nurseries and plantations in America and Asia. The larvae of these insects live up to 9 months underground. They have well-developed legs; they are white with a highly differentiated brown cephalic capsule and curl into a characteristic 'C' shape. The adults are brown beetles measuring 2–4 cm. They emerge massively with the first rains of the year and feed on foliage during the night (Waller *et al.*, 2007). Other members of Coleoptera are five species of beetles of the genus *Gonocephalum* (Tenebrionidae), whose larvae feed on roots, while the adults feed on the bark and collar of the coffee plants in Africa and Asia. In tropical America, the adults of several species of weevils of the genera *Diaprepes*, *Lachnopus* and *Pantomorus* (Curculionidae) feed on foliage, while the larvae eat roots (Waller *et al.*, 2007).

Chiromyza vittata Wiedemann (Diptera: Stratiomyiidae) is a fly that has been reported causing damage to roots of robusta coffee plantations in Brazil (Matiello, 1991; Fornazier *et al.*, 2000). The larvae of this insect appear in large quantities; they eat the smallest roots and make holes in the larger roots through which pathogenic microorganisms such as *Fusarium* can enter. Affected plantations exhibit chlorosis and production decreases.

Cutworms (Lepidoptera: Noctuidae) are significant pests of coffee nursery beds in many coffee-producing regions of the world. During the day, they are hidden under the soil surface, and at night, they leave their refuges to feed on the seedlings, cutting the collar at soil level. The most harmful species reported in Africa and Asia belong to the genera *Agrotis* and *Tycomarptes*, while in America they are of the genera *Agrotis*, *Feltia* and *Spodoptera* (Waller *et al.*, 2007).

Nymph and adult crickets (Gryllidae) and mole crickets (Gryllotalpidae) are insects of the order Orthoptera; they are polyphagous and nocturnal. They feed on roots and damage the collar of the coffee seedlings in nursery beds. During the day, these insects hide in burrows they make in the soil. The crickets *Anurogryllus abortivus* (Saussure), *Ceuthophilus* sp. and *Gryllus assimilis* (F.) have been reported in America; *Brachytrupes membranaceus* (Drury) and *Gryllus bimaculatus* De Geer in Africa; and *Brachytrupes portentosus* (Lichtenstein), *Gymnogryllus commodus* (Walker) and *Teleogryllus mitratus* (Burmeister) in Asia. A species of mole cricket, *Gryllotalpa africana* Palisot de Beauvois, has been reported attacking coffee in Africa and Asia, while another, *Scapteriscus didactylus* (Latreille), was observed in America (Crowe, 2004; Waller *et al.*, 2007).

Sixteen species of termites (Isoptera) belonging to the families Kalotermitidae (3), Nasutitermitidae (1), Rhinotermitidae (3) and Termitidae (9) have been seen infesting coffee plants in different regions of the world (Waller *et al.*, 2007). Termites are social insects that feed on wood, which they digest with the help of microorganisms that inhabit their digestive tract. In Africa, infestation begins in galleries abandoned by stem borer insects, such as *B. sierricola* (Crowe, 2004; Waller *et al.*, 2007).

Considerations for Pest Management in Coffee Plantations

Disruption of natural balance between pests and natural enemies

Pest management in coffee plantations has largely depended on the use of synthetic agrochemicals (Vayssière, 1964; Oduor and Simons, 2003). In fact, it has been erroneously assumed that the more technicized the plantation, the more it should use products of this type. Although the use of insecticides has permitted solving pest problems, their excessive and frequent use has promoted the development of resistant pests, as in the case of the coffee berry borer *H. hampei* in New Caledonia (Brun *et al.*, 1989) and of the leaf miner *L. coffeella* in Brazil (Fragoso *et al.*, 2002).

In many other documented cases, abuse in the use of insecticides has caused ecological imbalances (resurgence of pests and secondary pests) because they eliminate natural enemies. For example, *Lamprosema crocodora* (Meyrick) (Lepidoptera: Pyralidae) became a serious defoliator of *C. canephora* in the Democratic Republic of the Congo after use of persistent pesticides. Likewise, because of the effect of insecticides on its natural enemies, *Aspidiotus* sp. (Hemiptera: Diaspididae) became a serious pest in African coffee plantations (Waller *et al.*, 2007). Attacks on coffee plantations in Kenya and Tanzania by *Ascotis selenaria reciprocaria* Walker (Lepidoptera: Geometridae), a polyphagous caterpillar, occurred after intensive use of parathion to control leaf miners (Bigger, 1969). The dense populations of *Asterolecanium pustulans princeps* Castel Branco (Hemiptera: Asterolecaniidae) in coffee and cacao in São Tomé in the 1950s was attributed to elimination of the parasitoid *Encarsia citrina* (Crawford) (Hymenoptera: Aphelinidae) by insecticides used to control *Aspidiotus destructor* Signoret (Hemiptera: Diaspididae) (Waller *et al.*, 2007). In Mexico, *L. coffeella* infestations increased in coffee plantations that used parathion for control of *I. subquadratum* (Barrera *et al.*, 2003). In Brazil, the use of cupric fungicides to control the rust *H. vastatrix* caused increases in the leaf miner populations (Matiello, 1991). Other factors, such as droughts, can cause imbalances in regulation of populations of these insects. This was the case in Espírito Santo, Brazil, where long droughts have converted *Praelongorthezia praelonga* (Douglas) (Ortheziidae), a tropical insect important in citruses, into a serious pest that affects branches, leaves and berries of *C. canephora* (Matiello, 1991).

Complex interactions among organisms

The above examples illustrate the undesirable effects of insecticides when they are used frequently and as the first alternative for managing insect pests in coffee plantations, especially in plantations cultivated under shade trees, which are an important refuge for diverse organisms (Perfecto *et al.*, 1996; Jha *et al.*, 2014). Studies on trophic interactions associated with *C. viridis* in coffee plantations under shade in southern Mexico have discovered a highly complex network of relationships among different organisms, which results in natural control of the population of this scale (Vandermeer *et al.*, 2009, 2010; Jiménez-Soto *et al.*, 2013, Perfecto *et al.*, 2014). Some of these interactions occur around the scale *C. viridis*, a common phytophagous insect in the foliage of coffee plants. Among the several parasitoids and predators that attack the scale is *Azya orbigera* Mulsant (Coleoptera: Coccinellidae), a voracious ladybird beetle whose larvae and adults live near the scale colonies. On the other hand, the scale has a mutualistic relationship with the ant *Azteca sericeasur* Longino (Hymenoptera: Formicidae) (previously cited as *Azteca instabilis* (F. Smith)), which protects the scale from predators in exchange for the honeydew the scales excrete profusely. The ant constructs its nests in trees and forages over the coffee trees where the scale lives. The ladybird beetle has to take advantage of the ants' inattention to lay its eggs in the colony of scales, a very difficult task since the ant is highly aggressive. Nevertheless, other participants in the plot open up a window of opportunity for the ladybird beetle. At least three species of decapitating flies of the genus *Pseudacteon* (Diptera: Phoridae) parasitize ants. Indeed, studies show that the ant reduces its rhythm of activity up to 50% as a defensive behaviour. Some ants retreat to the nest for shelter and others remain motionless to avoid detection during attack by these flies. The ladybird beetles take advantage of the situation to oviposit on the branch where the scale colony is. Paradoxically, the ants will protect the ladybird beetle larvae from natural enemies as if they were members of the scale colony; in any case, they are incapable of harming the ladybird beetle larvae because of the waxy filaments that cover their bodies. Although at first a grower might see the presence of *A. sericeasur* on the coffee plant as negative because of their aggressiveness as well as the protection they give to *C. viridis*, less conspicuously, they also protect the coffee plant from other phytophagous pests. Studies have found that coffee plants with *A. sericeasur* have lower infestations of the seed borer *H. hampei*. Another interaction in this complex system is that of the fungus *Lecanicillium lecanii* (Zimm.) Zare and W. Gams, which is the most important natural enemy of the coffee rust *H. vastatrix* and attacks the scale *C. viridis*. A negative correlation exists between rust infection and the distance, on a scale of 15 m, to sites where a strong epizooty of *L. lecanii* on *C. viridis* occurred the previous year.

Because agricultural production is highly dependent on ecosystem services, such as pest control, pollination and fertilization to the soil, among others (Tscharntke *et al.*, 2012), pest management of these diversified systems raises the challenge of producing while conserving (Moguel and Toledo, 1999; Philpott and Dietsch, 2003; Garcia *et al.*, 2009). In particular, pest management in coffee plantations without synthetic agrochemicals is a huge challenge for growers who must comply with strict quality standards and economic and socio-environmental responsibility that organic coffee certifiers, among others, impose. Some of these standards limit the use of pesticides derived from the chemical industry and require that the crop should be cultivated in diversified shade systems managed with agronomic practices that promote conservation of biodiversity (Mas and Dietsch, 2004; Raynolds *et al.*, 2007; Méndez *et al.*, 2010).

Shade as a tool in pest management

Because shade trees greatly contribute to creating the microclimate in the coffee plantation (Velasco *et al.*, 2001), in these systems they are highly important for pest and disease management; an increase in shade can lower incidence of some pests but it can also favour the incidence of others (Beer *et al.*, 1998; Guharay *et al.*, 2001; Oduor and Simons, 2003). A now emblematic case of decreasing shade as a measure to control the coffee rust *H. vastatrix* and the effects on other pests and diseases of coffee occurred in Central America in the 1980s (Guharay *et al.*, 2001; Barrera, 2002; Mariño *et al.*, 2016). In many coffee plantations of this region, growers drastically reduced shade to allow better aeration and higher radiation. They believed that this would create unfavourable conditions for coffee rust, a fungus that requires high relative humidity for its development. However, the modification to the habitat created conditions favourable for the coffee leaf miner *L. coffeella* and brown eye spot (*Cercospora coffeicola*); both prosper with little shade. The increase in the incidence of these organisms motivated the growers to use pesticides, a measure that was practically non-existent before de-shading the coffee plantations to control coffee rust. Moreover, the growers were forced to use herbicides to control weeds that increased because of increased sunlight and nematicides for the control of nematodes, whose populations increased when their natural enemies decreased because of the smaller amount of organic matter from leaf litter from the shade trees incorporated into the soil. In addition, growers planted compact-growth varieties with short internodes and less vigorous roots and these were more vulnerable to nematodes than the long-internode varieties. Later, populations of the mealybug *P. citri* boomed, free of their natural enemies because of the frequent use of insecticides against the leaf miner. The lesions on the foliage caused by drifting drops of herbicide were exploited by anthracnosis (*Colletotrichum* spp.), increasing its incidence (Guharay *et al.*, 2001). The cascade of events triggered by the single action of reducing shade illustrates the need to understand the interaction among organisms that inhabit the coffee plantation, as well as the effects of agronomic practices on the plantation microclimate and its effect on organism interactions.

Studies on the effect of shade on production and quality of coffee are contradictory (Beer *et al.*, 1998; Guharay *et al.*, 2001), as are those that have studied pests, such as the coffee berry borer *H. hampei* (Soto *et al.*, 2002; Mariño *et al.*, 2016). For example, studies conducted in southern Chiapas, Mexico (Barrera and Covarrubias, 1984), Nicaragua (Matos *et al.*, 2004), Colombia (Bosselmann *et al.*, 2009) and Puerto Rico (Mariño *et al.*, 2016) indicate that an increase in shade favoured *H. hampei* infestation. In contrast, studies conducted in Kenya report that infestations of this pest were greater in coffee plantations exposed to the sun (Jaramillo *et al.*, 2013) or that shade had no effect on infestation, as occurs in the coffee plantations in northern Chiapas (Soto *et al.*, 2002). Concerning the leaf miner *L. coffeella*, the studies coincide in that infestation of this insect is higher in hot, dry conditions, which are those of coffee plantations at low altitudes exposed to the sun, especially in the dry season of the year, as occurs in Brazil (Souza *et al.*, 1998). In mountainous regions where shaded coffee plantations predominate, such as in Mexico, plant cover provided by shade trees does not appear to have great influence in leaf miner incidence (Lomelí-Flores *et al.*, 2010). In Cuba, it is recommended that plantation illumination above 70% be avoided; a measure to prevent favourable conditions for the floury mealybug complex integrated by the genera *Planococcoides*, *Pseudococcus* and *Planocuccus* (Martínez and Suris, 2000).

Because the upper limit accepted for coffee shade – a C3 plant – varies from 40% to 70% (Beer *et al.*, 1998), a level of 35–60% shade has been suggested for management of pests such as the coffee berry borer, leaf miner and mealybugs (Guharay *et al.*, 2001). It is recommended that shade be maintained at 60% in the dry season and at 45–50% at the onset of the rains. At the middle of the rainy season, an additional pruning of shade trees to a level of 35–40% is suggested, after which they should be left to grow to reach the 60% recommended for the dry season (Guharay *et al.*, 2001). However, before generalizing these recommendations, decisions on shade management should be made in relation to: (i) the incidence of pests; (ii) environmental characteristics of the plantations; and (iii) the benefits the shade trees provide to the growers, such as fruit, wood and other products (Beer *et al.*, 1998).

Other cultural control methods

In general, control through agronomic cultural practices is the most common way to control pests in coffee plantations. Besides shade management, keeping plants healthy and vigorous by pruning, balanced nutrition and weed control helps to prevent pest attacks and helps plants to tolerate them. One of the most used practices is elimination of infested plant material. In the case of *H. hampei*, regular harvesting of ripe berries and removal of fruits after harvest are the most important activities. Fruits infested by this pest should be burned, buried or boiled in water (Le Pelley, 1973; Wrigley, 1988; Barrera, 2008). Similar treatment is recommended for the control of high infestations of the berry moth *P. smaragdina*: manual removal of fruit clusters infested by the larvae. In this case, the material collected is put into holes in the ground that are afterwards covered with a fine mesh that allows parasitoids to exit, but impedes the adult pests from leaving (Gaitán *et al.*, 2015).

To control cerambycid borers, such as *B. sierricola*, *P. maculosus* and *X. quadripes*, infested plants should be pulled up and burned. Other recommended practices include manual collection of adults, interfering with oviposition by cleaning the stems or impregnating them with repellents and putting a wire into the perforations to kill the larvae that are inside. Control of branch borers, such as *X. compactus* and *X. morigerus*, should be done by cutting and burning sick branches. As soon as wilting is observed, the branch should be cut

5–8 cm below the perforation made by the insect. Recommended actions for the control of *X. compactus* are making sure that the coffee plants are not too heavily shaded and the plantation well drained. The black stem borer *Apate monachus* F. is controlled mainly by burning the coffee plants and other infested trees. It is also recommended that adults are picked off by hand very early in the morning; insects in the galleries can be killed with a wire. These practices may require quite a bit of manual labour (Le Pelley, 1973; Wrigley, 1988; Barrera, 2008; Gaitán *et al.*, 2015).

In the case of root pests, such as *Neorhizoecus coffeae* (Laing), which are very hard to detect and control, to prevent their dispersion and establishment in the plantation, roots of the coffee plants should be inspected before being transplanted (Gaitán *et al.*, 2015). Pruning the coffee plants to clear their central part is a practice suggested to dissuade the presence of *Antestiopsis* spp. bugs, since they prefer coffee plants with dense foliage (Wrigley, 1988).

Improving soil fertility by incorporating organic matter is an important practice in nature-friendly coffee plantations (Grossman, 2003). Generally, organic fertilizers (compost, vermicompost, bocashi, coffee pulp, animal manure, etc.) are rich in macro- and micronutrients and their interaction with soil flora and fauna improves plant health, vigour, tolerance and resilience to stress caused by pests and abiotic factors (Benzing, 2001; Altieri and Nicholls, 2003; Ayalew, 2014; Muschler, 2016). Soil attributes have an important role in managing coffee pests, such as the cerambycid *X. quadripes*; Thapa and Lantinga (2016) reported that plants that were grown with more nitrogen, higher pH and organic matter had lower infestations of this insect. Another common practice is the application of biofertilizers compatible with organic agriculture, such as 'Supermagro' (fresh bovine manure with micronutrients, bone meal, unrefined sugar and milk), 'calda Viçosa' (Bordeaux mixture supplemented with zinc, magnesium and copper sulfates, and boric acid) and calcium sulfate, which can also be used as a foliar fertilizer, can protect plants from attack by arthropod pests. An experiment conducted in a greenhouse with three products showed that they were up to 80% effective against the mite *Oligonychus ilicis* (McGregor) (Acari: Tetranychidae) 14 days after application, although the dosages of Supermagro and calda Viçosa were very high. According to this study, calcium sulfate, recognized for its effect against mites, had greater potential for mite control in organic coffee plantations (Tuelher *et al.*, 2014).

Other cultural practices aim to reduce crop exposure to certain pests. For example, to avoid the West African coffee borer *B. sierricola*, an insect that is believed to originate in secondary forests, planting a strip (30 m) of annual crops between the coffee plantation and the forest is suggested (Wrigley, 1988). Eliminating certain weeds that serve as hosts to mealybugs, such as *Dysmicoccus brevipes* (Cockerell), from the coffee plantation is also recommended (Gaitán *et al.*, 2015). In other cases, to prevent dust from roads falling on the coffee trees and reduce invasion of scales, such as *Cerococcus catenarius* (Fons.), live barriers of non-host plants can be established between roads and the coffee plantation (Wrigley, 1988).

Pest-resistant varieties

One of the most effective, economical and environmentally sound approaches for management of insect pests is growing cultivars that have improved levels of pest resistance (Goggin and Zhu-Salzman, 2015). However, in coffee very little research has been done regarding this topic; most has been aimed at developing disease-resistant varieties. As for pests, the leaf miner *L. coffeella* has been one of the most studied. Laboratory and field studies have found that several species of *Coffea* have different degrees of resistance to the miner. For example, because of the larval mortality they cause and, consequently, the reduced leaf area affected, Guerreiro-Filho (2006) considers *Coffea stenophylla*, *Coffea salvatrix*, *C. liberica* var. *liberica*, *Coffea brevipes*, *Coffea* sp. 'Moloundou', *Coffea jasminoides* and *Coffea farafaganensis* resistant. *Coffea kapakata*, *Coffea eugenioides*, *Coffea racemosa*, *C. liberica* var. *dewevrei*, *Coffea humilis*, *Coffea tetragona*, *Coffea tsirananae*, *Coffea resinosa*, *Coffea millotii*, *Coffea bertrandii*, *Coffea dolichophylla* and *Coffea bonnieri* are moderately resistant, while *Coffea congensis*, *Coffea sessiliflora*, *Psilanthus travancorensis* and *Coffea perrieri* are considered to be moderately susceptible. In Brazil, resistance genes were transferred from *C. racemosa* to highly productive susceptible *C. arabica* varieties (Guerreiro-Filho, 2006). Because it is still not possible sexually to propagate homogeneous, stable populations of resistant plants, clones were obtained from the mother plants and their agronomic performance was evaluated in the field. The results showed that clone '1059' – resistant to *L. coffeella* – had a 21% higher production than had 'Obatã IAC 1669-20', a susceptible *C. arabica* cultivar used as the control with and without chemical pest control (Mendonça *et al.*, 2016).

In the case of *H. hampei*, resistance has not been reported for commercial coffee varieties (Anthony *et al.*, 1999; Vega *et al.*, 2015). However, *C. arabica* is reported to be more susceptible than *C. canephora*, while *C. excelsa* and *C. liberica* are less attacked. This order can vary since for some authors robusta coffee is more susceptible than arabica (Le Pelley, 1973). Vayssière (1964) explained that coffees that have a thin pericarp (robustas and arabicas) were more susceptible to the coffee berry borer than were those with a thick pericarp (*C. liberica*, *C. excelsa*). Preliminary laboratory tests conducted in Brazil by Sera *et al.* (2007) indicate that *C. eugenioides* and *C. kapakata* may have volatile substances in the pulp that repel the coffee berry borer, while *Coffea bengalensis* has substances in the grain that make it resistant. In Colombia, recent laboratory studies conducted by Romero and Cortina-Guerrero (2004, 2007) to identify possible sources of resistance to the coffee berry borer, reported a certain level of antibiosis in *C. liberica* and in some *C. arabica* Ethiopian introductions, compared with the standard commercial *C. arabica* var. *Caturra*. Moreover, it is also known that caffeine, an alkaloid that plants use as defence against phytophagous organisms, does not affect attacks by the coffee berry borer (Guerreiro-Filho and Mazzafera, 2003). In general, there are few studies on *C. arabica* chemical resistance to this pest (Green *et al.*, 2015).

For several years, some research groups have conducted studies to develop transgenic coffee plants resistant to pests (Ferreira *et al.*, 2006; Mishra and Slater, 2012). For example, field tests of transgenic *C. canephora* plants implanted with the synthetic gene *cry*1Ac, which codes the Bt toxin (Leroy *et al.*, 2000), exhibited much less damage by the leaf miner *L. coffeella* than the control plants (Perthuis *et al.*, 2005). Genetically modified *C. arabica* plants that express the α-amylase (α-*AI1*) inhibitor-1 gene – a very active inhibitor of the digestive enzymes of *H. hampei* – under the control of phytohemagglutinin (PHA-L) – a promotor from *Phaseolus vulgaris* L. that is expressed specifically in the seed – delayed development of the pest larvae (Albuquerque *et al.*, 2015). There is still no commercial transgenic coffee with the above characteristics, and this new technology must overcome many ethical, socioeconomic, political, ecological and health issues (Baker, 2001), especially for application in nature-friendly productive systems.

Biological control

Conservation of natural enemies, mainly through judicious use of agrochemicals, has been the most common approach to biological control of coffee pests. In a few cases, exotic agents of biological control have been introduced. One of the most emblematic cases of the use of classical biological control was the collection of the parasitoid *Anagyrus* sp. nr. *kivuensis* Compere (Hymenoptera: Encyrtidae) in Uganda and its introduction to Kenya for control of the mealybug *Planococcus kenyae* (Le Pelley) (Hemiptera: Pseudococcidae) in 1938, resulting in a dramatic reduction of the pest by 1940 and an immense benefit (Le Pelley, 1943; Abasa 1975; Oduor and Simons, 2003).

A more recent case that resounded internationally was that of the coffee berry borer *H. hampei*. The first attempts at classical biological control of *H. hampei* date from 1923 and 1929 with the introduction of *Prorops nasuta* Waterston (Hymenoptera: Bethylidae) from Uganda to Java and Brazil, respectively (Le Pelley, 1973). However, the most important efforts began in the mid-1980s with the introduction of *P. nasuta* and *Cephalonomia stephanoderis* Betrem (Hymenoptera: Bethylidae) from several African countries such as Kenya, Togo and the Ivory Coast, to Ecuador (Klein-Koch *et al.*, 1988), Mexico (Barrera *et al.*, 1990a; Barrera, 1994) and Central America (Barrera *et al.*, 1990b). Later, programmes were started to introduce these parasitoids into Cuba (Peña-Marrero *et al.*, 2006), India (Balakrishnan *et al.*, 2000) and Colombia (Baker *et al.*, 2002). Colombia also coordinated the introduction of *Phymastichus coffea* LaSalle (Hymenoptera: Eulophidae) into several Latin American countries and India. *P. nasuta* and *C. stephanoderis* adult wasps feed on all stages of development, and the females parasitize completely developed *H. hampei* larvae, prepupae and pupae inside infested berries, while the eulophid *P. coffea* parasitizes the females of the coffee berry borer when they are on the coffee berries boring into the pulp (Vega *et al.*, 2015). With the exception of Colombia, where *P. nasuta* has more successfully established in the field (Maldonado and Benavides, 2007), *C. stephanoderis* was the most successful in India (Roobakkumar *et al.*, 2014), as well as in Cuba (Peña-Marrero *et al.*, 2006), Mexico and Central America (Barrera *et al.*, 2008a; Gómez *et al.*, 2010). Adult *P. coffea* parasitoids have not been able to establish in Colombia (Benavides *et al.*, 2012), Mexico or Central America (Barrera *et al.*, 2008a). Where they have been able to establish, the wasps contribute to natural control of *H. hampei*, with parasitism varying from 1% to 65% (Aristizábal *et al.*, 2016). The impact of the coffee berry borer's parasitoids has been improved by mass rearing in laboratories and farms, using berries, coffee parchment or semi-artificial diets infested with *H. hampei* (Barrera *et al.*, 2008a).

The use of entomopathogens is an alternative to chemical control when populations of certain insects are very high. The fungus *Beauveria bassiana* (Bals.) Vuill. is recommended for control of *H. hampei*, while for cicada nymphs that feed on coffee roots, the use of entomopathogenic nematodes is recommended. When the populations of defoliating caterpillars of the Geometridae family are out of control, spraying with the bacteria *Bacillus thuringiensis* Berliner (*Bt*) or the fungus *B. bassiana* is recommended (Gaitán *et al.*, 2015).

Insect trapping

Insect traps play an important role in monitoring and controlling some coffee pests. A mixture of methanol and ethanol in a 1:1 or 3:1 proportion has been shown to be a powerful attractant for capturing flying *H. hampei* females (Barrera *et al.*, 2004, 2007; Barrera, 2008). Better control of the coffee berry borer has been found by combining trapping with removal of infested fruits after harvest (Dufour *et al.*, 2007). In Brazil, a white delta trap baited with 300 μg of the racemic mixture of the synthetic pheromone 5,9-dimethylpentadecane was used over 4 ha to monitor the leaf miner *L. coffeella* (Bacca *et al.*, 2006). However, as a tool to confuse mating, this pheromone has not been successful (Ambrogi *et al.*, 2006). In India, sticky white cross-vane traps are baited with (*S*)-2-hydroxy-3-decanone, the main component of the male pheromone, to capture and control female *X. quadripes* (Venkatesha and Dinesh, 2012). In El Salvador and Mexico, the use of 'refuge-traps' (e.g. a piece of bamboo stalk, closed at one end and open at the other, placed among the coffee plants) is recommended to capture *I. subquadratum* nymphs and adults (Barrera *et al.*, 2002b; Barrera, 2008).

Botanical insecticides

Chemical control with pesticides from the chemical industry, especially those of higher toxicity and residuality and those that are not registered or are prohibited, are not permitted or strongly limited by companies that certify nature-friendly productive systems. Some less toxic and less residual products, such as toxic baits, highly refined mineral oils (white oils) and botanical insecticides have potential for use in this type of production system, particularly on small farms. Toxic bait, for example, made from a mixture of maize bran, sugarcane bagasse and 1% insecticide is effective against ants associated with scales such as *C. viridis* and mealybugs such as *P. citri* (Gaitán *et al.*, 2015). Spraying with white oils is also recommended for control of soft-bodied insects that inhabit foliage, such as *C. viridis*, *A. coffeae* and *Cerococcus catenarius* (Fons.) (Wrigley, 1988). Moreover, botanical insecticides obtained from locally cultivated plants are an alternative to conventional chemical insecticides (Green *et al.*, 2015). In the case of the coffee berry borer *H. hampei*, many laboratory studies have been conducted to evaluate the insecticidal effect of plants on adults. For example, laboratory tests conducted by Zorzetti *et al.* (2012) indicated that coffee leaves treated with ethanol extracts of *Tephrosia purpurea* (Linn.) Pers. (*Fabaceae*) leaves at concentrations of 10% caused as much mortality in adults as leaves treated with endosulfan, one of the most toxic products used against *H. hampei*. In this study, aqueous and ethanol extracts of *Moringa oleifera* Lam. (*Moringaceae*) seeds had similar effects. Another study by Santos *et al.* (2013a) found that ethanol extracts from the root of *Piper alatabaccum* Trel and Yuncker (*Piperaceae*) had insecticidal activity against *H. hampei* adults. Mendesil *et al.* (2012) achieved between 80% and 90% mortality of *H. hampei* adults treated with essential oils from *Thymus vulgaris* L. (*Lamiaceae*), *Aloysia* sp. (*Verbenaceae*), *Ruta chalepensis* L. (*Rutaceae*), *Chenopodium ambrosioides* L. (*Chenopodiaceae*) and *Cymbopogon nardus* (L.) Kuntze (*Poaceae*). Other plants reported to have insecticidal effects against *H. hampei* are *Aeollanthus pubescens* Benth. (*Lamiaceae*) (Mawussi *et al.*, 2009), *Ocimum canum* Sims (*Lamiaceae*) (Mawussi *et al.*, 2012), *Schinus terebinthifolius* (*Anacardiaceae*) Raddi (Santos *et al.*, 2013b) and *Ricinus communis* L. (*Euphorbiaceae*) (Celestino *et al.*, 2016). Besides toxic effects that cause mortality, the extracts of some plants are repellent: for example, in free selection tests, ethanol extracts of *T. purpurea* seeds and *M. oleifera* roots strongly repelled *H. hampei*; other plants reported to have repellent effects on *H. hampei* are *Azadirachta indica* A. Juss (*Meliaceae*) and *Nerium oleander* L. (*Apocynaceae*) (Zorzetti *et al.*, 2012). Reports of insecticidal effects of plants on other coffee pests, such as the leaf miner *L. coffeella* (Alvez *et al.*, 2011) and *Antestiopsis* bugs (Mendesil *et al.*, 2012) are less frequent. Nevertheless, they could be more interesting because, unlike *H. hampei*, which passes most of its life protected inside the coffee berries, these pests are found on the foliage where they can be sprayed with plant extracts (Green *et al.*, 2015).

Finally, it is important to say that the insect communities associated with nature-friendly coffee plantations, particularly in Latin America, are undergoing very strong pressure exerted by the coffee rust *H. vastatrix* and its management. Indeed, the serious infestation of this fungus that occurred from 2008 to 2013 in Colombia and Central America (Avelino *et al.*, 2015) defoliated susceptible coffee plants and caused early loss of berries and death of branches and plants with very strong impacts on yields and grower's incomes, leading to a crisis in the coffee industry of these countries. To deal with the disease, important changes have been made in agricultural practices. For example, in immediate response, applications of fungicides have increased and shade has been modified to protect susceptible *C. arabica* varieties. There has also been a gradual, but notable, process of substitution of susceptible varieties for *H. vastatrix*-resistant varieties. In addition, *C. arabica* is being substituted by *C. canephora* in the Soconusco region (Chiapas) of Mexico. These changes involve profound effects on management of the production systems, as well as on associated pests, natural enemies and biodiversity in general, especially in nature-friendly systems. The environmental, social and economic impacts have not been foreseen in their entire dimension. If management of the coffee rust is not handled with care, the cascade of ecological imbalances that occurred in Central America in the 1980s (Guharay *et al.*, 2001) could repeat itself. We hope that the mistakes of the past help to make better decisions today.

Acknowledgements

This work was partly supported by the project 'Diseño, construcción, equipamiento y puesta en marcha de un Centro Estatal de Innovación y Transferencia de Tecnología para el Desarrollo de la Caficultura Chiapaneca' (FOMIX #249930), and by the ECOSUR Multidisciplinary and Transversal project 'Innovación socioambiental en zonas cafetaleras para la reducción de la vulnerabilidad'.

References

Abasa, R.O. (1975) A review of the biological control of coffee insect pests in Kenya. *East African Agricultural and Forestry Journal* 40, 292–299.

Aguiar-Menezes, E.L., Souza, S.A.S., Santos, C.M.A., Resende, A.L.S., Strikis, P.C., Costa, J.R. and Ricci, M.S.F. (2007) Susceptibilidade de seis cultivares de café arábica às moscas-das-frutas (Diptera: Tephritoidea) em sistema orgânico com e sem arborização em Valença, RJ. *Neotropical Entomology* 36, 268–273.

Albuquerque, É.V.S., Bezerra, C.A., Romero, J.V., Valencia, J.W.A., Valencia-Jiménez, A., Pimenta, L.M., Barbosa, A.E.A.D., Silva, M.C.M., Meneguim, A.M., Sá, M.E.L., Engler, G., Almeida-Engler, J.de, Fernandez, D. and Grossi-de-Sá, M.F. (2015) Seed-specific stable expression of the α-*AI1* inhibitor in coffee grains and the *in vivo* implications for the development of the coffee berry borer. *Tropical Plant Biology* 8, 98–107.

Altieri, M.A. and Nicholls, C.I. (2003) Soil fertility management and insect pests: harmonizing soil and plant health in agroecosystems. *Soil and Tillage Research* 72, 203–211.

Alves, D.S., Oliveira, D.F., Carvalho, G.A., dos Santos, H.M. Jr, Carvalho, D.A., Santos, M.A.I. and de Carvalho, H.W.P. (2011) Plant extracts as an alternative to control *Leucoptera coffeella* (Guérin-Mèneville) (Lepidoptera: Lyonetiidae). *Neotropical Entomology* 40, 123–128.

Ambrogi, B.G., Lima, E.R. and Sousa-Souto, L. (2006) Efficacy of mating disruption for control of the coffee leaf miner *Leucoptera coffeella* (Guérin-Méneville) (Lepidoptera: Lyonetiidae). *BioAssay* 1, 1–5.

Anthony, F., Astorga, C. and Berthaud, J. (1999) Los recursos genéticos: las bases de una olución genética a los problemas de la caficultura latinoamericana. In: Bertrand, B. and Rapidel, B. (eds) *Desafíos de la Caficultura en Centroamérica*. IICA-PROMECAFÉ, CIRAD, IDR, CCCR, San José, Costa Rica, pp. 369–406.

Aranda-Delgado, E. (1986) Control natural del minador de la hoja del cafeto en México *Leucoptera coffella* (Guer.-Men. 1842) (Lep. Lyonetiidae). Undergraduate dissertation, Universidad Veracruzana, Xalapa, Veracruz, México, 217 pp.

Aristizábal, L.F., Bustillo, A.E. and Arthurs, S.P. (2016) Integrated pest management of coffee berry borer: strategies from Latin America that could be useful for coffee farmers in Hawaii. *Insects* 7(1), 6–24.

Asociación Nacional del Café (ANACAFÉ) (2014) Nueva plaga detectada en cafetales de Huehuetenango. *El Cafetal* 40, 8–9.

Avelino, J., Cristancho, M., Georgiou, S., Imbach, P., Aguilar, L., Bornemann, G., Läderach, P., Anzueto, F., Hruska, A.J. and Morales, C. (2015) The coffee rust crises in Colombia and Central America (2008–2013): impacts, plausible causes and proposed solutions. *Food Security* 7, 303–321.

Ayalew, T. (2014) Characterization of organic coffee production, certification and marketing systems: Ethiopia as a main indicator: a review. *Asian Journal of Agricultural Research* 8, 170–180.

Bacca, T., Lima, E.R., Picanço, M.C., Guedes, R.N.C. and Viana, J.H.M. (2006) Optimum spacing of pheromone traps for monitoring the coffee leaf miner *Leucoptera coffeella*. *Entomologia Experimentalis et Applicata* 119, 39–45.

Bacon, C.M., Méndez, V.E., Gliessman, S.R., Goodman, D. and Fox, J.A. (eds) (2008) *Confronting the Coffee Crisis: Fair Trade, Sustainable Livelihoods and Ecosystems in Mexico and Central America*. The MIT Press, Cambridge, Massachusetts, 390 pp.

Baker, P.S. (2001) GM coffee. In: Baker, P.S. (ed.) *Coffee Futures: A Source Book of Some Critical Issues Confronting the Coffee Industry*. CABI-FEDERECAFE, USDA-ICO. The Commodities Press, Chinchiná, Colombia, pp. 104–111.

Baker, P.S. and Barrera, J.F. (1993) A field study of a population of coffee berry borer, *Hypothenemus hampei* (Coleoptera: Scolytidae), in Chiapas, Mexico. *Tropical Agriculture* 70, 351–355.

Baker, P.S., Barrera, J.F. and Rivas, A. (1992) Life history studies of the coffee berry borer (*Hypothenemus hampei*, Scolytidae) on coffee trees in southern Mexico. *Journal of Applied Entomology* 29, 656–662.

Baker, P.S., Jackson, J.A.F. and Murphy, S.T. (2002) *Natural Enemies, Natural Allies*. Project completion report of the integrated management of the coffee berry borer project, CFC/ICO/02 (1998–2002). The Commodities Press, CABI Commodities, Egham, UK and Cenicafé, Chinchiná, Colombia.

Balakrishnan, M.M., Prakasan, C.B. and Sreedharan, K. (2000) Biocontrol of coffee berry borer in India. In: *International Scientific Symposium on Coffee*. Central Coffee Research Institute (CCRI), Bangalore, India, p. 111.

Barrera, G.J.F. (1994) Dynamique des populations du scolyte des fruits du caféier, *Hypothenemus hampei* (Coleoptera: Scolytidae), et lutte biologique avec le parasïtoide *Cephalonomia stephanoderis* (Hymenoptera: Bethylidae), au Chiapas, Mexique. PhD thesis, Université Paul Sabatier, Toulouse, France, p. 301.

Barrera, J.F. (2002) Principios agroecológicos para el manejo de plagas en cafetales. In: Pohlan, J. (ed.) *México y la Cafeticultura Chiapaneca; Reflexiones y Alternativas para los Caficultores*. Shaker Verlag, Herzogenrath, Germany, pp. 201–208.

Barrera, J.F. (2008) Coffee pests and their management. In: Capinera, J.L. (ed.) *Encyclopedia of Entomology*, 2nd edn. Springer, New York, pp. 961–998.

Barrera, J.F. and Covarrubias, M.L. (1984) Efecto de diferentes condiciones de sombra del cafetal sobre la intensidad del ataque de la broca del grano de café *Hypothenemus hampei* (Ferr.) (Coleoptera: Scolytidae) en el Soconusco. In: *Memorias del II Congreso Nacional de Manejo Integrado de Plagas*. AgMIP. Asociación Guatemalteca de Manejo Integrado de Plagas (AGMIP), Guatemala, Chiapas, México, pp. 208–218.

Barrera, J.F., Baker, P.S., Schwarz, A. and Valenzuela, J.E. (1990a) Introducción de dos especies de parasitoides africanos a México para el control biológico de la broca del café *Hypothenemus hampei* (Ferrari) (Coleoptera: Scolytidae). *Folia Entomologica Mexicana* 79, 245–247.

Barrera, J.F., Infante, F., Vega, M., González, O., Carrillo, E., Campos, O., Muñoz, R., Serrano, A., Osorto, J.J., Decazy, B. and Moore, D. (1990b) Introducción de *Cephalonomia stephanoderis* (Hymenoptera: Bethylidae) a Centroamérica para el control biológico de la broca del cafeto, *Hypothenemus hampei* (Coleoptera: Scolytidae). *Turrialba* 40, 570–574.

Barrera, J.F., López, G., Herrera, J., Ventura, S. and Nieto, G. (2002a) Bioecología y hábitos del Taladrador de las ramas del café robusta en el Soconusco, Chiapas. In: Barrera, J.F. (ed.) *Tres Plagas del Café en Chiapas*. El Colegio de la Frontera Sur, Mexico, pp. 85–94.

Barrera, J.F., Herrera, J., Zúñiga, J.A., Moreno, B. and Junghans, C. (2002b) Bioecología y hábitos del Chacuatete del café en Siltepec, Chiapas. In: Barrera, J.F. (ed.) *Tres Plagas del Café en Chiapas*. El Colegio de la Frontera Sur, Mexico, pp. 49–58.

Barrera, J.F., Herrera, J., Rabanales, R. and Pinson, E. (2003) Es *Leucoptera coffeella* (Lepidoptera: Lyonetiidae) plaga secundaria en cafetales donde *Idiarthron subquadratum* (Orthoptera: Tettigoniidae) es la plaga clave? In: Vázquez, M., Pérez, J.F., Ibarra, K.H., Balpuesta, C.I., Vázquez, J.R., Cervantes, J. and Ibarra, N. (eds) *Memorias del XXVI Congreso Nacional de Control Biológico*. Guadalajara, Jalisco, México, pp. 114–117.

Barrera, J.F., Villacorta, A. and Herrera, J. (2004) Fluctuación estacional de las capturas de 'La Broca del café' (*Hypothenemus hampei*) con trampas de etanol – metanol e implicaciones sobre el número de trampas. *Entomología Mexicana* 3, 540–544.

Barrera, J.F., Villacorta, A., Herrera, J., García, H. and Cruz, L. (2007) Aplicación de trampas para el monitoreo de la broca del café. In: Hohmann, C.L. (ed.) (organizer) *Anais – Manejo da Broca-do-Café: Workshop Internacional*, Workshop Internacional, Londrina, Paraná, Brazil. Instituto Agronômico do Paraná (IAPAR), Paraná, Brazil, pp. 95–112.

Barrera, J.F., Gómez, J., Castillo, A., López, E., Herrera, J. and González, G. (2008a) Broca del café, *Hypothenemus hampei* (Coleoptera: Curculionidae). In: Arredondo-Bernal, H.C. and Rodríguez-del-Bosque, L.A. (eds) *Casos de Control Biológico en México*. Mundi-Prensa, Mexico, pp. 101–120.

Barrera, J.F., Herrera, J. and Gómez, J. (2008b) Fluctuación de la población de adultos de *Rhabdopterus jansoni* (Jacoby) (Coleoptera: Chrysomelidae) en cafetales del Soconusco, Chiapas, Mexico. *Entomologica Mexicana* 7, 246–251.

Beer, J., Muschler, R., Kass, D. and Somarriba, E. (1998) Shade management in coffee and cacao plantations. *Agroforestry Systems* 38, 139–164.

Benavides, P., Góngora, C. and Bustillo, A. (2012) IPM program to control coffee berry borer *Hypothenemus hampei*, with emphasis on highly pathogenic mixed strains of *Beauveria bassiana*, to overcome insecticide resistance in Colombia. In: Perveen, F. (ed.) *Insecticides – Advances in Integrated Pest Management*. InTech Europe, Rijeka, Croatia, pp. 511–540.

Benzing, A. (2001) *Agricultura Orgánica: Fundamentos para la Región Andina*. Neckar-Verlag, Villengen-Schwenningen, Germany.

Bethke, J.A., Dreistadt, S.H. and Varela, L.G. (2014) *Thrips: Integrated Pest Management for Home Gardeners and Landscape Professionals*. Pest notes. University of California Publication 7429. University of California, Davis, California, p. 8.

Bigger, M. (1969) Giant looper *Ascotis selenaria reciprocaria* Walk. in Tanzania. *East African Agricultural and Forestry Journal* 35, 49–51.

Blackman, A. and Naranjo, M.A. (2012) Does eco-certification have environmental benefits? Organic coffee in Costa Rica. *Ecological Economics* 83, 58–66.

Bosselmann, A.S., Dons, K., Oberthur, T., Olsen, C.S., Ræbild, A. and Usma, H. (2009) The influence of shade trees on coffee quality in small holder coffee agroforestry systems in Southern Colombia. *Agriculture, Ecosystems & Environment* 129, 253–260.

Brun, L.O., Marcillaud, C., Gaudichon, V. and Suckling, D.M. (1989) Endosulfan resistance in *Hypothenemus hampei*. *Journal of Economic Entomology* 82, 1311–1316.

Campos, O.G., Decazy, B. and Carrillo, E. (1989) Dinámica poblacional del minador de la hoja del cafeto *Leucoptera coffeella* y sus enemigos naturales en la zona de Nuevo San Carlos, Retalhuleu, Guatemala. *Turrialba* 3, 393–399.

Cannell, M.G.R. (1985) Physiology of the coffee crop. In: Clifford, M.N. and Wilson, K.C. (eds) *Coffee: Botany, Biochemistry and Production of Beans and Beverage*. AVI Publishing Company, Westport, Connecticut, pp. 108–134.

Cárdenas-Murillo, R. and Posada-Flórez, F.J. (2001) *Los Insectos y Otros Habitantes de Cafetales y Platanales*. Cenicafé, Colombia, p. 259.

Celestino, F.V., Pratissoli, D., Machado, L.C., dos Santos Junior, H.J.G., de Queiroz, V.T. and Mardgan, L. (2016) Control of coffee berry borer, *Hypothenemus hampei* (Ferrari) (Coleoptera: Curculionidae: Scolytinae) with botanical insecticides and mineral oils. *Acta Scientiarum* 38, 1–8.

Constantino, L.M., Benavides, P.M. and Esteban-Durán, J.R. (2014) Description of a new species of coffee stem and root borer of the genus *Plagiohammus* Dillon and Dillon from Colombia (Coleoptera: Cerambycidae: Lamiinae), with a key to the Neotropical species. *Insecta Mundi* 0337, 1–21.

Coste, R. (1964) *Cafetos y Cafés del Mundo*. Los Cafetos. Tomo Primero. G.-P. Maisonneuve & Larose, Paris, p. 459.

Crowe, T.J. (2004) Coffee pests in Africa. In: Wintgens, J.N. (ed.) *Coffee: Growing, Processing, Sustainable Production*. Wiley-VCH Verlag GmbH & Co., Weinheim, Germany, pp. 421–458.

Culik, M.P., dos, D., Martins, S., Ventura, J.A. and Wolff, V.S. (2008) Diaspididae (Hemiptera: Coccoidea) of Espírito Santo, Brazil. *Journal of Insect Science* 8(17), 6.

David-Rueda, G., Constantino, L.M., Montoya, E.C., Ortega, O.E., Gil, Z.N. and Benavides-Machado, P. (2016) Diagnóstico de *Leucoptera coffeella* (Lepidoptera: Lyonetiidae) y sus parasitoides en el departamento de Antioquia. *Revista Colombiana de Entomología* 42, 4–11.

de Graaff, J. (1986) *The Economy of Coffee*. *Economics of Crops in Developing Countries* No. 1. Centre for Agricultural Publishing and Documentation, Wageningen, The Netherlands, p. 294.

Dufour, B.P., Franco, F.F. and Hernández, A. (2007) Evaluación del trampeo en el marco del manejo integrado de la broca del café. In: Barrera, J.F., García, A., Domínguez, V. and Luna, C. (eds) *La Broca del Café en América Tropical: Hallazgos y Enfoques*. Sociedad Mexicana de Entomología y El Colegio de la Frontera Sur, Mexico, pp. 89–99.

Eakin, H., Morales, H., Castellanos, E., Cruz-Bello, G. and Barrera, J.F. (2013) Coffee, disasters and social-ecological resilience in Guatemala and Chiapas, Mexico. In: Boulter, S., Palutikof, J., Karoly, D.J. and Guitart, D. (eds) *Natural Disasters and Adaptation to Climate Change*. Cambridge University Press, Cambridge, pp. 174–180.

Eakin, H., Tucker, C., Castellanos, E., Díaz, R., Barrera, J.F. and Morales, H. (2014) Adaptation in a multi-stressor environment: perceptions and responses to climatic and economic risks by coffee growers in Mesoamerica. *Environment, Development and Sustainability* 16, 123–139.

Ferreira, R.A.L.F. Protasio, P. and Vieira, L.G.E. (2006) Genetic transformation of coffee. *Brazilian Journal of Plant Physiology* 18, 83–94.

Fornazier, M.J., Krohling, B., Ambrozim, W. and da Rocha, A.C. (2000) Ocorrência das moscas-das-raízes em café arábica na região das Montanhas do Espírito Santo. In: *Simposio de Pesquisa dos Cafés do Brasil*. Poços de Caldas, Brasil, pp. 1171–1174.

Fragoso, D.B., Guedes, R.N.C., Picanço, M.C. and Zambolim, L. (2002) Insecticide use and organophosphate resistance in the coffee leaf miner *Leucoptera coffeella* (Lepidoptera: Lyonetiidae). *Bulletin of Entomological Research* 92, 203–212.

Gaitán, A.L., Cristancho, M.A., Castro-Caicedo, B.L., Rivillas, C.A. and Cadena-Gómez, G. (2015) *Compendium of Coffee Diseases and Pests*. The American Phytopathological Society (APS Press), Saint Paul, Minnesota, p. 79.

Garcia, C.A., Bhagwat, S.A., Ghazoul, J., Nath, C.D., Nanaya, K.M., Kushalappa, C.G., Raghuramulu, Y., Nasi, R. and Vaast, P. (2009) Biodiversity conservation in agricultural landscapes: challenges and opportunities of coffee agroforests in the Western Ghats, India. *Conservation Biology* 24, 479–488.

Giovannucci, D. and Koekoek, F.J. (2003) *The State of Sustainable Coffee: A Study of Twelve Major Markets*. International Institute for Sustainable Development (IISD), United Nations Conference on Trade and Development (UNCTAD) and International Coffee Organization (ICO), Cali, Valle del Cauca, Colombia, 198 pp.

Giraldo-Jaramillo, M., Benavides-Machado, P. and Villegas-García, C. (2010) Aspectos morfológicos y biológicos de *Monalonion velezangeli* Carvalho & Costa (Hemiptera: Miridae) en café. *Cenicafé* 61, 195–205.

Godoy, C. and Villalobos, W. (2006) Two new species of *Graphocephala* Van Duzee from Costa Rica (Cicadellidae: Cicadellinae). *Zootaxa* 1298, 61–68.

Godoy, C., Garita-Cambronero, J., Rivera, C. and Villalobos, W. (2006) Two new species of *Kapateira* Young from Costa Rica (Auchenorrhyncha: Cicadellidae: Cicadellinae). *Zootaxa* 1282, 29–38.

Goggin, F.L. and Zhu-Salzman, K. (2015) Pests and resistance: social networking – studying the web of plant–insect interactions to improve host plant resistance. [Editorial overview.] *Current Opinion in Insect Science* 9, v–viii.

Gómez, R.J.A., Santos, O., Valle-Mora, J. and Montoya, P.J.G. (2010) Determinación del establecimiento de parasitoides de la broca del café *Hypothenemus hampei* (Coleoptera: Curculionidae, Scolytinae) en cafetales del Soconusco, Chiapas, México. *Entomotropica* 25, 25–36.

Goodman, D. (2008) The international coffee crisis: a review of the issues. In: Bacon, C.M., Méndez, V.E., Gliessman, S.R., Goodman, D. and Fox, J.A. (eds) *Confronting the Coffee Crisis: Fair Trade, Sustainable Livelihoods and Ecosystems in Mexico and Central America*. The MIT Press, Cambridge, Massachusetts, pp. 3–25.

Green, P.W.C., Davis, A.P., Cossé, A.A. and Vega, F.E. (2015) Can coffee chemical compounds and insecticidal plants be harnessed for control of major coffee pests? *Journal of Agricultural and Food Chemistry* 63, 9427–9434.

Grossman, J.M. (2003) Exploring farmer knowledge of soil processes in organic coffee systems of Chiapas, Mexico. *Geoderma* 111, 267–287.

Guedes, R.N.C. and Oliveira, E.E. (2002) Resistência a inseticidas em insetos-pragas do cafeeiro: situação e perspectivas. In: Zambolim, L. (ed.) *O Estado da Arte de Tecnologias na Produção de Café*. Universidade Federal de Viçosa, Minas Gerais, Brazil, pp. 471–497.

Guerreiro Filho, O. (2006) Coffee leaf miner resistance. *Brazilian Journal of Plant Physiology* 18, 109–117.

Guerreiro-Filho, O. and Mazzafera, P. (2003) Caffeine and resistance of coffee to the berry borer *Hypothenemus hampei* (Coleoptera: Scolytidae). *Journal of Agricultural and Food Chemistry* 51, 6987–6991.

Guharay, F., Monterroso, D. and Staver, C. (2001) El diseño y manejo de la sombra para la supresión de plagas en cafetales de América Central. *Agroforestería en las Américas* 8, 22–29.

Haarer, A.E. (1964) *Producción Moderna de Café*. Cía. Editorial Continental, S.A. de C.V., Mexico, p. 652.

Herzog, E. (1994) Multipurpose shade trees in coffee and cocoa plantations in Côte d'Ivoire. *Agroforestry Systems* 27, 259–267.

International Coffee Organization (ICO) (2016a) Total Production by all Exporting Countries. Available at: http://www.ico.org/prices/po-production.pdf (accessed 30 June 2016).

International Coffee Organization (ICO) (2016b) Developing a Sustainable Coffee Economy. Available at: http://www.ico.org/sustaindev_e.asp?section=What_We_Do (accessed 15 November 2016).

Jacques, M.-A., Denancé, N., Legendre, B., Morel, E., Briand, M., Mississipi, S., Durand, K., Olivier, V., Portier, P., Poliakoff, F. and Crouzillat, D. (2016) New coffee plant infecting *Xylella fastidiosa* variants derived via homologous recombination. *Application of Environmental Microbiology* 82, 1556–1568.

Janse, J.D. and Obradovic, A. (2010) *Xylella fastidiosa*: its biology, diagnosis, control and risks. *Journal of Plant Pathology* 92(1), S1.35–S1.48.

Jaramillo, J., Setamou, M., Muchugu, E., Chabi-Olaye, A., Jaramillo, A., Mukabana, J., Maina, J., Gathara, S. and Borgemeister, C. (2013) Climate change or urbanization? Impacts on a traditional coffee production system in East Africa over the last 80 years. *PLoS One* 8, e51815.

Jha, S., Bacon, C.M., Philpott, S.M., Méndez, V.E., Läderach, P. and Rice, R.A. (2014) Shade coffee: update on a disappearing refuge for biodiversity. *BioScience* 64, 416–428.

Jiménez-Soto, M.E., Cruz-Rodríguez, J.A., Vandermeer, J. and Perfecto, I. (2013) *Hypothenemus hampei* (Coleoptera: Curculionidae) and its interactions with *Azteca instabilis* and *Pheidole synanthropica* (Hymenoptera: Formicidae) in a shade coffee agroecosystem. *Environment Entomology* 42, 915–924.

Klein-Koch, C., Espinoza, O., Tandazo, A., Cisneros, P. and Delgado, D. (1988) Factores naturales de regulación y control biológico de la broca del café (*Hypothenemus hampei*). *Sanidad Vegetal* 3, 5–30.

Kondo, T. (2013) A new species of *Toumeyella* Cockerell (Hemiptera: Coccoidea: Coccidae) on coffee roots, *Coffea arabica* L. (Rubiaceae), from Colombia and Venezuela. *Corpoica Ciencia y Tecnología Agropecuaria* 14, 39–51.

Kuesel, R., Gonthier, D.J., Cruz, M., Vaiyda, C., Iverson, A.L. and Perfecto, I. (2014) Local management and landscape use intensity associated with a coffee leaf-chewing beetle. *Agroecology and Sustainable Food Systems* 38, 532–540.

Kutywayo, D., Chemura, A., Kusena, W., Chidoko, P. and Mahoya, C. (2013) The impact of climate change on the potential distribution of agricultural pests: the case of the coffee white stem borer (*Monochamus leuconotus* P.) in Zimbabwe. *PLoS One* 8, e73432.

Lan, C.C. and Wintgens, J.N. (2004) Coffee pest in Africa. In: Wintgens, J.N. (ed.) *Coffee: Growing, Processing, Sustainable Production*. Wiley-VCH Verlag GmbH & Co., Weinheim, Germany, pp. 459–473.

Le Pelley, R.H. (1943) The biological control of a mealy bug on coffee and other crops in Kenya. *Empire Journal of Experimental Agriculture* 11, 78–88.

Le Pelley, R.H. (1973) *Las Plagas del Café*. Editorial Labor, S.A., Barcelona, Spain, p. 693.

Leroy, L., Henry, A.-M., Royer, M., Altosaar, I., Frutos, R., Duris, D. and Philippe, R. (2000) Genetically modified coffee plants expressing the *Bacillus thuringiensis cry*1Ac gene for resistance to leaf miner. *Plant Cell Reports* 19, 382–389.

Li, W.-B., Pria, W.D. Jr, Teixeira, D.C., Miranda, V.S., Ayres, A.J., Franco, C.F., Costa, M.G., He, C.-X., Costa, P.I. and Hartung, J.S. (2001) Coffee leaf scorch caused by a strain of *Xylella fastidiosa* from citrus. *Plant Disease* 85, 501–505.

Lomelí-Flores, J.R., Barrera, J.F. and Bernal, J.S. (2009) Impact of natural enemies on coffee leafminer *Leucoptera coffeella* (Lepidoptera: Lyonetiidae) population dynamics in Chiapas, Mexico. *Biological Control* 51, 51–60.

Lomelí-Flores, J.R., Barrera, J.F. and Bernal, J.S. (2010) Impacts of weather, shade cover and elevation on coffee leafminer *Leucoptera coffeella* (Lepidoptera: Lyonetiidae) population dynamics and natural enemies. *Crop Protection* 29, 1039–1048.

Maestri, M., Barros, R.S. and Rena, A.B. (2001) Coffee. In: Last, F.T. (ed.) *Tree Crop Ecosystems*. Elsevier, Amsterdam, pp. 339–360.

Maldonado, L.C.E. and Benavides, M.P. (2007) Evaluación del establecimiento de *Cephalonomia stephanoderis* y *Prorops nasuta*, controladores de *Hypothenemus hampei*, en Colombia. *Cenicafé* 58, 333–339.

Mariño, Y.A., Pérez, M.-E., Gallardo, F., Trifilio, M., Cruz, M. and Bayman, P. (2016) Sun vs. shade affects infestation, total population and sex ratio of the coffee berry borer (*Hypothenemus hampei*) in Puerto Rico. *Agriculture, Ecosystems & Environment* 222, 258–266.

Martinelli, N.M. and Zucchi, R.A. (1997) Cigarras (Hemiptera: Cicadidae: Tibicinidae) associadas ao cafeeiro: Distribuição, hospedeiros e chave para as espécies. *Anais da Sociedade Entomológica do Brasil* 26, 133–143.

Martínez, M.A. and Suris, M. (2000) Bases bioecológicas para el manejo de chinches harinosas en el cultivo del café en Cuba. *Manejo Integrado de Plagas* 57, 58–64.

Martínez-Torres, M.E. (2006) *Organic Coffee: Sustainable Development by Mayan Farmers*. Ohio University Press, Athens, p. 176.

Marucci, R.C., Lopes, J.R.S. and Cavichioli, R.R. (2008) Transmission efficiency of *Xylella fastidiosa* by sharpshooters (Hemiptera: Cicadellidae) in coffee and citrus. *Journal of Economic Entomology* 101, 1114–1121.

Mas, A.H. and Dietsch, T.V. (2004) Linking shade coffee certification to biodiversity conservation: Butterflies and birds in Chiapas, Mexico. *Ecological Applications* 14, 642–654.

Matiello, J.B. (1991) *O Café. Do Cultivo ao Consumo*. Publicações Globo Rural, Coleção do Agricultor, Grãos. Editora Globo, S.A., Brazil, 320 pp.

Matos, D.M., Guharay, F. and Beer, J. (2004) Incidencia de la broca (*Hypothenemus hampei*) en plantas de café a pleno sol y bajo sombra de *Eugenia jambos* y *Gliricidia sepium* en San Marcos, Nicaragua. *Agroforestería de las Américas* 41(42), 56–61.

Mawussi, G., Vilarem, G., Raynaud, C., Merlina, G., Gbongli, A.K., Wegbe, K. and Sanda, K. (2009) Chemical composition and insecticidal activity of *Aeollanthus pubescens* essential oil against coffee berry borer (*Hypothenemus hampei* Ferrari) (Coleoptera: Scolytidae). *Journal of Essential Oil Bearing Plants* 12, 327–332.

Mawussi, G., Tounou, A.K., Ayisah, K.D., Vilarem, G., Raynaud, C., Merlina, G., Wegbe, K. and Sanda, K. (2012) Chemical composition and insecticidal activity of *Ocimum canum* essential oil against coffee berry borer, *Hypothenemus hampei* (Ferrari) (Coleoptera: Scolytidae). *Journal of Essential Oil Bearing Plants* 15, 955–963.

Mendesil, E., Tadesse, M. and Negash, M. (2012) Efficacy of plant essential oils against two major insect pests of coffee (coffee berry borer, *Hypothenemus hampei*, and antestia bug, *Antestiopsis intricata*) and maize weevil, *Sitophilus zeamais*. *Archives of Phytopathology and Plant Protection* 45, 366–372.

Méndez, V.E., Bacon, C.M., Olson, M., Petchers, S., Herrador, D., Carranza, C., Trujillo, L., Guadarrama-Zugasti, C., Cordón, A. and Mendoza, A. (2010) Effects of Fair Trade and organic certifications on small-scale coffee farmer households in Central America and Mexico. *Renewable Agriculture and Food Systems* 25, 236–251.

Mendonça, A.P., Nonato, J.V.A., Andrade, V.T., Fatobene, B.J.D.R., Braghini, M.T., Prela-Pantano, A. and Guerreiro Filho, O. (2016) *Coffea arabica* clones resistant to coffee leaf miner. *Crop Breeding and Applied Biotechnology* 16, 42–47.

Mishra, M.K. and Slater, A. (2012) Recent advances in the genetic transformation of coffee. *Biotechnology Research International* 2012, 1–17.

Moguel, P. and Toledo, V.M. (1999) Biodiversity conservation in traditional coffee systems of Mexico. *Conservation Biology* 13, 11–21.

Morón, M.A. and Rodríguez-del-Bosque, L.A. (2010) Importancia, historia y retos. In: Rodríguez-del-Bosque, L.A. and Morón, M.A. (eds) *Plagas del Suelo*. Mundi-Prensa, Mexico, pp. 3–17.

Muñoz, R. (1990) *Manual de Plagas y Enfermedades del Café*. Instituto Hondureño del Café, Tegucigalpa, Honduras, p. 61.

Murphy, S.T. (1991) Insect natural enemies of coffee green scales (Hemiptera: Coccidae) in Kenya and their potential for biological control of *Coccus celatus* and *C. viridis* in Papua New Guinea. *Entomophaga* 36, 519–529.

Muschler, R.G. (2001) Shade improves coffee quality in a sub-optimal coffee-zone of Costa Rica. *Agroforestry Systems* 85, 131–139.

Muschler, R.G. (2016) Agroforestry: essential for sustainable and climate-smart land use? In: Pancel, L. and Köhl, M. (eds) *Tropical Forestry Handbook*. Springer, Heidelberg, Germany, pp. 2013–2116.

Nathanson, J.A. (1984) Caffeine and related methylxanthines: possible naturally occurring pesticides. *Science* 226, 184–187.

Osorio, N. (2002) The global coffee crisis: a threat to sustainable development. Submission to the World Summit on Sustainable Development, Johannesburg, August 2002 by Néstor Osorio, Executive Director of the International Coffee Organization. International Coffee Organization (ICO), Johannesburg, South Africa, p. 4.

Osorio, N. (2004) Lessons from the world coffee crisis: a serious problem for sustainable development. Submission to United Nations Conference on Trade and Development (UNCTAD) XI, São Paulo, Brazil, June 2004 by Néstor Osorio, Executive Director of the International Coffee Organization. International Coffee Organization (ICO), São Paulo, Brazil, p. 4.

Oduor, G.I. and Simons, S.A. (2003) Biological control in IPM for coffee. In: Neuenschwander, P., Borgemeister, C. and Langewald, J. (eds) *Biological Control in IPM in Africa*. CAB International, Wallingford, UK, pp. 347–362.

Padilla, M.R. and Rodríguez, H.W. (2000) Caracterización del grillo indiano del café (*Paroecanthus* spp. Sauss. Orthoptera: Gryllidae) y acciones para el manejo del insecto. In: *XIX Simposio Latinoamericano de Caficultura*. Instituto Interamericano de Cooperación para la Agricultura (IICA)/Programa Cooperativo Regional para el Desarrollo Tecnológico y Modernización de la Caficultura (PROMECAFE) and Instituto del Café de Costa Rica (ICAFÉ), San José, Costa Rica, pp. 423–432.

Peña-Marrero, E., García-Hernández, M., Blanco-Rodríguez, E. and Barrera-Gaytán, J.F. (2006) Introducción de la avispa de Costa de Marfil *Cephalonomia stephanoderis* Betrem (Hymenoptera: Bethylidae), parasitoide de la broca del fruto del cafeto *Hypothenemus hampei* Ferrari (Coleoptera: Scolytidae) en Cuba. *Fitosanidad* 10, 33–36.

Perfecto, I., Rice, R.A., Greenberg, R. and Van der Voort, M.E. (1996) Shade coffee: a disappearing refuge for biodiversity. *BioScience* 46, 598–608.

Perfecto, I., Vandermeer, J. and Philpott, S.M. (2014) Complex ecological interactions in the coffee agroecosystem. *Annual Review of Ecology, Evolution, and Systematics* 45, 137–158.

Perthuis, B., Pradon, J.L., Montagnon, C., Dufour, M. and Leroy, T. (2005) Stable resistance against the leaf miner *Leucoptera coffeella* expressed by genetically transformed *Coffea canephora* in a pluriannual field experiment in French Guiana. *Euphytica* 144, 321–329.

Philpott, S.M. and Dietsch, T. (2003) Coffee and conservation: a global context and the value of farmer involvement. *Conservation Biology* 17, 1844–1846.

Raga, A., Mineiro, J.L.C. and Wolff, V.R.S. (2003) Novos registros de hospedeiros de cochonilhas (Hemiptera: Diaspididae, Coccidae) no estado de São Paulo. *Arquivos do Instituto Biologico* (São Paulo) 70(Supplement 3), 57–60.

Raynolds, L.T., Murray, D. and Heller, A. (2007) Regulating sustainability in the coffee sector: a comparative analysis. *Agriculture and Human Values* 24, 147–163.

Reis, P.R. and de Souza, J.C. (1986) Pragas do cafeiro. In: Rena, A.B., Malavolta, E., Rocha, M. and Yamada, T. (eds) *Cultura do Cafeiro: Fatores que Afetam a Productividade*. Associação Brasileira para Pesquisa da Potassa e do Fosfato, Piracicaba, Brazil, pp. 323–338.

Rodríguez, C.M., Obando, J.J., Villalobos, W., Moreira, L. and Rivera, C. (2001) First report of *Xylella fastidiosa* infecting coffee in Costa Rica. *Plant Disease* 85, 1027.

Romero, J.V. and Cortina-Guerrero, H. (2004) Fecundidad y ciclo de vida de *Hypothenemus hampei* (Coleoptera: Curculionidae: Scolytinae) en introducciones silvestres de café. *Cenicafé* 55, 221–231.

Romero, J.V. and Cortina, H.A.G. (2007) Tablas de vida de *Hypothenemus hampei* (Coleoptera: Curculionidae: Scolytinae) sobre tres introducciones de café. *Revista Colombiana de Entomología* 33, 10–16.

Roobakkumar, A., Samuel, S.D., Balakrishnan, M.M. and Sreedharan, K. (2014) Release and establishment of the parasitoid *Cephalonomia stephanoderis* Betrem against the coffee berry borer *Hypothenemus hampei* Ferrari in Pulney Hills, Tamil Nadu, India. *Entomological News* 124, 221–223.

Santos, M.R.A., Lima, R.A., Silva, A.G., Teixeira, C.A.D., Alpirez, I.P.V. and Facundo, V.A. (2013a) Composiçãõ química e atividade inseticida do extrato acetônico de *Piper alatabaccum* Trel & Yuncker (Piperaceae) sobre *Hypothenemus hampei* Ferrari. *Revista Brasileira de Plantas Medicinais* 15, 332–336.

Santos, M.R.A., Lima, R.A., Silva, A.G., Lima, D.K.S., Sallet, L.A.P., Teixeira, C.D.A. and Facundo, V.A. (2013b) Composiçãõ química e atividade inseticida do óleo essencial de *Schinus terebinthifolius* Raddi (Anacardiaceae) sobre a broca-do-café (*Hypothenemus hampei*) Ferrari. *Revista Brasileira de Plantas Medicinais* 15, 757–762.

Sera, G.H., Sera, T., Ito, D.S., de Azevedo, J.A., Filho, C.R., da Mata, J.S., Cotarelli, V.M. and Doi, D.S. (2007) Resistênsia à broca em espécies e variedades de café. In: Hohmann, C.L. (ed.) (organizer) *Anais – Manejo da Broca-do-Café*, Workshop Internacional, Londrina, Paraná, Brazil. Instituto Agronômico do Paraná (IAPAR), Paraná, Brazil, pp. 263–272.

Smith, R.F. (1985) A history of coffee. In: Clifford, M.N. and Wilson, K.C. (eds) *Coffee: Botany, Biochemistry and Production of Beans and Beverage*. AVI Publishing Company, Westport, Connecticut, pp. 1–12.

Soto-Pinto, L., Perfecto, I. and Caballero-Nieto, J. (2002) Shade over coffee: its effects on berry borer, leaf rust and spontaneous herbs in Chiapas, Mexico. *Agroforestry Systems* 55, 37–45.

Souza, J.C., Reis, P.R. and Rigitano, R.L.O. (1998) *Bicho-mineiro do cafeeiro: biologia, danos e manejo integrado. Boletim Técnico* No. 54. EPAMIG (Agricultural Research Company of Minas Gerais), Belo Horizonte, Brazil, p. 48.

Thapa, S. and Lantinga, E.A. (2016) Infestation by coffee white stem borer, *Xylotrechus quadripes*, in relation to soil and plant nutrient content and associated quality aspects. *Southwestern Entomologist* 41, 331–335.

Tscharntke, T., Clough, Y., Wanger, T.C., Jackson, L., Motzke, I., Perfecto, I., Vandermeer, J. and Whitbread, A. (2012) Global food security, biodiversity conservation and the future of agricultural intensification. *Biological Conservation* 151, 53–59.

Tuelher, E.S., Venzon, M., Guedes, R.N.C. and Pallini, A. (2014) Toxicity of organic-coffee-approved products to the southern red mite *Oligonychus ilicis* and to its predator *Iphiseiodes zuluagai*. *Crop Protection* 55, 28–34.

Vandermeer, J., Perfecto, I. and Liere, H. (2009) Evidence for hyperparasitism of coffee rust (*Hemileia vastatrix*) by the entomogenous fungus, *Lecanicillium lecanii*, through a complex ecological web. *Plant Pathology* 58, 636–641.

Vandermeer, J., Perfecto, I. and Philpott, S.M. (2010) Ecological complexity and pest control in organic coffee production: uncovering an autonomous ecosystem service. *BioScience* 60, 527–537.

Vayssière, P. (1964) Los parásitos animales del cafeto. In: Coste, R. (ed.) *Cafetos y Cafés en el Mundo*. Los cafetos, Tomo Primero. G.-P. Maisonneuve & Larose, Paris, pp. 283–388.

Vega, F.E., Infante, F. and Johnson, A. (2015) The genus *Hypothenemus*, with emphasis on *H. hampei*, the coffee berry borer. In: Vega, F.E. and Hofstetter, R.W. (eds) *Bark Beetles: Biology and Ecology of Native and Invasive Species*. Academic Press, Elsevier, New York, pp. 427–494.

Velasco, E., Verdecia, J., Medina, R. and Rodríguez, L. (2001) Variaciones en el microclima de un cafetal en dependencia de la exposición a la radiación solar en las condiciones del macizo de la Sierra Maestra. *Cultivos Tropicales* 22, 53–59.

Vélez-Angel, R. (1972) Aguacate y sauce: nuevos hospederos del pasador del cafeto *Xylosandrus* (*Xyleborus*) *morigerus* Bland. *La Revista Facultad Nacional de Agronomía* 37, 78–81.

Venkatesha, M.G. and Dinesh, A.S. (2012) The coffee white stemborer *Xylotrechus quadripes* (Coleoptera: Cerambycidae): bioecology, status and management. *International Journal of Tropical Insect Science* 32, 177–188.

Waller, J.M., Bigger, M. and Hillocks, R.J. (2007) *Coffee Pests, Diseases and their Management*. Crop Protection Programme. CAB International, Wallingford, UK, 434 pp.

Watson, G.W. and Cox, J.M. (1990) Identity of the African coffee root mealybug, with descriptions of two new species of *Planococcus* (Homoptera: Pseudococcidae). *Bulletin of Entomological Research* 80, 99–105.

Wells, J.M., Raju, B.C., Hung, H.Y., Weisburg, W.G., Mandelco-Paul, L. and Brenner, D.J. (1987) *Xylella fastidiosa* gen. nov., sp. nov: Gram-negative, xylem-limited, fastidious plant bacteria related to *Xanthomonas* spp. *International Journal of Systematic Bacteriology* 37, 136–143.

Wood, S.L. (1982) The bark and ambrosia beetles of North and Central America (Coleoptera: Scolytidae), a taxonomic monograph. *Great Basin Naturalist Memoirs* 6, 1–1358.

Wrigley, G. (1988) *Coffee*. Longman Scientific & Technical, Harlow, Essex, UK.

Zavala-Olalde, J.A., Barrera, J.F., Morales, H. and Rojas, M. (2005) Design and evaluation of traps for *Idiarthron subquadratum* (Orthoptera: Tettigoniidae) with farmers participation in coffee plantations of Chiapas, Mexico. *Journal of Economic Entomology* 98, 821–835.

Zorzetti, J., Oliveira Janeiro Neves, P.M., Constanski, K.C., Santoro, P.H. and Batista Fonseca, I.C. (2012) Extratos vegetais sobre *Hypothenemus hampei* (Coleoptera: Curculionidae) e *Beauveria bassiana*. *Semina Ciencias Agrarias* 33, 2849–2861.

Zúñiga, J.A., Barrera, J.F., Williams, T. and Valle, J. (2002) Estimating population of *Idiarthron subquadratum* (Orthoptera: Tettigoniidae) using mark-recapture methods in coffee plantations in Chiapas, Mexico. *Environmental Entomology* 31, 515–522.

20 Pest Management in Organic Cacao

Régis Babin*
International Centre of Insect Physiology and Ecology (icipe), Nairobi, Kenya

Introduction

General information on cacao

Cacao, *Theobroma cacao*, is a small tree from the family *Malvaceae*, and originated in different forest areas of South and Central America (Wood, 1985). During the 20th century, the cacao-growing belt spread considerably over tropical areas of America, Africa and Asia, and is around 10 million ha today (FAOSTAT, 2014). Cocoa beans are produced for butter and powder that are used mainly in chocolate manufacture. In 2014, chocolate confectionery produced revenues of around US$120 bn, and these are expected to grow with the developing markets in countries with rising middle classes (Hawkins and Chen, 2014). At the same time, cocoa world production rose constantly for decades and reached 5 million t in 2012 (FAOSTAT, 2014). In 2012, Africa alone produced around 66% of total world production with four countries in the top five cocoa-producing nations, namely Ivory Coast (with 1.6 million t), Ghana, Nigeria and Cameroon. Asia produced around 19% of world production, with Indonesia being the world's second largest producer. The Americas produced around 14% of total world production of cocoa (FAOSTAT, 2014).

Cacao crop expansion in Africa and Asia came with the emergence of major pests and diseases, which have adapted to the crop from their local host plants. The most infamous examples are the cocoa mirids *Sahlbergella singularis* Hagl. and *Distantiella theobroma* Dist. (Hemiptera: Miridae), and the black pod disease due to *Phytophthora palmivora* Butler and *Phytophthora megakarya*, which became major threats for West African-producing countries in the 1960s and 1970s, respectively (Entwistle, 1985; Lass, 1985). In Latin America, witches' broom disease due to the basidiomycete fungus *Moniliophthora perniciosa* highly impacted production of cocoa in Brazil in the 1990s (Meinhardt *et al.*, 2008), while the frosty pod rot, due to *Moniliophthora roreri*, that is widely spread in Latin America, currently leads to low yield and crop abandonment (Phillips-Mora *et al.*, 2007). The cocoa pod borer *Conopomorpha cramerella* became a major pest of cacao in South-east Asia in the mid-1980s and is considered the main threat for cocoa production in most Asian-producing countries since the early 2000s (Posada *et al.*, 2011).

* regis.babin@cirad.fr

Cacao pests and diseases are numerous and no part of the plant is spared. Some insects are disease vectors, such as mealybugs of the Pseudococcidae family, which transmit the cacao swollen shoot virus (CSSV), a plant pathogenic virus that infects cacao trees in West Africa, affecting yields and often killing the trees within a few years (Domfeh *et al.*, 2011). Global crop losses due to these pests and diseases are usually assessed at 30–40% of the world cocoa production (ICCO, 2013). But pest and disease pressure is also responsible for higher costs of production, health and environmental issues due to the use of pesticides and farmer despondency leading to lower investment in cacao farming. Indeed, pest and disease management is usually problematic for cacao farmers, especially in Africa and Asia, because of inadequate farmer knowledge and practices, as well as limited access to resistant varieties and agrochemicals (ICCO, 2013).

General considerations of organic cocoa production

It is important to consider that cacao is not an industrial crop like other crops in developed countries. Cacao (90%) is grown by smallholders. Their number has been estimated at around 5 million worldwide (Hawkins and Chen, 2014). Although cacao is the main income source for most of them, farming practices still suffer from insufficient knowledge and capital investment (Hawkins and Chen, 2014). Thus cacao is often grown with little or no use of synthetic inputs, which are usually too expensive for farmers. In Africa, this situation is due to cocoa sector evolution in the last five decades. From the 1960s, governments promoted cacao cultivation and invested a lot of money in supporting farmers, especially for pest and disease management. In Cameroon, for instance, government subsidized spraying campaigns for cocoa mirids and fungicide distribution for black pod disease. Because of the 1990s' economic crisis, the pesticide and cocoa sectors were liberalized and subsidies dropped in a few years, leaving cacao farmers unprepared for facing the threats (Sonwa *et al.*, 2008). The private sector was not able to offer support to the farmers, and so pest and disease pressure worsened in the next three decades, contributing to low yields and cocoa beans of poor quality.

However, certified organic cacao exists, in low but growing proportions worldwide. Latest statistics from the International Cocoa Organization (ICCO) estimate production of certified organic cocoa at 15,000 t, less than 0.5% of the world production (ICCO, 2014). A recent world survey conducted by the Research Institute of Organic Agriculture (FiBL) and the International Federation of Organic Agriculture Movements (IFOAM) gives a more optimistic report with the total area under organic cacao assessed at 220,000 ha in 2011, around 2.3% of the world cacao-growing area. This is more than twice the proportion of the world's organic agricultural land, estimated at 0.9% in 2011 (Willer and Lernoud, 2013). The same survey indicates that the area under organic cacao increased fivefold since 2004, which is much more than most other crops (Willer and Lernoud, 2013). The fact is that countries with the highest volumes of certified organic cocoa are the minor producers of Latin America such as Bolivia, Mexico, Honduras and Peru, or other minor cocoa-producing countries like the Dominican Republic, the United Republic of Tanzania, Madagascar and São Tomé and Príncipe. By contrast, the largest cocoa producers produce low levels of certified organic cocoa, with Ivory Coast producing 0.0%, Indonesia 0.1%, Ghana 0.5% and Nigeria 0.4% (Willer and Lernoud, 2013).

Cacao pest management, organic by default

While the proportion of certified organic cocoa production is globally low worldwide, pest management on cacao is conducted for a significant part through ecologically sound practices, for two main reasons: (i) some pests have proven to be totally immune to chemical spraying because of their biology or because they develop resistance to insecticides; and (ii) the difficulties in accessing chemical inputs for smallholder farmers has resulted in use of more economic practices including the evaluation of the farm natural environment. In fact, a significant part of recommended (or traditional) practices currently used by farmers for insect pest management are fully compatible with organic production standards. However, it has to be noted that, currently, there is no single organic solution for the control of the major pests of cacao. Farmers are usually told to employ several practices to keep pest infestations under economic thresholds, and organic pest management practices are sometimes associated with inappropriate chemical spraying.

The aim of this chapter is to collate the existing knowledge of the pest management practices and tactics that could be used in organic cacao farming. The main pests of cacao are reviewed followed by details on the strategies compatible with organic farming that have been developed for their management.

Pests of Cacao

Major pests

Cacao crop development in tropical Africa and Asia came with the emergence of major insect pests, which adapted to the crop from their local host plants. Currently, in Latin America, its native continent, the crop is relatively unaffected by insect pests compared with Africa and Asia, where some major pests are widely distributed causing extensive damage to the crop.

Cocoa mirids

Mirid bugs are the most widespread and harmful insect pests of cacao worldwide. However, among the 40 species of the family Miridae damaging cacao, only a few have major economic impact on the world cocoa production. All the mirids injurious to cacao belong to the subfamily Bryocorinae and to two different tribes, the Odoniellini and the Monaloniini. The two tribes have very different morphological traits: mirids from tribe Odoniellini are usually robust insects and brownish in colour, while tribe Monaloniini are gracile and brightly coloured insects. Synthetic publications were devoted to cocoa mirid systematics, biogeography, biology, ecology and management, in the 1970s (Entwistle, 1972; Lavabre, 1977).

Sahlbergella singularis Hagl. and *Distantiella theobroma* Distant, of the tribe Odoniellini, are two closely related species native to the forest area of West and Central Africa. They were described for the first time on cacao around the beginning of the 20th century, and since then, have considerably spread with the crop throughout the current cocoa-producing countries of the region. *S. singularis* and *D. theobroma* are 10 mm long in the adult stage, and brown in colour. Their overall appearance mimics the bark of trees where they usually rest during the day. On cacao, eggs are inserted into pods and green shoots. Mirid population density is usually low in cacao plantations with seasonal maximum density of 2500 individuals/ha (around two individuals per tree). However, damage to the crop is considerable, due to mirid feeding behaviour. Like most Hemiptera, the mouthparts (stylets) are inserted in fruits at different developmental stages, as well as buds and green shoots. A large supply of saliva with hydrolytic enzymes is injected, leading to the liquefaction of plant tissues, which are finally ingested by the bug. Mirid feeding lesions on cacao pods and shoots appear as a black plug of dead tissue. On young pods, this damage may cause distortion during growth, sometimes leading to yellowing and fruit abortion. But the main damage is on vegetative parts of trees with the death of the terminal part of branches as well as many lesions. Mirid-damaged cacao plants are susceptible to fungal infection, resulting in cankering or bark roughening, destruction of the flower cushions, severe dieback of twigs and branches, and degradation of cacao farms. Economic losses attributed to African cocoa mirids have been assessed at 25–30% of the cocoa production of four of the five most important producing countries of the world, namely Ivory Coast, Ghana, Nigeria and Cameroon (Lavabre, 1977).

In Asia, Monaloniini mirids from the genus *Helopeltis* are numerous and widely distributed. Unlike mirids from tribe Odoniellini, *Helopeltis* spp. are gracile and coloured, and sometimes called cocoa mosquito bugs. Several *Helopeltis* species are major pests of important cash crops in Asia, such as black pepper (*Piper nigrum*), cashew (*Anacardium occidentale*), cinchona (*Cinchona* spp.), cacao and tea (*Camellia sinensis*) (Stonedahl, 1991). *Helopeltis antonii* Sign. was first observed on cacao in 1863 in Ceylon (present-day Sri Lanka). Nowadays, *H. antonii* and *Helopeltis theivora* (with *Helopeltis theobromae* as a subspecies) are widely distributed on cacao, with damage similar to that of African cocoa mirids, although economic loss is mainly due to damage on pods. They make characteristic necrotic lesions that kill young pods and shoots. In Malaysia, maximum damage to pods has been estimated at 85% during the fruiting months and yield losses of around 50% have been reported (Tong-Kwee *et al.*, 1989).

Mirids are usually well controlled with insecticides. The spraying campaigns implemented by West African governments from the 1960s to the 1980s led to a quick and significant increase of cocoa production in West Africa.

Cocoa pod borer

The cocoa pod borer, *Conopomorpha cramerella* Snellen is a small moth in the family Gracillariidae, endemic to South-east Asia, where it is known to affect different native fruit trees, such as rambutan (*Nephelium lappaceum*), pulasan (*Nephelium mutabile*), nam-nam (*Cynometra cauliflora*), kasai (*Pometia pinnata*) and different *Cola* species (Lim, 1992). *C. cramerella* was first reported attacking cacao in the 1860s in Sulawesi (present-day Indonesia) (Yen *et al.*, 2010). At the end of the 1980s, it was considered as the main pest of cacao in South-east Asia being widely distributed in Indonesia, the Philippines and Malaysia (Keane, 1992; Posada *et al.*, 2011). Cocoa pod borer has more recently reached New Guinea where it is now considered a major pest of cacao (Yen *et al.*, 2010). The adult female lays eggs on the cacao pod surface and the newly hatched nymph bores into the pod epidermis to reach the placenta (pulp) on which it feeds, disturbing the development of beans (Lim, 1992). Cocoa yield can be reduced by 60–84% in the case of severe infestation, and dry bean quality is also affected, which led to economic losses assessed at US$500 million/year in Asia in the early 2000s (Posada *et al.*, 2011). There is no single management strategy able to control this pest. As the nymph lives inside the pod, it is out of reach of insecticide spraying (Day, 1989; Shapiro *et al.*, 2008).

Mealybugs as vectors of CSSV

Hemiptera like aphids, scale insects and mealybugs are numerous on cacao but are usually not considered as major pests of the crop. However, some mealybugs of the Pseudococcidae family, such as *Planococcoides njalensis* (Laing), *Planococcus citri* (Risso), *Ferrisia virgata* (Cock.) and *Phenacoccus hargreavesi* (Laing) are vectors of a devastating disease in West Africa, the cocoa swollen shoot virus disease (CSSVD) (Bigger, 1981; Nguyen-Ban, 1984). Trees infected by the virus show swelling of stems and roots, mosaics, distortion of pods, as well as dieback, which lead to low yield and often to the short-term death of the trees (Lot *et al.*, 1991). The disease was first reported in 1936 in Ghana and is now affecting most parts of this country, resulting in the cutting of millions of cacao trees (Domfeh *et al.*, 2011). Nowadays, CSSVD is considered a serious threat to most of the cocoa-producing countries of West Africa, including Nigeria, Togo, Ghana and Ivory Coast.

Secondary pests

Secondary pests include insects that feed and develop on cacao but with infestation usually kept under the economic threshold by environmental factors. However, pest outbreaks can be observed in particular cropping conditions or in geographically limited areas, where these pests are then considered as major pests. In spite of this, they rarely cause significant economic losses at the world scale.

Cocoa mirids of secondary importance

Bryocoropsis laticollis Schum. (Odoniellini) has morphology very similar to *S. singularis* and *D. theobroma* and shares the same habitat. However, *B. laticollis* only feeds on pods and does not cause damage to the cacao canopy (Kumar and Ansari, 1974). *Boxiopsis madagascariensis* Lavabre is an Odoniellini endemic to the coast of Madagascar, where it was initially found on *Urena lobata* (Malvaceae). On cacao, *B. madagascariensis* causes damage similar to that of *S. singularis* and *D. theobroma* and has been considered as a major pest of cacao in Madagascar (Decazy, 1977). The bee bug *Platyngomiriodes apiformis* Ghauri is considered as an important pest of cacao in Sabah, Malaysia (Lim *et al.*, 1992). *Pseudodoniella laensis* Miller, *Pseudodoniella pacifica* China & Carvalho and *Pseudodoniella typica* (China & Carvalho) are known as important pests of cacao in New Guinea (Entwistle, 1972).

From the dozen species of Monaloniini of the genus *Afropeltis* recorded on cacao in Africa, two are commonly found in plantations: *Afropeltis lalandei* Carayon in West Africa and *Afropeltis corbisieri* Schmitz in Central Africa. Outbreaks of these two species have been locally noted, leading to significant production losses (Collingwood, 1977a). The genus *Afropeltis* is closely related to the genus *Helopeltis* with similar morphological and life history traits, and sometimes *Afropeltis* species are included in the *Helopeltis* genus.

The genus *Monalonion* is represented on cacao by seven species, distributed in Latin America from Mexico to Bolivia (de Abreu, 1977). These species feed almost exclusively on cacao pods. Heavy damage can lead to young fruit abortion and lowering of bean quality. These insects are sometimes considered as minor pests of cacao but the species *Monalonion dissimulatum*, which is the most widely distributed on cacao, was the cause of considerable production losses in Venezuela and Peru in the first half of the 20th century. Nowadays, *M. dissimulatum* is considered as a major pest in several countries of Latin America such as Bolivia and Ecuador (Ferrari *et al.*, 2014). In Brazil, the seven *Monalonion* species are present in cacao plantations but *Monalonion bondari* is the most common (Entwistle, 1972).

The shield bug Bathycoelia thalassina

The shield bug *Bathycoelia thalassina* (Herrich-Schaeffer) (Hemiptera: Pentatomidae) is a pest of cacao in most of the producing countries of West and Central Africa. Nymphs and adults feed on developing beans that they reach through the pod cortex with their long stylets, leading to bean abortion, pod distortion and premature ripening. Outbreaks of this pest in some localities of Ghana and Nigeria are responsible for significant production losses that were estimated at 18% in Ghana in the late 1970s (Owusu-Manu, 1976, 1990).

The cocoa borer Steirastoma breve

In the Neotropics, the longicorn beetles *Steirastoma* spp. (Coleoptera: Cerambycidae) are widely distributed on various host plants. *S. breve* damages young cacao trees: adult females feed on the bark and nymphs bore into the cacao stems making galleries, which opens the door for pathogenic microorganisms to colonize the plants. The beetle is considered a major pest of cacao in Venezuela, and in some areas of Brazil and the Caribbean islands, such as Trinidad (Liendo-Barandiaran *et al.*, 2010).

The cocoa fruit borer Carmenta theobromae

The cocoa fruit borer, *Carmenta theobromae* (Busck) (Lepidoptera: Sesiidae), is a small moth newly considered as a major pest of cacao in some areas of Venezuela and Colombia. Another species, *Carmenta foraseminis* increasingly affects cocoa production in Colombia. Similar to the cocoa pod borer, *Carmenta* spp. females lay eggs on the cacao pods and the larvae bore galleries inside, causing the fruit to rot following infections with pathogenic fungi (Morillo *et al.*, 2009).

The cocoa weevil Pantorhytes *spp.*

Weevils from the genus *Pantorhytes* (Coleoptera: Curculionidae) are major pests of cacao in New Guinea. Of the 11 species injurious to cacao plants, six have been found to be of economic importance. They are robust apterous insects about 1–5 cm long. Larvae tunnel into the cacao stem between the bark and the wood leading to weakened trees, infection by microbes and often sudden death of trees. Totally destroyed plantations have been reported in some areas of Papua New Guinea and the weevil has contributed to cocoa industry collapse in some important producing regions of the country (Moxon, 1992).

The cocoa stem borer Eulophonotus myrmeleon

Eulophonotus myrmeleon Fldr. is a moth from the family Cossidae, whose larvae feed on cacao wood, boring galleries in the stem. Initially considered as a minor pest, reports of *E. myrmeleon* on cacao became more numerous in the 1990s and early 2000s in West Africa. The pest has been recently recorded as serious in some areas of Nigeria and Ivory Coast. Infestation levels reaching around 5% of trees damaged by the pest have been reported in plantations near Ibadan in Nigeria (Anikwe, 2010).

Minor pests of cacao

Large numbers of insects can feed or breed on the plant, or both, without affecting production significantly. They are usually considered as minor pests of cacao although some of them can be major pests of other crops.

The cocoa psyllid *Tyora tessmani* (Aulmann) (Hemiptera: Psyllidae) feeds on cacao shoots leading to interference in leaf development. In cacao nurseries, large populations of the psyllid affect seedling development and should be controlled by chemical spraying (Igboekwe, 1983). Several Scolytidae species have been reported to damage cacao worldwide. Species of the beetle genera *Xyleborus* and *Xylosandrus* attack twigs, damaging the cacao canopy (Navarro and Liendo, 2010). Other species feed on pods and could be involved in pod infection by pathogens causing serious cacao pod diseases such as black pod (*Phytophthora palmivora*) (Konam and Guest, 2004). Leaf-feeding moths and beetles are numerous on cacao although rarely associated with damage of economic importance. The cacao armyworm *Tiracola plagiata*, a noctuid moth, is a pest of cacao in Asia and has been shown to be more abundant in plantations shaded with *Leucaena leucocephala* (Room and Smith, 1975). The cacao plume moth, *Michaelophorus nubilus*, is a moth of the family Pterophoridae damaging young cacao leaves in Latin America (Matthews and Miller, 2010).

In sub-Saharan Africa, two moth species of the family Nolidae are found on cacao: (i) the cacao pod borer *Characoma stictigrapta* Hmps.; and (ii) the spiny bollworm *Earias biplaga* Walk. The former species feeds on cacao leaves and pods while the larvae of *E. biplaga* attack the buds and young leaves, leading to serious damage on seedlings especially (Entwistle, 1972; Akotoye and Kumar, 1976). Some leaf-feeding beetles are reported as pests of cacao, among which is the Scarabaeidae *Adoretus versutus* Har., an Asian polyphagous chafer beetle, outbreaks of which caused serious defoliations in cacao plantations in Vanuatu in the 1980s (Beaudoin *et al.*, 1995). Chafer beetles from the genera *Apogonia*, *Anomala* and *Chaetadoretus* include leaf-feeding pests of cacao as well (Entwistle, 1985). Longhorned beetles (Cerambycidae) include several species of cocoa stem borers damaging cacao branches and stems worldwide. Genera *Phosphorus* and *Tragocephala* are commonly found boring galleries in cacao wood in West Africa, where they also damage coffee (Entwistle, 1972). Longhorned beetles from the genus *Glenea* include many pests of trees, some of which are found on cacao in different countries, notably in Papua New Guinea (Entwistle, 1972).

Pest Management Practices Compatible with Organic Cocoa Production

As noted above, certified organic cocoa represents a very small part of the world cocoa production and a tiny fraction of the cocoa crop in the biggest producing countries of West Africa and South-east Asia. Yet most of these countries have to face major insect pests, and taking into account the increasing consumer demand for organic cocoa, stakes are high for the development of organic means to control these pests. Hence, all the pest management strategies are being considered today – those made available to farmers after decades of research for alternatives to chemical control, as well as those traditionally implemented by farmers to protect their farms. The following paragraphs will present these solutions, with for each of them, an assessment of the degree of implementation.

Preventive solutions

The term 'preventive' here means those solutions implemented at the initial time of cacao planting or during routine maintenance work, to prevent infestation by insect pests and their damage. A significant part of these practices is based on farmer traditional knowledge and others come from scientific knowledge of the biology and ecology of insect pests. In any case, these solutions are currently highlighted as the engine of agroecological concepts implementation.

Planting resistant cacao varieties

An important consideration is that a large part of smallholder cacao farmers still use their own seeds or seeds collected from nearby farms for planting. Seeds come from pods usually collected on trees selected for their vigour, productivity and tolerance to pests and diseases. This traditional selection process, as well as improved variety dissemination by governments, has resulted in a large genetic variability within farms, which is now used by cacao selection programmes, incorporating farmers' perception within a participative approach (Eskes, 2011).

From the researchers' point of view, selection of resistant cacao varieties for pest and disease management is probably the strategy that has generated the most work, especially in the last three decades. However, it should be noted that if varietal solutions have been found and implemented for some cacao diseases, such as the witches' broom disease caused by *M. perniciosa* in Brazil, no definitive solution has been found for any cacao insect pest. For African cocoa mirids, ongoing research shows how the mechanisms involved in the resistance are complex. Resistance has been assessed through records of cumulative damage in selection trials, notably in Ivory Coast (Sounigo *et al.*, 2003). Antixenosis, antibiosis and tolerance of different genotypes have been tested through choice tests with cacao twigs in the laboratory and by enclosing mirids in sleeves on trees in Cameroon, Ivory Coast and Nigeria (Dibog *et al.*, 2008; N'Guessan *et al.*, 2008; Anikwe *et al.*, 2009). These studies allowed the selection of promising clones for further use in breeding programmes, but sometimes with inconsistent results, and much work remains to be done before cacao farmers can actually plant improved cacao varieties for pest control (Eskes, 2011). Major challenges are the identification of varieties combining resistance to mirids, the black pod disease and the cacao swollen shoot virus disease, as well as improved seed production and dissemination to farmers.

To a lesser extent, similar work has been done for the cocoa pod borer, resulting in similar challenges. Pod-surface smoothness, timing of pod development and pod hardness are factors affecting the breeding success of the moth and are pointed out as potential levers for cacao resistance to cocoa pod borer (Teh *et al.*, 2006). But for now, no totally resistant genotype exists and the strategy could be the planting of a mix of various genotypes, including a few susceptible ones. Such genetic diversity may force the moth to make a choice of cacao pods for egg laying, leading to lower global infestation of the plantation (McMahon *et al.*, 2009).

Some work on resistance of cacao to secondary or minor pests, such as *Steirastoma breve* (Morillo *et al.*, 2008), has been conducted and has revealed the potential of some varieties.

Cacao maintenance

Among good agricultural practices for the maintenance of the cacao tree, some are specifically recommended for pest management. For African cocoa mirids, pruning practices aim to prevent chupons on trees. Chupons are vertical shoots growing on the trunk, usually near the ground or below the tree crown. Chupons are particularly attractive to mirids, which feed and lay eggs on them, contributing to maintaining mirid populations on farms even when trees do not bear fruits. Since isolated cacao trees have been found to shelter more mirids than others, another good practice for mirid control is to maintain (while pruning) a continuous cacao canopy with branches touching each other in a thin continuous layer (Padi *et al.*, 2002a).

Given the gravity of the cocoa pod borer threat in some areas of South-east Asia, a radical cultural practice has been used since the early 20th century. Known as 'rampassen', the practice consists of removing all the fruits from cacao in a plantation in order to break the pest life cycle (Lim, 1992). Assessments of the impact of the practice on cocoa pod borer populations yielded an uneven picture, among which were major constraints such as labour costs and economic losses for cacao farmers (Lim, 1992). A conservative practice is to harvest ripening pods as frequently as possible and to break them open immediately to collect the cacao beans (Lim, 1992). After bean extraction, the pod husks can be used as a mulch to destroy immature stages of the pest.

Plant association

Cacao is an understorey crop traditionally grown under shade trees within agroforestry systems. Where possible, smallholder farmers establish cacao in the forest after having cleared the ground of understorey vegetation. Where there are no forests, farmers often shade cacao with trees they grow for fruits, firewood, timber or traditional medicine. Some of the practices recommended for cacao–tree association are specific to pest control. They involve tree species to be planted in association with the crop, and tree canopy management for shade.

REMOVAL OF ALTERNATIVE HOST PLANTS OF CACAO PESTS. A common recommendation in pest control is to remove alternative host plants of cacao pests from the crop or from the surrounding environment. In West Africa, cocoa mirids have adapted to cacao from native forest trees of the family *Malvaceae*. The most famous of them is the kola tree (genus *Cola*), which is grown for its nuts, and which contains seeds rich in caffeine. Due to the lack of farmer knowledge, kola trees are often used to shade cacao and some authors have suggested that the quick expansion of mirid dispersion on cacao in West Africa might be linked to kola tree–cacao associations (Entwistle, 1972). Similar recommendations exist for cocoa pod borer in Asia (Lim, 1992).

SHADE MANAGEMENT. Shade has proven to be a determining factor of cocoa mirid infestation and damage in West Africa. Unshaded plantations are usually more damaged by mirids than shaded ones and in shaded plantations mirid populations are usually sheltered by cacao trees exposed to direct sunlight through gaps in the shade-tree canopy layer (Babin *et al.*, 2010). A common shade recommendation for mirid control in West Africa is to maintain a regular shade level in cacao plantations (Padi *et al.*, 2002a). High forest trees have been shown to be more suitable than fruit trees because they provide a lighter and more uniform shade (Babin *et al.*, 2010).

PLANTING TREES TO FAVOUR PEST NATURAL ENEMIES. Some tree species are recommended in cacao plantations because they provide habitats for pest natural enemies. For example, in Malaysia, associations between cacao and coconut palms are recommended to improve the control of cacao pests, such as the mirid

Helopeltis theobromae by the generalist predator ants *Dolichoderus thoracicus* and *Oecophylla smaragdina*, through providing ants with nesting sites (Way and Khoo, 1991).

Biological control

Parasitoids

Old studies on natural enemies of the cocoa mirids *S. singularis* and *D. theobroma* in Africa report that parasitoids are few and do not lead to sufficient parasitism rates to be good candidates for biological control. Only one species of nymphal parasitoid has been recorded, *Leiophron (Euphorus) sahlbergellae* Wlk. (Braconidae, Euphorinae), with parasitism rates of 15–40% and 6–20% assessed in Ghana and Nigeria, respectively (Collingwood, 1977b). A hyperparasitoid, *Mesochorus melanothorax* Wlk. (Ichneumonidae) attacks *L. sahlbergellae* while feeding (Entwistle, 1972). Three other parasitoids from the genera *Telenomus* (Scelionidae), *Pediobus* (Eulophidae) and family Signiphoridae have been collected from *S. singularis* eggs, with parasitism rates lower than 10% (Entwistle, 1972).

By contrast, an indigenous egg parasitoid, *Trichogrammatoidea bactrae fumata* Ngaraja (Hymenoptera: Trichogrammatidae) was found to be associated with cocoa pod borer in Malaysia in 1982, and is now considered a good biological control agent for the pest on cacao (Lim, 1992). Parasitism rates ranging from 10% to 56% were observed in the 1980s and strong density dependence between the parasitoid and the pest was demonstrated (Lim, 1992). However, to the best of the author's knowledge, the literature does not give clear evidence of the use of large-scale releases of this parasitoid for the biological control of cocoa pod borer.

Generalist predators

Entwistle (1972) listed arthropod predators of African cocoa mirids. An old study by Williams (1954) revealed levels of predation of mirid nymphs as high as 16%, 19% and 21% for praying mantises (Mantidae), Reduviidae and ants, respectively. However, because they are generalist feeders, they are usually not considered as good candidates for biological control.

By contrast, some species of ants with very large polydomus colonies have been considered for biological control on cacao. As for most tropical ecosystems, ants are a key component of cacao agrosystems (Philpott and Armbrecht, 2006). In West Africa notably, ant communities have been well described in the past especially the arboreal species (Williams, 1954; Bigger, 1981). A high level of species diversity as well as strong spatial structuration of communities led authors to characterize them as ant mosaics (Tadu *et al.*, 2014a). In cacao agrosystems, ant mosaics are usually structured by highly dominant species such as *Oecophylla longinoda*, *Tetramorium aculeatum* and *Crematogaster* spp., which prey on a large range of invertebrates, including insect pests such as mirids and shield bugs. However, their actual impact on damage by mirids is still controversial and, to the best of our knowledge, there are no specific recommendations for the use of ants as biological control agents for any pest of cacao in Africa.

In that regard, farmers have set positive examples worldwide. In southern Cameroon for instance, in the 1960s, cacao farmers successfully used the little fire ant, *Wasmannia auropunctata* Roger, to get rid of insect pests on their plantations, building colonies in their farms by trapping ants with sweet baits (Bruneau de Miré, 1969). Unfortunately, by doing this, farmers have probably contributed to the expansion of this invasive tramp species, accidentally introduced in this area and now widely considered as a threat for Congo basin forest biodiversity.

In Asia, farmers have used ants for several centuries now, especially to protect fruits from insect pests. On cacao, the Asian weaver ant *O. smaragdina* is known as a beneficial predator of many pests such as the mirids *Helopeltis theobromae* and *Pseudodoniella laensis*, and the cocoa weevils (*Pantorhytes* spp.) in Papua New Guinea (Way and Khoo, 1992). But this species is aggressive to people and may hinder agricultural operations, in such a way that it is not always welcome in plantations. By contrast, the black ant *D. thoracicus* is not aggressive and Indonesian cacao farmers have used it to protect pods from mirid damage since the early 1900s (Way and Khoo, 1992). Since then, research has confirmed that *D. thoracicus* is a valuable biological control agent for the major pests *Helopeltis antonii* and *H. theivora* in Indonesia and *H. theobromae* in Malaysia (Saripah and Azhar, 2012). Research on *D. thoracicus* has led to recommendations to favour colony establishment in plantations, among which the destruction of antagonist ant species, the improvement of nesting conditions by planting coconut palms or introducing artificial nests, and artificial infestation of cacao with mealybugs, which are tended by black ants for honeydew (Way and Khoo, 1992).

The crazy ant, *Anoplolepsis longipes*, has shown promise for the control of *Pantorhytes* spp. in Papua New Guinea and methods have been developed for establishing colonies in cacao plantations (Moxon, 1992). But the use of crazy ants as a biological control agent is questioned due to the difficulty of maintaining large colonies in plantations over time.

Pesticides

Bacterial and fungal preparations

Several *Bacillus thuringiensis* (*Bt*) toxins have been tested for the control of cocoa pod borer. Eight of the 12 Cry1 proteins tested through laboratory bioassays were able to kill 50% of cocoa pod borer larvae maintained on an artificial diet (Santoso *et al.*, 2004). Field trials of *Bt* insecticide formulations in Indonesia have shown significant reduction of cocoa pod borer infestation and yield increase (Senewe *et al.*, 2013). Commercialized and local strains of *Beauveria bassiana* have been tested with promising results for *Monalonion dissimulatum* through field investigation in Bolivia (Ferrari *et al.*, 2014) and for *Pantorhytes plutus* in Papua New Guinea. Moreover, *B. bassiana* has been established as an endophyte of the cacao tree by spraying seedlings or flowers. In the latter case, the entomopathogenic fungus was re-isolated from pods, suggesting that the method could be used for major pest management of cocoa pod borer and mirids (Posada *et al.*, 2010). Suspensions of local strains of entomopathogenic fungi from *Paecilomyces* and *Lecanicillium* genera were tested on *Carmenta foraseminis* with success (Figueroa Medina *et al.*, 2013). Although these results are promising, to the best of our knowledge, there is no report on the use of commercialized bacterial and fungal preparations by cacao farmers for the control of insect pests.

Botanical pesticides

A curative practice that can be implemented in organic farming is the use of approved insecticides of biological and mineral origin. These are defined by the IFOAM *Basic Standards for Organic Production and Processing* (IFOAM, 2005). Neem (*Azadirachta indica*) extracts are one of the commonly used natural insecticides as they have shown real efficiency on several pests worldwide. On cacao, neem crude extracts at different concentrations, as well as commercial formulations, were tested on mirids in Ivory Coast and Ghana, and gave high levels of mortality in the laboratory and in the field (Padi *et al.*, 2002b). The repellent or deterrent effects of neem on mirids were also shown through attractiveness tests in the laboratory (N'Guessan *et al.*, 2006). Also, neem gave promising results on mealybugs, vectors of the CSSV, and on other pests such as *Helopeltis* spp. and the psyllid *Tyora tessmanni*.

At the present time, no data on the actual use of neem-based insecticides by cacao farmers is available. By contrast, the use of natural pesticides developed by farmers themselves in response to problems of agrochemical supply is reported. For example, a study conducted in Cameroon reveals that various herbal preparations of hemp (*Cannabis sativa*) are used alone or mixed with extracts from tobacco leaves, indigenous trees or with chemicals for the control of pests and diseases (including cocoa mirids and black pod) (Coulibaly *et al.*, 2002).

Mechanical control

Physical barriers

In Indonesia, cacao smallholder farmers have developed a physical method for controlling the cocoa pod borer. They use plastic bags to sleeve pods and prevent the moth from laying eggs on them. The issue of plastic bag pollution has led researchers to test the method using biodegradable plastic bags. A recent study conducted in Indonesia shows that only 50% of the pods were preserved this way from cocoa pod borer and discusses the importance of good timing of the pod sleeving that seems to depend on cacao variety and season (Rosmana *et al.*, 2010). Sleeving pods at an earlier developmental stage can reduce cocoa pod borer infestation by 85–100% but this also increases the risk of production losses due to physiological death of pods (wilt) and *Phytophthora* pod rot.

A few studies report assessments of other methods to physically protect pods from insect attack, by spraying kaolin and silicon-based products, but results should be confirmed with more investigations of these products (Ferrari *et al.*, 2014).

Hand picking and physical destruction of pests

Another mechanical practice developed by cacao growers worldwide for the control of stem borers is the poking of holes tunnelled by the pest with a wooden stick or a wire to kill the larvae. This practice has proved inefficient in controlling increasing populations of *Eulophonotus myrmeleon* in Nigeria when implemented alone, but is recommended in combination with well-targeted chemical control (Anikwe, 2010).

Hand picking may be a good strategy for some cacao pests that are easily seen on trees. This is true for the cocoa weevils *Pantorhytes* spp., whose adults are easily detected in the trees and destroyed by farmers in Papua New Guinea (Moxon, 1992).

Semiochemical control

Traps containing synthetic sex pheromones of cocoa mirids have been tested in Ghana and Cameroon (Padi *et al.*, 2004; Mahob *et al.*, 2011). Sticky traps baited with different blends of two components of the *Sahlbergella singularis* female sex pheromone, namely hexyl (R)-3-((E)-2-butenoyloxy)-butyrate and hexyl (R)-3-hydroxybutyrate, gave promising results for this pest, suggesting that the trap could be used for pest monitoring at least (Mahob *et al.*, 2011). There is still much work to be done, however, to measure its efficiency in reducing infestation and improving yield, before this can be considered.

Similar work has been conducted for the cocoa pod borer in Indonesia and Malaysia (Zhang *et al.*, 2008). Different blends of synthetic female sex pheromone of *C. cramerella*, including (E,Z,Z)- and (E,E,Z)-4,6,10-hexadecatrienyl acetates and the corresponding alcohols showed satisfactory attractiveness. Here again, experimentation on a larger scale is needed before including pheromone traps within strategies for biological control of cocoa pod borer.

Some pest control methods based on evaluation of semiochemicals have been tested for cacao pests of secondary importance. For example, the use of cacao brushwood piles has been tested for the control of the cocoa beetle *Steirastoma breve* in Venezuela (Liendo-Barandiaran *et al.*, 2010).

The Future of Biological Control of Pests on Cacao

Collective management strategies to be thought of in time and space

Pest management strategies compatible with certified organic cacao are numerous. Farmers are already implementing some, others have proven to be efficient although not yet widely used and some need more work for their efficiency to be demonstrated. However, if implemented alone, none of these practices has proven to be a complete and definitive solution, either because they failed to keep pest damage under economic thresholds or they caused other problems that overshadowed the benefits. Control of major pests, especially, requires a combination of practices implemented from the beginning of cacao plantation establishment, and taking into account recommendations for other major production constraints. This should be the case for some areas of West Africa where cacao production is threatened by damage on trees from both mirids and CSSV, and pod loss due to black pod disease.

Moreover, as a tree crop, the cacao development period can extend for several decades. Routine maintenance of a cacao farm is crucial for good productivity, and decisions need to be made at the time of plantation establishment, notably in terms of cacao varieties and plant association, which are of primary importance, with long-term consequences.

In many countries worldwide, cacao is grown continuously over wide areas, but by a large number of farmers, each owning a few acres. For pests with good dispersion ability, such as mirids and cocoa pod borer, pest management has to be organized at a larger scale than at individual farm level. This shows the importance of farmer organization and how a socio-economic approach plays a crucial role in pest management on cacao. This also shows how a landscape approach is important, taking into account spatial arrangements of cacao farms as well as the different components of the agricultural landscape.

Plant diversification as the main lever of agroecology

One of the aims of agroecology is to value ecological mechanisms for the design and management of sustainable agrosystems. Many studies on many crops worldwide have shown that enhancement of plant diversity in agrosystems is a good strategy for pest and disease regulation (Ratnadass *et al.*, 2012). Plant diversification helps to return the natural balance through re-establishment of trophic webs.

Since cacao is still grown in highly diversified agroforestry systems in various environments worldwide, cacao agrosystems offer excellent models for the study of ecological mechanisms involved in pest regulation. Recent studies showed how tree associations should be viewed in terms of composition and spatial structure for the regulation of cocoa mirids, through shade management and natural enemies' promotion (Gidoin *et al.*, 2014; Tadu *et al.*, 2014b).

Farmer knowledge as a cornerstone of agroecology

Because they suffered in the past and still suffer in some areas from financial constraints, smallholder farmers have developed their own management strategies for pest control worldwide. Most of these strategies are based on a better use of what is present on their farm or surrounding farms, as well as what the surrounding natural environment offers. The most persuasive evidence, already mentioned above, is the use of their own cacao varieties, the development of pesticides from local plants and enhancement of pest natural enemies. Farmers' technologies usually need improvement but accumulated knowledge is always of great interest and should contribute to an agroecological approach to pest control.

Cacao certification: a solution for organic pest management?

Due to the limited production of organic cacao worldwide, few studies have measured the impact of organic practices on cocoa production and environment. A study conducted in Bolivia showed that a certified organic cacao environment had greater plant diversity compared with a traditional agrosystem as well as providing better yields, leading to higher family income. This is explained by better organic farmer knowledge and practices, linked to self-organization and affiliation to farmers' organizations (Jacobi *et al.*, 2013).

By contrast, other certifications are exploding in growth, involving thousands of farmers, especially in the major producing countries of West Africa. As a matter of fact, the main cocoa industry companies, with the aim of improving their image, have set an ambitious target of 100% certified cacao for 2020. Certifications based on environmental and ethical standards as well as good farming practices are promoted by international non-governmental organizations (NGOs) such as The Rainforest Alliance and UTZ Certified. However, mass certification as practised currently in West Africa is not producing the desired results in terms of yield growth, farmer welfare and environmental protection (Ruf *et al.*, 2013). Regarding pest and disease control, a recent study conducted in Ivory Coast showed that certified farmers do not usually follow the chemical spraying recommendations (Ruf *et al.*, 2013).

This analysis, among others, clearly underlines that cacao certification should be carefully considered and planned from the beginning, when starting a cacao plantation and on a long-term basis, by incorporating farmers' expectations and constraints, as well as the knowledge and innovations that they have developed (Ayenor *et al.*, 2004, 2007).

References

Akotoye, N.A.K. and Kumar, R. (1976) Population dynamics of *Characoma stictigrapta* Hmps. (Lepidoptera: Noctuidae), on cocoa in Ghana. *Journal of Applied Ecology* 13, 753–773.

Anikwe, J.C. (2010) The seasonal occurrence and control of the cocoa stem borer, *Eulophonotus myrmeleon* Fldr. (Lepidoptera: Cossidae) on cocoa in Ibadan, Nigeria. *Libyan Agriculture Research Center Journal International* 1, 142–146.

Anikwe, J.C., Omoloye, A.A., Aikpokpodion, P.O., Okelana, F.A. and Eskes, A.B. (2009) Evaluation of resistance in selected cocoa genotypes to the brown cocoa mirid, *Sahlbergella singularis* Haglund in Nigeria. *Crop Protection* 28, 350–355.

Ayenor, G.K., Röling, N.G., Padi, B., Huis, A.V., Obeng-Ofori, D. and Atengdem, P.B. (2004) Converging farmers' and scientists' perspectives on researchable constraints on organic cocoa production in Ghana: results of a diagnostic study. *NJAS Wageningen Journal of Life Sciences* 52, 261–284.

Ayenor, G.K., van Huis, A., Obeng-Ofori, D., Padi, B. and Röling, N.G. (2007) Facilitating the use of alternative capsid control methods towards sustainable production of organic cocoa in Ghana. *International Journal of Tropical Insect Science* 27, 85–94.

Babin, R., ten Hoopen, G.M., Cilas, C., Enjalric, F., Yede Gendre, P. and Lumaret, J.-P. (2010) Impact of shade on the spatial distribution of *Sahlbergella singularis* in traditional cocoa agroforests. *Agricultural and Forest Entomology* 12, 69–79.

Beaudoin, L., Morin, J.P., Nguyen, C. and Decazy, B. (1995) Study of underground *Adoretus versutus* Har. (Col., Scarabaeidae) populations in Vanuatu: detection of cohabitation with other white grubs. *Journal of Applied Entomology* 119, 391–397.

Bigger, M. (1981) The relative abundance of the mealybug vectors (Hemiptera: Coccidae and Pseudococcidae) of cocoa swollen shoot disease in Ghana. *Bulletin of Entomological Research* 71, 435–448.

Bruneau de Miré, P. (1969) Une fourmi utilisée au Cameroun dans la lutte contre les mirides du cacaoyer: *Wasmannia auropunctata* Roger. *Café, Cacao, Thé* 13, 209–212.

Collingwood, C.A. (1977a) African mirids. In: Lavabre, E.M. (ed.) *Les Mirides du Cacaoyer*. G.P. Maisonneuve et Larose, Paris, pp. 71–83.

Collingwood, C.A. (1977b) Biological control and relations with other insects. In: Lavabre, E.M. (ed.) *Les Mirides du Cacaoyer*. G.P. Maisonneuve et Larose, Paris, pp. 237–255.

Coulibaly, O., Mbila, D., Sonwa, D.J., Adesina, A. and Bakala, J. (2002) Responding to economic crisis in sub-Saharan Africa: new farmer-developed pest management strategies in cocoa-based plantations in Southern Cameroon. *Integrated Pest Management Reviews* 7, 165–172.

Day, R.K. (1989) Effect of cocoa pod borer, *Conopomorpha cramerella*, on cocoa yield and quality in Sabah, Malaysia. *Crop Protection* 8, 332–339.

de Abreu, J.M. (1977) Mirideos neotropicais associados ao cacaueiro. In: Lavabre, E.M. (ed.) *Les Mirides du Cacaoyer*. G.P. Maisonneuve et Larose, Paris, pp. 85–106.

Decazy, B. (1977) Les mirides du cacaoyer à Madagascar: *Boxiopsis madagascariensis* Lavabre. In: Lavabre, E.M. (ed.) *Les Mirides du Cacaoyer*. G.P. Maisonneuve et Larose, Paris, pp. 123–137.

Dibog, L., Babin, R., Mbang, J.A., Decazy, B., Nyassé, S., Cilas, C. and Eskes, A.B. (2008) Effect of genotype of cocoa (*Theobroma cacao*) on attractiveness to the mirid *Sahlbergella singularis* (Hemiptera: Miridae) in the laboratory. *Pest Management Science* 64, 977–980.

Domfeh, O., Dzahini-Obiatey, H., Ameyaw, G.A., Abaka-Ewusie, K. and Opoku, G. (2011) Cocoa swollen shoot virus disease situation in Ghana: a review of current trends. *African Journal of Agricultural Research* 6, 5033–5039.

Entwistle, P.F. (1972) *Pests of Cocoa*. Longman Group Limited, London, p. 779.

Entwistle, P.F. (1985) Insects and cocoa. In: Wood, G.A.R. and Lass, R.A. (eds) *Cocoa*. Longman, London, pp. 366–443.

Eskes, A.B. (ed.) (2011) *Collaborative and Participatory Approaches to Cocoa Variety Improvement*. Final report of the CFC/ICCO/Bioversity project on 'Cocoa Productivity and Quality Improvement: A Participatory Approach' (2004-2010). CFC, Amsterdam, The Netherlands/ICCO, London, UK/Bioversity International, Rome, Italy, p. 205.

Ferrari, L., Flores, A., Velásquez, F., Schneider, M., Andres, C., Milz, J., Trujillo, G., Alcon, F. and Studer, C. (2014) Evaluation of organic pest management strategies to control the cocoa mirid (*Monalonion dissimulatum* Dist.), Alto Beni, Bolivia. In: *International Agriculture in a Changing World: Good News from the Field* 19 June 2014. HAFL, Zollikofen, Switzerland. Available at: http://www.systems-comparison.fibl.org/fileadmin/documents/en/syscom/Poster_exhibitions/Beitrag_Ferrari_etal_HAFL_symposium_2014.pdf (accessed 18 February 2015).

Figueroa Medina, W., Ramirez Sulvaran, J.A. and Sigarroa Rieche, A.K. (2013) Effect of native strains *Paecilomyces* sp. (Bainier) and *Lecanicillium* sp. (Zimm) on the control of *Carmenta foraseminis* Eichlin (Lepidoptera: Sesiidae) on cocoa (*Theobroma cacao* L.) crops. *Acta Agronómica* 62, 279–286.

FAOSTAT (2014) Food and Agriculture Organization of the United Nations (FAO) Corporate Statistical Database. Available at: http://faostat3.fao.org/home/E (accessed 25 September 2014).

Gidoin, C., Babin, R., Bagny Beilhe, L., Cilas, C., ten Hoopen, G.M. and Ngo Bieng, M.A. (2014) Tree spatial structure, host composition and resource availability influence mirid density or black pod prevalence in cacao agroforests in Cameroon. *PLoS One* 9, e109405.

Hawkins, D. and Chen, Y. (2014) *Giant on a Pinhead: A Profile of the Cocoa Sector*. Hardman & Co., London, 76 pp.

Igboekwe, A.D. (1983) Studies on the damage to young cocoa seedlings by the cocoa psyllid *Tyora tessmanni* (Aulmann) (Homoptera: Psyllidae). *Café, Cacao, Thé* 27, 67–70.

International Cocoa Organization (ICCO) (2013) *Annual Report 2012/2103*. ICCO, London, 64 pp.

International Cocoa Organization (ICCO) (2014) The Chocolate Industry. ICCO. Available at: http://www.icco.org/about-cocoa/chocolate-industry.html (accessed 25 September 2014).

International Federation of Organic Agriculture Movements (IFOAM) (2005) *The IFOAM Norms for Organic Production and Processing, including IFOAM Basic Standards for Organic Production and Processing, and IFOAM Accreditation Criteria for Bodies Certifying Organic Production and Processing*. IFOAM, Bonn, Germany, p. 126.

Jacobi, J., Andres, C., Schneider, M., Pillco, M., Calizaya, P. and Rist, S. (2013) Carbon stocks, tree diversity, and the role of organic certification in different cocoa production systems in Alto Beni, Bolivia. *Agroforestry Systems* 88, 1117–1132.

Keane, P.J. (1992) Diseases and pests of cocoa: an overview. In: Keane, P.J. and Putter, C.A.J. (eds) *Cocoa Pest and Disease Management in Southeast Asia and Australasia*. FAO Plant Production and Protection Paper No. 112. Food and Agriculture Organization of the United Nations (FAO), Rome, pp. 1–11.

Konam, J.K. and Guest, D.I. (2004) Role of beetles (Coleoptera: Scolytidae and Nitidulidae) in the spread of *Phytophthora palmivora* pod rot of cocoa in Papua New Guinea. *Australasian Plant Pathology* 33, 55–59.

Kumar, R. and Ansari, A.K. (1974) Biology, immature stages and rearing of cocoa-capsids (Miridae: Heteroptera). *Zoological Journal of the Linnean Society* 54, 1–29.

Lass, R.A. (1985) Diseases. In: Wood, G.A.R. and Lass, R.A. (eds) *Cocoa*. Longman, London, pp. 265–365.

Lavabre, E.M. (1977) *Les Mirides du Cacaoyer*. G.P. Maisonneuve et Larose, Paris, 366 pp.

Liendo-Barandiaran, C.V., Herrera-Malaver, B., Morillo, F., Sanchez, P. and Hernandez, J.V. (2010) Behavioral responses of *Steirastoma breve* (Sulzer) (Coleoptera: Cerambycidae) to host plant *Theobroma cacao* L., brushwood piles, under field conditions. *Applied Entomology and Zoology* 45, 489–496.

Lim, G.T. (1992) Biology, ecology and control of cocoa podborer *Conopomorpha cramerella* (Snellen). In: Keane, P.J. and Putter, C.A.J. (eds) *Cocoa Pest and Disease Management in Southeast Asia and Australasia*.

FAO Plant Production and Protection Paper No. 112. Food and Agriculture Organization of the United Nations (FAO), Rome, pp. 85–100.

Lim, G.T., Ooi, P.A.C., Lim, G.S. and Teng, P.S. (1992) Recent development of cocoa insect pests management in Sabah Malaysia. In: *Proceedings of the 3rd International Conference on Plant Protection in the Tropics, No. 4*. Malaysian Plant Protection Society, Pahang, Malaysia, pp. 36–53.

Lot, H., Djiekpor, E. and Jacquemond, M. (1991) Characterization of the genome of cacao swollen shoot virus. *Journal of General Virology* 72, 1735–1739.

Mahob, R.J., Babin, R., ten Hoopen, G.M., Dibog, L., Yede, Hall, D.R. and Bilong Bilong, C.F. (2011) Field evaluation of synthetic sex pheromone traps for the cocoa mirid *Sahlbergella singularis* (Hemiptera: Miridae). *Pest Management Science* 67, 672–676.

Matthews, D.L. and Miller, J.Y. (2010) Notes on the cacao plume moth in Honduras and description of the larvae and pupae (Lepidoptera: Pterophoridae). *Tropical Lepidoptera Research* 20, 28–34.

McMahon, P., Iswanto, A., Susilo, A.W., Sulistyowati, E., Wahab, A., Imron, M., Purwantara, A., Mufrihati, E., Dewi, V.S., Lambert, S., Guest, D. and Keane, P. (2009) On-farm selection for quality and resistance to pest/diseases of cocoa in Sulawesi: (i) performance of selections against cocoa pod borer, *Conopomorpha cramerella*. *International Journal of Pest Management* 55, 325–337.

Meinhardt, L.W., Rincones, J., Bailey, B.A., Aime, M.C., Griffith, G.W., Zhang, D. and Pereira, G.A.G. (2008) *Moniliophthora perniciosa*, the causal agent of witches' broom disease of cacao: what's new from this old foe? *Molecular Plant Pathology* 9, 577–588.

Morillo, F., Sánchez, P., Giron, C., Valera, A., Muñoz, W. and Guerra, J. (2008) Behavior of cacao hybrids (*Theobroma cacao*) to attack by *Steirastoma breve* (Coleoptera: Cerambycidae). *Revista Colombiana de Entomologia* 34, 151–155.

Morillo, F., Sánchez, P., Herrera, B., Liendo-Barandiaran, C., Muñoz, W. and Vicente Hernández, J. (2009) Pupal development, longevity and behavior of *Carmenta theobromae* (Lepidoptera: Sesiidae). *Florida Entomologist* 92, 355–361.

Moxon, J.E. (1992) Insect pests of cocoa in Papua New Guinea, importance and control. In: Keane, P.J. and Putter, C.A.J. (eds) *Cocoa Pest and Disease Management in Southeast Asia and Australasia*. FAO Plant Production and Protection Paper No. 112. Food and Agriculture Organization of the United Nations (FAO), Rome, pp. 129–144.

Navarro, R. and Liendo, R. (2010) Population fluctuation of Scolytidae (Insecta: Coleoptera) in cocoa of Aragua state, Venezuela. *Agronomia Tropical* 60, 255–261.

N'Guessan, F.K., Kouassi, A.F. and Atindehou, K. (2006) Study on the effect of the neem, *Azadiractha indica* Juss (Meliaceae) on *Sahlbergella singularis* (Hemiptera: Miridae), an important pest of cocoa. In: COPAL (ed.) *Proceedings of the 15th International Cocoa Research Conference*. Cocoa Producers' Alliance (COPAL), Lagos, Nigeria, pp. 1287–1296.

N'Guessan, K.F., N'Goran, J.A.K. and Eskes, A.B. (2008) Resistance of cacao (*Theobroma cacao* L.) to *Sahlbergella singularis* (Hemiptera: Miridae): investigation of antixenosis, antibiosis and tolerance. *International Journal of Tropical Insect Science* 28, 201–210.

Nguyen-Ban, J. (1984) Variations d'abondance des pseudococcines vectrices de la maladie du swollen shoot au Togo. *Café, Cacao, Thé* 28, 103–110.

Owusu-Manu, E. (1976) Estimation of cocoa pod losses caused by *Bathycoelia thalassina* (H.-S.) (Hemiptera, Pentatomidae). *Ghana Journal of Agricultural Science* 9, 81–83.

Owusu-Manu, E. (1990) Feeding behaviour and the damage caused by *Bathycoelia thalassina* (Herrich-Schaeffer) (Hemiptera: Pentatomidae). *Café, Cacao, Thé* 34, 97–104.

Padi, B., Ackonor, J.B. and Opoku, I.Y. (2002a) Cocoa IPM research and implementation in Ghana. In: Neuenschwander, P. and Vos, J.G.M. (eds) *West African Regional Cocoa IPM Workshop – Proceedings*. CPL Press, Newbury, UK, pp. 54–62.

Padi, B., Adu-Acheampong, R. and Nkansah, A. (2002b) Botanical pesticides for the control of cocoa capsids (Heteroptera: Miridae). In: COPAL (ed.), *Proceedings of the 13th International Cocoa Research Conference*. Cocoa Producers' Alliance (COPAL), Lagos, Nigeria, pp. 403–413.

Padi, B., Hall, D.R., Sarfo, J.E., Downham, M.C.A. and Farman, D.I. (2004) Development of sex pheromone traps for the monitoring and control of cocoa capsids in Ghana: update on field trials. In: Akrofi, A.Y., Ackonor, J.B. and Ollenu, L.A.A. (eds) *Proceedings of the 4th International Permanent Working Group for Cocoa Pests and Diseases (INCOPED) Seminar 'Dealing with Pressing Crop Protection Problems'*. Ghana Cocoa Board, Accra, Ghana, pp. 79–83.

Phillips-Mora, W., Aime, M.C. and Wilkinson, M.J. (2007) Biodiversity and biogeography of the cacao (*Theobroma cacao*) pathogen *Moniliophthora roreri* in tropical America. *Plant Pathology* 56, 911–922.

Philpott, S.M. and Armbrecht, I. (2006) Biodiversity in tropical agroforests and the ecological role of ants and ant diversity in predatory function. *Ecological Entomology* 31, 369–377.

Posada, F.J., Chaves, F.C., Gianfagna, T.J., Pava-Ripoll, M. and Hebbar, P. (2010) Establishement of the fungal entomopathogen *Beauveria bassiana* as an endophyte in cocoa pods (*Theobroma cacao* L.). *Revista UDCA Actualidad & Divulgación Científica* 13, 71–78.

Posada, F.J., Virdiana, I., Navies, M., Pava-Ripoll, M. and Hebbar, P. (2011) Sexual dimorphism of pupae and adults of the cocoa pod borer, *Conopomorpha cramerella*. *Journal of Insect Science* 11, 52.

Ratnadass, A., Fernandes, P., Avelino, J. and Habib, R. (2012) Plant species diversity for sustainable management of crop pests and diseases in agroecosystems: a review. *Agronomy for Sustainable Development* 32, 273–303.

Room, P.M. and Smith, E.S.C. (1975) Relative abundance and distribution of insect pests, ants and other components of the cocoa ecosystem in Papua New Guinea. *Journal of Applied Ecology* 12, 31–46.

Rosmana, A., Shepard, M., Hebbar, P. and Mustari, A. (2010) Control of cocoa pod borer and *Phytophthora* pod rot using degradable plastic pod sleeves and a nematode, *Steinernema carpocapsae*. *Indonesian Journal of Agricultural Science* 11, 41–47.

Ruf, F., N'Dao, Y. and Lemeilleur, S. (2013) Certification du cacao, stratégie à hauts risques. *Bulletin de veille Inter-réseaux Développement Rural* 217, 7.

Santoso, D., Chaidamsari, T., Wiryadiputra, S. and de Maagd, R.A. (2004) Activity of *Bacillus thuringiensis* toxins against cocoa pod borer larvae. *Pest Management Science* 60, 735–738.

Saripah, B. and Azhar, I. (2012) Five years of using cocoa black ants, to control cocoa pod borer at farmer plot – an epilogue. *Malaysian Cocoa Journal* 7, 8–14.

Senewe, R.E., Wagiman, F.X. and Wiryadiputra, S. (2013) Effectiveness of bioinsecticide *Bacillus thuringiensis* formulation against cocoa pod borer in field condition. *Pelita Perkebunan* 29, 108–119.

Shapiro, L.H., Scheffer, S.J., Maisin, N., Lambert, S., Purung, H.B., Sulistyowati, E., Vega, F.E., Gende, P., Laup, S., Rosmana, A., Djam, S. and Hebbar, P.K. (2008) *Conopomorpha cramerella* (Lepidoptera: Gracillariidae) in the Malay Archipelago: genetic signature of a bottlenecked population? *Annals of the Entomological Society of America* 101, 930–938.

Sonwa, D.J., Coulibaly, O., Weise, S.F., Akinwumi Adesina, A. and Janssens, M.J.J. (2008) Management of cocoa: constraints during acquisition and application of pesticides in the humid forest zones of southern Cameroon. *Crop Protection* 27, 1159–1164.

Sounigo, O., Coulibaly, N., Brun, L., N'Goran, J.K.A., Cilas, C. and Eskes, A. (2003) Evaluation of resistance of *Theobroma cacao* L. to mirids in Côte d'Ivoire: results of comparative progeny trials. *Crop Protection* 22, 615–621.

Stonedahl, G.M. (1991) The oriental species of *Helopeltis* (Heteroptera: Miridae): a review of economic literature and guide to identification. *Bulletin of Entomological Research* 81, 465–490.

Tadu, Z., Djiéto-Lordon, C., Yede Messop Youbi, E.B., Fomena, A. and Babin, R. (2014a) Ant diversity in different cocoa agroforest habitats in the Centre Region of Cameroon. *African Entomology* 22, 388–404.

Tadu, Z., Djiéto-Lordon, C., Yede Youbi, E.M., Aléné, C.D., Fomena, A. and Babin, R. (2014b) Ant mosaics in cocoa agroforestry systems of southern Cameroon: influence of shade on the occurrence and spatial distribution of dominant ants. *Agroforestry Systems* 88, 1067–1079.

Teh, C.-L., Pang, J.T.-Y. and Ho, C.-T. (2006) Variation of the response of clonal cocoa to attack by cocoa pod borer *Conopomorpha cramerella* (Lepidoptera: Gracillariidae) in Sabah. *Crop Protection* 25, 712–717.

Tong-Kwee, L., Muhamad, R., Fee, C.G. and Lan, C.C. (1989) Studies on *Beauveria bassiana* isolated from the cocoa mirid, *Helopeltis theobromae*. *Crop Protection* 8, 358–362.

Way, M. and Khoo, K.C. (1991) Colony dispersion and nesting habits of the ants, *Dolichoderus thoracicus* and *Oecophylla smaragdina* (Hymenoptera: Formicidae), in relation to their success as biological control agents on cocoa. *Bulletin of Entomological Research* 81, 341–350.

Way, M.J. and Khoo, K.C. (1992) Role of ants in pest management. *Annual Review of Entomology* 37, 479–503.

Willer, H. and Lernoud, J. (2013) Current statistics on organic agriculture worldwide: organic area, producers and market. In: Willer, H., Lernoud, J. and Kilcher, L. (eds) *The World of Organic Agriculture: Statistics and Emerging Trends 2013*. Research Institute of Organic Agriculture (FiBL), Frick, Switzerland and International Federation of Organic Agriculture Movements (IFOAM), Bonn, Germany, pp. 36–128.

Williams, G. (1954) Field observations on the cacao mirids, *Sahlbergella singularis* Hagl. and *Distantiella theobroma* (Dist.) in the Gold Coast. Part III. Population fluctuations. *Bulletin of Entomological Research* 45, 723–744.

Wood, G.A.R. (1985) History and development. In: Wood, G.A.R. and Lass, R.A. (eds) *Cocoa*. Longman, London, pp. 1–10.

Yen, J.D.L., Waters, E.K. and Hamilton, A.J. (2010) Cocoa pod borer (*Conopomorpha cramerella* Snellen) in Papua New Guinea: biosecurity models for New Ireland and the Autonomous Region of Bougainville. *Risk Analysis* 30, 293–309.

Zhang, A.J., Kuang, L.F., Maisin, N., Karumuru, B., Hall, D.R., Virdiana, I., Lambert, S., Purung, H.B., Wang, S. and Hebbar, P. (2008) Activity evaluation of cocoa pod borer sex pheromone in cacao fields. *Environmental Entomology* 37, 719–724.

21 Integrated Pest Management of Cassava Crops in South-east Asia

Ignazio Graziosi,[1]* Ngo Tien Dung,[2] Chamaiporn Buamas,[3] Prapit Wongtiem[4] and Kris A.G. Wyckhuys[1]
[1]International Center for Tropical Agriculture (CIAT) Asia regional office, Hanoi, Vietnam; [2]Ministry of Agriculture, Hanoi, Vietnam; [3]Department of Agriculture, Bangkok, Thailand; [4]Rayong Field Crops Research Center, Rayong, Thailand

Introduction

Cassava, *Manihot esculenta* Crantz (*Malpighiales*: *Euphorbiaceae*), is a perennial woody plant native to South America and cultivated for its starchy roots throughout the tropics (Olsen and Schaal, 1999). Approximately 57% of the world's cassava is grown in Africa, where it provides food security for more than 200 million people (Prudencio and Al-Hassan, 1994; Manyong, 2000; FAOSTAT, 2015). In South-east Asia, cassava is of increasing importance as a food, cash and bioenergy crop. Since the turn of the century, cassava productivity in South-east Asia has nearly doubled, and the crop has fast become one of the key agricultural commodities of the region. Local cassava provides industrial starches, dry chips for animal feed, biofuel and food for human consumption, for both domestic use and export (Hahn, 1989; Maziya-Dixon *et al.*, 2007). Indonesia and Thailand are the lead producers, with cassava of prime importance as a staple crop in the former country, and Thailand largely growing the crop for the billion-dollar starch industry (WITS, 2005; Ceballos, 2012). Cassava has a well-earned reputation as a hardy 'survivor' crop, able to sustain yields under relatively low soil fertility and proving exceptionally resilient to climate change, which is of particular value to the region's resource-poor farmers (Devendra and Thomas, 2002; Ceballos *et al.*, 2011). In fact, cassava continues to be grown primarily by smallholder farmers, who often operate on marginal lands and cultivate plots that are regularly smaller than 1 ha (Howeler, 2006; Valentin *et al.*, 2008). Hence, stabilization and augmentation of cassava yields can have important spillover benefits on local and regional food security (e.g. in Indonesia, southern Philippines), household and livelihood security, prosperity of rural economies and overall human welfare (Piya *et al.*, 2011; van Donge *et al.*, 2012; De Koninck and Rousseau, 2014).

Over the past decades, fast-accelerating genetic gain has permitted a steady growth of South-east Asian cassava yields, which reached approximately 21 t/ha by the early 2000s, primarily ascribed to fast-accelerating genetic gain (Henry and Gottret, 1996; Waddington *et al.*, 2010; FAOSTAT, 2015). This trend, however, is changing radically, following the accidental introduction of

* Corresponding author: i.graziosi@uky.edu

several deleterious non-native arthropods and pathogenic microorganisms, and the increased outbreaks of a number of long-time invaders and indigenous species (Bellotti *et al.*, 2012a; Alvarez *et al.*, 2013). In particular, multiple species of non-native mealybugs (Hemiptera: Pseudococcidae) have made their appearance in South-east Asia, and are spreading aggressively, aided by natural and anthropogenic means of dispersal and changing climates in many growing areas (Muniappan *et al.*, 2009; Parsa *et al.*, 2012). Some of these pests have caused substantial yield losses, and triggered the widespread adoption of (systemic) insecticides and acaracides. The (prophylactic) use of chemical insecticides could aggravate the economic burden for smallholders, and might have long-lasting detrimental impacts on the environmental integrity of local agroecosystems and human health (Kogan, 1998; Wilson and Tisdell, 2001; Goulson, 2013; Van der Sluijs *et al.*, 2015). In the meantime, as past work in Africa and South America has shown, cassava is a crop that is tailor-made for biological control. Hence, increased efforts should be made to formulate and promote integrated pest management (IPM) strategies, to further mitigate the impact of emerging biotic threats and secure the sustainability and resilience of local cropping systems (Edwards, 1989; Birch *et al.*, 2011).

In recent years, global interest in resource-conserving crop and pest management strategies has grown dramatically. In developing countries in particular, there is an urgent need to implement cultivation practices that can limit costs for small-scale farmers, reduce chemical inputs, promote agroecosystem health and mitigate the effects of climatic change (and climate-triggered pests or diseases) (Pretty *et al.*, 2003, 2006). On the one hand, organic agriculture has expanded as a sustainable and self-sufficient farming practice and is now contributing significantly to the world's food supply (Rigby and Cáceres, 2001; Badgley *et al.*, 2007). With over 3 million ha under organic production, Asia holds about 10% of the world's organic acreage. Organic farms are mostly located in China, India, Thailand, the Philippines and Vietnam, largely comprise horticultural crops, and are of increasing importance thanks to rising (regional) demand for organic produce (Willer and Kilcher, 2011). In spite of being grown for food and industrial purposes, South-east Asian cassava equally has ample potential as an organically managed crop and there is even a small but lucrative niche for organic tapioca starches and cassava flours (Anurugsa and Anuluxtipun, 2010; Andrade *et al.*, 2014). Given this scenario, organic crop management seems to be particularly well suited for a (semi-perennial) crop that is regularly cultivated with limited inputs, on small plots or degraded soils (Bouaguimbeck, 2011; Radhakrishnan *et al.*, 2015). On the other hand, non-chemical pest management, including organic options, clearly has a place in conventional cassava production systems. More so, there is an urgent need for applied research on environmentally sound tactics (and cropping systems adaptations) to address the complex of invasive, deleterious and fast-spreading pests and build broader systems resilience.

This chapter provides an updated account of the arthropod pest complex that is affecting South-east Asian cassava, describes current pest management options, and highlights future opportunities for development and scale-up of IPM and organic pest-control technologies.

The Cassava Pest Complex

The arthropod pest complex of South-east Asian cassava crops is primarily constituted by non-native, piercing, sucking herbivores, comprising both new and old-time invaders. Two newly arrived exotics, the cassava mealybug *Phenacoccus manihoti* Matile-Ferrero (Hemiptera: Pseudococcidae) and the green mite *Mononychellus mcgregori* (Flechtmann and Baker) (Acari: Tetranychidae) are cassava herbivores that have co-evolved with their host plant in its native South America. Other species are generalist herbivores that have gradually adapted to cassava and only in recent years are increasingly reported at heightened population levels (Bellotti *et al.*, 2012b). Among those are several non-native mealybugs and well-known pests of many crops like the papaya mealybug, various red spider mites (Acari: Tetranychidae) and multiple cosmopolitan whitefly species (Hemiptera: Aleyrodidae). Some non-natives, such as the globally important striped mealybug *Ferrisia virgata* (Cockerell), made it into South-east Asia over the past century, but only recently have been suspected to cause yield losses (Bayubay and Corpuz-Raros, 2006; Bellotti *et al.*, 2012a).

Among the endemic taxa are species within the red mite complex and herbivores often associated with cassava crops, such as white grubs or termites. Mealybugs, mites and whiteflies are the main taxonomic groups relevant to South-east Asian cassava. Presented here is an overview of the composition and bio-ecology for these groups, including their occurrence and local infestation pressure (Fig. 21.1) (Graziosi *et al.*, 2016).

The mealybug complex

Mealybugs (Hemiptera: Pseudococcidae) are soft-bodied sap-feeding insects with a marked sexual dimorphism and the ability to produce protective wax secretions (Gullan and Kosztarab, 1997; Hardy *et al.*, 2008). Mealybugs are among the world's most invasive groups of herbivorous insects (Miller and Miller, 2002; Wang *et al.*, 2010; Beltrà *et al.*, 2013). Their cryptic habit and minute size make mealybugs difficult to detect upon plant inspection, this facilitating their accidental introduction with plant stocks (Ben-Dov, 1994). Furthermore, the wide host range and virus-vectoring capability of a number of species significantly enhances their respective invasiveness and pest status (Minafra and Hadidi, 1998; Sether *et al.*, 1998; Wetten *et al.*, 2016). In cassava, mealybugs are typical dry-season pests, with quick population build-ups triggered by lowered precipitation and increased temperatures (Bellotti *et al.*, 1994; Calatayud *et al.*, 1994). Distinctive symptoms of mealybug damage include leaf yellowing and curling, defoliation and bud death (Ayanru, 1987; Bellotti, 2000), with detrimental effects on plant growth and root yield (Nwanze, 1982; Zeddies *et al.*, 2001).

Morphological features are used to organize the different mealybug taxa, with a distinction between short- and long-tailed species (Sanders, 1909). Short-tailed mealybugs have reduced waxy filaments that are equal or shorter than 75% of the body length, and produce ovisacs, wax-covered egg masses deposited on host plants. Long-tailed mealybugs are then characterized by conspicuous abdominal filaments (i.e. so-called 'tails'), production of live nymphs and absence of ovisacs. Among the 24 species of mealybugs known to feed on cassava worldwide, eight taxa are reported from South-east Asia (Parsa *et al.*, 2012). Five non-native short-tailed mealybugs are identified in the region (Table 21.1).

One of the most worrying recent invaders in South-east Asia is the cassava mealybug *P. manihoti*; the Neotropical herbivore that previously invaded Africa during the 1970s and devastated local cassava crops (Nwanze, 1982; Zeddies *et al.*, 2001). Compared with other (generalist) mealybugs occurring on South-east Asian cassava, *P. manihoti* is an oligophagous species, affecting as little as nine host plant genera (Ben-Dov, 1994). Parthenogenetic females lay hundreds of eggs in ovisacs, and newly hatched nymphs disperse both within the plant and to neighbouring plants, settling on growing tips and going through four instars prior to adult moulting (Nwanze *et al.*, 1979; Herrera *et al.*, 1989). Due to the release of toxins during mealybug feeding, young leaves are deformed and infested plants can be identified by the clumped appearance of the growing point (i.e. so-called bunchy-top). Following its first detection in eastern Thailand in 2008, this insect rapidly dispersed into Laos, Cambodia, Vietnam, Indonesia and Malaysia (Muniappan *et al.*, 2009; Winotai *et al.*, 2010; Parsa *et al.*, 2012; Sartiami *et al.*, 2015). Geographic spread and local proliferation were probably facilitated by unregulated trade of cassava planting material, particularly suitable local climatic conditions and absence of effective natural enemies. The fast-spreading insect heavily impacted cassava fields throughout the region, and yield drops up to 50% were reported from Thailand (Winotai *et al.*, 2010; Bellotti *et al.*, 2012a). The remaining four short-tailed

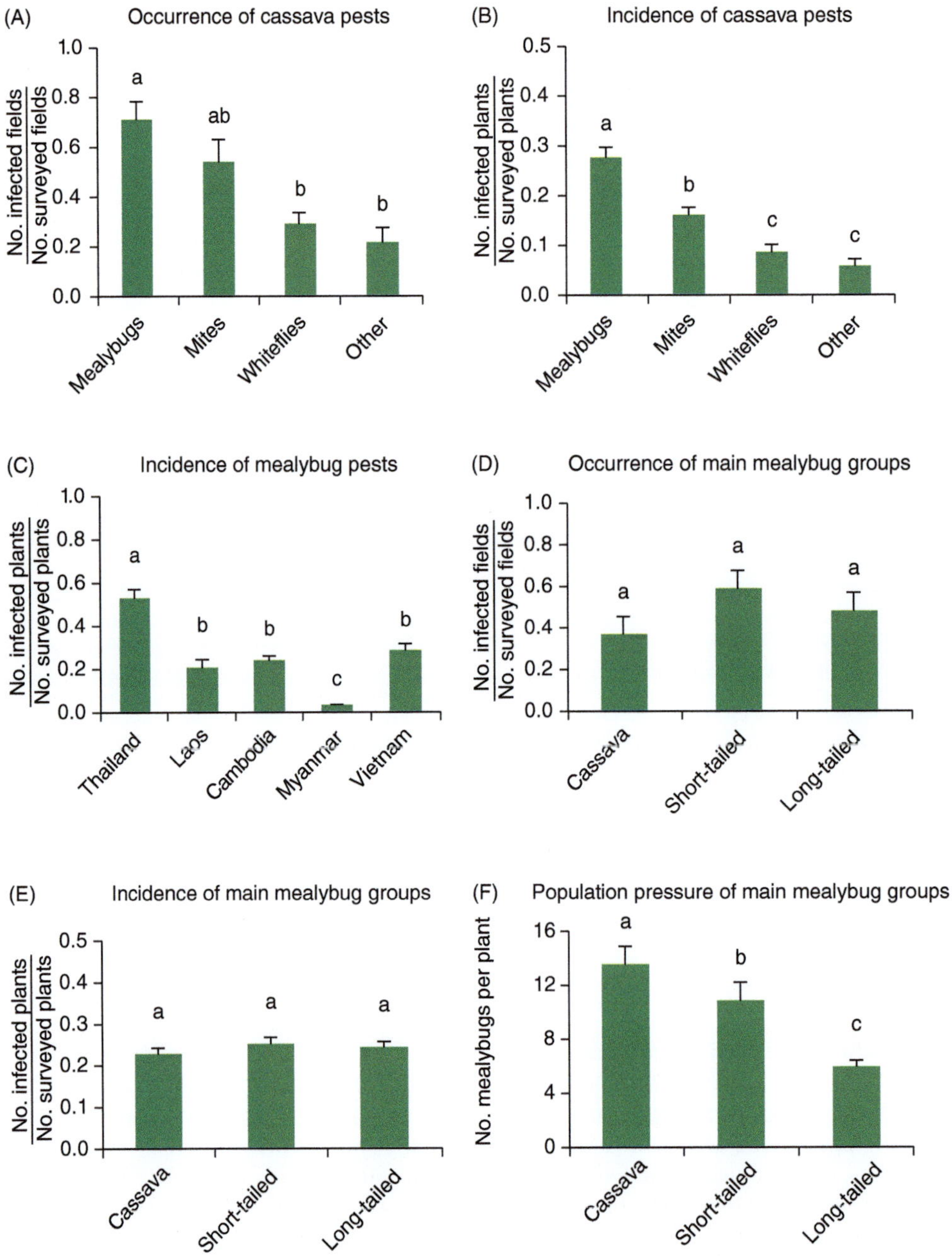

Fig. 21.1. Arthropod pests affecting cassava crops in South-east Asia. (A) Occurrence (proportion of affected fields per province) and (B) incidence (proportion of affected plants per affected field) of cassava pests. 'Other' arthropods include cassava scale, Caribbean black scale and termites. (C) Incidence of mealybug pests in different South-east Asian countries. (D) Occurrence, (E) incidence, and (F) population pressure of main mealybug species or species groups. 'Short-tailed' mealybug here refers to short-tailed mealybugs other than cassava mealybug. Means followed by the same letter do not differ (ANOVA, $\alpha = 0.05$); 429 fields in 26 provinces across five countries were monitored. (Modified from Graziosi *et al.*, 2016.)

Table 21.1. Main arthropod taxa reported on South-east Asian cassava. (Modified from Graziosi *et al.*, 2016.)

Taxa	Common name	Origin	Record in South-east Asia	Maximum yield loss[a]
Short-tailed mealybugs				
Phenacoccus manihoti	Cassava mealybug	Neotropics	2008	84%
Paracoccus marginatus	Papaya mealybug	Nearctic	2008	40%
Phenacoccus solenopsis	Cotton mealybug	Nearctic	2006	n.a.
Phenacoccus madeirensis	Madeira mealybug	Neotropics	2004	–
Maconellicoccus hirsutus	Hibiscus mealybug	Neotropics	1942	–
Long-tailed mealybug				
Pseudococcus jackbeardsleyi	Jack Beardsley mealybug	Neotropics	1996	n.a.
Ferrisia virgata	Striped mealybug	Neotropics	1942	–
Green mite complex				
Mononychellus mcgregori	Green mite	Neotropics	2009	60%
Red mite complex				
Neotetranychus lek	Red mite	n.a.	2013	53%[b]
Oligonychus thelytokus	Red mite	Afrotropical	1998	–
Tetranychus marianae	Red mite	Nearctic	1975	–
Eutetranychus africanus	African red mite	Neotropics	1975	–
Tetranychus yusti	Red mite	Palearctic	1975	–
Tetranychus urticae	Red mite	Afrotropical	1975	–
Oligonychus biharensis	Red mite	Oriental	1975	–
Tetranychus cinnabarinus	Red mite	Palearctic	1969	–
Tetranychus neocaledonicus	Red mite	Australian	1962	–
Tetranychus truncatus	Red mite	Palearctic	1962	–
Tetranychus kanzawai	Red mite	Palearctic	1962	–
Eutetranychus orientalis	Red mite	Palearctic	1962	–
Whiteflies				
Aleurodicus dispersus	Spiralling whitefly	Nearctic	1987	n.a.
Bemisia tabaci	Silverleaf whitefly	Palearctic	1933	–
Scale insects				
Aonidomytilus albus	Cassava scale	Nearctic	1935	n.a.
Parasaissetia nigra	Black scale	n.a.	n.a.	–
Saissetia miranda	Black scale	Nearctic	n.a.	–
Secondary pests				
Isoptera: Rhinotermitidae	Termites	Oriental	n.a.	n.a.
Coleoptera: Scarabedidae	Whitegrubs	Oriental	n.a.	–

[a]n.a., Not available.
[b]Value of 53% refers to the whole red mite complex.

mealybug species are well-known pests of other crops, and non-native generalist herbivores that adapted to *M. esculenta*. The papaya mealybug *Paracoccus marginatus* Williams and Granara de Willink is a global pest of papaya and numerous other crops, which invaded Asia in 2008, and has since spread to Indonesia, India, Cambodia, Thailand and the Philippines (Miller and Miller, 2002; Muniappan *et al.*, 2008; Muniappan, 2011). In India, yield loss due to *P. marginatus* reached 40% (Myrick *et al.*, 2014). The cotton mealybug *Phenacoccus solenopsis* Tinsley, and the Madeira mealybug *Phenacoccus madeirensis* Green are also recent invaders proliferating in climatically suitable areas of South-east Asia, while the hibiscus mealybug *Maconellicoccus hirsutus* Green has been present for a much a longer period of time (Takahashi, 1942). For most of the above species, little is known about their occurrence and impact on local cassava crops (Lu *et al.*, 2004; Muniappan *et al.*, 2009; Nagrare *et al.*, 2009; Wang *et al.*, 2010). Among the three long-tailed mealybugs, *Ferrisia virgata* is the oldest invader (Takahashi, 1942).

As a highly polyphagous and cosmopolitan herbivore, this species commonly reaches outbreaks in coffee, sweet potato, citrus and cotton (da Silva-Torres *et al.*, 2013). On cassava, *F. virgata* populations are thought to have increased substantially over the past decade, occasionally bringing about losses in root yield (Bellotti *et al.*, 2012a). Two other species, *Pseudococcus jackbeardsleyi* Gimpel & Miller (Fig. 21.1C), a major agricultural and horticultural pest (Shylesha, 2013), and *Pseudococcus longispinus* Targioni Tozzetti, a well-known issue in vineyard and fruit crops (Swirski *et al.*, 1980; Charles, 1981), are also inhabiting South-east Asian cassava fields, and their presence and yield impact on local crops is yet to be evaluated (Williams, 2004; Mani *et al.*, 2013).

In dry-season cassava plots, mealybugs are the most common group of arthropods and the pest complex with the highest incidence, occurring in 70% of fields (Fig. 21.1A) and infesting > 25% of plants on average (Fig. 21.1B, C). Within this complex, occurrence and field-level incidence of *P. manihoti* was comparable to those of other mealybugs (Fig 21.1D, E). In 2014 *P. manihoti* occurred in most of the region's key cassava growing areas. Overall, *P. manihoti* population levels were highest, followed by (other) short-tailed and long-tailed taxa (Fig. 21.1F), this highlighting the invasiveness of the former insect.

Mites

The global cassava mite complex counts approximately 50 taxa of red and green spider mites (Acari: Tetranychidae) (Byrne *et al.*, 1983). In Asian cassava crops, one species of green mite and at least 12 different red mite species have been reported (Table 21.1) (Byrne *et al.*, 1983). Mites can rapidly attain outbreak levels during the dry season, thanks to high reproductive rates and comparatively slow colonization (and build-up) of natural enemy populations, especially in larger monocultures (Bellotti, 2008). In Indonesia, red mites cause yield reductions up to 50%, while they are also noted as major pests in eastern Thailand and central Vietnam (Byrne *et al.*, 1983). Green mites are oligophagous herbivores and major pests of cassava in Africa (Yaninek and Hanna, 2003), while red mites are cosmopolitan and ubiquitous herbivores, with exceptional ability to adapt to new hosts (Agrawal, 2000). Most green mites preferentially infest young leaves and growing tips, and red mites usually feed on the underside of leaves located on the lower canopy, causing leaf yellowing and defoliation. When occurring at high population levels, red mites also affect upper parts of the plant (Byrne *et al.*, 1983; Yaninek *et al.*, 1989). At present, there is very little information on the composition and relative species abundance of the cassava mite complex in Asia, and novel species continue to be described (Flechtmann, 2013). Some of the most common species are *Tetranychus urticae* Koch, *Tetranychus truncatus* Ehara, *Tetranychus cinnabarinus* Boisd., *Tetranychus kanzawai* Kishida and the African red mite *Eutetranychus africanus* (Tucker) (Table 21.1) (Oei-Dharma, 1969; Baker, 1975a, b; Flechtmann, 1981; Byrne *et al.*, 1983; Sakunwarin *et al.*, 2003, 2004; Hinomoto *et al.*, 2007; Bellotti *et al.*, 2012b, c; Naing *et al.*, 2014).

In addition to red mites, the Neotropical green mite *M. mcgregori* has been identified in Vietnam and is thought to occur in China and Cambodia, but accurate information is needed regarding its status and occurrence in those countries (Bellotti *et al.*, 2012a; Parsa *et al.*, 2015). Furthermore, there is certain concern about an eventual presence of the cassava green mite *Mononychellus tanajoa*, a crucial pest of cassava capable of yield losses over 80% in Africa (Lu *et al.*, 2014; Parsa *et al.*, 2015).

During 2014 region-wide monitoring, mites were reported from over half of (dry-season) cassava fields (Fig. 21.1A), and on 16% of plants (Fig. 21.1B). Initial data show that the cassava mite complex was primarily composed of red mite species, with an absence of *M. mcgregori* outbreaks.

Whiteflies

Whiteflies (Hemiptera: Aleyrodidae) are global pests, damaging a wide range of agricultural and horticultural crops in open-field and greenhouse settings (Gerling, 1990; Bellotti and Arias, 2001). Plant damage is caused by direct feeding, virus transmission and proliferation of sooty moulds on leaves due to honeydew secretions (Jones, 2003). The most prominent whitefly-vectored viruses are the cassava mosaic virus (CMV) in Africa, India and Sri Lanka, and cassava brown streak disease in Africa. At present, there are no reports of whitefly-vectored plant pathogens in South-east Asia. Even though a highly diverse group of whitefly species impacts cassava in the Neotropics, only two whitefly species are found across the region (Bellotti, 2002; Bellotti *et al.*, 2012a), where they occasionally reach high infestation levels and cause crop damage and defoliation (Bellotti and Arias, 2001; Bellotti, 2008). The spiralling whitefly *Aleurodicus dispersus* Russel (Hemiptera Aleyrodidae) is a Neotropical species impacting cassava and various other crops worldwide (Neuenschwander, 2009). In 1987, *A. dispersus* made its initial appearance in the Philippines, and has since been found throughout the region (Kajita *et al.*, 1991; Mani and Krishnamoorthy, 2002; Muniappan *et al.*, 2009; Bellotti *et al.*, 2012a). While *A. dispersus* attains its highest population densities during the dry season, nymphal development is also facilitated by high humidity (D'Almeida *et al.*, 1998). The silverleaf whitefly *Bemisia tabaci* Gennadius is a cosmopolitan invader, able to develop on over 600 host plants, and thought to have invaded South-east Asia during the 1930s (Oliveira *et al.*, 2001). The insect is a rainy-season pest; numbers critically increase at the end of each rainy period, but background populations persist all year round (Bellotti and van Schoonhoven, 1978; Robertson, 1987). *B. tabaci* is a complex of morphologically identical species with distinct geographical distributions and genetic traits (Brown *et al.*, 1995; De Barro *et al.*, 2011). Multiple *B. tabaci* genotypes are described feeding on cassava, and different genetic populations are occurring in Vietnam, Thailand, Malaysia and Indonesia (Berry *et al.*, 2004; De Barro, 2005; Sseruwagi *et al.*, 2006). While a lot of information is available in the global literature on both *A. dispersus* and *B. tabaci*, the exact taxonomic composition, bio-ecology and incidence of whiteflies in South-east Asian cassava fields are yet to be determined.

In region-wide monitoring efforts during the 2014 dry season, *B. tabaci* and *A. dispersus* were observed in less than one-third of cassava fields, infesting on average 8% of plants (Fig. 21.1A, B). While no information is available on whitefly populations during the rainy season, we expect infestation pressure to be (somewhat) higher.

Secondary pests

A number of other arthropods are known to feed on Asian cassava, and localized yield losses may occur. The cassava or tapioca scale *Aonidomytilus albus* infests plants and 'stakes' (i.e. planting material, as cassava is vegetatively propagated) in India and Thailand (Subramaniam *et al.*, 1977; Wongkobrat *et al.*, 1987). Black or brown scales *Parasaissetia nigra* Nietner and *Saissetia miranda* Cockerell and Parell have also been reported from different cassava-growing areas (Peña and Waddill, 1982; Bellotti *et al.*, 2012b, c). Furthermore, white grubs (Coleoptera: Scarabeidae) and termites (Isoptera: Rhinotermitidae) occasionally infest the lower part of the stem and roots, and cause differing levels of yield loss (Ananda and Rasdiman, 1979; Miura *et al.*, 1990; Atu, 1993; Ranga Rao *et al.*, 2006; Bellotti *et al.*, 2012c). Scales or termites that are occasionally reported in local cassava plots occur in one-fifth of fields and infest only 6% of plants (Fig. 21.1A, B).

Guidelines for Non-Chemical Pest Management

With the recent arrival of several invasive and deleterious pests, chemical control has been widely promoted in a number of Asian countries and has fast gained a foothold in intensive cassava cropping systems (Franco *et al.*, 2009; Parsa *et al.*, 2012; L.T.P. Nghiem and K.A.G. Wyckhuys, 2017, unpublished data). For instance aqueous solutions of systemic neonicotinoids thiamethoxam, imidacloprid and dinotefuran are now extensively used in Thailand and actively promoted in neighbouring countries for stake disinfection and as prophylactics targeting mealybugs (Parsa *et al.*, 2012; K.A.G. Wyckhuys, 2014, unpublished data). There is mounting concern about the efficacy, economic sustainability and environmental impact of the preventative use of systemic insecticides, and its incompatibility with long-established principles of IPM (Goulson, 2013). Also, the extraordinary ability of the predominant pests (i.e. mealybugs, mites) to rapidly build up their populations, and their propensity to develop insecticide resistance can constrain the long-term sustainability of chemically based pest control in cassava fields (Herron *et al.*, 1998; Franco *et al.*, 2009; Koh *et al.*, 2009; Van Leeuwen *et al.*, 2010; Bellotti *et al.*, 2012a). Moreover, intensive insecticide applications can severely impact in-field abundance, activity and fitness of locally occurring or released natural enemies (Poletti *et al.*, 2007; Wang *et al.*, 2008; Bellotti *et al.*, 2012a). The recurrent, irrational use of systemic insecticides can lead to resistance development and natural enemy suppression, and ultimately trigger increased infestations and damage by tetranychid mites or mealybugs (Sclar *et al.*, 1998). The above also further exemplifies the increased value and need for sustainable and environmentally sound IPM packages. IPM aims to reduce pest damage to tolerable levels by adopting multiple compatible techniques including natural enemies, plant resistance, cultural practices, and (eventually) minimal, targeted pesticide use, while taking into account economic, ecological and sociological consequences (Bottrell, 1979; Kogan, 1998). Furthermore, IPM tactics combine pest suppression and mitigation strategies, with monitoring, risk assessment and decision-support tools (Ehler, 2006; Lemic *et al.*, 2015).

In cassava, a number of most-valuable cultural and biological control options are available, almost tailor-made to smallholder production systems. Several of those practices can be readily promoted, and could prove particularly effective to address emerging pest threats (Table 21.2). Biological control stands out as a cost-effective and environmentally sound option for cassava pest management. The introduction of parasitic and predatory arthropods facilitated long-term control of several key invaders, such as the cassava mealybug and the cassava green mite in Africa (Zeddies *et al.*, 2001; Yaninek and Hanna, 2003), while recent research is showing ample promise for entomopathogens (Maniania *et al.*, 2008; Franco *et al.*, 2009; Mascarin *et al.*, 2013). Crop diversification and intercropping tactics have also been under-researched, but could contribute substantially to pest control and broader cropping systems' resilience (Gold *et al.*, 1989; Trenbath, 1993) (Table 21.3). While intercropping tactics could be particularly well suited for smallholder producers, there is certain concern about increased labour demand for their establishment and maintenance, in light of dropping labour availability in many of

Table 21.2. Overview of integrated pest management (IPM) options for managing cassava pests in South-east Asia.

Pest complex	Biological control[a,b]	Cultural[a,b]	Host resistance	Other
Mealybugs	Paras[1,2,3], pred[1], entom fungi[3]	Intercrop[1,2], clean plant material, removal[1]	Low[3]	Pheromone traps[2]
Mites	Pred[1,2], entom fungi[3]	Intercrop[1,2]	Moderate[2]	Botanical extracts[2,3]
Whiteflies	paras[2], entom fungi[2]	Intercrop[1,2]	High levels[2]	–

[a]Paras, parasitic Hymenoptera; pred, predatory arthropods; entom fungi, fungal entomopathogens; intercrop, intercropping; removal, physical removal and destruction of infested plant parts.
[b]1, Implemented at some level in South-east Asia; 2, available and implemented in other areas and similar systems; 3, under evaluation.

Table 21.3. Effects of intercropping strategies on arthropod prevalence in cassava. All studies were conducted in Africa.

Target pest	Intercrop species	Effect (+/– variation)	No. of studies
Cassava mealybug (*Phenacoccus manihoti*)	Maize	–57–83% population pressure, no effect on percentage infested plant	2
Green mite (*Mononychellus tanajoa*)	Beans, maize, sweet potato	–7% incidence, no effect to –21% nymph, –40% adults on young leaves, no effects on predatory mite	2
Whiteflies (*Bemisia tabaci*)	Cotton and jatropha, maize and cowpea, bean maize and sweet potato	–79% eggs, –21–58% nymphs, –40–65% adults	3

South-east Asia's rural areas (Rigg, 2013, 2015). In the sections below, different non-chemical pest management options are presented for each of the pest complexes, emphasizing those already implemented on South-east Asian cassava and highlighting technologies that wait to be fine-tuned or transferred to local farmers.

Mealybugs

Natural enemies play a prominent role in the regulation of mealybug populations in a diverse range of cropping systems (Franco *et al.*, 2009; Daane *et al.*, 2012). More specifically, various species of parasitic or predatory insects and fungal entomopathogens have been reported for mealybug taxa relevant to cassava (Bellotti *et al.*, 2012a, b). Encyrtid wasps (Hymenoptera: Encyrtidae) are typical mealybug parasitoids (Noyes and Hayat, 1994), and have been effectively used in a number of biological control programmes such as the continent-wide management of *P. manihoti* in Africa, *P. marginatus* in the Pacific, *P. hirsutus* in the Caribbean and *Ps. longispinus* in Israel (Swirski *et al.*, 1980; Herren and Neuenschwander, 1991; Kairo *et al.*, 2000; Muniappan *et al.*, 2006). While biological control methods have been successfully wielded against pestiferous mealybugs, their effectiveness greatly depends upon local climate, crop management and soil conditions (Neuenschwander *et al.*, 1990; Schulthess *et al.*, 1997).

Initial yet deliberate steps have been taken to employ biological control against some of the mealybug invaders in Asia. Following the first outbreaks of the cassava mealybug in 2008, Thai institutions teamed up with the Food and Agriculture Organization of the United Nations (FAO) and global Consultative Group on International Agricultural Research (CGIAR) centres to introduce the Neotropical nymphal parasitoid *Anagyrus lopezi* (De Santis, 1964) (Hymenoptera: Encyrtidae). This minute wasp causes high insect mortality through host feeding, mutilation, oviposition probing and reproduction (Neuenschwander and Madojemu, 1986), and already earned its stripes in an earlier *P. manihoti* management campaign in the 1980s, in Africa. Wasps were moved from Benin to Thailand and mass reared, with releases conducted in late 2009 (Winotai *et al.*, 2010). Subsequently, successful establishment and lowered mealybug population levels were observed in most Thai cassava fields (Bellotti *et al.*, 2012a; Soysouvanh and Siri, 2013). Next, inoculative releases were conducted in Laos, Cambodia and Indonesia, and small rearing operations were established in southern Vietnam (FAO-IPM Asia Regional Programme, 2013, 2014). In some *P. manihoti*-infested areas in Thailand, local authorities and private sector actors continue to mass rear and release *A. lopezi*. These augmentative biological control schemes are very different from past practices in Africa, where no follow-up releases were conducted (Neuenschwander, 2001). Possibly, these schemes are still required, given the vastly different make-up of local cassava cropping systems, as defined by monocropping arrangements, increasing (broad-spectrum) pesticide use, staggered planting of fields, among others. In addition to *A. lopezi*, three non-native wasps, *Acerophagus papayae* Noyes and Schauff, *Pseudleptomastix mexicana* Noyes and Schauff and *Anagyrus loecki* Noyes and Menezes (Hymenoptera: Encyrtidae) are successfully controlling *P. marginatus* outbreaks in India, and wait to be employed in other countries (Myrick *et al.*, 2014). Also, a diverse community of naturally occurring predatory arthropods inflict mortality on mealybug taxa feeding on cassava, with ladybird beetles (Coleopotera: Coccinellidae) and green and brown lacewings (Neuroptera: Chrysopidae, Hemerobiidae) often being reported (Löhr *et al.*, 1990; Mani and Krishnamoorthy, 2008). Within South-east Asia, the coccinellid *Sasajiscymnus quinquepunctatus* (Weise) has been identified as a candidate for biological control of *P. marginatus* in Thailand (Saengyot and Burikam, 2012a), and augmentative releases of green lacewings *Plesiochrysa ramburi* (Schneider) and *Mallada basalis* (Walker) (Neuroptera: Chrysopidae) target the Thai cassava mealybugs complex (Suasa-ard, 2010). Lastly, augmentative biological control has also been considered using the endemic apefly *Spalgis epius* (Westwood) (Lepidoptera: Lycaenidae) (Saengyot and Burikam, 2012b).

The use of fungal entomopathogens also carries ample promise, with *Neozygites fumosa* (Entomophthorales) actively promoted for *P. manihoti* control in African cassava crops (Le Rü *et al.*, 1985; Le Rü, 1986). Other entomopathogenic fungi such as *Beauveria bassiana* have been used to control *P. marginatus*, *P. solenopsis* and *M. hirsutus* on cotton in Asia (Banu and Gopalakrishanan, 2012; Surulivelu *et al.*, 2012; Kumar *et al.*, 2014; Ujjan *et al.*, 2015).

Sex pheromones are also increasingly employed for mealybug management (Millar *et al.*, 2002; Franco *et al.*, 2009). Pheromone lures are readily used to monitor mealybug populations, and thus guide pest management decisions. In the meantime, they could also be used for mass trapping, mating disruption or volatile-mediated recruitment of natural enemies (Branco *et al.*, 2006; Franco *et al.*, 2011). To our knowledge, these info-chemicals remain under-researched and unused for mealybug pest control in cassava.

While the above tactics represent immediate, curative options to tackle pest outbreaks, an appropriate planning and implementation of a set of 'good practices' for cassava crop management can contribute substantially to preventing pest problems. For example, the use of good-quality, pest-free planting material is crucial and non-chemical tactics should be used for on-farm disinfection of cassava stakes. Also, cultural control and proper field sanitation remain under-used, but could readily prevent localized mealybug outbreaks to a limited extent. In Africa, the use of maize as an intercrop drastically reduces population pressure of the cassava mealybug (Agbobli, 1987; Schulthess *et al.*, 2004) (Table 21.3), but data from Asia and for other intercrop and mealybug species appear to be missing. Also, work in the 1990s showed that biological control efficacy is directly tied to soil fertility (Schulthess *et al.*, 1997), but Asian cassava researchers (and farmers) all too often forget that a healthy and well-fertilized soil is prerequisite for effective pest suppression and crop health (Schulthess *et al.*, 1997). At present, some of these crop and pest management practices are being promoted through novel extension tools, such as farmer-to-farmer educational videos (L.T.P. Nghiem and K.A.G. Wyckhuys, 2017, unpublished data).

The above arsenal of (curative and preventative) control tactics can diversify and improve some of the management tactics that South-east Asian farmers currently use to address mealybug invaders. While some technologies, such as sampling protocols, economic thresholds or other decision-support tools, still wait to be developed and additional knowledge needs to be gained on biocontrol options for other mealybugs, the authors of this chapter are confident that these can further advance environmentally sound and low-cost pest management for Asian cassava crops.

Mites

An array of options are available for the management of cassava mite infestations, ranging from biological control and cultural practices to recently validated botanical acaricides (Byrne *et al.*, 1983; Corbett *et al.*, 1991; Bellotti *et al.*, 1999; Isman, 2000).

Natural enemies can effectively regulate mite populations, and successful biological control of mites has been achieved using predatory mites (Acari: Phytoseiidae) (Hamamura, 1986; McMurtry and Croft, 1997; Parsa *et al.*, 2015). Cassava green mite outbreaks in Africa were effectively managed using the Neotropical phytoseiid *Typhlodromalus aripo* (De Leon) (Yaninek and Hanna, 2003; Mutisya *et al.*, 2015). In South-east Asia, the phytoseiid predator *Amblyseius* (*Neoseiulus*) *longispinosus* (Evans) is deployed for control of *T. kanzawai*, *T. truncatus*, *T. urticae* and *E. africanus* on cassava and other crops in Thailand and the Philippines. However, this species is highly susceptible to common insecticides, which greatly impede survival, development and ultimate effectiveness (Hamamura, 1986; Vasquez, 1994; Nusartlert *et al.*, 2010; Attia *et al.*, 2013). Myriad insect and arachnid predators are known to prey upon cassava mites worldwide: coccinellid *Stethorus gilvifrons* Mulsant and *Stethorus siphonulus* Kapar, *Pullus* sp., *Serangium* sp., rove beetle (Colepotera Staphylinidae) *Oligota* sp. and spiders *Poecilochroa* sp. (family Gnaphosidae) and *Olios* sp. (family Sparassidae) are described from Indonesia and India (Byrne *et al.*, 1983). Some of those predators could be used in augmentative biological control schemes, as exemplified by the successful use of predatory lacewings to manage *T. urticae* and *T. kanzawai* populations on Taiwanese strawberry crops (Chang and Huang, 1995).

Entomopathogens are increasingly considered to be good to outstanding candidates for integrated management of spider mites (Irigaray *et al.*, 2003). In cassava, *Neozygites* (Entomophthorales) strains can inflict high mortality to *M. tanajoa* in both native South America and Africa (Delalibera *et al.*, 2004). Researchers in Indonesia and the Philippines are evaluating different entomopathogenic fungi to control *T. kanzawai* (Sanjaya *et al.*, 2013a, b). It might be advisable to bank on these promising developments and test those fungal strains for potential use against mites in local cassava crops.

Cultural control options are still limited, and further research is needed to evaluate different crop management tools that can help mitigate mite outbreaks. One such option is through cropping systems diversification, as intercropping with bean, maize and sweet potato has shown to drastically lower densities of the cassava green mite in Africa (Toko *et al.*, 1996; Night *et al.*, 2011) (Table 21.3). While cassava varieties with high mite resistance (i.e. immunity) are not available, the combined use of biological control and moderately resistant varieties effectively controls *M. tanajoa* populations (Bellotti, 2008; Bellotti *et al.*, 2012a). In general, varieties with pronounced leaf pubescence are less susceptible to green mite infestations (Byrne, 1980). While research has been conducted in Africa and South America on cultural control or varietal resistance against mites, much remains to be done in terms of varietal screening for resistance to *Tetranychidae* in Asia.

Last but not least, recent work reveals highly effective repellent or acaricidal properties of different plant extracts, against red mites *T. urticae*, *T. cinnabarinus* and *Oligonychus biharensis*, this offering new options for an integrated or organic approach to cassava mite management (Zhang *et al.*, 2005, 2008; Wei *et al.*, 2011; Attia *et al.*, 2013; Roh *et al.*, 2013). Also, further work may be merited to explore opportunities for conservation biological control, eventually through provision of non-prey (alternative) food resources such as maize pollen, for predatory phytoseiid mites (e.g. Onzo *et al.*, 2005). In conclusion, recent research advances point at clear opportunities for non-chemical (organic) management of mites in cassava crops, and also stress the need for continued applied research to develop, evaluate and adapt IPM technologies for some of South-east Asia's most prominent cassava mite species.

Whiteflies

In the absence of recurrent population outbreaks, cassava whiteflies usually don't require control. However, this situation is starkly different in areas that are affected by whitefly-borne viral diseases (Thresh and Otim-Nape, 1994). Although no whitefly-transmitted viruses are reported from South-east Asia and local farmers generally don't control whiteflies, it is valuable to highlight a number of non-chemical management options.

Whitefly suppression can be achieved using locally occurring or introduced parasitic Hymenoptera. In West Africa, outbreaks of spiralling whitefly were effectively suppressed following fortuitous introduction and establishment of Neotropical parasitoids *Encarsia haitiensis* Dozier and *Encarsia guadeloupe* Viggiani (Hymenoptera: Aphelinindae) (D'Almeida *et al.*, 1998). The latter species also effectively controls *A. dispersus* in India and Taiwan (Mani *et al.*, 2004). While predators were largely thought to be of minor importance in controlling cassava whitefly, Lundgren *et al.* (2014) reported a surprisingly diverse community of predators, including a netwing beetle (Lycidae), spiders, coccinellid beetles and lacewings, to be attacking whitefly in Colombian cassava fields. On another front, the research on fungal entomopathogens is quickly advancing: different *Paecilomyces* species are associated with *B. tabaci* in India, and *Fusarium* strains are under evaluation against spiralling whitefly on vegetable crops in Asia (Faria and Wraight, 2001; Aiswariya *et al.*, 2007). None of the above fungal strains have been evaluated against whitefly in Asia, and (very) limited work has been done describing the natural enemy communities associated with local whitefly populations.

CGIAR institutions such as the International Center for Tropical Agriculture (CIAT) have spearheaded whitefly resistance screening and breeding, and have made available a number of cultivars highly resistant to *Aleurotrachelus socialis* and *B. tabaci* to cassava growers in South America and Africa (Bellotti and Arias, 2001; Ceballos *et al.*, 2004; Omongo *et al.*, 2012). Whitefly-resistant varieties are also present in germplasm collections in countries such as Thailand, China or Vietnam, but they are not deployed in the field. In addition to host plant resistance, cropping systems diversification tactics can contribute to pest control by inferring associational resistance or directly favouring resident natural enemy communities (Bellotti *et al.*, 2012a; Lundgren *et al.*, 2014). African trials show how various intercropping schemes can drastically lower in-field abundance of different whitefly development stages (Table 21.3) (Fondong *et al.*, 2002; Ewusie, 2008; Night *et al.*, 2011). Hence, diversification schemes and the incorporation of (moderately) resistant varieties constitute clear mechanisms for whitefly population suppression, and are on-the-shelf technologies in case whitefly control becomes a necessity in South-east Asia. In the meantime, biological control may carry considerable potential as well, but here critical insights still need to be gained.

Discussion and Conclusion

The work of the authors provides one of the first, comprehensive overviews of the magnitude of biotic invasions in South-east Asian cassava crops. This 4 million ha crop, which directly underpins the livelihoods of 8 million (smallholder) farmers and provides food security to entire nations, has recently been hit by a number of (fast-spreading) pest invaders, and increasing outbreaks of long-time insect colonists. While these mounting invasive pest problems can be ascribed to globalization (and the associated, cross-boundary movement of biota) and linked to climate change (Hobbs, 2000; Meyerson and Mooney, 2007; Peh, 2010), they also elucidate inherent, critical weaknesses and deficiencies of South-east Asia's cassava cropping systems. Grown primarily under monoculture arrangements, with complete abandonment of crop rotation schemes and deficient soil management and fertilization, local cassava crops have proven particularly susceptible to pest attack (see also Brown, 1963; Nweke,

1994; Hambäck and Englund, 2005). More so, with local farmers massively embracing the prophylactic use of (systemic) insecticides to address invasive pest problems, natural defences of these cropping systems may even be further undermined. This chapter has presented a number of options for non-chemical (or organic) management of some of these pest invaders, ranging from biological control to host plant resistance or the development of IPM decision-support tools. In the meantime, the authors also point at immediate opportunities (e.g. cropping systems diversification) to enhance or restore the resilience of local cassava cropping systems (see Aye and Howeler, 2012; Brooker *et al.*, 2014), and optimally enable natural plant defence mechanisms.

This chapter thus provides clear and immediate opportunities for low-cost, non-chemical pest management in local cassava crops, and for approaches that are fully in line with long-established principles of IPM. Only by safeguarding those valuable practices, and by recognizing, valuing and building upon past work from other continents, can the long-term profitability and sustainability of cassava production systems throughout South-east Asia be ensured.

Acknowledgements

We would like to thank Tin Maung Aye, Sok Sophearith and Lao Thao for logistical support to monitoring efforts. Phanuwat Moonjuntha, Sirilak Lankaew and Department of Agricultural Extension officers facilitated monitoring activities in Thailand. We are grateful to Khanxay Somchanda and Plant Protection Center, Lao PDR colleagues for assistance in identifying mealybugs, and to Mr Chhun Hy Heng (General Directorate of Agriculture) for help in orchestrating countrywide sampling efforts in Cambodia. Monitoring work in Vietnam, Laos and Cambodia was made possible through respective provincial and district-level Plant Protection Sub-Department, Provincial Agriculture and Forestry Office/District Agriculture and Forestry Office and Provincial Department of Agriculture offices. We thank Linh Giang and Louis Parker for developing maps on pest and cassava distribution, and Erik Delaquis for providing knowledge on intercropping included in Table 21.3. This initiative was conducted as part of a European Community-funded, International Fund for Agricultural Development-managed, CIAT-executed programme (CIAT-EGC-CO 1000001285).

References

Agbobli, C.A. (1987) *Incidence des Dates de Plantation sur l'infestation du Manioc par les Cochenilles et la Mosaïque Africaine en Culture Associée (manioc-maïs)*. Orstom-Institut Francais de Recherche Scientifique pour le Developpement en Cooperation, Lomé, Togo, 10 pp.

Agrawal, A.A. (2000) Host-range evolution: adaptation and trade-offs in fitness of mites on alternative hosts. *Ecology* 81, 500–508.

Aiswariya, K.K., Manjunatha, M. and Naik, M.I. (2007) Evaluation of fungi *Fusarium semitectum* Berk. and Ravenel and *Verticillium lecanii* (Zimm.) Viegas against spiraling whitefly *Aleurodicus dispersus* Russell on guava. *Karnataka Journal of Agricultural Sciences* 20, 283–287.

Alvarez, E., Pardo, J.M., Mejía, J.F., Bertaccini, A., Thanh, N.D. and Hoat, T.X. (2013) Detection and identification of '*Candidatus* Phytoplasma asteris'-related phytoplasmas associated with a witches' broom disease of cassava in Vietnam. *Phytopathogenic Mollicutes* 3, 77–81.

Ananda, K. and Rasdiman, S. (1979) White grubs identification on cassava at Kalasan region, Yogyakarta (Central Java, Indonesia). Paper presented at Kongres Entomologi, 9–11 January 1979, Jakarta, Indonesia.

Andrade, M.M.P., de Oliveira, C.S., Colman, T.A.D., da Costa, F.J.O.G. and Schnitzler, E. (2014) Effects of heat–moisture treatment on organic cassava starch. *The Journal of Thermal Analysis Calorimetry* 115, 2115–2122.

Anurugsa, B. and Anuluxtipun, Y. (2010) *Organic Tapioca and Paddy Rice*. Technical manual, Thai Land Development Department, Thailand. Available at: http://www.ldd.go.th (accessed 26 February 2016).

Attia, S., Grissa, K.L., Lognay, G., Bitume, E., Hance, T. and Mailleux, A.C. (2013) A review of the major biological approaches to control the worldwide pest *Tetranychus urticae* (Acari: Tetranychidae) with special reference to natural pesticides. *Journal of Pest Science* 86, 361–386.

Atu, U.G. (1993) Cultural practices for the control of termite (Isoptera) damage to yams and cassava in south-eastern Nigeria. *International Journal of Pest Management* 39, 462–466.

Ayanru, D.K.G. (1987) Effects of mealybug (*Phenacoccus manihoti*) infestation on cassava yield components and plant tissue quality. *Journal of Agricultural and Rural Development in the Tropics and Subtropics* 88, 5–10.

Aye, T.M. and Howeler, R. (2012) Cassava agronomy: intercropping systems. In: Howeler, R.H. (ed.) *The Cassava Handbook*. CIAT, Cali, Colombia, pp. 613–625.

Badgley, C., Moghtader, J., Quintero, E., Zakem, E., Chappell, M.J., Aviles-Vazquez, K. and Perfecto, I. (2007) Organic agriculture and the global food supply. *Renewable Agricultural and Food Systems* 22, 86–108.

Baker, E.W. (1975a) Plant-feeding mites of Thailand (Tetranychidae, Tenuipalpidae, and Tuckerellidae). *Plant Protection Service Technical Bulletin*. Plant Protection Service, Bangkok.

Baker, E.W. (1975b) Spider mites (Tetranychidae: Acarina) from Southeast Asia and Japan. *Cooperative Economic Insect Report* 25, 911–921.

Banu, J.G. and Gopalakrishanan, N. (2012) Development of formulations of a native entomopathogenic fungus, *Lecanicillium lecanii* and testing virulence against mealybug, *Paracoccus marginatus* infesting cotton. *Indian Journal of Plant Protection* 40, 182–186.

Bayubay, A.L.G. and Corpuz-Raros, L.A. (2006) Taxonomic survey of mites and sap-sucking insects associated with cassava especially in commercial plantations in Isabela Province, Philippines. *Philippine Entomologist* 20, 102–125.

Bellotti, A.C. (2000) Las plagas principales del cultivo de la yuca: un panorama global. In: *Simposio Avances en el Manejo de Plagas. Memorias del XXVII Congreso de SOCOLEN*, July 2000, Medellín, Colombia. Sociedad Colombiana de Entomologia (SOCOLEN), Bogotá, pp. 189–217.

Bellotti, A.C. (2002) Arthropod pests. In: Hillocks, R.J., Wydra, K. and Bellotti, A. (eds) *Cassava: Biology, Production and Utilization*. CAB International, Wallingford, UK, pp. 209–235.

Bellotti, A.C. (2008) Cassava pests and their management. In: Capnera, J.L. (ed.) *Encyclopedia of Entomology*, 2nd edn. Springer, Dordrecht, The Netherlands.

Bellotti, A.C. and Arias, B. (2001) Host plant resistance to whiteflies with emphasis on cassava as a case study. *Crop Protection* 20, 813–823.

Bellotti, A.C. and van Schoonhoven, A. (1978) Mite and insect pests of cassava. *Annual Review of Entomology* 23, 39–67.

Bellotti, A.C., Braun, A.R., Arias, B., Castillo, J.A. and Guerrero, J.M. (1994) Origin and management of neotropical cassava arthropod pests. *African Crop Science Journal* 4, 407–417.

Bellotti, A.C., Smith, L. and Lapointe, S.L. (1999) Recent advances in cassava pest management. *Annual Review of Entomology* 44, 343–370.

Bellotti, A.C., Campo, B.V.H. and Hyman, G. (2012a) Cassava production and pest management: present and potential threats in a changing environment. *Tropical Plant Biology* 5, 39–72.

Bellotti, A.C., Arias, B.V., Vargas, O.H., Rejes, J.A.Q. and Guerrero, G.M. (2012b) Insects and mites that attack cassava, and their control. In: Ospina, B. and Veballos, H. (eds) *Cassava in the Third Millennium: Modern Production, Processing, Use, and Marketing Systems*. CIAT, Cali, Colombia, pp. 213–250.

Bellotti, A.C., Herrera, C.J., del Pilar Hernandez, M., Arias, B., Guerrero, G.M. and Melo, E.L. (2012c) Cassava pests in Latin America, Africa and Asia, In: Howeler, R.H. (ed.) *The Cassava Handbook*. CIAT, Cali, Colombia, pp.199–257.

Beltrà, A., Tena, A. and Soto, A. (2013) Fortuitous biological control of the invasive mealybug *Phenacoccus peruvianus* in southern Europe. *BioControl* 58, 309–317.

Ben-Dov, Y. (1994) *A Systematic Catalogue of the Mealybugs of the World (Insecta: Homoptera: Coccoidea: Pseudococcidae and Putoidae) with Data on Geographical Distribution, Host Plants*, Biology and Economic Importance. Intercept Limited, Andover, UK.

Berry, S.D., Fondong, V.N., Rey, C., Rogan, D., Fauquet, C.M. and Brown, J.K. (2004) Molecular evidence for five distinct *Bemisia tabaci* (Homoptera: Aleyrodidae) geographic haplotypes associated with cassava plants in sub-Saharan Africa. *Annals of the Entomological Society of America* 97, 852–859.

Birch, A.N.E., Begg, G.S. and Squire, G.R. (2011) How agro-ecological research helps to address food security issues under new IPM and pesticide reduction policies for global crop production systems. *Phyton, International Journal of Experimental Botany* 62, 3251–3261.

Bottrell, D.R. (1979) *Integrated Pest Management*. US Government Printing Office, Washington, DC, 120 pp.

Bouaguimbeck, H. (2011) *African Organic Agriculture Training Manual*. A Resource Manual for Trainers. FiBL, Research Institute of Organic Agriculture, Frick, Switzerland.

Branco, M., Jactel, H., Franco, J.C. and Mendel, Z. (2006) Modelling response of insect trap captures to pheromone dose. *Ecological Modelling* 197, 247–257.

Brooker, R.W., Bennett, A.E., Cong, W.F., Daniell, T.J., George, T.S., Hallett, P.D., Hawes, C., Iannetta, P.P.M., Jones, H.G., Karley, A.J., Li, L., McKenzie, B.M., Pakeman, R.J., Paterson, E., Schob, C., Shen, J., Squire, G., Watson, C.A., Zhang, C., Zhang, F., Zhang, J. and White, P.J. (2014) Improving intercropping: a synthesis of research in agronomy, plant physiology and ecology. *New Phytologist* 206, 107–117.

Brown, J.K., Frohlich, D.R. and Rosell, R.C. (1995) The sweetpotato or silverleaf whiteflies: biotypes of *Bemisia tabaci* or a species complex? *Annual Review of Entomology* 40, 511–534.

Brown, L.R. (1963) Agricultural diversification and economic development in Thailand: a case study. *Economic Research Service Foreign Agriculture Economic Report* 8, 34. United States Department of Agriculture, Foreign Agricultural Report, Washington, DC.

Byrne, D.H. (1980) Studies of resistance to the mites *Mononychellus tanajoa* (Bondar) and *Mononychellus caribbeanae* (McGregor) in cassava, *Manihot esculenta* (Crantz). *Dissertation Abstracts International* B(41), 750.

Byrne, D.H., Bellotti, A.C. and Guerrero, J.M. (1983) The cassava mites. *International Journal of Pest Management* 29, 378–394.

Calatayud, P.A., Tertuliano, M. and Le Rü, B. (1994) Seasonal changes in secondary compounds in the phloem sap of cassava in relation to plant genotype and infestation by *Phenacoccus manihoti* (Homoptera: Pseudococcidae). *Bulletin of Entomological Research* 84, 453–459.

Ceballos, H. (2012) Cassava in Colombia and the world. New prospects for a millennial crop. In: Ospina, B. and Veballos, H. (eds) *Cassava in the Third Millennium: Modern Production, Processing, Use, and Marketing Systems*. CIAT, Cali, Colombia, pp. 1–11.

Ceballos, H., Iglesias, C.A., Pérez, J.C. and Dixon, A.G. (2004) Cassava breeding: opportunities and challenges. *Plant Molecular Biology* 56, 503–516.

Ceballos, H., Ramirez, J., Bellotti, A.C., Jarvis, A. and Alvarez, E. (2011) Adaptation of cassava to changing climates. In: Yadav, S.S., Redden, R.J., Hatfield, J.L., Lotze-Campen, H. and Hall, A.E. (eds) *Crop Adaptation to Climate Change*. Blackwell Publishing, Oxford, pp. 411–425.

Chang, C.P. and Huang, S.C. (1995) Evaluation of the effectiveness of releasing green lacewing, *Mallada basalis* (Walker) for the control of tetranychid mites on strawberry. *Plant Protection Bulletin (Taipei)* 37, 41–58.

Charles, J.G. (1981) Distribution and life history of the longtailed mealy bug, *Pseudococcus longispinus* (Homoptera: Pseudococcidae), in Auckland vineyards. *New Zealand Journal of Zoology* 8, 285–293.

Corbett, A., Leigh, T.F. and Wilson, L.T. (1991). Interplanting alfalfa as a source of *Metaseiulus occidentalis* (Acari: Phytoseiidae) for managing spider mites in cotton. *Biological Control* 1, 188–196

Crowder, D.W. and Harwood, J.D. (2014) Promoting biological control in a rapidly changing world. *Biological Control* 75, 1–7.

Daane, K.M., Almeida, R.P., Bell, V.A., Walker, J.T., Botton, M., Fallahzadeh, M., Mani, M., Miano, J.L., Sforza, R., Walton, V.M. and Zaviezo, T. (2012) Biology and management of mealybugs in vineyards. In: Bostanian, N.J., Vincent, C. and Isaacs, R. (eds) *Arthropod Management in Vineyards*. Springer, Dordrecht, The Netherlands, pp. 271–307.

D'Almeida, Y.A., Lys, J.A., Neuenschwander, P. and Ajuonu, O. (1998) Impact of two accidentally introduced *Encarsia* species (Hymenoptera: Aphelinidae) and other biotic and abiotic factors on the spiralling whitefly *Aleurodicus dispersus* (Russell) (Homoptera: Aleyrodidae), in Benin, West Africa. *Biocontrol Science and Technology* 8, 163–173.

da Silva-Torres, C.S.A., de Oliveira, M.D. and Torres, J.B. (2013) Host selection and establishment of striped mealybug, *Ferrisia virgata*, on cotton cultivars. *Phytoparasitica* 41, 31–40.

De Barro, P.J. (2005) Genetic structure of the whitefly *Bemisia tabaci* in the Asia-Pacific region revealed using microsatellite markers. *Molecular Ecology* 14, 3695–3718.

De Barro, P.J., Liu, S.S., Boykin, L.M. and Dinsdale, A.B. (2011) *Bemisia tabaci*: a statement of species status. *Annual Review of Entomology* 56, 1–19.

De Koninck, R. and Rousseau, J.F.G. (2014) *Gambling with the Land: The Contemporary Evolution of Southeast Asian Agriculture*. Nus Press, Singapore.

Delalibera, I., Hajek, A.E. and Humber, R.A. (2004) *Neozygites tanajoae* sp. nov., a pathogen of the cassava green mite. *Mycologia* 96, 1002–1009.

Devendra, C. and Thomas, D. (2002) Smallholder farming systems in Asia. *Agricultural Systems* 71, 17–25.

Edwards, C.A. (1989) The importance of integration in sustainable agricultural systems. *Agriculture, Ecosystems & Environment* 27, 25–35.

Ehler, L.E. (2006) Integrated pest management (IPM): definition, historical development and implementation, and the other IPM. *Pest Management Science* 62, 787–789.

Ewusie, E. (2008) *Potentials and Prospect of Strip Cropping in the Management of Cassava Whitefly*. Texas Tech University, Lubbock, Texas.

FAO-IPM Asia Regional Programme (2013) Vietnam Proceeds with Mass Rearing and Field Releases of the Parasitoid, *Anagyrus lopezi*, for the Control of Pink Cassava Mealybug Tay Ninh. Vietnam Food and Agriculture Organization of the United Nations (FAO) Integrated Pest Management (IPM) online news. Available at: http://www.vegetableipmasia.org (accessed 5 October 2015).

FAO-IPM Asia Regional Programme (2014) FAO and CIAT Assist Indonesia with Biological Control for Pink Cassava Mealybug Bogor, Indonesia. Vietnam Food and Agriculture Organization of the United Nations (FAO) Integrated Pest Management (IPM) online news. Available at: http://www.vegetableipmasia.org/news/view/108 (accessed 5 October 2015).

FAOSTAT (2015) Food and Agriculture Organization of the United Nations, Corporate Statistical Database. Available at: http://faostat3.fao.org/home (accessed 5 October 2015).

Faria, M. and Wraight, S.P. (2001) Biological control of *Bemisia tabaci* with fungi. *Crop Protection* 20, 767–778.

Flechtmann, C.H.W. (1981) The cassava mite complex. II. New records and description of two new species in the genus *Tetranychus* from Asia. *International Journal of Acarology* 7, 81–86.

Flechtmann, C.H.W. (2013) A new species of *Neotetranychus trägårdh* (Acari, Prostigmata, Tetranychidae) from Thailand with a key to world species. *Persian Journal of Acarology* 2, 35–40.

Fondong, V.N., Thresh, J.M. and Zok, S. (2002) Spatial and temporal spread of cassava mosaic virus disease in cassava grown alone and when intercropped with maize and/or cowpea. *Journal of Phytopathology* 150, 365–374.

Franco, J.C., Zada, A. and Mendel, Z. (2009) Novel approaches for the management of mealybug pests. In: *Biorational Control of Arthropod Pests*. Springer, Dordrecht, The Netherlands, pp. 233–278.

Franco, J.C., Da Silva, E.B., Fortuna, T., Cortegano, E., Branco, M., Suma, P., Torre, I.L., Russo, A., Elyahu, M., Protasov, A., Levi-Zada, A. and Mendel, Z. (2011) Vine mealybug sex pheromone increases citrus mealybug parasitism by *Anagyrus* sp. nr. *pseudococci* (Girault). *Biological Control* 58, 230–238.

Gerling, D. (1990) Natural enemies of whiteflies: predators and parasitoids. In: Gerling, D. (ed.) *Whiteflies: Their Bionomics, Pest Status and Management*. Intercept Limited, Andover, UK.

Gold, C.S., Altieri, M.A. and Bellotti, A.C. (1989) Cassava intercropping and pest incidence: a review illustrated with a case study from Colombia. *International Journal of Pest Management* 35, 339–344.

Goulson, D. (2013) An overview of the environmental risks posed by neonicotinoid insecticides. *Journal of Applied Ecology* 50, 977–987.

Graziosi, I., Minato, N., Alvarez, E., Ngo, D.T., Hoat, T.X., Aye, T.M., Pardo, J.M., Wongtiem, P. and Wyckhuys, K.A. (2016) Emerging pests and diseases of South-east Asian cassava: a comprehensive evaluation of geographic priorities, management options and research needs. *Pest Management Science* 72, 1071–1089.

Gullan, P.J. and Kosztarab, M. (1997) Adaptations in scale insects. *Annual Review of Entomology* 42, 23–50.

Hahn, S.K. (1989) An overview of African traditional cassava processing and utilization. *Outlook on Agriculture* 18, 110–118.

Hamamura, T. (1986) Studies on the biological control of Kanzawa spider mite, *Tetranychus kanzawai* Kishida by the chemical resistant predacious mite, *Amblyseius longispinosus* (Evans) in tea fields (Acarina: Tetranychidae, Phytoseiidae). *Bulletin of the National Research Institute of Tea (Japan)* 21, 121–201. (in Japanese)

Hambäck, P.A. and Englund, G. (2005) Patch area, population density and the scaling of migration rates: the resource concentration hypothesis revisited. *Ecology Letters* 8, 1057–1065.

Hardy, N.B., Gullan, P.J. and Hodgson, C.J. (2008) A subfamily-level classification of mealybugs (Hemiptera: Pseudococcidae) based on integrated molecular and morphological data. *Systematic Entomology* 33, 51–71.

Henry, G. and Gottret, V. (1996) Global cassava trends. Reassessing the crop's future. CIAT Working document No. 157. CIAT, Cali, Colombia, 45 pp.

Herren, H.R. and Neuenschwander, P. (1991) Biological control of cassava pests in Africa. *Annual Review of Entomology* 36, 257–283.

Herrera, C.J., Driesche, R.V. and Bellotti, A.C. (1989) Temperature-dependent growth rates for the cassava mealybug, *Phenacoccus herreni*, and two of its encyrtid parasitoids, *Epidinocarsis diversicornis* and *Acerophagus coccois* in Colombia. *Entomologia Experimentalis et Applicata* 50, 21–27.

Herron, G.A., Edge, V.E., Wilson, L.J. and Rophail, J. (1998) Organophosphate resistance in spider mites (Acari: Tetranychidae) from cotton in Australia. *Experimental and Applied Acarology* 22, 17–30.

Hinomoto, N., Tran, D.P., Pham, A.T., Ngoc Le, T.B., Tajima, R., Ohashi, K., Osakabe, M. and Takafuji, A. (2007) Identification of spider mites (Acari: Tetranychidae) by DNA sequences: a case study in Northern Vietnam. *International Journal of Acarology* 33, 53–60.

Hobbs, R.J. (2000) Land-use changes and invasions. In: Mooney, H.A. and Hobbs, R.J. (eds) *Invasive Species in a Changing World*. Island Press, Washington, DC, pp. 55–64.

Howeler, R.H. (2006) Cassava in Asia: trends in cassava production, processing and marketing. Paper presented at Workshop on Partnership in Modern Science to Develop a Strong Cassava Commercial Sector in Africa and Appropriate Varieties by 2020, May 2006, Bellagio, Italy.

Irigaray, F.J.S.D.C., Marco-Mancebón, V. and Pérez-Moreno, I. (2003) The entomopathogenic fungus *Beauveria bassiana* and its compatibility with triflumuron: effects on the two-spotted spider mite *Tetranychus urticae*. *Biological Control* 26, 168–173.

Isman, M.B. (2000) Plant essential oils for pest and disease management. *Crop Protection* 19, 603–608.

Jones, D.R. (2003) Plant viruses transmitted by whiteflies. *European Journal of Plant Pathology* 109, 195–219.

Kairo, M.T., Pollard, G.V., Peterkin, D.D. and Lopez, V.F. (2000) Biological control of the hibiscus mealybug, *Maconellicoccus hirsutus* Green (Hemiptera: Pseudococcidae) in the Caribbean. *Integrated Pest Management Reviews* 5, 241–254.

Kajita, H., Samudra, I.M. and Naito, A. (1991) Discovery of the spiraling whitefly *Aleurodicus dispersus* Russell (Homoptera: Aleyrodidae) from Indonesia, with notes on its host plants and natural enemies. *Applied Entomology and Zoology* 26, 397–400.

Kogan, M. (1998) Integrated pest management: historical perspectives and contemporary developments. *Annual Review of Entomology* 43, 243–270.

Koh, S.H., Ahn, J., Im, J.S., Jung, C., Lee, S.H. and Lee, J.H. (2009) Monitoring of acaricide resistance of *Tetranychus urticae* (Acari: Tetranychidae) from Korean apple orchards. *Journal of Asia-Pacific Entomology* 12, 15–21.

Kumar, K.I., Kennedy, J.S. and Suresh, S. (2014) Laboratory evaluation of *Beauveria bassiana* (Balsamo) Vuillemin against crawler stage of *Paracoccus marginatus* (Williams and Granara de Willink). *Trends in Biosciences* 7, 246–249.

Lemic, D., Mikac, K.M., Kozina, A., Benitez, H.A., McLean, C.M. and Bažok, R. (2015) Monitoring techniques of the western corn rootworm are the precursor to effective IPM strategies. *Pest Management Science* 72, 405–417.

Le Rü, B. (1986) Étude de l'évolution d'une mycose a *Neozygites fumosa* [Zygomycetes, Entomophthorales] dans une population de la cochenille du manioc, *Phenacoccus manihoti* [Hom.: Pseudococcidae]. *Entomophaga* 31, 79–89.

Le Rü, B., Silvie, P. and Papierok, B. (1985) L'entomophthorale *Neozygites fumosa* pathogène de la Cochenille du manioc, *Phenacoccus manihoti* [Hom.: Pseudococcidae], en Républiquc populairc du Congo. *Entomophaga* 30, 23–29.

Löhr, B., Varela, A.M. and Santos, B. (1990) Exploration for natural enemies of the cassava mealybug, *Phenacoccus manihoti* (Homoptera: Pseudococcidae), in South America for the biological control of this introduced pest in Africa. *Bulletin of Entomological Research* 80, 417–425.

Lu, H., Lu, F.P., Xu, X.L. and Chen, Q. (2014) Potential geographic distribution areas of *Mononychellus mcgregori* in Guangxi province. *Applied Mechanics and Materials* 522, 1051–1054.

Lundgren, J.G., López-Lavalle, L.A.B., Parsa, S. and Wyckhuys, K.A. (2014) Molecular determination of the predator community of a cassava whitefly in Colombia: pest-specific primer development and field validation. *Journal of Pest Science* 87, 125–131.

Mani, M. and Krishnamoorthy, A. (2002) Classical biological control of the spiralling whitefly, *Aleurodicus dispersus* Russell – an appraisal. *International Journal of Tropical Insect Science* 22, 263–273.

Mani, M. and Krishnamoorthy, A. (2008) Biological suppression of the mealybugs *Planococcus citri* (Risso), *Ferrisia virgata* (Cockerell) and *Nipaecoccus viridis* (Newstead) on pummelo with *Cryptolaemus montrouzieri* Mulsant in India. *Journal of Biological Control* 22, 169–172.

Mani, M., Joshi, S., Kalyanasundaram, M., Shivaraju, C., Krishnamoorthy, A., Asokan, R. and Rebijith, K.B. (2013) A new invasive Jack Beardsley mealybug, *Pseudococcus jackbeardsleyi* (Hemiptera: Pseudococcidae) on papaya in India. *Florida Entomologist* 96, 242–245.

Maniania, N.K., Bugeme, D.M., Wekesa, V.W., Delalibera, I. Jr and Knapp, M. (2008) Role of entomopathogenic fungi in the control of *Tetranychus evansi* and *Tetranychus urticae* (Acari: Tetranychidae), pests of horticultural crops. *Experimental and Applied Acarology* 46, 259–274.

Manyong, V.M. (2000) *Impact: The Contribution of IITA-Improved Cassava to Food Security in Sub-Saharan Africa*. International Institute of Tropical Agriculture (IITA), Ibadan, Nigeria.

Mascarin, G.M., Kobori, N.N., Quintela, E.D. and Delalibera, I. (2013) The virulence of entomopathogenic fungi against *Bemisia tabaci* biotype B (Hemiptera: Aleyrodidae) and their conidial production using solid substrate fermentation. *Biological Control* 66, 209–218.

Maziya-Dixon, B., Dixon, A.G. and Adebowale, A.R.A. (2007) Targeting different end uses of cassava: genotypic variations for cyanogenic potentials and pasting properties. *International Journal of Food Science and Technology* 42, 969–976.

McMurtry, J.A. and Croft, B.A. (1997) Life styles of phytoseiid mites and their roles in biological control. *Annual Review of Entomology* 42, 291–321.

Meyerson, L.A. and Mooney, H.A. (2007) Invasive alien species in an era of globalization. *Frontiers in Ecology and the Environment* 5, 199–208.

Millar, J.G., Daane, K.M., Mcelfresh, J.S., Moreira, J.A., Malakar-Kuenen, R., Guillén, M. and Bentley, W.J. (2002) Development and optimization of methods for using sex pheromone for monitoring the mealybug *Planococcus ficus* (Homoptera: Pseudococcidae) in California vineyards. *Journal of Economic Entomology* 95, 706–714.

Miller, D.R. and Miller, G.L. (2002) Redescription of *Paracoccus marginatus* Williams and Granara de Willink (Hemiptera: Coccoidea: Pseudococcidae), including descriptions of the immature stages and adult male. *Proceedings of the Entomological Society of Washington* 104, 1–159.

Minafra, A. and Hadidi, A. (1994) Sensitive detection of grapevine virus A, B, or leafroll-associated III from viruliferous mealybugs and infected tissue by cDNA amplification. *Journal of Virological Methods* 47, 175–187.

Miura, K., Subhasaram, T., Tawinthung, N., Noochan, N. and Shiraishi K. (1990) Effects of termite activity on soils in northeast Thailand. *Japanese Journal of Tropical Agriculture* 34, 40–47.

Muniappan, R. (2011) Recent invasive hemipterans and their biological control in Asia. Paper presented at the 5th Meeting of the Asian Cotton Research and Development Network, 23–25 February 2011, Lahore, Pakistan.

Muniappan, R., Meyerdirk, D.E., Sengebau, F.M., Berringer, D.D. and Reddy, G.V.P. (2006) Classical biological control of the papaya mealybug, *Paracoccus marginatus* (Hemiptera: Pseudococcidae) in the Republic of Palau. *Florida Entomologist* 89, 212–217.

Muniappan, R., Shepard, B.M., Watson, G.W., Carner, G.R., Sartiami, D., Rauf, A. and Hammig, M.D. (2008) First report of the papaya mealybug, *Paracoccus marginatus* (Hemiptera: Pseudococcidae), in Indonesia and India. *Journal of Agricultural and Urban Entomology* 25, 37–40.

Muniappan, R., Shepard, B.M., Watson, G.W., Carner, G.R., Rauf, A., Sartiami, D., Hidayat, P., Afun, J.V.K., Goergen, G. and Rahman, A.Z. (2009) New records of invasive insects (Hemiptera: Sternorrhyncha) in Southeast Asia and West Africa. *Journal of Agricultural and Urban Entomology* 26, 167–174.

Mutisya, D.L., El-Banhawy, E.M., Khamala, C.P.M. and Kariuki, C.W. (2015) Management of cassava green mite *Mononychellus progresivus* (Acari: Tetranychidae) in different agro-ecological zones of Kenya. *Systematic and Applied Acarology* 30, 39–50.

Myrick, S., Norton, G.W., Selvaraj, K.N., Natarajan, K. and Muniappan, R. (2014) Economic impact of classical biological control of papaya mealybug in India. *Crop Protection* 56, 82–86.

Nagrare, V.S., Kranthi, S., Biradar, V.K., Zade, N.N., Sangode, V., Kakde, G., Shukla, R.M., Shivare, D., Khadi, B.M. and Kranthi, K.R. (2009) Widespread infestation of the exotic mealybug species, *Phenacoccus solenopsis* (Tinsley) (Hemiptera: Pseudococcidae), on cotton in India. *Bulletin of Entomological Research* 99, 537–541.

Naing, H.H., Chandrapatya, A., Navajas, M. and Auger, P. (2014) Know more about spider mites (Acari: Tetranychidae) in Myanmar. Paper presented at Conference at Yezin Agricultural University, Yezin, Myanmar, December 2014.

Neuenschwander, P. (2001) Biological control of the cassava mealybug in Africa: a review. *Biological Control* 21, 214–229.

Neuenschwander, P. (2009) Spiralling whitefly. *Aleurodicus dispersus*, a recent invader and new cassava pest. *African Crop Science Journal* 2, 419–421.

Neuenschwander, P. and Madojemu, E. (1986) Mortality of the cassava mealybug, *Phenacoccus manihoti* Mat.-Ferr.(Hom., Pseudococcidae), associated with an attack by *Epidinocarsis lopezi* (Hym., Encyrtidae). *Mitteilungen der Schweizerishen Entomologische Gesellschaft* 59, 57–62.

Neuenschwander, P., Hammond, W.N.O., Ajuonu, O., Gado, A., Echendu, N., Bokonon-Ganta, A.H., Allomasso, R. and Okon, I. (1990) Biological control of the cassava mealybug, *Phenacoccus manihoti* (Hom., Pseudococcidae) by *Epidinocarsis lopezi* (Hym., Encyrtidae) in West Africa, as influenced by climate and soil. *Agriculture, Ecosystems & Environment* 32, 39–55.

Night, G., Asiimwe, P., Gashaka, G., Nkezabahizi, D., Legg, J.P., Okao-Okuja, G., Obonyo, R., Nyirahorana, C., Mukakanyana, C., Mukase, F., Munyabarenzi, I. and Mutumwinka, M. (2011) Occurrence and distribution of cassava pests and diseases in Rwanda. *Agriculture Ecosystems & Environment* 140, 492–497.

Noyes, J.S. and Hayat, M. (1994) *Oriental Mealybug Parasitoids of the Anagyrini* (Hymenoptera: Encyrtidae). CAB International, Wallingford, UK.

Nusartlert, N., Vichitbandha, P., Baker, G.T. and Chandrapatya, A. (2010) Pesticide-induced mortality and prey-dependent life history of the predatory mite *Neoseiulus longispinosus* (Acari: Phytoseiidae). In: *Trends in Acarology Proceedings of the 12th International Congress*. Springer, Dordrecht, The Netherlands, pp. 495–498.

Nwanze, K.F. (1982) Relationships between cassava root yields and crop infestations by the mealybug, *Phenacoccus manihoti*. *International Journal of Pest Management* 28, 27–32.

Nwanze, K.F., Leuschner, K. and Ezumah, H.C. (1979) The cassava mealybug, *Phenacoccus* sp. in the Republic of Zaire. *Proceedings of the National Academy of Sciences of the United States of America* 25, 125–130.

Nweke, F.I. (1994) Farm level practices relevant to cassava plant protection. *African Crop Science Journal* 2, 563–582.

Oei-Dharma, H.P. (1969) Use of pesticides and control of economic pests and diseases in Indonesia. In: Brill, E.J. (ed.) *Use of Pesticides and Control of Economic Pests and Diseases in Indonesia*. Leiden, The Netherlands, p. 199.

Oliveira, M.R.V., Henneberry, T.J. and Anderson, P. (2001) History, current status, and collaborative research projects for *Bemisia tabaci*. *Crop Protection* 20, 709–723.

Olsen, K.M. and Schaal, B.A. (1999) Evidence on the origin of cassava: phylogeography of *Manihot esculenta*. *Proceedings of the National Academy of Sciences of the United States of America* 96, 5586–5591.

Omongo, C.A., Kawuki, R., Bellotti, A.C., Alicai, T., Baguma, Y., Maruthi, M.N., Bua, A. and Colvin, J. (2012) African cassava whitefly, *Bemisia tabaci*, resistance in African and South American cassava genotypes. *Journal of Integrative Agriculture* 11, 327–336.

Onzo, A., Hanna, R., Negloh, K., Toko, M. and Sabelis, M.W. (2005) Biological control of cassava green mite with exotic and indigenous phytoseiid predators – effects of intraguild predation and supplementary food. *Biological Control* 33, 143–152.

Parsa, S., Kondo, T. and Winotai, A. (2012) The cassava mealybug (*Phenacoccus manihoti*) in Asia: first records, potential distribution, and an identification key. *PLoS One* 7, e47675.

Parsa, S., Hazzi, N.A., Chen, Q., Lu, F., Campo, B.V.H., Yaninek, J.S. and Vásquez-Ordóñez, A.A. (2015) Potential geographic distribution of two invasive cassava green mites. *Experimental and Applied Acarology* 65, 195–204.

Peh, K.S.H. (2010) Invasive species in Southeast Asia: the knowledge so far. *Biodiversity and Conservation* 19, 1083–1099.

Peña, J.E. and Waddill, V. (1982) Pests of cassava in South Florida. *Florida Entomologist* 65, 143–149.

Piya, S., Kiminami, A. and Hironori, Y. (2011) Sources of agricultural productivity in South and Southeast Asia. *Trends in Agricultural Economics* 4, 18–29.

Poletti, M., Maia, A.H.N. and Omoto, C. (2007) Toxicity of neonicotinoid insecticides to *Neoseiulus californicus* and *Phytoseiulus macropilis* (Acari: Phytoseiidae) and their impact on functional response to *Tetranychus urticae* (Acari: Tetranychidae). *Biological Control* 40, 30–36.

Pretty, J.N., Morison, J.I. and Hine, R.E. (2003) Reducing food poverty by increasing agricultural sustainability in developing countries. *Agricultural Ecosystems & Environment* 95, 217–234.

Pretty, J.N., Noble, A.D., Bossio, D., Dixon, J., Hine, R.E., Penning de Vries, F.W. and Morison, J.I. (2006) Resource-conserving agriculture increases yields in developing countries. *Environmental Science & Technology* 40, 1114–1119.

Prudencio, Y.C. and Al-Hassan, R. (1994) The food security stabilization roles of cassava in Africa. *Food Policy* 19, 57–64.

Radhakrishnan, A.S., Girija, S. and Anil, A.T. (2015) Organic vs conventional management in cassava: growth dynamics, yield and soil properties. *Journal of Root Crops* 39, 93–99.

Ranga Rao, G.V., Giang, N.T.L., Lieu, P. and Tram, N.T.H. (2006) Occurrence of white grubs in groundnut crop in uplands of South Vietnam: a new report. *International Arachis Newsletter* 26, 29–31.

Rigby, D. and Cáceres, D. (2001) Organic farming and the sustainability of agricultural systems. *Agricultural Systems* 68, 21–40.

Rigg, J. (2005) Poverty and livelihoods after full-time farming: a South-east Asian view. *Asia-Pacific Viewpoint* 46, 173.

Rigg, J. (2013) From rural to urban: a geography of boundary crossing in Southeast Asia. *Trans-Regional and National Studies of Southeast Asia* 1, 5–26.

Robertson, I.A.D. (1987) The whitefly, *Bemisia tabaci* (Gennadius) as a vector of African cassava mosaic virus at the Kenya coast and ways in which the yield losses in cassava, *Manihot esculenta* Crantz caused by the virus can be reduced. *International Journal of Tropical Insect Science* 8, 797–801.

Roh, H.S., Lee, B.H. and Park, C.G. (2013) Acaricidal and repellent effects of myrtacean essential oils and their major constituents against *Tetranychus urticae* (Tetranychidae). *Journal of Asia-Pacific Entomology* 16, 245–249.

Saengyot, S. and Burikam, I. (2012a) Development and growth ratio of predaceous coccinellid, *Sasajiscymnus quinquepunctatus* (Weise) on papaya mealybug, *Paracoccus marginatus* Williams & Granara de Willink. *Kasetsart Journal (Natural Science)* 46, 418–426.

Saengyot, S. and Burikam, I. (2012b) Bionomics of the apefly, *Spalgis epius* (Lepidoptera: Lycaenidae), predatory on the papaya mealybug, *Paracoccus marginatus* (Hemiptera: Pseudococcidae), in Thailand. *Songklanakarin Journal of Science and Technology* 34, 1.

Sakunwarin, S., Chandrapatya, A. and Baker, G. (2003) Biology and life table of the cassava mite, *Tetranychus truncatus* Ehara (Acari: Tetranychidae). *Systematic and Applied Acarology* 8, 13–24.

Sakunwarin, S., Chandrapatya, A. and Visetson, S. (2004) Synergism and detoxification mechanism of crude sugar apple seed extract in *Tetranychus truncatus* Ehara (Prostigmata: Tetranychidae). *Kasetsart Journal (Natural Science)* 38, 340–348.

Sanders, J.G. (1909) The identity and synonymy of some of our soft scale-insects. *Journal of Economic Entomology* 2, 428–448.

Sanjaya, Y., Ocampo, V.R. and Caoili, B.L. (2013a) Selection of entomopathogenic fungi against the red spider mite *Tetranychus kanzawai* (Kishida) (Tetranychidae: Acarina). *Arthropods* 2, 208–215.

Sanjaya, Y., Ocampo, V.R. and Caoili, B.L. (2013b) Transmission effect of entomopathogenic fungi on population of *Tetanychus kanzawai* (Kishida) (Tetranychidae: Acarina). *Arthropods* 2, 36–41.

Sartiami, D., Watson, G.W., Roff, M.M.N., Hanifah, M.Y. and Idris, A.B. (2015) First record of cassava mealybug, *Phenacoccus manihoti* (Hemiptera: Pseudococcidae), in Malaysia. *Zootaxa* 3957, 235–238.

Schulthess, F., Neuenschwander, P. and Gounou, S. (1997) Multi-trophic interactions in cassava, *Manihot esculenta*, cropping systems in the subhumid tropics of West Africa. *Agriculture Ecosystems & Environment* 66, 211–222.

Schulthess, F., Chabi-Olaye, A. and Gounou, S. (2004) Multi-trophic level interactions in a cassava–maize mixed cropping system in the humid tropics of West Africa. *Bulletin of Entomological Research* 94, 261–272.

Sclar, D.C., Gerace, D. and Cranshaw, W.S. (1998) Observations of population increases and injury by spider mites (Acari: Tetranychidae) on ornamental plants treated with imidacloprid. *Journal of Economic Entomology* 91, 250–255.

Sether, D.M., Ullman, D.E. and Hu, J.S. (1998) Transmission of pineapple mealybug wilt-associated virus by two species of mealybug (*Dysmicoccus* spp.). *Phytopathology* 88, 1224–1230.

Shylesha, A.N. (2013) Host range of invasive Jack Beardsley mealybug, *Pseudococcus jackbeardsleyi* Gimpel and Miller in Karnataka. *Pest Management in Horticultural Ecosystems* 19, 106–107.

Soysouvanh, P. and Siri, N. (2013) Population abundance of pink mealybug, *Phenacoccus manihoti* on four cassava varieties. *Khon Kaen Agriculture Journal* 41, 149–153.

Sseruwagi, P., Maruthi, M.N., Colvin, J., Rey, M.E.C., Brown, J.K. and Legg, J.P. (2006) Colonization of non-cassava plant species by cassava whiteflies (*Bemisia tabaci*) in Uganda. *Entomologia Experimentalis et Applicata* 119, 145–153.

Suasa-ard, W. (2010) Natural enemies of important insect pests of field crops and utilization as biological control agents in Thailand. Presented at the International Seminar on Enhancement of Functional Biodiversity Relevant to Sustainable Food Production in ASPAC (Asia Pacific Network of Science and Technology Centres), 9–11 November, Tsukuba, Japan.

Subramaniam, T.R., David, B.V., Thangavel, P. and Abraham, E.V. (1977) Insect pest problems of tuber crops in Tamil Nadu. *Journal of Root Crops* 3, 43–50.

Surulivelu, T., Banu, J.G., Rajan, T.S., Dharajothi, B. and Amutha, M. (2012) Evaluation of fungal pathogens for the management of mealybugs in Bt cotton. *Journal of Biological Control* 26, 92–96.

Swirski, E., Izhar, Y., Wysoki, M., Gurevitz, E. and Greenberg, S. (1980) Integrated control of the long-tailed mealybug, *Pseudococcus longispinus* [Hom.: Pseudococcidae], in avocado plantations in Israel. *Entomophaga* 25, 415–426

Takahashi, R. (1942) Some injurious insects of agricultural plants and forest trees in Thailand and Indo-China. II. Coccidae. Formosa Government Research Institute Department of Agriculture Report 81. Formosa Government Research Institute, Taihoku, Taiwan, pp. 1–56.

Thresh, J.M. and Otim-Nape, G.W. (1994) Strategies for controlling African cassava mosaic geminivirus. In: Harris, K.F. (ed.) *Advances in Disease Vector Research*. Springer, New York, pp. 215–236.

Toko, M., Yaninek, J.S. and O'Neil, R.J. (1996) Response of *Mononychellus tanajoa* (Acari: Tetranychidae) to cropping systems, cultivars, and pest interventions. *Environmental Entomology* 25, 237–249.

Trenbath, B.R. (1993) Intercropping for the management of pests and diseases. *Field Crops Research* 34, 381–405.

Ujjan, A.A., Khanzada, M.A. and Shahzad, S. (2015) Efficiency of *Metarhizium* spp. (Sorokīn) strains and insecticides against cotton mealybug *Phenacoccus solenopsis* (Tinsley). *Pakistan Journal of Zoology* 47, 351–360.

Valentin, C., Agus, F., Alamban, R., Boosaner, A., Bricquet, J.P., Chaplot, V., de Guzman, T., de Rouw, A., Janeau, J.L., Orange, D., Phachomphonh, K., Phai, D.D., Podwojewski, P., Ribolzi, O., Silvera, N., Subagyono, K., Thiebaux, J.P., Toan, T.D. and Vadari, T. (2008) Runoff and sediment losses from 27 upland catchments in Southeast Asia: impact of rapid land use changes and conservation practices. *Agricultural Ecosystems & Environment* 128, 225–238.

Van der Sluijs, J.P., Amaral-Rogers, V., Belzunces, L.P., van Lexmond, M.B., Bonmatin, J.M., Chagnon, M., Downs, C.S., Furlan, L., Gibbons, D.W., Giorio, C., Girolami, V., Goulson, D., Kreutzweiser, D.P., Krupke, C., Liess, M., Long, E., McField, M., Mineau, P., Mitchell, E.A.D., Morissey, C.S., Noome, D.A., Pisa, L., Settele, J., Simon-Delso, N., Stark, J.D., Tapparo, A., Van Dyck, H., van Praagh, J., Whitehorn, P.R. and Wiemers, M. (2015) Conclusions of the Worldwide Integrated Assessment on the risks of neonicotinoids and fipronil to biodiversity and ecosystem functioning. *Environmental Science and Pollution Research* 22, 148–154.

Van Donge, J.K., Henley, D. and Lewis, P. (2012) Tracking development in South-east Asia and sub-Saharan Africa: the primacy of policy. *Development Policy Review* 30, s5–s24.

Van Leeuwen, T., Vontas, J., Tsagkarakou, A., Dermauw, W. and Tirry, L. (2010) Acaricide resistance mechanisms in the two-spotted spider mite *Tetranychus urticae* and other important Acari: a review. *Insect Biochemistry and Molecular Biology* 40, 563–572.

Vasquez, E.A. (1994) *Biological Control of Cassava Red Spider Mite* Tetranychus kanzawai *Kishida*. Philippine Rootcrop Research and Training Center, Baybay, the Philippines.

Waddington, S.R., Li, X., Dixon, J., Hyman, G. and De Vicente, M.C. (2010) Getting the focus right: production constraints for six major food crops in Asian and African farming systems. *Food Security* 2, 27–48.

Wang, H.Y., Yang, Y., Su, J.Y., Shen, J.L., Gao, C.F. and Zhu, Y.C. (2008) Assessment of the impact of insecticides on *Anagrus nilaparvatae* (Pang et Wang) (Hymenoptera: Mymanidae), an egg parasitoid of the rice planthopper, *Nilaparvata lugens* (Hemiptera: Delphacidae). *Crop Protection* 27, 514–522.

Wang, Y., Watson, G.W. and Zhang, R. (2010) The potential distribution of an invasive mealybug *Phenacoccus solenopsis* and its threat to cotton in Asia. *Agricultural and Forest Entomology* 12, 403–416.

Wei, J., Ding, W., Zhao, Y.G. and Vanichpakorn, P. (2011) Acaricidal activity of *Aloe vera* L. leaf extracts against *Tetranychus cinnabarinus* (Boisduval) (Acarina: Tetranychidae). *Journal of Asia-Pacific Entomology* 14, 353–356.

Wetten, A., Campbell, C. and Allainguillaume, J. (2016) High-resolution melt and morphological analyses of mealybugs (Hemiptera: Pseudococcidae) from cacao: tools for the control of cacao swollen shoot virus spread. *Pest Management Science* 72, 527–533.

Willer, H. and Kilcher, L. (eds) (2011) *The World of Organic Agriculture: Statistics and Emerging Trends 2011*. Research Institute of Organic Agriculture (FiBL), Frick, Switzerland and International Federation of Organic Agriculture Movements (IFOAM), Bonn, Germany.

Williams, D.J. (2004) *Mealybugs of Southern Asia*. The Natural History Museum, Kualu Lumpur, Malaysia, 896 pp.

Wilson, C. and Tisdell, C. (2001) Why farmers continue to use pesticides despite environmental, health and sustainability costs. *Ecological Economics* 39, 449–462.

Winotai, A., Goergen, G., Tamò, M. and Neuenschwander, P. (2010) Cassava mealybug has reached Asia. *Biocontrol News and Information* 31, 10N–11N.

Wongkobrat, A., Weerawut, T. and Prachuabmoh, O. (1987) Effect of white scale insect [*Aonidomytilus albus*] infestation on cassava yield and quality. AgrisFAO online record. Available at: http://agris.fao.org/aos/records/TH9121467 (accessed 5 October 2015).

World Integrated Trade Solution software (WITS) (2005) The World Bank. Available at: http://wits.worldbank.org (accessed 5 October 2015).

Yaninek, S. and Hanna, R. (2003) Cassava green mite in Africa – a unique example of successful classical biological control of a mite pest on a continental scale, In: Neuenschwander, P., Borgemeister, C. and Langewald, J. (eds) *Biological Control in IPM systems in Africa*. CAB International, Wallingford, UK, pp. 61–75.

Yaninek, J.S., Moraes, G.J. and Markham, R.H. (1989) *Handbook on the Cassava Green Mite (*Mononychellus tanajoa*) in Africa: a Guide to its Biology and Procedures for Implementing Classical Biological Control.* International Institute of Tropical Agriculture (IITA), Ibadan, Nigeria.

Zeddies, J., Schaab, R.P., Neuenschwander, P. and Herren, H.R. (2001) Economics of biological control of cassava mealybug in Africa. *Agricultural Economics* 24, 209–219.

Zhang, F., Liu, H., Luo, Y., Jin, Q. and Fu, Y. (2005) Repellent effects of the alcohol extracts of 25 tropical plants to *Oligonychus biharensis* Hirst. *Plant Protection* 32, 57–60.

Zhang, Y.Q., Wei, D., Zhao, Z.M., Jing, W.U. and Fan, Y.H. (2008) Studies on acaricidal bioactivities of *Artemisia annua* L. extracts against *Tetranychus cinnabarinus* Bois. (Acari: Tetranychidae). *Agricultural Sciences in China* 7, 577–584.

Index